Handbook of Modern Physics

Handbook of Modern Physics

Editor: Kate Fellows

NY RESEARCH PRESS
New York

Published by NY Research Press
118-35 Queens Blvd., Suite 400,
Forest Hills, NY 11375, USA
www.nyresearchpress.com

Handbook of Modern Physics
Edited by Kate Fellows

International Standard Book Number: 978-1-63238-604-5 (Hardback)

Cataloging-in-Publication Data

Handbook of modern physics / edited by Kate Fellows.
p. cm.
Includes bibliographical references and index.
ISBN 978-1-63238-604-5
1. Physics. I. Fellows, Kate.
QC21.3 .H36 2018
530--dc23

Printed in China.

Contents

Preface

This book aims to highlight the current researches and provides a platform to further the scope of innovations in this area. This book is a product of the combined efforts of many researchers and scientists, after going through thorough studies and analysis from different parts of the world. The objective of this book is to provide the readers with the latest information of the field.

Physics is a vast field of study. It primarily studies the concepts of motion, energy and force with respect to matter. Modern physics expands into interdisciplinary fields such as quantum chemistry, biophysics, particle physics, etc. This book attempts to understand the multiple branches that fall under the discipline of modern physics and how such concepts have practical applications. The various studies that are constantly contributing towards advancing technologies and evolution of this field are examined in detail. Scientists and students actively engaged in this field will find this book full of crucial and unexplored concepts.

I would like to express my sincere thanks to the authors for their dedicated efforts in the completion of this book. I acknowledge the efforts of the publisher for providing constant support. Lastly, I would like to thank my family for their support in all academic endeavors.

Editor

1

Uncovering the spatial structure of mobility networks

Thomas Louail[1,2], Maxime Lenormand[3], Miguel Picornell[4], Oliva García Cantú[4], Ricardo Herranz[4], Enrique Frias-Martinez[5], José J. Ramasco[3] & Marc Barthelemy[1,6]

The extraction of a clear and simple footprint of the structure of large, weighted and directed networks is a general problem that has relevance for many applications. An important example is seen in origin-destination matrices, which contain the complete information on commuting flows, but are difficult to analyze and compare. We propose here a versatile method, which extracts a coarse-grained signature of mobility networks, under the form of a 2×2 matrix that separates the flows into four categories. We apply this method to origin-destination matrices extracted from mobile phone data recorded in 31 Spanish cities. We show that these cities essentially differ by their proportion of two types of flows: integrated (between residential and employment hotspots) and random flows, whose importance increases with city size. Finally, the method allows the determination of categories of networks, and in the mobility case, the classification of cities according to their commuting structure.

[1] Institut de Physique Théorique, CEA-CNRS (URA 2306), Orme-des-Merisiers Batiment 774, F-91191 Paris, France. [2] Géographie-Cités, CNRS-Paris 1-Paris 7 (UMR 8504), 13 rue du four, FR-75006 Paris, France. [3] IFISC, Instituto de Física Interdisciplinar y Sistemas Complejos (CSIC-UIB), Campus Universitat de les Illes Balears, Palma de Mallorca E-07122, Spain. [4] Nommon Solutions and Technologies, calle Cañas 8, Madrid E-28043, Spain. [5] Telefonica Research, Madrid E-28050, Spain. [6] Centre d'Analyse et de Mathématique Sociales, EHESS-CNRS (UMR 8557), 190-198 avenue de France, FR-75013 Paris, France. Correspondence and requests for materials should be addressed to M.B. (email: marc.barthelemy@cea.fr).

The increasing availability of pervasive data in various fields has opened up exciting possibilities of renewed quantitative approaches to many phenomena. This is particularly true for cities and urban systems for which different devices at different scales produce a very large amount of data potentially useful to construct a 'new science of cities'[1].

A new problem we have to solve is to then extract useful information from these huge data sets. In particular, we are interested in extracting coarse-grained information and stylized facts that encode the essence of a phenomenon, and that any reasonable model should reproduce. Such meso-scale information helps us to understand the system, to compare different systems and also to propose models. This issue is particularly striking in the study of commuting in urban systems. In transportation research and urban planning, individuals' daily mobility is usually captured in origin-destination (OD) matrices, which contain the flows of individuals going from a point to another (see refs 2,3). An OD matrix thus encapsulates the complete information about the flow of individuals in a city, at a given spatial scale and for a specific purpose. It is a large network, and as such does not provide clear, synthetic and useful information about the structure of the mobility in the city. More generally, it is very difficult to extract high-level, synthetic information from large networks and methods such as community detection[4] and stochastic block modelling (see for example, refs 5–7) were recently proposed. Both these methods group nodes in clusters according to certain criteria, and nodes in a given cluster have similar properties (for example, in the stochastic block modelling nodes in a given group have similar neighbourhood). These methods are very interesting when one wants to extract meso-scale information from a network, but they are unable to construct expressive categories of links and propose a classification of weighted (directed) networks. This is particularly true in the case of commuting networks in cities, where edges represent flows of individuals that travel daily from their residential neighbourhood to their main activity area. Several types of links can be distinguished in these mobility networks; some constitute the backbone of the city by connecting major residential neighbourhoods to employment centres, while other flows converge from smaller residential areas to important employment centres, or diverge from major residential neighbourhoods to smaller activity areas. In addition, the spatial properties of these commuting flows are fundamental in cities and a relevant method should be able to take this aspect into account.

There is important literature in quantitative geography and transportation research that focuses on the morphological comparison of cities[8–11] and notably on multiple aspects of polycentrism, ranging from schematic pictures proposed by urban planners and architects[12] to quantitative case studies and contextualized comparisons of cities[13–15]. So far, most comparisons of large sets of cities have been based on morphological indicators[8,9] (built-up areas, residential density, number of sub-centres and so on) and aggregated mobility indicators[10,11] (motorization rate, average number of trips per day, energy consumption *per capita* per transport mode and so on), and have focused on the spatial organization of residences and employment centres. But these previous studies did not propose generic methods to take into account the spatial structure of commuting trips, which consist of both an origin and a destination. Such comparisons based on aggregated indicators thus fail to give an idea of the morphology of the city in terms of daily commuting flows. We still need some generic methods that are expressive in a urban context, and that could constitute the quantitative equivalent of the schematic pictures of city forms that have been pictured for so long by urban planners[12].

In this paper, we propose a simple and versatile method designed to compare the structure of large, weighted and directed networks. In the next section, we describe this method in detail. The guiding idea is that a simple and clear picture can be provided by considering the distribution of flows between different types of nodes. We then apply the method to commuting (journey to work) OD matrices of 31 cities extracted from a large mobile phone data set. We discuss the urban spatial patterns that our method reveals, and we compare these patterns observed in empirical data to those obtained with a reasonable null model that generates random commuting networks. Finally, the method allows determining categories of networks with respect to their structure, and here to classify cities according to their commuting structure. This classification highlights a clear relation between commuting structure and city size.

Results

Extracting coarse-grained information from OD matrices. For the sake of clarity, we will use here the language of OD matrices, but the method could easily be applied to any weighted and directed network from which we want to extract high-level information.

We assume that for a given city, we have the $n \times n$ matrix F_{ij} where n is the number of spatial units that compose the city at the spatial aggregation level considered (for example, a grid composed of square cells of size a, see Methods). This OD matrix F_{ij} represents the number of individuals living in location i and commuting to location j where they have their main, regular activity (work or school for most people). By convention, when computing the number of inhabitants and workers in each cell, we do not consider the diagonal of the OD matrix. This means that we omit the individuals who live and work in the same cell (considered as 'immobile' at this spatial scale).

To extract a simple signature of the OD matrix, we proceed in two steps. We first extract both the residential and the work locations with a large density—the so called 'hotspots' (see ref. 16). The number of residents of cell i is given by $\sum_{j\neq i} F_{ij}$ and its number of workers is given by $\sum_{j\neq i} F_{ji}$. The hotspots then correspond to local maxima of these quantities. It is important to note that the method is general, and does not depend on how we determine these hotspots.

Once we have determined the cells that are the residential and the work hotspots (some cells can possibly be both), we proceed to the second and main step of the method. We reorder the rows and columns of the OD matrix to separate hotspots from non-hotspots. We put the m residential hotspots on the top lines, and do the same for columns by putting the p work hotspots on the left columns. The OD matrix then becomes a four-quadrant matrix, where the flows F_{ij} are spatially positioned in the matrix with respect to their nature: on the top left the individuals that live in hotspots and work in hotspots; at the top right the individuals that live in hotspots and do not work in hotspots; at the bottom left individuals that do not live in hotspots but work in hotspots and finally in the bottom right corner the individuals that neither live or work in hotspots. For each quadrant we sum the number of commuters and normalize it by the total number of commuters in the OD matrix, which gives the proportion of individuals in each of the four categories of flows. In other words, for a given city, we reduce the OD matrix to a 2×2 matrix

$$\Lambda = \begin{pmatrix} I & D \\ C & R \end{pmatrix} \tag{1}$$

where

$$I = \sum_{i\in 1..m, j\in 1..p} F_{ij} \Big/ \sum_{i,j\in 1..n} F_{ij}$$

is the proportion of integrated flows that go from residential hotspots to work hotspots;

$$C = \sum_{i \in m+1..n, j \in 1..p} F_{ij} / \sum_{i,j \in 1..n} F_{ij}$$

is the proportion of convergent flows that go from random residential places to work hotspots;

$$D = \sum_{i \in 1..m, j \in p+1..n} F_{ij} / \sum_{i,j \in 1..n} F_{ij}$$

is the proportion of divergent flows that go from residential hotspots to random activity places;

$$R = \sum_{i \in m+1..n, j \in p+1..n} F_{ij} / \sum_{i,j \in 1..n} F_{ij}$$

is the proportion of random flows that occur 'at random' in the city, that is, that are going from and to places that are not hotspots.

By construction, we have $I, C, D, R \in [0,1]$ and $I+C+D+R=1$. This matrix Λ is thus a very simple footprint of the OD matrix that gives an expressive picture of the structure of commuting in the city, as illustrated by Fig. 1.

Commuting data, hotspots and ICDR values. Large scale individual mobility networks are nowadays extracted from pervasive geolocated data, such as mobile phone, GPS, public transport cards or social apps data[17–21]. In particular, if an individual's mobile phone geolocated activity is available during a sufficiently long period of time, it is possible—under certain regularity conditions—to infer the most likely locations of her home and her workplace, and by aggregation to construct OD matrices[22–24]. Several parameters, however, impact the construction of OD matrices such as the nature of the data source (survey or user-generated geolocated data), or the spatial scale at which the OD matrix is built, which can be dictated by administrative units (divisions in wards, counties, municipalities and so on) or by technical reasons such as the density of antennas in the case of mobile phone data. Given this variety of data collection protocols,

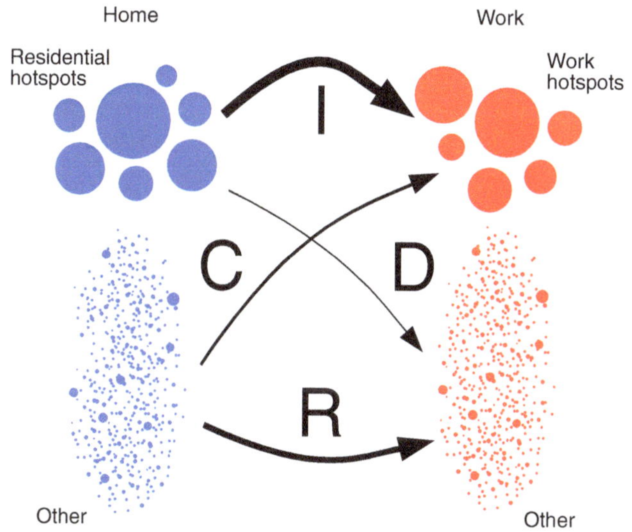

Figure 1 | Illustration of the ICDR method. The method decomposes the commuting flows in the city in four categories: the integrated flows (*I*) from hotspot to hotspot, the convergent flows (*C*) to hotspots, the divergent flows (*D*) originating at hotspots and finally the random flows (*R*) which are neither starting nor ending at hotspots. For each city with its origin-destination matrix, we can compute the importance of each commuting flow category and get a simple picture of the mobility structure in the city.

it is thus particularly remarkable that when considering the commuting flows at a city scale, various sources of pervasive data provide a very similar mobility information when compared with the OD matrices built from surveys[24]. This result needs to be confirmed for other cities and countries, but it already opens the door to a systematic use of pervasive, geolocated data as a relevant substitute to traditional transport surveys.

In the following, we apply our ICDR method to OD matrices that have been extracted from mobile phone records in 31 Spanish urban areas during a 5-week period (see the Methods section for details on the data set and the calculation of the OD matrices).

As described above, the first step consists in determining the hotspots. Several possible methods have been proposed in the literature[15,25,26], and we use here a parameter-free method based on the Lorenz curve of the densities that we have proposed in a recent study (see ref. 16 and the Methods section). Once we have determined both the origin (residential) hotspots and the destination (work) hotspots in each city, we first observe how their number scale with the population size of the city. Both these numbers for residential and employment hotspots scale sublinearly with the population size (see Supplementary Figs 4 and 5). The number of work hotspots grows significantly slower than the number of residential hotspots, showing that residential areas are (i) more dispersed in the city, and (ii) are more numerous than activity centres, as intuitively expected (see Supplementary Fig. 6 that displays the locations of home and work hotspots in four cities that exhibit different spatial organizations). We also note here that the sublinear scaling of the work hotspots confirms previous results obtained with a totally different data set (the number of employment centres in US cities)[27].

We now apply the second part of the method to calculate the *I*, *C*, *D* and *R* values for each OD matrix. For the 31 Spanish urban areas under study (see Supplementary Fig. 1), we obtain the values shown in Fig. 2. In Fig. 2a, we plot these values versus the population size of these cities. For this sample of cities, we see that globally the proportion *I* of individuals that commute from hotspot to hotspot decreases as the population size increases, while the proportion *R* of 'random' flows increases and the proportions *C* and *D* of convergent and divergent flows, respectively, seem surprisingly constant whatever the city size. In Fig. 2b we plot the same values but sorted by decreasing values of *I*, which shows clearly that the *I* and *R* values are the relevant parameters for distinguishing cities from each other.

We also notice that the values obtained for another spatial scale of data aggregation confirm this trend (see Supplementary Fig. 7). The decay of *I* flows ('integrated') flows in favour of *R* flows ('random') when *P* increases shows that the population growth among Spanish cities goes with a decentralization of both activity places and residences. As cities get bigger, their numbers of residential and employment hotspots grow (sublinearly), but these hotspots catch a smaller part of the commuting flows.

A null model. To evaluate to what extent the ICDR signatures of cities are characteristic of their commuting structure, we compare these values to the ones returned by a null model of commuting flows. For each city, we generate random OD matrices of the same size than the reference OD matrix but with random flows of individuals that preserve the in and out degree of each node (see Supplementary Note 5). Figure 2c shows the average values and s.d. obtained for 100 replications. In Fig. 2d, we plot the *Z*-scores of the *I*, *C*, *D*, and *R* values of each city when compared with the values I^*, C^*, D^* and R^* returned by our null model (for example, for the quantity *R* of the city *i*, the *Z*-score is given by $Z(R_i) = (R_i - \overline{R_i^*})/\sigma(R_i^*)$). Essentially, we observe that the

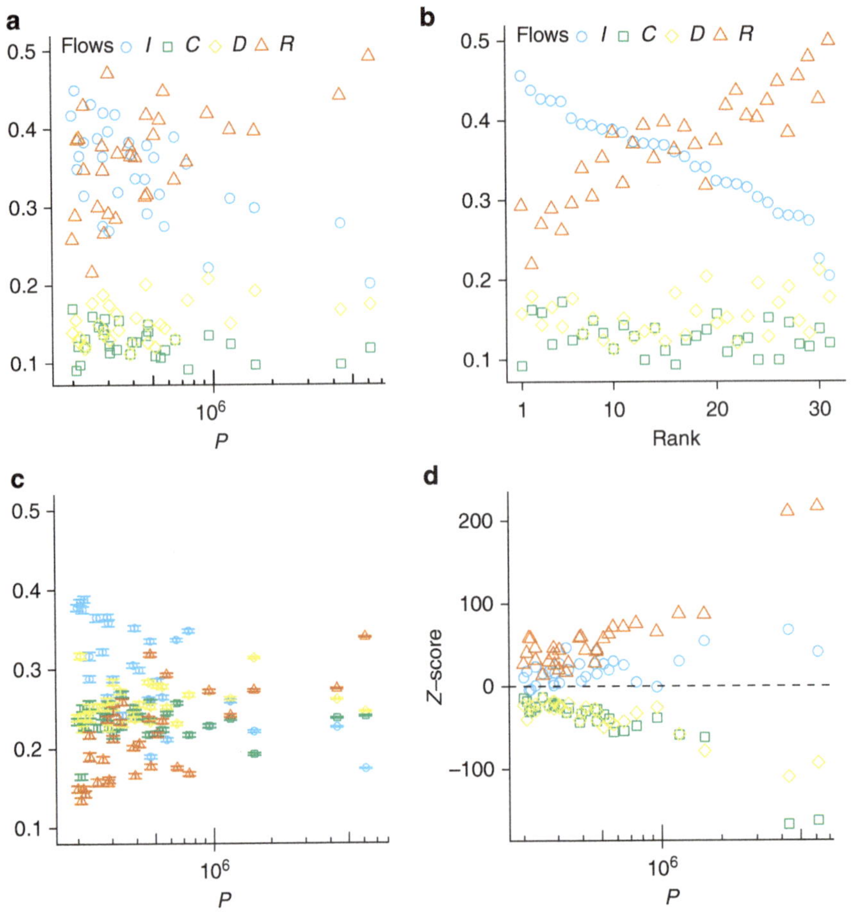

Figure 2 | Results for 31 Spanish cities. (**a**) *I* (integrated), *C* (convergent), *D* (divergent) and *R* (random) values versus population size for 31 Spanish urban areas. (**b**) Same ICDR values as in **a** but sorted by decreasing order of *I* (note that by definition, we have for each city $I+C+D+R=1$). It is remarkable that *I* and *R* dominate and seem almost sufficient to distinguish cities, while *C* and *D* are almost constant whatever the city size (see Supplementary Fig. 7 for the values obtained with another size *a* of grid cells). (**c**) *I*, *C*, *D* and *R* average values and s.d. obtained for 100 replications of a null model, where the inflow and outflow at each node are kept constant while flows are randomly distributed at random between nodes. (**d**) *Z*-scores obtained by comparing the empirical data and the values returned by the null model. Large values of *Z*-scores show that the actual commuting networks cannot be considered as resulting from connecting the nodes at random. The *I*, *C*, *D* and *R* values of a specific city are then a signature of its structure.

Z-scores of *I* and *R* are positive and large, while those of *C* and *D* are negative (and large in absolute value). Also as cities grow, the *Z*-scores of *I* and *R* increase while those of *C* and *D* decrease. These results demonstrate that the larger a city, and the less random it appears. This is in contrast with the naive expectation that the larger a city, the more disordered is the structure of individuals' mobility. For large cities, there seems to be a commuting backbone, which cannot result from purely random movements of individuals. This backbone is the footprint of the city's structure and history, and probably results from strong constraints and efficiency considerations.

Robustness. Since our method first requires to determine origin and destination hotspots, one could argue that the interpretation of the *I*, *C*, *D* and *R* values will crucially depend on the particular method chosen to define these hotspots. The identification of hotspots is a problem that has been broadly discussed in urban economics (see Supplementary Note 3). Roughly speaking, starting from a spatial distribution of densities, the goal is to identify the local maxima and amounts to choose a threshold ρ^* for the density ρ of individuals: a cell i is a hotspot if the local density of people is such that $\rho(i)>\rho^*$. To test the impact of the choice of a particular threshold ρ^* on the resulting ICDR values, we measure the sensitivity of these values to the density threshold ρ^* (see Supplementary Fig. 8). As expected, the lower the density threshold, the larger the number of origin and destination hotspots, and consequently the larger *I* and the smaller *R*. In contrast, changing the density threshold has little impact on the *C* and *D* terms. More importantly, the conclusions drawn from the comparison of the ICDR values across cities remain the same; whatever the density threshold ρ^* chosen to define residential/employment hotspots, we still observe the same qualitative behaviour: a decay of integrated flows (*I*) in favour of 'random' flows (*R*), when the population size increases.

Distance of each type of flows. We now want to characterize spatially these different flows and the relation between city size and the commuting distances travelled by individuals. In each city, we compute the average distance travelled by individuals per type of flows *I*, *C*, *D* and *R*. The resulting average distances measured in data are plotted in red on Fig. 3a. We observe that the average distance for all categories of flows increases with population size, an expected effect as the city's area also grows with population size. We also observe that the average distance of convergent flows *C* increases faster than for other types of flows (we note that the average distance associated to convergent flows increases more than the distance associated to divergent flows *D*, showing that these flows are not symmetric as one could have

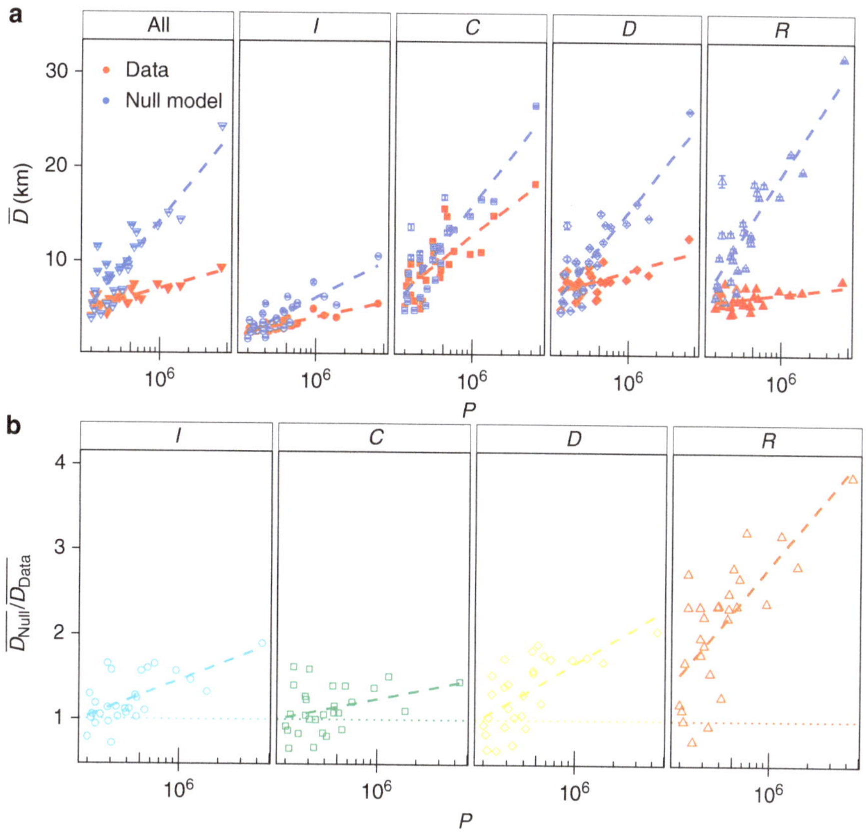

Figure 3 | Distance per type of flow and population size. (**a**) Average commuting distance versus population size, per type of flows; (**b**) ratio $\overline{D_{Null}}/\overline{D_{Data}}$ per type of flows. The LouBar criteria[16] is used here to define residential and employment hotspots.

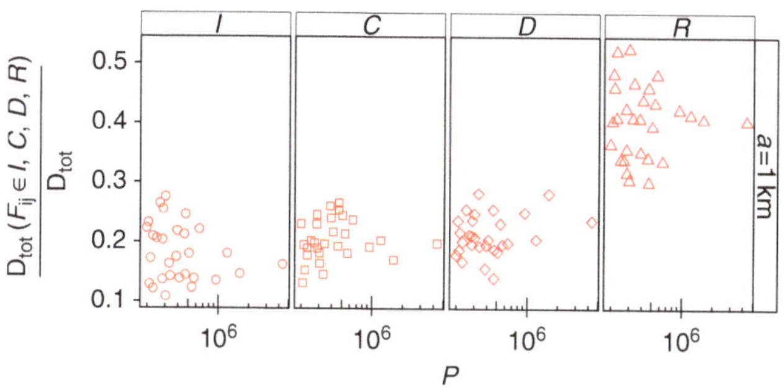

Figure 4 | Total commuting distance. Contributions to the total commuting distance of each type of flows versus population size. We observe that the variations are small and that the largest contribution comes from the *R* flows.

naively expected). This result means that for this set of Spanish cities, commuters from small residential areas to important activity centres travel on average a longer distance than all other individuals. This observation could be an indication that for our set of cities, residential areas have expanded while activity centres remained at their location, leading to longer commuting distances.

Another interesting information is provided by the comparison of distances measured in the data with average distances measured from the random OD matrices generated by the null model. The average distances associated to the null model are plotted in blue in Fig. 3a. We see that for all types of flows the distances measured in the empirical data are shorter than those generated by the null model. This is another clear indication of the spatial organization of individual flows in cities. It also highlights the importance of the travel time budget in the residential locations choice. Remarkably enough, the distance of convergent flows (*C*) is both the largest and the one that increases the fastest with population, indicating a low degree of efficiency.

The comparison of this behaviour with the null model leads to interesting results. In Fig. 3b, we plot the ratio $\overline{D_{Null}}/\overline{D_{Data}}$ for the four types of flow. Values <1 indicate that the average commuting distance generated by the null model is shorter than the distance observed in the city. Surprisingly, we observe that small cities display a value <1, indicating the lesser importance of space at this short scale. We also see that this ratio increases faster for random flows (*R*) than for the others (*D*, *C*, *I*), suggesting a remarkable spatial structure of these *R* flows.

We also consider the fraction of total commuting distance by type of flow (Fig. 4). We see that for each type of flows, their respective fraction is constant and independent of city size. With the LouBar hotspots detection method[16] (see Supplementary Fig. 3) and with a grid of 1 km^2 cells, we measure that roughly 40% of the total commuting distance is made on random flows while the other types represent each about 20%. This result shows that the method is able to identify where most of the commuting distance is travelled. In particular, we see that the natural, obvious flows (*I*) from residential centres to activity centres are not the most important ones, and that the decentralization of commuting flows seems to be the rule for the Spanish cities in our sample (see Supplementary Figs 10 and 11 for various aggregation scales).

Classification of cities. Finally, the ICDR signature of their OD matrix allows to cluster cities with respect to the structure of their commuting patterns. We measure the euclidian distance between the cities' ICDR signatures and we then perform a hierarchical cluster analysis. Figure 5 shows the dendrogram resulting from the classification. Four well-separated clusters are identified on this dendrogram, and Table 1 gives the average value of each term along with the average population of the cities composing the cluster. Remarkably these summary statistics show that largest cities are clustered together and are characterized by a larger proportion of 'random' flows (*R*) of individuals both living and working in parts of the city that are not the dominant residential and activity centres. This can be interpreted as an increased facility in bigger urban areas to commute from any part of the city to any other part. Further studies on other cities and countries are needed at this stage to discuss the relevance of the proposed classification.

It is also important to test the robustness of this classification and we show that introducing a reasonable amount of noise in the OD matrices does not change the classification (see Methods and Supplementary Fig. 12). This sensitivity test confirms that the clustering is robust against possible errors in the data source and in the extraction of the mobility networks. The classification of cities based on their *ICDR* values is also robust to a change of the method used to define residential and employment hotspots (see Supplementary Fig. 13).

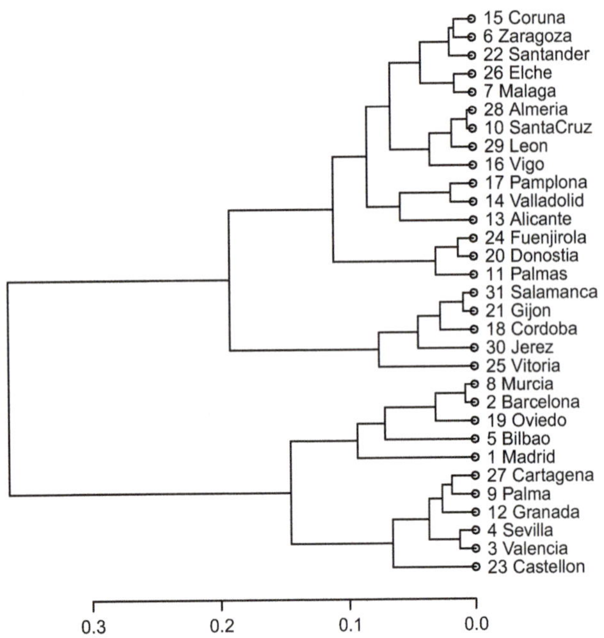

Figure 5 | Classification of cities. Dendrogram resulting from the hierarchical clustering on cities based on their ICDR values. In front of each city name, we indicate its rank in the hierarchy of population sizes. The largest cities are clustered together. As cities get bigger, the 'random' component (*R*) of their commuting flows increases, which signals that it is easier to commute from any place to any other in large cities.

Discussion

We have proposed a method to extract high-level information from large, weighted and directed networks, such as OD matrices. This method relies on the identification of origin and destination hotspots, and this first step can be performed with any reasonable criterion. The important second step consists of aggregating flows of four different types, depending whether they start and end from/to a hotspot or not.

We have applied this method to commuting networks extracted from mobile phone data available in 31 Spanish cities. The method has allowed us to highlight several remarkable patterns in the data:

- Independently of the density threshold chosen to determine hotspots, the proportion of integrated flows (*I*) decreases with city size, while the proportion of random flows (*R*) increases;
- On average and for all cities considered here, individuals who live in residential main hubs and who work in employment main hubs (*I* flows) travel shorter distances than the others (*C*, *D*, *R* flows);
- When the city size increases, the largest impact is on convergent flows (*C*) of individuals living in smaller residential areas (typically in the suburbs) and commuting to important employment centres;
- The classification of cities based on the ICDR values leads to groups with consistent population size, highlighting a clear relationship between the population size of cities and their commuting structure.

In addition, the comparison with a null model led to interesting conclusions. Flows in cities display a high level of spatial organization and as the population size of the city grows, the increase of the *Z*-scores of *I*, *C*, *D*, *R* shows that the structure of the mobility is more and more specific, and far from a random organization. We note that an interesting direction for future research would be to find some analytical arguments using simple models of city organization for estimating these flows, and how they vary with population size.

Our coarse-graining method provides a large scale picture of individual flows in the city. In this respect it could be particularly useful for validating synthetic results of urban mobility models (such as ref. 28), and for comparing different models. An accurate modelling of mobility is indeed crucial in a large number of applications, including the important case of epidemic spreading, which needs to be better understood, in particular at the intra-urban level[29,30].

It would also be interesting to apply the proposed method to other mobility data sets, with different time and spatial scales, to test the robustness of these commuting patterns. Another direction for future studies could be to inspect the time evolution of the *I*, *C*, *D* and *R* values for data sets describing travel to work journeys over several decades (such as national travel surveys). The method could also be applied at larger spatial scales, for

Table 1 | Classification of cities.

Cluster	Cities	$\bar{P}$	$\bar{I}$	$\bar{R}$	$\bar{D}$	$\bar{C}$
Orange	Salamanca, Gijon, Cordoba,...	255,330	0.43	0.27	0.16	0.14
Dark blue	La Coruña, Zaragoza, Santander, Elche,...	392,970	0.37	0.36	0.15	0.13
Green	Cartagena, Palma, Granada,...	732,992	0.31	0.41	0.16	0.13
Light blue	Murcia, Barcelona, Bilbao, Madrid,...	2,463,551	0.25	0.46	0.17	0.12

Average ICDR values and average population sizes of the cities composing each of the four clusters represented in Fig. 5. As the population grows the proportion of random flows increases while the proportion of integrated decreases. The weights of convergent and divergent flows stay constant among the groups.

example, to capture dominant effects in international migration flows. More generally, we believe that an important feature of the ICDR method is its versatility, as it could be applied to any type of data that is naturally represented by a weighted and directed network.

Methods

Data. The data set used for our analysis comprises 55 days of aggregated and anonymized records for 31 urban areas of more than 200,000 inhabitants. No individual information or records were available for this study. The records included the set of base transceiver stations (communication antennas, BTS) used for the communications as presented in the call detail records (CDR). A CDR is produced for each active phone event, including call/sms sending or reception. The number of anonymized users represents on average 2% of the total population and at most 5% of the total population. These percentages are almost the same for all the urban areas. From the CDR data obtained for 20 weekdays (from mondays to thursdays only), we extracted home and work places for all the anonymized mobile phone users in the data set. The output of this processing phase is an OD commuting matrix for each urban area, at the scale of the BTS point pattern. To facilitate the calculations and the comparison of the results between different cities, the OD matrices are then transposed on regular square-cell grids of varying size a (see Supplementary Note 2).

Extraction of OD matrices from mobile phone data. In order to extract OD matrices from phone calls, we select a subset of users with a mobility displaying a sufficient level of statistical regularity. For this analysis, we considered commuting patterns during workdays only. The users' home and work locations are identified as the Voronoi cells, which are the most frequently visited on weekdays by each user between 2000 and 0700 h (home) and between 0900 and 1700 h (work). We assume that there must be a daily travel between the home and work locations of each individual. Users with a call activity $>40\%$ of the days under study at home or work are considered as valid. We then aggregate the complete flow of users and construct the OD matrix with the flows between a Voronoi cell classified as home and another cell classified as a workplace. Since the Voronoi areas do not exactly match the grid cells (see for example Supplementary Fig. 2), we use a transition matrix to change the spatial scale of the OD matrix, that is to transform the F_{ij} values of the OD matrix where i and j are Voronoi cells into $F'_{i'j'}$ values where i' and j' are the cells of a regular grid (see Supplementary Note 2).

Spatial scale of the OD matrix. The OD matrix is the standard object in mobility studies and transport planning[2], and contains information about movement of individuals in a given area. More precisely, an OD matrix is a $n \times m$ matrix, where n is the number of different 'origin' zones, m is the number of 'destination' zones and F_{ij} is the number of people commuting from place i to place j during a given period of time. In transport surveys, the size of the OD matrix depends on the spatial scale at which the mobility data has been collected. Traditionally, the zones that are used to partition the city are the administrative units, whose size can vary from census and electoral units to whole departments or states, depending on the purpose for building the OD matrix.

In this study, we applied our ICDR method to cities divided in square cells that are smaller than administrative units, allowing for a better spatial resolution. In the case of OD matrices extracted from CDR mobile phone data, the maximal resolution corresponds to the BTS (antennas) point pattern. The ICDR method proposed in this paper does, however, depend on a particular spatial scale and can be applied on OD matrices available at coarser spatial resolutions as well. For a given territory, the results obtained with the ICDR method—the I, C, D and R values in the first place—will obviously depend on the spatial resolution and will also depend on the method used to define hotspots. It is important to note that when the ICDR method is used for comparing cities, the spatial resolution and the hotspots identification method should be the same for all cities (see Supplementary Notes 4 and 6 and Supplementary Figs 8 and 9 for results obtained with another hotspots delimitation method, and see Supplementary Fig. 7 for results obtained when considering another spatial scale of aggregation).

Robustness of the classification of cities. To ensure that the classification of cities based on their ICDR matrices is robust, we introduce a noise in the flows F_{ij} (for all the 31 cities). We focus on the case where the workplace of an individual can be modified and where the number of individuals living in each cell i is kept constant. More precisely, the noise is introduced as follows:

1. We pick up a uniform random positive integer g, the number of individuals whose workplace is reshuffled. This number g varies from 1 to $N=\sum F_{ij}$ the total number of commuters in the city.
2. We repeat g times the following operation: we pick up randomly a residence and a workplace (a couple of values (i, j)) and move one individual from her workplace j to put another randomly chosen workplace j': $F_{ij} \to F_{ij}-1$ and $F_{ij'} \to F_{ij'}+1$

The parameter of this workplace reshuffling is then $f=g/N$. To evaluate how much the classification of cities is affected by this noise, we compute the Jaccard index J_I between the reference classification of cities in k groups and the classification in k groups obtained for the noisy OD matrices. The Jaccard index measures the similarity of two partitions P and P' of same size:

$$J_I = \frac{a}{a+b+c} \quad (2)$$

where

- a: number of city pairs that are in the same group for both P and P'
- b: number of city pairs that are in different groups in P but in the same one in P'
- c: number of city pairs that are in the same group in P but not in P' (or conversely)

The Jaccard index J_I is in the range [0, 1] and the closest it is to 1, the larger the similarity between P and P'.

We generate 100 noisy matrices for each value of f and compute the average value $\bar{J}_I$ of the Jaccard index. This average value encodes the distance between the reference partition P of cities in k groups and the partitions of cities in k groups obtained for the noisy OD matrices. Supplementary Fig. 12 shows the values of $\bar{J}_I$ versus the proportion f of reshuffled individuals, for different number of groups k. The red shaded rectangle on each panel corresponds to the mean value ± s.d. obtained for 1,000 replications of a null model, in which permutations of cities among the k groups are randomly performed. We observe here that for up to 20% of reshuffled individuals, the average value $\bar{J}_I$ obtained is always significantly larger than the one obtained for the null model, indicating that the classification is robust even for important values of noise.

References

1. Batty, M. *The New Science of Cities* (MIT Press, 2013).
2. Ortuzar, J. D. & Willumsen, L. G. Modelling transport 4th edn (Wiley, 2011).
3. Weiner, E. *Urban Transportation Planning in the United States* (US Department of Transportation, 1986).
4. Fortunato, S. Community detection in graphs. *Phys. Rep.* **486,** 75–174 (2010).
5. Faust, F. & Wasserman, S. *Social Network Analysis* (Cambridge Univ. Press, 1992).
6. Karrer, B. & Newman, M. E. Stochastic blockmodels and community structure in networks. *Phys. Rev. E* **83,** 016107 (2011).
7. Aicher, C., Jacobs, A. Z. & Clauset, A. Learning Latent Block Structure in Weighted Networks. Preprint at http://arXiv.org/abs/1404.0431 (2014).
8. Tsai, Y. H. Quantifying urban form: compactness versus sprawl. *Urban. Stud.* **42,** 141–161 (2005).
9. Guérois, M. & Pumain, D. Built-up encroachment and the urban field: a comparison of forty european cities. *Env. Plan. A* **40,** 2186–2203 (2008).
10. Schwarz, N. Urban form revisited - selecting indicators for characterising european cities. *Landscape Urban Plan.* **96,** 29–47 (2010).
11. Le Néchet, F. Urban spatial structure, daily mobility and energy consumption: a study of 34 european cities. *Cybergeo* 580 (2012).
12. Bertaud, A. & Malpezzi, S. The Spatial Distributio of Population in 48 World Cities: implications for Economies in Transition (World Bank Report, 2003).
13. Salomon, I., Bovy, P. & Orfeuil, J. P. *A Billion Trips a Day: Tradition and Transition in European Travel Patterns* (Kluwer Academic Publishers, 1993).
14. Cattan, N. *Cities and Networks in Europe: a Critical Approach of Polycentrism* (John Libbey Eurotext, 2007).
15. Berroir, S., Mathian, H., Saint-Julien, T. & Sanders, L. in *Modelling Urban Dynamics* (eds Desrosiers, F. & Thériault, M.) 1–25 (ISTE-Wiley, 2011).
16. Louail, T. *et al.* From mobile phone data to the spatial structure of cities. *Sci. Rep.* **4,** 5276 (2014).
17. Schneider, C. M., Belik, V., Couronné, T., Smoreda, Z. & González, M. C. Unravelling daily human mobility motifs. *J. R. Soc. Interface* **10,** 20130246 (2013).
18. Roth, C., Kang, S. M., Batty, M. & Barthelemy, M. Structure of urban movements: polycentric activity and entangled hierarchical flows. *PLoS ONE* **6,** e15923 (2011).
19. Noulas, A., Scellato, S., Lambiotte, R., Pontil, M. & Mascolo, C. A tale of many cities: universal patterns in human urban mobility. *PLoS ONE* **7,** e37027 (2012).
20. Wu, L., Zhi, Y., Sui, Z. & Liu, Y. Intra-urban human mobility and activity transition: evidence from social media check-in data. *PLoS ONE* **9,** e97010 (2014).
21. Zhong, C., Arisona, S. M., Huang, X., Batty, M. & Schmitt, G. Detecting the dynamics of urban structure through spatial network analysis. *Int. J. Geogr. Inf. Sci.* **28,** 2178–2199 (2014).
22. Kung, K. S., Sobolevsky, S. & Ratti, C. Exploring universal patterns in human home/work commuting from mobile phone data. *PLoS ONE* **9,** e96180 (2014).
23. Jiang, S. *et al.* in Proceedings of the 2nd ACM SIGKDD (International Workshop on Urban Computing, 2013).

24. Lenormand, M. *et al.* Cross-checking different sources of mobility information. *PLoS ONE* **9,** e105184 (2014).
25. McMillen, D. P. Nonparametric employment subcenter identification. *J. Urban Econ.* **50,** 448–473 (2001).
26. Griffith, D. A. Modelling urban population density in a multi-centered city. *J. Urban Econ.* **9,** 298–310 (1981).
27. Louf, R. & Barthelemy, M. Modeling the polycentric transition of cities. *Phys. Rev. Lett.* **111,** 198702 (2013).
28. Eubank, S. *et al.* Modelling disease outbreaks in realistic urban social networks. *Nature* **429,** 180–184 (2004).
29. Balcan, D. *et al.* Multiscale mobility networks and the spatial spreading of infectious diseases. *Proc. Natl Acad. Sci. USA* **106,** 21484–21489 (2009).
30. Dalziel, B. D., Pourbohloul, B. & Ellner, S. P. Human mobility patterns predict divergent epidemic dynamics among cities. *Proc. R. Soc. B* **280,** 20130763 (2013).

Acknowledgements

We thank the anonymous referees for interesting and constructive comments. The research leading to these results has received funding from the European Union Seventh Framework Programme FP7/2007-2013 under grant agreement 318367 (EUNOIA project). M.L. and J.J.R. received partial financial support from the Spanish Ministry of Economy (MINECO) and FEDER (EU) under projects MODASS (FIS2011-24785) and INTENSE@COSYP (FIS2012-30634). The work of M.L. has been funded under the PD/004/2013 project, from the Conselleria de Educacion, Cultura y Universidades of the Government of the Balearic Islands and from the European Social Fund through the Balearic Islands ESF operational program for 2013–2017.

Author contributions

T.L. designed the study, processed and analyzed the data and wrote the manuscript; M.L. processed and analyzed the data; O.G.C. and M.P. processed the data; R.H. and J.J.R. coordinated the study; E.F.-M. obtained and processed the data; M.B. coordinated and designed the study, and wrote the manuscript. All authors read, commented and approved the final version of the manuscript.

Additional information

Competing financial interests: The authors declare no competing financial interests.

2

Efficient entanglement distillation without quantum memory

Daniela Abdelkhalek[1,2], Mareike Syllwasschy[2], Nicolas J. Cerf[3], Jaromír Fiurášek[4] & Roman Schnabel[1,2]

Entanglement distribution between distant parties is an essential component to most quantum communication protocols. Unfortunately, decoherence effects such as phase noise in optical fibres are known to demolish entanglement. Iterative (multistep) entanglement distillation protocols have long been proposed to overcome decoherence, but their probabilistic nature makes them inefficient since the success probability decays exponentially with the number of steps. Quantum memories have been contemplated to make entanglement distillation practical, but suitable quantum memories are not realised to date. Here, we present the theory for an efficient iterative entanglement distillation protocol without quantum memories and provide a proof-of-principle experimental demonstration. The scheme is applied to phase-diffused two-mode-squeezed states and proven to distil entanglement for up to three iteration steps. The data are indistinguishable from those that an efficient scheme using quantum memories would produce. Since our protocol includes the final measurement it is particularly promising for enhancing continuous-variable quantum key distribution.

[1] Institut für Laserphysik, Universität Hamburg, Hamburg 22761, Germany. [2] Institut für Gravitationsphysik, Leibniz Universität Hannover and Max-Planck-Institut für Gravitationsphysik (Albert-Einstein-Institut), Hannover 30167, Germany. [3] Quantum Information and Communication, Ecole Polytechnique de Bruxelles, CP 165, Université libre de Bruxelles, Brussels 1050, Belgium. [4] Department of Optics, Palacký University, Olomouc 77146, Czech Republic. Correspondence and requests for materials should be addressed to R.S. (email: roman.schnabel@physnet.uni-hamburg.de).

Light is the most suitable carrier of quantum information over long distances. Distribution of entangled states of light among distant nodes of a quantum communication network[1,2] may be used for various purposes, such as quantum cryptography[3,4] or quantum teleportation[5–7]. However, the transmission of quantum states of light is, in practice, unavoidably affected by losses and other decoherence effects that usually grow with distance. To fight these detrimental effects, entanglement distillation can be used, which extracts from a large number of noisy and weakly entangled states a smaller number of copies with increased entanglement and purity[8–16]. Crucially, entanglement distillation only requires local quantum operations and classical communication between the spatially separated parties holding parts of the entangled state. A canonical iterative entanglement distillation protocol[8,9] is illustrated in Fig. 1a. At each step, two copies of a decohered entangled state are consumed and, with some probability, one copy of a distilled state with improved properties is produced, which in turn serves as the input for the next round of the protocol. An efficient implementation of this iterative protocol requires a quantum memory[17], which motivates the extensive current effort for harnessing the quantum light-matter interface, ultimately leading to quantum repeaters[18]. Without quantum memory, all elementary two-copy distillation steps need to succeed simultaneously, which imposes an exponential overhead in terms of required resources, and hence drastically reduces the success rate of the protocol. Let us consider N iterations of the protocol and suppose, for simplicity, that the success probability of each elementary distillation step is the same and equal to P. A single attempt to perform N iterations requires 2^N input states, and all 2^N-1 elementary distillation steps must succeed simultaneously. To obtain 1 distilled copy, one thus needs to consume on average $2^N/P^{2^N-1}$ input states. With quantum memories, the distillation becomes far more efficient because the successfully distilled states after each step of the protocol can be stored and used in the subsequent step as required. Ideally, this reduces the required number of input states to $2^N/P^N$. Note that the hardware-efficient pumping Gaussifier proposed in ref. 19 requires only a single quantum memory unit on each side, and enables sequential processing of the individual copies of distilled quantum state but achieves similarly inefficient distillation rate as a purely optical scheme without quantum memory.

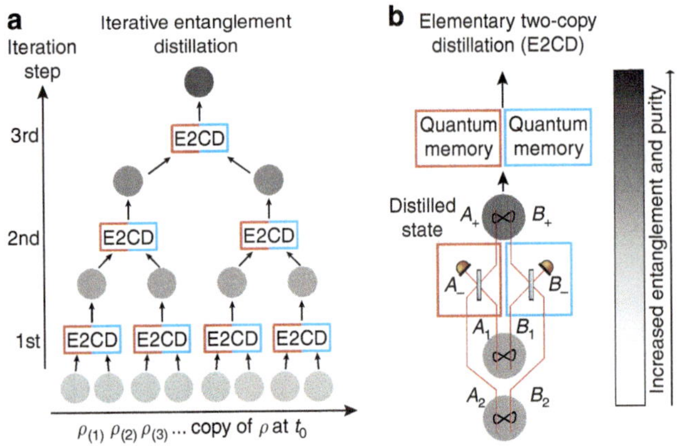

Figure 1 | Iterative entanglement distillation. (**a**) At each elementary step of the protocol, two copies of the input state, described by the density matrices ρ, are combined to produce, with some probability, one output copy with better properties. (**b**) Elementary step of iterative Gaussification of two-mode continuous-variable quantum states. Alice's and Bob's modes locally interfere on balanced beam splitters and one output mode on each side is measured with a photodetector. The procedure succeeds if both modes are projected onto vacuum. A quantum memory was previously deemed necessary for an efficient implementation of such an iterative protocol as one must store the successfully distilled states in a quantum memory to make them available for subsequent distillation steps.

To date, single-copy entanglement concentration[10,13,14,16,20], and elementary two-copy entanglement distillation[11,12] have been demonstrated for both discrete- and continuous-variable quantum states of light. A collective three-copy distillation of continuous-variable entangled states could even be implemented[15], but it remained inefficient in the absence of a quantum memory. Aside from this, an efficient realization of the full iterative multicopy entanglement distillation protocol could never be done because it is pending on operating an efficient quantum memory.

Here, we address this challenge from an opposite viewpoint and cancel out the need for quantum memories by exploiting the recently proposed concept of emulation of a quantum protocol[21,22]. The emulation replaces the actual physical implementation of a certain quantum operation by suitable postprocessing of measurement data and postselection, which offers a high potential to circumvent hardware implementation problems. In particular, the emulation of noiseless quantum amplification and attenuation by processing the data resulting from (eight-port) homodyne detection was proposed[21,22], and a proof-of-principle experimental emulation of single-mode noiseless quantum amplification was reported[23].

In this article, we apply such a strategy to multicopy continuous-variable entanglement distillation. Our scheme is solely based on the measurement data taken on a single beam of light that carries a continuous stream of copies of a decohered entangled state. Remarkably, due to the specific nature of the measurement and postprocessing, neither quantum memories nor the simultaneous physical realization of many copies are required, in contrast to the common knowledge on entanglement distillation. In principle, an arbitrary number of elementary (two-copy) distillation steps can be emulated by postprocessing the experimental data, solely limited by the length of the measured data stream. Specifically, we demonstrate the distillation of phase-diffused two-mode-squeezed states of light by iterative Gaussification[24–27] based on the interference of two copies of the state on balanced beam splitters, followed by projecting one output port on each side onto a vacuum state, see Fig. 1b. Here, we show that this scheme can be emulated by suitable processing of data obtained by eight-port homodyne detection on each mode of each phase-diffused state. Crucially, such emulation is completely indistinguishable from a full physical implementation with even ideal quantum memories to anyone outside Alice and Bob's labs. The only difference is that the distilled states are already detected and not physically available for further processing.

Nevertheless, the data may be used to fully characterize the distillation protocol and, for example, to extract a secret key from the distilled states. Our procedure is therefore particularly suitable for quantum cryptography, where it can convert seemingly useless highly noisy states into states that allow for the extraction of a secret key. It is similarly applicable to all related quantum communication protocols provided they terminate with suitable measurements.

Results

Emulation of iterative entanglement distillation. The starting point of our work is the iterative Gaussification scheme[24,25] illustrated in Fig. 1, which can serve for entanglement distillation of non-Gaussian quantum states of light. Each elementary step of

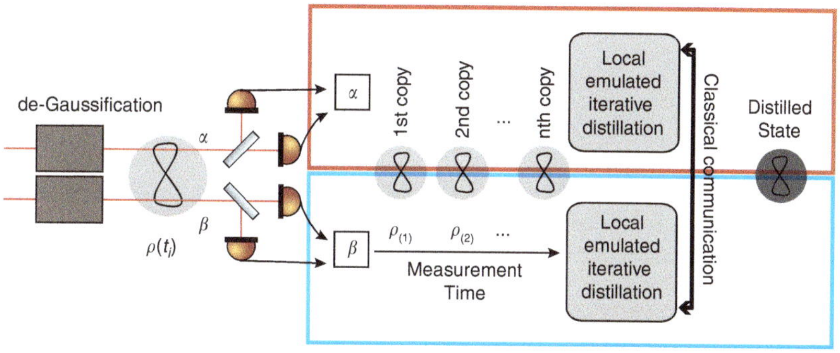

Figure 2 | Emulation of iterative Gaussification protocol. A single, continuously operated source distributes copies of the entangled and decohered states ρ to Alice and Bob. The states' coherent amplitudes are locally measured with two out-of-phase BHDs each, the so-called eight-port homodyne detectors. By prolonging the measurement time any quantity of copies may be generated. An emulated iterative distillation protocol based on the local operations and classical communication is realized as described in the main text. In the end, Alice and Bob share data that are equivalent to that from measurements on distilled entangled states.

the Gaussification protocol requires two copies of a two-mode entangled state, whose modes are labelled as A_1, B_1, and A_2, B_2, respectively. Modes A_1 and A_2 belong to Alice, and modes B_1 and B_2 are held by Bob. Both Alice and Bob combine their pairs of modes on a balanced beam splitter and obtain output modes A_+, A_- and B_+, B_-, where $+$ and $-$ refer to constructive and destructive interference, respectively. Subsequently, Alice and Bob perform photon number measurement on the modes A_- and B_- and exchange the measurement outcomes. This elementary distillation step is successful if both modes A_- and B_- are projected onto a vacuum state, and the distilled state in modes A_+ and B_+ provides the input for another round of the iterative protocol. Suppose now that Alice and Bob decide to measure the final distilled state with eight-port homodyne detectors, which perform projections onto coherent states. The generally impracticable implementation of the full iterative distillation protocol can be replaced by the following emulation procedure as depicted in Fig. 2. Alice and Bob perform the eight-port homodyne detection directly on the distributed decohered entangled state, and they repeat this measurement many times to collect a sufficiently large data set. Let α_j and β_j denote the measurement outcomes (complex amplitudes of coherent states) for the jth copy of the distributed state. Alice and Bob then combine the measurement results into pairs, say $(\alpha_{2n}, \alpha_{2n+1})$ and $(\beta_{2n}, \beta_{2n+1})$, and emulate the interference on beam splitters making the first layer of the protocol by performing simple additions and subtractions of the measurement results, which faithfully mimics interference of coherent states,

$$\begin{aligned}\alpha_{n,+} &= \tfrac{1}{\sqrt{2}}(\alpha_{2n}+\alpha_{2n+1}), \quad \alpha_{n,-} = \tfrac{1}{\sqrt{2}}(\alpha_{2n}-\alpha_{2n+1}),\\ \beta_{n,+} &= \tfrac{1}{\sqrt{2}}(\beta_{2n}+\beta_{2n+1}), \quad \beta_{n,-} = \tfrac{1}{\sqrt{2}}(\beta_{2n}-\beta_{2n+1}).\end{aligned} \tag{1}$$

Projection of the output modes A_- and B_- onto vacuum is equivalent to conditioning on $\alpha_{n,-}=\beta_{n,-}=0$. After the conditioning we obtain a reduced set of pairs $\alpha_j^{(1)}=\alpha_{n_j,+}$ and $\beta_j^{(1)}=\beta_{n_j,+}$ where n_j represent the values of n for which the conditioning was successful. These pairs of data represent effective measurement outcomes of eight-port homodyne detections on modes A and B of the distilled two-mode state after the first iteration of the distillation protocol. This procedure can be repeated, and outputs of the kth round of the protocol, $\alpha_j^{(k)}$ and $\beta_j^{(k)}$, can be used as inputs of the next round of the protocol, resulting in $\alpha_j^{(k+1)}$ and $\beta_j^{(k+1)}$.

Since the probability to obtain the specific outcomes $\alpha_{n,-}=\beta_{n,-}=0$ in eight-port homodyning vanishes, we need to modify the conditioning to make the protocol practicable. A theoretically appealing modification is to impose an acceptance probability that is a Gaussian function of the complex amplitudes,

$$P_{\mathrm{acc}}(\alpha_-,\beta_-) = \exp\left(-\frac{|\alpha_-|^2}{\bar{n}}\right)\exp\left(-\frac{|\beta_-|^2}{\bar{n}}\right). \tag{2}$$

Physically, this corresponds to a modified conditioning in Fig. 1b, where modes A_- and B_- are projected onto thermal states with a mean number of thermal photons equal to $\bar{n}$. The choice of Gaussian P_{acc} guarantees that the iterative Gaussification protocol indeed converges to a Gaussian state[25,27], which greatly simplifies its theoretical treatment and allows us to derive analytically the asymptotic state to which the protocol converges.

To be more specific, let us consider as an example the distillation of phase-diffused two-mode-squeezed states, as used in our experiment. A covariance matrix of the initial Gaussian symmetric two-mode-squeezed state before phase diffusion reads[1]

$$\gamma_{AB} = \begin{pmatrix} a & 0 & b & 0\\ 0 & a & 0 & -b\\ b & 0 & a & 0\\ 0 & -b & 0 & a\end{pmatrix}. \tag{3}$$

Here, a denotes the (symmetric) variance of phase space projections of the complex amplitude (quadratures) of the individual modes, and b represents the correlations between quadratures of modes A and B. Suppose now that modes A and B are sent through noisy channels where random phase shifts ϕ_A and ϕ_B are imposed. For the sake of simplicity, we shall assume that the phase diffusions are independent and have the same statistics for both modes (although our results can be easily extended to more general scenarios). Note that the phase-noise does not modify the value of parameter a, since each mode is locally in a thermal state that is invariant under phase-shift. So the covariance matrix of the phase-diffused non-Gaussian state at the output of the noisy channels preserves the form (3), but the phase diffusion reduces the intermodal correlations,

$$a_{\mathrm{PD}} = a, \quad b_{\mathrm{PD}} = qb, \tag{4}$$

where $q=\langle\cos\phi_A\ \cos\phi_B\rangle$ quantifies the phase diffusion for uncorrelated noise. The stronger the phase diffusion, the smaller is $|q|$. We assume here that the phase noise is symmetric, $\langle\sin\phi_A\rangle=\langle\sin\phi_B\rangle=0$. If no phase diffusion is present (ϕ_A and ϕ_B always zero) q equals unity. If the phases are fully random q equals zero. For given fixed values of the random phase shifts ϕ_A

and ϕ_B, the covariance matrix γ_{AB} of an input two-mode state is transformed according to $R\gamma_{AB}R^T$ with

$$R(\phi) = \begin{pmatrix} \cos\phi_A & \sin\phi_A & 0 & 0 \\ -\sin\phi_A & \cos\phi_A & 0 & 0 \\ 0 & 0 & \cos\phi_B & \sin\phi_B \\ 0 & 0 & -\sin\phi_B & \cos\phi_B \end{pmatrix} \quad (5)$$

describing the action of the particular phase-shift values ϕ_A and ϕ_B at the level of the covariance matrix. This expression has to be averaged over the random phase shifts to obtain the resulting covariance matrix of the de-phased state. Note that in this last averaging step we assume that the mean values of quadratures of the input state vanish, which is satisfied in our experiment where we utilize squeezed vacuum state. If non-zero, the mean values can be eliminated and set to zero by suitable local coherent displacements, which in our emulation-based scheme can be implemented simply as displacements of the measured data. With the help of the general theory of iterative Gaussification protocols presented in ref. 27, one can derive the following analytical expression for the covariance matrix of the asymptotic Gaussian state,

$$\gamma_{AB,\infty} = \left\langle R(\phi)[\gamma_{AB} + (2\bar{n}+1)I]^{-1}R^T(\phi)\right\rangle_\phi^{-1} - (2\bar{n}+1)I. \quad (6)$$

Where, I denotes the identity matrix and $\langle . \rangle_\phi$ represents statistical averaging over random phase diffusions.

The performance of the distillation protocol and its usefulness for quantum key distribution is illustrated in Fig. 3. The squeezing properties of the two-mode state can be characterized by the squeezed variance V_{sq} defined as the minimum eigenvalue of the covariance matrix, $V_{sq} = \min(\mathrm{eig}(\gamma_{AB}))$, while entanglement can be detected using the Duan–Simon criterion[28,29]. Technically, one has to determine the minimum symplectic eigenvalue μ of a covariance matrix corresponding to the partially transposed state, and entanglement is witnessed if $\mu < 1$. For a symmetric two-mode-squeezed state as in equation (3), we find that $V_{sq} = \mu = a - |b|$, hence the presence of squeezing, $V_{sq} < 1$, also indicates that the state is entangled. In Fig. 3a, the dependence of μ on the dephasing parameter q is plotted for the phase-diffused state (blue solid line) and the asymptotic distilled state for different $\bar{n}$ (red lines). The enhancement of entanglement by distillation is clearly visible, and the closer we choose $\bar{n}$ to zero (which corresponds to projection on vacuum) the stronger is the effect. Importantly, the distillation protocol can even recover seemingly lost quadrature entanglement as the distilled state can exhibit $\mu < 1$ even if the initial phase-diffused state exhibited $\mu > 1$. Distillation of entanglement necessarily implies that the input state was (weakly) entangled too. However, its initial entanglement is not visible via the covariance matrix (which is consistent with a separable Gaussian state) but hidden in higher-order correlations of the quadratures.

We can determine the maximum tolerable phase noise (minimum $|q|$) for which we can still distil Gaussian entanglement. For the *pure* symmetric two-mode Gaussian state with covariance matrix given by equation (3), and $a = \cosh(2r)$ and $b = \sinh(2r)$ we find that the Gaussian entanglement is asymptotically distilled provided that

$$\frac{\bar{n}}{\bar{n}+1}\tanh r < |q|. \quad (7)$$

Here, r denotes the squeezing parameter of the squeezed input states. If we project on vacuum ($\bar{n} = 0$) then entanglement is distilled for any $r > 0$ and $|q| > 0$. For $\bar{n} > 0$, however, entanglement is only distilled in the asymptotic limit if the dephasing is not too large ($|q|$ not too small). For mixed states, the entanglement might not be distillable even if $\bar{n} = 0$, c.f. the red solid line in Fig. 3a which crosses the horizontal line $\mu = 1$ at non-zero q.

We can go further and calculate the distillable secret key in continuous-variable quantum key distribution $K = I_{AB} - \chi_{AE}$ (Fig. 3b), where I_{AB} is the mutual information between Alice's and Bob's data and χ_{AE} is Eve's classically accessible information on Alice's quantum state. The key rate can be calculated analytically using the optimality of Gaussian attacks[30–32] and the well-known formulas for entropies of Gaussian probability distributions and Gaussian quantum states (see refs 1,21,33 for more details). As shown in Fig. 3b, the distillation protocol can significantly enlarge the range of q values for which a secret key can be distilled. Crucially, the key rate for the distilled state can be positive even if the initial phase-diffused state does not exhibit squeezing, $V_{sq} > 1$, that is, if its covariance matrix is compatible with that of a Gaussian separable state. In such a case, no secure key could be obtained from the decohered state by any protocol involving Gaussian measurements and a security analysis based on the covariance matrix of that state. The emulation of iterative entanglement distillation can thus efficiently convert seemingly useless noisy data into data from which a non-zero secret key could be extracted.

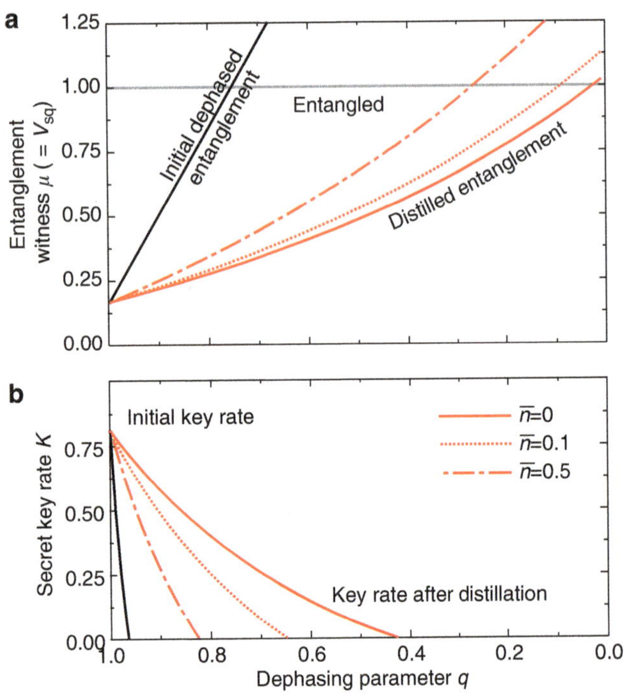

Figure 3 | Entanglement distillation of phase-diffused two-mode-squeezed states. Squeezing variance V_{sq} (**a**) and distillable secret key K (**b**) are plotted for the phase-diffused state (blue) and for the asymptotic distilled state for different $\bar{n}$ (red lines). $\bar{n}$ gives here the mean number of thermal photons of the state we project onto. The parameter q quantifies the strength of the phase diffusion: $q = 1$ indicates that no phase noise is present, whereas $q = 0$ corresponds to a completely randomized phase. The parameters of the input state before phase-diffusion read $a = 3.583$ and $b = 3.417$, which corresponds to variances of squeezed and anti-squeezed quadratures $V_{sq} = 1/6$ and $V_{antisq} = 7$.

Experimental setup. A schematic picture of the experimental setup is shown in Fig. 4. In a first step, we generated two Gaussian entangled light fields, which was achieved by superimposing two squeezed vacuum states on a balanced beam splitter with a 90° phase-shift (refs 6,34). Previous realizations of multicopy continuous-variable entanglement distillation protocols used the

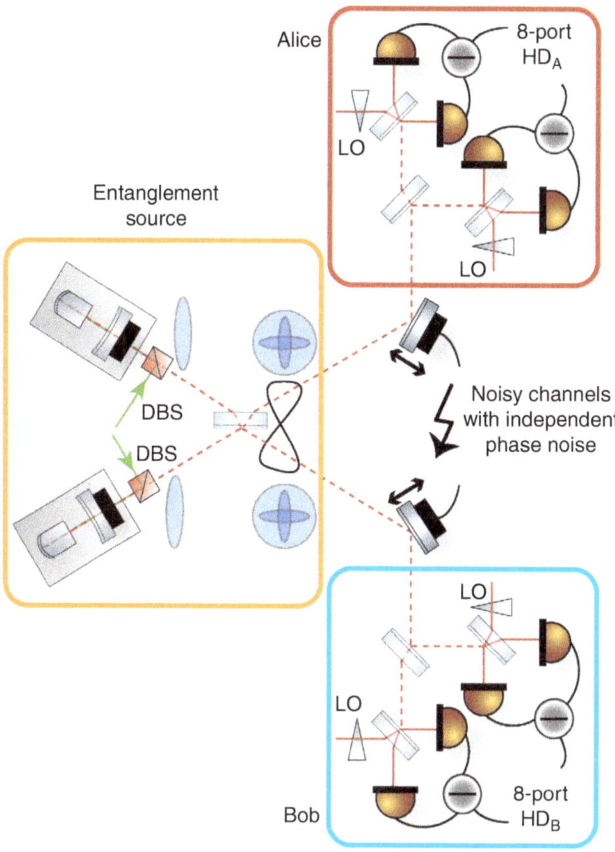

Figure 4 | Experimental setup. Two propagating squeezed-light modes are superimposed on a balanced beam splitter. The entangled outputs are sent to two separate sites. During transmission the modes are exposed to phase noise, which produces de-Gaussified mixed quantum states. For strong phase noise the two-mode squeezing, that is, the entanglement in the second moments of the quadratures, is completely destroyed. The decohered modes are continuously and unconditionally detected by eight-port homodyne detectors, which have negligible intrinsic phase noise and detection efficiencies of about 87 ± 6%. DBS, dichroic beam splitter.

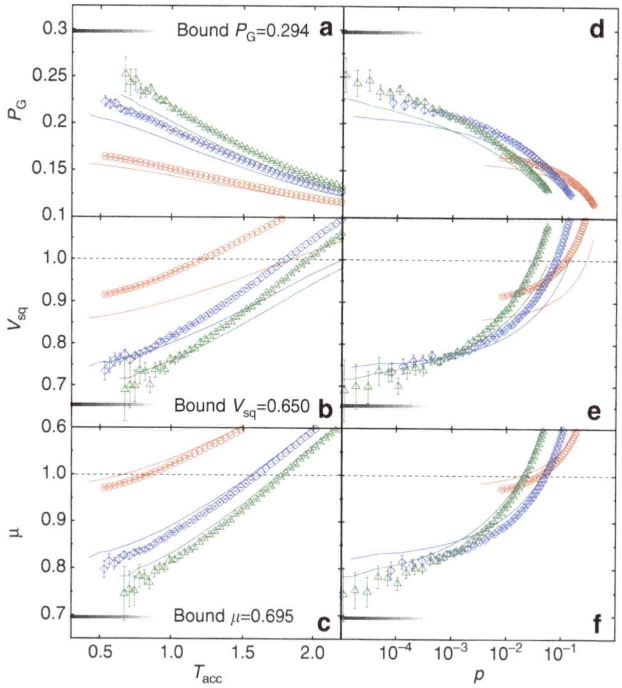

Figure 5 | Experimental distillation results. The panels **a-f** show the dependences of V_{sq}, μ and P_G on the threshold T_{acc} and success probability p for one (red, circle), two (blue, square) and three (green, triangle) iterations of the distillation protocol as discussed in the main text. The error bars give a 2σ confidence interval and are obtained by bootstrapping the samples 100 times where the length of each bootstrapping sample equals the length of the analysed data set. The solid lines give theoretical predictions obtained by Monte Carlo simulations for 10^{10} samples. These simulations use the measured undiffused covariance matrix (equation 8) and the experimentally determined phase-noise factor $q = 0.78$. The model takes into account additional losses induced by the implementation of our de-phasing channel, and, for simplicity, we assume an uncorrelated Gaussian distribution of the random phase shifts with equal variances. The general good agreement between simulation and experiment suggests that our model represents a good approximation. The discrepancy for V_{sq} can be explained by the fact that the actual phase noise does not precisely meet our assumptions. By a slight change of q we could get a good fit for V_{sq} at the expense of a worse fit for μ. The short fading horizontal lines represent the asymptotic limit for projection onto vacuum ($\bar{n} = 0$).

so-called v-class entangled states[12,15], which are simply achieved by superimposing a squeezed-vacuum mode and an ordinary vacuum mode on a balanced beam splitter. Replacing the vacuum mode by a second squeezed light field increases the experimental effort, but also allows, in principle, stronger entanglement, which is needed for QKD. Our scheme has the advantage that the (efficient) multistep iterative distillation of entanglement only requires at-most two squeezed light sources, whereas an actual hardware implementation of a (still inefficient) iterative distillation without quantum memories requires doubling the number of squeezed light sources with each iteration step. In our proof-of-principle experiment the squeezed-vacuum sources produced slightly different squeezing values, around 3 dB. Due to this asymmetry the covariance matrix of the entangled state did not hold the form given by equation (3) but was given by

$$\gamma_{AB,exp} = \begin{pmatrix} 3.20 & -0.13 & -2.90 & -0.04 \\ -0.13 & 6.24 & -0.03 & 6.08 \\ -2.90 & -0.03 & 3.70 & -0.06 \\ -0.04 & 6.08 & -0.06 & 6.83 \end{pmatrix}. \quad (8)$$

We reconstructed this covariance matrix by measuring the prepared state without additional phase diffusion with the eight-port homodyne detectors and using the formula $\gamma_{AB,exp} = 2\gamma_{EHD} - I$, where γ_{EHD} is a covariance matrix calculated directly from the eight-port homodyne data, and I denotes the identity matrix. The small values at the off-diagonal elements are not exactly zero due to imprecisions in the quadrature phases. The different values of the diagonal elements are mainly caused by the different squeezing values. The matrix $\gamma_{AB,exp}$ was reconstructed without correcting for detection inefficiencies and detector dark noise. For this reason, the covariance matrix is a good description of the actual measurement data (but a less good description of the quantum state before detection). We note that two-mode squeezing of >10 dB has already been demonstrated[34] and our scheme would work equally well with these higher squeezing strengths.

Our entangled modes were distributed between Alice and Bob's sites. During transmission, the modes were exposed to independent phase noise, which models a noisy transmission through, for example, optical fibres. The noise was applied by varying the position of steering mirrors in the path, driven by piezo electric transducers. We varied the strength of the noise by amplifying or attenuating the voltage given to the actuators to examine the behaviour of the protocol for different noise strengths. For all measurements shown in the next section we used the same phase diffusion which corresponded to $q = 0.78$. At the sites, the received light beams were absorbed in eight-port homodyne detectors, consisting of a balanced beam splitter and two conventional balanced homodyne detectors (BHDs), which measure the amplitude and phase quadrature on the first and the second beam, respectively. This measurement corresponds to a projection onto coherent states. The recorded data provided the basis to execute the emulated distillation protocol.

Experimental results. In the experiment, we collected up to 5×10^8 data points for a fixed strength q of phase noise, and we emulated up to three rounds of the iterative entanglement distillation protocol. To simplify data processing, we used a deterministic conditioning rule and the elementary entanglement distillation step was taken successful if $|\alpha_-| < T_{acc}$ and $|\beta_-| < T_{acc}$, where T_{acc} is a tunable threshold. While such conditioning may lead to some residual non-Gaussianity of the asymptotic state, it does not modify the qualitative properties of the distillation protocol. We have used the resulting data to reconstruct the covariance matrix γ of the two-mode state after each iteration of the distillation protocol, and we have used the covariance matrix to determine various properties of the distilled state. The results are plotted in Fig. 5, which shows the two-mode squeezing variance V_{sq}, the symplectic eigenvalue μ witnessing entanglement, and the quantity $P_G = 1/\sqrt{\det \gamma}$, which for Gaussian states coincides with the state purity.

Figure 5a–c illustrate that each iteration of the protocol increases squeezing and (Gaussian) entanglement of the distilled state for a fixed threshold T_{acc}, as indicated by the reduction of V_{sq} and μ. Furthermore, P_G increases with the number of iterations, which is a strong indication that the purity of the state is increased by the protocol. The absolute bounds presented in all panels are the asymptotic bounds for projection onto vacuum ($\bar{n} = 0$) calculated with equation (6). Statistical properties of the random phase shifts required for evaluation of the phase average in equation (6) were determined by comparing the covariance matrices of the initial and the de-phased states. It is remarkable how close the data come to the absolute bound with only three iteration steps and a finite threshold T_{acc}. Figure 5d–f show the practically more relevant dependence of V_{sq}, μ and P_G on the total success probability of the protocol p, which was determined as the ratio of the number of distilled copies of the state versus the total number of input copies of the state. These plots fully demonstrate the usefulness of an iterative distillation scheme. For a fixed total success probability, more iterations may be advantageous as they lead to better squeezing, entanglement and purity. The crossover between one and two iterations is clearly and unambiguously demonstrated by the data, while the crossover between two and three iterations can be seen in the region of $10^{-4} < P < 10^{-3}$, albeit the results in this region are already affected by statistical uncertainties. We have repeated the experiment for several different strengths of phase noise, and the results were in all cases qualitatively similar to those shown in Fig. 5.

We emphasize that the emulated iterative entanglement distillation presented here is inherently as efficient as if the protocol had been implemented with quantum memories, which would store successfully distilled states from previous rounds of the protocol to be utilized in the next rounds of the protocol. In the emulation, we naturally use in the subsequent rounds of the protocol only the successfully distilled states from previous rounds, which ensures that the average number of input copies that are consumed to produce a single distilled copy scales exactly as if quantum memories had been used.

Discussion

Our proof-of-principle experiment paves the way to efficient quantum communication protocols, where a specific data processing mimicks quantum memories (hence, provides the same advantage in terms of resources) without actually requiring them. This strategy can be applied at the end points of any entanglement distribution scheme where the end users perform eight-port homodyne detection. Importantly, our protocol is not limited to schemes where entanglement is physically distributed, but is also applicable to a prepare-and-measure quantum key distribution schemes where Alice prepares Gaussian-modulated coherent states and sends them to Bob who performs eight-port homodyne detection[35]. Indeed, since the preparation of coherent states by Alice is indistinguishable from the preparation of two-mode-squeezed vacuum followed by eight-port homodyne detection on Alice's mode, the preparation can be equivalently interpreted as a measurement on a (virtual) shared entangled state. Similarly, our procedure is also applicable to the recently proposed and demonstrated measurement-device independent continuous-variable QKD protocol[36,37], where Alice and Bob both prepare randomly modulated coherent states and send them to an untrusted relay which performs a continuous-variable Bell measurement on the two modes and publicly announces the results of the measurement. Given its wide range of potential applications, our proposal represents a promising tool for improving the performance of various quantum communication systems.

Methods

Squeezed-light preparation. The main light source was a Nd:YAG laser that produced a continuous-wave field at a wavelength of 1064 nm, as well as frequency-doubled light at 532 nm. The infrared light provided the control fields for the active-length stabilization of two squeezed-light resonators containing 7% magnesium-oxid-doped lithium-niobate crystals (MgO:LiNbO3), as well as the optical local oscillators for homodyne detection. The green light field was mode-matched into the squeezed-light resonators to pump degenerate type I parametric-downconversion processes, which provided resonator output modes at 1064 nm in squeezed-vacuum states. The squeezed-light resonators were singly resonant for 1064 nm, had a standing-wave and half-monolithic design, and a length of about 40 mm. The resonators' coupling mirrors were attached to piezo electric transducers to stabilize the cavity lengths on resonance using the Pound–Drever–Hall locking scheme[38,39]. An active temperature control stabilized the crystal temperatures at phase matching of the fundamental and the harmonic fields at about 60 °C.

Light detection and data acquisition. The two-mode (squeezed) state was detected with two eight-port homodyne detectors incorporating altogether four conventional BHDs. Each BHD consisted of a balanced beam splitter, two high-quantum efficiency PIN photodiodes and used a homodyne local oscillator power of about 3 mW at 1064 nm. The difference photo-electric current of the two photodiodes was amplified and transferred to a voltage by a trans-impedance amplifier. The voltage signal was then mixed with a 6.4 MHz electronic local oscillator, anti-alias filtered with a corner frequency of 400 kHz and finally synchronously sampled at a frequency of 1 MHz.

References

1. Weedbrook, C. *et al.* Gaussian quantum information. *Rev. Mod. Phys.* **84,** 621–669 (2012).
2. Kimble, H. J. The quantum internet. *Nature* **453,** 1023–1030 (2008).
3. Ekert, A. K. Quantum cryptography based on Bell's theorem. *Phys. Rev. Lett.* **67,** 661–663 (1991).
4. Scarani, V. *et al.* The security of practical quantum key distribution. *Rev. Mod. Phys.* **81,** 1301–1350 (2009).
5. Bouwmeester, D. *et al.* Experimental quantum teleportation. *Nature* **390,** 575–579 (1997).
6. Furusawa, A. *et al.* Unconditional quantum teleportation. *Science* **282,** 706–709 (1998).

7. Bowen, W. P *et al.* Experimental investigation of continuous variable quantum teleportation. *Phys. Rev. A* **67,** 032302 (2003).
8. Bennett, C. H. *et al.* Purification of noisy entanglement and faithful teleportation via noisy channels. *Phys. Rev. Lett.* **76,** 722–725 (1996).
9. Deutsch, D. *et al.* Quantum privacy amplification and the security of quantum cryptography over noisy channels. *Phys. Rev. Lett.* 77, 2818–2821 (1996).
10. Kwiat, P. G., Barraza-Lopez, S., Stefanov, A. & Gisin, N. Experimental entanglement distillation and 'hidden' non-locality. *Nature* **409,** 1014–1017 (2001).
11. Pan, J.-W., Gasparoni, S., Ursin, R., Weihs, G. & Zeilinger, A. Experimental entanglement purification of arbitrary unknown states. *Nature* **423,** 417–422 (2003).
12. Hage, B. *et al.* Preparation of distilled and purified continuous-variable entangled states. *Nat. Phys.* **4,** 915–918 (2008).
13. Dong, R. *et al.* Experimental entanglement distillation of mesoscopic quantum states. *Nat. Phys.* **4,** 919–923 (2008).
14. Takahashi, H. *et al.* Entanglement distillation from Gaussian input states. *Nat. Photon.* **4,** 178–181 (2010).
15. Hage, B., Samblowski, A., Diguglielmo, J., Fiurášek, J. & Schnabel, R. Iterative entanglement distillation: approaching the elimination of decoherence. *Phys. Rev. Lett.* **105,** 230502 (2010).
16. Kurochkin, Y., Prasad, A. S. & Lvovsky, A. I. Distillation of the two-mode squeezed state. *Phys. Rev. Lett.* **112,** 070402 (2014).
17. Simon, C. *et al.* Quantum memories. *Eur. Phys. J. D* **58,** 1–22 (2010).
18. Duan, L.-M., Lukin, M. D., Cirac, J. I. & Zoller, P. Long-distance quantum communication with atomic ensembles and linear optics. *Nature* **414,** 413–418 (2001).
19. Campbell, E. T., Genoni, M. G. & Eisert, J. Continuous-variable entanglement distillation and noncommutative central limit theorems. *Phys. Rev. A* **87,** 42330 (2013).
20. Ulanov, A. E. *et al.* Undoing the effect of loss on quantum entanglement. *Nat. Photon.* **9,** 764–768 (2015).
21. Fiurášek, J. & Cerf, N. J. Gaussian postselection and virtual noiseless amplification in continuous-variable quantum key distribution. *Phys. Rev. A* **86,** 060302 (2012).
22. Walk, N., Ralph, T. C., Symul, T. & Lam, P. K. Security of continuous-variable quantum cryptography with Gaussian postselection. *Phys. Rev. A* **87,** 020303 (2013).
23. Chrzanowski, H. M. *et al.* Measurement-based noiseless linear amplification for quantum communication. *Nat. Photon.* **8,** 333–338 (2014).
24. Browne, D., Eisert, J., Scheel, S. & Plenio, M. Driving non-Gaussian to Gaussian states with linear optics. *Phys. Rev. A* **67,** 062320 (2003).
25. Eisert, J., Browne, D. E., Scheel, S. & Plenio, M. B. Distillation of continuous-variable entanglement with optical means. *Ann. Phys.* **311,** 431–458 (2004).
26. Datta, A. *et al.* Compact continuous-variable entanglement distillation. *Phys. Rev. Lett.* **108,** 060502 (2012).
27. Campbell, E. T. & Eisert, J. Gaussification and entanglement distillation of continuous-variable systems: a unifying picture. *Phys. Rev. Lett.* **108,** 020501 (2012).
28. Duan, L., Giedke, G., Cirac, J. I. & Zoller, P. Inseparability criterion for continuous variable systems. *Phys. Rev. Lett.* **84,** 2722–2725 (2000).
29. Simon, R. Peres-Horodecki separability criterion for continuous variable systems. *Phys. Rev. Lett.* **84,** 2726–2729 (2000).
30. Grosshans, F. & Cerf, N. J. Continuous-variable quantum cryptography is secure against non-gaussian attacks. *Phys. Rev. Lett.* **92,** 047905 (2004).
31. Navascues, M., Grosshans, F. & Acin, A. Optimality of Gaussian attacks in continuous-variable quantum cryptography. *Phys. Rev. Lett.* **97,** 190502 (2006).
32. Garcia-Patron, R. & Cerf, N. J. Unconditional optimality of Gaussian attacks against continuous-variable quantum key distribution. *Phys. Rev. Lett.* **97,** 190503 (2006).
33. Lodewyck, J. *et al.* Quantum key distribution over 25km with an all-fiber continuous-variable system. *Phys. Rev. A* **76,** 042305 (2007).
34. Eberle, T., Händchen, V. & Schnabel, R. Stable control of 10 dB two-mode squeezed vacuum states of light. *Opt. Express* **21,** 11546–11553 (2013).
35. Weedbrook, C. *et al.* Quantum cryptography without switching. *Phys. Rev. Lett.* **93,** 170504 (2004).
36. Zhang, Y.-C. *et al.* Continuous-variable measurement-device-independent quantum key distribution using squeezed states. *Phys. Rev. A* **90,** 052325 (2014).
37. Pirandola, S. *et al.* High-rate measurement-device-independent quantum cryptography. *Nat. Photon.* **9,** 397–402 (2015).
38. Drever, R. *et al.* Laser phase and frequency stabilization using an optical resonator. *Appl. Phys. B* **31,** 97–105 (1983).
39. Black, E. D. An introduction to Pound-Drever-Hall laser frequency stabilization. *Am. J. Phys.* **69,** 79 (2001).

Acknowledgements

This research was partially supported by the Deutsche Forschungsgemeinschaft (project SCHN 757/5-1). J.F. acknowledges financial support from the EU FP7 under Grant Agreement No. 308803 (project BRISQ2) cofinanced by MSMT CR (7E13032). N.J.C. acknowledges financial support from the Fonds de la Recherche Scientifique (F.R.S.-FNRS) under grant T.0199.13 and from the Belgian Science Policy Office (BELSPO) under grant IAP P7-35 Photonics@be. D.A. acknowledges financial support from the DFG Graduate School 'Quantum mechanical noise in complex systems' (RTG1991). We thank Tobias Gehring and Aiko Samblowski for helpful discussions.

Author contributions

J.F. and N.J.C. developed the theory, D.A. and M.S. performed the experiment and all measurements under the supervision of R.S. The data were analysed by J.F. and D.A., and J.F., D.A., R.S. and N.J.C. wrote the paper.

Additional information

Competing financial interests: The authors declare no competing financial interests.

3

Evolutionary games of condensates in coupled birth–death processes

Johannes Knebel[1,*], Markus F. Weber[1,*], Torben Krüger[2,*] & Erwin Frey[1]

Condensation phenomena arise through a collective behaviour of particles. They are observed in both classical and quantum systems, ranging from the formation of traffic jams in mass transport models to the macroscopic occupation of the energetic ground state in ultra-cold bosonic gases (Bose–Einstein condensation). Recently, it has been shown that a driven and dissipative system of bosons may form multiple condensates. Which states become the condensates has, however, remained elusive thus far. The dynamics of this condensation are described by coupled birth–death processes, which also occur in evolutionary game theory. Here we apply concepts from evolutionary game theory to explain the formation of multiple condensates in such driven-dissipative bosonic systems. We show that the vanishing of relative entropy production determines their selection. The condensation proceeds exponentially fast, but the system never comes to rest. Instead, the occupation numbers of condensates may oscillate, as we demonstrate for a rock–paper–scissors game of condensates.

[1]Arnold Sommerfeld Center for Theoretical Physics and Center for NanoScience, Department of Physics, Ludwig-Maximilians-Universität München, Theresienstraße 37, 80333 München, Germany. [2]IST Austria, Am Campus 1, 3400 Klosterneuburg, Austria. * These authors contributed equally to this work. Correspondence and requests for materials should be addressed to E.F. (email: frey@lmu.de).

Condensation phenomena occur in a broad range of contexts in both classical and quantum systems. Networks such as the World Wide Web or the citation network perpetually grow by the addition of nodes or links and they evolve by rewiring. Over time, a finite fraction of the links of a network may be attached to particular nodes. These nodes become hubs and thereby dominate the dynamics of the whole network; they become condensate nodes[1–3]. Condensation also occurs in models for the jamming of traffic[4–7] and in related mass transport models in which particles hop between sites on a lattice[3,8,9]. A condensate forms when a finite fraction of all particles aggregates into a cluster that dominates the total particle flow. Bose–Einstein condensation, on the other hand, is a quintessentially quantum mechanical phenomenon. When an equilibrated, dilute gas of bosonic particles is cooled to a temperature near absolute zero, a finite fraction of bosons may condense into the energetic ground state[10–12]. Long-range phase coherence builds up and quantum physics becomes manifest on the macroscopic scale[13,14].

In both the classical and the quantum mechanical context, condensation occurs when one or multiple states become macroscopically occupied (they become condensates), whereas the other states become depleted[15,16]. However, the physical origins of condensation in the above examples differ from each other. Why and how condensation arises in a particular system remains a topic of general interest and vivid research.

Here we study condensation in two systems from different fields of research: incoherently driven-dissipative systems of non-interacting bosons and evolutionary games of competing agents. As we show below, the physical principle of vanishing entropy production governs the formation of condensates in both of these systems. The entities that constitute the respective system shall be called particles. They may be quantum or classical particles (bosons or agents). The dynamics of these particles eventually lead to the condensation into particular states (quantum states or strategies). Before describing the above two systems, we now introduce the mathematical framework of our study.

On an abstract level, we consider a system of S (non-degenerate) states $E_i, i = 1, \ldots, S$, each of which is occupied by $N_i \geq 0$ indistinguishable particles, see Fig. 1a. The configuration of the system at time t is fully characterized by the occupation numbers $\mathbf{N} = (N_1, N_2, \ldots, N_S)$. This configuration changes continuously in time due to the transition of particles between states. The total number of particles in this coupled birth–death process is conserved ($N = \sum_i N_i$). We are interested in the probability $P(\mathbf{N}, t)$ of finding the system in configuration $\mathbf{N}$ at time t. The temporal evolution of the probability distribution $P(\mathbf{N}, t)$ is governed by the classical master equation[17,18]:

$$\partial_t P(\mathbf{N}, t) = \sum_{\substack{i,j=1 \\ j \neq i}}^{S} \left(\Gamma_{i \leftarrow j}(N_i - 1, N_j + 1) P(\mathbf{N} - \mathbf{e}_i + \mathbf{e}_j, t) - \Gamma_{i \leftarrow j}(N_i, N_j) P(\mathbf{N}, t) \right), \tag{1}$$

where $\mathbf{e}_i \in \mathbb{Z}^S$ denotes the unit vector in direction i (equal to 1 at index i, otherwise 0). The rate for the transition of particles from state E_j to E_i depends linearly on the number of particles in the departure and in the arrival state:

$$\Gamma_{i \leftarrow j} = r_{ij}(N_i + s_{ij}) N_j\ , \tag{2}$$

with rate constant $r_{ij} \geq 0$ and constant $s_{ij} \geq 0$.

Condensation in this framework is understood as the macroscopic occupation of one or multiple states[15,16]: We consider a state E_i as a condensate when the long-time average of the number of particles in this state scales linearly with the system size ($\langle N_i \rangle_t \sim \mathcal{O}(N)$ for large t). Hence, a condensate

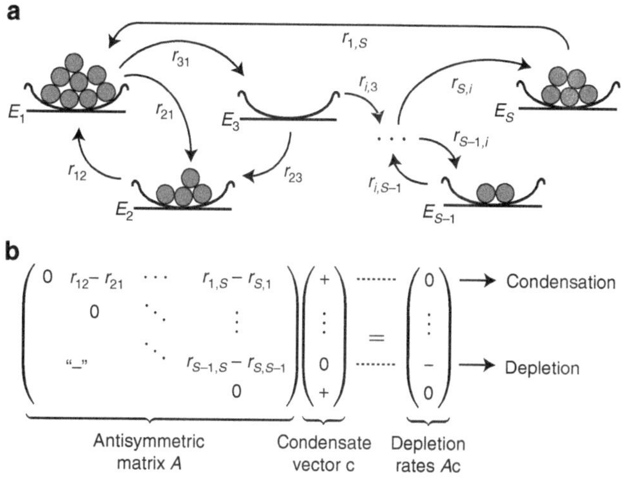

Figure 1 | Condensation into multiple states due to particle transitions between states and mathematics of condensate selection. (**a**) With respect to condensation in an incoherently driven-dissipative quantum system, each bowl represents a state E_i that is occupied by N_i non-interacting bosons (filled circles). If indicated by an arrow, bosons may undergo transitions from state E_j to state E_i at a rate $\Gamma_{i \leftarrow j} = r_{ij}(N_i + 1)N_j$, with rate constant r_{ij}. In the language of evolutionary game theory, the figure depicts the interaction of N_i agents (filled circles) playing strategies E_i (bowls) for $i = 1,\ldots,S$. An agent playing strategy E_j adopts strategy E_i at a rate $\Gamma_{i \leftarrow j} = r_{ij}N_iN_j$. The above rate of bosonic condensate selection is recovered if agents may also spontaneously mutate from E_j to E_i at a rate r_{ij}. (**b**) A condensate vector c for an antisymmetric matrix A has two properties: its entries are positive for indices for which Ac is zero, and they are zero for indices for which Ac is negative ("-" signifies the antisymmetry of matrix A). Temporal evolution of the relative entropy of the condensate vector to the state concentrations under the ALVE (3) relates positive entries of the condensate vector to condensates, and its zero entries to depleted states. Generically, positive entries of c represent the asymptotic temporal average of oscillating condensate concentrations according to the ALVE (3), and negative entries of Ac represent depletion rates.

harbours a finite fraction of the total number of particles for large systems ($N \gg 1$). We refer to a state as depleted when its average occupation number scales less than linearly with the system size. Therefore, the fraction of particles in a depleted state vanishes in the limit of large systems.

Depending on the values of the rate constants r_{ij}, numerical simulation of equations (1) with rates (2) reveals that all states, multiple states or only one state become condensates when the particle density N/S is large enough to detect condensation[19]. Thus far, various questions about condensation have remained elusive for the coupled birth–death process defined by equation (1): Which of the states become condensates? How does this selection of condensates proceed? Is it possible to construct systems that condense into a specific set of condensates?

In the following, we answer these questions by illuminating the physical principle that governs the formation of multiple condensates on the leading-order timescale. We show that the vanishing of relative entropy production determines the selection of condensates (see equations (3) and (4) below). We elaborate how condensate selection is determined by the rate constants r_{ij}. The condensation proceeds exponentially fast into a dynamic, metastable steady state within which the occupation numbers of condensates may oscillate. By applying our general results to systems with many states, we show that the interplay between critical properties of such networks of states[20] and dynamically stable network motifs[21] determines the selection of condensates. The results of our analysis apply to any system whose dynamics are described by the coupled birth–death processes (1) with rates (2). Before proceeding to the mathematical and numerical analysis of condensation in these processes, we now give a brief overview of such systems.

Results

Non-interacting bosons in driven-dissipative systems. The classical master equation (1) has recently been derived by Vorberg *et al.*[19] in the study of bosonic systems that are dissipative and driven by external sources. For a system of non-interacting bosons that is weakly coupled to a reservoir and driven by an external time-periodic force (a so-called Floquet system)[22–24], one can eliminate the reservoir degrees of freedom (Born and Markov approximation)[25,26] and the density matrix of the system becomes diagonal (see the Supplement of the work of Vorberg *et al.*[19]). The effective dynamics of the bosons become incoherent and are captured on a macroscopic level in terms of the coupled birth–death processes (1) with rates $\Gamma_{i\leftarrow j}=r_{ij}(N_i+1)N_j$ (that is all $s_{ij}=1$ in the rates (2)). These non-equilibrium set-ups may not only lead the bosons into a single, but also into multiple condensates[19].

For the incoherently driven-dissipative systems described above, the state E_i denotes a time-dependent Floquet state[22–24]. The total rate $\Gamma_{i\leftarrow j}$ for the transition of a boson from state E_j to E_i depends linearly on the number of bosons in the departure state (N_j) and the arrival state (N_i+1). The latter factor stems from the indistinguishability of bosons and reflects their tendency to congregate. Although we refer to equation (1) as a classical master equation and coherence does not build up, the quantum statistics of bosons is still encoded in the functional form of $\Gamma_{i\leftarrow j}$. The rate constant r_{ij} is determined by the microscopic properties of the system and the reservoir.

Condensation in the above set-up is to be distinguished from Bose–Einstein condensation. Typically, studies on Bose–Einstein condensation focus on the existence of long-range phase coherence in thermal equilibrium[10–15], its kinetic formation[13,14,27–32] and the fragmentation of a coherent condensate into multiple condensates (for example, when the equilibrium ground state is degenerate)[13,16]. In contrast, the classical birth–death processes (1) with rates (2) describe condensation in bosonic systems that are externally driven by a continuing supply of energy, dissipate into the environment and exhibit decoherence.

Equations of type (1) may also arise in atomic physics and quantum optics and are known as Pauli master equations[33–35]. They describe how the population of S non-degenerate energy levels changes over time when a system harbours N indistinguishable, non-interacting bosonic atoms. Such changes may occur by interactions with a radiation field that induces transitions between energy levels. A theoretical description of these transitions in terms of a Pauli master equation is appropriate if coherence is negligible. As in the previous example, the system then approaches a state in which some of the energy levels are macroscopically occupied (condensates), whereas the others are depleted. More generally, whenever a rate constant r_{ij} governs the transition of a single boson from a state E_j to E_i, the rates (2) with $s_{ij}=1$ for all i and j apply if N non-interacting bosons are brought into the system[19].

Strategy selection in evolutionary game theory. The classical master equation (1) also occurs in evolutionary game theory (EGT). Historically, EGT was developed to study the evolutionary processes that are driven by selection and mutation[36,37] and seeks to identify optimal strategies for competitive interactions. For example, EGT has been applied in the study of the prominent 'rock–paper–scissors' (RPS) game, which was proposed as a facilitator of species coexistence and has inspired both experimental and theoretical research[38–42]. Furthermore, the 'prisoner's dilemma' game serves as a paradigmatic model to explore the evolution and maintenance of cooperation[43,44]. The interplay between non-linear and stochastic effects underlies the dynamics of such evolutionary games[45–50].

In EGT, one typically considers a system of N interacting agents (classical particles) who repeatedly play one fixed strategy E_i out of the S possible choices $E_1, E_2, \ldots, E_S$. In each succeeding interaction, the defeated agent adopts the strategy of its opponent. Since N_j agents playing strategy E_j can potentially be defeated by one of the N_i agents playing strategy E_i, the rate of change is $\Gamma_{i\leftarrow j}=r_{ij}N_iN_j$. If an agent who plays E_j can also spontaneously mutate into an agent who plays E_i (with rate $\mu_{ij}=r_{ij}s_{ij}$), one recovers the classical master equation (1) with rates (2).

Thus, there exists a correspondence between condensation in incoherently driven-dissipative bosonic systems and strategy selection in EGT—the transition of bosons between states can be interpreted in terms of the interaction and mutation of agents employing evolutionary strategies. In effect, the states in an incoherently driven-dissipative set-up play an evolutionary game and the winning states form the condensates.

After having introduced the above examples, we now proceed with the mathematical and numerical analysis of the classical master equation (1). We show that the dynamics of condensation change on two distinct timescales. At the leading-order timescale, the dynamics are described by a set of non-linearly coupled, ordinary differential equations (see equation (3) below), which determine the states that become condensates. We identify these states by applying concepts from EGT. After an exposition of the physical principles that underlie the condensation dynamics, implications of our general results for incoherently driven-dissipative systems are discussed.

The antisymmetric Lotka–Volterra equation. The total number of particles needed for condensation phenomena to occur is large ($N \gg 1$). To detect macroscopic occupancies, it is also assumed that the particle density N/S is large. Therefore, one may approximate the classical master equation (1) by a Langevin equation for the state concentrations $x_i(t)=N_i(t)/N$ (details of the derivation are provided in Supplementary Note 1). Originally proposed for Brownian particles suspended in a liquid, the Langevin equation decomposes the dynamics of a sample trajectory of the random process into two contributions—into a deterministic drift and into noise stemming from the discreteness of particle numbers ('demographic fluctuations'). Both the demographic fluctuations and the contribution to the deterministic drift that corresponds to mutations in the EGT setting are suppressed by a small prefactor $1/N$. Therefore, these terms change the dynamics only slowly. The deterministic drift that corresponds to interactions between agents is, however, not suppressed. It thus governs the dynamics to leading order.

Hence, we find that the leading-order dynamics of the condensation process (1)–(2) are described by the differential equations:

$$\frac{\mathrm{d}}{\mathrm{d}t}x_i = x_i(A\mathbf{x})_i\,. \tag{3}$$

The matrix A is antisymmetric and encodes the effective transition rates between states ($a_{ij}=r_{ij}-r_{ji}$). The constants s_{ij} that occur in the definition of the rates (2) do not change the leading-order dynamics, but they become relevant on subleading-order timescales.

We refer to equation (3) as the antisymmetric Lotka–Volterra equation (ALVE). It provides a description of pairwise interactions that preserve the total number of particles. Therefore, the

Table 1 | Condensation processes described by the ALVE in different fields of research.

Field of research	Entity	State	Process	Dynamics
Quantum physics[19]	Boson	Quantum state	Incoherent transition	Condensation/depletion
Evolutionary game theory[59,60]	Agent	Strategy	Game (+mutation)	Win/loss
Population dynamics[47,51-53]	Individual	Species	Competition (+mutation)	Survival/extinction
Chemical kinetics[56-58]	Molecule	Chemical species	Reaction (+conversion)	Production/consumption
Plasma physics[54,55]	Plasmon	Jet	Scattering	Increase/decrease

ALVE, antisymmetric Lotka-Volterra equation.
The ALVE (3) governs condensation processes in diverse fields of research. For example, for incoherently driven-dissipative bosonic systems, the ALVE describes condensation and depletion of states by incoherent transitions of non-interacting bosons. In EGT, the ALVE occurs in the context of winning and losing strategies played by agents.

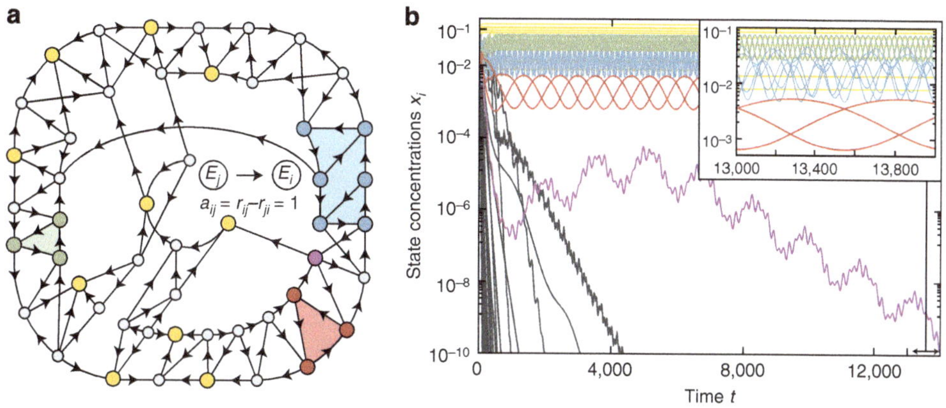

Figure 2 | Fragmentation of an exemplary system into multiple condensates with oscillating state concentrations. (**a**) Randomly sampled network of 50 states. Disks represent states. An arrow from state E_j to state E_i represents an effective rate constant $a_{ij} = r_{ij} - r_{ji} = 1$ (a missing arrow indicates a forbidden transition with $a_{ij} = 0$). Computation of a condensate vector c predicted relaxation into 10 isolated condensates (yellow), one interacting subsystem with six condensates (blue) and two RPS cycles (red and green). All other states become depleted. The complete network also comprises RPS cycles of which some states become depleted. Knowledge of the network topology alone is thus insufficient to determine condensates. (**b**) Temporal evolution of state concentrations x_i (logarithmic scale). Colours in accordance with **a**. Numerical integration of the ALVE (3) confirmed the selection of states based on the condensate vector c. Subsystems with six (blue) and three (red and green) condensates exhibit oscillations of concentrations with non-vanishing particle flow. Depletion of states occurs exponentially fast. Identifying condensates from a condensate vector c is more reliable than through numerical integration: The concentration of the state associated to the purple disk in **a** decays exponentially to a concentration of 1.5×10^{-7} before recovering transiently. Numerical integration cannot rule out permanent recovery at later times. Supplementary Fig. 2 demonstrates such a case.

ALVE finds a broad range of applications in diverse fields of research, in addition to the aforementioned condensation of bosons far from equilibrium. It was first studied by Volterra[51] in the context of predator–prey oscillations in population biology[47,52,53]. In plasma physics, the ALVE describes the spectra of plasma oscillations (Langmuir waves)[54,55], and in chemical kinetics it captures the dynamics of bimolecular autocatalytic reactions[18,56–58]. In EGT, the ALVE is known as the replicator equation of zero-sum games such as the RPS game[47,59–61]. Table 1 summarizes all of the above analogies.

Despite the simple structure of the ALVE, it exhibits a rich and complex behaviour. In the following, we show how the mathematical analysis of the ALVE explains condensation into multiple states (condensate selection). To this end, we extend an approach for the analysis of the ALVE that was introduced in the context of EGT[59,60].

Production of relative entropy and condensate selection. Our analysis starts from a theorem in linear programming theory[62]. Given an antisymmetric matrix A, it is always possible to find a vector **c** that fulfils the following conditions: its entries are positive for indices in $I \subseteq \{1, \ldots, S\}$ and zero for indices in $\bar{I} = \{1, \ldots, S\} - I$, whereas the entries of $A\mathbf{c}$ are zero for indices in I and negative for indices in $\bar{I}$ (Fig. 1b). Although several vectors **c** with these properties may exist, the index set I is unique and, thus, determined by the antisymmetric matrix A. Finding such a 'condensate vector' **c** is crucial for the understanding of condensate selection and of the condensation dynamics. The condensate vector has the following physical interpretations.

All condensate vectors yield fixed points of the ALVE (3). Because of the antisymmetry of matrix A, a linear stability analysis of these fixed points does not yield insight into the global dynamics (Supplementary Note 1). However, the global stability properties can be inferred by showing that the relative entropy of a condensate vector to the state concentrations,

$$D(\mathbf{c}||\mathbf{x}) = \sum_{i \in I} c_i \log(c_i/x_i), \qquad (4)$$

is a Lyapunov function (note that we do not consider the relative entropy of the state concentrations to the condensate vector, but define the relative entropy vice versa). The relative entropy (4) decreases with time and is bounded from below (see Methods and Supplementary Fig. 1). Therefore, the dynamics relax to a subsystem in which the relative entropy production is zero. The relaxation of relative entropy production is reminiscent of Prigogine's study of open systems in non-equilibrium thermodynamics. Indeed, we find that the system, to cite Prigogine's phrase, 'settles down to the state of least dissipation'[63].

This state of least dissipation is characterized even further by the condensate vector **c**. Considering the definition of the relative entropy (4) and its boundedness, it follows that every concentration x_i with $i \in I$ remains larger than a positive constant. On the

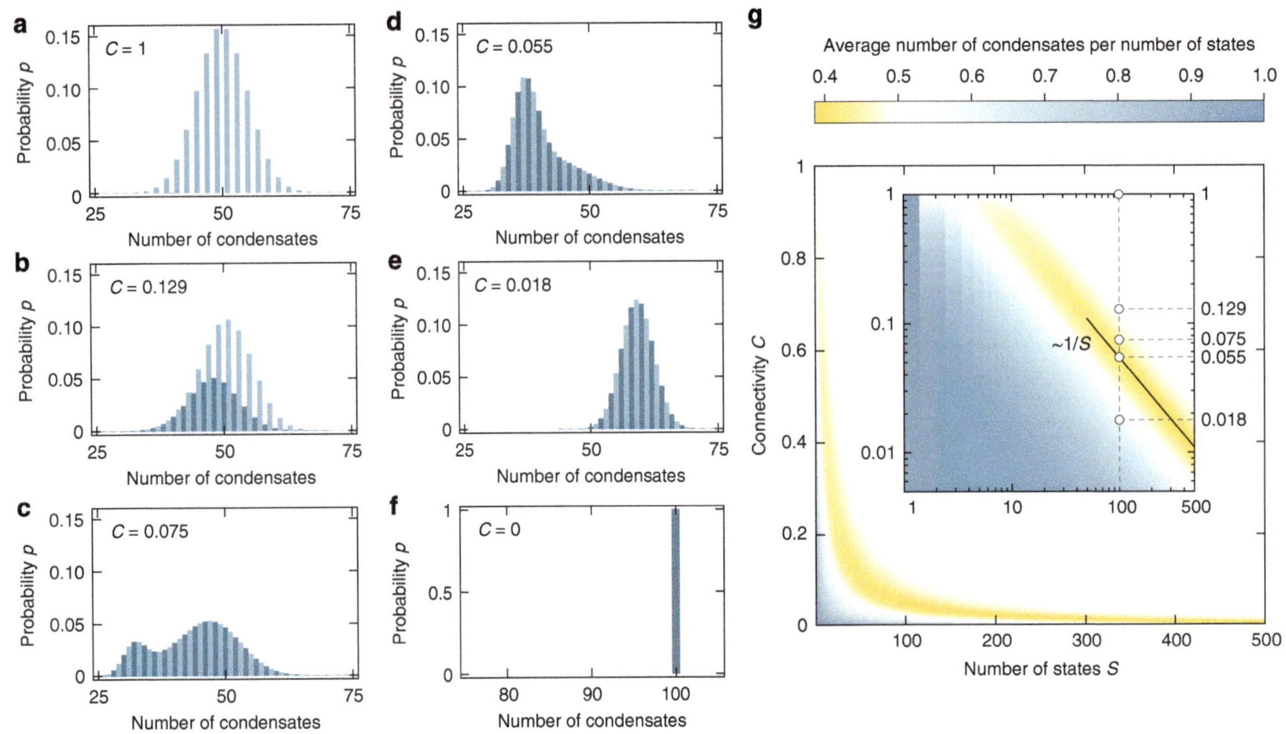

Figure 3 | Dependence of the number of condensates on the connectivity of states in random networks. (**a-f**) Measured probability p of finding a particular number of condensates for a system with $S=100$ states and with connectivity C (5×10^6 systems analysed per histogram). The connectivity specifies the percentage of states between which transitions of particles occur with a non-zero effective rate constant $a_{ij}=r_{ij}-r_{ji}$. Effective rate constants a_{ij} were sampled from a Gaussian distribution (zero mean, unit variance). (**a**) At full connectivity, the distribution is pseudobinomial with only odd numbers of condensates ($C=1$; light blue bars)[19,60,66]. (**b**) As the connectivity is reduced, even numbers of condensates become possible when systems decouple into even numbers of subsystems ($C=0.129$; dark blue bars). (**c**) The distribution exhibits bimodality ($C=0.075$) and (**d**) approaches a minimal average number of 40.2 condensates ($C=0.055$). (**e**) This average subsequently increases ($C=0.018$) because isolated states are trivially selected as condensates ($C=0$) as shown in **f**. (**g**) Average number of condensates per number of states (colour coded) plotted against the number of states S and the connectivity C (log-log graph in inset; $\geq 10^4$ systems per data point, see Supplementary Fig. 4 for the reliability of the linear programming algorithm). White circles correspond to distributions shown in (**a-e**). The minimal relative number of condensates conforms to the power law $C \sim 1/S^{\gamma}$ with $\gamma=0.998 \pm 0.008$ (s.e.m.) and can be related to the criticality of random networks[20].

other hand, states with indices in $\bar{I}$ become depleted for long times (see Methods). Therefore, we find that the condensate vector determines the selection of condensates. Positive entries of **c** correspond to states that become condensates, whereas zero entries of **c** correspond to states that become depleted. Both the set of condensates and the set of depleted states are unique (Fig. 1b) and independent of the initial conditions. Generically, the entries of the condensate vector are also unique upon normalization (its entries sum up to one) and yield the rate $|(A\mathbf{c})_i|$ at which a state E_i becomes depleted. The condensate selection occurs exponentially fast (see Fig. 2 and Methods).

After relaxation, the dynamics of the system are restricted to the condensates. In other words, the condensates form the attractor of the dynamics. However, the dynamics in this subsystem do not come to rest. The state of least dissipation is a dynamic state with a perpetually changing number of particles in the condensates—periodic, quasiperiodic and non-periodic oscillations are observed (Fig. 2b and Supplementary Fig. 1). In the generic case, the entries of the condensate vector represent the temporal average of condensate concentrations according to the ALVE (3). After condensate selection, the dynamics of these active condensates take place on a high-dimensional, deformed sphere[61].

An algebraic algorithm to find the condensates. Numerical integration of the ALVE (3) is neither a feasible nor a reliable method for identifying condensates (Fig. 2, Supplementary Figs 1 and 2). Instead, we determine these states by numerically searching for a condensate vector **c**. To this end, we reformulate the above conditions on **c** in terms of two linear inequalities[62]:

$$A\mathbf{c} \leq 0 \text{ and } \mathbf{c} - A\mathbf{c} > 0 \,. \tag{5}$$

We solve these inequalities with a linear programming algorithm that is both reliable and efficient. The time to find a condensate vector scales only polynomially with the number of states S (see Supplementary Fig. 3 and Methods for details).

Condensation in large random networks of states. We used our combined analytical and numerical approach to study how the connectivity of a random network of states affects the selection of condensates under the dynamics of the ALVE (3). The connectivity specifies the percentage of states between which particle transitions occur[20,64]. After having generated a network with a given connectivity, the strength and direction of an allowed transition between states E_j and E_i were determined by randomly sampling the corresponding effective rate constant $a_{ij}=r_{ij}-r_{ji}$.

Our results for condensation in large random networks of states are summarized in Fig. 3. When the connectivity of a network is zero, all of its states are isolated. Particles are not exchanged between states and none of the states becomes depleted. For an increased connectivity, isolated pairs of states are sampled in a random network. One state in an isolated pair is always depleted and the average number of condensates decreases

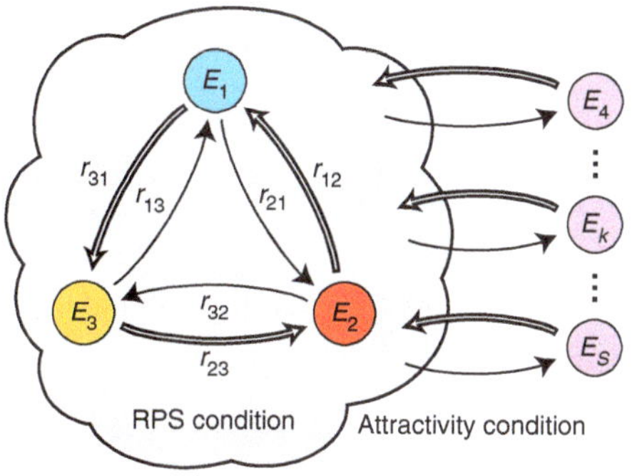

Figure 4 | Conditions for the emergence of a RPS cycle of condensates. Three particular states E_1, E_2 and E_3 (blue, red and yellow disks) of a network condense into a RPS cycle if, and only if, two conditions are fulfilled: First, the 'RPS condition' requires that the rate constants r_{ij} between the three states form a RPS network: $r_{i-1,i+1} > r_{i+1,i-1}$ (indices are counted modulo 3, for example, $r_{42} = r_{12}$ (framed arrows denote rate constants that are larger than rate constants for the respective reverse direction). Differences between these rate constants define the entries $c_i = r_{i-1,i+1} - r_{i+1,i-1}$ of an admissible condensate vector c. Second, the 'attractivity condition' requires that the weighted sum of rates from any exterior state E_k (purple disks) into the RPS cycle, $\sum_{j=1}^{3} c_j r_{jk}$ (framed arrows), is larger than the weighted sum of outbound rates, $\sum_{j=1}^{3} c_j r_{kj}$ (black arrows). In other words, the inflow of particles into the RPS cycle from any exterior state needs to be greater than the outflow to that state.

rapidly. On approaching a critical connectivity, cycles and trees of all orders become embedded in a random network. This critical connectivity scales inversely with the number of states[20]. We observe that, under the dynamics of the ALVE (3), the average number of condensates becomes minimal for a connectivity that also scales inversely with the number of states (see Fig. 3g). We attribute this minimum to the interplay between the criticality of random networks and condensate selection on connected components of the network[61,65]. Embedded directed cycles are a recurring motif[21] in the remaining network of condensates after condensate selection. Above the critical connectivity, a single giant cluster is formed. On average, half the number of states in this giant cluster become condensates once the network is fully connected ($C = 1$; Fig. 3a)[19,60,66]. Thus, our analysis emphasizes the importance of critical properties of random networks for condensate selection.

Design of active condensates. Our understanding of the condensate selection can be used to design systems that condense into a particular network of states, a game of condensates. We exemplify this procedure by formulating conditions under which a system relaxes into a RPS game of condensates[40]. Three particular states E_1, E_2 and E_3 in a system become a RPS cycle of condensates if, and only if, the following two conditions are fulfilled (Fig. 4). First, the 'RPS condition' requires that the rate constants between the three states form a RPS network (for example, $r_{12} > r_{21}$, $r_{23} > r_{32}$ and $r_{31} > r_{13}$). Second, the 'attractivity condition' requires that the inflow of particles into the RPS cycle from any other state E_k is greater than the outflow to that state E_k (for all $k = 4, \ldots, S$). The values of the rate constants between the states that become depleted are irrelevant. More complex games of condensates can be designed by formulating similar conditions on the rate constants. These conditions are formulated as inequalities that depend on Pfaffians of the antisymmetric matrix A and its submatrices (see Methods)[61]. The flow of particles between states in these systems causes condensate concentrations to oscillate (Fig. 2b).

Discussion

Our findings thus suggest intriguing dynamics of condensates in systems whose temporal evolutions are captured by the classical master equation (1) with rates (2), for example, in driven-dissipative systems of non-interacting bosons. Condensates observed on the leading-order timescale are metastable. For longer times, relaxation into a steady state occurs[19,67]. When detailed balance is broken in the system of condensates, the net probability current between at least two states does not vanish and a non-equilibrium steady state is approached[68,69]. The simplest way of designing such condensates is illustrated by the above RPS game. In this game, detailed balance is broken, for example, when the transition of particles is unidirectional (with totally asymmetric rate constants $r_{12} > r_{21} = 0$, $r_{23} > r_{32} = 0$, and $r_{31} > r_{13} = 0$). For non-interacting bosons in driven-dissipative systems, the continuing supply with energy through the external time-periodic driving force (Floquet system) and the dissipation of energy into the environment may, therefore, prevent the system from reaching equilibrium. How such systems may be realized in an experiment poses an interesting question for future research.

The transition of particles between condensates in the here-studied coupled birth–death processes parallels the interaction and mutation of winning agents in evolutionary game theory, reflecting an 'evolutionary game of condensates'. Our results suggest the possibility of creating novel bosonic systems with an oscillating occupation of condensates. Non-interacting bosons in incoherently driven-dissipative systems are promising candidates. Since the antisymmetric Lotka-Volterra equation also arises in population biology, chemical kinetics and plasma physics, all of our mathematical results apply to these fields as well.

Methods

Asymptotics of the ALVE. The asymptotic behaviour of the ALVE (3) can be characterized as follows: for every antisymmetric matrix A there exists a unique subset of states $I \subseteq \{1, \ldots, S\}$ whose concentrations stay away from zero for all times, that is,

$$x_i(t) \geq Const(A, \mathbf{x}_0) > 0 \text{ for all } t \geq 0 \text{ and for every } i \in I\,. \tag{6}$$

The set I is the set of condensates. All of the other states with indices in $\bar{I} = \{1, \ldots, S\} - I$ become depleted as $t \to \infty$, that is,

$$x_i(t) \to 0 \text{ as } t \to \infty \text{ for every } i \in \bar{I}\,. \tag{7}$$

The set of condensates can be determined algebraically from the antisymmetric matrix A and does not depend on the initial conditions $\mathbf{x}_0 \in \Delta_{S-1} = \{\mathbf{x} \in \mathbb{R}^S \mid x_i > 0 \text{ for all } i, \sum_{i=1}^{S} x_i = 1\}$.

To show this result, the time-dependent entropy $D(\mathbf{c}||\mathbf{x})(t)$ of a condensate vector $\mathbf{c} = (c_1, \ldots, c_n) \in \bar{\Delta}_{S-1}$ ($c_i \geq 0$ for all i and $\sum_i c_i = 1$) relative to the trajectory $\mathbf{x}(t)$ is considered (that is, the Kullback-Leibler divergence of $\mathbf{x}(t)$ from $\mathbf{c}$), see equation (4). A condensate vector is defined via the properties (see Fig. 1b):

$$c_i > 0 \text{ and } (A\mathbf{c})_i = 0 \quad \text{for } i \in I, \text{ and} \tag{8}$$

$$c_i = 0 \text{ and } (A\mathbf{c})_i < 0 \quad \text{for } i \in \bar{I}\,. \tag{9}$$

Such a vector can always be found for an antisymmetric matrix[62]. Notably, the index set I is unique although more than one condensate vector may exist.

Considering the time derivative of the relative entropy $D(\mathbf{c}||\mathbf{x})(t)$ and employing equations (3) and (8) yields:

$$\begin{aligned}\frac{\mathrm{d}}{\mathrm{d}t} D(\mathbf{c}||\mathbf{x})(t) &= -\sum_{i=1}^{S} c_i \frac{\partial_t x_i}{x_i} = -\sum_{i=1}^{S} c_i (A\mathbf{x})_i = \sum_{i=1}^{S} (A\mathbf{c})_i x_i \\ &= \sum_{i \in \bar{I}} (A\mathbf{c})_i x_i\,. \end{aligned} \tag{10}$$

Since $(A\mathbf{c})_{\bar{I}} < 0$ and $\mathbf{x} > 0$, it follows that $\partial_t D(\mathbf{c}||\mathbf{x})(t) < 0$ (please note the overbars in subscripts, which may be lost when read in low resolution). Therefore, the relative entropy $D(\mathbf{c}||\mathbf{x})$ is a Lyapunov function if $\mathbf{c}$ is chosen in accordance with equations (8) and (9). Moreover, $D(\mathbf{c}||\mathbf{x})$ is bounded from above by $D(\mathbf{c}||\mathbf{x})(0)$ and from below by zero for all times. This can be seen from the definition of D, and from the integration of equation (10) (using that $(A\mathbf{c})_{\bar{I}}<0$ and $\mathbf{x}>0$):

$$0\leq D(\mathbf{c}||\mathbf{x})(t)=D(\mathbf{c}||\mathbf{x})(0)+\int_0^t ds\sum_{i\in\bar{I}}(A\mathbf{c})_i x_i(s)\leq D(\mathbf{c}||\mathbf{x})(0)\,. \quad (11)$$

From the definition of the relative entropy in equation (4), it follows that every concentration x_i with $i\in I$ remains larger than a positive constant, that is, $x_i(t)\geq Const(A,\mathbf{x}_0)>0$ for all times t (if $x_i(t)\to 0$ for $i\in I$, it follows that $D\to\infty$, which contradicts the boundedness of D).

Furthermore, equation (11) implies that,

$$-\int_0^t \mathrm{d}s\,(A\mathbf{c})_i x_i(s)\leq -\int_0^t \mathrm{d}s\sum_{i\in\bar{I}}(A\mathbf{c})_i x_i(s)\leq D(\mathbf{c}||\mathbf{x})(0)\,, \quad (12)$$

for every $i\in\bar{I}$ and for all t. Therefore, concentration x_i is integrable for every $i\in\bar{I}$ ($x_i\in L^1(0,\infty)$) with the bound:

$$0<\int_0^\infty \mathrm{d}s\,x_i(s)\leq\frac{D(\mathbf{c}||\mathbf{x})(0)}{-(A\mathbf{c})_i}=Const(A,\mathbf{x}_0)\quad\text{for every } i\in\bar{I}\,. \quad (13)$$

Since the derivative of the concentrations is bounded from above and below, $|\partial_t x_i|=|x_i(A\mathbf{x})_i|\leq\|(A\mathbf{x})\|_\infty\leq\|A\|_{\infty\to\infty}\leq\mathrm{Const}(A)$, one concludes that x_i is uniformly continuous ($\|A\|_{\infty\to\infty}$ denotes the operator norm of A induced by the maximum norm on $\mathbb{R}^S$). Together with the integrability (13), it follows that the states with indices in $\bar{I}$ become depleted as $t\to\infty$, that is, $x_i(t)\to 0$ for $i\in\bar{I}$.

In conclusion, given an antisymmetric matrix $A=R-R^T$ via a rate constant matrix $R=\{r_{ij}\}_{i,j}$, one finds a condensate vector $\mathbf{c}$ that satisfies inequalities (8)–(9). The index set I, for which the entries of $\mathbf{c}$ are positive, represents condensates. The index set $\bar{I}$, for which entries of $\mathbf{c}$ are zero, represents states that become depleted. Moreover, equation (10) implies that the relative entropy becomes a conserved quantity in the subsystem of condensates (Supplementary Fig. 1).

Temporal average of condensate concentrations. The ALVE (3) is solved implicitly by,

$$x_i(t)=x_i(0)e^{t\cdot(A\langle\mathbf{x}\rangle_t)_i}\,, \quad (14)$$

with the time average of the trajectory $\langle\mathbf{x}\rangle_t$ defined as:

$$\langle\mathbf{x}\rangle_t=\frac{1}{t}\int_0^t \mathrm{d}s\,\mathbf{x}(s)\,. \quad (15)$$

It is shown above that $0<Const(A,\mathbf{x}_0)\leq x_i(t)\leq 1$ holds for the states that become condensates ($i\in I$). By rearranging equation (14), one thus obtains:

$$|(A\langle\mathbf{x}\rangle_t)_i|\leq\frac{1}{t}\left|\log\left(\frac{x_i(t)}{x_i(0)}\right)\right|\leq\frac{Const(A,\mathbf{x}_0)}{t}\quad\text{for all } i\in I\,. \quad (16)$$

Note that $Const$ is used to denote arbitrary positive, time-independent constants. Therefore, the right-hand side of equation (16) vanishes for $t\to\infty$. On the other hand, x_i is integrable for $i\in\bar{I}$ (equation (13)). Thus, the corresponding component of the time average converges to zero,

$$\langle x_i\rangle_t\leq\frac{Const(A,\mathbf{x}_0)}{t}\to 0\text{ as } t\to\infty\text{ for every } i\in\bar{I}\,. \quad (17)$$

Hence, the distance of the time average $\langle\mathbf{x}\rangle_t$ to the kernel of the antisymmetric submatrix A^I converges to zero (the submatrix A^I corresponds to the system of condensates with indices in I).

Structure of a generic antisymmetric matrix. For systems with an even number of states S, the antisymmetric matrix $A=R-R^T$ generically has a trivial kernel, whereas for systems with an odd number of states, the kernel of A is generically one dimensional. A higher dimensional kernel of A only occurs if the matrix entries are tuned[52,61,70]. As a consequence, when all of the entries above the diagonal of A are, for example, randomly drawn from a continuous probability distribution (for example from a Gaussian distribution), all 2^S submatrices of A have a kernel with dimension of less than or equal to one.

The projection of $\mathbf{x}\in\mathbb{R}^S$ to the subspace $\mathbb{R}^{(J)}\subseteq\mathbb{R}^S$ for an arbitrary index set $J\subseteq\{1,\ldots,S\}$ is defined as $\mathbf{x}_J:=P_J\mathbf{x}:=(x_j)_{j\in J}$. In other words, entries of $\mathbf{x}_J$ are zero for indices in the complement $\bar{J}$. In the following, the short notation $A^J:=P_JAP_J$ is also used (see above). Furthermore, the set of antisymmetric matrices whose submatrices have a kernel with dimension ≤ 1 is defined:

$$\Omega:=\left\{A\in\mathbb{R}^{S\times S}\mid A\text{ is antisymmetric and }\dim\ker A^J\leq 1\text{ for all } J\subseteq\{1,\ldots,S\}\right\}. \quad (18)$$

The complement $\bar{\Omega}$ has measure zero with respect to the flat measure dA on antisymmetric matrices (the translation invariant measure, which is sigma-finite and not trivial).

For antisymmetric matrices $A\in\Omega$, the kernel can be characterized as follows[61,70]. If the number of states S is even, the kernel of A is trivial: $\ker A=\{\mathbf{0}\}$. If the number of states S is odd, the kernel is one-dimensional: $\ker A=\{\mathbf{v}\}$. This kernel element can be computed analytically in terms of Pfaffians of submatrices of A:

$$\mathbf{v}=\left(\mathbf{Pf}(A_{\hat{1}}),-\mathbf{Pf}(A_{\hat{2}}),\ldots,\mathbf{Pf}(A_{\hat{S}})\right)\,. \quad (19)$$

The submatrix $A_{\hat{k}}\in\mathbb{R}^{(S-1)\times(S-1)}$ denotes the matrix for which the k-th column and row are removed from A.

For antisymmetric matrices $A\in\Omega$, the normalized condensate vector $\mathbf{c}$ with $\sum_i c_i=1$ and with properties (8), (9) is unique. The latter follows from $A^I\mathbf{c}=0$ (equations (8) and (18)). Therefore, the condensate vector is the unique kernel vector of the subsystem of condensates whose interactions are characterized by the matrix A^I. Furthermore, I contains an odd number of elements. To determine the condensate vector for $A\in\Omega$, one can proceed as follows. For each odd-dimensional submatrix A^I with $I\subseteq\{1,\ldots,S\}$, one computes the kernel element $\mathbf{v}$ according to equation (19) and defines the vector $\mathbf{w}\in\mathbb{R}^S$ by setting $\mathbf{w}_I=\mathbf{v}$ and $\mathbf{w}_{\bar{I}}=0$. There exists exactly one set I for which $(A\mathbf{w})_{\bar{I}}<0$. The corresponding vector $\mathbf{w}$ is the unique condensate vector upon normalization.

Temporal average of condensate concentrations (generic case). It was shown above that the temporal average of condensate concentrations $\langle\mathbf{x}\rangle_t$ converges to a non-negative kernel element of the antisymmetric matrix A^I. In the generic case, the condensate vector $\mathbf{c}$ is the unique kernel element of A^I upon normalization. Therefore, positive entries of $\mathbf{c}$ represent the asymptotic temporal average of condensate concentrations,

$$\|\langle\mathbf{x}\rangle_t-\mathbf{c}\|_\infty\leq\frac{Const(A,\mathbf{x}_0)}{t}\to 0\text{ as } t\to\infty. \quad (20)$$

Exponentially fast depletion of states (generic case). On inserting equation (20) into the implicit solution (14) of the ALVE, the exponentially fast depletion of states with $i\in\bar{I}$ can be seen as follows (note that $(A\mathbf{c})_i<0$ according to the choice of the condensate vector in equations (8) and (9)):

$$x_i(t)=x_i(0)e^{t\cdot(A\langle\mathbf{x}\rangle_t)_i} \quad (21)$$

$$\leq x_i(0)e^{t\cdot\left((A\mathbf{c})_i+\|A(\langle\mathbf{x}\rangle_t-\mathbf{c})\|_\infty\right)} \quad (22)$$

$$\leq x_i(0)e^{t\cdot(A\mathbf{c})_i+Const(A,\mathbf{x}_0)} \quad (23)$$

$$=Const(A,\mathbf{x}_0)e^{t\cdot(A\mathbf{c})_i}\,, \quad (24)$$

and analogously,

$$x_i(t)=x_i(0)e^{t\cdot(A\langle\mathbf{x}\rangle_t)_i}\geq Const(A,\mathbf{x}_0)e^{t\cdot(A\mathbf{c})_i}\,. \quad (25)$$

Therefore, condensate selection occurs exponentially fast at depletion rate $|(A\mathbf{c})_i|$. The dynamics of cases for non-generic antisymmetric matrices are discussed in Supplementary Note 2.

Linear programming algorithm. For the numeric computation of condensate vectors $\mathbf{c}$, a finite threshold $\delta>0$ was introduced into the inequalities (5): $A\mathbf{c}\leq 0$ and $\mathbf{c}-A\mathbf{c}\geq\delta>0$. Its value was set to $\delta=1$ by rescaling of $\mathbf{c}$. Numerical solution of the inequalities was performed by using the IBM ILOG CPLEX Optimization Studio 12.5 and its interface to the C++ language. The software Mathematica 9.0 from Wolfram Research was also found to be applicable. Further information on the calibration of the linear programming algorithm and a simplified Mathematica algorithm are provided in Supplementary Note 3.

References

1. Krapivsky, P. L., Redner, S. & Leyvraz, F. Connectivity of growing random networks. *Phys. Rev. Lett.* **85,** 4629–4632 (2000).
2. Bianconi, G. & Barabási, A.-L. Bose-Einstein condensation in complex networks. *Phys. Rev. Lett.* **86,** 5632–5635 (2001).
3. Evans, M. R. & Hanney, T. Nonequilibrium statistical mechanics of the zero-range process and related models. *J. Phys. A: Math. Gen.* **38,** R195–R240 (2005).
4. Evans, M. R. Bose-Einstein condensation in disordered exclusion models and relation to traffic flow. *Europhys. Lett.* **36,** 13–18 (1996).
5. Krug, J. & Ferrari, P. A. Phase transitions in driven diffusive systems with random rates. *J. Phys. A: Math. Gen.* **29,** L465–L471 (1996).
6. Chowdhury, D., Santen, L. & Schadschneider, A. Statistical physics of vehicular traffic and some related systems. *Phys. Rep.* **329,** 199–329 (2000).
7. Kaupužs, J., Mahnke, R. & Harris, R. J. Zero-range model of traffic flow. *Phys. Rev. E* **72,** 056125 (2005).
8. Spitzer, F. Interaction of Markov processes. *Adv. Math.* **5,** 246–290 (1970).
9. Evans, M. R. & Waclaw, B. Condensation in stochastic mass transport models: beyond the zero-range process. *J. Phys. A: Math. Theor.* **47,** 095001 (2014).
10. Bose, S. N. Plancks Gesetz und Lichtquantenhypothese. *Z. Phys.* **26,** 178–181 (1924).
11. Einstein, A. Quantentheorie des einatomigen idealen Gases. *Sitzb. d. Preuss. Akad. d. Wiss* 261–267 (1924).

12. Einstein, A. Quantentheorie des einatomigen idealen Gases. Zweite Abhandlung. *Sitzb. d. Preuss. Akad. d. Wiss* 3–14 (1925).
13. Griffin, A., Snoke, D. & Stringari, G. *Bose Einstein Condensation* (Cambridge Univ. Press, 1995).
14. Anglin, J. R. & Ketterle, W. Bose-Einstein condensation of atomic gases. *Nature* **416,** 211–218 (2002).
15. Penrose, O. & Onsager, L. Bose-Einstein condensation and liquid helium. *Phys. Rev.* **104,** 576–584 (1956).
16. Mueller, E. J., Ho, T.-L., Ueda, M. & Baym, G. Fragmentation of Bose-Einstein condensates. *Phys. Rev. A* **74,** 033612 (2006).
17. Gardiner, C. *Stochastic Methods: A Handbook for the Natural and Social Sciences* (Springer, 2009).
18. Van Kampen, N. G. *Stochastic Processes in Physics and Chemistry* (Elsevier, 2007).
19. Vorberg, D., Wustmann, W., Ketzmerick, R. & Eckardt, A. Generalized Bose-Einstein condensation into multiple states in driven-dissipative systems. *Phys. Rev. Lett.* **111,** 240405 (2013).
20. Albert, R. & Barabási, A.-L. Statistical mechanics of complex networks. *Rev. Mod. Phys.* **74,** 47–97 (2002).
21. Milo, R. *et al.* Network motifs: simple building blocks of complex networks. *Science* **298,** 824–827 (2002).
22. Blümel, R. *et al.* Dynamical localization in the microwave interaction of Rydberg atoms: the influence of noise. *Phys. Rev. A* **44,** 4521–4540 (1991).
23. Kohler, S., Dittrich, T. & Hänggi, P. Floquet-Markovian description of the parametrically driven, dissipative harmonic quantum oscillator. *Phys. Rev. E* **55,** 300–313 (1997).
24. Breuer, H.-P., Huber, W. & Petruccione, F. Quasistationary distributions of dissipative nonlinear quantum oscillators in strong periodic driving fields. *Phys. Rev. E* **61,** 4883–4889 (2000).
25. Grifoni, M. & Hänggi, P. Driven quantum tunneling. *Phys. Rep.* **304,** 229–354 (1998).
26. Breuer, H.-P. & Petruccione, F. *The Theory of Open Quantum Systems* (Oxford Univ. Press, 2002).
27. Gardiner, C. W. & Zoller, P. Quantum kinetic theory: a quantum kinetic master equation for condensation of a weakly interacting Bose gas without a trapping potential. *Phys. Rev. A* **55,** 2902–2921 (1997).
28. Kagan, Y. & Svistunov, B. V. Evolution of correlation properties and appearance of broken symmetry in the process of Bose-Einstein condensation. *Phys. Rev. Lett.* **79,** 3331–3334 (1997).
29. Bijlsma, M. J., Zaremba, E. & Stoof, H. T. C. Condensate growth in trapped Bose gases. *Phys. Rev. A* **62,** 063609 (2000).
30. Gardiner, C. W., Lee, M. D., Ballagh, R. J., Davis, M. J. & Zoller, P. Quantum kinetic theory of condensate growth: comparison of experiment and theory. *Phys. Rev. Lett.* **81,** 5266–5269 (1998).
31. Walser, R., Williams, J., Cooper, J. & Holland, M. Quantum kinetic theory for a condensed bosonic gas. *Phys. Rev. A* **59,** 3878–3889 (1999).
32. Kocharovsky, V. V., Scully, M. O., Zhu, S.-Y. & Suhail Zubairy, M. Condensation of N bosons. II. Nonequilibrium analysis of an ideal Bose gas and the laser phase-transition analogy. *Phys. Rev. A* **61,** 023609 (2000).
33. Pauli, W. *Festschrift zum 60. Geburtstage A. Sommerfeld* (Hirzel, 1928).
34. Mandel, L. & Wolf, E. *Optical Coherence and Quantum Optics* (Cambridge Univ. Press, 1995).
35. Gardiner, C. W. & Zoller, P. *Quantum Noise* (Springer, 2004).
36. Maynard Smith, J. *Evolution and the Theory of Games* (Cambridge Univ. Press, 1982).
37. Nowak, M. A. & Sigmund, K. Evolutionary dynamics of biological games. *Science* **303,** 793–799 (2004).
38. Sinervo, B. & Lively, C. M. The rock-paper-scissors game and the evolution of alternative male strategies. *Nature* **380,** 240–243 (1996).
39. Kerr, B., Riley, M., Feldman, M. & Bohannan, B. Local dispersal promotes biodiversity in a real-life game of rock-paper-scissors. *Nature* **418,** 171–174 (2002).
40. Reichenbach, T., Mobilia, M. & Frey, E. Mobility promotes and jeopardizes biodiversity in rock-paper-scissors games. *Nature* **448,** 1046–1049 (2007).
41. Weber, M. F., Poxleitner, G., Hebisch, E., Frey, E. & Opitz, M. Chemical warfare and survival strategies in bacterial range expansions. *J. R. Soc. Interface* **11,** 20140172 (2014).
42. Szolnoki, A. *et al.* Cyclic dominance in evolutionary games: a review. *J. R. Soc. Interface* **11,** 20140735 (2014).
43. Nowak, M. A., Sasaki, A., Taylor, C. & Fudenberg, D. Emergence of cooperation and evolutionary stability in finite populations. *Nature* **428,** 646–650 (2004).
44. Szolnoki, A., Antonioni, A., Tomassini, M. & Perc, M. Binary birth-death dynamics and the expansion of cooperation by means of self-organized growth. *EPL* **105,** 48001 (2014).
45. McKane, A. J. & Newman, T. J. Predator-prey cycles from resonant amplification of demographic stochasticity. *Phys. Rev. Lett.* **94,** 218102 (2005).
46. Traulsen, A., Claussen, J. C. & Hauert, C. Coevolutionary dynamics: from finite to infinite populations. *Phys. Rev. Lett.* **95,** 238701 (2005).
47. Reichenbach, T., Mobilia, M. & Frey, E. Coexistence versus extinction in the stochastic cyclic Lotka-Volterra model. *Phys. Rev. E* **74,** 51907 (2006).
48. Melbinger, A., Cremer, J. & Frey, E. Evolutionary game theory in growing populations. *Phys. Rev. Lett.* **105,** 178101 (2010).
49. Biancalani, T., Dyson, L. & McKane, A. J. Noise-induced bistable states and their mean switching time in foraging colonies. *Phys. Rev. Lett.* **112,** 038101 (2014).
50. Rulands, S., Jahn, D. & Frey, E. Specialization and bet hedging in heterogeneous populations. *Phys. Rev. Lett.* **113,** 108102 (2014).
51. Volterra, V. *Leçons sur la Théorie Mathématique de la Lutte pour la Vie* (Gauthier-Villars, 1931).
52. Goel, N. S., Maitra, S. C. & Montroll, E. W. On the Volterra and other nonlinear models of interacting populations. *Rev. Mod. Phys.* **43,** 231–276 (1971).
53. May, R. M. *Stability and Complexity in Model Ecosystems* (Princeton Univ. Press, 1973).
54. Zakharov, V., Musher, S. & Rubenchik, A. Nonlinear stage of parametric wave excitation in a plasma. *JETP Lett.* **19,** 151–152 (1974).
55. Manakov, S. Complete integrability and stochastization of discrete dynamical systems. *Sov. Phys.-JETP* **40,** 269–274 (1975).
56. Itoh, Y. Boltzmann equation on some algebraic structure concerning struggle for existence. *Proc. Jpn Acad.* **47,** 854–858 (1971).
57. Di Cera, E., Phillipson, P. E. & Wyman, J. Chemical oscillations in closed macromolecular systems. *Proc. Natl Acad. Sci. USA* **85,** 5923–5926 (1988).
58. Di Cera, E., Phillipson, P. E. & Wyman, J. Limit-cycle oscillations and chaos in reaction networks subject to conservation of mass. *Proc. Natl Acad. Sci. USA* **86,** 142–146 (1989).
59. Akin, E. & Losert, V. Evolutionary dynamics of zero-sum games. *J. Math. Biol.* **20,** 231–258 (1984).
60. Chawanya, T. & Tokita, K. Large-dimensional replicator equations with antisymmetric random interactions. *J. Phys. Soc. Jpn* **71,** 429–431 (2002).
61. Knebel, J., Krüger, T., Weber, M. F. & Frey, E. Coexistence and survival in conservative Lotka-Volterra networks. *Phys. Rev. Lett.* **110,** 168106 (2013).
62. Kuhn, H. & Tucker, A. *Linear Inequalities and Related Systems* (Princeton Univ. Press, 1956).
63. Prigogine, I. Time, structure, and fluctuations. *Science* **201,** 777–785 (1978).
64. Gardner, M. R. & Ashby, W. R. Connectance of large dynamic (cybernetic) systems: critical values for stability. *Nature* **228,** 784 (1970).
65. Durney, C. H., Case, S. O., Pleimling, M. & Zia, R. K. P. Saddles, arrows, and spirals: deterministic trajectories in cyclic competition of four species. *Phys. Rev. E* **83,** 051108 (2011).
66. Allesina, S. & Levine, J. M. A competitive network theory of species diversity. *Proc. Natl Acad. Sci. USA* **108,** 5638–5642 (2011).
67. Eisert, J., Friesdorf, M. & Gogolin, C. Quantum many-body systems out of equilibrium. *Nat. Phys.* **11,** 124–130 (2015).
68. Zia, R. K. P. & Schmittmann, B. Probability currents as principal characteristics in the statistical mechanics of non-equilibrium steady states. *J. Stat. Mech* **2007,** P07012 (2007).
69. Kriecherbauer, T. & Krug, J. A pedestrian's view on interacting particle systems, KPZ universality and random matrices. *J. Phys. A: Math. Theor.* **43,** 403001 (2010).
70. Cullis, C. E. *Matrices and Determinoids* vol. I and II (Cambridge Univ. Press, 1913).

Acknowledgements

We are grateful for fruitful discussions with Peter Zoller, Ulrich Schollwöck, Immanuel Bloch, Wilhelm Zwerger, André Eckardt, Daniel Vorberg, Alexander Schnell, Marianne Bauer, Brendan Osberg, Jacob Halatek, Matthias Bauer and Sebastian Koch. This work was supported by the Deutsche Forschungsgemeinschaft as project A7 of the SFB TR 12 'Symmetry and Universality in Mesoscopic Systems', and by the German Excellence Initiative via the programme 'Nanosystems Initiative Munich' (NIM). J.K. acknowledges funding from the Studienstiftung des Deutschen Volkes.

Author contributions

J.K., M.F.W., T.K. and E.F. designed, discussed and planned the study. T.K., J.K. and M.F.W. developed the analytical results. M.F.W., T.K. and J.K. developed the numerical algorithms and generated the data. J.K., M.F.W., T.K. and E.F. interpreted the results and wrote the manuscript.

Additional information

Competing financial interests: The authors declare no competing financial interests.

4

Efficient quantum walk on a quantum processor

Xiaogang Qiang[1,*], Thomas Loke[2,*], Ashley Montanaro[3], Kanin Aungskunsiri[1], Xiaoqi Zhou[1,4], Jeremy L. O'Brien[1], Jingbo B. Wang[2] & Jonathan C.F. Matthews[1]

The random walk formalism is used across a wide range of applications, from modelling share prices to predicting population genetics. Likewise, quantum walks have shown much potential as a framework for developing new quantum algorithms. Here we present explicit efficient quantum circuits for implementing continuous-time quantum walks on the circulant class of graphs. These circuits allow us to sample from the output probability distributions of quantum walks on circulant graphs efficiently. We also show that solving the same sampling problem for arbitrary circulant quantum circuits is intractable for a classical computer, assuming conjectures from computational complexity theory. This is a new link between continuous-time quantum walks and computational complexity theory and it indicates a family of tasks that could ultimately demonstrate quantum supremacy over classical computers. As a proof of principle, we experimentally implement the proposed quantum circuit on an example circulant graph using a two-qubit photonics quantum processor.

[1] Centre for Quantum Photonics, H.H. Wills Physics Laboratory and Department of Electrical and Electronic Engineering, University of Bristol, Bristol BS8 1UB, UK. [2] School of Physics, The University of Western Australia, Crawley, WA 6009, Australia. [3] School of Mathematics, University of Bristol, Bristol BS8 1TW, UK. [4] State Key Laboratory of Optoelectronic Materials and Technologies and School of Physics, Sun Yat-sen University, Guangzhou 510275, China. * These authors contributed equally to this work. Correspondence and requests for materials should be addressed to J.B.W. (email: jingbo.wang@uwa.edu.au) or to J.C.F.M. (email: jonathan.matthews@bristol.ac.uk).

Quantum walks are the quantum mechanical analogue of the well-known classical random walk and they have established roles in quantum information processing[1–3]. In particular, they are central to quantum algorithms created to tackle database search[4], graph isomorphism[5–7], network analysis and navigation[8,9], and quantum simulation[10–12], as well as modelling biological processes[13,14]. Meanwhile, physical properties of quantum walks have been demonstrated in a variety of systems, such as nuclear magnetic resonance[15,16], bulk[17] and fibre[18] optics, trapped ions[19–21], trapped neutral atoms[22] and photonics[23,24]. Almost all physical implementations of quantum walk so far followed an analogue approach as for quantum simulation[25], whereby the apparatus is dedicated to implement specific instances of Hamiltonians without translation onto quantum logic. However, there is no existing method to implement analogue quantum simulations with error correction or fault tolerance, and they do not scale efficiently in resources when simulating broad classes of large graphs. Some exceptions of demonstrations of quantum walks, such as ref. 15, adopted the qubit model, but did not discuss potentially efficient implementation of quantum walks.

Efficient quantum circuit implementations of continuous-time quantum walks (CTQWs) have been presented for sparse and efficiently row-computable graphs[26,27], and specific non-sparse graphs[28,29]. However, the design of quantum circuits for implementing CTQWs is in general difficult, since the time-evolution operator is time dependent and non-local[1]. A subset of circulant graphs have the property that their eigenvalues and eigenvectors can be classically computed efficiently[30,31]. This enables construction of a scheme that efficiently outputs the quantum state $|\psi(t)\rangle$, which corresponds to the time-evolution state of a CTQW on corresponding graphs. One can then either implement further quantum circuit operations or perform direct measurements on $|\psi(t)\rangle$ to extract physically meaningful information. For example the 'SWAP test'[32] can be used to estimate the similarity of dynamical behaviours of two circulant Hamiltonians operating on two different initial states, as shown in Fig. 1a. This procedure can also be adapted to study the stability of quantum dynamics of circulant molecules (for example, the DNA Möbius strips[33]) in a perturbational environment[34,35]. When measuring $|\psi(t)\rangle$ in the computational basis we can sample the probability distribution

$$p(x) := |\langle x|\psi(t)\rangle|^2 \tag{1}$$

that describes the probability of observing the quantum walker at position $x\in\{0,\ 1\}^n$—an n-bit string, labelling one of the 2^n vertices of the given graph, as shown in Fig. 1b. Sampling of this form is sufficient to solve various search and characterization problems[4,9], and can be used to deduce critical parameters of the quantum walk, such as mixing time[2].

Here we present efficient quantum circuits for implementing CTQWs on circulant graphs with an eigenvalue spectrum that can be classically computed efficiently. These quantum circuits provide the time-evolution states of CTQWs on circulant graphs exponentially faster than best previously known methods[30]. We report a proof-of-principle experiment, where we implement CTQWs on an example circulant graph (namely the complete graph of four vertices) using a two-qubit photonics quantum processor to sample the probability distributions and perform state tomography on the output state of a CTQW. We also provide evidence from computational complexity theory that the probability distributions $p(x)$ that are output from the circuits of this circulant form are in general hard to sample from using a classical computer, implying our scheme also provides an exponential speedup for sampling. We adapt the methodology of refs 36–38 to show that if there did exist a classical sampler for a somewhat more general class of circuits, then this would have the following unlikely complexity-theoretic implication: the infinite tower of complexity classes known as the polynomial hierarchy would collapse. This evidence of hardness exists despite the classical efficiency with which properties of the CTQW, such as the eigenvalues of circulant graphs, can be computed on a classical machine.

Results

Quantum circuit for CTQW on circulant graph. For an undirected graph G of N vertices, a quantum particle (or 'quantum walker') placed on G evolves into a superposition $|\psi(t)\rangle$ of states in the orthonormal basis $\{|1\rangle, |2\rangle, \ldots, |N\rangle\}$ that correspond to vertices of G. The exact evolution of the CTQW is governed by connections between the vertices of G: $|\psi(t)\rangle = \exp(-itH)|\psi(0)\rangle$ where the Hamiltonian is given by $H = \gamma A$ for hopping rate per edge per unit time γ and where A is the N-by-N symmetric adjacency matrix, whose entries are $A_{jk} = 1$, if vertices j and k are connected by an edge in G, and $A_{jk} = 0$ otherwise[1]. The dynamics of a CTQW on a graph with N vertices can be evaluated in time poly(N) on a classical computer. When a CTQW takes place on a graph G of exponential size, that is, $N = 2^n$ for an input of size n, it becomes interesting to use quantum processors to simulate dynamics.

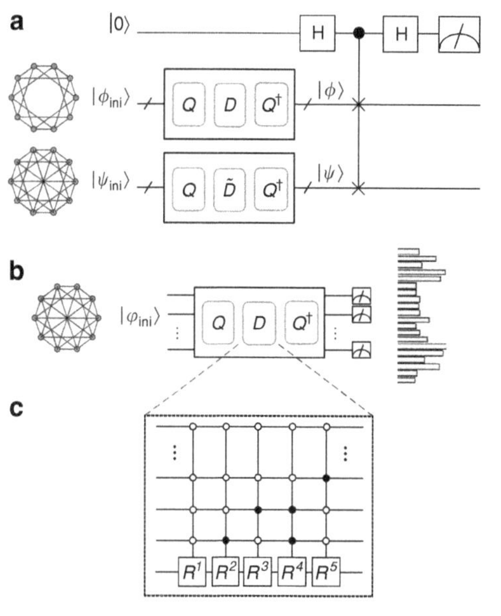

Figure 1 | Applications for generating the time-evolution state of circulant Hamiltonians. (**a**) The SWAP test[32] can be used to estimate the similarity of two evolution states of two similar circulant systems, or when one of the Hamiltonians is non-circulant but efficiently implementable. In brief, an ancillary qubit is entangled with the output states ψ and ϕ of two compared processes according to $\frac{1}{2}|0\rangle[|\phi\rangle|\psi\rangle + |\psi\rangle|\phi\rangle] + \frac{1}{2}|1\rangle[|\phi\rangle|\psi\rangle - |\psi\rangle|\phi\rangle]$. On measuring the ancillary qubit we obtain outcome '1' with probability $\frac{1}{2}(1 - |\langle\phi|\psi\rangle|^2)$—the probability of observing '1' indicates the similarity of dynamical behaviours of the two processes. See its complexity analysis in Supplementary Note 1. (**b**) Probability distributions are sampled by measuring the evolution state in a complete basis, such as the computational basis. (**c**) An example of the quantum circuit for implementing diagonal unitary operator $D = \exp(-it\Lambda)$, where the circulant Hamiltonian has 5 non-zero eigenvalues. The open and solid circles represent the control qubits as 'if $|0\rangle$' and 'if $|1\rangle$', respectively. $R^i = [1, 0; 0, \exp(-it\lambda_i)] (i = 1, \cdots, 5)$, where λ_i is the corresponding eigenvalue.

Circulant graphs are defined by symmetric circulant adjacency matrices for which each row j when right rotated by one element, equals the next row $j+1$—for example, complete graphs, cycle graphs and Mobius ladder graphs are all subclasses of circulant graphs, and further examples are shown in Supplementary Note 2. It follows that Hamiltonians for CTQWs on any circulant graph have a symmetric circulant matrix representation, which can be diagonalized by the unitary Fourier transform[31], that is, $H = Q^{\dagger}\Lambda Q$, where

$$Q_{jk} = \frac{1}{\sqrt{N}}\omega^{jk},\ \omega = \exp(2\pi i/N) \qquad (2)$$

and Λ is a diagonal matrix containing eigenvalues of H, which are all real and whose order is determined by the order of the eigenvectors in Q. Consequently, we have $\exp(-itH) = Q^{\dagger}\exp(-it\Lambda)Q$, where the time dependence of $\exp(-itH)$ is confined to the diagonal unitary operator $D = \exp(-it\Lambda)$.

The Fourier transformation Q can be implemented efficiently by the well-known QFT quantum circuit[39]. For a circulant graph that has $N = 2^n$ vertices, the required QFT of N dimensions can be implemented with $O((\log N)^2) = O(n^2)$ quantum gates acting on $O(n)$ qubits. To implement the inverse QFT, the same circuit is used in reverse order with phase gates of opposite sign. D can in general be implemented using at most $N = 2^n$ controlled-phase gates with phase values being a linear function of t, because an arbitrary phase can be applied to an arbitrary basis state, conditional on at most n-1 qubits. However, given a circulant graph that has O(poly(n)) non-zero eigenvalues, only O(poly(n)) controlled-phase gates are needed to implement D. If the given circulant graph has $O(2^n)$ distinct eigenvalues, which can be characterized efficiently (such as the cycle graphs and Mobius ladder graphs), then we are still able to implement the diagonal unitary operator D using polynomial quantum resources. A general construction of efficient quantum circuits for D was given by Childs[40], and is shown in Supplementary Fig. 1 and Supplementary Note 3 for completeness. Thus, the quantum circuit implementations of CTQWs on circulant graphs can be constructed, which have an overall complexity of O(poly(n)), and act on at most $O(n)$ qubits. Compared with the best-known classical algorithm based on fast Fourier transform, that has the computational complexity of $O(n2^n)$ (ref. 30), the proposed quantum circuit implementation generates the evolution state $|\psi(t)\rangle$ with an exponential advantage in speed.

Experimental demonstration. To demonstrate implementation of our scheme with two qubits, we have built photonic quantum logic to simulate CTQWs on the K_4 graph—a complete graph with self loops on four vertices (Fig. 2a). The family of complete graphs K_N are a special kind of circulant graph, with an adjacency matrix A where $A_{jk} = 1$ for all j, k. Their Hamiltonian has only 2 distinct eigenvalues, 0 and $N\gamma$. Therefore, the diagonal matrix of eigenvalues of K_4 is $\Lambda = \mathrm{diag}(\{4\gamma, 0, 0, 0\})$. We can readily construct the quantum circuit for implementing CTQWs on K_4 based on diagonalization, using the QFT matrix. However, the choice of using the QFT matrix as the eigenbasis of Hamiltonian is not strictly necessary—any equivalent eigenbasis can be selected. Through the diagonalization using Hadamard eigenbasis, an alternative efficient quantum circuit for implementing CTQWs on K_4 is shown in Fig. 2b, which can be easily extended to K_N.

We built a configurable two-qubit photonics quantum processor (Fig. 2c), adapting the entanglement-based technique presented in ref. 41, and implemented CTQWs on K_4 graph with various evolving times and initial states. Specifically, we prepared two different initial states $|\varphi_{\mathrm{ini}}\rangle_1 = [1, 0, 0, 0]'$ and $|\varphi_{\mathrm{ini}}\rangle_2 = \frac{1}{\sqrt{2}}[1, 1, 0, 0]'$, which represent the quantum walker starting from vertex 1, and the superposition of vertices 1 and 2, respectively. We chose the evolution time following the list $\{0, \frac{1}{8}\pi, \frac{2}{8}\pi, \frac{3}{8}\pi, \frac{4}{8}\pi, \frac{5}{8}\pi, \frac{6}{8}\pi, \frac{7}{8}\pi, \pi\}$, which covers the whole

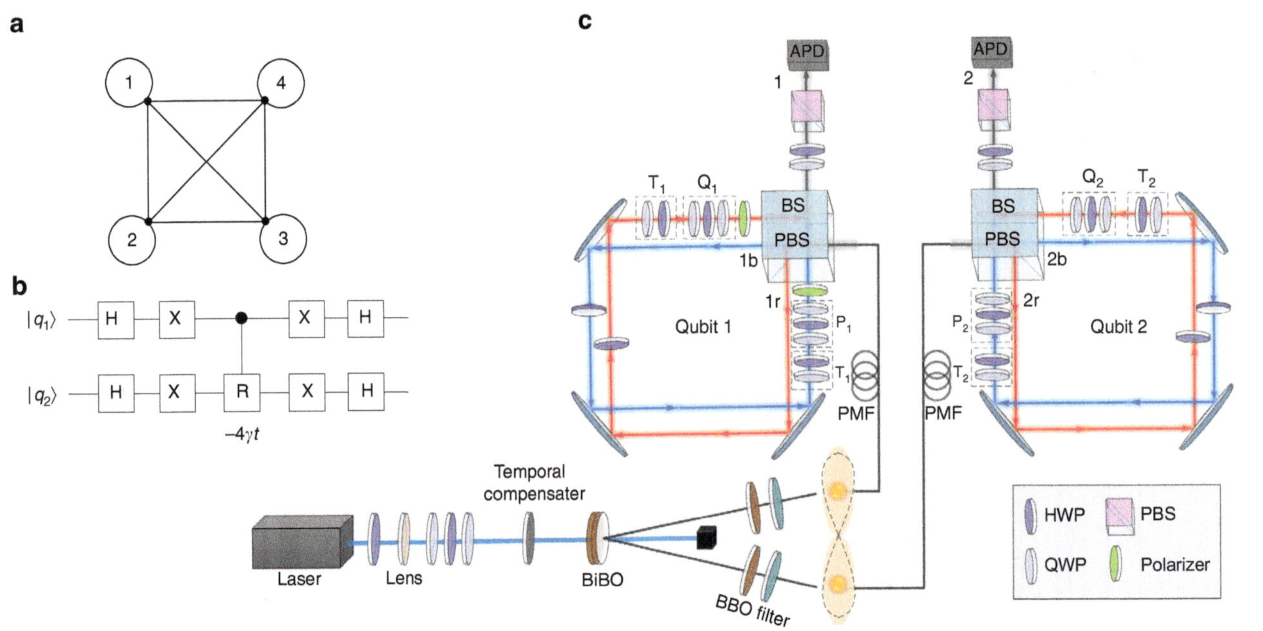

Figure 2 | The schematic diagram and set-up of experimental demonstration. **(a)** The K_4 graph. **(b)** The quantum circuit for implementing CTQW on the K_4 graph. This can also be used to implement CTQW on the K_4 graph without self-loops, up to a global phase factor $\exp(i\gamma t)$. H and X represent the Hadamard and Pauli-X gate, respectively. $R = (1, 0; 0, \exp(-i4\gamma t))$ is a phase gate. **(c)** The experimental set-up for a reconfigurable two-qubit photonics quantum processor, consisting of a polarization-entangled photon source using paired type-I BiBO crystal in the sandwich configuration and displaced Sagnac interferometers. See further details in Methods.

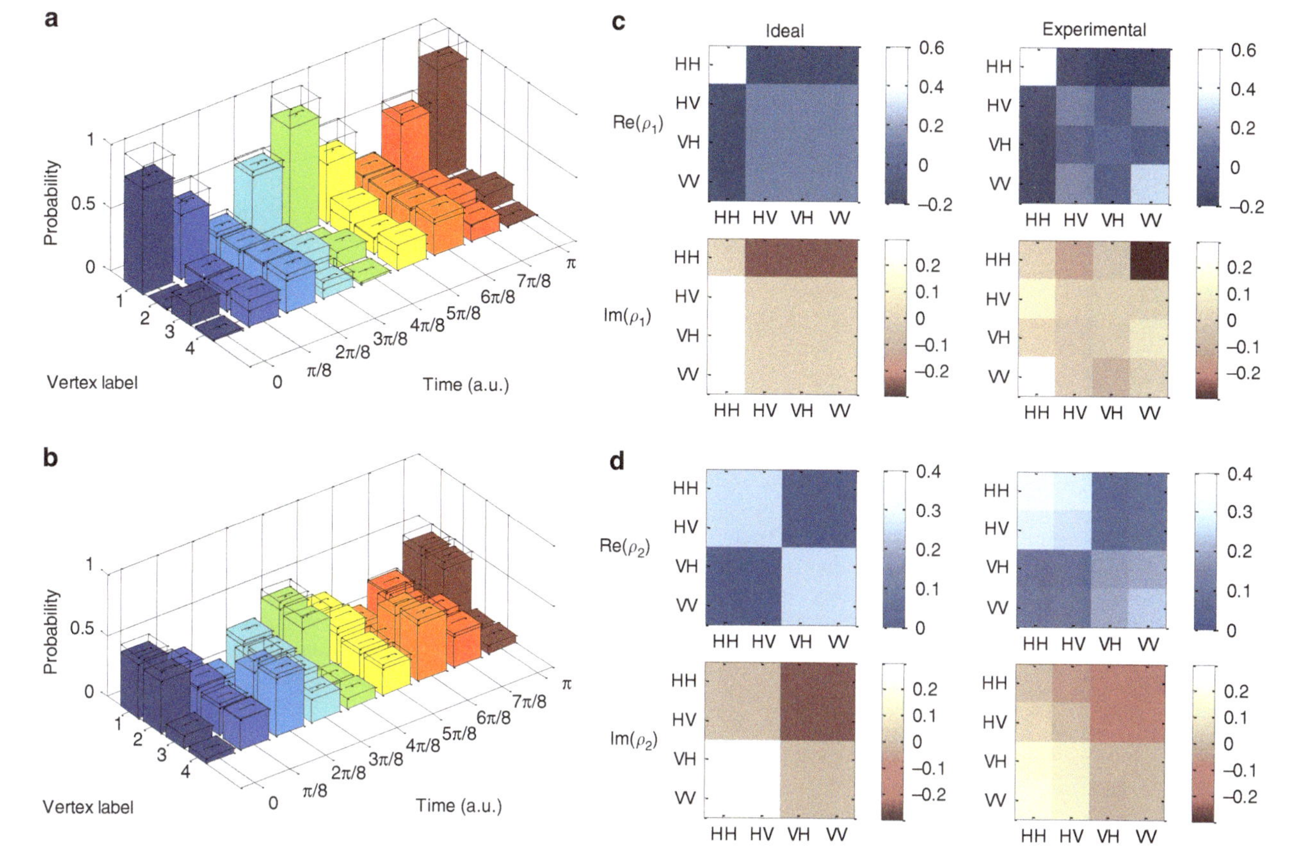

Figure 3 | Experimental results for simulating CTQWs on K_4. (**a**,**b**) The experimental sampled probability distributions with ideal theoretical distributions overlaid, for CTQWs on K_4 graph with initial states $|\varphi_{\rm ini}\rangle_1 = [1, 0, 0, 0]'$ and $|\varphi_{\rm ini}\rangle_2 = \frac{1}{\sqrt{2}}[1, 1, 0, 0]'$. The s.d. of each individual probability is also plotted, which is calculated by propagating error assuming Poissonian statistics. (**c**,**d**) The ideal theoretical and experimentally reconstructed density matrices for the states $|\varphi_{\rm out}\rangle_1 = [0.75 + 0.25i, -0.25 + 0.25i, -0.25 + 0.25i, -0.25 + 0.25i]'$ (corresponding to ρ_1) and $|\varphi_{\rm out}\rangle_2 = [0.3536 + 0.3536i, 0.3536 + 0.3536i, -0.3536 + 0.3536i, -0.3536 + 0.3536i]'$ (corresponding to ρ_2). Both of the real and imaginary parts of the density matrices are obtained through the maximum likelihood estimation technique, and is shown as Re(ρ) and Im(ρ), respectively. Further results are shown in Supplementary Table 1, Supplementary Fig. 2 and Supplementary Note 4.

periodical characteristics of CTQWs on K_4 graph. For each evolution, we sampled the corresponding probability distribution with fixed integration time, shown in Fig. 3a,b. To measure how close the experimental and ideal probability distributions are, we calculated the average fidelities defined as $F_{\rm average} = \frac{1}{9}\sum_{n=1}^{9}\sum_{i=1}^{4}\sqrt{P_{\rm ideal,n}(i)P_{\rm exp,n}(t)}$. The achieved average fidelities for the samplings with two distinct initial states are $96.68 \pm 0.27\%$ and $95.82 \pm 0.25\%$, respectively. Through the proposed circuit implementation, we are also able to examine the evolution states using quantum state tomography, which is generally difficult for the analogue simulations. For two specific evolution states $|\varphi_{\rm out}\rangle_1 = \exp(-iH\frac{7}{8}\pi)|\varphi_{\rm ini}\rangle_1$ and $|\varphi_{\rm out}\rangle_2 = \exp(-iH\frac{7}{8}\pi)|\varphi_{\rm ini}\rangle_2$, we performed quantum state tomography and reconstructed the density matrices using the maximum likelihood estimation technique. The two reconstructed density matrices achieve fidelities of $85.81 \pm 1.08\%$ and $88.44 \pm 0.97\%$, respectively, shown in Fig. 3c,d.

Here we have chosen to use K_4 in our experiment because it is simple enough to be implementable with state of the art photonics capability, while it provides an example to demonstrate our protocol for simulating CTQW on a circulant graph with controlled quantum logic. As the size of graph increases, the simplicity of K_N implies that CTQWs on this family of graphs can easily be simulated classically for arbitrary N—for CTQW on a complete graph of size N, an arbitrary output probability amplitude $\langle y|\exp(-itH)|x\rangle$ can be readily obtained as $(N - 1 + \exp(-itN\gamma))N^{-1}$ if $x = y$, and $(-1 + \exp(-itN\gamma))N^{-1}$ otherwise, where $|x\rangle$ and $|y\rangle$ represent the initial state and evolution state, respectively. However, our outlined quantum circuit implementation (Fig. 1) extends to implement CTQW on far more complicated circulant graphs.

Hardness of the sampling problem. To provide evidence that simulating CTQW on general circulant graphs is likely to be hard classically, we consider a circuit of the form $Q^{\dagger}DQ$, where D is a diagonal matrix made up of poly(n) controlled-phase gates and Q is the quantum Fourier transform. Define $p_{\rm D}$ to be the probability of measuring all qubits to be 0 in the computational basis after $Q^{\dagger}DQ$ is applied to the input state $|0\rangle^{\otimes n}$. It is readily shown that

$$\begin{aligned} p_{\rm D} &= \left|\langle 0|^{\otimes n} Q^{\dagger} D Q |0\rangle^{\otimes n}\right|^2 \\ &= \left|\langle +|^{\otimes n} D |+\rangle^{\otimes n}\right|^2 \\ &= \left|\langle 0|^{\otimes n} H^{\otimes n} D H^{\otimes n} |0\rangle^{\otimes n}\right|^2. \end{aligned} \quad (3)$$

This implies that $p_{\rm D}$ can also be obtained through a circuit of form $H^{\otimes n}DH^{\otimes n}$ with D unchanged—this represents a class of circuits known as instantaneous quantum polynomial time (IQP), which has the following structure: each qubit line begins and ends with a Hadamard (H) gate, and, in between, every gate is diagonal in the computational basis[37,42]. As such, p_D is a probability that is classically hard to compute—it is known that computing p_D for arbitrary diagonal unitaries D made up of circuits of poly(n) gates, even if each acts on $O(1)$ qubits, is #P-hard[38,43,44]. This hardness result even holds for approximating p_D up to any relative error strictly less than 1/2 (refs 38,43,44), where $\widetilde{p_D}$ is said to approximate p_D up to relative error ϵ if

$$\left|\widetilde{p_D} - p_D\right| \leq \epsilon p_D. \tag{4}$$

Note that other output probabilities $p(x)$ cannot be achieved using IQP circuits since a general circulant graph cannot be diagonalized by Hadamard matrices but rather by more heterogeneous Fourier matrices.

Towards a contradiction, assume that there exists a polynomial-time randomized classical algorithm, which samples from p, as defined in equation (1). Then a classic result of Stockmeyer[45] states that there is an algorithm in the complexity class FBPP^{NP}, which can approximate any desired probability $p(x)$ to within relative error $O(1/\text{poly}(n))$. This complexity class FBPP^{NP}—described as polynomial-time randomized classical computation equipped with an oracle to solve arbitrary NP problems—sits within the infinite tower of complexity classes known as the polynomial(-time) hierarchy[46]. Combining with the above hardness result of approximating p_D, we find that the assumption implies that an FBPP^{NP} algorithm solves a #P-hard problem, so P$^{\#P}$ would be contained within FBPP^{NP}, and therefore the polynomial hierarchy would collapse to its third level. This consequence is considered very unlikely in computational complexity theory[46]. A similar methodology has been used to prove the hardness of IQP and boson sampling[36–38].

We therefore conclude that, in general, a polynomial-time randomized classical sampler from the distribution p is unlikely to exist. Further, this even holds for classical algorithms which sample from any distribution $\widetilde{p}$ which approximates p up to relative error strictly $<1/2$ in each probability $p(x)$. It is worth noting that if the output distribution results from measurements on only $O(\text{poly}(\log n))$ qubits[47], or obeys the sparsity promise that only a poly(n)-sized, and *a priori* unknown, subset of the measurement probabilities are non-zero[48], it could be classically efficiently sampled. It was shown in ref. 38 that assuming certain conjectures in complexity theory, it is classically hard to sample from distributions that are close in total variation distance to arbitrary IQP probability distributions. The differences between circulant and IQP circuits imply that this result does not go through immediately in our setting. Therefore, it remains open to prove hardness of approximate simulation of CTQWs on circulant graphs, which specifically requires to show that computing most of the output probabilities of circulant circuits is hard, assuming some conjectures in complexity theory.

Discussion

In this paper, we have described how CTQWs on circulant graphs can be efficiently implemented on a quantum computer, if the eigenvalues of the graphs can be characterized efficiently classically. In fact, we can construct an efficient quantum circuit to implement CTQWs on any graph whose adjacency matrix is efficiently diagonalisable, in other words, as long as the matrix of column eigenvectors Q and the diagonal matrix of the eigenvalue exponentials D can be implemented efficiently. To demonstrate our implementation scheme, we simulated CTQWs on an example 4-vertex circulant graph, K_4, using a two-qubit photonic quantum logic circuit. We have shown that the problem of sampling from the output probability distributions of quantum circuits of the form $Q^{\dagger}DQ$ is hard for classical computers, based on a highly plausible conjecture that the polynomial hierarchy does not collapse. This observation is particularly interesting from both perspectives of CTQW and computational complexity theory, as it provides new insights into the CTQW framework and also helps to classify and identify new problems in computational complexity theory. For the CTQWs on the circulant graphs of poly(n) non-zero eigenvalues, the proposed quantum circuit implementations do not need a fully universal quantum computer, and thus can be viewed as an intermediate model of quantum computation. Meanwhile, the evidence we provided for hardness of the sampling problem indicates a promising candidate for experimentally establishing quantum supremacy over classical computers, and further evidence against the extended Church–Turing thesis. To claim in an experiment super-classical performance based on the conjecture outlined in this work, future demonstrations would need to consider circulant graphs that are more general than K_N and that are of sufficient size to be outside the capabilities of a classical computer. For photonics, the biggest challenges remain increasing the number of indistinguishable photons and controlled gate operations. For any platform, quantum circuit implementation of CTQWs could be more appealing due to available methods in fault tolerance and error correction, which are difficult to implement for other intermediate models like boson sampling[49] and for analogue quantum simulation. Our results may also lead to other practical applications through the use of CTQWs for quantum algorithm design.

Methods

Experimental set-up. A diagonally polarized, 120 mW, continuous-wave laser beam with central wavelength of 404 nm is focused at the centre of paired type-I BiBO crystals with their optical axes orthogonally aligned to each other, to create the polarization entangled photon-pairs[50]. Through the spontaneous parametric downconversion process, the photon pairs are generated in the state of $\frac{1}{\sqrt{2}}(|H_1H_2\rangle + |V_1V_2\rangle)$, where H and V represent horizontal and vertical polarization, respectively. The photons pass through the polarization beam-splitter (PBS) part of the dual PBS/beam-splitter cubes on both arms to generate two-photon four-mode state of the form $\frac{1}{\sqrt{2}}(|H_{1b}H_{2b}\rangle + |V_{1r}V_{2r}\rangle)$ (where r and b labels the red and blue paths shown in Fig. 2c, respectively). Rotations T_1 and T_2 on each path, consisting of half wave-plate (HWP) and quarter wave plate (QWP), convert the state into $\frac{1}{\sqrt{2}}(|\phi_{1b}\phi_{2b}\rangle + |\phi_{1r}\phi_{2r}\rangle)$, where $|\phi_1\rangle$ and $|\phi_2\rangle$ can be arbitrary single-qubit states. The four spatial modes 1b, 2b, 1r and 2r pass through four single-qubit quantum gates P_1, P_2, Q_1 and Q_2, respectively, where each of the four gates is implemented through three wave plates: QWP, HWP and QWP. The spatial modes 1b and 1r (2b and 2r) are then mixed on the beam-splitter part of the cube. By post-selecting the case where the two photons exit at ports 1 and 2, we obtain the state $(P_1 \otimes P_2 + Q_1 \otimes Q_2)|\phi_1\phi_2\rangle$. In this way, we implement a two-qubit quantum operation of the form $P_1 \otimes P_2 + Q_1 \otimes Q_2$ on the initialized state $|\phi_1\phi_2\rangle$.

As shown in Fig. 2b, the quantum circuit for implementing CTQW on the K_4 graph consists of Hadamard gates (H), Pauli-X gates (X) and controlled-phase gate (CP). CP is implemented by configuring $P_1 = |H\rangle\langle H|$, $P_2 = I$, $Q_1 = |V\rangle\langle V|$, $Q_2 = R(= [1,0;0,e^{-i4\gamma t}])$, where P_1 and Q_1 are implemented by polarizers. Altogether with combining the operation (H·X)⊗(H·X) before CP with state preparation and the operation (X·H)⊗X·H after CP with measurement setting, we implement the whole-quantum circuit on the experimental set-up. The evolution time of CTQW is controlled by the phase value of R, which is determined by setting the three wave plates of Q_2 in Fig. 2c to QWP$(\frac{\pi}{4})$, HWP(ω), QWP$(\frac{\pi}{4})$, where the angle ω of HWP equals to the phase of R: -4γ t. The evolution time t is then given by $t = -\omega/(4\gamma)$.

References

1. Farhi, E. & Gutmann, S. Quantum computation and decision trees. *Phys. Rev. A* **58,** 915 (1998).
2. Kempe, J. Quantum random walks: an introductory overview. *Contemp. Phys.* **44,** 307–327 (2003).
3. Childs, A. M., Gosset, D. & Webb, Z. Universal computation by multiparticle quantum walk. *Science* **339,** 791–794 (2013).
4. Childs, A. M. & Goldstone, J. Spatial search by quantum walk. *Phys. Rev. A* **70,** 022314 (2004).
5. Douglas, B. L. & Wang, J. B. A classical approach to the graph isomorphism problem using quantum walks. *J. Phys. A* **41,** 075303 (2008).

6. Gamble, J. K., Friesen, M., Zhou, D., Joynt, R. & Coppersmith, S. N. Two-particle quantum walks applied to the graph isomorphism problem. *Phys. Rev. A* **81,** 052313 (2010).
7. Berry, S. D. & Wang, J. B. Two-particle quantum walks: entanglement and graph isomorphism testing. *Phys. Rev. A* **83,** 042317 (2011).
8. Berry, S. D. & Wang, J. B. Quantum-walk-based search and centrality. *Phys. Rev. A* **82,** 042333 (2010).
9. Sánchez-Burillo, E., Duch, J., Gómez-Gardeñes, J. & Zueco, D. Quantum navigation and ranking in complex networks. *Sci. Rep.* **2,** 605 (2012).
10. Lloyd, S. Universal quantum simulators. *Science* **273,** 1073–1078 (1996).
11. Berry, D. W. & Childs, A. M. Black-box hamiltonian simulation and unitary implementation. *Quantum Inf. Comput.* **12,** 29–62 (2012).
12. Schreiber, A. *et al.* A 2D quantum walk simulation of two-particle dynamics. *Science* **336,** 55–58 (2012).
13. Engel, G. S. *et al.* Evidence for wavelike energy transfer through quantum coherence in photosynthetic systems. *Nature* **446,** 782–786 (2007).
14. Rebentrost, P. *et al.* Environment-assisted quantum transport. *New J. Phys.* **11,** 033003 (2009).
15. Du, J. *et al.* Experimental implementation of the quantum random-walk algorithm. *Phys. Rev. A* **67,** 042316 (2003).
16. Ryan, C. A. *et al.* Experimental implementation of a discrete-time quantum random walk on an NMR quantum-information processor. *Phys. Rev. A* **72,** 062317 (2005).
17. Do, B. *et al.* Experimental realization of a quantum quincunx by use of linear optical elements. *J. Opt. Soc. Am. B* **22,** 499–504 (2005).
18. Schreiber, A. *et al.* Photons walking the line: a quantum walk with adjustable coin operations. *Phys. Rev. Lett.* **104,** 050502 (2010).
19. Xue, P., Sanders, B. C. & Leibfried, D. Quantum walk on a line for a trapped ion. *Phys. Rev. Lett.* **103,** 183602 (2009).
20. Schmitz, H. *et al.* Quantum walk of a trapped ion in phase space. *Phys. Rev. Lett.* **103,** 090504 (2009).
21. Zähringer, F. *et al.* Realization of a quantum walk with one and two trapped ions. *Phys. Rev. Lett.* **104,** 100503 (2010).
22. Karski, M. *et al.* Quantum walk in position space with single optically trapped atoms. *Science* **325,** 174–177 (2009).
23. Perets, H. B. *et al.* Realization of quantum walks with negligible decoherence in waveguide lattices. *Phys. Rev. Lett.* **100,** 170506 (2008).
24. Carolan, J. *et al.* On the experimental verification of quantum complexity in linear optics. *Nat. Photon.* **8,** 621 (2014).
25. Manouchehri, K. & Wang, J. B. *Physical Implementation of Quantum Walks* (Springer-Verlag, 2014).
26. Aharonov, D. & Ta-Shma, A. in *Proceedings of the 35th Annual ACM Symposium on Theory of Computing* 20–29 (ACM, New York, NY, USA, 2003).
27. Berry, D. W., Ahokas, G., Cleve, R. & Sanders, B. C. Efficient quantum algorithms for simulating sparse hamiltonians. *Commun. Math. Phys.* **270,** 359–371 (2007).
28. Childs, A. M. & Kothari, R. Limitations on the simulation of non-sparse hamiltonians. *Quantum Inf. Comput.* **10,** 669–684 (2009).
29. Childs, A. M. On the relationship between continuous-and discrete-time quantum walk. *Commun. Math. Phys.* **294,** 581–603 (2010).
30. Ng, M. K. *Iterative Methods for Toeplitz Systems* (Oxford Univ. Press, 2004).
31. Gray, R. M. *Toeplitz and Circulant Matrices: A Review* (Now Publishers Inc., 2006).
32. Buhrman, H., Cleve, R., Watrous, J. & De Wolf, R. Quantum fingerprinting. *Phys. Rev. Lett.* **87,** 167902 (2001).
33. Han, D. *et al.* Folding and cutting DNA into reconfigurable topological nanostructures. *Nat. Nanotechnol.* **5,** 712–717 (2010).
34. Peres, A. Stability of quantum motion in chaotic and regular systems. *Phys. Rev. A* **30,** 1610 (1984).
35. Prosen, T. & Znidaric, M. Stability of quantum motion and correlation decay. *J. Phys. A Math. Gen.* **35,** 1455 (2002).
36. Aaronson, S. & Arkhipov, A. in *Proceedings of the 43rd Annual ACM Symposium on Theory of Computing* 333-342 (ACM Press, New York, NY, USA, 2011).
37. Bremner, M. J., Jozsa, R. & Shepherd, D. J. Classical simulation of commuting quantum computations implies collapse of the polynomial hierarchy. *Proc. R. Soc. A Math. Phys. Engineering Science* **467,** 459–472 (2010).
38. Bremner, M. J., Montanaro, A. & Shepherd, D. J. Average-case complexity versus approximate simulation of commuting quantum computations. Preprint at http://arxiv.org/abs/1504.07999 (2015).
39. Nielsen, N. A. & Chuang, I. L. *Quantum Computation and Quantum Information* (Cambridge Univ. Press, 2010).
40. Childs, A. M. *Quantum Information Processing in Continuous Time* (PhD thesis, Massachusetts Institute of Technology, 2004).
41. Zhou, X. Q. *et al.* Adding control to arbitrary unknown quantum operations. *Nat. Commun.* **2,** 413 (2011).
42. Shepherd, D. & Bremner, M. J. Temporally unstructured quantum computation. *Proceedings of the Royal Society A: Mathematical, Physical and Engineering Science* **465,** 1413–1439 (2009).
43. Fujii, K. & Morimae, T. Quantum commuting circuits and complexity of Ising partition functions. Preprint at http://arxiv.org/abs/1311.2128 (2013).
44. Goldberg, L. A. & Guo, H. The complexity of approximating complex-valued Ising and Tutte partition functions. Preprint at http://arxiv.org/abs/1409.5627 (2014).
45. Stockmeyer, L. J. On approximation algorithms for #P. *SIAM J. Comput.* **14,** 849–861 (1985).
46. Papadimitriou, C. *Computational Complexity* (Addison-Wesley, 1994).
47. Nest, M. Simulating quantum computers with probabilistic methods. *Quantum Inf. Comput.* **11,** 784–812 (2011).
48. Schwarz, M. & Nest, M. Simulating quantum circuits with sparse output distributions. Preprint at http://arxiv.org/abs/1310.6749 (2013).
49. Rohde, P. P. & Ralph, T. C. Error tolerance of the boson-sampling model for linear optics quantum computing. *Phys. Rev. A* **85,** 022332 (2012).
50. Rangarajan, R., Goggin, M. & Kwiat, P. Optimizing type-I polarization-entangled photons. *Opt. Express* **17,** 18920–18933 (2009).

Acknowledgements

We thank Anthony Laing and Peter J. Shadbolt for helpful discussions. We acknowledge support from the Engineering and Physical Sciences Research Council (EPSRC), the European Research Council (ERC)—including BBOI, QUCHIP (H2020-FETPROACT-3-2014: Quantum simulation) and PIQUE (FP7-PEOPLE-2013-ITN)—the Centre for Nanoscience and Quantum Information (NSQI), the US Army Research Office (ARO) grant W911NF-14-1-0133. X.Q. acknowledges support from University of Bristol and China Scholarship Council. T.L. is supported by an International Postgraduate Research Scholarship, Australian Postgraduate Award and the Bruce and Betty Green Postgraduate Research Top-Up Scholarship. J.L.O.B. acknowledges a Royal Society Wolfson Merit Award and a Royal Academy of Engineering Chair in Emerging Technologies. J.B.W. acknowledges the Benjamin Meaker visiting professorship provided by IAS at University of Bristol. J.C.F.M. was supported by a Leverhulme Trust Early Career Fellowship and an EPSRC Early Career Fellowship.

Author contributions

X.Q., T.L., J.L.O.B., J.B.W. and J.C.F.M. conceived and designed the project. T.L. and J.B.W. proposed the circuit construction scheme. X.Q., K.A. and X.Z. built the experiment set-up. X.Q. performed the experiments and analysed the data. A.M. provided the complexity theory proofs. X.Q., T.L., A.M., J.L.O.B., J.B.W. and J.C.F.M. wrote the manuscript. J.L.O.B., J.B.W. and J.C.F.M. supervised the project.

Additional information

Competing financial interests: The authors declare no competing financial interests.

5

Mapping quantum state dynamics in spontaneous emission

M. Naghiloo[1], N. Foroozani[1], D. Tan[1], A. Jadbabaie[1] & K.W. Murch[1,2]

The evolution of a quantum state undergoing radiative decay depends on how its emission is detected. If the emission is detected in the form of energy quanta, the evolution is characterized by a quantum jump to a lower energy state. In contrast, detection of the wave nature of the emitted radiation leads to different dynamics. Here, we investigate the diffusive dynamics of a superconducting artificial atom under continuous homodyne detection of its spontaneous emission. Using quantum state tomography, we characterize the correlation between the detected homodyne signal and the emitter's state, and map out the conditional back-action of homodyne measurement. By tracking the diffusive quantum trajectories of the state as it decays, we characterize selective stochastic excitation induced by the choice of measurement basis. Our results demonstrate dramatic differences from the quantum jump evolution associated with photodetection and highlight how continuous field detection can be harnessed to control quantum evolution.

[1]Department of Physics, Washington University, St Louis, Missouri 63130, USA. [2]Institute for Materials Science and Engineering, St Louis, Missouri 63130, USA. Correspondence and requests for materials should be addressed to K.W.M. (email: murch@physics.wustl.edu).

In spontaneous emission, an emitter decays from an excited state by releasing radiation into a quantized mode of the electromagnetic field. From the point of view of quantum measurement theory, the light–matter interaction entangles the quantum state of the emitter with its electromagnetic environment[1,2]. The emission field may therefore serve as a pointer state used to indirectly monitor the emitter's evolution. Subsequent measurements of the field convey information about the state of the emitter and consequently cause back-action[3]. Typically, spontaneous emission is detected in the form of energy quanta, resulting in an instantaneous jump of the emitter to a lower energy state. However, if the emission is measured with a detector that is not sensitive to quanta, but rather to the amplitude of the field, the emitter's state undergoes different dynamics over finite timescales. In this case, the emitter is predicted to diffuse through its state space rather than abruptly decay to the ground state in a quantum jump, and the choice of measurement on the pointer will affect the dynamics of diffusion[4–7].

In this work, we perform homodyne measurements of the spontaneous emission from an effective two-level system, formed by the strong light–matter interaction between a superconducting circuit and a microwave cavity. By performing phase-sensitive amplification, we selectively amplify and de-amplify orthogonal quadratures of the emission field, enforcing a choice of measurement basis on the pointer state and limiting the measurement back-action experienced by the emitter[8,9]. We use quantum state tomography, in addition to weak measurements, to study the back-action of homodyne measurements and track the time evolution of the emitter's state under radiative decay. Contrary to the evolution expected for photodetection, we observe quantum trajectories[3,8,10] that stochastically diffuse through the state space of the emitter. For certain phases of homodyne detection we observe that the emitter evolves towards its excited state under radiative decay[5]. This stochastic excitation demonstrates how phase-sensitive amplification of spontaneous emission can be utilized to control the emitter's evolution.

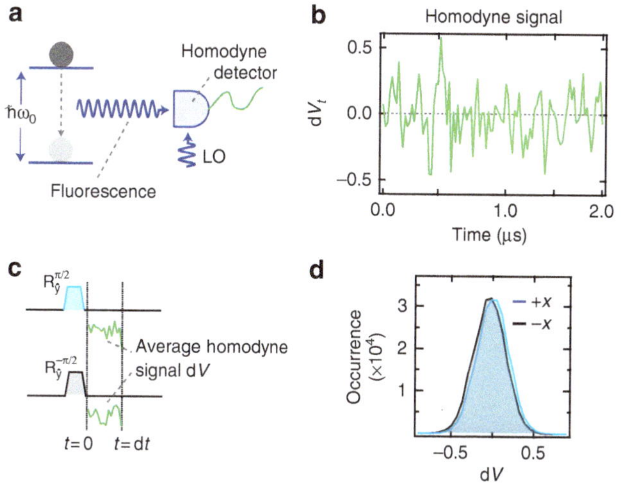

Figure 1 | Experimental set-up. (**a**) The experiment uses a near-quantum-limited Josephson parametric amplifier, labelled here as a homodyne detector, to perform homodyne measurements of the fluorescence emitted by an effective two-level system. (**b**) A sample of the dimensionless homodyne signal (denoted dV_t at time step t). The noisy signal reflects the quantum fluctuations of the measured electromagnetic mode and is normalized such that its variance is γdt. (**c**) To calibrate the measurement, we use $\pi/2$ rotations to prepare the emitter in the states $\pm x$ and average the ensuing homodyne signal for a time $dt = 20$ ns. (**d**) Histograms of the averaged homodyne signals **c** show how the measurement carries partial information about the σ_x quadrature of the emitter's dipole. The strength of measurement, set by the emitter's decay rate γ, is inversely proportional to the overlap of the histograms.

Results

Experimental set-up. Our system (Fig. 1a) consists of an effective two-level emitter formed by the resonant interaction of a transmon circuit[11] and a three-dimensional waveguide cavity[12]. The strong light–matter interaction between the circuit and the cavity strips them of their individual character and gives rise to hybrid circuit-cavity states. We use the lowest energy transition ($\omega_0/2\pi = 6.83$ GHz) as an effective two-level system. Deliberate coupling to a 50 Ω transmission line results in a radiative decay rate $\gamma = 2.3 \times 10^6\,\mathrm{s}^{-1}$. The process of emission is described by the interaction Hamiltonian, $H_{\mathrm{int}} = \gamma(a^\dagger\sigma_- + a\sigma_+)$, where $a^\dagger$ (a) is the creation (annihilation) operator for a photon in the transmission line, and σ_+ (σ_-) is the pseudo-spin raising (lowering) operator. This interaction couples an arbitrary field quadrature $a^\dagger e^{i\phi} + ae^{-i\phi}$ to a corresponding emitter dipole $\sigma_- e^{i\phi} + \sigma_+ e^{-i\phi}$. Due to the Heisenberg uncertainty relations, the emitted radiation exhibits quantum fluctuations in its quadrature amplitudes. If these fluctuations are measured, they provide information on the emitter state and drive its stochastic evolution. Conversely, if the fluctuations are de-amplified, their information is no longer available, eliminating the corresponding stochastic back-action on the emitter state.

To accurately detect these quantum fluctuations, we perform phase-sensitive amplification[13] of outgoing signals near the emission frequency using a near-quantum-limited Josephson parametric amplifier[14,15]. In this mode of operation, the amplifier squeezes the outgoing light along an axis in quadrature space given by the phase of the amplifier pump ϕ. This constitutes a homodyne measurement of the amplified field quadrature $a^\dagger e^{i\phi} + ae^{-i\phi}$. Due to the emitter-field interaction, the choice of ϕ effectively enforces a choice of measurement basis on the emitter. In our experiment, we choose the amplifier phase $\phi = 0$; the corresponding noisy homodyne signal (denoted dV_t, Fig. 1b) is then sensitive to the emitter dipole $\sigma_- + \sigma_+ = \sigma_x$.

The variance of the homodyne signal originates not only from the quantum fluctuations of the detected mode, but also from losses and added noise in the amplification chain. We account for this loss of information into the incoherent and dissipative classical environment with the quantum efficiency η. Meanwhile, we treat the quantum noise as a Weiner process; the fluctuations of the measurement signal dV_t in an infinitesimal time step dt are described by stochastic noise increments dW_t. Known as Weiner increments[3], these are zero-mean, Gaussian random variables with variance dt. To accurately reflect this stochastic and dimensionless nature of the homodyne signal, we scale dV_t such that it has a variance $\sigma^2 = \gamma dt$, with the full measurement record given by $dV_t = \sqrt{\eta}\gamma\langle\sigma_x\rangle dt + \sqrt{\gamma}dW_t$.

To experimentally demonstrate that our homodyne detection scheme is sensitive to a single quadrature of the emitter's dipole, we prepare the emitter in a specific state, perform homodyne measurement with $\phi = 0$, and integrate the resulting signal (Fig. 1c,d). By repeating the measurement for several iterations, we can create histograms of the homodyne signal. We compare the resulting distributions for two state preparations, $\pm x$ (the positive or negative eigenstates of the σ_x Pauli operator). The observed separation of the two histograms, $\Delta V = 2\sqrt{\eta}\gamma dt$, gives the quantum efficiency of our detection set-up as $\eta = 0.3$.

Conditional dynamics of radiative decay. We now study the conditional dynamics of the emitter's state under radiative decay.

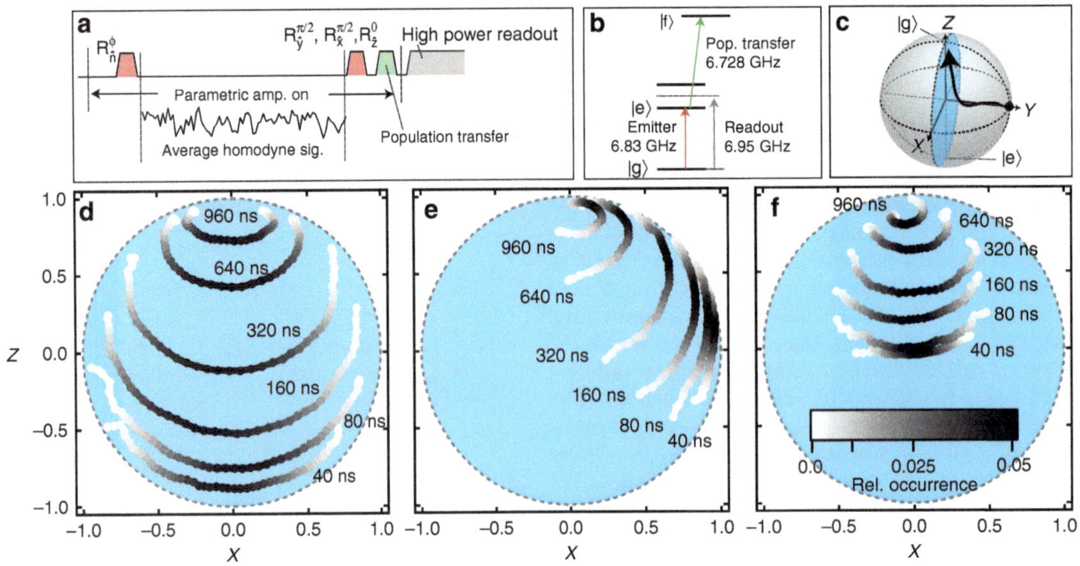

Figure 2 | Mapping spontaneous decay. (**a**) The experimental sequence prepares the emitter in an initial state, then homodyne detection is used to record the emitted radiation. Following a variable period of time, further rotations are applied to the emitter before state readout to perform quantum state tomography. To enhance the readout contrast, a pulse is applied to transfer the excited state population to a higher state of the system. (**b**) The level structure of the system and frequencies of the three microwave drives. The emitter is given by the lowest two energy levels. (**c**) We average the state tomography to determine $x \equiv \langle\sigma_x\rangle|_{\bar{V}}$, and $z \equiv \langle\sigma_z\rangle|_{\bar{V}}$ conditioned on the outcome of the homodyne measurement. These correlated tomography results are displayed on the *X*-*Z* plane of the Bloch sphere for three different initial states: $-z$ (**d**) $+x$ (**e**) and $+y$ (**f**). The colour scale indicates the relative occurrence of each measurement value. Note the different back-action dynamics between **e** and **f**, a result of phase-sensitive amplification.

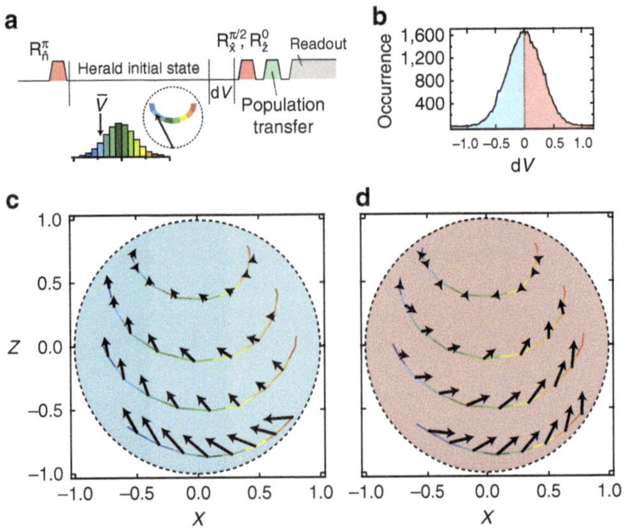

Figure 3 | Back-action vector maps. (**a**) An arbitrary initial state in the *X*-*Z* plane of the Bloch sphere is prepared by heralding on the average homodyne signal $\bar{V}$. After heralding, we digitize the signal for an additional 40 ns to obtain dV. Finally, we use quantum state tomography to examine the conditional back-action based on the value of dV. (**b**) Histogram of the signals dV which we separate into positive or negative dV. The back-action imparted on the emitter for negative (**c**) or positive (**d**) values of dV is depicted by an arrow at different locations in the *X*-*Z* plane of the Bloch sphere.

We conduct the experimental sequence depicted in Fig. 2a,b; we first use a resonant rotation to prepare an initial state, then obtain the average homodyne signal $\bar{V}$ by integrating the detected homodyne signal for a variable period of time, and finally perform projective measurements to conduct quantum state tomography as described in the Methods section. The results of these projective measurements are averaged conditionally on the integrated homodyne signal. This yields the conditional Pauli averages, $\langle\sigma_x\rangle|_{\bar{V}}$, $\langle\sigma_y\rangle|_{\bar{V}}$, $\langle\sigma_z\rangle|_{\bar{V}}$. In Fig. 2d–f we plot $\langle\sigma_z\rangle|_{\bar{V}}$ and $\langle\sigma_x\rangle|_{\bar{V}}$ parametrically on the *X*-*Z* plane of the Bloch sphere for different integration times. We study the conditional evolution for three different state preparations.

When the emitter is prepared in the excited state (Fig. 2d), the *x*-component of the state develops a correlation with the average homodyne signal. This highlights how our homodyne measurement provides an indirect signature[6] of only the real part of $\sigma_- = (\sigma_x + i\sigma_y)/2$. As the state is allowed more time to decay, it evolves to different deterministic arcs in the interior of the Bloch sphere.

Under phase-sensitive amplification, the choice of homodyne phase can vary the stochastic back-action on the emitter's state. To study this, we compare two different state preparations, $+x$ and $+y$. When the emitter is prepared in the state $+x$ (Fig. 2e), we observe that some of the conditioned states evolve towards the excited state[5]. This stochastic excitation is unique to amplitude measurements of the field quadrature, since such excitation is not possible under photodetection[6]. In contrast, when the emitter is prepared in the state $+y$, an eigenstate of the imaginary part of our measured operator $\sigma_- = (\sigma_x + i\sigma_y)/2$, the emitter dipole corresponds to the de-amplified quadrature of the emission field, and no stochastic excitation is observed (Fig. 2f). This different state preparation is equivalent to preparing the emitter in the same state $+x$ (as depicted in Fig. 2e) and instead changing the homodyne phase by $\pi/2$. This demonstrates how the choice of homodyne measurement phase can be used to control the evolution of the emitter.

We take advantage of the deterministic evolution of the emitter, conditioned on the integrated homodyne signal, to characterize the back-action at different points in the Bloch sphere. Figure 3 shows a vector map of the state evolution due to a specific detected homodyne signal dV at various points. By preparing the emitter in the excited state and averaging the homodyne signal for various periods of time, we can prepare a

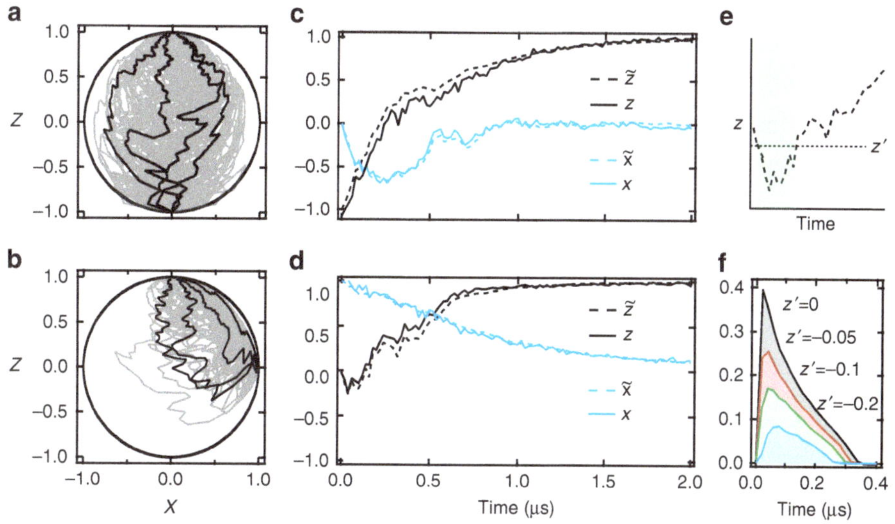

Figure 4 | Quantum trajectories. (**a**,**b**) Quantum trajectories of spontaneous decay calculated by the stochastic master equation, initialized at $-z$ (**a**) and $+x$ (**b**). Several trajectories are depicted in grey, and a few individual trajectories are highlighted in black. (**c**,**d**) Individual trajectories $(\tilde{x}, \tilde{z})$ that originate from $-z$ (**c**) and $+x$ (**d**) are shown as dashed lines and the tomographic reconstructions (Methods section) based on projective measurements are shown as solid lines. (**e**) For trajectories that are initialized along $+x$ some are excited (attaining values below a threshold z'). (**f**) Fraction of the trajectories that are excited below the threshold z' versus time.

nearly arbitrary mixed state through heralding. After selecting a decay time and a specific initial state (x_i, z_i), based on an average signal $\bar{V}$, we digitize the homodyne signal for an additional time $\mathrm{d}t = 40$ ns to obtain $\mathrm{d}V$. We then use quantum state tomography to determine the final state (x_f, z_f), conditioned on the detection of $\mathrm{d}V$ within a specified range. The back-action at a specific location in state space, associated with the detection of a given value of $\mathrm{d}V$, is provided by the vector connecting (x_i, z_i) and (x_f, z_f). The back-action vector maps demonstrate how positive (negative) measurement results push the state towards $+x$ $(-x)$. Furthermore, the maps show that the back-action is stronger near the state $-z$, indicating that the measurement strength is proportional to the emitter's excitation.

Quantum trajectories. The back-action maps that we present in Fig. 3 allow us to calculate the evolution of the emitter's state conditioned on a sequence of homodyne measurement results. Formally, this evolution is described by the stochastic master equation[5],

$$\mathrm{d}\rho = \gamma\mathcal{D}[\sigma_-]\rho\mathrm{d}t + \sqrt{\eta\gamma}\mathcal{H}[\sigma_-\mathrm{d}W_t]\rho. \quad (1)$$

Where $\mathcal{D}[\sigma_-]\rho = \sigma_-\rho\sigma_+ - \frac{1}{2}(\sigma_+\sigma_-\rho + \rho\sigma_+\sigma_-)$ and $\mathcal{H}[O]\rho = O\rho + \rho O^\dagger - \mathrm{tr}[(O+O^\dagger)\rho]\rho$ are the dissipation and jump superoperators, respectively. When we ignore the results of homodyne monitoring (for example by setting $\eta = 0$), the state follows deterministic evolution from an initial state to the ground state, as described by the first term of equation (1). The second term accounts for information conveyed by the homodyne measurement through stochastic noise increments $\mathrm{d}W_t$. We can recast this stochastic master equation in terms of the Bloch vector components x, z, y,

$$\mathrm{d}x = -\frac{\gamma}{2}x\mathrm{d}t + \sqrt{\eta}(1 - z - x^2)(\mathrm{d}V_t - \gamma\sqrt{\eta}x\mathrm{d}t), \quad (2)$$

$$\mathrm{d}z = \gamma(1 - z)\mathrm{d}t + \sqrt{\eta}x(1 - z)(\mathrm{d}V_t - \gamma\sqrt{\eta}x\mathrm{d}t), \quad (3)$$

$$\mathrm{d}y = -\frac{\gamma}{2}y\mathrm{d}t - \sqrt{\eta}xy(\mathrm{d}V_t - \gamma\sqrt{\eta}x\mathrm{d}t). \quad (4)$$

We now turn to calculating individual quantum trajectories for the emitter's state. In Fig. 4, we prepare the emitter in the excited state and then digitize the detected homodyne signal for 2 μs. Based on this signal, we use equations (2–4) to calculate the emitter's trajectory using time steps of $\mathrm{d}t = 20$ ns. Instead of taking a straight path to the ground state, the trajectory diffuses through the Bloch sphere, subject to back-action from the measured quantum fluctuations of the emission field.

We also study quantum trajectories originating from the state $+x$. In this case, the stochastic back-action causes some of the trajectories to become more excited as they decay under homodyne detection. In Fig. 4f, we quantify this feature by extracting the probability of excitation above a certain threshold at different times. By examining the measurement term in equation (3), proportional to $\sqrt{\eta}$, we see that the state at $+x$ will be stochastically excited if the Weiner increment $\mathrm{d}W_t$, obtained from the detected signal $\mathrm{d}V_t$, is less than $-\sqrt{\gamma/\eta}\mathrm{d}t$, predicting that ~35% of the trajectories should be excited in the first time step.

Discussion

In recent years, several experiments have demonstrated control over the emission process by either altering vacuum fluctuations[16] or engineering the electromagnetic environment[17,18], allowing spontaneous emission to be used as a resource for quantum information[19,20]. In addition, the entanglement between a quantum emitter and its spontaneous emission field has been studied in experiments using natural atoms[1] and solid state systems[2], and can be used to herald entanglement between spatially separated systems[21]. Our work highlights how spontaneous emission can also be used as a resource for quantum measurement, where the emission field serves as a pointer system for indirect measurements of the emitter state. Contemporary experiments[7–10,22–26] that harness Bayesian statistics or use quantum optics to track the evolution of quantum states have yielded a deeper understanding of quantum measurement evolution. Here, we have shown how specific quadrature measurements of the fluorescence from a quantum emitter result in a rich conditional evolution of the state. We have harnessed this evolution to map out the back-action associated with such measurements, and we have tracked the individual quantum trajectories an emitter takes when decaying through

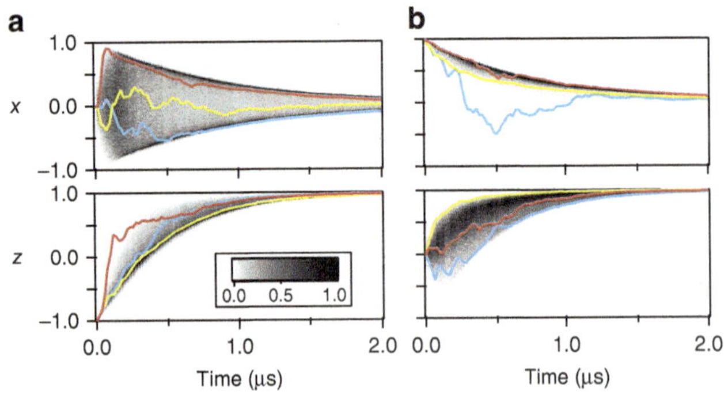

Figure 5 | State histograms. Greyscale histograms represent the distribution for values of x and z at each time point. The greyscale shading is normalized such that the most frequent value is 1 at each time point. (**a**) Histograms of the state when the emitter is initialized in the state $-z$ with a few sample trajectories shown in colour. (**b**) Histograms associated with decay from the state $+x$.

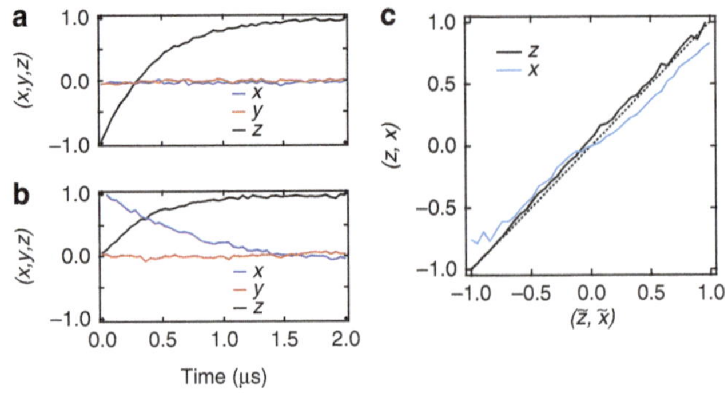

Figure 6 | Tomography calibrations. The ensemble decay as determined by projective measurements for initial states $-z$ (**a**) and $+x$ (**b**). (**c**) Tomographic validation for the ensemble of trajectories shows the average tomography values (x, z) versus the values obtained from individual trajectories.

fluorescence. In contrast to the instantaneous dynamics of emission due to measurements of quanta, here we show that spontaneous emission may also occur over finite timescales.

Measurements, and more broadly, control over a quantum environment, can in principle be used to steer quantum evolution[27,28]. Though the physical mechanism of energy decay remains constant between homodyne detection and photodetection, by changing the basis for the measurement of the emission field, we have significantly altered the emitter state's dynamics. In this work, quantum noise of the pointer state drives the evolution of the emitter, demonstrated clearly by our use of a phase-sensitive amplifier. By squeezing the pointer state, we change the nature of its quantum fluctuations, and therefore cause selective measurement back-action on the emitter. Such control over the quantum light–matter interaction has the potential to advance techniques in fluorescence based imaging, and will be essential in quantum feedback control[3,9,29] of quantum systems.

Methods

Device fabrication and parameters. The emitter system consists of a transmon circuit characterized by charging energy $E_C/h = 270$ MHz and Josephson energy $E_J/h = 24.6$ GHz. The circuit was fabricated by double-angle evaporation of aluminium on a high resistivity silicon substrate. The circuit was then placed at the centre of a waveguide cavity (dimensions 34.15 × 27.9 × 5.25 mm) machined from 6061 aluminium. The cavity geometry was chosen to be resonant with the lowest energy transition of the transmon circuit. The resonant interaction between the circuit and the cavity (characterized by coupling rate $g/2\pi = 136$ MHz) results in hybrid states, as described by the Jaynes–Cummings Hamiltonian. The cavity is deliberately coupled to two 50 Ω cables: one weakly coupled port, characterized by coupling quality factor $Q_c \simeq 10^5$, is used to drive the system, while a more strongly coupled port $Q_c \simeq 10^4$ sets the total radiative decay time of the system. This configuration results in an effectively 'one dimensional atom', where all of the radiative decay is captured by the strongly coupled cable[16]. Spontaneous emission from this 'artificial atom' is amplified by a near-quantum-limited Josephson parametric amplifier, consisting of a 1.5 pF capacitor, shunted by a superconducting quantum interference device (SQUID) composed of two $I_0 = 1$ μA Josephson junctions. The amplifier is operated with negligible flux threading the SQUID loop and produces 20 dB of gain with an instantaneous 3-dB-bandwidth of 20 MHz.

We used standard techniques to measure the energy decay time $T_1 = 430$ ns and Ramsey decay time $T_2^* = 830$ ns, indicating that the emitter experiences a negligibly small amount of pure dephasing. We also examined the equilibrium state populations of the emitter using a Rabi driving technique[30], and found the excited state population to be less than 3%.

State tracking. We use a master equation (equivalent to equations (2–4)) to propagate the density matrix for the emitter's state conditioned on the detected homodyne signal. The signal is digitized in 20 ns steps, and scaled such that its variance is γdt. At each time step, we update the density matrix components $\rho_{11}[i]$ and $\rho_{01}[i]$ based on the detected measurement signal $dV[i]$, where $z \equiv 1 - 2\rho_{11}$ and $x \equiv 2\mathrm{Re}[\rho_{01}]$. Our state update is consistent with the Itô formulation of stochastic calculus.

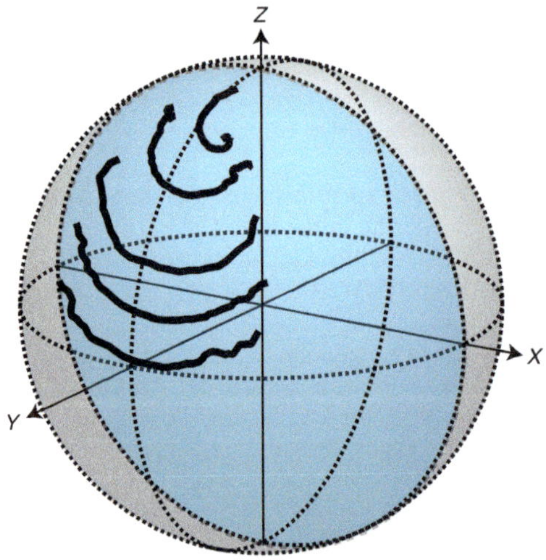

Figure 7 | Spontaneous decay from the state $+y$. The emitter's state at different times conditioned on the integrated homodyne measurement signal. The decay times are 80, 160, 320, 640 and 960 ns, and the data correspond to what is depicted in Fig. 2f. The X–Z plane plotted in Fig. 2f is highlighted in blue.

$$\rho_{11}[i+1] = \rho_{11}[i] - \gamma\rho_{11}[i]dt - \sqrt{\eta}\left(dV[i] - \sqrt{\eta\gamma}2\rho_{01}[i]dt\right) \times \left(2\rho_{01}[i]\rho_{11}[i]\right) \quad (5)$$

$$\rho_{01}[i+1] = \rho_{01}[i] - \gamma\rho_{01}[i]/2dt + \sqrt{\eta}\left(dV[i] - \sqrt{\eta\gamma}2\rho_{01}[i]dt\right) \times \left(\rho_{11}[i] - 2\rho_{01}[i]\rho_{01}[i]\right) \quad (6)$$

Ensemble dynamics. Based on 9×10^5 repetitions of the experiment and associated quantum trajectories, we can examine ensemble dynamics of the paths on the Bloch sphere taken by our decaying emitter. The behaviour of single trajectories characterizes the dynamics of spontaneous decay subject to homodyne detection, and is distinctly different than the full ensemble behaviour that decays deterministically towards the ground state.

Figure 5 displays greyscale histograms of the state at different points in time for two different initial conditions. For trajectories initialized in $-z$ (Fig. 5a), these histograms demonstrate how the decay paths are restricted to a deterministic arc in the Bloch sphere. Curiously enough, a state prepared in a traditional eigenstate of spontaneous emission will develop some quantum coherence when monitored under homodyne detection. The x-components of such trajectories may be pinned to the edges of this arc on the X-axis, or instead may oscillate about the central

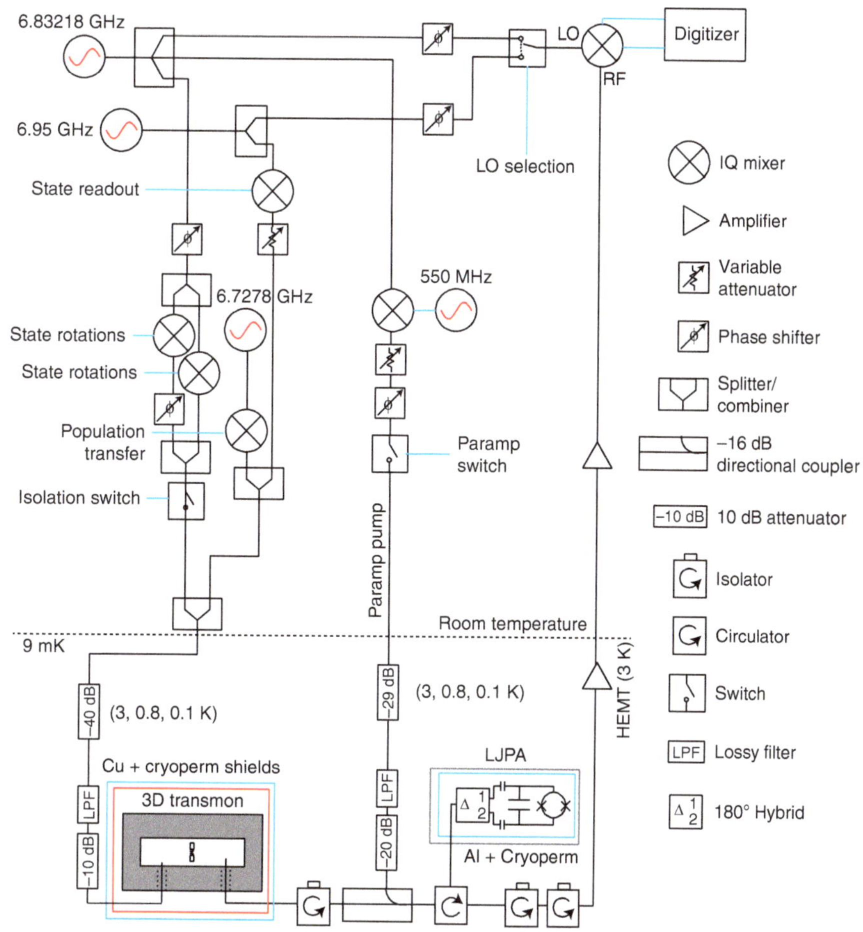

Figure 8 | Experimental set-up. A single microwave frequency generator is used for the qubit rotations, the amplifier pump and the local oscillator for demodulation. A second generator, operating at the high-power resonance frequency of the cavity is used for state readout for tomography. The third generator is used to transfer state population to a higher excited state of the circuit-cavity system to enhance the readout contrast.

value of $x = 0$. We note that though the trajectories exhibit an immediate diffusive behaviour for short timescales, the decay of coherence takes over at longer timescales, indicated by a decreasing upper bound on the stochastically acquired coherence. Examining behaviour along the Z-axis, we see that though some trajectories may decay by more quickly approaching the ground state, no trajectory may decay more slowly in z than a specific lower bound at each time step.

On the other hand, when the emitter is initialized along $+x$ in a superposition of its excited and ground states, the histograms of the Bloch sphere coordinates show different behaviour (Fig. 5b). The x-component of the trajectory encounters a decreasing upper bound on its maximum value, once more illustrating motion along a shrinking deterministic arc. The z-component, however, can exhibit extremely varied behaviour. In addition to following the average decay path, the state may also stochastically excite, or it may rapidly decay in z while approaching the surface of the Bloch sphere. Currently, it is these states that rapidly decay that have the highest purity on average, retaining the most information about the state. In comparison, due to our limited measurement efficiency, stochastically excited trajectories become more mixed as they diffuse towards the excited state. We note that for $\eta = 1$, all of our trajectories, regardless of dynamics, would describe pure states confined to move only on the surface on the Bloch sphere.

In fact, we expect the ensemble ratio of stochastically excited trajectories to increase with increasing η. As mentioned in the main text, trajectories experience $dz < 0$ when the Weiner increment obtained from the measurement record satisfies $dW_t < -\sqrt{\gamma}dt/\sqrt{\eta}x$. Recall that dW_t is a zero-mean random variable distributed with variance dt, and consider the back-action experienced by trajectories initialized with $x = 1$. Naively, the probability of stochastic excitation is then given by the integral,

$$\int_{-\infty}^{-\sqrt{\gamma/\eta}dt} dW_t (2\pi dt)^{-1/2} e^{-dW_t^2/2dt}. \quad (7)$$

As η increases, so does the value of this integral. For $\eta = 1$ and a time step $dt = 20$ ns, the probability for stochastic excitation for our system reaches a maximum value of ~41.5%. For our measured quantum efficiency of $\eta = 0.3$, we expect ~35% of trajectories to excite in the first time step.

Tomography and readout calibration. All tomography results are corrected for imperfect state preparation and readout fidelities. We perform state readout by first applying a resonant pulse at 6.73 GHz to transfer the excited state population to a higher excited state, and then proceed to drive the bare cavity resonance at 6.95 GHz at high power to conduct the Jaynes–Cummings high-power readout technique[31]. Tomography for y and x is achieved by first applying a 40 ns $\pi/2$ rotation about the X or Y axes. The combined state preparation and readout fidelity (80%) was determined from the contrast of resonant Rabi oscillations. Each experimental sequence includes separate calibration measurements used to determine the readout level of the ground state and the prepared excited state. These levels are used to scale the tomography results. Figure 6a,b shows the ensemble decay curves for the state preparations $-z$ and $+x$.

The emitter's state is characterized by expectation values (x, z). To characterize accuracy of the state tracking, we compare the expectation values that are calculated for a single iteration of the experiment to the values obtained from an ensemble of projective measurements. In Fig. 4 we show this comparison to reconstruct an individual trajectory. To accomplish this, we denote an individual trajectory $(\tilde{x}(t), \tilde{y}(t), \tilde{z}(t))$ (Note that $\tilde{y}(t)=0$). At each time point, we perform several experiments of total duration t', followed by one of three tomography and readout sequences. For each of these experiments, we calculate $(x(t'), z(t'))$; if $x(t')$ and $z(t')$ are within ± 0.12 of $\tilde{x}(t')$ and $\tilde{z}(t')$, then the subsequent tomography result is included in the tomographic validation at t'. We follow this process for each t' along the trajectory, resulting in a tomographic reconstruction of the trajectory.

We can further test the predictions given by the individual trajectories for all runs of the experiment at all times. Figure 6c displays the average projective measurement outcomes conditioned on the values of $\tilde{x}(t')$ or $\tilde{z}(t')$ compared with the values $\tilde{x}(t')$ or $\tilde{z}(t')$ showing good agreement between the individual trajectories and the projective measurements.

Phase-sensitive back-action. When the emitter is initialized in $+y$ the state dynamics are not confined to the X–Z plane. Figure 7 displays the state conditioned on the integrated homodyne signal and shows how the y-component does not acquire a correlation with the measurement signal. This may be understood as a result of phase-sensitive amplification with $\phi = 0$. When we perform our homodyne measurement of the real part of σ_-, we de-amplify the quadrature containing information on the imaginary part of σ_-, corresponding to σ_y on the Bloch sphere. The de-amplification of this orthogonal signal suppresses the magnitude of its quantum fluctuations, effectively eliminating the information associated with the σ_y quadrature of the emitter's dipole. Therefore we do not perform weak measurements of σ_y, and we do not observe quantum dynamics such as stochastic excitation.

We may also understand this phenomenon by examining the dz and dy segments of the stochastic master equation provided in the main text. The presence of an xy coefficient on the measurement term in equation (4), means the stochastic back-action has no effect on the state when it is in an eigenstate of σ_x or σ_y, limiting dynamics to a deterministic reduction in y. Meanwhile, if we examine equation (3) after factoring out a common factor of $(1 - z)$, which serves to push the trajectory towards the ground state, we see the measurement term is proportional only to x. Therefore, for a state prepared with $y = \pm 1$, there will be no initial stochastic excitation, and the state will begin its decay by deterministically approaching the ground state. However, once fluctuations in the measurement signal cause the state to acquire a nonzero x value, the trajectory's dynamics will cease to be trivial.

Experimental set-up. Figure 8 displays a simplified schematic of the experimental set-up. A single generator is used for qubit rotations, the amplifier pump, and demodulation of the amplified signal. The parametric amplifier is pumped by two sidebands that are equally separated from the carrier by 550 MHz, allowing for phase-sensitive amplification without leakage at the emitter's transition frequency. The experimental repetition rate is 8 kHz.

References

1. Blinov, B. B., Moehring, D. L., Duan, L.-M. & Monroe, C. Observation of entanglement between a single trapped atom and a single photon. *Nature* **428,** 153–157 (2004).
2. Eichler, C. *et al.* Observation of entanglement between itinerant microwave photons and a superconducting qubit. *Phys. Rev. Lett.* **109,** 240501 (2012).
3. Wiseman, H. & Milburn, G. *Quantum Measurement and Control* (Cambridge University Press, 2010).
4. Wiseman, H. M. & Gambetta, J. M. Are dynamical quantum jumps detector dependent? *Phys. Rev. Lett.* **108,** 220402 (2012).
5. Bolund, A. & Mølmer, K. Stochastic excitation during the decay of a two-level emitter subject to homodyne and heterodyne detection. *Phys. Rev. A* **89,** 023827 (2014).
6. Jordan, A. N., Chantasri, A., Rouchon, P. & Huard, B. Anatomy of fluorescence: quantum trajectory statistics from continuously measuring spontaneous emission. Preprint at http://arxiv.org/abs/1511.06677 (2015).
7. Campagne-Ibarcq, P. *et al.* Observing quantum state diffusion by heterodyne detection of fluorescence. *Phys. Rev. X* **6,** 011002 (2016).
8. Murch, K. W., Weber, S. J., Macklin, C. & Siddiqi, I. Observing single quantum trajectories of a superconducting qubit. *Nature* **502,** 211–214 (2013).
9. de Lange, G. *et al.* Reversing quantum trajectories with analog feedback. *Phys. Rev. Lett.* **112,** 080501 (2014).
10. Weber, S. J. *et al.* Mapping the optimal route between two quantum states. *Nature* **511,** 570–573 (2014).
11. Koch, J. *et al.* Charge-insensitive qubit design derived from the cooper pair box. *Phys. Rev. A* **76,** 042319 (2007).
12. Paik, H. *et al.* Observation of high coherence in josephson junction qubits measured in a three-dimensional circuit qed architecture. *Phys. Rev. Lett.* **107,** 240501 (2011).
13. Clerk, A. A., Devoret, M. H., Girvin, S. M., Marquardt, F. & Schoelkopf, R. J. Introduction to quantum noise, measurement, and amplification. *Rev. Mod. Phys.* **82,** 1155–1208 (2010).
14. Castellanos-Beltran, M. A., Irwin, K. D., Hilton, G. C., Vale, L. R. & Lehnert, K. W. Amplification and squeezing of quantum noise with a tunable josephson metamaterial. *Nat. Phys.* **4,** 929–931 (2008).
15. Hatridge, M., Vijay, R., Slichter, D. H., John Clarke & Siddiqi, I. Dispersive magnetometry with a quantum limited squid parametric amplifier. *Phys. Rev. B* **83,** 134501 (2011).
16. Murch, K. W., Weber, S. J., Beck, K. M., Ginossar, E. & Siddiqi, I. Reduction of the radiative decay of atomic coherence in squeezed vacuum. *Nature* **499,** 62–65 (2013).
17. Purcell, E. M. Spontaneous emission probabilities at radio frequencies. *Phys. Rev.* **69,** 681 (1946).
18. Houck, A. A. *et al.* Generating single microwave photons in a circuit. *Nature* **449,** 328–331 (2007).
19. Houck, A. A. *et al.* Controlling the spontaneous emission of a superconducting transmon qubit. *Phys. Rev. Lett.* **101,** 080502 (2008).
20. Hoi, I.-C. *et al.* Probing the quantum vacuum with an artificial atom in front of a mirror. *Nat. Phys.* **11,** 1045–1049 (2015).
21. Bernien, H. *et al.* Heralded entanglement between solid-state qubits separated by three metres. *Nature* **497,** 86–90 (2013).
22. Katz, N. *et al.* Coherent state evolution in a superconducting qubit from partial-collapse measurement. *Science* **312,** 1498–1500 (2006).
23. Guerlin, C. *et al.* Progressive field-state collapse and quantum non-demolition photon counting. *Nature* **448,** 889–893 (2007).
24. Hatridge, M. *et al.* Quantum back-action of an individual variable-strength measurement. *Science* **339,** 178–181 (2013).
25. Groen, J. P. *et al.* Partial-measurement backaction and nonclassical weak values in a superconducting circuit. *Phys. Rev. Lett.* **111,** 090506 (2013).
26. Roch, N. *et al.* Observation of measurement-induced entanglement and quantum trajectories of remote superconducting qubits. *Phys. Rev. Lett.* **112,** 170501 (2014).
27. Warren, W. S., Rabitz, H. & Dahleh, M. Coherent control of quantum dynamics: the dream is alive. *Science* **259,** 1581–1589 (1993).
28. Shapiro, M. & Brumer, P. *Quantum Control of Molecular Processes* (Wiley-VCH Verlag GmbH & Co. KGaA, 2011).
29. Vijay, R. *et al.* Stabilizing rabi oscillations in a superconducting qubit using quantum feedback. *Nature* **490,** 77–80 (2012).
30. Geerlings, K. *et al.* Demonstrating a driven reset protocol for a superconducting qubit. *Phys. Rev. Lett.* **110,** 120501 (2013).
31. Reed, M. D. *et al.* High-fidelity readout in circuit quantum electrodynamics using the jaynes-cummings nonlinearity. *Phys. Rev. Lett.* **105,** 173601 (2010).

Acknowledgements

We thank A.N. Jordan, K. Mølmer and P. Harrington for discussions. This research was supported in part by the John Templeton Foundation and the Sloan Foundation, and used facilities at the Institute of Materials Science and Engineering at Washington University.

Author contributions

M.N., N.F. and K.W.M. performed the experiments and D.T. fabricated the samples. All authors contributed to the data analysis and writing the manuscript.

Additional information

Competing financial interests: The authors declare no competing financial interests.

6

Coherent quantum depletion of an interacting atom condensate

M. Kira[1]

Sufficiently strong interactions promote coherent quantum transitions in spite of thermalization and losses, which are the adversaries of delicate effects such as reversibility and correlations. In atomic Bose-Einstein condensates (BECs), strong atom–atom interactions can eject atoms from the BEC to the normal component, yielding quantum depletion instead of temperature depletion. A recent experiment has already been verified to overcome losses. Here I show that it also achieves coherent quantum-depletion dynamics in a BEC swept fast enough from weak to strong atom–atom interactions. The elementary coherent process first excites the normal component into a liquid state that evolves into a spherical shell state, where the atom occupation peaks at a finite momentum to shield 50% of the BEC atoms from annihilation. The identified coherent processes resemble ultrafast semiconductor excitations expanding the scope of BEC explorations to many-body non-equilibrium studies.

[1] Department of Physics, Philipps-University Marburg, Renthof 5, D-35032 Marburg, Germany. Correspondence and requests for materials should be addressed to M.K. (email: mackillo.kira@physik.uni-marburg.de).

Strong interactions[1–3] are prerequisite for efficient quantum-state manipulations hindered by dissipation and thermalization. For ultracold atoms, thermalization[4] itself is advantageous because it creates the Bose–Einstein condensate (BEC) from the normal component, whereas quantum depletion[5,6] ejects atoms from the BEC via interactions. However, coherent quantum depletion is reversible because it is driven by a distinct coherent transition amplitude between the BEC and normal-component atoms, unlike losses and thermal depletion[4]. Atom–atom interactions may also bind atoms to clusters within a range defined by the atomic scattering length[7], a_{scatt}, that can be substantially larger than the atom size; a large (small) $|a_{scatt}|$ implies strong (weak) interactions. The a_{scatt} can be systematically changed by applying an external magnetic field B to tune the atom coupling in the vicinity of a Feshbach resonance[8]. A fast B sweep can then jump the system from weak to strong interactions on a timescale much faster than the three-atom loss[9] and thermalization. Figure 1 illustrates how an abrupt change in a_{scatt} changes atoms' sphere of influence, with radius a_{scatt}, from weak to strong interaction at time $t=0$. The strongest interaction is found at so-called unitarity[7,10] where a_{scatt} diverges. Unitarity can be detrimental for the BEC because atoms can then scatter over large distances, as indicated by the overlapping spheres. Fast changes in a_{scatt} have been shown to induce, for example, Sakharov oscillations[11] and Bogoliubov excitations[12–14].

An important milestone was reached in the fast-sweep experiment by Makotyn *et al.*[15] (f-s experiment) with a ^{85}Rb BEC swept to unitarity in just 5 μs. Beating the expectations[16,17], the BEC survived the sudden jump to unitarity by evolving into a new quasiequilibrium in roughly 100 μs, which was much faster than the three-body loss rate. Earlier analyses of the f-s experiment identified a two[18]- and three[19]-body contact in the high-momentum tail of an atom distribution, which explained[13,20] the momentum-dependent relaxation of the distribution tail, and assigned[9] the unexpectedly long lifetime of the BEC at unitarity to a long-lived three-body state. In addition, the possibility of observing coherent oscillations has been predicted[21] in an interacting Bose gas. Using full many-body computations[22–25], I show that the f-s experiment has also established another landmark—coherent quantum depletion dominated by atom-cluster excitations resembling ultrafast excitations in semiconductors[23,26,27].

Results

Atom-cluster dynamics in quantum depletion. Conceptually, an abrupt change in a_{scatt} brings the BEC to a non-equilibrium in the same manner as an ultrafast laser excites solids[23,26] or

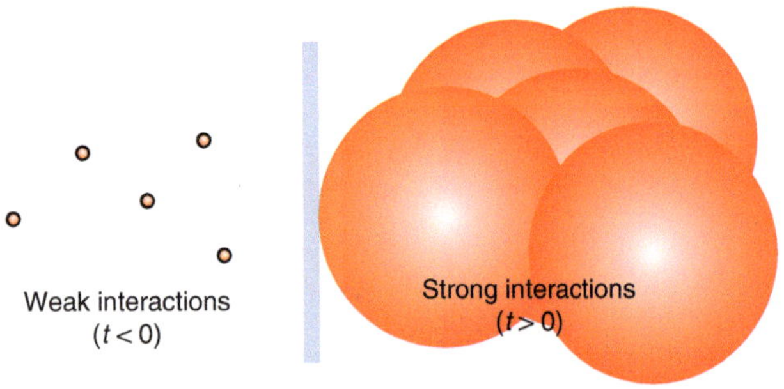

Figure 1 | Principle of fast-sweep experiments. Schematic illustration of the range of a_{scatt} change in fast-sweep experiments switched at time $t=0$ from weak (left) to strong (right) interactions; the sphere of influence has radius a_{scatt} around each atom.

molecules[28] to a non-equilibrium state. Ultrafast spectroscopy is renown for exciting both transient coherences[22,26,29] and relaxation oscillations[27] towards equilibrium. Therefore, one may expect that the f-s experiment accessed very similar effects. In semiconductors, ultrafast lasers can also excite a family of quasiparticles ranging from individual electrons and holes to correlated particle clusters[25] such as the dropleton[30]. Analogously, it clearly is interesting to explore what kind of many-body states the f-s experiment has excited.

Solving fast-sweep generated excitations is challenging because a full quantum-kinetic description must resolve how interactions break up the BEC atoms, clustered to all orders[25], into normal-component atoms that typically form independent clusters containing only few atoms. I apply a cluster-expansion approach[22,23] in the excitation picture[25]. The BEC is then presented completely and exactly[25] in terms of few atom clusters generated into the normal component; the approach is summarized in Supplementary Notes 1–4. Especially, both quantum-depletion and ultrafast semiconductor excitations share exactly the same cluster-dynamics structure where clusters are generated sequentially[23,24] from small to large ones. I utilize this fundamental connection to systematically determine both the formally exact atom-cluster dynamics and a systematic truncation scheme, presented in Supplementary Note 1.

The coherence itself is identified by the transition amplitude $s^\star_{\mathbf{k}} \equiv \langle \hat{L}_0^2 B^\dagger_{\mathbf{k}} B^\dagger_{-\mathbf{k}} \rangle$ to move two BEC atoms into the normal component. Microscopically, $\hat{L}_0^2$ removes two atoms from the BEC and $B^\dagger_{\mathbf{k}}$ is the usual boson-creation operator for a normal-component atom with a wave vector $\mathbf{k}\neq 0$. For large atom traps, the momentum of the BEC atoms vanishes such that $s^\star_{\mathbf{k}}$ conserves the momentum by creating the atoms to momenta $+\hbar\mathbf{k}$ and $-\hbar\mathbf{k}$. The accumulation of the normal component is determined by the atom occupation $f_{\mathbf{k}} \equiv \langle B^\dagger_{\mathbf{k}} B_{\mathbf{k}} \rangle$.

In the excitation picture[25,24], $f_{\mathbf{k}}$ and $s_{\mathbf{k}}$ define the relevant two-atom clusters excited by the coherent quantum depletion. To focus the analysis on the possible coherent effects, I assume that the atom trap is closed, preventing thermalization because energy and particle exchange out of the trap is excluded. Consequently, the BEC atom number $N_C \equiv \mathcal{N} - \sum_{\mathbf{k}\neq 0} f_{\mathbf{k}}$ is the difference of total atom number $\mathcal{N}$ and the number of atoms in the normal component. The resulting $(s_{\mathbf{k}}, f_{\mathbf{k}}, N_C)$ group forms a closed set of two-particle cluster dynamics, constituting the hyperbolic Bloch equations (HBEs)[24]. As demonstrated in Supplementary Note 4, the HBEs describe gapless BEC excitations also at unitarity and the total atom distribution becomes

$$n_{\mathbf{k}} \equiv N_C \delta_{\mathbf{k},0} + f_{\mathbf{k}} (1 - \delta_{\mathbf{k},0}), \qquad (1)$$

when no distinction is made between the BEC ($\mathbf{k}=0$) and normal-component atoms.

Connection of ultrafast semiconductor and BEC excitations. The HBEs resemble the semiconductor Bloch equations[27,31] (SBEs) describing how ultrafast lasers drive coherent semiconductor polarization and the quantum kinetics[22,23] that follows it, cf. Supplementary Note 2. In the HBEs, the coherent $s_{\mathbf{k}}$ is analogously driven by a $V_{\mathbf{k}}\,N_C$ source containing the Fourier transformation $V_{\mathbf{k}}$ of the pairwise atom–atom interaction. In analogy to the SBEs, the $V_{\mathbf{k}}\,N_C$ source is renormalized by a $\sum_{\mathbf{k}'} V_{\mathbf{k}'-\mathbf{k}} s_{\mathbf{k}'}$ sum that describes the possibility of forming bound atom pairs called dimers[7]. Furthermore, $V_{\mathbf{k}}$ yields a hierarchical coupling to three- and four-atom-cluster dynamics, which introduces, for example, dephasing and energy shift to $s_{\mathbf{k}}$ (ref. 24). Three-atom-cluster dynamics also contains the possibility of forming Efimov trimers[24,32].

Since atomic BECs and semiconductors are physically very different, it is clear that HBEs and SBEs yield also distinct differences. For example, the SBEs describe the dynamics of two kinds of fermions—electrons and holes—with different effective masses, which often produces different dynamics for the respective distributions; the HBEs used contain only one class of density distributions. The fermion character itself appears in the SBEs as Pauli blocking of transitions, contrasted by the bosonic stimulation in the HBEs. At the same time, $V_{\mathbf{k}}$ is a long-range interaction in the SBEs, whereas it has a short range in an atomic Bose gas. The long range and magnitude of $V_{\mathbf{k}}$ drives the relevant interaction timescales of the SBEs to be much faster (ultrafast sub-ps scale) than in the HBEs (μs scale).

In general, the semiconductors exhibit more possibilities of binding clusters due to stronger interactions and participation of more particles classes. Nevertheless, the cluster classes and phenomenology are often strikingly similar. For example, recently found dropletons[30] emerge close to the pair-ionization threshold (corresponding to unitarity), have quantized energy and are spatially extended clusters with more than three electron—hole pairs. Similar to Efimov trimers, the dropleton is significantly larger (roughly 20 times) than a bound electron—hole pair, while its energy scaling is different.

Most important, the $s_{\mathbf{k}}$ drives coherent quantum depletion in the same way as the polarization embodies the coherences of the ultrafast semiconductor excitations. In semiconductors, the polarization determines the extent, symmetries and timescales of ultrafast excitations, whereas higher-order clusters, such as the dropletons, modify 'only' the quantitative details[33]. Since including additional hierarchical levels in the SBEs is numerically tedious, systematic and accurate methods have been developed to reduce it to few-parameter scattering models[22] that satisfy the essential symmetries. Analogously, $s_{\mathbf{k}}$ dynamics must determine central aspects of the coherent quantum depletion, whereas the formation of, for example, Efimov trimers should yield 'only' quantitative modifications. When the hierarchical coupling is reduced to a dephasing γ and an energy-renormalization constant μ for $s_{\mathbf{k}}$, it must[24] be accompanied by an excitation-dependent relaxation for the $f_{\mathbf{k}}$, based on symmetry; the explicit systematic model is presented in Supplementary Note 2.

The capabilities of this level HBEs is thoroughly tested by matching the computational parameters exactly with the conditions of the f-s experiment as given in ref. 15. The atom mass is that of ^{85}Rb, the spherical trap has $\mathcal{N} = 6\times10^4$ atoms and the average density is $\langle\rho\rangle = 5.5\,\mu\text{m}^{-3}$. The interaction is ramped from weak ($a_{\text{scatt}} = 150\ a_0$, a_0 being the hydrogen Bohr radius) to unitarity ($a_{\text{scatt}} = \infty$) within 5 μs, which leaves γ and μ as the only free parameters chosen to match frequencies 0.761 and 0.475 kHz, respectively. The f-s experiment recorded the column distribution,

$$\tilde{n}_k \equiv \frac{\mathcal{N}}{\rho} \int_{-\infty}^{\infty} \mathrm{d}k_z\, n_{\sqrt{k^2+k_z^2}} \qquad (2)$$

which is the marginal distribution of all atoms. In the f-s experiment, $\tilde{n}_k$ was broadened[15] below $k = 3\,\mu\text{m}^{-1}$ and included in the computations by using a broadened BEC distribution $N_C \frac{\rho}{\mathcal{N}} \left(\frac{2\pi}{\sqrt{\pi} k_{\text{lim}}}\right)^3 e^{-\mathbf{k}^2/k_{\text{lim}}^2}$ to replace $N_C\ \delta_{\mathbf{k},0}$ in equation (1) with $k_{\text{lim}} = 1.1342\,\mu\text{m}^{-1}$.

Quantum kinetics of quantum depletion. Figure 2a presents $\tilde{n}_k$ of the HBE computations for the same times as those recorded in the f-s experiment (Fig. 2 in ref. 15) where t is the evolution time at unitarity. At $t = 0$ (shaded area), $\tilde{n}_k$ peaks at the origin because the system is still dominated by the BEC formed at the regime of weak interactions. Already $t = 15\,\mu\text{s}$ evolution at unitarity (dashed line) produces an extended tail in $\tilde{n}_k$. After that, the tail becomes steeper and evolves towards a steady state, while the BEC peak drops significantly. One also observes that the steady state is reached faster at high momenta (roughly above $k = 15\,\mu\text{m}^{-1}$) while residual deviations remain in the vicinity of $k = 5\,\mu\text{m}^{-1}$, even 170 μs after the switch is on. As shown in Supplementary Note 5, this computation explains quantitatively (γ and μ being the only free parameters) the f-s experiment timescales, overall excitation levels and distribution shapes, which confirms that the f-s experiment is indeed dominated by the coherent quantum depletion. To explain more quantitative details, one also needs to include the excitation-induced effects[22,33–37] beyond a constant μ, as shown later.

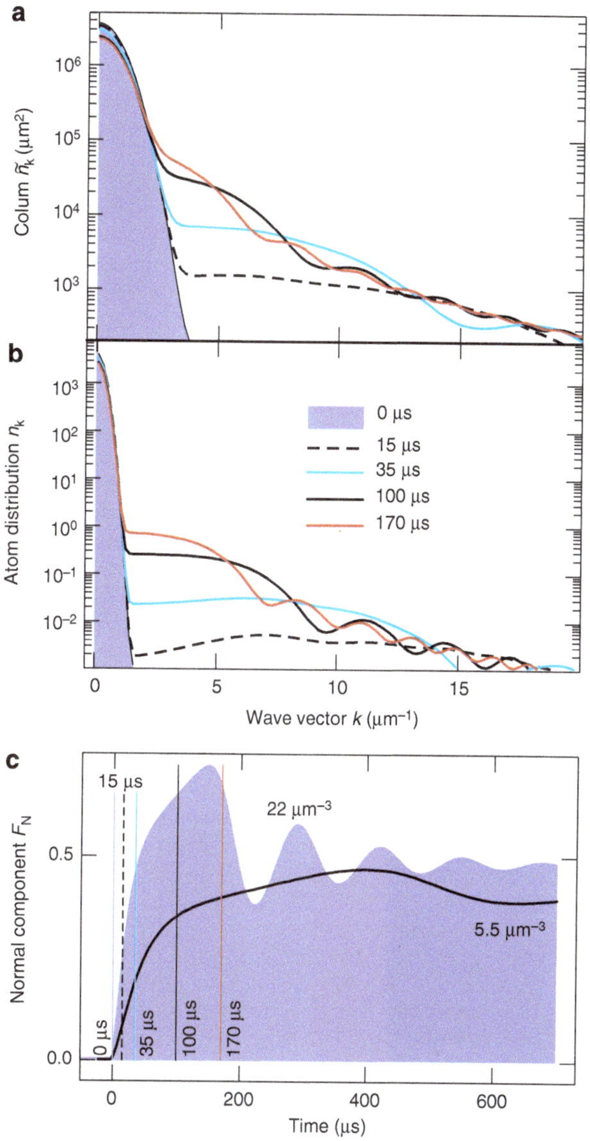

Figure 2 | Fast-sweep analysis of interacting Bose gas. (**a**) Computed column distributions (2) for five representative t. (**b**) The corresponding actual $n_{\mathbf{k}}$ distributions. The interactions are switched to unitarity at $t = 0$ and homogeneous ^{85}Rb-atom density is $\rho = 5.5\,\mu\text{m}^{-3}$. (**c**) Normal-component fraction for two representative ρ. The vertical lines identify the snapshots analysed in **a**,**b** with matching colour coding.

The computed $\tilde{n}_k$ also exhibits slight oscillations for 100 μs (black line) and 170 μs (red line), which seem to be only indicative in the f-s experiment; this minor difference is probably related to the experimental resolution. The actual $n_{\mathbf{k}}$ distribution is plotted as in Fig. 2b using a very narrow $k_{\mathrm{lim}} = 0.3781\,\mu\mathrm{m}^{-1}$ for the BEC. A clear BEC peak is observed in $n_{\mathbf{k}}$ for all times, which shows that the BEC indeed exists even at unitarity, and that $n_{\mathbf{k}}$ and $\tilde{n}_k$ resemble each other. The projection in equation (2) simply smoothens the oscillatory features.

The level of quantum depletion is defined by the fraction of normal-component atoms $F_{\mathrm{N}} \equiv \frac{1}{N}\sum_{\mathbf{k}\neq 0} f_{\mathbf{k}}$, presented in Fig. 2b as function of time for $\rho = 5.5\,\mu\mathrm{m}^{-3}$ (line); the vertical lines identify the $\tilde{n}_k$ snapshots of Fig. 2a. For comparison, F_{N} is also shown for $\rho = 22\,\mu\mathrm{m}^{-3}$ (shaded area). As a general trend, F_{N} increases as ρ becomes larger. However, the coherent quantum depletion cannot completely exhaust the BEC ($N_{\mathrm{C}} \to 0$) because the $V_{\mathbf{k}}\, N_{\mathrm{C}}$ source would become zero, which would reverse the coherent quantum depletion not sustained by its source. Instead, the steady-state F_{N} saturates towards 50% for large atom densities, for example, $\rho = 22\,\mu\mathrm{m}^{-3}$ yields pronounced relaxation oscillations towards 50% after peaking F_{N} to 72%. Also the $\rho = 5.5\,\mu\mathrm{m}^{-3}$ case shows some relaxation oscillations, but with a very long oscillation period.

The ρ dependence of relaxation oscillations is studied in Fig. 3a by comparing the peak F_{N} (dashed line) with the final F_{N} (red-solid line) evaluated at $t = 700\,\mu\mathrm{s}$; the vertical lines identify the densities of Fig. 2b. The final F_{N} saturates towards 50% from below at high densities, which shows that the coherent quantum depletion cannot collapse the BEC at the unitarity. For low densities, the peak and final F_{N} are identical while they deviate significantly above $\rho = 10\,\mu\mathrm{m}^{-3}$ due to the relaxation oscillations. Even though the f-s experiment does not show pronounced relaxation oscillations, they should become visible when ρ is increased by a factor of two.

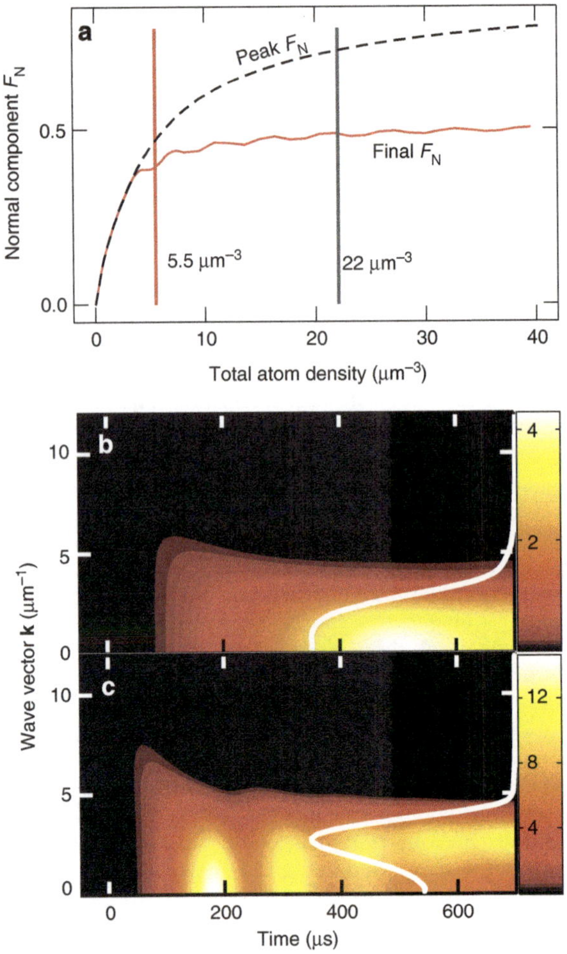

Figure 3 | Shell-state formation in quantum depletion. (**a**) Peak (dashed line) and final (solid line, $t = 700\,\mu\mathrm{s}$) fraction F_{N} of normal-component atoms generated by quantum depletion as function of density. The vertical lines indicate the densities analysed in **b**,**c** and Fig. 2. (**b**) Quantum kinetics of normal-component distribution for (**b**) $\rho = 5.5\,\mu\mathrm{m}^{-3}$ and (**c**) $\rho = 22\,\mu\mathrm{m}^{-3}$; the white line shows the shape of the final $f_{\mathbf{k}}$ at $t = 700\,\mu\mathrm{s}$.

Shell-state formation. Quantum kinetics of normal-component $f_{\mathbf{k}}$ is presented in Fig. 3b,c as function of t for $\rho = 5.5$ and $22\,\mu\mathrm{m}^{-3}$, respectively. The final distribution is indicated by the white line. Just after the switch is on ($t = 0$), $f_{\mathbf{k}}$ is excited to large momenta and then focuses towards the origin, as in Fig. 2a. For $\rho = 5.5\,\mu\mathrm{m}^{-3}$, $f_{\mathbf{k}}$ evolves smoothly towards a monotonically decaying steady-state distribution. For $\rho = 22\,\mu\mathrm{m}^{-3}$, $f_{\mathbf{k}}$ dynamics does not only show the relaxation oscillations but it approaches an unexpected form; $f_{\mathbf{k}}$ peaks roughly at $k = 2.8\,\mu\mathrm{m}^{-1}$ and not at the origin. In other words, the normal-component atoms prefer to occupy a shell around the BEC due to the coherent $s_{\mathbf{k}}$; I will call this many-body configuration a shell state.

As validity checks, I show in Supplementary Note 4 that the shell state survives three-body loss, the excitation energies are indeed gapless and the results depend only slightly on the atom-trap density profile when the excitation-induced μ-shift moves the position of the shell state as function of density. Supplementary Note 3 also shows analytically that the shell state limits the coherently driven F_{N} to 50%. Therefore, the coherent quantum depletion alone cannot annihilate the BEC—one needs additional losses[38,39] or thermalization to execute that. The unexpected survival of the BEC in the f-s experiment[15] was assigned to build-up of long-living (1 ms) Efimov states[9]. The presented results also demonstrate that the BEC is additionally shielded by the shell state. The presence of the identified coherences is also expected to reduce the losses, as shown for semiconductors[33].

The spatial extent of quantum depletion can be determined by analysing the atom–atom pair correlation function $g(\mathbf{r})$ that defines the normalized probability of finding an atom at position $\mathbf{r}$ when another one is held at the origin. The normalization is chosen to produce $g(\mathbf{r}) = 1$ for uncorrelated atoms, and $r^2(g(\mathbf{r}) - 1)$ determines deviations with a radial weight, as discussed in Supplementary Note 3. Figure 4 shows the $r^2(g(\mathbf{r}) - 1)$ dynamics for $\rho = 5.5\,\mu\mathrm{m}^{-3}$ in Fig. 4a and $\rho = 22\,\mu\mathrm{m}^{-3}$ in Fig. 4b, studied also in Fig. 3b and Fig. 3c, respectively. Switching to unitarity at $t = 0$ excites a long-range order extending beyond $4\,\mu\mathrm{m}$ and shows rapid oscillations that are typical for a liquid[40] where atoms are arranged into shells. At later times, the atom–shell separation increases, indicated by the fan-like expansion of the peaks and dips. In addition, the height of large-distance shells diminishes leaving only one dominant peak and dip for $\rho = 5.5\,\mu\mathrm{m}^{-3}$, indicating transition to a short-range order. For $\rho = 22\,\mu\mathrm{m}^{-3}$, the steady-state $r^2(g(\mathbf{r}) - 1)$ has about three clear peaks and dips separated roughly by $2.1\,\mu\mathrm{m}$ matching well with the 2.2-μm wave length predicted to be the shell-state peak in $f_{\mathbf{k}}$. Therefore, the relaxation observed in Figs 2 and 3 is connected with coherent quantum kinetics from long-range-order liquid to a short-range-order configuration whose range can be extended by forming the shell state.

Excitation-induced effects. To determine how quantitatively the computations already reproduce the f-s experiment, Fig. 5

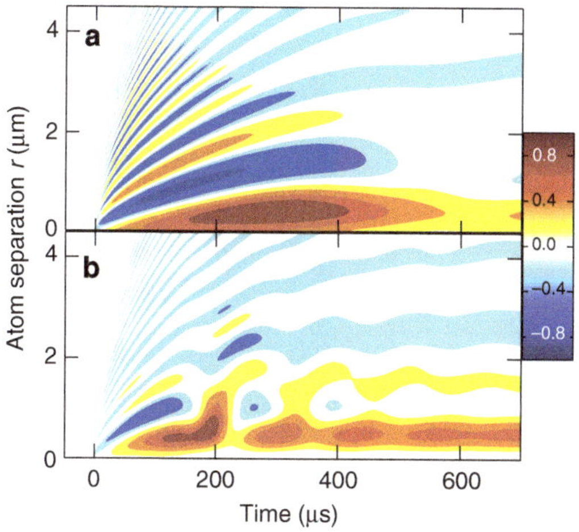

Figure 4 | Long-range order in quantum depletion. Quantum kinetics of $r^2(g(\mathbf{r})-1)$ is shown for (**a**) $\rho=5.5\,\mu\text{m}^{-3}$ (peak $r^2(g(\mathbf{r})-1)=15.8\,\mu\text{m}^2$) and (**b**) $\rho=22\,\mu\text{m}^{-3}$ (peak $r^2(g(\mathbf{r})-1)=22.3\,\mu\text{m}^2$) as function of atom separation. The colour coding is scaled with respect to peak $r^2(g(\mathbf{r})-1)$.

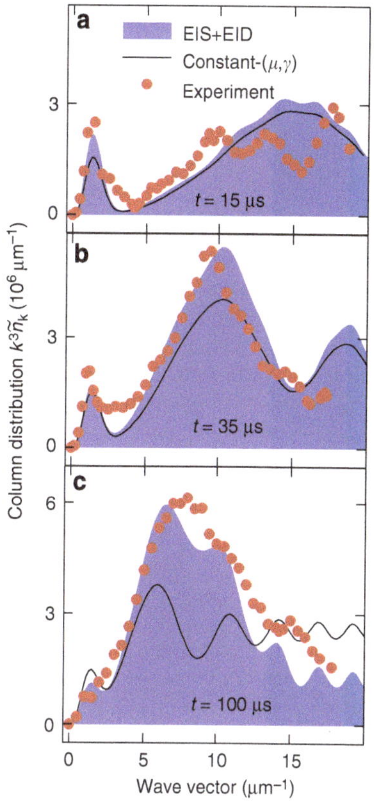

Figure 5 | Quantitative experiment-theory comparison. Experimental (filled circles) $k^3\tilde{n}_k$ is compared with constant-(μ, γ) (solid line) and full EIS/EID (shaded area) HBE computations for (**a**) $t=15\,\mu$s, (**b**) $t=35\,\mu$s and (**c**) $t=100\,\mu$s after switching to unitarity. The experimental points are read from the inset to Fig. 2 in ref. 15.

compares $k^3\tilde{n}_k$ of the f-s experiment (filled circles) with the constant-(μ,γ) computation (solid line). When the column distribution is multiplied by k^3, the so-called two-body contact[19,41,42] is observed as an asymptotically constant $k^3\tilde{n}_k$. A similar asymptotic behaviour is expected in the actual $k^4 n_{\mathbf{k}}$ distribution when multiplier k^4 is used instead of k^3 due to dimensionality difference of $n_{\mathbf{k}}$ and $\tilde{n}_k$.

Figure 5 shows that the constant-(μ,γ) computation already explains the experiment in great detail for the early times. However, it slightly underestimates n_k and $\tilde{n}_k$ at late times, close to $k=7\,\mu\text{m}^{-1}$. Such quantitative differences are not unexpected because the excitation-induced shift (EIS) in μ and the excitation-induced dephasing (EID) in γ should modify the quantitative distribution/spectral shapes, as in semiconductors[33].

To verify this explanation, I have also performed EIS computations that include the three-body loss, main features of EID/EIS and calculated the distributions using a local density approximation (LDA) average; details are given in Supplementary Note 5. Both EIS and EID are generated by the hierarchical coupling to three- and four-atom clusters that are induced by nonlinear source terms containing products of $f_{\mathbf{k}}$ and $s_{\mathbf{k}}$. As shown in ref. 24, such sources yield at least a quadratic density dependence for both γ and μ, in analogy to the three-body loss. To capture these excitation-induced aspects, I use the simplest functional form $\mu=211.4\,\text{Hz}+628.9(\rho_n/\mu\text{m}^{-3})^2\text{Hz}$ and $\gamma=21.14\,\text{Hz}+10.48(\rho_n/\mu\text{m}^{-3})^2\text{Hz}$, where ρ_n is the density of normal-component atoms. Many combinations of constant and quadratic coefficients produce qualitatively similar $\tilde{n}_k$ results; the used specific values yield roughly the best quantitative agreement with the experiment. The three-body loss model is defined in Supplementary Note 4 and its time constant is chosen to produce a 630-μs decay at $\langle\rho\rangle=5.5\,\mu\text{m}^{-3}$, as in the experiment[15]. Owing to density dependence of the three-atom loss, μ and γ, the shape of $\tilde{n}_k$ depends so strongly on the local density that the LDA average of Supplementary Note 4 needs to be implemented for a precise comparison.

The shaded areas in Fig. 5 present the LDA-averaged $k^3\tilde{n}_k$ of the EIS/EID computation. At early times ($t=15\,\mu$s), the EIS/EID and constant-(μ, γ) computation (solid line) distributions are essentially the same, both explaining the experiment very well. However, only the EIS/EID computation agrees with the experiment quantitatively at $t=100\,\mu$s. In other words, this analysis verifies that the quantitative deviations between the experiment and constant-(μ, γ) computation stem from the excitation-induced effects. Both computations explain the presence of weak oscillations in the experiment as relaxation oscillations. While the oscillation period agrees well, there are some differences in the experimental versus theoretical oscillation amplitude. However, the deviation is close to the experimental noise level such that one would need to measure these oscillations with a finer resolution and probably compute the excitation-induced effects fully quantum kinetically to explain all quantitative details; both studies are beyond the scope of the present investigation.

Discussion

Quantum depletion can be partially described in terms of the Bogoliubov excitations[24], but they alone destroy the BEC at high densities[13] because one needs the shell state to shield the BEC. The HBEs include the Bogoliubov excitations as a substructure[24], while the shell state as well as dimers result from the additional coupling among them. In fact, the coherent quantum depletion drives many dimer resonances, especially just after the jump to unitarity, based on the time-energy uncertainty. All of the initially generated dimer components in $s_{\mathbf{k}}$ evolve with their own eigen frequency, which generates quantum beats as the system evolves further, explaining the observed $n_{\mathbf{k}}$ oscillations and the transient long-range order. The relaxation oscillations appear through the nonlinear modulation in the quantum-depletion source.

The presented quantum-kinetic theory explains how an atomic BEC generates the normal component via sequential atom clustering when the atom–atom interactions are switched fast to unitarity. The resulting kinetics is initiated by a transition amplitude describing a coherent ejection of two BEC atoms to the normal component. This process is shown to explain the recent f-s experiment[15] in great detail, which connects quantum depletion with coherent non-equilibrium processes in semiconductors.

The coherent quantum depletion identified should exhibit partial reversibility when switched back to weak interactions. Coherent control should also become possible when a sequence of Feshbach-resonance modulations is implemented. For atom densities that are roughly two times higher than in the Makotyn *et al.*[15] experiment, the coherent quantum-depletion process is shown to form a shell state faster than 50 μs even when three-body losses are present, which prevents quantum depletion beyond 50% normal-component fraction at quasiequilibrium.

The identification of the shell state is simplest if one can experimentally access the normal-component distribution $f_{\mathbf{k}}$ of a homogeneous density. However, current measurements detect the overall distribution $n_{\mathbf{k}}$ including both the BEC and the normal component, and measure its density average. In a realistic experiment, the BEC contribution of $n_{\mathbf{k}}$ spreads around $\mathbf{k}=0$ due to finite-size effects[15]. The visibility of the shell state can, therefore, be enhanced by decreasing the BEC spreading and/or by having a flatter atom trap, which makes the density profile more homogeneous. Therefore, further experiments and/or analyses are needed for an unambiguous identification of the shell state. The computations suggest that the shell state should become appreciable at atom densities that are roughly two times higher than in ref. 15, and that the excitation-induced effects can be accessed by comparing experiment and theory $k^3\bar{n}_k$ distributions quantitatively; an analogous analysis has delivered very precise insights into the many-body states involved in semiconductor excitations[33].

The predicted non-equilibrium dynamics as well as shell-state formation can be experimentally accessed by measuring the evolution of atom distribution $n_{\mathbf{k}}$, which also provides direct information of the normal-component fraction and its shape. I also expect that the effect of higher-order clusters (such as Efimov trimers) can be identified through quantitative shape changes in the $n_{\mathbf{k}}$, in analogy to the cluster-based studies in semiconductors[30,33]. Experiments with multiple switch-on/off sequences may access analogous coherent non-equilibrium quantum kinetics as in ultrafast semiconductor excitations. Through this connection, one may extend the future fast-sweep experiments to utilize novel concepts such as coherent control[43] and quantum spectroscopy[30,44,45] to access unexpected many-body states.

Methods

General aspects. Coherent quantum depletion is solved with the HBEs[24] that establish a new method to study an interacting Bose gas. The HBEs are presented in Supplementary Note 1 while the actual cluster-expansion[22,23] derivation is performed in ref. 24. The HBEs are analogous to the SBEs that govern the physics of ultrafast excitations in semiconductors[22,23,27]. Using the formal HBE–SBE connection, the quantum-kinetics insights of semiconductor studies are applied to reduce the hierarchical many-body coupling to a dephasing and energy shift that conserve the energy and maintain the BEC excitation as a minimum uncertainty state of the Heisenberg uncertainty relation. At this level, the method studies exclusively coherent quantum depletion because it deliberately omits atom loss, energy loss and thermalization. Atom losses can be included straightforwardly as explained in Supplementary Note 4.

Numerical aspects. The HBEs couple the coherent transition amplitude $s_{\mathbf{k}}$ with the normal component and BEC occupations $f_{\mathbf{k}}$ and N_C, respectively. Owing to the radial symmetry, one needs to consider only the radial k dependency for both $s_{\mathbf{k}}$ and $f_{\mathbf{k}}$. The resulting HBE computations are performed by discretizing k into 800 intervals and the atom–atom interactions are modelled with a realistic Morse potential. In order to resolve both the short- and long-range quantum kinetics of the interacting atoms, the k grid is divided into two parts: the low-k grid extends up to $76\,\mu m^{-1}$ and the high-k grid reaches $6.0\times10^4\,\mu m^{-1}$, both containing 400 points. This grid covers k values for over 5 orders of magnitude, which is sufficient to reveal how the atom–atom interactions exhibit repulsion <100 pm and unitarity interaction extending to macroscopic distances. The HBE dynamics is then solved with a fourth-order Runge–Kutta method.

References

1. Raimond, J. M., Brune, M. & Haroche, S. Manipulating quantum entanglement with atoms and photons in a cavity. *Rev. Mod. Phys.* **73,** 565–582 (2001).
2. Khitrova, G., Gibbs, H. M., Kira, M., Koch, S. W. & Scherer, A. Vacuum Rabi splitting in semiconductors. *Nat. Phys.* **2,** 81–90 (2006).
3. Bloch, I., Dalibard, J. & Zwerger, W. Many-body physics with ultracold gases. *Rev. Mod. Phys.* **80,** 885–964 (2008).
4. Gardiner, C. W., Lee, M. D., Ballagh, R. J., Davis, M. J. & Zoller, P. Quantum kinetic theory of condensate growth: comparison of experiment and theory. *Phys. Rev. Lett.* **81,** 5266–5269 (1998).
5. Xu, K. *et al.* Observation of strong quantum depletion in a gaseous Bose—Einstein condensate. *Phys. Rev. Lett.* **96,** 180405 (2006).
6. Wüster, S. *et al.* Quantum depletion of collapsing Bose—Einstein condensates. *Phys. Rev. A* **75,** 043611 (2007).
7. Giorgini, S., Pitaevskii, L. & Stringari, S. Theory of ultracold atomic Fermi gases. *Rev. Mod. Phys.* **80,** 1215–1274 (2008).
8. Chin, C., Grimm, R., Julienne, P. & Tiesinga, E. Feshbach resonances in ultracold gases. *Rev. Mod. Phys.* **82,** 1225–1286 (2010).
9. Sykes, A. G. *et al.* Quenching to unitarity: quantum dynamics in a three-dimensional Bose gas. *Phys. Rev. A* **89,** 021601 (2014).
10. Braaten, E. & Hammer, H.-W. Universality in few-body systems with large scattering length. *Phys. Rep.* **428,** 259–390 (2006).
11. Hung, V., Gurarie, C.-L. & Chin, C. From cosmology to cold atoms: observation of Sakharov oscillations in a quenched atomic superfluid. *Science* **341,** 1213–1215 (2013).
12. Natu, S. & Mueller, E. Dynamics of correlations in a dilute Bose gas following an interaction quench. *Phys. Rev. A* **87,** 053607 (2013).
13. Yin, X. & Radzihovsky, L. Quench dynamics of a strongly interacting resonant Bose gas. *Phys. Rev. A* **88,** 063611 (2013).
14. Kain, B. & Ling, H. Y. Nonequilibrium states of a quenched Bose gas. *Phys. Rev. A* **90,** 063626 (2014).
15. Makotyn, P., Klauss, C., Goldberger, D., Cornell, E. & Jin, D. Universal dynamics of a degenerate unitary Bose gas. *Nat. Phys.* **10,** 116–119 (2014).
16. Papp, S. B. *et al.* Bragg spectroscopy of a strongly interacting Rb85 Bose-Einstein condensate. *Phys. Rev. Lett.* **101,** 135301 (2008).
17. Navon, N. *et al.* Dynamics and thermodynamics of the low-temperature strongly interacting Bose gas. *Phys. Rev. Lett.* **107,** 135301 (2011).
18. Corson, J. & Bohn, J. Bound-state signatures in quenched Bose—Einstein condensates. *Phys. Rev. A* **91,** 013616 (2014).
19. Smith, D., Braaten, E., Kang, D. & Platter, L. Two-body and three-body contacts for identical bosons near unitarity. *Phys. Rev. Lett.* **112,** 110402 (2014).
20. Rançon, A. & Levin, K. Equilibrating dynamics in quenched Bose gases: characterizing multiple time regimes. *Phys. Rev. A* **90,** 021602 (2014).
21. Holland, M., Park, J. & Walser, R. Formation of pairing fields in resonantly coupled atomic and molecular Bose—Einstein condensates. *Phys. Rev. Lett.* **86,** 1915–1918 (2001).
22. Kira, M. & Koch, S. W. Many-body correlations and excitonic effects in semiconductor spectroscopy. *Prog. Quant. Electron.* **30,** 155–296 (2006).
23. Kira, M. & Koch, S. W. *Semiconductor Quantum Optics* 1st edn (Cambridge University Press, 2011).
24. Kira, M. Hyperbolic Bloch equations: atom-cluster kinetics of an interacting Bose gas. Preprint at http://arxiv.org/abs/1407.4927 (2014).
25. Kira, M. Excitation picture of an interacting Bose gas. *Ann. Phys.* **351,** 200–249 (2014).
26. Shah, J. *Ultrafast Spectroscopy of Semiconductors and Semiconductor Nanostructures,* 2nd edn (Springer, 1999).
27. Haug, H. & Koch, S. *Quantum Theory of the Optical and Electronic Properties of Semiconductors* 5th edn (World Scientific, 2009).
28. Zewail, A. Laser femtochemistry. *Science* **242,** 1645–1653 (1988).
29. Chemla, D. & Shah, J. Many-body and correlation effects in semiconductors. *Nature* **411,** 549–557 (2001).
30. Almand-Hunter, A. *et al.* Quantum droplets of electrons and holes. *Nature* **506,** 471–475 (2014).
31. Lindberg, M. & Koch, S. W. Effective bloch equations for semiconductors. *Phys. Rev. B* **38,** 3342–3350 (1988).
32. Hammer, H.-W., Nogga, A. & Schwenk, A. Colloquium: three-body forces: from cold atoms to nuclei. *Rev. Mod. Phys.* **85,** 197–217 (2013).
33. Smith, R. P. *et al.* Extraction of many-body configurations from nonlinear absorption in semiconductor quantum wells. *Phys. Rev. Lett.* **104,** 247401 (2010).
34. Peyghambarian, N. *et al.* Blue shift of the exciton resonance due to exciton-exciton interactions in a multiple-quantum-well structure. *Phys. Rev. Lett.* **53,** 2433–2436 (1984).

35. Wang, H. *et al.* Transient nonlinear optical response from excitation induced dephasing in gaas. *Phys. Rev. Lett.* **71,** 1261–1264 (1993).
36. Wang, H. *et al.* Transient four-wave-mixing line shapes: effects of excitation-induced dephasing. *Phys. Rev. A* **49,** R1551–R1554 (1994).
37. Shacklette, J. M. & Cundiff, S. T. Role of excitation-induced shift in the coherent optical response of semiconductors. *Phys. Rev. B* **66,** 045309 (2002).
38. Burt, E. A. *et al.* Coherence, correlations, and collisions: what one learns about Bose—Einstein condensates from their decay. *Phys. Rev. Lett.* **79,** 337–340 (1997).
39. Donley, E. *et al.* Dynamics of collapsing and exploding Bose—Einstein condensates. *Nature* **412,** 295–299 (2001).
40. Jorgensen, W. L., Chandrasekhar, J., Madura, J. D., Impey, R. W. & Klein, M. L. Comparison of simple potential functions for simulating liquid water. *J. Chem. Phys.* **79,** 926–935 (1983).
41. Tan, S. Energetics of a strongly correlated Fermi gas. *Ann. Phys.* **323,** 2952–2970 (2008).
42. Tan, S. Large momentum part of a strongly correlated Fermi gas. *Ann. Phys.* **323,** 2971–2986 (2008).
43. Warren, W., Rabitz, H. & Dahleh, M. Coherent control of quantum dynamics: the dream is alive. *Science* **259,** 1581–1589 (1993).
44. Kira, M. & Koch, S. W. Quantum-optical spectroscopy of semiconductors. *Phys. Rev. A* **73,** 013813 (2006).
45. Kira, M., Koch, S. W., Smith, R. P., Hunter, A. E. & Cundiff, S. T. Quantum spectroscopy with Schrödinger-cat states. *Nat. Phys.* **7,** 799–804 (2011).

Acknowledgements
I thank Deborah Jin, JILA, for fruitful discussions about the fast-switch experiment and Renate Schmid for careful proof reading of the manuscript.

Author contributions
M.K. is solely responsible for the scientific content of this work.

Additional information

Competing financial interests: The authors declare no competing financial interests.

Engineering skyrmions in transition-metal multilayers for spintronics

B. Dupé[1,2], G. Bihlmayer[2], M. Böttcher[1], S. Blügel[2] & S. Heinze[1]

Magnetic skyrmions are localized, topologically protected spin structures that have been proposed for storing or processing information due to their intriguing dynamical and transport properties. Important in terms of applications is the recent discovery of interface stabilized skyrmions as evidenced in ultra-thin transition-metal films. However, so far only skyrmions at interfaces with a single atomic layer of a magnetic material were reported, which greatly limits their potential for application in devices. Here we predict the emergence of skyrmions in $[4d/Fe_2/5d]_n$ multilayers, that is, structures composed of Fe biatomic layers sandwiched between $4d$ and $5d$ transition-metal layers. In these composite structures, the exchange and the Dzyaloshinskii-Moriya interactions that control skyrmion formation can be tuned separately by the two interfaces. This allows engineering skyrmions as shown based on density functional theory and spin dynamics simulations.

[1] Institute of Theoretical Physics and Astrophysics, University of Kiel, Leibnizstrasse 15, 24098 Kiel, Germany. [2] Peter Grünberg Institut (PGI-1) and Institute for Advanced Simulation (IAS-1), Forschungszentrum Jülich and JARA, 52425 Jülich, Germany. Correspondence and requests for materials should be addressed to B.D. (email: dupe@theo-physik.uni-kiel.de).

Topologically protected spin structures, in particular magnetic skyrmions[1–3], have recently received much attention due to their promising applications in spintronics[4–6]. Among the different mechanisms that can induce a topologically non-trivial magnetic structure, the relativistic Dzyaloshinskii–Moriya interaction (DMI)[7,8] arising in non-centrosymmetric systems is arguably the most prominent one[9]. Magnetic skyrmions have been first revealed in a few non-centrosymmetric bulk magnets, such as the chiral MnSi[10,11] or FeCoSi[12]. In two-dimensional systems, the DMI arises due to the broken inversion symmetry at the interface between a magnetic film and a substrate[13–15]. For example, an Fe monolayer on an Ir(111) substrate exhibits strong DMI and on this system a nanoscale skyrmion lattice was first observed by scanning tunnelling microscopy[16]. This skyrmion lattice is stabilized by the short-range four-spin interaction and slight modifications of the film composition, for example, adding an atomic Pd overlayer, can push this system into a regime where individual skyrmions can exist, which can be manipulated by spin-polarized currents in scanning tunnelling microscopy[17].

The manipulation of ultra-thin films such as Fe/Ir(111) is a versatile route to tailor properties such as the size or magnetic field needed to induce a magnetic skyrmion[18,19]. For practical applications, however, there are several difficulties that have to be overcome: Manipulation of skyrmions by electric currents[6,20–24] and measurement of the topological Hall effect[25–27] requires that a substantial fraction of the current runs through the magnetic structure. In a single Fe layer on a metallic substrate, this is difficult to realize. Creating large-area single Fe layers on a substrate requires specialized preparation methods that hamper large-scale production of these structures. Temperature stability and optimization of the size of these magnetic structures requires a tuning of different parameters such as exchange interactions, the DMI and the magnetic anisotropy.

Here we propose multilayers composed of Fe bilayers epitaxially sandwiched between 4*d* and 5*d* transition-metal (TM) layers as promising systems towards solving these issues. The repetition of layered structures is a powerful method in spintronics to engineer materials properties[28–30]: besides increasing the magnetic material in $[4d/\text{Fe}/5d]_n$ multilayers, such structures have the key advantage that the essential magnetic interactions, the exchange and the DMI, can be controlled by two separate interfaces. Thereby, the requirement of pseudomorphic growth is limited to the interface, which induces the DMI, whereas intermixing at the other interface is not crucial. Based on density functional theory (DFT), we show a route to engineer the skyrmion systems by varying the interface composition. We focus on an Fe bilayer sandwiched between Ir and Rh/Pd layers; however, other combinations of 4*d* and 5*d* TMs are also feasible. By changing the 4*d* band filling—here by alloying Rh with Pd—the exchange interaction in the adjacent Fe layers can be tuned due to strong hybridization at the interface[31]. The 5*d* TM Ir, on the other hand, ensures a strong DMI. Using spin dynamics simulations, we find the emergence of skyrmions in the Fe bilayers and study their properties.

Results

First-principles calculations. To access the magnetic interactions in TM multilayers, we have performed DFT calculations for various non-collinear spin structures including spin–orbit coupling[32,33]. By mapping the total energies of the different magnetic configurations to an extended Heisenberg model, we can extract parameters for the interactions. We consider homogeneous flat spin spirals with a period λ, in which the magnetization rotates along a given crystallographic direction that lies in the plane of the interface by an angle of $\phi = (2\pi/\lambda)a$ from atom to atom, where a is the spacing between adjacent atoms (see Methods and Supplementary Note 1). This approach has been successfully used in the past to study complex spin textures in a wide range of systems such as biatomic chains at surfaces[34], ultra-thin film skyrmion systems[16,18] and magnetic mono-[15,35,36] and bilayers[37].

A characteristic quantity of the exchange interaction in the considered TM multilayers is the effective nearest-neighbour exchange constant J_{eff}. It is a measure of how quickly the energy rises for long-wave length spin spirals with respect to the ferromagnetic (FM) state. J_{eff} was obtained by fitting the dispersion curve $E(\mathbf{q})$ of spin spirals characterized by the wave vector $\mathbf{q}$ in the vicinity of the FM state ($|\mathbf{q}| < 0.1 \times 2\pi/a$) with a parabola $J_{\text{eff}}\frac{3a^2}{2}|q|^2$ where a is the nearest-neighbour distance within the layers (for energy dispersions, see Supplementary Fig. 1). Therefore, a negative value of J_{eff} indicates a spin spiral ground state driven by exchange interaction. As a second characteristic quantity, we include the energy difference between a spin spiral state with a wavelength $\lambda = 2\pi/|q| = 2.7$ nm and the FM state, denoted by E_{SS}.

J_{eff} and E_{SS} are displayed for various systems in Fig. 1. We start with the monolayer films on substrates, which can host single skyrmions as shown in experiment[17] and explained by theory[18]. For both fcc and hcp stacking of the Pd overlayer on Fe/Ir(111), we find a rather small value of the effective exchange constant, that is, $J_{\text{eff}} = -2.3$ and $+4.4$ meV, respectively. The DMI, which is of similar strength in both systems[18], can stabilize skyrmions in these films in an applied magnetic field. It is noteworthy that from recent experiments on the field-dependent diameter of skyrmions in Pd/Fe/Ir(111), Romming *et al.*[38] determined an exchange stiffness that corresponds approximately to a J_{eff} of 1.6 meV, which is very close to the values we find here. For two layers of Pd on Fe/Ir(111), J_{eff} is negative and slightly stronger than for Pd(fcc)/Fe/Ir(111) and a skyrmion phase also occurs in a magnetic field as found in our simulations. In these two systems, the spin spiral ground state is already enforced by the exchange interaction and becomes even more favourable due to the DMI. We can also conclude the spin spiral ground state from the negative value of E_{SS}, which closely follows the trend of J_{eff}.

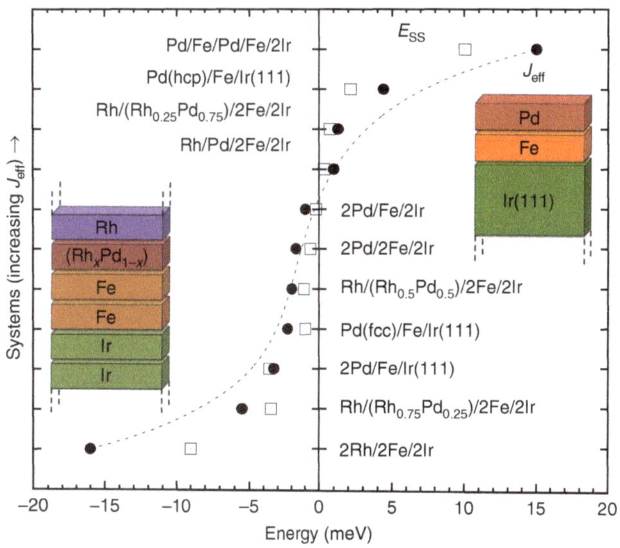

Figure 1 | Effective exchange interaction in multilayers. J_{eff} (full circles) is given for TM multilayers (black/blue fonts) and ultra-thin films on surfaces (grey fonts). Insets show sketches of the multilayer and thin film geometry. The energy difference between a spin spiral with a wavelength of 2.7 nm with respect to the FM state, E_{SS}, is given by empty squares. The dashed line is a guide for the eye.

Now we turn to the multilayers that are built by an epitaxial fcc stacking of (111) layers with a geometry as in an Ir bulk crystal but allowing for structural relaxation in the direction normal to the layers (see Methods and Supplementary Table 1). We first consider a multilayer, which is a repetition of a sandwich consisting of one atomic Fe layer between two Pd and two Ir layers denoted as 2Pd/Fe/2Ir. As shown in Fig. 1, the effective exchange interaction is very similar to that of the ultra-thin film system Pd(fcc)/Fe/Ir(111), which demonstrates that J_{eff} is mainly controlled by the interfaces. A first naive solution to increase the amount of magnetic material, that is, Fe layers in the sandwich, would be to repeat the Pd/Fe interface sequence as in Pd/Fe/Pd/Fe/2Ir. However, J_{eff} becomes much larger (≈ 15 meV), indicating strong FM exchange interaction, which prevents skyrmion formation. If, on the other hand, we consider the multilayer with a basic unit consisting of two layers of Pd, Fe and Ir, that is, 2Pd/2Fe/2Ir, the effective exchange interaction is of the same order as in the original single Fe-layer system.

We can understand the role of Pd by replacing it with Rh, which has one electron less in the 4*d* shell. For a multilayer consisting of 2Rh/2Fe/2Ir, we obtain a value of $J_{\mathrm{eff}} \approx -16$ meV, that is, negative and of large absolute value. In this system, we find an exchange-driven spin spiral ground state and no skyrmion formation in a reasonable external magnetic field. By choosing an intermediate interface composition, for example, $Rh/Rh_{0.75}Pd_{0.25}/2Fe/2Ir$, we can increase the 4*d* band filling quasi continuously and thereby increase the value of the effective exchange coupling J_{eff}. For a large range of interface alloy compositions Rh_xPd_{1-x}, we find multilayers with a small J_{eff}, which can host skyrmionic ground states as we show below using spin dynamics simulations.

Spin dynamics simulations. After having revealed the feasibility to tune the magnetic interactions in TM multilayers, we demonstrate the formation of skyrmions in such systems. To explore the magnetic phase space in an external magnetic field, we employ the spin Hamiltonian

$$\begin{aligned} H \;=\; & -\sum_{ij} J_{ij}\left(\mathbf{m}_i\cdot\mathbf{m}_j\right) - \sum_{ij}\mathbf{D}_{ij}\cdot\left(\mathbf{m}_i\times\mathbf{m}_j\right) \\ & -\sum_i K\left(m_i^z\right)^2 - \sum_i \mu_s \mathbf{B}\cdot\mathbf{m}_i, \end{aligned} \tag{1}$$

which describes the magnetic interactions between the magnetic moments $\mathbf{M}_i$ of atoms at sites $\mathbf{R}_i$ where $\mathbf{m}_i = \mathbf{M}_i/M_i$. The parameters for the exchange interaction (J_{ij}), the DMI ($\mathbf{D}_{ij}$), the magnetic moments (μ_s), as well as an uniaxial magnetocrystalline anisotropy (K) were obtained from DFT.

In a magnetic bilayer, it is convenient to split the exchange interaction into an intralayer contribution, $J_\delta^{||}$, which corresponds to the exchange coupling to an atom of the δ-th nearest neighbour shell within an Fe layer, and the interlayer part $J_\delta^{\perp}$, which corresponds to the coupling between the two Fe layers (see Supplementary Note 2 and Supplementary Table 2 for details). For the multilayer Rh/Pd/2Fe/2Ir, on which we focus below as a generic example, the interlayer coupling between nearest neighbours is very strong, $J_1^{\perp} = 24.73$ meV, and favours FM alignment of magnetic moments of the two Fe layers. However, the nearest-neighbour exchange coupling within each Fe layer, $J_1^{||} = 3.86$ meV, is much smaller and on the order of antiferromagnetic (AFM) exchange with third nearest neighbours, $J_3^{||} = -3.04$ meV. In total, the competition between intra- and interlayer exchange leads to the small effective exchange constant $J_{\mathrm{eff}} = 1$ meV (see Fig. 1), which indicates a very flat dispersion curve close to the FM state (see Supplementary Fig. 2 and Supplementary Note 3). It is noteworthy that taking into account separate intralayer exchange coupling constants within the two Fe layers does not lead to any qualitative changes (see Supplementary Note 4, Supplementary Figs 3–6 and Supplementary Table 3).

The DMI is of primary importance to stabilize skyrmions and has to be carefully evaluated for the bilayer case. Although there are contributions from the two interfaces, we find from our DFT calculations that the DMI is controlled by the Fe interface to the heavy 5*d* material (see Supplementary Note 3 and Supplementary Fig. 2). Within our spin Hamiltonian, we can describe it by an effective nearest-neighbour **D**-vector acting on the Fe moments at the Fe/Ir interface only. We obtain a value of $D = 1.3$ meV for Rh/Pd/2Fe/2Ir with very little variation on changing the composition of the Rh_xPd_{1-x} interface (on the order of 0.1 meV). This value is very similar to those obtained for the ultra-thin film system Pd/Fe/Ir(111), that is, $D = 1.0$ and 1.35 meV for fcc and hcp stacking of Pd, respectively[18]. The strength of the DMI is sufficient to create a spin spiral ground state for Rh/Pd/2Fe/2Ir with a right-hand rotation sense and a period of $\lambda = 2.25$ nm (*cf.* Supplementary Note 3). The obtained uniaxial anisotropy, K, is 0.6 meV per Fe atom and favours an out-of-plane magnetization.

We show the formation of skyrmions in multilayers by studying our spin Hamiltonian parametrized from first-principles using spin dynamics simulations (see Methods). In Fig. 2, the obtained low-temperature phase diagram is shown as a function of an external magnetic field applied perpendicular to the layers. At $B = 0$ T, the lowest energy is obtained for a spin spiral with a period of $\lambda = 2.25$ nm. As the magnetic field increases the spin spiral is destabilized and at $B = 2.75$ T a transition to a skyrmion lattice occurs. Fields of this order of magnitude have been used, for example, in ref. 17, to induce skyrmions in monolayers experimentally. Further fine-tuning of the materials in the multilayer can be used to get to even lower critical fields. At $B = 5.8$ T, the magnetic field becomes large enough to stabilize the FM state.

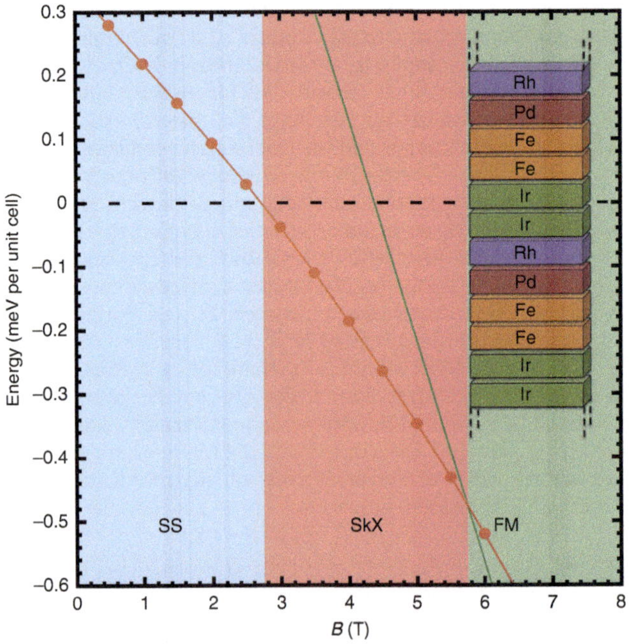

Figure 2 | Low-temperature magnetic phase diagram of the multilayer Rh/Pd/2Fe/2Ir. The total energy of the FM state (green line) and of the skyrmion lattice (SkX, red line) is shown with respect to the spin spiral (SS) state (dashed black line). The regimes of the SS, SkX and FM phase are indicated by blue, red and green colour, respectively. The multilayer structure is shown in the inset.

As the DMI does not vary much on changing the composition x of the $Rh/Rh_xPd_{1-x}/2Fe/2Ir$ multilayer and the influence of the magnetocrystalline anisotropy energy is rather moderate, the magnetic field strengths at the phase transitions depend mainly on the effective exchange interaction J_{eff}. Although this interaction is influenced by factors such as alloying at the interfaces, it can be tuned as shown in Fig. 1, allowing a control over the phase diagram. (Limiting scenarios with respect to variation of the exchange interaction and magnetocrystalline anisotropy are discussed in the Supplementary Note 5 and shown in Supplementary Figs 7 and 8.)

In Fig. 3, we present the profile of an isolated skyrmion in the bilayer at $B = 3.5$ T, that is, the polar angle θ of the local magnetization as a function of the distance from the centre. The profile was obtained by imposing the theoretical profile[39] and relaxing this spin structure within our spin dynamics simulation. The curves for both Fe layers show that at the centre of the skyrmion the magnetic moments are antiparallel to the surrounding FM background. We have obtained a skyrmion diameter defined as in ref. 39 of ~2.2 nm, consistent with the spin spiral minimum.

To understand the stabilization mechanism of a skyrmion in the Fe bilayer, we study the contributions of the different interactions to the site-resolved energy with respect to the FM background in each Fe layer for the isolated skyrmion at $B = 3.5$ T. The site-resolved energy due to exchange (Fig. 4a,b) is very similar for the two Fe layers apart from a rotation by 60° due to the different position of the first nearest-neighbour atoms in the adjacent layer, which are coupled by $J_1^{\perp}$ giving the main contribution to the exchange. The site-resolved exchange energy is minimal in the skyrmion centre where a moderate rotation of the magnetization (*cf.* Fig. 3) corresponding to a long wavelength spin spiral is energetically favoured. As we move away from the centre of the skyrmion, the canting between nearest-neighbour spins and the exchange energy increases, reaches a maximum at a distance of 1 nm from the core and decreases again when reaching the FM background.

The site-resolved total energy reveals the difference between the two Fe layers. For the Fe layer at the Pd interface (Fig. 4d) the DM interaction vanishes and the total energy is only increased by the Zeeman term and the anisotropy contribution. However, the Fe layer at the Ir interface (Fig. 4c) is subject to a significant DMI and its energy contribution outweighs the anisotropy and the Zeeman term, leading to a stabilizing (negative) energy within the entire skyrmion. The total site-resolved energy of the Fe bilayer,

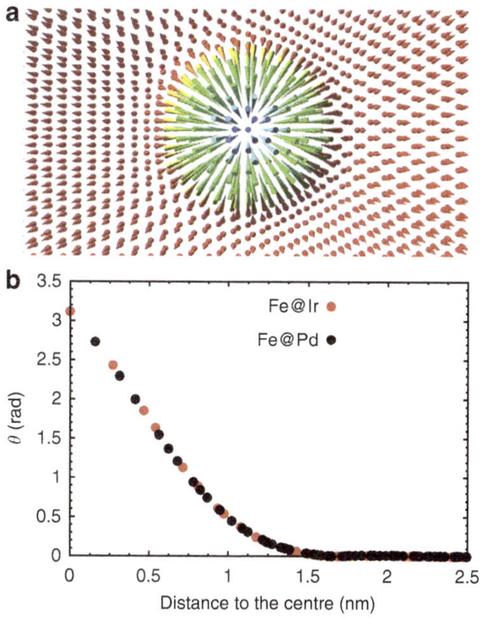

Figure 3 | Skyrmion profile in the multilayer Rh/Pd/2Fe/2Ir. (**a**) Top view of the magnetization of the skyrmion with red arrows pointing up and green/blue arrows pointing down. (**b**) Magnetization profile of an isolated skyrmion at a magnetic field of $B = 3.5$ T averaged at a given radial distance r from the centre. The polar angle θ of the magnetization vector is shown as a function of r. Red dots correspond to the Fe atoms at the Fe/Ir interface and the black dots correspond to the Fe atoms on the Pd side.

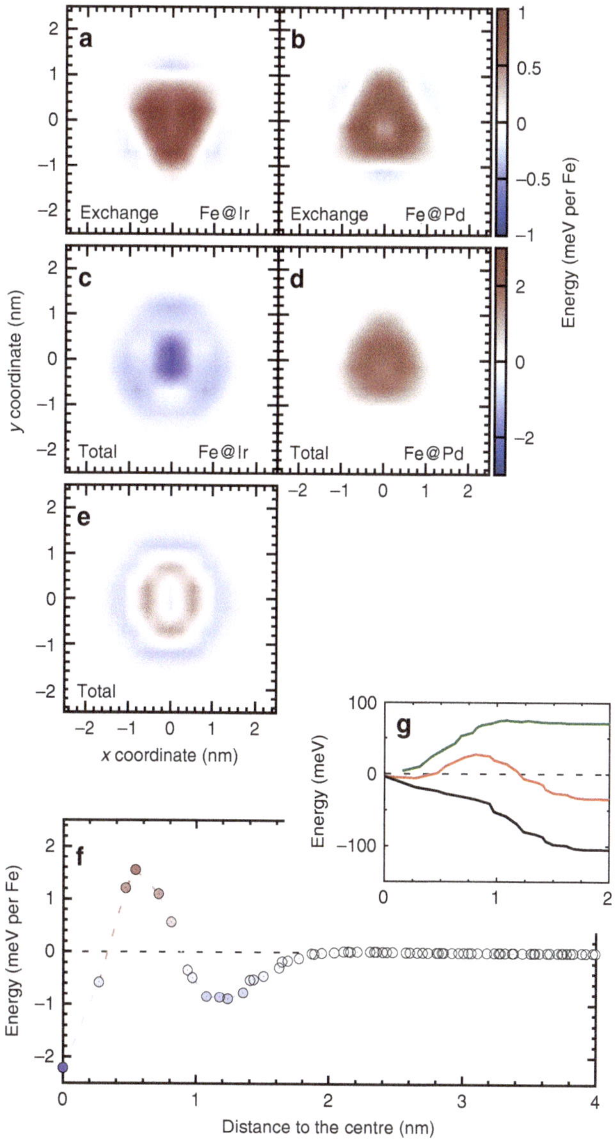

Figure 4 | Site-resolved energy contributions for a skyrmion in the multilayer Rh/Pd/2Fe/2Ir. The isolated skyrmion has been obtained at a magnetic field of $B = 3.5$ T (*cf.* skyrmion profile in Fig. 3). (**a**,**b**) The exchange energy in the Fe layer at the Ir interface (Fe@Ir) and at the Pd interface (Fe@Pd), respectively. (**c**,**d**) The total energy (exchange, DMI, Zeeman and anisotropy) for the two Fe layers and (**e**) for the complete Fe bilayer. The colour scale bar of the energy is given to the right of the panels, indicating stabilizing ($E < 0$) and destabilizing ($E > 0$) contributions. For (**e**), the scale bar is the same as for **c**,**d**. (**f**) Total energy per Fe atom averaged over the δth neighbour shell for **e**. The inset (**g**) shows the total energy summed over the shells from the skyrmion centre up to the radius r of the Fe bilayer (red), Fe@Pd (green) and Fe@Ir (black).

that is, the sum of the two layers, shown in Fig. 4e,f, exhibits a small local minimum at the centre of the skyrmion due to the compensation of the DMI from the Fe/Ir interface and the Zeeman and anisotropy terms. As we move away from the centre, the total energy first increases due to the strong exchange and Zeeman contributions and the rise of the anisotropy term, and decreases beyond ~ 0.5 nm due to the DM term (*cf.* Fig. 4f).

The site-resolved energy summed over all shells within the radius r around the skyrmion centre is shown for the Fe bilayer and the two Fe layers in Fig. 4g. For the Fe layer at the Pd interface (green curve), the energy is always positive as expected from the site-resolved energy (*cf.* Fig. 4d) and saturates at $\sim r = 1.25$ nm. For the Fe layer at the Ir interface (black curve), the energy is negative, indicating a skyrmion stabilization due to the DMI. The competition of these two contributions leads to the profile for the Fe bilayer (red curve), which displays a positive maximum at small radius but converges to a negative value of about -32 meV, indicating that the skyrmion is energetically favourable due to the DM contribution at the rim of the skyrmion.

Discussion

We have focused on the system composed of an Fe bilayer sandwiched between Rh/Pd and Ir layers. For Fe monolayers on 4*d* and 5*d* TM surfaces, we have demonstrated before that the exchange interaction can be tuned by the substrate band filling from FM to AFM[31,40,41]. Therefore, other TM multilayers, for example, Ru_xPd_{1-x}/2Fe/Pt, allow a similar tuning of J_{eff} by varying the composition x. The strength of the DMI, on the other hand, will be determined by the interface with the 5*d*-TM, for example, Ir or Pt[42,43].

We have not considered to increase the number of Fe layers, for example, to trilayers, as this would increase the effective exchange interaction and reduce the effect of the two interfaces on the magnetic properties. The increase of magnetic material is ensured in our setup by the repetitions of the sandwich layers and strong coupling between skyrmions in adjacent bilayers. This leads to enhanced transition temperatures as shown in Supplementary Note 6 and Supplementary Fig. 9. TM multilayers consisting of a repetition of a few atomic layers as proposed here have recently been realized experimentally to stabilize chiral domain walls[29].

In conclusion, we have shown that magnetic skyrmions can emerge in TM multilayers consisting of Fe bilayers sandwiched between 4*d* and 5*d* TMs. Such systems provide a rich field for the formation of skyrmions and allow to go beyond the monolayer systems discussed so far[16–18] to a thickness range where the unique transport properties of these systems can be studied and exploited in devices. Recent success in tailoring the DMI at the interface to the 5*d* material[28] can be combined with our concept to tune the exchange interaction by the choice of the 4*d* component of the multilayers, thus opening a route to engineer skyrmion properties for spintronic applications.

Methods

First-principles calculations. We have explored the multilayer structures from first-principles based on the full-potential linearized augmented plane wave method as implemented in the FLEUR code (www.flapw.de). Within this approach we can calculate the total energy of non-collinear magnetic structures such as spin spirals[32] including the DMI in first-order perturbation theory with respect to the spin–orbit coupling[33]. We have used a two-dimensional hexagonal p(1 × 1) unit cell within each layer and the in-plane lattice parameter of the Ir(111) surface as obtained from DFT in ref. 18. The distances between the different layers were relaxed using the mixed functional suggested in ref. 44, which is ideally suited for interfaces of 3*d* and 5*d* TMs. The magnetic properties were obtained within the local density approximation[45]. To study alloying within the RhPd layers, we have used the virtual crystal approximation[46], which has been used successfully for various types of systems from solid solutions[47] to magnetic ultra-thin films[40]. Further computational details can be found in Supplementary Note 1.

Spin dynamics simulations. To calculate the different energy differences between the FM, SkX and SS phases, we have relaxed a spin lattice of the bilayer with (100 × 100) spins in each layer according to the Landau–Lifshitz equation of spin dynamics:

$$\hbar \frac{d\mathbf{m}_i}{dt} = \frac{\partial H_i}{\partial \mathbf{m}_i} \times \mathbf{m}_i - \beta \left(\frac{\partial H_i}{\partial \mathbf{m}_i} \times \mathbf{m}_i \right) \times \mathbf{m}_i \qquad (2)$$

where β is the damping term and H_i is the Hamiltonian of spin i and is expressed by the Hamiltonian given in equation (1) if we omit the sum over i. To achieve fast relaxation, we have used $\beta = 0.95$. We have carried out simulations on a nanosecond timescale with time steps ranging from 0.1 to 10 fs. The equation of motion was integrated with a simple Euler, a Heun and two semi-implicit integrators[48] leading to the same results. To ensure that we reached the ground state, we have simultaneously run Monte Carlo simulations with a Metropolis algorithm on these configurations at $T = 0$ K. Depending on the system, both techniques give similar results with a precision on the order of a few μeV per spin, sufficient to discriminate between the different phases in our cases.

References

1. Bogdanov, A. & Yablonskii, D. A. Thermodynamically stable vortices in magnetically ordered crystals. The mixed state of magnets. *Sov. Phys. JETP* **68,** 101–103 (1989).
2. Bogdanov, A. & Hubert, A. Thermodynamically stable magnetic vortex states in magnetic crystals. *J. Mag. Mag. Mater.* **138,** 255–269 (1994).
3. Rößler, U. K., Bogdanov, A. N. & Pfleiderer, C. Spontaneous skyrmion ground states in magnetic metals. *Nature* **442,** 797–801 (2006).
4. Fert, A., Cros, V. & Sampaio, J. Skyrmions on the track. *Nat. Nanotechnol.* **8,** 152–156 (2013).
5. Kiselev, N., Bogdanov, A. N., Schäfer, R. & Rößler, U. K. Chiral skyrmions in thin magnetic films: new objects for magnetic storage technology? *J. Phys. D* **44,** 392001 (2011).
6. Sampaio, J., Cros, V., Rohart, S., Thiaville, A. & Fert, A. Nucleation, stability and current-induced motion of isolated magnetic skyrmions in nanostructures. *Nat. Nanotechnol.* **8,** 839–844 (2013).
7. Dzyaloshinskii, I. E. Thermodynamic theory of weak ferromagnetism in antiferromagnetic substances. *Sov. Phys. JETP* **5,** 1259–1262 (1957).
8. Moriya, T. Anisotropic superexchange interaction and weak ferromagnetism. *Phys. Rev.* **120,** 91–98 (1960).
9. Nagaosa, N. & Tokura, Y. Topological properties and dynamics of magnetic skyrmions. *Nat. Nanotechnol.* **8,** 899–911 (2013).
10. Mühlbauer, S. *et al.* Skyrmion lattice in a chiral magnet. *Science* **323,** 915–919 (2009).
11. Pappas, C. *et al.* Chiral paramagnetic skyrmion-like phase in MnSi. *Phys. Rev. Lett.* **102,** 197202 (2009).
12. Yu, X. *et al.* Real-space observation of a two-dimensional skyrmion crystal. *Nature* **465,** 901–904 (2010).
13. Crépieux, A. & Lacroix, C. Dzyaloshinskii-Moriya interactions induced by symmetry breaking at a surface. *J. Mag. Mag. Mater.* **182,** 341–349 (1998).
14. Bogdanov, A. N. & Rößler, U. K. Chiral symmetry breaking in magnetic thin films and multilayers. *Phys. Rev. Lett.* **87,** 037203 (2001).
15. Bode, M. *et al.* Chiral magnetic order at surfaces driven by inversion asymmetry. *Nature* **447,** 190–193 (2007).
16. Heinze, S. *et al.* Spontaneous atomic-scale magnetic skyrmion lattice in two dimensions. *Nat. Phys.* **7,** 713–718 (2011).
17. Romming, N. *et al.* Writing and deleting single magnetic skyrmions. *Science* **341,** 636–639 (2013).
18. Dupé, B., Hoffmann, M., Paillard, C. & Heinze, S. Tailoring magnetic skyrmions in ultra-thin transition metal films. *Nat. Commun.* **5,** 4030 (2014).
19. Simon, E., Palotás, K., Rózsa, L., Udvardi, L. & Szunyogh, L. Formation of magnetic skyrmions with tunable properties in PdFe bilayer deposited on Ir(111). *Phys. Rev. B* **90,** 094410 (2014).
20. Jonietz, F. *et al.* Spin transfer torques in MnSi at ultralow current densities. *Science* **330,** 1648–1651 (2010).
21. Yu, X. Z. *et al.* Skyrmion flow near room temperature in an ultralow current density. *Nat. Commun.* **3,** 988 (2012).
22. Everschor, K. *et al.* Rotating skyrmion lattices by spin torques and field or temperature gradients. *Phys. Rev. B* **86,** 054432 (2012).
23. Iwasaki, J., Mochizuki, M. & Nagaosa, N. Universal current-velocity relation of skyrmion motion in chiral magnets. *Nat. Commun.* **4,** 1463 (2013).

24. Iwasaki, J., Mochizuki, M. & Nagaosa, N. Current-induced skyrmion dynamics in constricted geometries. *Nat. Nanotechnol.* **8,** 742–747 (2013).
25. Lee, M., Kang, W., Onose, Y., Tokura, Y. & Ong, N. P. Unusual Hall effect anomaly in MnSi under pressure. *Phys. Rev. Lett.* **102,** 186601 (2009).
26. Neubauer, A. *et al.* Topological Hall effect in the a phase of MnSi. *Phys. Rev. Lett.* **102,** 186602 (2009).
27. Franz, C. *et al.* Real-space and reciprocal-space berry phases in the Hall effect of $Mn_{1-x}Fe_xSi$. *Phys. Rev. Lett.* **112,** 186601 (2014).
28. Moreau-Luchaire, C. *et al.* Additive interfacial chiral interaction in multilayers for stabilization of small individual skyrmions at room temperature. *Nat. Nanotechnol.* **11,** 444–448 (2016).
29. Chen, G., N'Diaye, A. T., Wu, Y. & Schmid, A. K. Ternary superlattice boosting interface-stabilized magnetic chirality. *App. Phys. Lett.* **106,** 062402 (2015).
30. Chen, G., Mascaraque, A., N'Diaye, A. T. & Schmid, A. K. Room temperature skyrmion ground state stabilized through interlayer exchange coupling. *Appl. Phys. Lett.* **106,** 242404 (2015).
31. Hardrat, B. *et al.* Complex magnetism of Fe monolayers on hexagonal transition-metal surfaces from first principles. *Phys. Rev. B* **79,** 094411 (2009).
32. Kurz, P., Förster, F., Nordström, L., Bihlmayer, G. & Blügel, S. Ab initio treatment of noncollinear magnets with the full-potential linearized augmented plane wave method. *Phys. Rev. B* **69,** 024415 (2004).
33. Heide, M., Bihlmayer, G. & Blügel, S. Describing Dzyaloshinskii-Moriya spirals from first principles. *Phys. B Condens. Matter* **404,** 2678–2683 (2009).
34. Menzel, M. *et al.* Information transfer by vector spin chirality in finite magnetic chains. *Phys. Rev. Lett.* **108,** 197204 (2012).
35. Ferriani, P. *et al.* Atomic-scale spin spiral with a unique rotational sense: Mn monolayer on W(001). *Phys. Rev. Lett.* **101,** 027201 (2008).
36. Zimmermann, B., Heide, M., Bihlmayer, G. & Blügel, S. First-principles analysis of a homochiral cycloidal magnetic structure in a monolayer Cr on W(110). *Phys. Rev. B* **90,** 115427 (2014).
37. Yoshida, Y. *et al.* Conical spin-spiral state in an ultrathin film driven by higher-order spin interactions. *Phys. Rev. Lett.* **108,** 087205 (2012).
38. Romming, N., Kubetzka, A., Hanneken, C., von Bergmann, K. & Wiesendanger, R. Field-dependent size and shape of single magnetic skyrmions. *Phys. Rev. Lett.* **114,** 177203 (2015).
39. Bogdanov, A. N. & Hubert, A. The properties of isolated magnetic vortices. *Phys. Stat. Solid.* **186,** 527–543 (1994).
40. Ferriani, P., Turek, I., Heinze, S., Bihlmayer, G. & Blügel, S. Magnetic phase control in monolayer films by substrate tuning. *Phys. Rev. Lett.* **99,** 187203 (2007).
41. Al-Zubi, A., Bihlmayer, G. & Blügel, S. Modelling magnetism of hexagonal Fe monolayers on 4d substrates. *Phys. Stat. Solid. B* **248,** 2242–2247 (2011).
42. Chen, G. *et al.* Tailoring the chirality of magnetic domain walls by interface engineering. *Nat. Commun.* **4,** 2671 (2013).
43. Hrabec, A. *et al.* Measuring and tailoring the Dzyaloshinskii-Moriya interaction in perpendicularly magnetized thin films. *Phys. Rev. B* **90,** 020402 (R) (2014).
44. De Santis, M. *et al.* Structure and magnetic properties of MnPt(110) – (1 × 2): a joint x-ray diffraction and theoretical study. *Phys. Rev. B* **75,** 205432 (2007).
45. Vosko, S. H., Wilk, L. & Nusair, M. Accurate spin-dependent electron liquid correlation energies for local spin density calculations: a critical analysis. *Can. J. Phys.* **58,** 1200–1211 (1980).
46. Turchi, P. E. A., Gonis, A., Drchal, V. & Kudrnovský, J. First-principles study of stability and local order in substitutional Ta-W alloys. *Phys. Rev. B* **64,** 085112 (2001).
47. Bellaiche, L. & Vanderbilt, D. Virtual crystal approximation revisited: application to dielectric and piezoelectric properties of perovskites. *Phys. Rev. B* **61,** 7877–7882 (2000).
48. Mentink, J. H., Tretyakov, M. V., Fasolino, A., Katsnelson, M. I. & Rasing, T. Stable and fast semi-implicit integration of the stochastic Landau-Lifshitz equation. *J. Phys. C* **22,** 176001 (2010).

Acknowledgements

We thank Stanislas Rohart for helpful discussions. Research leading to these results has received funding from the Deutsche Forschungsgemeinschaft via project DU1489/2-1. We gratefully acknowledge computing time at the JUROPA supercomputer from the Jülich Supercomputing Centre and the supercomputer of the North-German Supercomputing Alliance (HLRN). This project has received funding from the European Union Horizon 2020 research and innovation programme under grant agreement number 665095 (FET-Open project MAGicSky).

Author contributions

B.D. performed the spin dynamics simulations. S.H. and G.B. performed the first-principles calculations. M.B. and B.D. performed the Monte-Carlo simulations. B.D., G.B. and S.H. wrote the manuscript. All authors discussed the data and contributed to the paper.

Additional information

Competing financial interests: The authors declare no competing financial interests.

8

Quantum annealing for the number-partitioning problem using a tunable spin glass of ions

Tobias Graß[1], David Raventós[1], Bruno Juliá-Díaz[1,2,3], Christian Gogolin[1,4] & Maciej Lewenstein[1,5]

Exploiting quantum properties to outperform classical ways of information processing is an outstanding goal of modern physics. A promising route is quantum simulation, which aims at implementing relevant and computationally hard problems in controllable quantum systems. Here we demonstrate that in a trapped ion setup, with present day technology, it is possible to realize a spin model of the Mattis-type that exhibits spin glass phases. Our method produces the glassy behaviour without the need for any disorder potential, just by controlling the detuning of the spin-phonon coupling. Applying a transverse field, the system can be used to benchmark quantum annealing strategies which aim at reaching the ground state of the spin glass starting from the paramagnetic phase. In the vicinity of a phonon resonance, the problem maps onto number partitioning, and instances which are difficult to address classically can be implemented.

[1]ICFO-Institut de Ciències Fotòniques, The Barcelona Institute of Science and Technology, Avinguda Carl-Friedrich Gauss 2, Castelldefels 08860, Spain. [2]Departament de Física Quàntica i Astrofísica, Facultat de Física, Universitat de Barcelona, Barcelona 08028, Spain. [3]Institut de Ciències del Cosmos, Universitat de Barcelona, ICCUB, Martí i Franquès 1, Barcelona 08028, Spain. [4]Max-Planck-Institut für Quantenoptik, Hans-Kopfermann-Straße 1, Garching 85748, Germany. [5]ICREA-Institució Catalana de Recerca i Estudis Avançats, Lluís Companys 23, Barcelona 08010, Spain. Correspondence and requests for materials should be addressed to T.G. (email: tobias.grass@icfo.es).

Spin models are paradigms of multidisciplinary science: They find several applications in various fields of physics, from condensed matter to high-energy physics, but also beyond the physical sciences. In neuroscience, the famous Hopfield model describes brain functions such as associative memory by an interacting spin system[1]. This directly relates to computer and information sciences, where pattern recognition or error-free coding can be achieved using spin models[2]. Importantly, many optimization problems, like number partitioning or the travelling salesman problem, belonging to the class of NP-hard problems, can be mapped onto the problem of finding the ground state of a specific spin model[3,4]. This implies that solving spin models is a task for which no general efficient classical algorithm is known to exist. In physics, analytic replica methods have been developed in the context of spin glasses[5,6]. A controversial development, supposed to provide also an exact numerical understanding of spin glasses, regards the D-Wave machine. Recently introduced in the market, this device solves classical spin glass models, but the underlying mechanisms are not clear, and it remains an open question whether it provides a speed-up over the best classical algorithms[7–9].

This triggers interest in alternative quantum systems designed to solve the general spin models via quantum simulation. A noteworthy physical system for this goal are trapped ions: Nowadays, spin systems of trapped ions are available in many laboratories[10–15], and adiabatic state preparation, similar to quantum annealing, is experimental state-of-art. Moreover, such system can exhibit long-range spin–spin interactions[16] mediated by phonon modes, leading to a highly connected spin model. Here we demonstrate how to profit from these properties, using trapped ions as a quantum annealer of a classical spin glass model.

We consider a setup described by a time-dependent Dicke-like model:

$$H_0(t) = \sum_m \hbar\omega_m a_m^\dagger a_m + \sum_{i,m} \hbar\Omega_i \sqrt{\frac{\omega_{\text{recoil}}}{\omega_m}} \xi_{im} \sin(\omega_\text{L} t) \times \sigma_x^i \left(a_m + a_m^\dagger\right), \tag{1}$$

with a_m annihilating a phonon in mode m with frequency ω_m and characterized by the normalized collective coordinates ξ_{im}. The second term in H_0 couples the motion of the ions to an internal degree of freedom (spin) through a Raman beam which induces a spin flip on site i, described by σ_x^i, and (de)excites a phonon in mode m. Here, Ω_i is the Rabi frequency, $\hbar\omega_\text{recoil}$ the recoil energy, and ω_L the beatnote frequency of the Raman lasers. Before also studying the full model, we consider a much simpler effective Hamiltonian, derived from equation (1) by integrating out the phonons[16–19]. The model then reduces to a time-independent Ising-type spin Hamiltonian

$$H_J = -\hbar \sum_{ij} J_{ij} \sigma_x^i \sigma_x^j. \tag{2}$$

Each phonon mode contributes to the effective coupling J_{ij} in a factorizable way, proportional to $\xi_{im}\xi_{jm}$, and weighted by the inverse of the detuning from the mode $\Delta_m = \omega_m - \omega_\text{L}$:

$$J_{ij} = \Omega_i \Omega_j \frac{\omega_\text{recoil}}{2\omega_\text{L}} \sum_m \frac{\xi_{im}\xi_{jm}}{\Delta_m}. \tag{3}$$

The ξ_{im} imprint a pattern to the spin configuration, similar to the associative memory in a neural network[1,20]. The connection between multi-mode Dicke models with random couplings, the Hopfield model, and spin glass physics has been the subject of recent research[21–23], and the possibility of addressing number partitioning was mentioned in this context[23].

Before proceeding, we remind the reader that the concept of a spin glass used in the literature may have different and controversial meanings (*cf.*, refs 24–27): (i) long-range spin glass models[5] and neural networks[28,29], believed to be captured by the Parisi picture[30,31] and replica symmetry breaking. These lead to hierarchical organization of the exponentially degenerated free-energy landscape, breakdown of ergodicity and aging, slow dynamics due to a continuous splitting of the metastable states with decreasing temperatures (*cf.*, ref. 32). (ii) Short-range spin glass models believed to be captured by the Fisher–Huse[33,34] droplet/scaling model with a single ground state (up to a total spin flip), but a complex structure of the domain walls. For these models, aging, rejuvenation and memory, if any, have different nature and occurrence[32,35]; (iii) Mattis glasses[36], where the huge ground-state degeneracy becomes an exponential quasi-degeneracy, for which finding the ground state becomes computationally hard[37] (see the subsection 'Increasing complexity' in the 'Results' section). Note that exponential (quasi)degeneracy of the ground states (or the free-energy minima) characterizes also other interesting states: certain kinds of spin liquids or spin ice, and so on.

Here we analyse the trapped ion setup. Even without explicit randomness ($\Omega_i = \Omega = \text{const.}$), the coupling to a large number of phonon modes suggests the presence of glassy behaviour. This intuition comes from the fact that the associative memory of the related Hopfield model works correctly if the number of patterns is at most $0.138N$, where N is the number of spins[28]. For a larger number of patterns, the Hopfield model exhibits glassy behaviour since many patterns have similar energy and the dynamics gets stuck in local minima. However, it is not clear *a priori* how the weighting of each pattern, present in equation (3), modifies the behaviour of the spin model. In certain regimes the detuning suggests to neglect the contributions from all but one mode, leading to a Mattis-type model with factorizable couplings[36], $J_{ij} \propto \xi_{im}\xi_{jm}$. The possibility of negative detuning, that is, antiferromagnetic coupling to a pattern, drives such system into a glassy phase, characterized by a huge low-energy Hilbert space. The antiferromagnetic Mattis model is directly connected to the potentially NP-hard task of number partitioning[37,38]. Its solution can then be found via quantum annealing, that is, via adiabatic ramping down of a transverse magnetic field, see also the proposal of ref. 39 for a frustrated ion chain.

We start by giving analytical arguments to demonstrate the glassy behaviour in the classical spin chains. Using exact numerics, we then focus on the quantum Mattis model. By calculating the magnetic susceptibility, and an Edward–Anderson-like order parameter, we distinguish between glassy, paramagnetic and ferromagnetic regimes. The annealing dynamics is investigated using exact numerics and a semi-classical approximation. We demonstrate the feasibility of annealing for glassy instances. Finally, we show that the memory in the quantum Hopfield model is real-valued rather than binary. This property might be useful for various applications of quantum neural networks such as pattern-recognition schemes.

Results

Tunable spin–spin interactions. We start our analysis by inspecting the phonon modes. For a sufficiently anisotropic trapping potential, the ions self-arrange in a line along the z-axis[40,41]. The phonon Hamiltonian H_ph is obtained by a second-order expansion of Coulomb and trap potentials around these equilibrium positions z_i: $H_\text{ph} = (M/2)\sum_{ij} V_{ij} q_i q_j$, with q_i the displacement of the ith ion from equilibrium. For the transverse phonon branch, V_{ij} is given by[16]

$$V_{ij} = \begin{cases} \omega_\perp^2 - \dfrac{e^2/M}{4\pi\epsilon_0} \displaystyle\sum_{i''(\neq i)} \dfrac{1}{\left|z_i - z_{i''}\right|^3}, & i = j \\ \dfrac{e^2/M}{4\pi\epsilon_0} \dfrac{1}{\left|z_i - z_j\right|^3}. & i \neq j \end{cases} \tag{4}$$

Our exact numerical simulations have been performed for six $^{40}Ca^+$ ions in a trap of frequency $\omega_\perp = 2\pi \times 2.655$ MHz, as used in a recent experiment[14]. To maximize the bandwidth of the phonon spectrum and thereby facilitate the annealing process, we choose the radial trap frequency ω_z as large as allowed to avoid zig-zag transitions, $\omega_z \lesssim 1.37\omega_\perp \cdot N^{-0.86}$. Diagonalizing the matrix V_{ij} leads to the previously introduced mode vectors $\boldsymbol{\xi}_m = (\xi_{1m}, \ldots, \xi_{Nm})$, which are normalized to one, and ordered according to their frequency ω_m:

$$\boldsymbol{\xi}_{m'}^T V \boldsymbol{\xi}_m = \omega_m^2 \delta_{m,m'}. \quad (5)$$

The mode $\boldsymbol{\xi}_N$ with largest frequency, $\omega_N = \omega_\perp$, is the centre-of-mass mode, $\xi_{iN} = N^{-1/2}$. Parity symmetry of V_{ij} is reflected by the modes, $\xi_{im} = \pm \xi_{(N+1-i)m}$, and even $(+)$ and odd $(-)$ modes alternate in the phonon spectrum. We focus on even N, for which all components ξ_{im} are non-zero. Except for the centre-of-mass mode, all modes fulfil $\sum_i \xi_{im} = 0$.

Previous experiments[10–14] have mostly been performed with a beatnote frequency ω_L several kHz above ω_N, leading to an antiferromagnetic coupling $J_{ij} < 0$ with power-law decay, see Fig. 1a. Despite the presence of many modes, the couplings J_{ij} then take an ordered structure. This work, in contrast, focuses on the regime $\omega_L < \omega_N$, where modes with both positive and negative ξ_{im} contribute, and ferro and antiferromagnetic couplings coexist, see Fig. 1b. This reminds of the disordered scenario of common spin glass models like the Sherrington–Kirkpatrick model[5]. In the following we study the properties of the time-independent effective spin model, before considering the full time-depedent problem involving spin-phonon coupling.

Classical Mattis model. Close to a resonance with one phonon mode, simple arguments allow for deducing the spin configurations of the ground states. In this limit, we can neglect the other modes, leading to a Mattis model. A single pattern $\boldsymbol{\xi}_m$ then determines the coupling, $J_{ij} \propto \xi_{im}\xi_{jm}$. The sign of J_{ij} depends on the sign of the detuning: Below the resonance, we have $\text{sign}(J_{ij}) = \text{sign}(\xi_{im}\xi_{jm})$, and accordingly the energy $-\hbar J_{ij}\sigma_x^i\sigma_x^j$ is minimized if σ_x^i and σ_x^j are either both aligned or both anti-aligned with ξ_{im} and ξ_{jm}. Thus, we have a twofold degenerate ground state given by the patterns $\pm[\text{sign}(\xi_{1m}), \ldots, \text{sign}(\xi_{Nm})]$. We refer to this scenario as the ferromagnetic side of a resonance.

Crossing the resonance, the overall sign of the Hamiltonian changes. Naively, one might assume that this should not qualitatively affect the physics, since we have $\frac{1}{N}\sum_{i \neq j} \xi_{im}\xi_{jm} = -\frac{1}{N} \to 0$, that is, there is an equal balance between positive and negative J_{ij}. This expectation, however, turns out to be false. Recalling the relation between the Mattis model and number partitioning[4,37,38], the antiferromagnetic model maps onto an optimization problem in which the task is to find the optimal bi-partition of a given sequence of numbers $(\xi_i)_i$, such that the cost function $E = (\sum_{i\in\uparrow} \xi_i - \sum_{j\in\downarrow} \xi_j)^2$ is minimized. Here, the two partitions are denoted by $\uparrow$ and $\downarrow$. For a Hamiltonian of the form $H = \sum_{ij} \xi_i\sigma_x^i\xi_j\sigma_x^j = (\sum_i \xi_i\sigma_x^i)^2$, eigenvectors of σ_x^i are Hamiltonian eigenstates with an energy precisely given by the cost function E. Thus, in the limit of just one antiferromagnetic resonance, the ground state of H is exactly the configuration that minimizes the cost function.

With this insight, the ground states of the spin model are easily found exploiting the system's parity symmetry: For even modes, $\xi_{im} = \xi_{(N+1-i)m}$, and we simply have to choose $\langle\sigma_x^i\rangle = -\langle\sigma_x^{N+1-i}\rangle$ to minimize the cost function. For odd modes, $\xi_{im} = -\xi_{(N+1-i)m}$, and we must choose $\langle\sigma_x^i\rangle = \langle\sigma_x^{N+1-i}\rangle$. In both cases, this implies that we can choose half of the spins arbitrarily, leading to at least $2^{N/2}$ ground states.

The important observation is that an exponentially large number of ground states exists in the limit of being arbitrarily close above a resonance. This is a characteristic feature of spin glasses, yet it does not lead to computational hardness. In fact, as pointed out in ref. 38, the number-partitioning problem with exponentially many perfect partitions belongs to the 'easy phase'. How to reach hard instances will be explained in the section below.

Pushing our arguments further we consider the influence of a second resonance: In between two resonances, the exponential degeneracy of the antiferromagnetic coupling on one side is lifted by the influence of the ferromagnetically coupled mode on the other side. Interestingly, this does not lead to frustration, since even- and odd-parity modes alternate in the phonon spectrum, and the pattern favoured by the ferromagnetic coupling is always contained in the ground state manifold of the antiferromagnetic coupling. Accordingly, between two modes the ground-state pattern is uniquely defined by the upper mode.

Beyond this two-mode approximation, we rely on numerical results. Taking into account all modes, exact diagonalization of a small system ($N \leq 10$) shows that the two-mode model captures the behaviour correctly: At any detuning, the degeneracy due to the nearest antiferromagnetic coupling is lifted in favour of the pattern of the next ferromagnetic coupling. In Fig. 2a, we plot the cumulative density of states $\rho_{\text{cum}}(E)$, that is, the number of states with an energy below E. The corresponding phonon resonances ω_m are marked in Fig. 2b. The curves clearly reflect the very different behaviour in the red- and blue-detuned regimes: Fig. 2 illustrates the quick increase of $\rho_{\text{cum}}(E)$ at low energies, when the

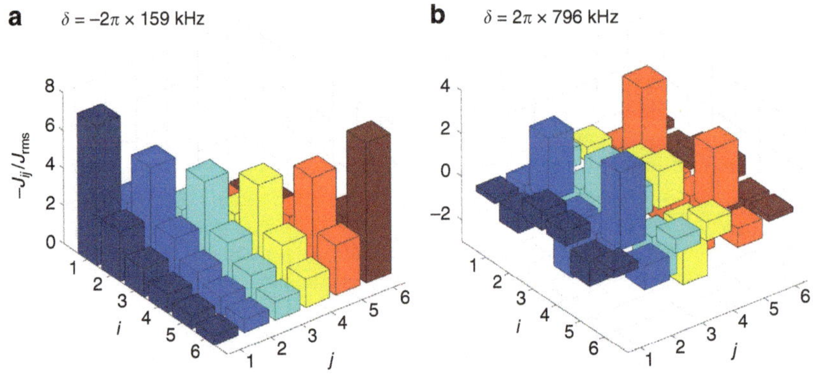

Figure 1 | Power-law decay against quasi-random patterns. We plot the coupling constants J_{ij} (in units of $J_{\text{rms}} \equiv \frac{1}{N(N-1)}\sum_{i\neq j}|J_{ij}|^2$), for a system of six ions at different detunings $\delta = \omega_N - \omega_L$. Rabi frequencies are taken as constants $\Omega_i = \text{const.}$ **(a)** For negative detuning, $\delta = -2\pi \times 159$ kHz, interactions have a power-law decay. **(b)** For positive detuning, $\delta = 2\pi \times 796$ kHz, the coupling constants resemble a spin glass.

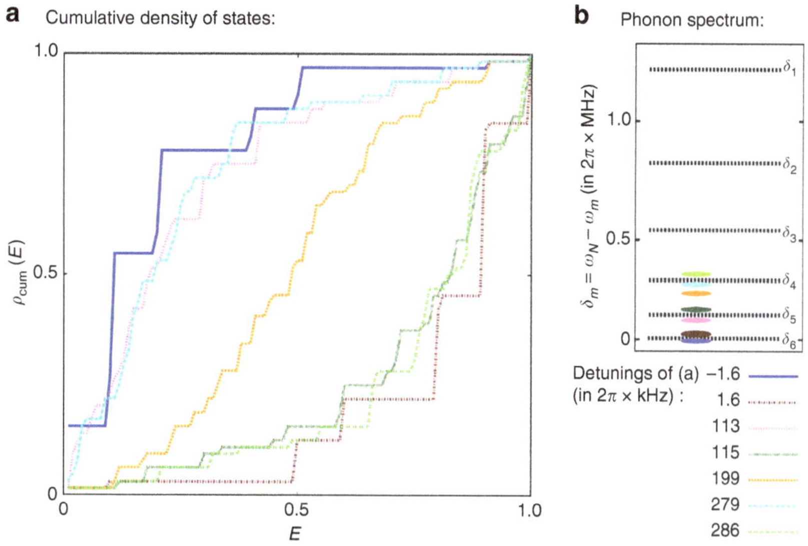

Figure 2 | Cumulative density of states. (**a**) For six ions and at trap frequency $\omega_z = 2\pi \times 770$ kHz, we plot the number of states (divided by the total Hilbert space dimension) below the normalized energy treshold (0: ground state energy, 1: energy of highest state), for different detunings from the centre-of-mass mode, $\delta = \omega_N - \omega_L$. The density of states at low energies is seen to strongly increase when the detuning is slightly above a phonon resonance. In **b** the position of the resonances (measured from the centre-of-mass mode at $\delta_N = 0$) are shown.

laser detuning is chosen on the antiferromagnetic side of a phonon resonance. In contrast, a low density of states characterizes the system on the ferromagnetic side of a resonance. In intermediate regimes, as shown for $\delta = 2\pi \times 199$ kHz, the spectrum is symmetric.

A breakdown of the two-mode approximation is expected for large numbers of spins: Since the distance between neighbouring resonances approximately scales with $1/N$ (at fixed trap frequencies), the influence of additional modes grows with the system size. The combined contribution of all antiferromagnetically coupled modes tries to select the fully polarized configurations as the true ground state, while all ferromagnetically coupled modes, except for the centre-of-mass mode, favour fully unpolarized configurations. As a consequence, it is *a priori* unclear which pattern will be selected in the presence of many modes.

This observation is crucial from the point of view of a complexity theory. In the presence of parity symmetry neither the one-mode problem (that is, the number-partitioning problem), nor the two-mode approximation are hard problems, as they can be solved by simple analytic arguments. However, when many modes lift the degeneracy of the exponentially large low-energy manifold in an *a priori* unknown way, one faces the situation where a true but unknown ground state is separated only by a very small gap. Identifying this state then usually requires scanning an exponentially large number of low-energy states, and classical annealing algorithms can easily get stuck in a wrong minimum. Below, we discuss how a transverse magnetic field opens up a way of finding the ground state via quantum annealing. Moreover, we will discuss strategies to make also the one-mode model, that is, the number-partitioning problem, computationally complex.

Increasing complexity. As discussed above, the instances of the number-partitioning problem realized in the ion chain are simple to solve due to parity symmetry. This is a convenient feature when testing the correct functioning of the quantum simulation, but our goal is the implementation of computationally complex and selectable instances of the problem in the device. One strategy is the use of microtraps to hold the ions[42]. The equilibrium positions of the ions can then be chosen at will, opening up the possibility to control the values of the ξ_{im}. The computational complexity of the number-partitioning problem then depends on the precision with which the ξ_{im} can be tuned. If the number of digits can be scaled with the number of spins, one enters the regime where number partitioning is proven to be NP-hard[38]. Thus, the number of digits must at least be of order $\log_{10} N$, which poses no problem for realistic systems involving tens of ions.

Another way of enhancing complexity even within a parity-symmetric trap would be to 'deactivate' some spins by a fast pump laser. For example, if all spins on the left half of the chain are forced to oscillate, $\sigma_x^j \rightarrow \sigma_x^j e^{i\omega_{\text{pump}} t}$, the part of the Hamiltonian which remains time-independent poses a number-partitioning problem of N/2 different numbers.

Another promising approach was recently suggested in ref. 43: operating on the antiferromagnetic side of the centre-of-mass resonance, single-site addressing allows one to use the Rabi frequency for defining the instance of the number-partitioning problem. This could indeed be the step to turn the trapped ions setup into a universal number-partitioning solver, where arbitrary user-defined instances can be implemented.

If one is not interested in the number-partitioning problem itself, one might also increase the system's complexity via resonant coupling to more than one mode. Equipping the Raman laser with several beatnote frequencies $\omega_L^{(\mu)}$ and Rabi frequencies $\Omega_i^{(\mu)}$, it is possible to engineer couplings of the form[44]:

$$J_{ij} \propto \sum_{\mu=1}^{\mu_{\max}} \Omega_i^{(\mu)} \Omega_j^{(\mu)} \sum_{m=1}^{N} \frac{\xi_{im}\xi_{jm}}{\omega_m - \omega_L^{(\mu)}}. \quad (6)$$

With an appropriate choice of Rabi frequencies and detunings, this allows for realizing the Hopfield model, $J_{ij} \propto \sum_{\mu=1}^{\mu_{\max}} \xi_{im_\mu} \xi_{jm_\mu}$, where each coupling μ is assumed to be in resonance with one mode m_μ. For ferromagnetic couplings, the low-energy states again are determined by the signs of the ξ_{im_μ}, but in general the different low-energy patterns are not degenerate, and a glassy regime is expected for large $\mu_{\max}$ (ref. 28).

Quantum phases. So far, we have considered classical spin chains lacking any non-commuting terms in the Hamiltonian. Quantum properties come into play if we either add an additional coupling $\sum_{ij}\sigma_y^i\sigma_y^j$, or a transverse magnetic field:

$$H_B = \hbar B \sum_i \sigma_z^i. \tag{7}$$

The latter has been realized in several experiments[10,12,14], and is convenient for our purposes, as the field strength B, if decaying with time, provides an annealing parameter: For large B, all spins are polarized along the z-direction, whereas for vanishing B one obtains the ground state of the classical Ising chain. As argued above, the latter exhibits spin glass phases with an exponentially large low-energy subspace. Even in those cases where the true ground state is known theoretically, finding it experimentally remains a difficult task. Our system hence provides an ideal test ground for experimenting with different annealing strategies.

Before presenting results for the simulated quantum annealing, let us first discuss the different phases expected for the effective Hamiltonian $H_{\text{eff}}=H_J+H_B+\hbar\epsilon\sigma_x^1$. The last term is a bias introduced to break the Z_2 symmetry. Our distinction between phases is based on certain quantities which combine thermal and quantum averages

$$\langle\langle\cdot\rangle^\alpha\rangle_T \equiv \frac{\sum_\lambda \langle\lambda|\cdot|\lambda\rangle^\alpha \exp(-E_\lambda/k_B T)}{\sum_\lambda \exp(-E_\lambda/k_B T)}, \tag{8}$$

with $|\lambda\rangle$ denoting the Hamiltonian eigenstates at energy E_λ. For $\alpha=1$, $\langle\langle\cdot\rangle^\alpha\rangle_T$ reduces to the normal thermal average. We will use low, but non-zero temperatures T of the order of the coupling constant, accounting in this way for the huge quasi-degeneracy in the glassy regime. The thermal average $\langle\cdot\rangle_T$ plays a role somewhat similar to the disorder average, as it averages over various quasi-ground states (pure thermodynamic phases). We therefore expect $\langle\langle\sigma_x^i\rangle\rangle_T$ to go to zero in the glassy phase. In contrast, a non-zero average $\langle\langle\sigma_x^i\rangle\rangle_T$ detects the ferromagnetic phase of the Mattis model, while it vanishes in the paramagnetic state. Taking its square to get rid of the sign, we obtain a global ferromagnetic order parameter by summing over all spins:

$$\nu_{\text{FM}} = \frac{1}{N}\sum_i \langle\langle\sigma_x^i\rangle\rangle_T^2. \tag{9}$$

On the other hand, in the spirit of an Edwards–Anderson-like parameter, we consider thermal averages of squared quantum averages, that is,

$$\nu_{\text{EA}} = \frac{1}{N}\sum_i \left\langle\langle\sigma_x^i\rangle^2\right\rangle_T. \tag{10}$$

At a sufficiently low temperature this average would still be zero for a paramagnetic system, but now it remains non-zero for both ferromagnetic and glassy systems. Accordingly, a parameter which is 'large' only for glassy systems is given by the ratio $\nu_{\text{SG}}=\frac{\nu_{\text{EA}}}{\nu_{\text{FM}}}$, used in Fig. 3 to detect the glassy regions.

Thermal averages are difficult to measure, but the contained information is also present in linear response functions at zero temperature. Therefore, we have calculated the longitudinal magnetic susceptibility χ, that is, the response of the system to a small local magnetic field h_x^j along the σ_x direction (in units J_{rms}^{-2}):

$$\chi = \frac{1}{N}\sum_{ij}\left(\frac{\partial\langle\sigma_x^i\rangle}{\partial h_x^j}\right)^2. \tag{11}$$

Due to the (quasi)degeneracy in the glass, one expects a huge response even from a weak field, and thus a divergent susceptibility.

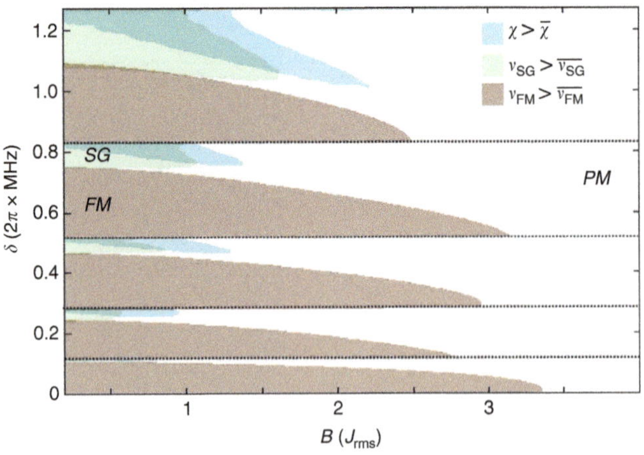

Figure 3 | Magnetic phases. On varying the laser detuning δ and the transverse magnetic field B, we mark, for $N=6$ ions and $\omega_z=2\pi\times 770$ kHz, those regions in configuration space where the ferromagnetic order parameter ν_{FM}, the spin glass order parameter ν_{SG}, or the longitudinal magnetic susceptibility χ take larger than average values ($\overline{\nu_{\text{FM}}}=0.13$, $\overline{\nu_{\text{SG}}}=5$, $\overline{\chi}=1.6$). Below each phonon resonance (marked by the dashed horizontal lines), there is a regime where large values of ν_{SG} and χ indicate spin glass behaviour for sufficiently weak transverse field B. To break the Z_2 symmetry of the Hamiltonian, all quantities were calculated in the presence of a biasing field $\epsilon\sigma_x^1$ (with $\epsilon=-J_{\text{rms}}$).

We have calculated ν_{FM}, ν_{SG} and χ, for $N=6$ between $0\leq 2\pi\times\delta\leq 8$ MHz, and $0\leq B\leq 4J_{\text{rms}}$. For the thermal averaging, we have chosen a temperature $k_B T=J_{\text{rms}}$. The results are summarized in Fig. 3, indicating the regions where these quantities take larger values than their configurational averages $\overline{\nu_{\text{FM}}}$, $\overline{\nu_{\text{SG}}}$ and $\overline{\chi}$, defined as $\overline{f} \equiv \int dB \int d\delta\, f(B,\delta)/(B_{\max}\delta_{\max})$. In this way, we identify and distinguish ferromagnetic behaviour below, and glassy behaviour above each resonance. Regions of large susceptibility χ overlap with regions of large ν_{SG}, attaining numerical values which are three orders of magnitude larger than the corresponding averages. For sufficiently strong field B, in contrast, none of these parameters is large, indicating paramagnetic behaviour.

Note that the existence of the purported glassy phase in the quantum case is an open problem. We provide here the evidence only for small systems, since it is numerically feasible and corresponds directly to current or near-future experiments. If we increase the complexity of our system by resonant coupling to many phonon modes, as discussed in the last paragraph of the previous subsection, the glassy behaviour will result from the interplay of contributions of many modes similarly as in the Hopfield model with Hebbian rule and random memory patterns. Here the beautiful results by Strack and Sachdev[21]—the quantum analogue of the Amit *et al.*[29] machinery—can be applied directly to obtain the phase diagram for large N. If, however, we increase the complexity by random positioning of the ion traps, then the resonance condition will pick up the contribution from one dominant (random) mode, and the Hebbian picture will apply.

Simulated quantum annealing. We will now turn to the more realistic description of the system in terms of a time-dependent Dicke model, described by the Hamiltonian $H_0(t)$ in equation (1) with an additional transverse field $H_B(t)$ from equation (7). We assume an exponential annealing protocol $B(t)=B_{\max}\exp(-t/\tau)$. Again we apply a bias field $h_{\text{bias}}=\hbar\epsilon\sigma_x^1$, lifting the Z_2 degeneracy of the classical ground state. We study the exact time evolution under the Hamiltonian $H(t)=H_0(t)+H_B(t)+h_{\text{bias}}$, using a

Krylov subspace method[45], and truncating the phonon number to a maximum of two phonons per mode.

Initially, the system is cooled to the motional ground state, and spins are polarized along σ_z. Choosing $B_{\max} \gg \epsilon, J_{\mathrm{rms}}$, this configuration is close to the ground state of the effective model $H_J + H_B + h_{\mathrm{bias}}$ at $t=0$. If the decay of $B(t)$ is slow enough, and if the entanglement between spins and phonons remains sufficiently low, the system stays close to the ground state for all times, and finally reaches the ground state of H_J.

We have simulated this process for $N=6$ ions, as shown in Fig. 4. As a result of the annealing, we are not interested in the final quantum state, but only in the signs of $\langle \sigma_x^i \rangle$, which fully determine the system in the classical configuration. This provides some robustness. We find that for a successful annealing procedure, yielding the correct sign for all i, the number of phonons produced during the evolution should not be larger than 1. At fixed detuning, we can reduce the number of phonons by decreasing the Rabi frequency, at the expense of increasing time scales. As a realistic choice[14,15], we demand that annealing is achieved within tens of miliseconds.

Figure 4a shows that one can operate at a detuning $\delta = 2\pi \times 239$ kHz, that is, at the onset of a glassy phase according to Fig. 3. The mode vector which selects the ground state is $\xi_5 = (0.61, 0.34, 0.11, -0.11, -0.34, -0.61)$, and the corresponding pattern can be read out after an annealing time $t \geq 15$ ms. On the other hand, even for long times, $\langle \sigma_x^{i=4} \rangle$ saturates only at -0.15, which is far from the classical value -1. As shown in Fig. 4b, a slower annealing protocol leads to more robust results $(|\langle \sigma_x^i \rangle| > 0.52\ \forall\ i)$. In Fig. 4c, a much simpler instance in the ferromagnetic regime is considered. Good results $(|\langle \sigma_x^i \rangle| > 0.36\ \forall\ i)$ can then be obtained within only a few ms.

In addition, dephasing due to instabilities of applied fields and spontaneous emission processes disturb the dynamics of the spins. In ref. 43 a master equation was derived that takes into account such noisy environment. To study the evolution of the system in this open scenario we have applied the Monte Carlo wave-function method[46]. As quantum jump operators are hermitean, σ_x^i for dephasing and σ_z^i for spontaneous emission, the evolution remains unitary, but is randomly interrupted by quantum jumps. Each jump has equal probability Γ, and the average number of jumps within the annealing time T is given by $\overline{n_{\mathrm{jumps}}} = 2N\Gamma T$, which we chose close to 1.

Since a faithful description requires statistics over many runs, we restrict ourselves to a small system, $N=4$, with short annealing times. In a sample of 100 runs, we noted 94 jumps (42 σ_x and 52 σ_z jumps). In 39 runs, no jump occured. Among the 61 runs in which at least one jump occured, 26 runs still produced the correct sign for all spin averages $\langle \sigma_x^i \rangle$. The full time evolution, averaged over all runs, is shown in Fig. 5. On average, the final result is (0.65, 0.50, -0.41, -0.68), that is, our annealing with noise still produces the correct answer, but with lower fidelity.

Whether an individual jump harms the evolution crucially depends on the time at which it occurs: While a spin flip (σ_z noise) is harmless in the beginning of the annealing, a dephasing event (σ_x noise) at an early stage of the evolution leads to wrong results. Oppositely, at the end of the annealing procedure, dephasing becomes harmless while spontaneous emission falsifies the result. An optimal annealing protocol has to balance the effect of different noise sources against non-adiabatic effects in the unitary evolution.

Scalability. Above we have demonstrated the feasibility of the proposed quantum annealing scheme in small systems. The usefulness of the approach, however, depends crucially on its behaviour on increasing the system size. While the exact treatment of the dynamics becomes intractable for longer chains, an efficient description can be derived from the Heisenberg equations:

$$i\hbar \frac{\mathrm{d}}{\mathrm{d}t}\langle a_m \rangle = \langle [a_m, H(t)] \rangle,$$
$$i\hbar \frac{\mathrm{d}}{\mathrm{d}t}\langle \sigma_\alpha^j \rangle = \langle [\sigma_\alpha^j, H(t)] \rangle, \tag{12}$$

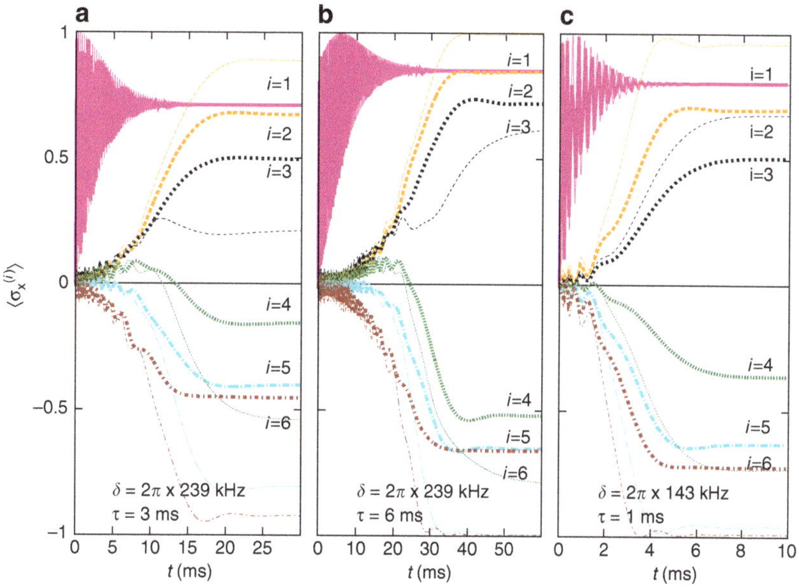

Figure 4 | Simulated quantum annealing in the closed system. Unitary time evolution of the full system ($N=6$) for different laser detunings δ between the fourth and fifth resonance. Thick lines show the result of the exact evolution, while the thin lines have been obtained from the semi-classical approximation. In all cases, the desired pattern $(+++---)$ can be read out after sufficiently long annealing times. While in **a**,**b** we operate at the onset of glassiness, $\delta = 2\pi \times 239$ kHz, **c** considers a ferromagnetic instance, $\delta = 2\pi \times 143$ kHz. The annealing protocol in **a** is defined by $B_{\max} = 50 J_{\mathrm{rms}}$ and $\tau = 3$ ms. In **b** the same instance is solved with higher fidelity by choosing $B_{\max} = 80\, J_{\mathrm{rms}}$ and $\tau = 6$ ms. Fast annealing, with $B_{\max} = 50\, J_{\mathrm{rms}}$ and $\tau = 1$ ms, is possible in the ferromagnetic instance in (**c**). In all simulations, we have chosen $\Omega = 2\pi \times 50$ kHz, $\omega_{\mathrm{recoil}} = 2\pi \times 15$ kHz, and $\epsilon = -10$ kHz.

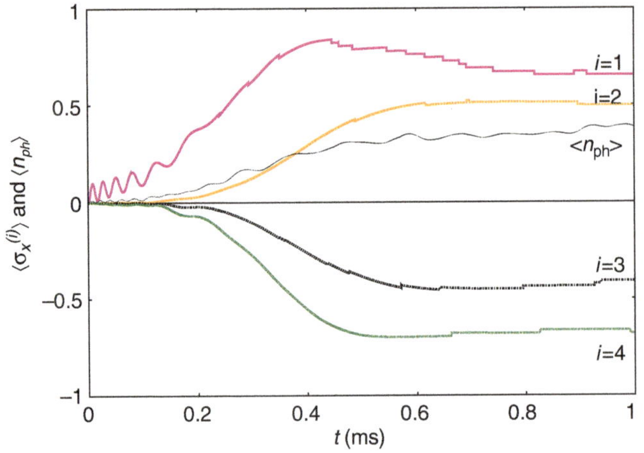

Figure 5 | Simulated quantum annealing in the open system. Using a Monte Carlo wave-function description averaged over 100 runs, we simulate the annealing process with $N=4$ ions in the presence of dephasing and spontaneous emission. With a noise rate $\Gamma=0.03\,J_{\mathrm{rms}}$ $\approx$120 Hz and a total annealing time $T=1$ ms, we have on average one jump per run, $\overline{n_{\mathrm{jumps}}}$=0.96. We plot expectation values of the spins and total number of phonons, produced by the coupling, as a function of time. Here, we have operated in the ferromagnetic regime between the second and third phonon resonance, $\delta=2\pi\times159$ kHz. Choosing $\Omega=2\pi\times50$ kHz, $\omega_{\mathrm{recoil}}=2\pi\times15$ kHz, $\omega_z=2\pi\times876$ kHz, $B_{\max}=50\,J_{\mathrm{rms}}\approx200$ kHz, and $\epsilon=-10$ kHz, we are able to perform very fast annealing, τ=0.1 ms, with high fidelity. After a time $T=1$ ms, the time evolution without noise has converged to $(\langle\sigma_x^1\rangle,\langle\sigma_x^2\rangle,\langle\sigma_x^3\rangle,\langle\sigma_x^4\rangle)=(0.97,\ 0.70,\ -0.65,\ -0.85)$, correctly reproducing the classical pattern $(++--)$.

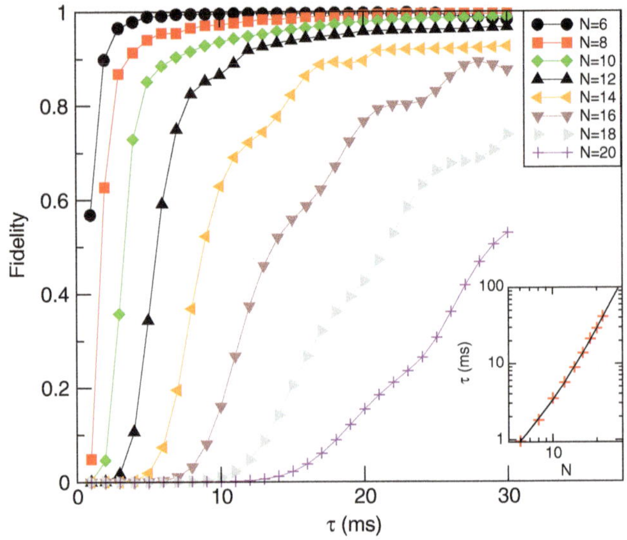

Figure 6 | Fidelity estimates of the quantum annealing and scaling of annealing times. We solve the equations of motion, equation (12), for a glassy instance at different system sizes N, and plot the fidelity of the outcome as a function of the annealing time τ. In the inset, we investigate the scaling behaviour by plotting (in double-logarithmic scale) the value of τ which is needed for a fidelity $F=0.5$ as a function of N. A fourth-order polynomial fit agrees very well with the data (black dashed line). The fit parameters, as defined by equation (13), are τ_0=(90 ± 40) ms and $\gamma=12.0\pm1.2$. For all calculations, we have chosen a beatnote frequency between the two lowest resonances, $\omega_{\mathrm{L}}=0.8\omega_1+0.2\omega_2$, a trap frequency $\omega_z=2\pi\times270$ kHz, and a bias potential $\epsilon=-1$ kHz. The initial value of the transverse field was $B_{\max}=10$ kHz.

with $H=H_0(t)+H_B(t)+h_{\mathrm{bias}}$. To solve this set of $5N$ first-order differential equations, we make a semi-classical approximation $\langle a_m\sigma_x^i\rangle\approx\langle a_m\rangle\langle\sigma_x^i\rangle$, and then proceed numerically using a fourth-order Runge–Kutta algorithm. The semi-classical approximation is justified as long as the system remains close to the phonon vacuum. A direct comparison with exact results for six ions (see Fig. 4) shows that the semi-classical approach, while slightly overestimating fidelities, accurately reproduces all relevant time scales.

This approach allows us to extend our simulations up to $N=22$ ions, at trap frequency $\omega_z=2\pi\times270$ kHz. We operate between the first and second resonance where the level spacing is largest, at a beatnote frequency $\omega_{\mathrm{L}}=\omega_1+0.2(\omega_2-\omega_1)$, that is with a fixed relative detuning between the two modes. This choice corresponds to $\delta=2\pi\times1.2$ MHz in Fig. 3, characterized as a glassy instance of the system.

Our aim is to find the relation of annealing time, measured by the decay parameter τ, and system size while the fidelity F is kept constant. For a practical definition we demand that F is zero when the annealing fails, that is when the sign of the spin averages $\langle\sigma_x^i\rangle$ does not agree with the classical target state for all i. If the annealing finds the correct signs, the robustness still depends on the absolute values of the spin averages. The fidelity is then defined as the smallest absolute value, $F=\min_i\left|\langle\sigma_x^i\rangle\right|$. Our results are summarized in Fig. 6: First, this figure shows that for all sizes $N\leq20$ large fidelities $F\geq0.5$ can be produced within experimentally feasible time scales, $\tau\leq30$ ms. Second, the time scale τ needed for a fidelity $F=0.5$ fits well to a fourth-order polynomial in N (with subleading terms of the order $\exp(1/N)$):

$$\tau(N)=N^4\tau_0\exp(\gamma/N), \qquad (13)$$

with τ_0 and γ being free fit parameters. Although the sample of 22 ions is too small to draw strong conclusions, it is noteworthy that the polynomial fit is more accurate than an exponential one. This suggests that the proposed quantum simulation is indeed an efficient way of solving a complex computational problem. One should also keep in mind that our estimates, based on a semi-classical approximation, neglect certain quantum fluctuations which could further speed-up the annealing process.

To study the scaling of dissipative effects, we have extendend the Monte Carlo wave-function approach to larger systems, which is feasible if the phonon dynamics is neglected. The unitary part of the evolution is then described by the effective Ising Hamiltonian $H_{\mathrm{eff}}=H_J+H_B(t)+h_{\mathrm{bias}}$. The dissipative part consists of random quantum jumps described by σ_x^i and σ_z^i. The results for a glassy instance ($\delta=2\pi\times198$ kHz at $\omega_z=2\pi\times700$ kHz) are summarized in Table 1 for $N=4$, 6, 8. The noise rate is chosen such that on average one quantum jump occurs in the system with four ions, while accordingly the system with eight ions suffers on average from two such events. In all cases, the annealing produces the correct pattern, $F>0$. As expected, F decreases for larger systems, but fortunately rather slowly (from $F=0.25$ at $N=4$ to $F=0.16$ at $N=8$). If the total amount of noise is kept constant, that is, $\Gamma\propto1/N$, the annealing is found to profit from larger system sizes, since a quantum jump at spin i is unlikely to affect the sign of $\langle\sigma_x^j\rangle$ for $j\neq i$. We note that the spin values produced by the Monte Carlo wave-function method cannot be described by a normal distribution. Importantly, the peak of each distribution, roughly coinciding with its median, is barely affected by the noise. Thus, larger fidelities can be obtained from the median rather than from the arithmetic mean of $\langle\sigma_x^i\rangle$.

Spin pattern in the quantum Mattis model. The quantum annealing discussed above exploits quantum effects to extract information encoded in the classical model. Now we search for

Table 1 | Annealing in effective spin model with and without noise.

	Closed system	(1.0, 0.97, −0.39, −0.42)
$N=4$	Open system, mean values	(0.79, 0.70, −0.25, −0.32)
	Open system, median values	(1.0, 0.97, −0.39, −0.42)
	Closed system	(1.0, 0.99, 0.76, −0.40, −0.64, −0.66)
$N=6$	Open system, mean values	(0.82, 0.68, 0.55, −0.23, −0.43, −0.46)
	Open system, median values	(1.0, 0.98, 0.76, −0.39, −0.64, −0.66)
	Closed system	(1.0, 0.99, 0.96, 0.58, −0.31, −0.70, −0.75, −0.77)
$N=8$	Open system, mean values	(0.83, 0.69, 0.69, 0.40, −0.16, −0.43, −0.55, −0.53)
	Open system, median values	(1.0, 0.98, 0.94, 0.58, −0.30, −0.69, −0.75, −0.76)

We perform quantum annealing ($\tau=3$ ms and $T=30$ ms) for a glassy instance ($\delta=2\pi\times198$ kHz at $\omega_z=2\pi\times700$ kHz, $\epsilon=-1$ kHz) at different system sizes N, using an effective spin model description. The results $(\langle\sigma_x^1\rangle, \ldots, \langle\sigma_x^N\rangle)$, are shown for the closed-system dynamics and for a noisy system, with mean and median over a sample of 2,000 runs. The average number of noisy event scales with the system size, and is adjusted to $4/N$. In all cases, the signs of $\langle\sigma_x^i\rangle$ reproduce correctly the mode pattern, and the fidelity decreases with the system size. In contrast to the arithmetic mean values, the median values in the noisy sample are barely affected by the noise.

information which is encoded in the quantum, but not in the classical model. Therefore, we focus on the ferromagnetic Mattis model, which in the classical case keeps a binary memory of a spin pattern, that is, of N bits. Our considerations can also be generalized to the Hopfield model[1], which memorizes multiple patterns. We will show how quantum effects can increase the amount of information encoded by these models.

Recall that in the classical case, the spin pattern was defined by a resonant mode in terms of the sign of each component. In the quantum case, however, one cannot simply replace classical spins by quantum averages, $\mathrm{sign}(\langle\sigma_x^i\rangle)$. Even in a weak transverse field B, this quantity vanishes due to the Z_2 symmetry, $\sigma_x\to-\sigma_x$. Instead, the pattern is reflected by $\lambda_i=\langle\Psi_1|\sigma_x^i|\Psi_2\rangle$, where $|\Psi_1\rangle$ and $|\Psi_2\rangle$ are the ground and first excited state. For small B, we find numerically $\mathrm{sign}(\lambda_i)=\mathrm{sign}(\xi_{im})$. For large B, the stronger relation $\lambda_i=\xi_{im}$ holds approximately, see Fig. 7. Thus, the former binary memory has become real-valued.

To show this behaviour, we note that for strong B, the ground state $|\Psi_1\rangle$ is fully polarized along z, and the first excited state $|\Psi_2\rangle$ is restricted to the N-dimensional subspace with one spin flipped, that is, $S_z=\sum_i\sigma_z^i=N-2$. Within this subspace the Hamiltonian H_J is given by an $N\times N$ matrix approximately proportional to $\tilde{J}_{ij}=-\xi_{im}\xi_{jm}$ for $i\neq j$, and $\tilde{J}_{ii}=$ constant. Here we neglect all but the m-th mode close to resonance.

It is easy to see that the vector $\boldsymbol{\xi}_m$ is a ground state of the matrix $-\xi_{im}\xi_{jm}$, which differs from $\tilde{J}_{ij}$ only by the diagonal elements, which approach unity for large N. Thus, the first excited state reads $|\Psi_2\rangle=\sum_{i=1}^N\xi_{im}|i\rangle$, where $|i\rangle$ denotes the state in which spin i is flipped relatively to all others (in the σ_z basis). This shows that $\lambda_i\approx\xi_{im}$, and the small deviations decrease quickly with N.

Measuring λ_i experimentally is possible by full state tomography. The absolute value of λ_i can be obtained via a simple σ_z^i measurement. In the limit of strong B-fields we have $\lambda_i=\left[(1-\langle\sigma_z^i\rangle)/2\right]^{1/2}$.

Many applications are known for the classical spin system with couplings defined by spin patterns, reaching from pattern recognition and associative memory in the Hopfield model[1] to noise-free coding[2,47]. Our analysis suggests that patterns given by real numbers could replace patterns of binary variables by exploiting the quantum character of the spins.

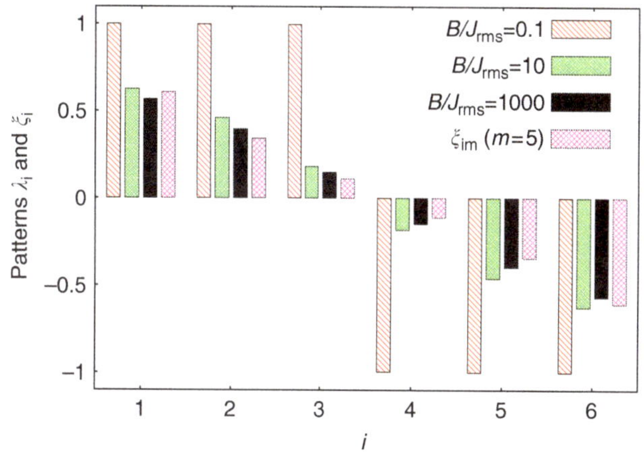

Figure 7 | From binary to real-valued patterns. The spin expectation values λ_i approach the real values of the mode vector ξ_{im} when a sufficiently strong transverse magnetic field B is present. The shown data was obtained from the effective spin Hamiltonian for $N=6$ ions in the ferromagnetic regime, $\delta=2\pi\times143$ kHz. Deviations from the equality $\lambda_i=\xi_{im}$ are smaller than 0.04.

Discussion

In summary, our work demonstrates the occurrence of Mattis glass behaviour in spin chains of trapped ions, if the detuning of the spin-phonon coupling is chosen between two resonances. In these regimes, the effective spin system has an exponentially large number of low-energy states, and finding its ground state corresponds to solving a number-partitioning problem. This establishes a direct connection between the properties of a physical system and the solution of a potentially NP-hard problem of computer science. Given the state-of-art in experiments with trapped ions, the physical implementation is feasible: In comparison with the previous experiments with trapped ions[10–14], only the detuning of the spin-phonon coupling needs to be adjusted. Differently from other approaches to spin glass physics, our scheme does not require any disorder. In its most natural implementation, parity symmetry allows one to analytically determine the ground state. Different ways to break this symmetry can be implemented to increase the complexity of the problem.

The ion chain then becomes an ideal test ground for applying quantum simulation strategies to solve computationally complex problems. By applying a transverse field to the ions, quantum annealing from a paramagnet to the glassy ground state is possible. The ionic system may be used to benchmark quantum annealing, which has become a subject of very lively and controversial debate since the launch of the D-Wave computers[7,8]. Exact calculations for small systems ($N=6$) and approximative calculations for larger system ($N=22$) demonstrate the feasibility of the proposed quantum annealing, and suggest a polynomial scaling of the annealing time. Accordingly, this approach may offer the sought-after quantum speed-up. In view of sizes of 30 and more ions already trapped in recent experiments (*cf.*, ref. 48), a realization of our proposal could not only confirm our semi-classical results, but also go beyond the sizes considered here.

Finally, resonant coupling to multiple modes opens an avenue to neural network models, where a finite number of patterns is memorized by couplings to different phonon modes. Quantum features can increase the memory of such networks from binary to real-valued numbers. It will be subject of future studies to work out the possible benefits which quantum neural networks may establish for information processing purposes.

References

1. Hopfield, J. J. Neural networks and physical systems with emergent collective computational abilities. *Proc. Natl Acad. Sci. USA* **79,** 2554–2558 (1982).
2. Nishimori, H. *International Series of Monographs on Physics* (Clarendon Press, 2001).
3. Barahona, F. On the computational complexity of Ising spin glass models. *J. Phys. A: Math. Gen.* **15,** 3241–3253 (1982).
4. Lucas, A. Ising formulations of many NP problems. *Front. Phys.* **2,** 5 (2014).
5. Sherrington, D. & Kirkpatrick, S. Solvable model of a spin-glass. *Phys. Rev. Lett.* **35,** 1792–1796 (1975).
6. Mézard, M., Parisi, G., Sourlas, N., Toulouse, G. & Virasoro, M. Replica symmetry breaking and the nature of the spin glass phase. *J. Phys. France* **45,** 843–854 (1984).
7. Rønnow, T. F. *et al.* Defining and detecting quantum speedup. *Science* **345,** 420–424 (2014).
8. Katzgraber, H. G., Hamze, F. & Andrist, R. S. Glassy chimeras could be blind to quantum speedup: designing better benchmarks for quantum annealing machines. *Phys. Rev. X* **4,** 021008 (2014).
9. Denchev, V. S. *et al.* What is the Computational Value of Finite Range Tunneling? Available at: http://arxiv.org/abs/1512.02206 (2015).
10. Friedenauer, A., Schmitz, H., Glueckert, J. T., Porras, D. & Schaetz, T. Simulating a quantum magnet with trapped ions. *Nat. Phys.* **4,** 757–761 (2008).
11. Kim, K. *et al.* Entanglement and tunable spin-spin couplings between trapped ions using multiple transverse modes. *Phys. Rev. Lett.* **103,** 120502 (2009).
12. Kim, K. *et al.* Quantum simulation of frustrated Ising spins with trapped ions. *Nature* **465,** 590–593 (2010).
13. Britton, J. W. *et al.* Engineered two-dimensional Ising interactions in a trapped-ion quantum simulator with hundreds of spins. *Nature* **484,** 489–492 (2012).
14. Jurcevic, P. *et al.* Quasiparticle engineering and entanglement propagation in a quantum many-body system. *Nature* **511,** 202–205 (2014).
15. Richerme, P. *et al.* Non-local propagation of correlations in quantum systems with long-range interactions. *Nature* **511,** 198–201 (2014).
16. Porras, D. & Cirac, J. I. Effective quantum spin systems with trapped ions. *Phys. Rev. Lett.* **92,** 207901 (2004).
17. Mintert, F. & Wunderlich, C. Ion-Trap quantum logic using long-wavelength radiation. *Phys. Rev. Lett.* **87,** 257904 (2001).
18. Schmitz, H. *et al.* The 'arch' of simulating quantum spin systems with trapped ions. *Appl. Phys. B* **95,** 195–203 (2009).
19. Joseph Wang, C.-C. & Freericks, J. K. Intrinsic phonon effects on analog quantum simulators with ultracold trapped ions. *Phys. Rev. A* **86,** 032329 (2012).
20. Pons, M. *et al.* Trapped ion chain as a neural network: error resistant quantum computation. *Phys. Rev. Lett.* **98,** 023003 (2007).
21. Strack, P. & Sachdev, S. Dicke quantum spin glass of atoms and photons. *Phys. Rev. Lett.* **107,** 277202 (2011).
22. Gopalakrishnan, S., Lev, B. L. & Goldbart, P. M. Frustration and glassiness in spin models with cavity-mediated interactions. *Phys. Rev. Lett.* **107,** 277201 (2011).
23. Rotondo, P., Lagomarsino, M. C. & Viola, G. Dicke simulators with emergent collective quantum computational abilities. *Phys. Rev. Lett.* **114,** 143601 (2015).
24. Stein, D. L. & Newman, C. M. Spin glasses: old and new complexity. *Complex Syst.* **20,** 115–126 (2011).
25. Stein, D. L. & Newman, C. M. *Spin Glasses and Complexity* (Princeton University Press, 2013).
26. Moore, M. A., Bokil, H. & Drossel, B. Evidence for the droplet picture of spin glasses. *Phys. Rev. Lett.* **81,** 4252 (1998).
27. Aspelmeier, T. *et al.* Finite-size critical scaling in Ising spin glasses in the mean-field regime. *Phys. Rev. E* **93,** 032123 (2016).
28. Amit, D. J., Gutfreund, H. & Sompolinsky, H. Storing infinite numbers of patterns in a spin-glass model of neural networks. *Phys. Rev. Lett.* **55,** 1530–1533 (1985).
29. Amit, D. J., Gutfreund, H. & Sompolinsky, H. Statistical mechanics of neural networks near saturation. *Ann. Phys.* **173,** 30–67 (1987).
30. Parisi, G. The order parameter for spin glasses: a function on the interval 0-1. *J. Phys. A (Math. Gen.)* **13,** 1101–1112 (1980).
31. Mezard, M., Parisi, G. & Virasoro, M. A. *Spin glass theory and beyond* (World Scientific, 1987).
32. Vincent, E. *Lecture Notes in Physics* (Springer, 2007).
33. Fisher, D. S. & Huse, D. A. Ordered phase of short-range ising spin-glasses. *Phys. Rev. Lett* **56,** 1601 (1986).
34. Daniel, S. F. & David, A. H. Equilibrium behavior of the spin-glass ordered phase. *Phys. Rev. B* **38,** 386 (1988).
35. Lefloch, F., Hammann, J., Ocio, M. & Vincent, E. Can aging phenomena discriminate between the droplet model and a hierarchical description in spin glasses? *EPL* **18,** 647 (1992).
36. Mattis, D. Solvable spin systems with random interactions. *Phys. Lett. A* **56,** 421–422 (1976).
37. Mertens, S. Phase transition in the number partitioning problem. *Phys. Rev. Lett.* **81,** 4281 (1998).
38. Mertens, S. The Easiest Hard Problem: Number Partitioning. Available at: http://arxiv.org/abs/condmat/0310317 (2003).
39. Nevado, P. & Porras, D. Hidden frustrated interactions and quantum annealing in trapped ion spin-phonon chains. *Phys. Rev. A* **93,** 013625 (2016).
40. Steane, A. M. The ion trap quantum information processor. *Appl. Phys. B* **64,** 623–643 (1997).
41. James, D. F. V. Quantum dynamics of cold trapped ions with application to quantum computation. *Appl. Phys. B* **66,** 181–190 (1998).
42. Bermudez, A., Schaetz, T. & Porras, D. Photon-assisted-tunneling toolbox for quantum simulations in ion traps. *New J. Phys.* **14,** 053049 (2012).
43. Hauke, P., Bonnes, L., Heyl, M. & Lechner, W. Probing entanglement in adiabatic quantum optimization with trapped ions. *Front. Phys.* **3,** 21 (2015).
44. Korenblit, S. *et al.* Quantum simulation of spin models on an arbitrary lattice with trapped ions. *New J. Phys.* **14,** 095024 (2012).
45. Park, J. H. & Light, J. C. Unitary quantum time evolution by iterative lanczos reduction. *J. Chem. Phys.* **86,** 5870–5876 (1986).
46. Mølmer, K., Castin, Y. & Dalibard, J. Monte Carlo wave-function method in quantum optics. *J. Opt. Soc. Am. B* **10,** 524–538 (1993).
47. Sourlas, N. Spin-glass models as error-correcting codes. *Nature* **339,** 693–695 (1989).
48. Ramm, M., Pruttivarasin, T. & Häffner, H. Energy transport in trapped ion chains. *New J. Phys.* **16,** 063062 (2014).

Acknowledgements

We thank A. Acín, R. Augusiak, A. Bermudez, J. Tura, P. Rotondo, L. Santos and P. Wittek for discussions. We acknowledge financial support from EU grants OSYRIS (ERC-2013-AdG Grant No. 339106), SIQS (FP7-ICT-2011-9 No. 600645), QUIC (H2020-FETPROACT-2014 No. 641122), EQuaM (FP7/2007-2013 Grant No. 323714), Spanish MINECO grants (FOQUS FIS2013-46768-P, FIS2014-54672-P), Severo Ochoa grant (SEV-2015-0522), Generalitat de Catalunya (2014 SGR 401, 2014 SGR 874 and SGR 875), the Maria de Maeztu grant (MDM-2014-0369) and Fundació Cellex. B.J.-D. is supported by the Ramon y Cajal program, and C.G. by MPQ-ICFO, ICFOnest+ (FP7-PEOPLE-2013-COFUND) and by the European Union's Marie Sklodowska-Curie Individual Fellowships (IF-EF) programme (GA 700140).

Author contributions

All authors contributed to the conceptual design of the quantum simulation, discussion and interpretation of results, and the writing of the manuscript. Exact numerical calculations were performed by T.G. Semi-classical calculations were done by B.J.-D. and D.R.

Additional information

Competing financial interests: The authors declare no competing financial interests.

9

An experimental test of noncontextuality without unphysical idealizations

Michael D. Mazurek[1,2], Matthew F. Pusey[3], Ravi Kunjwal[4], Kevin J. Resch[1,2] & Robert W. Spekkens[3]

To make precise the sense in which nature fails to respect classical physics, one requires a formal notion of classicality. Ideally, such a notion should be defined operationally, so that it can be subject to direct experimental test, and it should be applicable in a wide variety of experimental scenarios so that it can cover the breadth of phenomena thought to defy classical understanding. Bell's notion of local causality fulfils the first criterion but not the second. The notion of noncontextuality fulfils the second criterion, but it is a long-standing question whether it can be made to fulfil the first. Previous attempts to test noncontextuality have all assumed idealizations that real experiments cannot achieve, namely noiseless measurements and exact operational equivalences. Here we show how to devise tests that are free of these idealizations. We perform a photonic implementation of one such test, ruling out noncontextual models with high confidence.

[1] Institute for Quantum Computing, University of Waterloo, Waterloo, Ontario Canada N2L 3G1. [2] Department of Physics & Astronomy, University of Waterloo, Waterloo, Ontario Canada N2L 3G1. [3] Perimeter Institute for Theoretical Physics, 31 Caroline Street North, Waterloo, Ontario Canada N2L 2Y5. [4] Optics & Quantum Information Group, The Institute of Mathematical Sciences, C.I.T Campus, Taramani, Chennai 600 113, India. Correspondence and requests for materials should be addressed to R.W.S. (email: rspekkens@perimeterinstitute.ca).

Making precise the manner in which a quantum world differs from a classical one is a surprisingly difficult task. The most successful attempt, due to Bell[1], shows a conflict between quantum theory and a feature of classical relativistic theories termed local causality, which asserts that no causal influences propagate faster than light. But the latter assumption can only be tested for scenarios wherein there are two or more systems that are space-like separated. And yet few believe that this highly specialized situation is the only point where the quantum departs from the classical. A leading candidate for a notion of nonclassicality with a broader scope is the failure of quantum theory to admit of a noncontextual model, as proven by Kochen and Specker[2]. Recent work has highlighted how this notion lies at the heart of many phenomena that are taken to be distinctly quantum: the fact that quasi-probability representations go negative[3,4], the existence of quantum advantages for cryptography[5] and for computation[6–8], and the possibility of anomalous weak values[9]. Consequently, the study of noncontextuality has not only foundational significance but practical applications as well.

An experimental refutation of noncontextuality would demonstrate that the conflict with noncontextual models is not only a feature of quantum theory, but of nature itself, and hence also of any successor to quantum theory. The requirements for such an experimental test, however, have been a subject of much controversy[10–16].

A fundamental problem with most proposals for testing noncontextuality[17–24], and experiments performed to date[25–32], is that they seek to test a notion of noncontextuality which posits that measurements have a deterministic response in the noncontextual model. It has been shown that such determinism is only justified under the idealization that measurements are noiseless[33], which is never satisfied precisely by any real experiment. We refer to this issue as the problem of noisy measurements.

Another critical problem with previous proposals is the fact that the assumption of noncontextuality can only be brought to bear when two measurement events (an event is a measurement and an outcome) are operationally equivalent, which occurs when the two events are assigned exactly the same probability by all preparation procedures[34]; in this case they are said to differ only by the measurement context. In a real experiment, however, one never achieves the ideal of precise operational equivalence. Previous work on testing noncontextuality—including the only experiment to have circumvented the problem of noisy measurements (by focusing on preparations)[5]—has failed to provide a satisfactory account of how the deviation from strict operational equivalence should be accounted for in the interpretation of the results. We term this problem the problem of inexact operational equivalence.

In this work, we solve both of the above problems. We contend with the problem of noisy measurements by devising a test of a generalised notion of noncontextuality, proposed in ref. 34, that allows general measurements to have an indeterministic response while reducing to the traditional notion in the idealized case of projective quantum measurements. For the problem of inexact operational equivalence, whereas some have been led to consider modifying the definition of noncontextuality so that it applies to pairs of procedures that are merely *close* to operationally equivalent[35,36], we circumvent the problem by demonstrating a general technique that appeals to equivalences not among the procedures themselves, but certain convex mixtures thereof. Of course, any judgment of operational equivalence of measurements (preparations) rests on an assumption about which sets of preparations (measurements) are sufficient to establish such equivalence, that is, which sets are tomographically complete. We here assume that the cardinality of a tomographically-complete set of measurements (preparations) for a photon's polarization is three (four), as it is in quantum theory. We collect some experimental evidence for this assumption—another improvement over previous experiments—but the possibility of its failure is the most significant remaining loophole for tests of noncontextuality. For Bell's notion of local causality, the theoretical work of Clauser *et al.*[37] was critical to enabling an experimental test without unphysical idealizations, e.g., without the perfect anti-correlations presumed in Bell's original proof[1]. Similarly, the theoretical innovations we introduce here make it possible for the first time to subject noncontextuality to an experimental test without the idealizations described above. We report on a quantum-optical experiment of this kind, the results of which rule out noncontextual models with high confidence.

Results

A noncontexuality inequality. According to the operational approach proposed in ref. 34, to assume noncontextuality is to assume a constraint on model-construction, namely, that if procedures are statistically equivalent at the operational level then they ought to be statistically equivalent in the underlying model.

Operationally, a system is associated with a set $\mathcal{M}$ (resp. $\mathcal{P}$) of physically possible measurement (resp. preparation) procedures. An operational theory specifies the possibilities for the conditional probabilities $\{p(X|P,M) : P \in \mathcal{P}, M \in \mathcal{M}\}$ where X ranges over the outcomes of measurement M. In an ontological model of such a theory, the causal influence of the preparation on the measurement outcome is mediated by the ontic state of the system, that is, a full specification of the system's physical properties. We denote the space of ontic states by Λ. It is presumed that when the preparation P is implemented, the ontic state of the system, $\lambda \in \Lambda$, is sampled from a probability distribution $\mu(\lambda|P)$, and when the system is subjected to the measurement M, the outcome X is distributed as $\xi(X|M, \lambda)$. Finally, for the model to reproduce the experimental statistics, we require that

$$\sum_{\lambda \in \Lambda} \xi(X|M, \lambda)\mu(\lambda|P) = p(X|M, P). \qquad (1)$$

A general discussion of this notion of noncontextuality is provided in Supplementary Note 1, where we also explain how it differs from the notion used in previous experimental tests and why the latter makes unphysical idealizations. This notion can also be understood through the concrete example we consider here (which is based on a construction from Section V of ref. 34).

Suppose there is a measurement procedure, M_*, that is operationally indistinguishable from a fair coin flip: it always gives a uniformly random outcome regardless of the preparation procedure,

$$p(X = 0, 1|M_*, P) = \frac{1}{2}, \forall P \in \mathcal{P}. \qquad (2)$$

In this case, noncontextuality dictates that in the underlying model, the measurement should also give a uniformly random outcome regardless of the ontic state of the system,

$$\xi(X = 0, 1|M_*, \lambda) = \frac{1}{2}, \forall \lambda \in \Lambda. \qquad (3)$$

In other words, because M_* appears operationally to be just like a coin flip, noncontextuality dictates that physically it must be just like a coin flip.

The second application of noncontextuality is essentially a time-reversed version of the first. Suppose there is a triple of preparation procedures, P_1, P_2 and P_3, that are operationally indistinguishable from one another: no measurement reveals any information about which of these preparations was implemented,

$$\forall M \in \mathcal{M} : p(X|M, P_1) = p(X|M, P_2) = p(X|M, P_3). \quad (4)$$

In this case, noncontextuality dictates that in the underlying model, the ontic state of the system does not contain any information about which of these preparation procedures was implemented,

$$\forall \lambda \in \Lambda : \mu(\lambda|P_1) = \mu(\lambda|P_2) = \mu(\lambda|P_3). \quad (5)$$

In other words, because it is impossible, operationally, to extract such information, noncontextuality dictates that physically, the information is not present in the system.

Suppose that M_* can be realized as a uniform mixture of three other binary-outcome measurements, denoted M_1, M_2 and M_3. That is, one implements M_* by uniformly sampling $t \in \{1, 2, 3\}$, implementing M_t, then outputting its outcome as the outcome of M_* (ignoring t thereafter). Finally, suppose that each preparation P_t can be realized as the equal mixture of two other preparation procedures, denoted $P_{t,0}$ and $P_{t,1}$.

Consider implementing M_t on $P_{t,b}$, and consider the average degree of correlation between the measurement outcome X and the preparation variable b:

$$A \equiv \frac{1}{6} \sum_{t\in\{1,2,3\}} \sum_{b\in\{0,1\}} p(X = b|M_t, P_{t,b}). \quad (6)$$

We now show that noncontextuality implies a nontrivial bound on A.

The proof is by contradiction. In order to have perfect correlation on average, we require perfect correlation in each term, which implies that for all ontic states λ assigned nonzero probability by $P_{t,b}$, the measurement M_t must respond deterministically with the $X = b$ outcome. Given that P_t is an equal mixture of $P_{t,0}$ and $P_{t,1}$, it follows that for all ontic states λ assigned nonzero probability by P_t, the measurement M_t must have a deterministic response, i.e., $\xi(X = b|M_t, \lambda) \in \{0, 1\}$.

But equation (5) (which follows from the assumption of noncontextuality) asserts that the preparations P_1, P_2 and P_3 must assign nonzero probability to precisely the same set of ontic states. Therefore, to achieve perfect correlation on average, each measurement must respond deterministically to all the ontic states in this set.

Now note that by the definition of M_*, the probability of its outcome $X = b$ is $\xi(X=b|M_*, \lambda) = \frac{1}{3}\sum_{t\in\{1,2,3\}} \xi(X=b|M_t, \lambda)$. But then equation (3) (which follows from the assumption of noncontextuality) says

$$\frac{1}{3} \sum_{t\in\{1,2,3\}} \xi(X = b|M_t, \lambda) = \frac{1}{2}. \quad (7)$$

For each deterministic assignment of values, $(\xi(X = b|M_1, \lambda), \xi(X = b|M_2, \lambda), \xi(X = b|M_3, \lambda)) \in \{(0, 0, 0), (0, 0, 1), \ldots, (1, 1, 1)\}$, the constraint of equation (7) is violated. It follows, therefore, that for a given λ, one of M_1, M_2 or M_3 must fail to have a deterministic response, contradicting the requirement for perfect correlation on average. This concludes the proof.

The precise (i.e., tight) bound is

$$A \le \frac{5}{6}, \quad (8)$$

as we demonstrate in Supplementary Figs 1–3, Supplementary Tables 1 and 2, and Supplementary Note 2. This is our noncontextuality inequality.

Quantum violation of the inequality. Quantum theory predicts there is a set of preparations and measurements on a qubit having the supposed properties and achieving $A = 1$, the logical maximum. Take the M_t to be represented by the observables $\boldsymbol{\sigma} \cdot \boldsymbol{n}_t$ where $\boldsymbol{\sigma}$ is the vector of Pauli operators and the unit vectors $\{\boldsymbol{n}_1, \boldsymbol{n}_2, \boldsymbol{n}_3\}$ are separated by 120° in the $\boldsymbol{x} - \boldsymbol{z}$ plane of the Bloch sphere of qubit states[38]. The $P_{t,b}$ are the eigenstates of these observables, where we associate the positive eigenstate $|+\boldsymbol{n}_t\rangle\langle+\boldsymbol{n}_t|$ with $b = 0$. To see that the statistical equivalence of equation (2) is satisfied, it suffices to note that

$$\frac{1}{3}|+\boldsymbol{n}_1\rangle\langle+\boldsymbol{n}_1| + \frac{1}{3}|+\boldsymbol{n}_2\rangle\langle+\boldsymbol{n}_2| + \frac{1}{3}|+\boldsymbol{n}_3\rangle\langle+\boldsymbol{n}_3| = \frac{1}{2}\mathbb{I}, \quad (9)$$

and to recall that for any density operator ρ, $\mathrm{tr}(\rho\frac{1}{2}\mathbb{I}) = \frac{1}{2}$. To see that the statistical equivalence of equation (4) is satisfied, it suffices to note that for all pairs $t, t' \in \{1, 2, 3\}$,

$$\frac{1}{2}|+\boldsymbol{n}_t\rangle\langle+\boldsymbol{n}_t| + \frac{1}{2}|-\boldsymbol{n}_t\rangle\langle-\boldsymbol{n}_t| = \frac{1}{2}|+\boldsymbol{n}_{t'}\rangle\langle+\boldsymbol{n}_{t'}| + \frac{1}{2}|-\boldsymbol{n}_{t'}\rangle\langle-\boldsymbol{n}_{t'}|, \quad (10)$$

which asserts that the average density operator for each value of t is the same, and therefore leads to precisely the same statistics for all measurements. Finally, it is clear that the outcome of the measurement of $\boldsymbol{\sigma} \cdot \boldsymbol{n}_t$ is necessarily perfectly correlated with whether the state was $|+\boldsymbol{n}_t\rangle\langle+\boldsymbol{n}_t|$ or $|-\boldsymbol{n}_t\rangle\langle-\boldsymbol{n}_t|$, so that $A = 1$.

These quantum measurements and preparations are what we seek to implement experimentally, so we refer to them as ideal, and denote them by M_t^{i} and $P_{t,b}^{\mathrm{i}}$.

Note that our noncontextuality inequality can accommodate noise in both the measurements and the preparations, up to the point where the average of $p(X = b|M_t, P_{t,b})$ drops below $\frac{5}{6}$. It is in this sense that our inequality does not presume the idealization of noiseless measurements.

Contending with the lack of exact operational equivalence. The actual preparations and measurements in the experiment, which we call the primary procedures and denote by $P_{1,0}^{\mathrm{p}}, P_{1,1}^{\mathrm{p}}, P_{2,0}^{\mathrm{p}}, P_{2,1}^{\mathrm{p}}, P_{3,0}^{\mathrm{p}}, P_{3,1}^{\mathrm{p}}$ and $M_1^{\mathrm{p}}, M_2^{\mathrm{p}}, M_3^{\mathrm{p}}$, almost certainly deviate from the ideal versions and consequently their mixtures, that is, $P_1^{\mathrm{p}}, P_2^{\mathrm{p}}, P_3^{\mathrm{p}}$ and M_*^{p}, fail to achieve strict equality in equations (2) and (4).

We solve this problem as follows. From the outcome probabilities on the six primary preparations, one can infer the outcome probabilities on the entire family of probabilistic mixtures of these. It is possible to find within this family many sets of six preparations, $P_{1,0}^{\mathrm{s}}, P_{1,1}^{\mathrm{s}}, P_{2,0}^{\mathrm{s}}, P_{2,1}^{\mathrm{s}}, P_{3,0}^{\mathrm{s}}, P_{3,1}^{\mathrm{s}}$, which define mixed preparations $P_1^{\mathrm{s}}, P_2^{\mathrm{s}}, P_3^{\mathrm{s}}$ that satisfy the operational equivalences of equation (4) exactly. We call the $P_{t,b}^{\mathrm{s}}$ secondary preparations. We can define secondary measurements $M_1^{\mathrm{s}}, M_2^{\mathrm{s}}, M_3^{\mathrm{s}}$ and their uniform mixture M_*^{s} in a similar fashion. The essence of our approach, then, is to identify such secondary sets of procedures and use these to calculate A. If quantum theory correctly models our experiment, then we expect to get a value of A close to 1 if and only if we can find suitable secondary procedures that are close to the ideal versions.

To test the hypothesis of noncontextuality, one must allow for the possibility that the experimental procedures do not admit of a quantum model. Nonetheless, for pedagogical purposes, we will first provide the details of how one would construct the secondary sets under the assumption that all the experimental procedures do admit of a quantum model.

In Fig. 1, we describe the construction of secondary preparations in a simplified example of six density operators that deviate from the ideal states only within the $\boldsymbol{x} - \boldsymbol{z}$ plane of the Bloch sphere.

In practice, the six density operators realized in the experiment will not quite lie in a plane. We use the same idea to contend with

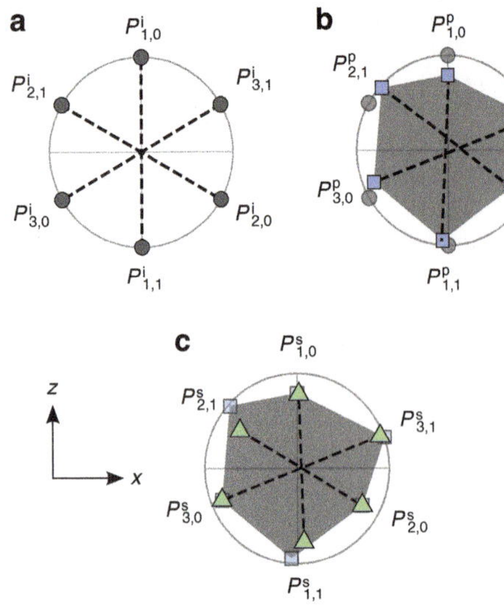

Figure 1 | Solution to the problem of inexact operational equivalences. Here, we illustrate our solution for the case of preparations under the simplifying assumption that these are confined to the $\boldsymbol{x}-\boldsymbol{z}$ plane of the Bloch sphere. For a given pair, $P_{t,0}$ and $P_{t,1}$, the midpoint along the line connecting the corresponding points represents their equal mixture, P_t. (**a**) The target preparations $P^i_{t,b}$, with the coincidence of the midpoints of the three lines illustrating that they satisfy the operational equivalence (4) exactly. (**b**) Illustration of how errors in the experiment (exaggerated in magnitude) will imply that the realized preparations $P^p_{t,b}$ (termed primary) will deviate from the ideal. The lines indicate that not only do these preparations fail to satify the operational equivalence (4), but since the three lines do not all meet at the same point, no mixtures of the $P^p_{t,0}$ and $P^p_{t,1}$ can be found at a single point independent of t. The set of preparations corresponding to probabilistic mixtures of the $P^p_{t,b}$ are depicted by the grey region. (**c**) Secondary preparations $P^s_{t,b}$ have been chosen from this grey region, with the coincidence of the midpoints of the three lines indicating that the operational equivalence (4) has been restored. Note that we require only that the mixtures of the three pairs of preparations be the same, not that they correspond to the completely mixed state.

this, but with one refinement: we supplement our set of ideal preparations with two additional ones, denoted $P^i_{4,0}$ and $P^i_{4,1}$ corresponding to the two eigenstates of $\boldsymbol{\sigma}\cdot\boldsymbol{y}$. The two procedures that are actually realized in the experiment are denoted $P^p_{4,0}$ and $P^p_{4,1}$ and are considered supplements to the primary set. We then search for our six secondary preparations among the probabilistic mixtures of this supplemented set of primaries rather than among the probabilistic mixtures of the original set. Without this refinement, it can happen that one cannot find six secondary preparations that are close to the ideal versions, as we explain in Supplementary Note 3.

The scheme for defining secondary measurement procedures is also described in Supplementary Fig. 4 and Supplementary Note 3. Analogously to the case of preparations, one contends with deviations from the plane by supplementing the ideal set with the observable $\boldsymbol{\sigma}\cdot\boldsymbol{y}$.

Note that in order to identify which density operators have been realized in an experiment, the set of measurements must be complete for state tomography[39]. Similarly, to identify which sets of effects have been realized, the set of preparations must be complete for measurement tomography[40]. However, the original ideal sets fail to be tomographically complete because they are restricted to a plane of the Bloch sphere, and an effective way to complete them is to add the observable $\boldsymbol{\sigma}\cdot\boldsymbol{y}$ to the measurements and its eigenstates to the preparations. Therefore, even if we did not already need to supplement these ideal sets for the purpose of providing greater leeway in the construction of the secondary procedures, we would be forced to do so in order to ensure that one can achieve full tomography.

The relevant procedure here is not quite state tomography in the usual sense, since we want to allow for systematic errors in the measurements as well as the preparations. Hence the task[41,42] is to find a set of qubit density operators, $\rho_{t,b}$, and POVMs, $\{E_{X|t}\}$, that together make the measured data as likely as possible (we cannot expect $\mathrm{tr}(\rho_{t,b}E_{X|t})$ to match the measured relative frequencies exactly due to the finite number of experimental runs).

To analyze our data in a manner that does not prejudice which model—noncontextual, quantum, or otherwise—does justice to it, we must search for representations of the preparations and measurements not amongst density operators and sets of effects, but rather their more abstract counterparts in the formalism of generalised probabilistic theories[43,44] (GPTs), called generalised states and effects. The assumption that the system is a qubit is replaced by the strictly weaker assumption that three two-outcome measurements are tomographically complete. (In GPTs, a set of measurements are called tomographically complete if their statistics suffice to determine the state.) We take these states and effects as estimates of our primary preparations and measurements, and we define our estimate of the secondary procedures in terms of these, which in turn are used to calculate our estimate for A. We explain how the raw data is fit to a set of generalised states and effects in Supplementary Note 4. We characterize the quality of this fit with a χ^2 test.

Experiment. We use the polarization of single photons to test our noncontextuality inequality. The set-up, shown in Fig. 2, consists of a heralded single-photon source[45–47], polarization-state preparation and polarization measurement. We generate photons using spontaneous parametric downconversion and prepare eight polarization states using a polarizer followed by a quarter-wave plate (QWP) and half-wave plate (HWP). The four polarization measurements are performed using a HWP, QWP and polarizing beamsplitter. Photons are counted after the beamsplitter and the counts are taken to be fair samples of the true probabilities for obtaining each outcome for every preparation-measurement pair. Since the orientations of the preparation waveplates lead to small deflections of the beam, some information about the preparation gets encoded spatially, and similarly the measurement waveplates create sensitivity to spatial information; coupling the beam into the single-mode fibre connecting the state-preparation and measurement stages of the experiment removes sensitivity to these effects. For a single experimental run we implement each preparation-measurement pair for 4 s (approximately 10^5 counts). We performed 100 such runs.

Preparations are represented by vectors of raw data specifying the relative frequencies of outcomes for each measurement, uncertainties on which are calculated assuming Poissonian uncertainty in the photon counts. For each run, the raw data is fit to a set of states and effects in a GPT in which three binary-outcome measurements are tomographically complete. This is done using a total weighted least-squares method[48,49]. The average χ^2 over the 100 runs is 3.9 ± 0.3, agreeing with the expected value of 4, and indicating that the model fits the data well (see Supplementary Note 4, Supplementary Data 1 and 2, and Supplementary Software 1). The fit returns a 4×8 matrix that serves to define the 8 GPT states and 4 GPT effects, which are our estimates of the primary preparations and measurements. The column of this matrix associated to the t,b preparation, which we denote $\mathbf{P}^p_{t,b}$, specifies our estimate of the probabilities assigned by the primary preparation $P^p_{t,b}$ to outcome '0' of each of the

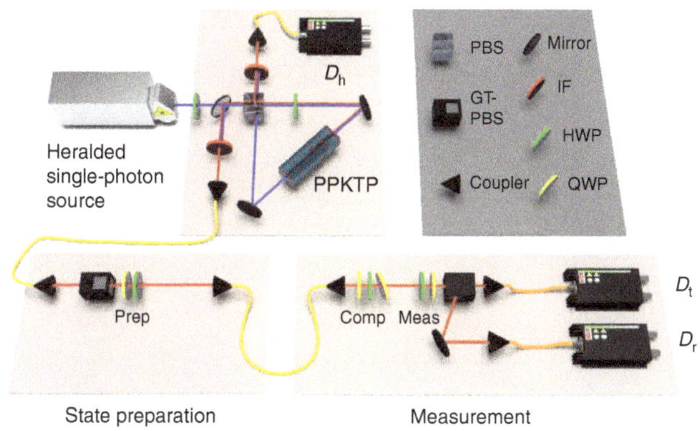

Figure 2 | The experimental setup. Polarization-separable photon pairs are created via parametric downconversion, and detection of a photon at detector D_h heralds the presence of a single photon. The polarization state of this photon is prepared with a polarizer and two waveplates (Prep). A single-mode fibre is a spatial filter that decouples beam deflections caused by the state-preparation and measurement waveplates from the coupling efficiency into the detectors. Three waveplates (Comp) are set to undo the polarization rotation caused by the fibre. Two waveplates (Meas), a polarizing beamsplitter, and detectors D_r and D_t perform a two-outcome measurement on the state. PPKTP, periodically-poled potassium titanyl phosphate; PBS, polarizing beamsplitter; GT-PBS, Glan-Taylor polarizing beamsplitter; IF, interference filter; HWP, half-waveplate; QWP, quarter-waveplate.

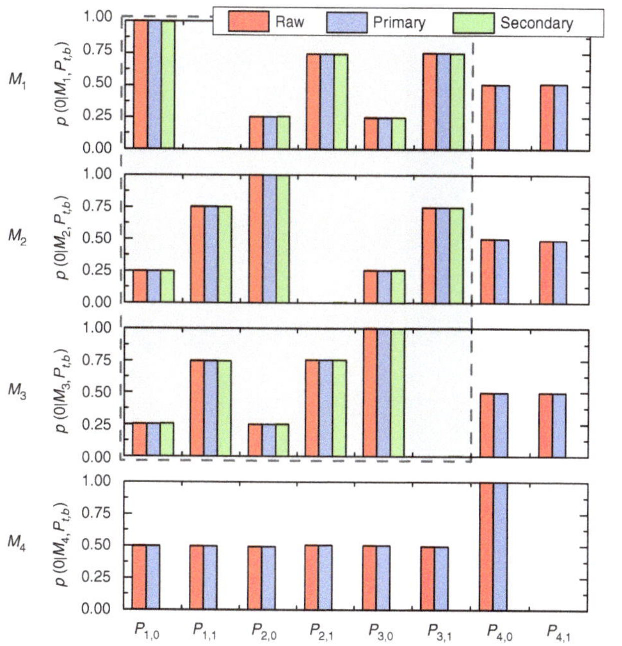

Figure 3 | Outcome probabilities for each measurement-preparation pair. For each such pair, we report the probability of obtaining outcome 0 in the measurement. Red bars are relative frequencies calculated from the raw counts, blue bars are our estimates of the outcome probabilities of the primary measurements on the primary preparations obtained from a best-fit of the raw data, and green bars are our estimates of the outcome probabilities of the secondary measurements on the secondary preparations. The shaded grey background highlights the measurements and preparations for which secondary procedures were found. Error bars are not visible on this scale, neither are discrepancies between the obtained probabilities and the ideal values thereof, which are at most 0.013; statistical error due to Poissonian count statistics is at most 0.002.

primary measurements. The raw and primary data are compared in Fig. 3. The probabilities are indistinguishable on this scale. We plot the probabilities for P_1, P_2, and P_3 in Fig. 4a on a much finer scale. We then see that the primary data are within error of the raw data, as expected given the high quality of the fit to the GPT. However, the operational equivalences of equations (2) and (4) are not satisfied by our estimates of the primary preparations and measurements, illustrating the need for secondary procedures.

We define the six secondary preparations as probabilistic mixtures of the eight primaries: $\mathbf{P}^{s}_{t,b}=\sum_{t'=1}^{4}\sum_{b'=0}^{1} u^{t,b}_{t',b'}\mathbf{P}^{p}_{t',b'}$, where the $u^{t,b}_{t',b'}$ are the weights in the mixture. We maximize $C_P=\frac{1}{6}\sum_{t=1}^{3}\sum_{b=0}^{1} u^{t,b}_{t,b}$ over valid $u^{t,b}_{t',b'}$ subject to the constraint of equation (4), that is, $\frac{1}{2}\sum_b \mathbf{P}^{s}_{1,b}=\frac{1}{2}\sum_b \mathbf{P}^{s}_{2,b}=\frac{1}{2}\sum_b \mathbf{P}^{s}_{3,b}$ (a linear program). A high value of C_P ensures that each of the six secondary preparations is close to its corresponding primary. Averaging over 100 runs, we find $C_P = 0.9969 \pm 0.0001$, close to the maximum of 1. An analogous linear program to select secondary measurements yields similar results. Supplementary Tables 3 and 4 display the weights that define each secondary preparation and measurement, averaged over 100 runs. Figure 3 also displays the outcome probabilities for the secondary procedures, confirming that they are close to ideal. Figure 4 demonstrates how our construction enforces the operational equivalences.

We analyzed each experimental run separately and found the degree of correlation $p(X=b|M^{s}_{t}, P^{s}_{t,b}\cdot)$ for each value of t and b. The averages over the 100 runs are shown in Fig. 5a and are all in excess of 0.995. Averaging over t and b yields an experimental value $A = 0.99709 \pm 0.00007$, which violates the noncontextual bound of $5/6 \approx 0.833$ by 2300σ (Fig. 5b).

Discussion

Using the techniques described here, it is possible to convert proofs of the failure of noncontextuality in quantum theory into experimental tests of noncontextuality that are robust to noise and experimental imprecisions[50,51]. For any phenomenon, therefore, one can determine which of its operational features are genuinely nonclassical. This is likely to have applications for scientific fields wherein quantum effects are important and for developing novel quantum technologies. The definition of operational equivalence of preparations (measurements) required them to be statistically equivalent relative to a tomographically complete set of measurements (preparations). There are two examples of how the assumption of tomographic completeness is expected not to hold exactly in our experiment, even if one grants the correctness of quantum theory. First, our source produces a small multi-photon component. We measure the $g^{(2)}(0)$ of our source[52] to be 0.0105 ± 0.0001 and from this we estimate the ratio of heralded detection events caused by multiple photons to those caused by single photons to be 1:4,000. Regardless of the value of A that one presumes for multi-photon events, one can infer that the value of A we would have achieved had the source been purely single-photon could have been less than the value given above by at most 10^{-6}, a difference that does not affect our conclusions. We also expect the assumption to not hold exactly because of the inevitable coupling of the polarization into the spatial degree of freedom of the photon, which could be caused, for example, by a wedge in a waveplate. Indeed, we found that if the spatial filter was omitted from the experiment, our fitting routine returned large χ^2 values, which we attributed to the fact that different angles of the waveplates led to different deflections of the beam. A more abstract worry is that nature might conflict with the assumption (and prediction of quantum theory) that three independent binary-outcome measurements are tomographically

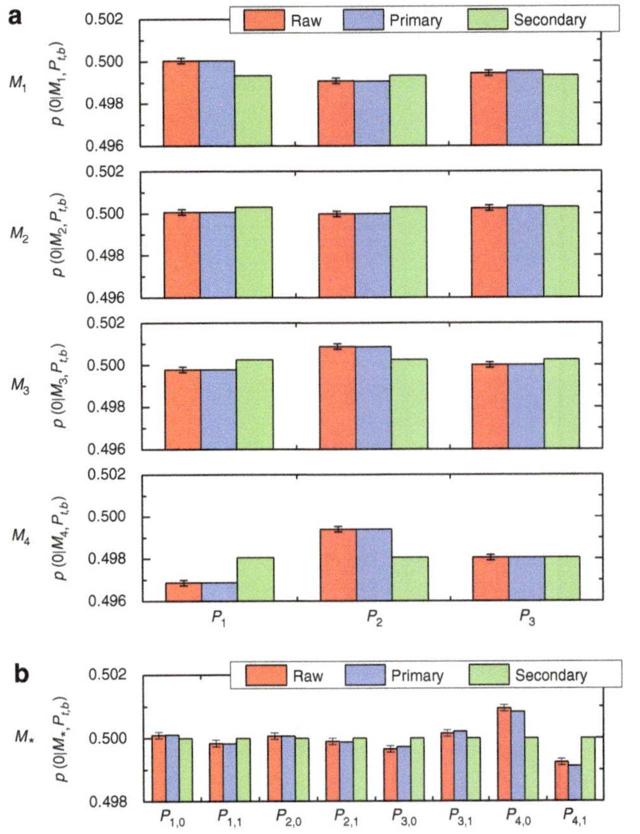

Figure 4 | Verifying the requisite operational equivalences. Operational statistics for raw, primary, and secondary preparations and measurements, averaged over 100 experimental runs. (**a**) The probabilities of the primary measurements (blue bars) differ depending on which of the three mixed preparations P_1^p, P_2^p, and P_3^p are measured. These probabilities are within error of the raw data (red bars), indicating that a generalised probabilistic theory in which three two-outcome measurements are tomographically complete fits the data well. Probabilities for primary measurements on the secondary preparations (green bars) are independent of the preparation, hence the secondary preparations satisfy equation (4). Note that one expects these probabilities to deviate from 0.5. In the example of Fig. 1c, this corresponds to the fact that the intersection of the lines is not the completely mixed state. (**b**) Outcome probabilities of measurement M_* on the eight preparations. Red bars are raw data, blue bars are the measurement M_*^p on the primary preparations, and green bars are M_*^s on the primary preparations. Regardless of the input state, M_*^s returns outcome 0 with probability 0.5, hence it is operationally indistinguishable from a fair-coin flip (equation (2)). Error bars in all plots are calculated assuming Poissonian count statistics.

complete for the polarization of a photon. Our experiment has provided evidence in favour of the assumption insofar as we have fit data from four measurements to a theory where three are tomographically complete and found a good χ^2 value for the fit. One can imagine accumulating much more evidence of this sort, but it is difficult to see how any experiment could conclusively vindicate the assumption, given that one can never test all possible measurements. This, therefore, represents the most significant loophole in experimental tests of noncontextuality, and new ideas for how one might seal it or circumvent it represent the new frontier for improving such tests.

Methods

Preparation procedure. A 20-mW diode laser with a wavelength of 404.7 nm produces photon pairs, one horizontally polarized and the other vertically polarized, via spontaneous parametric down-conversion in a 20-mm type-II PPKTP crystal.

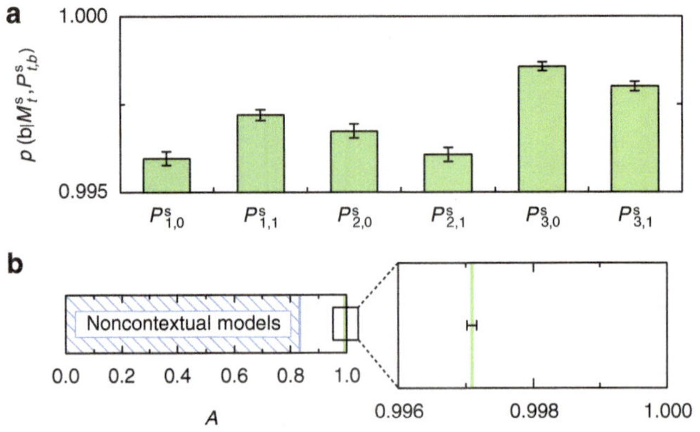

Figure 5 | Violation of the noncontextuality inequality. (**a**) Values of the six degrees of correlation in equation (8), averaged over 100 experimental runs. (**b**) Average measured value for A contrasted with the noncontextual bound $A = 5/6$. We find $A = 0.99709 \pm 0.00007$, which violates the noncontextual bound by 2300σ. Error bars in both plots represent the standard deviation in the average of the measured values over the 100 experimental runs.

The downconversion crystal is inside a Sagnac loop and the pump laser is polarized vertically to ensure it only travels counter-clockwise around the loop. Photon pairs are separated at a polarizing beamsplitter and coupled into two single-mode fibres (SMFs). Vertically-polarized photons are detected immediately at detector D_h, heralding the presence of the horizontally-polarized signal photons which emerge from the SMF and pass through a state-preparation stage before they are measured. Herald photons were detected at a rate of 400 kHz. Signal photons emerge from the fibre and pass through a Glan-Taylor polarizing beamsplitter (GT-PBS) which transmits vertically-polarized light. Polarization controllers in the fibre maximize the number of photons which pass through the beamsplitter. A quarter- and half-waveplate set the polarization of the signal photons to one of eight states.

Spatial mode filter. A single-mode fibre acts as a spatial mode filter. This filter ensures that information about the angles of the state-preparation waveplates cannot be encoded in the spatial mode of the photons, and that our measurement procedures do not have a response that depends on the spatial mode, but only on polarization as intended. The SMF induces a fixed polarization rotation, so a set of three compensation waveplates are included after the SMF to undo this rotation. It follows that the preparation-measurement pairs implemented in our experiment are in fact a rotated version of the ideal preparation and a similarly-rotated version of the ideal measurement. Such a fixed rotation, however, does not impact any of our analysis.

Measurement procedure. Measurements are performed in four bases, set by a half- and quarter-waveplate. A second GT-PBS splits the light, and both output ports are detected. Due to differences in the coupling and detection efficiencies in each path after the beamsplitter, each measurement consists of two parts. First, the waveplates are aligned such that states corresponding to outcome '0' are transmitted by the GT-PBS, and the number of heralded photons detected in a two-second window is recorded for each port. Second, the waveplate angles are changed in such a way as to invert the outcomes, so the detector in the reflected port corresponds to outcome '0' and heralded photons are detected for another two seconds. The counts are added together and the probability for outcome '0' is calculated by dividing the number of detections corresponding to outcome '0' by the total number of detection events in the four-second window. The single-photon detection rate at detectors D_r and D_t depends on the measurement settings. In the transmissive and reflective ports of the measurement GT-PBS photons were detected at maximum rates of 330 and 250 kHz, respectively. Coincident detection events between herald photons and the transmissive and reflective ports of the measurement GT-PBS were up to 22 and 16 kHz, respectively.

Code availability. The authors declare that the data-analysis code supporting the findings of this study are available within the article's Supplementary Information files (Supplementary Software 1).

References

1. Bell, J. S. On the Einstein-Podolsky-Rosen paradox. *Physics* **1**, 195–200 (1964).
2. Kochen, S. & Specker, E. The problem of hidden variables in quantum mechanics. *Indiana Univ. Math. J.* **17**, 59–87 (1968).
3. Spekkens, R. W. Negativity and contextuality are equivalent notions of nonclassicality. *Phys. Rev. Lett.* **101**, 020401 (2008).
4. Ferrie, C. & Emerson, J. Frame representations of quantum mechanics and the necessity of negativity in quasi-probability representations. *J. Phys. A: Math. Theor.* **41**, 352001 (2008).
5. Spekkens, R. W., Buzacott, D. H., Keehn, A. J., Toner, B. & Pryde, G. J. Preparation contextuality powers parity-oblivious multiplexing. *Phys. Rev. Lett.* **102**, 010401 (2009).
6. Howard, M., Wallman, J., Veitch, V. & Emerson, J. Contextuality supplies the 'magic' for quantum computation. *Nature* **510**, 351–355 (2014).
7. Raussendorf, R. Contextuality in measurement-based quantum computation. *Phys. Rev. A* **88**, 022322 (2013).
8. Hoban, M. J., Campbell, E. T., Loukopoulos, K. & Browne, D. E. Non-adaptive measurement-based quantum computation and multi-party bell inequalities. *New J. Phys.* **13**, 023014 (2011).
9. Pusey, M. F. Anomalous weak values are proofs of contextuality. *Phys. Rev. Lett.* **113**, 200401 (2014).
10. Meyer, D. A. Finite precision measurement nullifies the Kochen-Specker theorem. *Phys. Rev. Lett.* **83**, 3751–3754 (1999).
11. Kent, A. Noncontextual hidden variables and physical measurements. *Phys. Rev. Lett.* **83**, 3755–3757 (1999).
12. Clifton, R. & Kent, A. Simulating quantum mechanics by non-contextual hidden variables. *Proc. R. Soc. Lond. A* **456**, 2101–2114 (2000).
13. Mermin, N. D. A Kochen-Specker theorem for imprecisely specified measurement. Preprint at http://arxiv.org/abs/quant-ph/9912081 (1999).
14. Simon, C., Brukner, Č. & Zeilinger, A. Hidden-variable theorems for real experiments. *Phys. Rev. Lett.* **86**, 4427–4430 (2001).
15. Larsson, J.-Å. A Kochen-Specker inequality. *Europhys. Lett.* **58**, 799–805 (2002).
16. Barrett, J. & Kent, A. Non-contextuality, finite precision measurement and the Kochen-Specker theorem. *Stud. Hist. Phil. Mod. Phys.* **35**, 151–176 (2004).
17. Cabello, A. & Garcia-Alcaine, G. Proposed experimental tests of the Bell-Kochen-Specker theorem. *Phys. Rev. Lett.* **80**, 1797 (1998).
18. Cabello, A. Kochen-Specker theorem and experimental test on hidden variables. *Int. J. Mod. Phys. A* **15**, 2813–2820 (2000).
19. Simon, C., Żukowski, M., Weinfurter, H. & Zeilinger, A. Feasible 'Kochen-Specker' experiment with single particles. *Phys. Rev. Lett.* **85**, 1783–1786 (2000).
20. Cabello, A., Filipp, S., Rauch, H. & Hasegawa, Y. Proposed experiment for testing quantum contextuality with neutrons. *Phys. Rev. Lett.* **100**, 130404 (2008).
21. Klyachko, A. A., Can, M. A., Binicioğlu, S. & Shumovsky, A. S. Simple test for hidden variables in spin-1 systems. *Phys. Rev. Lett.* **101**, 020403 (2008).
22. Cabello, A. Experimentally testable state-independent quantum contextuality. *Phys. Rev. Lett.* **101**, 210401 (2008).
23. Badziag, P., Bengtsson, I., Cabello, A. & Pitowsky, I. Universality of state-independent violation of correlation inequalities for noncontextual theories. *Phys. Rev. Lett.* **103**, 050401 (2009).
24. Gühne, O. *et al.* Compatibility and noncontextuality for sequential measurements. *Phys. Rev. A* **81**, 022121 (2010).
25. Michler, M., Weinfurter, H. & Żukowski, M. Experiments towards falsification of noncontextual hidden variable theories. *Phys. Rev. Lett.* **84**, 5457–5461 (2000).
26. Huang, Y.-F., Li, C.-F., Zhang, Y.-S., Pan, J.-W. & Guo, G.-C. Experimental test of the Kochen-Specker theorem with single photons. *Phys. Rev. Lett.* **90**, 250401 (2003).
27. Hasegawa, Y., Loidl, R., Badurek, G., Baron, M. & Rauch, H. Quantum contextuality in a single-neutron optical experiment. *Phys. Rev. Lett.* **97**, 230401 (2006).
28. Kirchmair, G. *et al.* State-independent experimental test of quantum contextuality. *Nature* **460**, 494–497 (2009).
29. Bartosik, H. *et al.* Experimental test of quantum contextuality in neutron interferometry. *Phys. Rev. Lett.* **103**, 040403 (2009).
30. Amselem, E., Rådmark, M., Bourennane, M. & Cabello, A. State-independent quantum contextuality with single photons. *Phys. Rev. Lett.* **103**, 160405 (2009).
31. Moussa, O., Ryan, C. A., Cory, D. G. & Laflamme, R. Testing contextuality on quantum ensembles with one clean qubit. *Phys. Rev. Lett.* **104**, 160501 (2010).
32. Lapkiewicz, R. *et al.* Experimental non-classicality of an indivisible quantum system. *Nature* **474**, 490–493 (2011).
33. Spekkens, R. W. The status of determinism in proofs of the impossibility of a noncontextual model of quantum theory. *Found. Phys.* **44**, 1125–1155 (2014).
34. Spekkens, R. W. Contextuality for preparations, transformations, and unsharp measurements. *Phys. Rev. A* **71**, 052108 (2005).
35. Winter, A. What does an experimental test of quantum contextuality prove or disprove? *J. Phys. A: Math. Theor.* **47**, 424031 (2014).
36. Kujala, J. V., Dzhafarov, E. N. & Larsson, J.-A. Necessary and sufficient conditions for an extended noncontextuality in a broad class of quantum mechanical systems. *Phys. Rev. Lett.* **115**, 150401 (2015).
37. Clauser, J. F., Horne, M. A., Shimony, A. & Holt, R. A. Proposed experiment to test local hidden-variable theories. *Phys. Rev. Lett.* **23**, 880–884 (1969).
38. Nielsen, M. A. & Chuang, I. L. *Quantum Computation and Quantum Information: 10th Anniversary Edition* (Cambridge University Press, 2011).
39. James, D. F. V., Kwiat, P. G., Munro, W. J. & White, A. G. Measurement of qubits. *Phys. Rev. A* **64**, 052312 (2001).
40. Lundeen, J. S. *et al.* Tomography of quantum detectors. *Nature Phys.* **5**, 27–30 (2009).
41. Stark, C. Self-consistent tomography of the state-measurement gram matrix. *Phys. Rev. A* **89**, 052109 (2014).
42. Stark, C. J. & Harrow, A. W. Compressibility of positive semidefinite factorizations and quantum models. *IEEE Trans. Inf. Theory* **62**, 2867–2880 (2016).
43. Hardy, L. Quantum theory from five reasonable axioms. Preprint at http://arxiv.org/abs/quant-ph/0101012 (2001).
44. Barrett, J. Information processing in generalized probabilistic theories. *Phys. Rev. A* **75**, 032304 (2007).
45. Kim, T., Fiorentino, M. & Wong, F. N. C. Phase-stable source of polarization-entangled photons using a polarization sagnac interferometer. *Phys. Rev. A* **73**, 012316 (2006).
46. Fedrizzi, A., Herbst, T., Poppe, A., Jennewein, T. & Zeilinger, A. A wavelength-tunable fiber-coupled source of narrowband entangled photons. *Opt. Express* **15**, 15377–15386 (2007).
47. Biggerstaff, D. N. *et al.* Cluster-state quantum computing enhanced by high-fidelity generalized measurements. *Phys. Rev. Lett.* **103**, 240504 (2009).
48. Krystek, M. & Anton, M. A weighted total least-squares algorithm for fitting a straight line. *Meas. Sci. Technol.* **18**, 3438 (2007).
49. Press, W. H., Teukolsky, S. A., Vetterling, W. T. & Flannery, B. P. *Numerical Recipes: The Art of Scientific Computing* 3rd edn (Cambridge University Press, 2007).
50. Kunjwal, R. & Spekkens, R. W. From the Kochen-Specker theorem to noncontextuality inequalities without assuming determinism. *Phys. Rev. Lett.* **115**, 110403 (2015).
51. Pusey, M. F. The robust noncontextuality inequalities in the simplest scenario. Preprint at http://arxiv.org/abs/1506.04178 (2015).
52. Grangier, P., Roger, G. & Aspect, A. Experimental evidence for a photon anticorrelation effect on a beam splitter: A new light on single-photon interferences. *Europhys. Lett.* **1**, 173–179 (1986).

Acknowledgements

The authors thank Megan Agnew for assistance with data acquisition software. This research was supported in part by the Natural Sciences and Engineering Research Council of Canada (NSERC), Canada Research Chairs, Ontario Centres of Excellence, Industry Canada, and the Canada Foundation for Innovation (CFI). M.D.M. acknowledges support from the Ontario Ministry of Training, Colleges, and Universities. R.K. thanks the Perimeter Institute for hospitality during his visit there, which was made possible in part through the support of a grant from the John Templeton Foundation and through the support of the Institute of Mathematical Sciences, Chennai. Research at Perimeter Institute is supported by the Government of Canada through Industry Canada and by the Province of Ontario through the Ministry of Research and Innovation.

Author contributions

R.K., M.D.M., M.F.P. and R.W.S. developed the theory. M.D.M. and K.J.R. designed the experiment. M.D.M. performed the experiment and implemented the numerical analysis of the data. All authors contributed to writing the article.

Additional information

Competing financial interests: The authors declare no competing financial interests.

10

Stabilization of active matter by flow-vortex lattices and defect ordering

Amin Doostmohammadi[1], Michael F. Adamer[1], Sumesh P. Thampi[1] & Julia M. Yeomans[1]

Active systems, from bacterial suspensions to cellular monolayers, are continuously driven out of equilibrium by local injection of energy from their constituent elements and exhibit turbulent-like and chaotic patterns. Here we demonstrate both theoretically and through numerical simulations, that the crossover between wet active systems, whose behaviour is dominated by hydrodynamics, and dry active matter where any flow is screened, can be achieved by using friction as a control parameter. Moreover, we discover unexpected vortex ordering at this wet-dry crossover. We show that the self organization of vortices into lattices is accompanied by the spatial ordering of topological defects leading to active crystal-like structures. The emergence of vortex lattices, which leads to the positional ordering of topological defects, suggests potential applications in the design and control of active materials.

[1]The Rudolf Peierls Centre for Theoretical Physics, 1 Keble Road, Oxford OX1 3NP, UK. Correspondence and requests for materials should be addressed to A.D. (email: amin.doostmohammadi@physics.ox.ac.uk).

Whether it is biological matter such as cytoskeletal networks, cellular colonies and suspensions of bacteria or synthetic systems such as Janus catalysts and vibrating granular rods, continuous injection of energy from the constituent elements leads to exotic behaviour such as collective motion[1–3], pattern formation[4–7], topological defects[8–10] and active turbulence[11–13]. Active systems where hydrodynamic interactions are key, such as suspensions of swimming bacteria, are referred to as wet. The energy input to such systems is dissipated by the viscosity of the fluid. However, in many active materials there are alternative forms of dissipation, such as wall friction between the active particles and a substrate[14]. In the limit where the frictional damping dominates, it screens out any velocity fields that are generated by the activity. For instance, the Vicsek model for migrating animal herds[15], and assemblies of vibrated granular rods[16,17] lack long-range hydrodynamic interactions and are classified as dry active materials. In general, a given active material will fall between these limits with the relative magnitudes of frictional and dissipative damping controlling its position on the wet–dry spectrum. Examples of experimental two-dimensional active nematics where friction could be varied to control the crossover between the wet and dry limits include cells moving and dividing on substrates[18,19], microswimmers confined between parallel walls[11] and suspensions of microtubules and molecular motors moving at an interface between two fluids[6].

Here we unify two different classes of active matter by using friction as a control parameter to interpolate between wet and dry active matter, and at the wet–dry crossover we discover an unexpected regime, where otherwise disordered vortices self-organize into lattices interleaved with ordered arrays of topological defects. The vortex lattices and their corresponding network of ordered defects arise from the competition between friction and viscous dissipation and stablize the active system. Our results contribute to understanding the physics of matter operating out of equilibrium, with its potential in the design of active micro- and nano-machines.

Results

Wet–dry crossover. The behaviour of dry and wet active nematics is summarized pictorially at the right- and left-hand sides of Fig. 1. In the absence of momentum conservation, dry active matter is commonly described in terms of the concentration of active components and their orientational order[20–22]. The existence of any curvature in the orientational order induces an instability that leads to the formation of bands of concentration[21,22]. The concentration bands are, however, always unstable and eventually break into filaments that, in turn, coalesce and form new bands. The process is repeated and a chaotic regime is established[23].

Introducing hydrodynamics leads to extra complexity due to the coupling of the density and orientational order of active particles to the fluid flow, and distinct dynamical features are manifest in the models of wet active nematics. In particular, there is a hydrodynamic instability of nematic regions that leads to the formation of walls, lines of high distortions in the director field, where elastic energy is stored. The walls are continuously broken up by the creation and annihilation of topological defects, a process that gives rise once again to unstable nematic regions. The formation and removal of the walls is maintained by the active forcing, and a state termed active turbulence is established, characterized by a chaotic velocity field with regions of high vorticity[9,24].

Theory. We build on the nematohydrodynamic equations of liquid crystals to describe a natural route from wet to dry active nematics. To this end, the continuum description of the dynamics of passive liquid crystals is modified to account for the active stresses generated by constituent elements[25]. This continuum approach, which allows for coarse graining over the microscopic details, has proven successful in reproducing several experimental observations including the flow behaviour and defect dynamics observed in experiments on microtuble bundles driven by motor proteins[6,9,24], the spatial organization of bacterial cultures in confined environments[26], tissue growth[27,28] and fluidization[29], and the flow fields associated with cell division in cellular monolayers[30].

The variables needed to describe the hydrodynamics of a wet active nematic are: ϕ the relative concentration of active and passive particles, ρ the total density, $\mathbf{u}$ the velocity vector, and $\mathbf{Q} = \frac{3q}{2}(\mathbf{nn} - \mathbf{I}/3)$, which is the nematic tensor, with q the magnitude of the orientational order, $\mathbf{n}$ the director, and $\mathbf{I}$ the identity tensor. The four coupled continuum equations describing the time evolution of these quantities are[31]

$$\partial_t \phi + \partial_i (u_i \phi) = \Gamma_\phi \nabla^2 \mu, \tag{1}$$

$$(\partial_t + u_k \partial_k) Q_{ij} - S_{ij} = \Gamma_Q H_{ij}, \tag{2}$$

$$\partial_t \rho + \partial_i (\rho u_i) = 0, \tag{3}$$

$$\rho(\partial_t + u_k \partial_k) u_i = \partial_j \Pi_{ij} - \gamma u_i, \tag{4}$$

where the mobility and the rotational diffusivity are denoted by Γ_ϕ and Γ_Q, respectively. The co-rotation term,

$$\begin{aligned} S_{ij} =& [\lambda E_{ik} + \Omega_{ik}]\left(Q_{kj} + \frac{\delta_{kj}}{3}\right) + \left(Q_{ik} + \frac{\delta_{ik}}{3}\right)[\lambda E_{kj} - \Omega_{kj}] \\ & - 2\lambda\left(Q_{ij} + \frac{\delta_{ij}}{3}\right)(Q_{kl}\partial_k u_l), \end{aligned} \tag{5}$$

accounts for the response of the orientational order to the flow gradients characterized by the strain rate tensor $E_{ij} = (\partial_i u_j + \partial_j u_i)/2$ and vorticity tensor $\Omega_{ij} = (\partial_j u_i - \partial_i u_j)/2$. The alignment parameter λ, which describes the responses of particles to the strain and vorticity takes different values for different particle shapes[32] with $\lambda > 0$, $\lambda < 0$ and $\lambda = 0$ for rod-like, disk-like and spherical particles, respectively. The chemical potential $\mu = \delta\mathcal{F}/\delta\phi$ and molecular field $H_{ij} = -\delta\mathcal{F}/\delta Q_{ij} + (\delta_{ij}/3)\mathrm{Tr}(\delta\mathcal{F}/\delta Q_{kl})$ represent the relaxation of the concentration and the orientational order to equilibrium, where

$$\begin{aligned} \mathcal{F} =& \frac{a}{2}\phi^2 + \frac{A}{2}\phi\mathbf{Q}^2 + \frac{C}{4}\mathbf{Q}^4 \\ & + \frac{K_\phi}{2}(\nabla\phi)^2 + \frac{K_Q}{2}(\nabla\mathbf{Q})^2 + K_c\mathbf{Q} : \nabla\nabla\phi, \end{aligned} \tag{6}$$

is the free energy functional and a, A, C, K_ϕ, K_Q, K_c are material constants.

The terms on the left-hand side of equation (4) are the usual terms in the Navier–Stokes equation describing the inertia of the fluid. These are small at low Reynolds number, the limit relevant to active particles on the micron-scale or smaller. Contributions to the stress tensor, Π_{ij}, are the viscous dissipation $\Pi_{ij}^{\text{viscous}} = 2\eta E_{ij}$, where η is viscosity, the elastic stresses that generate backflow and encode mechanical forces on the fluid due to the relaxational motion of the active entities

$$\begin{aligned} \Pi_{ij}^{\text{elastic}} =& -P\delta_{ij} + (\mathcal{F} - \mu\phi)\delta_{ij} \\ & + 2\lambda\left(Q_{ij} + \frac{\delta_{ij}}{3}\right)(Q_{kl}H_{lk}) \\ & - \lambda H_{ik}\left(Q_{kj} + \frac{\delta_{kj}}{3}\right) - \lambda\left(Q_{ik} + \frac{\delta_{ik}}{3}\right)H_{kj} \\ & - \partial_i Q_{kl}\frac{\delta\mathcal{F}}{\delta\partial_j Q_{lk}} - \partial_i\phi\frac{\delta\mathcal{F}}{\delta\partial_j\phi} + Q_{ik}H_{kj} - H_{ik}Q_{kj}, \end{aligned} \tag{7}$$

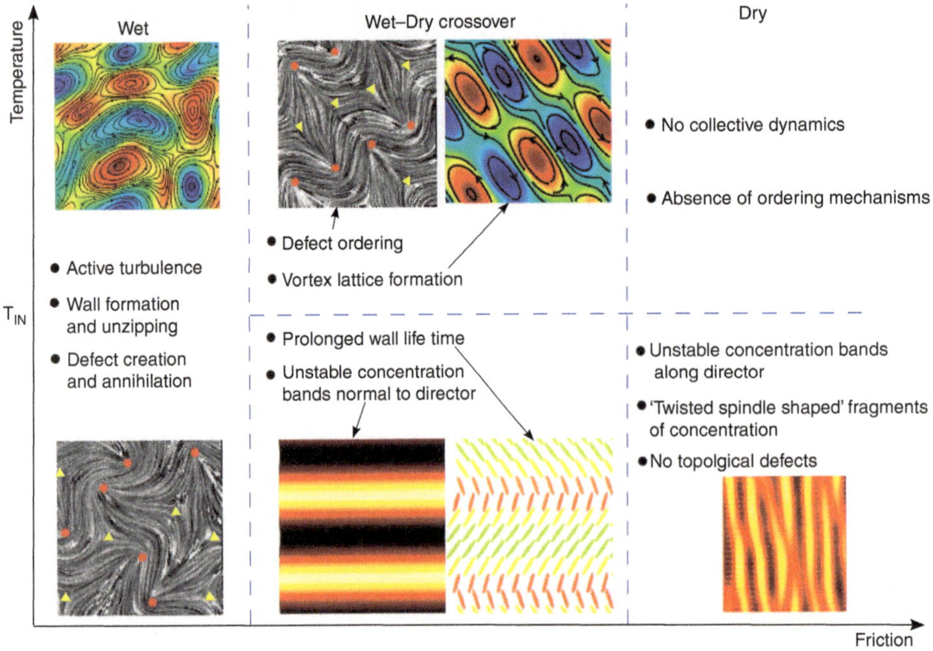

Figure 1 | Dynamical behaviours of active nematics in the temperature-friction phase space. T_{IN} denotes the isotropic-nematic transition temperature: above T_{IN} at intermediate frictions we find a novel vortex lattice that entangles an ordered defect array. The blue-red colourmaps represent vorticity fields superimposed by streamlines (black solid lines). The concentration fields are depicted by red-yellow colourmaps. In the mid-lower section, the director field is illustrated by ellipsoids coloured by their orientations. Grey panels show director fields superimposed by topological defects (red circles and yellow triangles correspond to $+1/2$ and $-1/2$ defects).

where P denotes the pressure, and, of primary relevance to our current argument, the active stress $\Pi_{ij}^{active} = -\zeta Q_{ij}$, which is proportional to the orientational order parameter and distinguishes active from passive nematics[25]. The strength of the activity is set by ζ. The final term in equation (4) accounts for the friction, for example, with a substrate, with γ, the associated friction coefficient. Details of this model in the context of passive and active liquid crystals can be found in refs 31,33–39.

In a wet active nematic, the momentum propagation is more effectively suppressed as the friction coefficient is increased. In the limit of sufficiently high friction, the viscous and elastic stresses become small and the friction forces balance with the forces generated by the active term. In this regime the velocity can be approximated in terms of the nematic tensor as

$$u_i \approx -\frac{\zeta}{\gamma}\partial_j Q_{ij}. \tag{8}$$

In writing equation (8), we are neglecting the fluid inertia, which is dominated by the viscosity for bacterial suspensions and cytoskeletal filaments. We also neglect the elastic terms in the stress tensor that generate backflow. Non-dimensionalizing the momentum equation (see Methods section for more details) shows that the viscous stress dominates the backflow by a factor controlled by the Ericksen number $Er = \eta UL/K$, where U, L are a characteristic velocity and length. Er is $\sim 10^2$ in our simulations.

We chose parameters that are in a range that has been previously used to reproduce experimental results measuring the velocity correlations of microtuble bundles driven by molecular motors[6,24] and the flow fields of dividing Madine Darby Canine Kidney cells[30]. Moreover, the scaling suggested by equation (8), that velocity increases proportionately with activity, has been observed in both simulations[24] and experiments[6].

Considering equation (8) as an equality and substituting in the equations (1) and (2) for the concentration and orientational order evolution, we reproduce the standard equations for dry active matter,

$$\partial_t \phi = \Gamma_\phi \nabla^2 \mu + \frac{\zeta\phi}{\gamma}\partial_i\partial_j Q_{ij} + \ \ldots\ , \tag{9}$$

$$\partial_t Q_{ij} = \Gamma_Q H_{ij} + \ \ldots \tag{10}$$

The second term on the right-hand side of equation (9) is identical to that commonly introduced to represent the current due to curvature in the director field, which drives ordering in dry active nematics. Thus, we demonstrate that the equations of dry active nematics can be generically reproduced from the nematohydrodynamic equations, showing that the dry limit arises naturally when friction dominates viscosity in wet active materials. We use $+\ \ldots$ to denote that flow-driven terms that are non-linear in ϕ and $\mathbf{Q}$ also appear in equation (9) and (10). These are listed explicitly in the Methods section and their appearance can be expected from symmetry arguments. In addition, we have also used a kinetic approach[21] to show that our analytical derivation of the dry active nematic equations from equations of wet active nematics, matches the microscopic derivation at the kinetic equation level. We show that even higher order terms predicted by our derivation can be expected from the expansion of the kinetic equations to higher order (Methods section).

Simulation. To demonstrate the wet to dry crossover, we present numerical solutions of the active nematic equations. We first illustrate the effect of increasing the friction on the evolution of the concentration and order parameter in a strongly ordered nematic (Fig. 2). Figure 2a shows the time evolution of the concentration field (top row) and director field (bottom row) for an active nematic with an intermediate friction coefficient. As for zero-friction the nematic state is hydrodynamically unstable and walls form in the director field. However, both wall creation, and subsequent destruction by topological defects, are significantly slowed by the friction, which leads to a reduction in the root-mean-square (RMS) velocity by a factor ~10 (Fig. 2c). Surprisingly, the slower dynamics is accompanied by an increase in the number of topological defects (Fig. 2d), a consequence of the larger number of walls at higher friction. Similar trends have been recently observed in active nematics without any concentration variation[40]. However, as described earlier, concentration variation is necessary to approach the dry limit where activity manifests in concentration phase separation.

Linked to the wall formation is the emergence of coincident concentration bands, with a higher number of active particles at the walls, where the magnitude of the nematic ordering is reduced. The concentration ordering is driven by an advective flux of active particles towards the walls, down the gradient in **Q**. In the steady state this is balanced by diffusion of the active particles from high to low concentrations.

Figure 2c,d show that the active system behaves very differently for higher values of the friction. The RMS velocity is very low and topological defects are not formed. We note that although, topological defects do not appear in the dry limit, such defects have been observed in shaken rod experiments and in a simulation of active rods. As pointed out in ref. 41 this occurs, in their model at least, because the collisions include a relative rotation of the colliding particles as well as nematic alignment. Figure 2b indicates that concentration bands still appear in the dry limit but, in the absence of an advective flux, the way in which they are formed must be very different. The relevant mechanism is an instability driven by curvature in the nematic order, described by the final term in equation (9). The coupling between the concentration and the nematic order is established by the molecular potential resulting in a strong (weak) ordering at high

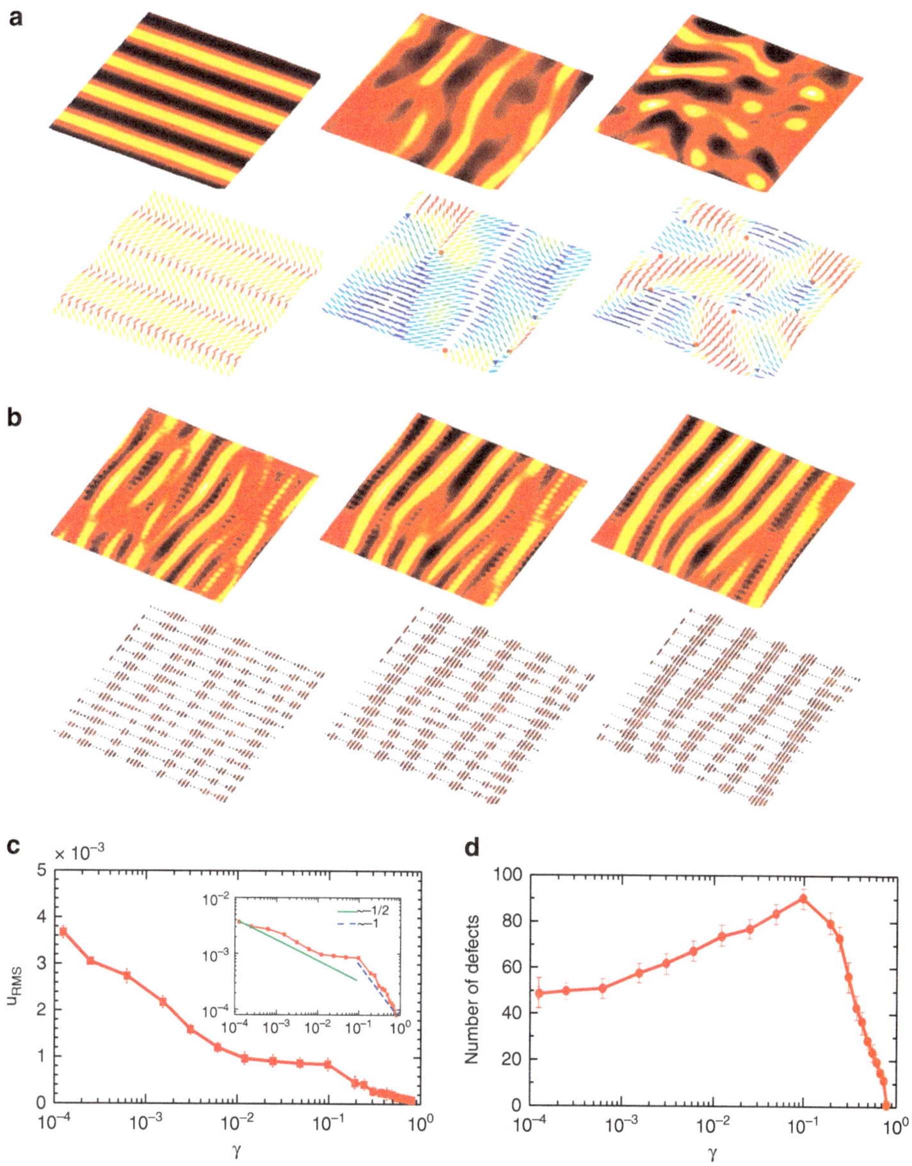

Figure 2 | Increasing friction drives the crossover from wet to dry active nematics. (**a**,**b**) Temporal evolution of the concentration field and nematic director field for (**a**) $\gamma = 0.08$ and (**b**) $\gamma = 0.8$ with $t = 10^5$, 5×10^5, 8×10^5 (in LB units) for left, middle and right columns, respectively. For each value of the friction the top row is a colour map indicating variations in concentration and the bottom row is the corresponding director field coloured by the orientation of nematic directors. +1/2 and −1/2 defects are denoted by red circles and blue triangles. (**c**) The RMS velocity is reduced by increasing friction. The inset illustrates the variation in log-log plot, showing the existence of different exponents. (**d**) The total number of topological defects initially increases, but drops sharply at $\gamma \sim 0.1$ and disappears at the dry limit.

(low) concentrations. Note the striking difference between the wet and dry limits: in the wet limit, concentration bands are initially formed perpendicular to the director, and the concentration and the nematic order are anti-correlated. In the dry limit, the bands form parallel to the director, and the concentration and nematic fields are correlated. The variation of the RMS velocity suggests the existence of different regimes (Fig. 2c, inset). From equation (8), the magnitude of the velocity can be estimated as $u_{\mathrm{RMS}} \approx \zeta q/\gamma L_q$, where L_q is the characteristic length scale of the nematic order variation. When the frictional damping is strong, L_q is set by the distance between the walls, which is approximately proportional to the screening length $L_{sc} \sim \sqrt{\eta/\gamma\rho}$ and, therefore, the velocity is expected to vary as $u_{\mathrm{RMS}} \sim \gamma^{-1/2}$ (Fig. 2c, inset: green line). At even higher frictions, there is not enough energy for hydrodynamic wall formation and the active flux-induced instabilities of dry matter dominate. Therefore, L_q does not depend on the screening length and $u_{\mathrm{RMS}} \sim \gamma^{-1}$ (Fig. 2c, inset: blue line).

Flow-vortex lattices and ordered topological defects

We next focus on the dynamical effects of friction at a temperature corresponding to the (passive) isotropic–nematic transition (Fig. 3). In a passive nematic there would be no orientational order. However, order can be stabilized as a result of extensional flows generated by the active forcing for $\zeta\gamma > 0$, that is for an extensile, rod-like system ($\zeta > 0$, $\lambda > 0$) or a contractile, disc-like system ($\zeta < 0$, $\lambda < 0$)[42].

Considering first extensile rods, for the smallest value of friction considered active turbulence is established. Flow-vortices move around chaotically, decay and re-form (Fig. 3a and Supplementary Movies 1 and 2). As the friction coefficient is increased there is a striking change in behaviour. As is apparent in Fig. 3b, the velocity vortices form ordered, stationary arrays (Supplementary Movies 3 and 4). Figure 3c indicates how the vortices become narrower at higher friction.

The vortex lattice is interleaved with an ordered network of topological defects (Fig. 3e,f). The defect lattice is directly linked to the velocity pattern: $+1/2$ defects are generated between pairs of counter-rotating vortices due to the change in the sign of the vorticity[24], while $-1/2$ defects are created between two pairs of co-rotating vortices due to distinct flow-induced reorientation of nematic directors (Fig. 3g). The shear flow in the middle of a vortex turns a nematogen towards a 45° orientation relative to the vortex line, while extensional and compressional flows at the edges of a vortex induce vertical and horizontal reorientations, respectively.

Similar behaviour is seen for a system of contractile, disk-like particles. In the absence of friction, the activity-induced ordering results in the generation of turbulent-like flow patterns (Fig. 4a,b). As the friction coefficient is increased, vortices arrange to form a lattice (Fig. 4c–e). The long-range order of the vortex lattice is evident from measurements of the vorticity–vorticity correlation functions (Fig. 4f), which are defined as $C_{\omega\omega}(r) = \langle \boldsymbol{\omega}(r,t) \cdot \boldsymbol{\omega}(0,t)\rangle / \langle \boldsymbol{\omega}(0,t)^2 \rangle$, where $\boldsymbol{\omega}$ is the vorticity. A similar ordered array of vortices has recently been reported in experiments on dividing endothelial cells[19] where the vortex structure was associated with the division-induced flow field.

In the absence of friction, the vorticity length scale is set by the activity and the elastic constant[13,37]. However, for both extensile rod-like and contractile disk-like particles, on increasing friction, the vorticity length scale drops. When it becomes comparable to the screening length vortices are no longer disturbed by the flow and the vortex lattice is established (Fig. 5a).

The positional ordering of defects for extensile and contractile systems can be compared by calculating the structure factor $S_{\mathbf{k}} = \sum_{n=1}^{N_d} e^{2\pi i(\mathbf{k}\cdot\mathbf{x}_n)}$, where $\mathbf{x}_n$ denotes the position of the nth defect[43]. At low friction, positions of topological defects are uncorrelated (Figs 3d and 5b,c; low friction), while positional ordering is observed at high friction values (Figs 3e,f and 5b,c; high friction). The mechanism of the ordering is different in extensile and contractile systems as the primary mode of instability in contractile systems is a splay distortion of the director field, while in extensile systems, bend distortions dominate[44]. This manifests itself in different structure factors of defect ordering for extensile rod-like and contractile disk-like particles, which correspond to rectangular and centred rectangular Bravais crystal structures[43], respectively (Fig. 5b,c).

Discussion

In addition to the vortex lattice surrounding epithelial cells, a similar positional ordering of defects has been observed in active polar systems in the form of aster arrays[45–47]. Moreover, the vorticity pattern we report at high friction resembles self-organized vortices of sperm cells on a surface[48] and periodic vortex arrays reported for a motility assay of microtubules with short-range attractions[7]. Recently, orientational, but not positional, ordering of topological defects has been observed in experiments on microtubule bundles driven by molecular motors[49], but the mechanism for this is not yet understood.

Friction is present in many active systems and can be tuned, providing mechanistic insights into pattern formation in active materials. For example, in the experiments reported in ref. 6, the microtubule bundles and kinesin molecular motors move on an oil–water interface. The active layer experiences friction due to the momentum transfer to the surrounding fluid and thus the friction can be varied by changing the viscosity difference between the upper and lower fluid layers[50]. In addition, experimental measurements of the flow fields of endothelial cells[19] could be extended by exploring the flow patterns on substrates with different coatings to alter the frictional properties. Indeed recent studies have demonstrated the important role of friction in altering the self-propulsion mechanism of a single cell in confined migration[51] and have shown that the cortical friction can stabilize actin patterns in epithelial tubes[52]. The latter study shows that in the absence of orientational order, the combined effects of friction, actin treadmilling and myosin contractility control the formation of the actin ring on the cell cortex[52].

The ability to control flow vorticity and defect structure will be important in the design and operation of biological and biomimetic materials. Scientists are just starting to try to create active machines that mimic nature on the microscopic scale[10] and active matter is being considered by those traditionally interested in device physics and novel materials[53]. The formation of the vortex lattice leads to the positional ordering of topological defects that could provide an early step towards the design and control of active materials producing well defined velocity fields. For example, one might envisage that such a vortex lattice has the potential to drive an array of microscopic gears.

Methods

Non-dimensionalization. To characterize the relative importance of different stress contributions, we turn to the non-dimensional form of the momentum equation (equation (4)) introducing the characterisitic length and characterisitc velocity as L, U, respectively, we have $\mathbf{x}^* = \mathbf{x}L$, $\mathbf{u}^* = \mathbf{u}U$ and $p^* = p/(\rho U^2)$, where * denotes dimensionless variables. Equation (4) is then written in

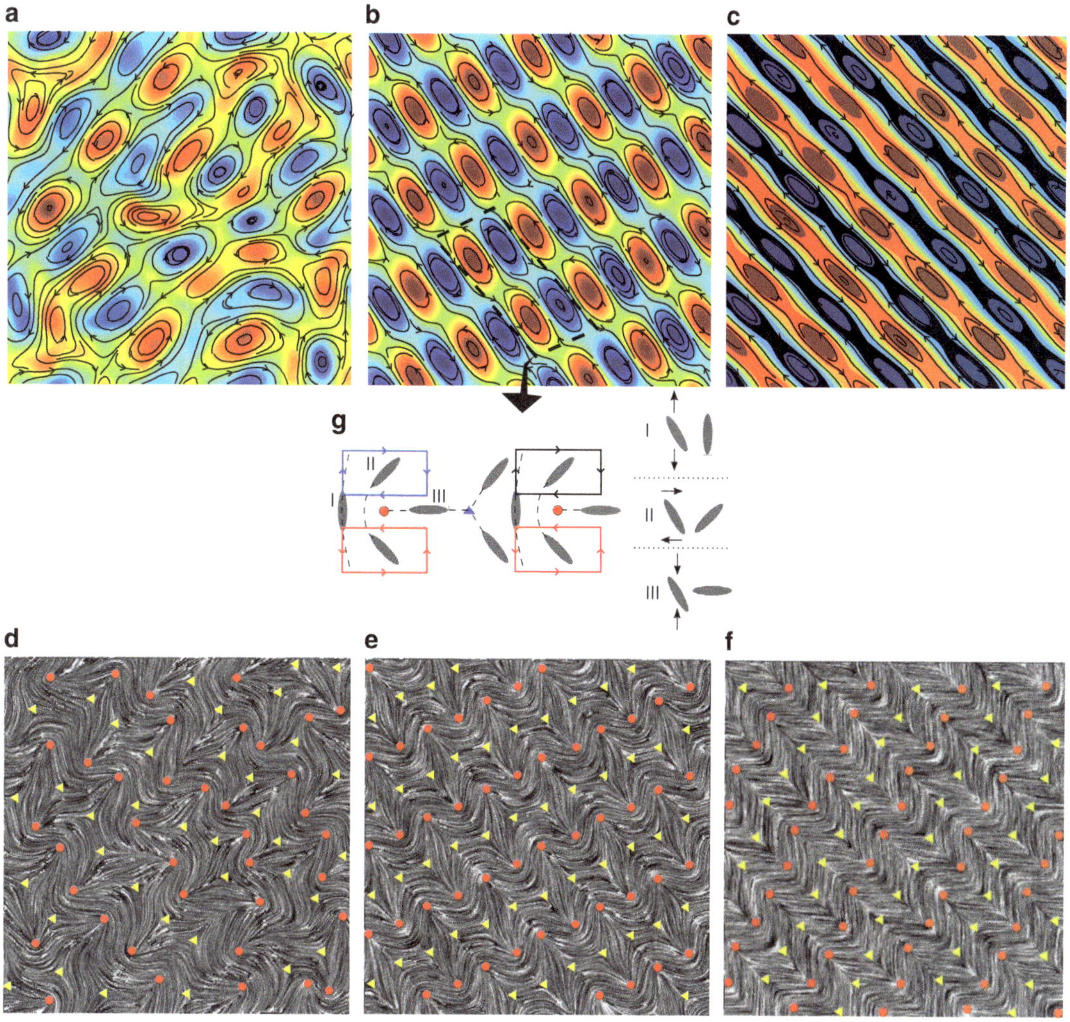

Figure 3 | Emergence of a vortex lattice and defect ordering for extensile rod-like particles. (**a–c**) Velocity field coloured by the magnitude of the vorticity. (**d–f**) Director fields visualized by Line Integral Convolution and superimposed by topological defects (with $+1/2$ and $-1/2$ defects denoted by red circles and yellow triangles). The hydrodynamic screening length is $L_{sc} = 15.30$, 7.51, 5.10, (lattice units) for the left, middle, and right columns, respectively. (**g**) Orientation of a defect is determined by its position relative to neighbouring vortices. A director responds differently to either extensional (I) or compressional (III) flow at the edges of vortices while it experiences a shear (II) inside a vortex.

non-dimensional form as

$$\partial_t^* \mathbf{u}^* + \mathbf{u}^*.(\nabla^* \mathbf{u}^*) = -\nabla^* p^* + \nabla^*.\left(\frac{1}{Re}\mathbf{\Pi}^{\text{viscous}^*} + \frac{1}{ReEr}\mathbf{\Pi}^{\text{elastic}^*} + \frac{1}{Re}\frac{\zeta L}{\eta U}\mathbf{\Pi}^{\text{active}^*} + \frac{1}{Re}\frac{\rho\gamma L^2}{\eta}\mathbf{\Pi}^{\text{friction}^*}\right). \tag{11}$$

It shuold be noted that $1/Re$ where $Re = \rho UL/\eta$, is the Reynolds number, is the common prefactor for all the stress terms. For the parameters that we use in our simulations, both active stress ($\zeta L/\eta U$) and frictional stress ($\rho\gamma L^2/\eta$) are of the same order for large enough friction ($> O(0.01)$), and much larger than unity. The viscous stresses are on the order of $1/Re$ and the elastic terms are on the order of $1/(ReEr)$, where $Er = \eta UL/K$ is the Ericksen number. The Ericksen number characterizes the ratio of viscous to elastic forces and is on the order of $O(10^2)$ in our simulations. Thus both viscous and elastic stresses are not important for large values of friction coefficient.

The high friction limit. When the energy generated by the active particles is primarily dissipated as friction we may write $\gamma\mathbf{u} \approx -\zeta\nabla\cdot\mathbf{Q}$, which gives an expression for the velocity field in terms of distortions in the nematic director field, $\mathbf{u} \approx -(\zeta/\gamma)\nabla\cdot\mathbf{Q}$. Considering this as an equality and substituting into the equation (1) for the concentration evolution gives

$$\partial_t\phi = \Gamma_\phi a\nabla^2\phi + \frac{\zeta\phi}{\gamma}\partial_i\partial_j Q_{ij} + \frac{\zeta}{\gamma}(\partial_j Q_{ij})(\partial_i\phi) + \Gamma_\phi\nabla^2\left[\frac{A}{2}\mathbf{Q}^2 - K_\phi\nabla^2\phi + K_c\partial_i\partial_j Q_{ij}\right]. \tag{12}$$

In this equation, the terms in Γ_ϕ represent diffusive dynamics arising from free energy gradients. The second term on the right-hand side is the current of active particles due to self-generated flow. This is usually introduced as a phenomenological, curvature-driven current in models of dry active nematics[3,21,23]. The third term is a non-linear term arising again from advection of concentration but rarely used in models of dry active systems.

Similarly, substituting for the velocity field in the equation (2) for the nematic order parameter we obtain

$$\begin{aligned}\partial_t Q_{ij} = &\Gamma_Q\left[(-A\phi - CQ_{pq}Q_{qp})Q_{ij} + K_Q\partial_\gamma^2 Q_{ij} + K_c\left(\frac{1}{2}\delta_{ij}\partial_k^2 - \partial_i\partial_j\right)\phi\right] \\ &+ \frac{\zeta}{\gamma}[\partial_m Q_{km}\partial_k Q_{ij}] - \frac{\zeta}{2\gamma}[\lambda\partial_m(\partial_i Q_{km} + \partial_k Q_{im}) + \partial_m(\partial_k Q_{im} - \partial_i Q_{km})]\left[Q_{kj} + \frac{\delta_{kj}}{3}\right] \\ &- \frac{\zeta}{2\gamma}\left[Q_{ik} + \frac{\delta_{ik}}{3}\right][\lambda\partial_m(\partial_k Q_{jm} + \partial_j Q_{km}) - \partial_m(\partial_j Q_{km} - \partial_k Q_{jm})] \\ &+ \frac{2\lambda\zeta}{\gamma}\left[Q_{ij} + \frac{\delta_{ij}}{3}\right]Q_{kl}\partial_k\partial_m Q_{ml}.\end{aligned} \tag{13}$$

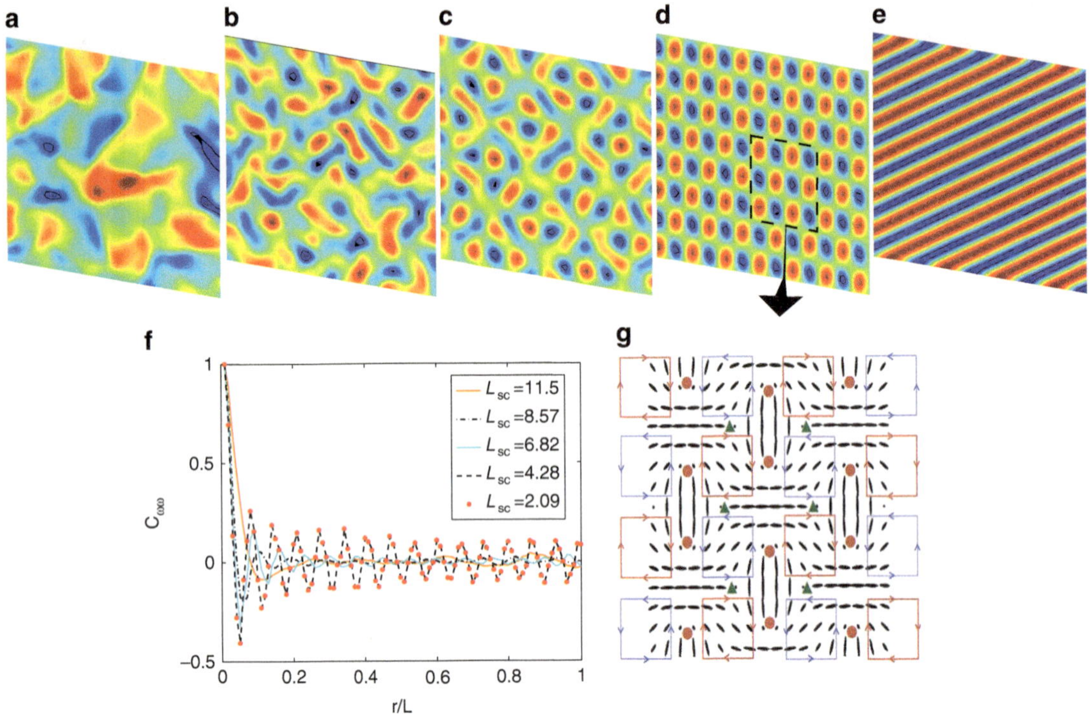

Figure 4 | Emergence of a vortex lattice and defect ordering for contractile disk-like particles. (**a-e**) Velocity field coloured by the magnitude of the vorticity. The hydrodynamic screening length is $L_{sc} = 11.5$, 8.57, 6.82, 4.28, 2.09 (lattice units) for (**a-e**) respectively. (**f**) Vorticity-vorticity correlations function $C_{\omega\omega}$ demonstrates the transition from active turbulence state to the vortex lattice configuration. (**g**) The stable structure of nematic directors and topological defects in the vortex lattice. Solid red and blue lines illustrate the clockwise and counterclockwise vortices. Topological defects with charges $+1/2$, $-1/2$, are shown by red circles and green triangles, respectively.

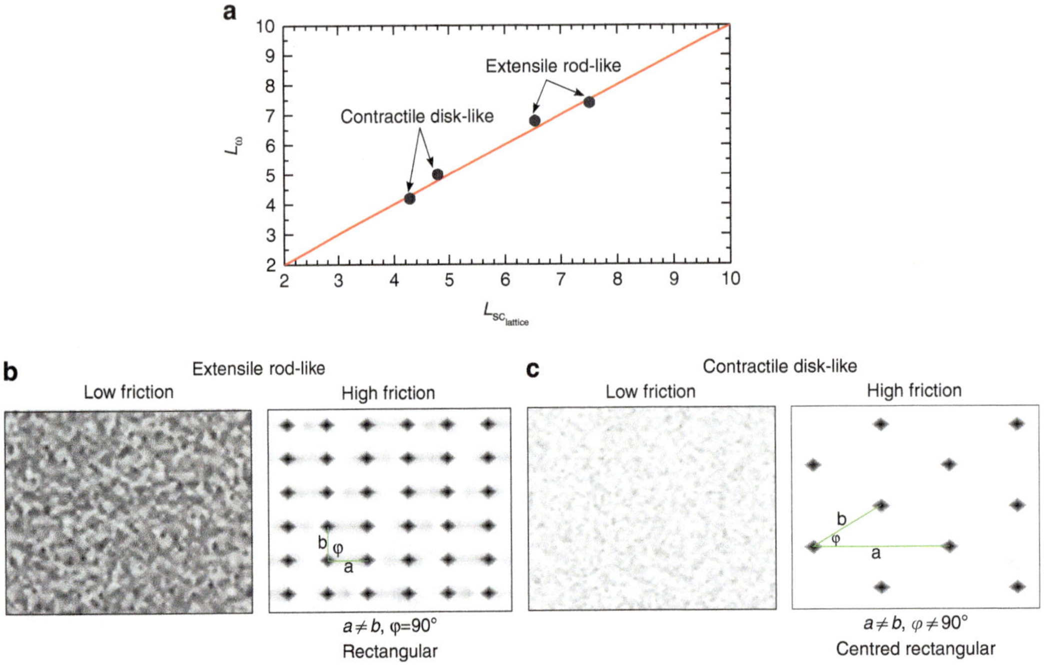

Figure 5 | Properties of the vortex lattice. (**a**) Dots indicate the characteristic vorticity length scale and the screening length at which the vortex lattice emerges for extensile and contractile systems, each for two different values of the activity. The red line indicates $L_\omega = L_{sc}$. (**b**,**c**) Colormaps of the structure factor for defects showing the emergence of positional ordering of defects at high friction for (**b**) extensile rod-like and (**c**) contractile disk-like particles.

Here, the terms written on the first line are generally used to describe the dynamics of dry active nematics. Terms on the other lines, that are proportional to ζ/γ, are generated by the presence of active flow and flow gradients, just like the curvature-driven term that appeared in equation (6) in the main text. These contributions are not usually considered in dry active nematics [3,21], because they represent non-linear contributions which are assumed to be sub-dominant in defining the dynamics. We now follow the approach of Bertin *et al.* [21] to show explicitly that such flow coupling terms can arise in general descriptions of dry systems.

Microscopic derivation of the flow-driven terms. We briefly review the microscopic approach used by Bertin *et al.* [21] to derive the equations of motion of dry active nematics. (For consistency, we use the relative concentration ϕ rather than the number density used in ref. 21.) Bertin *et al.* consider N particles which are characterized by position $\mathbf{x}$ and orientation θ. The alignment interactions are similar to those in a Vicsek model[20]; however there is no net motion along the orientation vector of each particle, thus imposing nematic symmetry. Considering the motion of the particles, taking into account their angular diffusion and inter-particle collisions, Bertin *et al.* showed that the evolution of the Fourier transform components of the single particle probability distribution $\hat{f}_k(\mathbf{x}, \theta, t)$ follow the master equations

$$\begin{aligned}\partial_t\hat{f}_k(\mathbf{x},t) =& \frac{1}{2}\tilde{\nabla}\tilde{\nabla}^*\hat{f}_k(\mathbf{x},t)+\frac{1}{4}\left(\tilde{\nabla}^{*2}\hat{f}_{k+1}+\tilde{\nabla}^2\hat{f}_{k-1}\right)+\left[\hat{P}_k-1\right]\hat{f}_k(\mathbf{x},t)\\ &+\frac{1}{\pi}\sum_q J\left(\hat{P}_k,k,q\right)\hat{f}_q(\mathbf{x},t)\hat{f}_{k-q}(\mathbf{x},t),\end{aligned} \tag{14}$$

where $\hat{P}_k$ is the Fourier transform of the noise distribution for angular displacements and $J\left(\hat{P}_k, k, q\right) = 4\hat{P}_k\frac{1+w\sqrt{2}(2q-k)(-q)^q\sin(k\pi/2)}{1-4(2q-k)^2} - \frac{4}{1-16q^2}$. Also, $\tilde{\nabla} = \partial_x + i\partial_y$ and $\tilde{\nabla}^*$ is the complex conjugate of $\tilde{\nabla}$.

To close this infinite hierarchy of equations, a small parameter ϵ is introduced that characterizes the distance to the isotropic–nematic transition. Assuming that $\hat{f}_k \sim |\epsilon|^k, \partial_t \sim \epsilon^2, \partial_x \sim \epsilon$, the first two modes ($k=0$, 1) are given by

$$\partial_t\hat{f}_0(\mathbf{x},t) = \frac{1}{2}\tilde{\nabla}\tilde{\nabla}^*\hat{f}_0(\mathbf{x},t)+\frac{1}{4}\left[\tilde{\nabla}^{*2}\hat{f}_1+\tilde{\nabla}^2\hat{f}_1^*\right], \tag{15}$$

$$\begin{aligned}\partial_t\hat{f}_1(\mathbf{x},t) =& \frac{1}{2}\tilde{\nabla}\tilde{\nabla}^*\hat{f}_1(\mathbf{x},t)+\frac{1}{4}\tilde{\nabla}\tilde{\nabla}\hat{f}_0\\ &+A'\hat{f}_1+C'\hat{f}_1^*\hat{f}_1^2+D'\tilde{\nabla}^*\tilde{\nabla}^*\hat{f}_1^2+E'\hat{f}_1\tilde{\nabla}\tilde{\nabla}\hat{f}_1,\end{aligned} \tag{16}$$

where the equations are truncated $O(\varepsilon^4)$ and the $k=2$ mode is used to close the equations. Coefficients A', C', D' and E' represent material parameters, that may also depend on ρ. Identifying the zeroth mode as the concentration ($\hat{f}_0 = \phi$) and first mode as the concentration weighted nematic field ($\hat{f}_1 = 2\left(\phi Q_{xx} + i\phi Q_{xy}\right)$), Bertin *et al.*[21] obtained the following equations $O(\epsilon^3)$:

$$\partial_t\phi = \frac{1}{2}\nabla^2\phi + \partial_i\partial_j\left(\phi Q_{ij}\right), \tag{17}$$

$$\begin{aligned}\partial_t\left(\phi Q_{ij}\right) =& A'\left(\phi Q_{ij}\right)+\frac{C'}{2}\left[\left(\phi Q_{pq}\right)\left(\phi Q_{qp}\right)\right]\left(\phi Q_{ij}\right)\\ &+\frac{1}{2}\nabla^2\left(\phi Q_{ij}\right)-\frac{1}{4}\left(\frac{\delta_{ij}}{2}\partial_k^2-\partial_i\partial_j\right)\phi.\end{aligned} \tag{18}$$

These are the familiar equations describing dry active nematics, see eg equations (6,7) in the main text or the first lines of equations (12,13) (A minor difference may be attributed to the approaches adopted: as is usual in kinetic descriptions the equations describe the evolution of $\phi\mathbf{Q}$ rather than $\mathbf{Q}$, while continuum approaches are generally not written in terms of concentration weighted fields.)

The two additional terms that appear $O(\epsilon^4)$ in the analysis are $D'\tilde{\nabla}^*\tilde{\nabla}^*\hat{f}_1^2 + E'\hat{f}_1\tilde{\nabla}\tilde{\nabla}\hat{f}_1$, the last two terms of equation (16). These terms, that are not used in Bertin *et al.*[21] have the role of advecting and co-rotating $\mathbf{Q}$ with the flow field. They give rise to all the non-linear terms in the last three lines of equation (13) except those proportional to the alignment parameter λ. Thus we establish that the flow-driven terms in equation (13) can be present in the kinetic description of dry active systems, but appear only at higher orders.

It is interesting to note that kinetic theory treats the case $\lambda = 0$, that is, the role of the symmetric part of the velocity gradient tensor is not taken into consideration. Further physics will need to be incorporated in the kinetic approach to capture the flow aligning behaviour of active nematics (terms in the equations of motion that depend on λ). Similarly modifying the interactions between the particles may also yield additional coupling and the higher order diffusive terms that appear on the second line of equation (12).

Numerics. The governing equations of active nematohydrodynamics (equations (1–4)) describe a coupled system of partial differential equations. We employ a hybrid lattice Boltzmann technique to solve them. In this method, the equations describing the evolution of density and momentum (equations (3,4)) are solved in tandem using a discretized version of the Boltzmann equation where the first and second moments of the particle distribution function give the density and momentum respectively. The Bhatnagar–Gross–Krook approximation with a single relaxation time is used in the collision operator. Equations (1,2) which respectively describe the evolution of the concentration and orientational order parameter are solved using the method of lines. All spatial differentials are discretized using the second order central difference scheme and time integration is performed using an Euler method. The time step for the method of lines is chosen as 1/10th of that for the lattice Boltzmann updates. The flow field used to evolve the order parameters is updated after every lattice Boltzmann time step while the active and passive stress are determined using the updated order parameter fields in every lattice Boltzmann step, hence ensuring the coupling between the equations in the algorithm. As usual in the lattice Boltzmann technique, discrete space and time steps are chosen as unity. Details can be found in refs 24,31. Simulations are performed on a 100×100 lattice. Unless otherwise stated, the parameters used are $\Gamma_\phi = 1.0$, $\Gamma_Q = 0.1$, $a = 0$, $A = -1.0$, $C = 0.6$, $K_\phi = 0.1$, $K_Q = 0.01$, $K_c = 0.0$, and $\eta = 0.05$, in lattice units. We use $\zeta = 0.008$, $\lambda = 0.7$ for extensile, rod-like particles and $\zeta = -0.02$, $\lambda = -0.3$ for contractile, disk-like active systems.

It is noteworthy that material parameters in active systems are not known experimentally, and the research in this direction is in progress. In our simulations, we have used the usual lattice Boltzmann units[31], and therefore an experimentalist may suitably dimensionalize the system of interest by choosing a length, time and mass scale[24,54].

References

1. Ramaswamy, S. The mechanics and statistics of active matter. *Annu. Rev. Cond. Mat. Phys.* **1,** 323–345 (2010).
2. Koch, D. L. & Subramanian, G. Collective hydrodynamics of swimming microorganisms: living fluids. *Annu. Rev. Fluid Mech.* **43,** 637–659 (2011).
3. Marchetti, M. C. *et al.* Hydrodynamics of soft active matter. *Rev. Mod. Phys.* **85,** 1143–1189 (2013).
4. Dombrowski, C., Cisneros, L., Chatkaew, S., Goldstein, R. E. & Kessler, J. O. Self-concentration and large-scale coherence in bacterial dynamics. *Phys. Rev. Lett.* **93,** 098103 (2004).
5. Voituriez, R., Joanny, J.-F. & Prost, J. Generic phase diagram of active polar films. *Phys. Rev. Lett.* **96,** 028102 (2006).
6. Sanchez, T., Chen, D. T. N., DeCamp, S. J., Heymann, M. & Dogic, Z. Spontaneous motion in hierarchically assembled active matter. *Nature* **491,** 431–434 (2012).
7. Sumino, Y. *et al.* Large-scale vortex lattice emerging from collectively moving microtubules. *Nature* **483,** 448–452 (2012).
8. Narayan, V., Ramaswamy, S. & Menon, N. Long-lived giant number fluctuations in a swarming granular nematic. *Science* **317,** 105–108 (2007).
9. Giomi, L., Bowick, M. J., Ma, X. & Marchetti, M. C. Defect annihilation and proliferation in active nematics. *Phys. Rev. Lett.* **110,** 228101 (2013).
10. Keber, F. C. *et al.* Topology and dynamics of active nematic vesicles. *Science* **345,** 1135–1139 (2014).
11. Wensink, H. H. *et al.* Meso-scale turbulence in living fluids. *Proc. Natl Acad. Sci. USA* **109,** 14308–14313 (2012).
12. Dunkel, J. *et al.* Fluid dynamics of bacterial turbulence. *Phys. Rev. Lett.* **110,** 228102 (2013).
13. Giomi, L. Geometry and topology of turbulence in active nematics. *Phys. Rev. X* **5,** 031003 (2015).
14. Vedula, S. K. *et al.* Emerging modes of collective cell migration induced by geometrical constraints. *Proc. Natl Acad. Sci. USA* **109,** 12974–12979 (2012).
15. Vicsek, T. & Zafeiris, A. Collective motion. *Phys. Rep.* **517,** 71–140 (2012).
16. Deseigne, J., Dauchot, O. & Chaté, H. Collective motion of vibrated polar disks. *Phys. Rev. Lett.* **105,** 098001 (2010).
17. Kumar, N., Soni, H., Ramaswamy, S. & Sood, A. K. Flocking at a distance in active granular matter. *Nat. Commun.* **5,** 4688 (2014).
18. Gruler, H., Dewald, U. & Eberhardt, M. Nematic liquid crystals formed by living amoeboid cells. *Eur. Phys. J.* **11,** 187–192 (1999).
19. Rossen, N. S., Tarp, J. M., Mathiesen, J., Jensen, M. H. & Oddershede, L. B. Long-range ordered vorticity patterns in living tissue induced by cell division. *Nat. Commun.* **5,** 5720 (2014).
20. Vicsek, T., Czirók, A., Ben-Jacob, E., Cohen, I. & Shochet, O. Novel type of phase transition in a system of self-driven particles. *Phys. Rev. Lett.* **75,** 1226 (1995).
21. Bertin, E. *et al.* Mesoscopic theory for fluctuating active nematics. *New J. Phys.* **15,** 085032 (2013).
22. Ngo, S. *et al.* Large-scale chaos and fluctuations in active nematics. *Phys. Rev. Lett.* **113,** 038302 (2014).
23. Shi, X., Chaté, H. & Ma, Y. Instabilities and chaos in a kinetic equation for active nematics. *New J. Phys.* **16,** 035003 (2014).
24. Thampi, S. P., Golestanian, R. & Yeomans, J. M. Velocity correlations in an active nematic. *Phys. Rev. Lett.* **111,** 118101 (2013).
25. Simha, R. A. & Ramaswamy, S. Hydrodynamic fluctuations and instabilities in ordered suspensions of self-propelled particles. *Phys. Rev. Lett.* **89,** 058101 (2002).
26. Volfson, D., Cookson, S., Hasty, J. & Tsimring, L. S. Biomechanical ordering of dense cell populations. *Proc. Natl Acad. Sci. USA* **105,** 15346–15351 (2008).
27. Bittig, T., Wartlick, O., Kicheva, A., González-Gaitán, M.s. & Jülicher, F. Dynamics of anisotropic tissue growth. *New J. Phys.* **10,** 063001 (2008).

28. Bittig, T., Wartlick, O., González-Gaitán, M. & Jülicher, F. Quantification of growth asymmetries in developing epithelia. *Eur. Phys. J* **30,** 93–99 (2009).
29. Ranft, J. *et al.* Fluidization of tissues by cell division and apoptosis. *Proc. Natl Acad. Sci. USA* **107,** 20863–20868 (2010).
30. Doostmohammadi, A. *et al.* Celebrating Soft Matter's 10th Anniversary: cell division: a source of active stress in cellular monolayers. *Soft Matter* **11,** 7328 (2015).
31. Marenduzzo, D., Orlandini, E., Cates, M. E. & Yeomans, J. M. Steady-state hydrodynamic instabilities of active liquid crystals: hybrid lattice Boltzmann simulations. *Phys. Rev. E* **76,** 031921 (2007).
32. Larson., R. G. *The Structure and Rheology of Complex Fluids,* volume 33 (Oxford University Press, 1999).
33. de Gennes, P. G. & Prost, J. *The Physics of Liquid Crystals* (Oxford University Press, 1995).
34. Beris, A. N. & Edwards, B. J. *Thermodynamics of Flowing Systems* (Oxford University Press, 1994).
35. Denniston, C., Marenduzzo, D., Orlandini, E. & Yeomans, J. M. Lattice Boltzmann algorithm for three dimensional liquid crystal hydrodynamics. *Phil. Trans. R. Soc. Lond. A* **362,** 1745–1754 (2004).
36. Fielding, S. M., Marenduzzo, D. & Cates, M. E. Nonlinear dynamics and rheology of active fluids: Simulations in two dimensions. *Phys. Rev. E* **83,** 041910 (2011).
37. Thampi, S. P., Golestanian, R. & Yeomans, J. M. Vorticity, defects and correlations in active turbulence. *Phil. Trans. R. Soc. A* **372,** 20130366 (2014).
38. Sulaiman, N., Marenduzzo, D. & Yeomans, J. M. Lattice boltzmann algorithm to simulate isotropic-nematic emulsions. *Phys. Rev. E* **74,** 041708 (2006).
39. Blow, M. L., Thampi, S. P. & Yeomans, J. M. Lyotropic active nematics. *Phys. Rev. Lett.* **113,** 248303 (2014).
40. Thampi, S. P., Golestanian, R. & Yeomans, J. M. Active nematic materials with substrate friction. *Phys. Rev. E* **90,** 062307 (2014).
41. Shi, X. & Ma, Y. Topological structure dynamics revealing collective evolution in active nematics. *Nat. Commun.* **4,** 3013 (2013).
42. Thampi, S. P., Doostmohammadi, A., Golestanian, R. & Yeomans, J. M. Intrinsic free energy in active nematics. *Europhys. Lett.* **112,** 28004 (2015).
43. Chaikin, P. M. & Lubensky., T. C. *Principles of Condensed Matter Physics* (Cambridge University Press, 2000).
44. Edwards, S. & Yeomans, J. M. Spontaneous flow states in active nematics: A unified picture. *Europhys. Lett.* **85,** 18008 (2009).
45. Ziebert, F. & Zimmermann, W. Nonlinear competition between asters and stripes in filament-motor systems. *Euro. Phys. J. E* **18,** 41–54 (2005).
46. Gowrishankar, K. & Rao, M. Nonequilibrium phase transitions in active contractile polar filaments. Preprint at http://arxiv.org/abs/1201.3938 (2012).
47. Menzel, A. M. & Löwen., H. Traveling and resting crystals in active systems. *Phys. Rev. Lett.* **110,** 055702 (2013).
48. Riedel, I. H., Kruse, K. & Howard, J. A self-organized vortex array of hydrodynamically entrained sperm cells. *Science* **309,** 300–303 (2005).
49. DeCamp, S. J., Redner, G. S., Baskaran, A., Hagan, M. & Dogic, Z. F. Orientational order of motile defects in active nematics. *Nat. Mater.* **14,** 1110–1115 (2015).
50. Guillamat, P., Ignés-Mullol, J. & Sagués, F. Patterning active materials with addressable soft interfaces. Preprint at http://arxiv.org/abs/1511.03880 (2015).
51. Bergert, M. *et al.* Paluch. Force transmission during adhesion-independent migration. *Nat. Cell Biol.* **17,** 524–529 (2015).
52. Hannezo, E., Dong, B., Recho, P., Joanny, J. & Hayashi, S. Cortical instability drives periodic supracellular actin pattern formation in epithelial tubes. *Proc. Natl Acad. Sci. USA* **112,** 8620–8625 (2015).
53. Zhou, S., Sokolov, A., Lavrentovich, O. D. & Aranson, I. S. Living liquid crystals. *Proc. Natl. Acad. Sci. USA* **111,** 1265–1270 (2014).
54. Cates, M. E., Fielding, S. M., Marenduzzo, D., Orlandini, E. & Yeomans, J. M. Shearing active gels close to the isotropic-nematic transition. *Phys. Rev. Lett.* **101,** 068102 (2008).

Acknowledgements

We acknowledge funding from the ERC Advanced Grant MiCE. We thank Tyler Shendruk and Ramin Golestanian for helpful conversations.

Author contributions

S.P.T., A.D. and J.M.Y. conceived the research and J.M.Y. and S.P.T. supervised the project. A.D. carried out the simulations. All authors contributed to performing analytical calculations, interpreting the results and writing the paper.

Additional information

Competing financial interests: The authors declare no competing financial interests.

11

Topological crystalline metal in orthorhombic perovskite iridates

Yige Chen[1], Yuan-Ming Lu[2,3] & Hae-Young Kee[1,4]

Since topological insulators were theoretically predicted and experimentally observed in semiconductors with strong spin-orbit coupling, increasing attention has been drawn to topological materials that host exotic surface states. These surface excitations are stable against perturbations since they are protected by global or spatial/lattice symmetries. Following the success in achieving various topological insulators, a tempting challenge now is to search for metallic materials with novel topological properties. Here we predict that orthorhombic perovskite iridates realize a new class of metals dubbed topological crystalline metals, which support zero-energy surface states protected by certain lattice symmetry. These surface states can be probed by photoemission and tunnelling experiments. Furthermore, we show that by applying magnetic fields, the topological crystalline metal can be driven into other topological metallic phases, with different topological properties and surface states.

[1] Department of Physics, University of Toronto, Toronto, Ontario M5S 1A7, Canada. [2] Department of Physics, University of California, Berkeley, California 94720, USA. [3] Materials Sciences Division, Lawrence Berkeley National Laboratory, Berkeley, California 94720, USA. [4] Canadian Institute for Advanced Research, Toronto, Ontario M5G 1Z8, Canada. Correspondence and requests for materials should be addressed to H.-Y.K. (email: hykee@physics.utoronto.ca).

Recent discovery of topological insulators (TIs) reveals a large class of new materials that, despite an insulating bulk, host robust metallic surface states[1–9]. Unlike conventional ordered phases characterized by their symmetries, these quantum phases are featured by non-trivial topology of their band structures, and remarkably they harbour conducting surface states protected by global symmetries such as time-reversal and charge conservation. More recently, it was realized that certain insulators can support surface states protected by crystal symmetry, and they are named topological crystalline insulators[10–12]. The rich topology of insulators in the presence of symmetries lead to a natural question: are there similar 'topological metals' hosting protected surface states? After the proposal of Weyl semimetal[13], a large class of topological metals are classified[14], which harbours surface flat bands protected by global symmetries such as charge conservation. However, an experimental confirmation of these phases is still lacking[15].

In this work, we propose that orthorhombic perovskite iridates $AIrO_3$, where A is an alkaline-earth metal, with strong spin–orbit coupling (SOC) and Pbnm structure, can realize a new class of metal dubbed topological crystalline metal (TCM). Topological properties of such a TCM phase include zero-energy surface states protected by the mirror-reflection symmetry, and gapless helical modes located at the core of lattice dislocations. Photoemission and tunnelling spectroscopy are natural experimental probes for these topological surface states. We further show how this TCM phase can be driven to metallic phases with different topology by applying magnetic fields, which breaks the mirror symmetry. All these results will be supported by topological classification in the framework of K-theory, as well as numerical calculations of topological invariants and surface/dislocation spectra, as presented in the Methods section.

Results

Crystal structure and lattice symmetry in $SrIrO_3$. Iridates have attracted much attention due to the strong SOC in 5d-Iridium (Ir) and a variety of crystal structures ranging from layered perovskites to pyrochlore lattices[16,17]. Despite structural differences, a common ingredient of these iridates with Ir^{4+} is $J_{\text{eff}}=1/2$ states governing low-energy physics, resulted from a combination of strong SOC and crystal field splitting. Among them, orthorhombic perovskite iridates $AIrO_3$ (where A is an alkaline-earth metal) belongs to Pbnm space group and can be tuned into a TI[18].

A unit cell of $AIrO_3$ contains four Ir atoms as shown in Fig. 1a, and there are three types of symmetry plane: b-glide, n-glide and mirror plane perpendicular to $\hat{\mathbf{c}}$ axis. Each of them can be assigned to the symmetry operators as follows: Π_b, Π_n and Π_m, as listed in the Methods section. Introducing three Pauli matrices corresponding to the in-plane sublattice ($\boldsymbol{\tau}$), layer ($\boldsymbol{\nu}$) and pseudo-spin $J_{\text{eff}}=1/2$ ($\boldsymbol{\sigma}$), the tight-binding Hamiltonian $H(\mathbf{k})$[18] has a relatively simple form shown in equation (6).

The band structure of tight-binding Hamiltonian $H(\mathbf{k})$ exhibits a ring-shaped one-dimensional (1D) Fermi surface (FS) close to the Fermi level as shown in Fig. 1b, which we call the nodal ring, as the energy dispersion is linear in two perpendicular directions. This nodal ring FS was confirmed by first-principle *ab initio* calculations[18,19], and it remains intact in the presence of Hubbard U up to 2.5 eV. Therefore, the semimetallic character of $SrIrO_3$ with U around 2 eV is consistent with the previous experimental results[20,21]. It was further shown that the size of nodal ring is determined by rotation and tilting angles of the oxygen octahedra around each Ir atom. In the case of $SrIrO_3$, which has a minimal distortion of octahedra, the nodal ring is centred around the $U \equiv (k_a=0, k_b=\frac{\pi}{a}, k_c=\frac{\pi}{c})$ point, and extends in two-dimensional (2D) U–R–S–X plane (perpendicular to $\hat{\mathbf{b}}$ axis) in three-dimensional (3D) Brillouin zone (BZ) as shown in Fig. 1b.

Zero-energy surface states and dislocation helical modes. Here we show that the nodal ring FS exhibits non-trivial topology that leads to localized surface zero modes protected by the mirror symmetry, thus coined TCM. To demonstrate the existence of zero-energy surface states, the band structure calculation for the $\hat{\mathbf{a}}\hat{\mathbf{c}}$-side plane was carried out with open boundary, that is, $[1\bar{1}0]$ surface perpendicular to $\hat{\mathbf{b}}$ axis. The energy dispersion at $k_c=\frac{\pi}{c}$, as displayed in Fig. 1c, reveals a dispersionless zero-energy flat band marked by red colour for all k_a on $[1\bar{1}0]$ surface of the sample. On the other hand, the slab spectrum at $k_a=0$ shows that the surface states are gapped except at $k_c=\frac{\pi}{c}$, as shown in Fig. 1d. Similar calculations for $\hat{\mathbf{b}}\hat{\mathbf{c}}$-side plane show that [110] surface (perpendicular to $\hat{\mathbf{a}}$ axis) also supports localized surface zero modes at $k_c=\frac{\pi}{c}$, while [010] surface (perpendicular to $\hat{\mathbf{y}}$ axis) does not harbour any zero-energy states. As elaborated in the Methods section, these surface states manifest a mirror-symmetry-protected weak index labelled by a vector[22] $\mathbf{M}=\hat{\mathbf{a}}+\hat{\mathbf{b}}//\hat{\mathbf{y}}$, where $\hat{\mathbf{a}}$ and $\hat{\mathbf{b}}$ are Bravais lattice primitive vectors. A direct consequence of this weak index is the existence of $k_c=\frac{\pi}{c}$ zero-energy states for any side surface, except for [010] surface perpendicular to vector $\mathbf{M}$.

Another consequence of weak index $\mathbf{M}$ is the existence of pairs of counter-propagating zero modes ('helical modes') localized in a dislocation line, which respects mirror symmetry Π_m. The number of zero-mode pairs N_0 in each dislocation line is determined by its Burgers vector $\mathbf{B}$ by $N_0=\mathbf{B}\cdot\mathbf{M}/2\pi$ (ref. 23). We have performed numerical calculations that demonstrate a pair of gapless helical modes in a dislocation line along $\hat{\mathbf{c}}$ axis with Burgers vector $\mathbf{B}=\pm\hat{\mathbf{a}}$. Detailed results are presented in the Methods section.

Classification and topological invariants. To understand the topological nature of the zero-energy surface states, let us first clarify the symmetry of tight-binding model $H(\mathbf{k})$. It turns out in the mirror-reflection-symmetric $k_c=\frac{\pi}{c}$ plane, $H(\mathbf{k})$ has an emergent chiral symmetry, that is, $\{H(\mathbf{k}), \mathcal{C}\}=0$, where $\mathcal{C}=\sigma_z\nu_x\tau_z$. Here $\boldsymbol{\sigma}$, $\boldsymbol{\nu}$ and $\boldsymbol{\tau}$ represent $J_{\text{eff}}=\frac{1}{2}$ pesudo-spin space, inter-plane and in-plane sublattice, respectively. A chiral symmetry can be understood as the combination of time-reversal symmetry and certain particle-hole symmetry[24], hence switching the sign of Hamiltonian. The presence of chiral symmetry $\mathcal{C}$ enforces the energy spectrum of $H(\mathbf{k})$ to be symmetrical with respect to the zero energy. It is known that chiral symmetry can protect zero-energy surface modes[25,26]. On the other hand, various crystal symmetries such as mirror reflection Π_m bring extra non-trivial topological properties into the system we studied. Starting from the $k_c=\frac{\pi}{c}$ plane with mirror reflection $\Pi_m=\sigma_z\nu_x$ and chiral symmetry $\mathcal{C}$, we can classify possible surface flat bands in the mathematical framework of K-theory[27], as summarized in Table 1 (see Methods section for details).

In particular with both mirror and chiral symmetries, the classification is $\mathbb{Z}\times\mathbb{Z}$ characterized by a pair of integer topological invariants (W^+, W^-). Since the Hilbert space can be decomposed into two subspaces with different mirror eigenvalues $\Pi_m=\pm 1$, in each subspace we can obtain a 1D winding number[14]

$$W^{\pm}_{[lmn]}(\mathbf{k}_\parallel)=\frac{1}{2\pi i}\int dk_\perp \mathrm{Tr}\left[Ch^{-1}_{\pm}(\mathbf{k})\frac{\partial}{\partial k_\perp}h_\pm(\mathbf{k})\right], \quad (1)$$

where $k_\perp$ is the crystal momentum along $[lmn]$ direction and $h_\pm(\mathbf{k})=H^{\pm}_{\mathbf{k}_\parallel}(k_\perp)$ is the 1D Hamiltonian parametrized by $[lmn]$ surface momentum $\mathbf{k}_\parallel$ in $\Pi_m=\pm 1$ subspace. For both $\hat{\mathbf{b}}\hat{\mathbf{c}}$ plane ([110] surface) and $\hat{\mathbf{a}}\hat{\mathbf{c}}$ plane ($[1\bar{1}0]$ surface), we have $(W^+,$

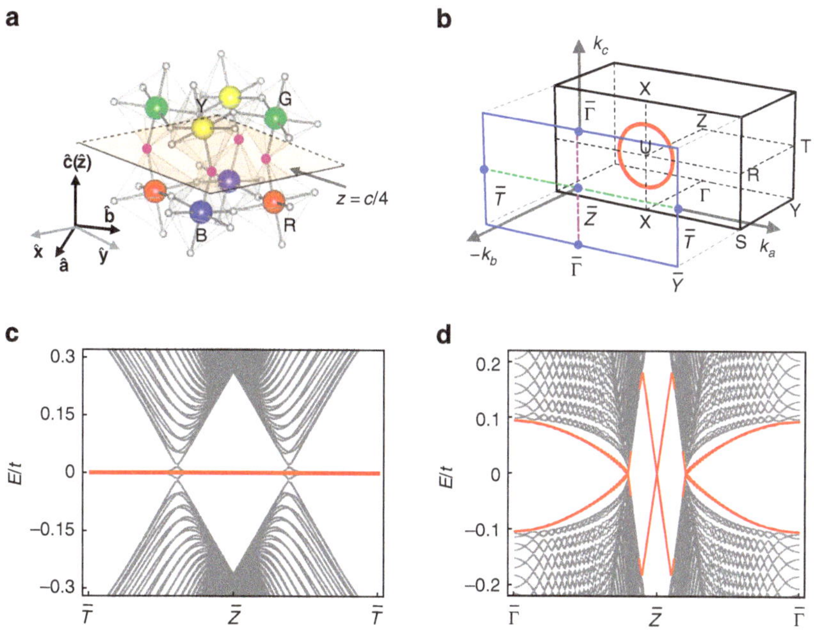

Figure 1 | Crystal structure and surface states on [1$\bar{1}$0] side plane. (**a**) Crystal structure of $AIrO_3$ with $\hat{\mathbf{a}}$ axis along [110], $\hat{\mathbf{b}}$ axis along [1$\bar{1}$0] and $\hat{\mathbf{c}}$ axis along $\hat{\mathbf{z}}$ direction, respectively. The unit cell contains four Ir atoms: blue (B), red (R), yellow (Y) and green (G) represent different oxygen octahedra environment. The length of Bravais lattice unit vectors $\hat{\mathbf{a}}/\hat{\mathbf{b}}$ and $\hat{\mathbf{c}}$ axis are a/a and c, respectively. A mirror symmetry plane (coloured with light orange), mapping z to $-z$, locates at $z=\frac{c}{4}$, where four oxygen atoms (purple solid circles) are within the same plane. (**b**) Special k-points in the 3D bulk BZ and 2D surface BZ for [1$\bar{1}$0] surface. The location of the nodal ring depict as red circles on U-R-S-X plane in 3D BZ. Slab-geometry surface spectrum (**c**) for [1$\bar{1}$0] surface at $k_c=\frac{\pi}{c}$, plotted along the high-symmetry line (green dashed line) $\bar{T}\rightarrow\bar{Z}\rightarrow\bar{T}$, and (**d**) for [1$\bar{1}$0] surface at $k_a=0$ by varying k_c to follow $\bar{\Gamma}\rightarrow\bar{Z}\rightarrow\bar{\Gamma}$ (purple dashed line), where red lines represent surface states.

Table 1 | Classification of topological crystalline metals and possible surface states on a generic side surface parallel to $\hat{\mathbf{z}}$-axis.

Mirror Π_m	Chiral $\mathcal{C}$	Classification	Topological invariants	Surface zero modes
Yes	Yes	$\mathbb{Z}\times\mathbb{Z}$	$(W^+, W^-)=(1,-1)$	Yes $(k_c=\frac{\pi}{c})$
No	Yes	$\mathbb{Z}$	$W\equiv W^+ + W^-$	Yes $(k_c=\frac{\pi}{c}+\delta,\ \lvert\delta\rvert\ll 1)$
Yes	No	0	—	No
No	No	0	—	No

Note that $[\Pi_m,\mathcal{C}]=0$. (W^+, W^-) is the pair of winding numbers obtained in $\Pi_m=\pm1$ subspaces of block-diagonalized Hamiltonian.

$W^-)=(1,-1)$. These quantized winding numbers correspond to a pair of zero modes for each surface momentum with $k_c=\frac{\pi}{c}$, and they cannot hybridize due to opposite mirror eigenvalues. Meanwhile, for [010] surface both $W\pm$ vanish, indicating a weak index[22] $\mathbf{M}=\hat{\mathbf{a}}+\hat{\mathbf{b}}//\hat{\mathbf{y}}$. Intuitively, the system can be considered as a stacked array of $\hat{\mathbf{x}}\hat{\mathbf{z}}$ planes, each plane (perpendicular to $\hat{\mathbf{y}}$ axis) with a pair of mirror-protected zero-energy edge modes at $k_c=\frac{\pi}{c}$.

Once the mirror symmetry Π_m is broken, the classification becomes $\mathbb{Z}$ characterized by one integer topological invariant: the total winding number $W\equiv W^+ + W^-$ associated with 1D TI in symmetry class AIII[24,27]. This total winding number vanishes for all surface momenta at $k_c=\frac{\pi}{c}$ though.

Topological metal/semimetal induced by magnetic fields. Time-reversal (TR) breaking perturbations such as a magnetic field can drive the system from the TCM phase to other metallic phases with different topological properties. In the presence of Zeeman coupling $\mu_B\mathbf{h}\cdot\boldsymbol{\sigma}$ introduced by magnetic field $\mathbf{h}$, clearly the chiral symmetry $\mathcal{C}$ is still preserved as long as the field is in $\hat{\mathbf{x}}\hat{\mathbf{y}}$ plane ($\mathbf{h}\perp\hat{\mathbf{z}}$). Magnetic field parallels to $\hat{\mathbf{z}}$ axis breaks chiral symmetry, which gaps the nodal ring, making the system trivial. Meanwhile, mirror symmetry Π_m will be broken unless the magnetic field is along $\hat{\mathbf{z}}$ direction. Thus, our focus below will be magnetic field in the $\hat{\mathbf{x}}\hat{\mathbf{y}}$ plane. The FS topology and associated surface states with a magnetic field along different directions are listed in Table 2.

Chiral topological metal protected by chiral symmetry. Due to TR and inversion symmetry, the nodal ring in TCM always has twofold Kramers degeneracy. After applying [1$\bar{1}$0] direction magnetic field $\mathbf{h}//\hat{\mathbf{b}}$, the doubly degenerate nodal ring in Fig. 1b splits into two rings in Fig. 2a shifted along $\hat{\mathbf{c}}(\hat{\mathbf{z}})$ axis on U-R-S-X plane. Although the mirror symmetry is broken by the [1$\bar{1}$0] magnetic field, chiral symmetry $\mathcal{C}$ is still preserved. Therefore, the topological properties of the two nodal rings are captured by an integer winding number W in the symmetry class AIII[14]. Consider [1$\bar{1}$0] surface for instance, depending on the surface momentum $\mathbf{k}_{||}$, the winding number $W_{\mathbf{k}_{||}}$ is plotted in Fig. 2b. It vanishes in the region where the two nodal rings (blue and red) overlap, but becomes ±1 in other regions within the two nodal rings.

The energy spectra on a slab geometry with open [1$\bar{1}$0] surfaces are displayed in Fig. 2c,d, plotted as a function of k_a with $k_c=\frac{\pi}{c}$ and $k_c=\frac{\pi}{c}-\delta$, respectively. There is no zero modes at $k_c=\frac{\pi}{c}$, corresponding to trivial winding number. Meanwhile, zero-energy flat bands highlighted by red colour in Fig. 2d exist inside the nodal rings. It confirms the non-trivial topology of the bulk nodal rings with quantized winding number $W_{[1\bar{1}0]} = \pm 1$ shown in Fig. 2b. It turns out this chiral topological metal supports localized flat band protected by chiral symmetry on any surface, as long as its normal vector $\hat{\mathbf{n}}$ is not perpendicular to $\hat{\mathbf{b}}$ axis. Meanwhile, the two nodal rings are stable against any perturbations preserving chiral symmetry, since the winding number changes when we cross each nodal ring[14].

Weyl semimetal. Once we apply a magnetic field along [110] direction (or $\hat{\mathbf{a}}$ axis), both mirror (Π_m) and n-glide (Π_n) symmetries are broken. Consequently, the nodal ring is replaced by a pair of 3D Dirac nodes, appearing at momenta $\left(\pm k_0, \frac{\pi}{a}, \frac{\pi}{c}\right)$ along the path R→U→R BZ line. However, these Dirac nodes are not symmetry protected, since a sublattice potential mv_z alternating by layers would further split each Dirac point into a pair of Weyl nodes. And this mv_z term has the same symmetry as a Zeeman field $h_{[110]}$ along $\hat{\mathbf{a}}$ axis.

In the presence of both magnetic field $h_{[110]}$ and layer-alternating potential mv_z, the system still preserves chiral symmetry C, b-glide Π_b and inversion symmetry. Two pairs of Weyl nodes emerge at $\pm\mathbf{k}_1 = \left(\pm k_1, \frac{\pi}{a}, \frac{\pi}{c}\right)$ and $\pm\mathbf{k}_2 = \left(\pm k_2, \frac{\pi}{a}, \frac{\pi}{c}\right)$ along R→U→ R line in Fig. 3a. The low-energy Hamiltonian around $\mathbf{k}_1$ has linear dispersion along all $\mathbf{k}$ directions

$$H_{\text{eff}}(\delta\mathbf{k}) = \mathbf{p}(\delta\mathbf{k}) \cdot \sigma, \tag{2}$$

$$\mathbf{p}(\delta\mathbf{k}) \equiv (p_x, p_y, p_z) = (A_1\delta k_b - A_2\delta k_c, B_1\delta k_c, D_1\delta k_a),$$

with $\delta\mathbf{k}\equiv\mathbf{k}\text{-}\mathbf{k}_1$. Various coefficients can be expressed in terms of the tight-binding hopping parameters (see Methods).

There exists a 'jump' for the Chern number C for all occupied bands from $C=0$ to $C=1$ when k_a crosses k_1, and similarly an opposite 'jump' from $C=1$ to $C=0$ after k_a passes k_2. The different signs of jumps in Chern number indicate that the Weyl fermions at $\pm\mathbf{k}_1$ and $\pm\mathbf{k}_2$ have opposite topological charge $+1$ (blue) and -1 (red), respectively, as shown in Fig. 3a.

The surface states on $[1\bar{1}0]$ surface at $k_a = \frac{0.7}{a}$, between the two Weyl nodes with opposite chirality, are plotted along k_c in Fig. 3b. There is a single dispersing zero mode coloured with red in Fig. 3b, which is localized on each surface of the sample. A series of one-way-dispersing zero modes for all surface momenta between $\mathbf{k}_1$ and $\mathbf{k}_2$ form a 'chiral Fermi arc'[13,28] on $[1\bar{1}0]$ surface, as shown by the green lines in Fig. 3a.

Table 2 | Various TR-breaking perturbations and consequences on the nodal ring and surface states (for example, on $[1\bar{1}0]$ side surface).

B field	b-glide Π_b	n-glide Π_n	Mirror Π_m	Chiral C	Fermi surfaces	Zero-energy surface states
$[1\bar{1}0]$	No	Yes	No	Yes	Nodal rings	Flat bands at $k_c = \frac{\pi}{c} \pm \delta$
[110]	Yes	No	No	Yes	Nodal points	Fermi arcs between Weyl points
[001]	No	No	Yes	No	Gapped	No

When the magnetic field is along [110] direction, the nodal ring splits into a pair of Dirac nodes along U→R. On the other hand, the nodal ring is completely gapped if the magnetic field has a nonzero $\hat{\mathbf{z}}(\hat{\mathbf{c}})$ direction component.

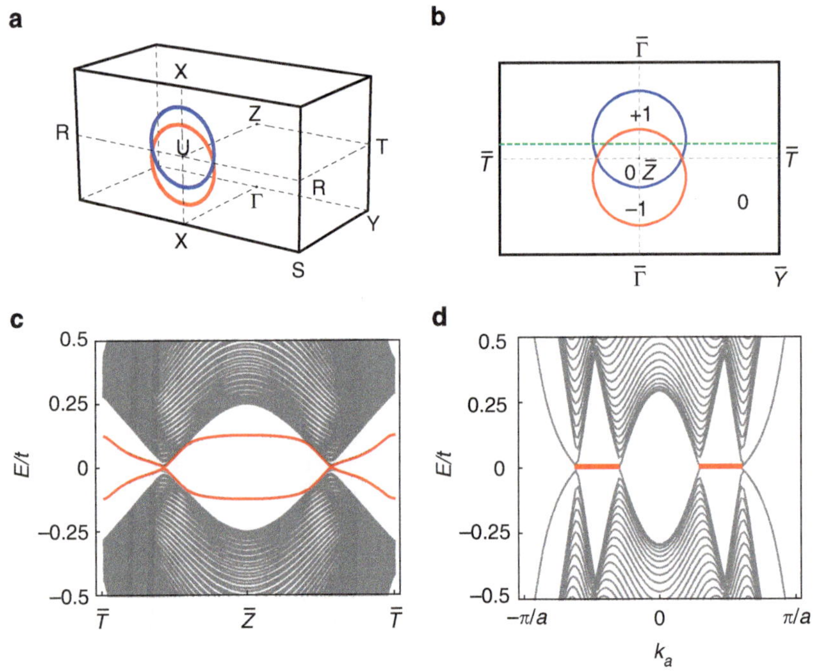

Figure 2 | Evolution of the nodal ring under various TR-breaking terms and corresponding surface modes. (**a**) When the magnetic field $\hat{\mathbf{h}} \parallel \hat{\mathbf{b}}$, a pair of nodal rings contain one ring (blue) shifted upwards along $\hat{\mathbf{c}}(\hat{\mathbf{z}})$ axis and the other one (red) shifted downwards. (**b**) The corresponding winding number $W_{[1\bar{1}0]}$ distribution on $\bar{Z}-\bar{T}-\bar{Y}-\bar{\Gamma}$ plane. Slab spectrum (**c**) when $k_c = \frac{\pi}{c}$ for $[1\bar{1}0]$ surface plotted as a function of k_a along the high-symmetry line $\bar{T} \rightarrow \bar{Z} \rightarrow \bar{T}$, and (**d**) when $k_c = \frac{\pi}{c} + \delta$ indicated by the green dashed line in (**b**).

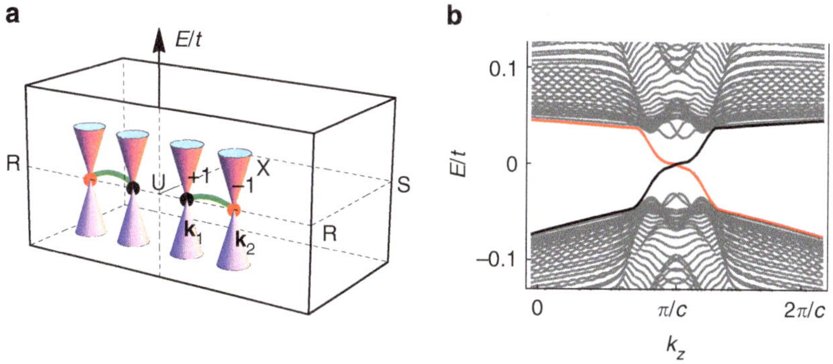

Figure 3 | Emergence of Weyl fermions on [$1\bar{1}0$] side surface. (**a**) Two pairs of Weyl node located along the high-symmetry line U→R at $\mathbf{k}_1$ and $\mathbf{k}_2$, respectively. One of the Weyl fermion (blue) has +1 chirality but the other Weyl node (red) has opposite chirality. The Fermi arc connecting those two Weyl nodes is coloured by green. (**b**)The edge states spectra when $k_a = \frac{0.7}{a}$, which is in between two Weyl nodes plotted as a function of k_c.

Discussion

The existence of surface zero modes in our TCM phase originates from the chiral and mirror-reflection symmetry of $AIrO_3$ with Pbnm structure. Any side surface other than [010] plane should exhibit robust zero-energy surface states independent of the details. In a generic band structure of $SrIrO_3$, this nodal ring does not sit exactly at the Fermi level E_F, but slightly below E_F (unless SOC is stronger than an atomic SOC used in the first-principle calculation), and a hole-like pocket FS occurs around Γ point in Fig. 1b[19]. In other words, the nodal ring occurs around $k_c = \frac{\pi}{c}$, while the small bulk Fermi pocket is located near $k_c = 0$. Therefore, the zero-energy surface modes are well separated from the bulk FS pockets in momentum, and a momentum-resolved probe is required to detect the surface states. Angle-resolved photoemission spectroscopy (ARPES) would be the best tool to observe the momentum-resolved surface states shown in Fig. 1c,d on a side plane of $AIrO_3$. Notice that, ARPES has successfully detected the topological surface states in Dirac semimetal material[29]. Due to the presence of extremely small orbital overlap amplitudes between further Ir sites, the surface states acquire a slight dispersion. However, as we emphasized above, the mirror symmetry is a crucial ingredient to support such non-trivial surface states detectable by ARPES, despite the 'weak breaking' of chiral symmetry.

These surface states also contribute finite surface density of states (SDOSs) near zero energy. In contrast, the semimetallic bulk band contribution to SDOS vanishes around zero energy due to the presence of bulk nodal ring. Therefore, protected surface modes can be detected as a zero bias hump (finite SDOSs) in the dI/dV curve of scanning tunnelling microscopy. However, in real materials, it will be difficult to separate the contribution of the surface states to SDOS from the bulk part. On the other hand, there exists protected propagating fermion modes in dislocation lines[23] that preserves mirror symmetry. One advantage is that these topological helical modes are protected by mirror and chiral symmetries, and hence will not be destroyed by hybridization with bulk gapless excitations. In particular, counter-propagating gapless fermions show up in pairs in the dislocation core, and the number of gapless fermion pairs is given by $\mathbf{M}\cdot\mathbf{B}/2\pi$, where $\mathbf{B}$ is the Burgers vector of dislocation as shown in the Methods section. Unlike other gapless and dispersive bulk excitations that are extended in space, these helical fermion zero modes are localized near the dislocation core, which in hence can be detected by scanning tunnelling microscopy.

Since a bulk sample of $AIrO_3$ such as $SrIrO_3$ requires high pressure to achieve the Pbnm crystal structure[30,31], it is desirable to grow a film of $AIrO_3$. Recently, superlattices of atomically thin slices of $SrIrO_3$ was made by pulsed laser deposition along [001] plane[32]. This work is a first step towards possible topological phases in iridates. Furthermore, a successful growth of film along [111] plane was also reported[33]. Thus, growing a film of AIrO3 along [110] (or [$1\bar{1}0$]) is plausible. To confirm the proposed TCM, an ARPES study should be performed on a film of AIrO3 grown along [110] (or [$1\bar{1}0$]), where the mirror symmetry of Pbnm structure is kept. This analysis should reveal a flat surface band near $k_c = \frac{\pi}{c}$ below E_F.

Methods

Symmetry operators and tight-binding Hamiltonian. The tight-binding Hamiltonian is defined in the basis of eight component spinor ψ, organized as[18]

$$\psi \equiv \left(c_{B\uparrow}, c_{R\uparrow}, c_{Y\uparrow}, c_{G\uparrow}, c_{B\downarrow}, c_{R\downarrow}, c_{Y\downarrow}, c_{G\downarrow}\right)^T, \tag{3}$$

where B, R, R, G correspond to four sublattices. We define Pauli matrices $\boldsymbol{\tau}$ and $\boldsymbol{\nu}$ in terms of following sublattice rotations

$$\begin{aligned} B \overset{\tau_x}{\longleftrightarrow} R, \quad Y \overset{\tau_x}{\longleftrightarrow} G; \\ B \overset{\nu_x}{\longleftrightarrow} Y, \quad R \overset{\nu_x}{\longleftrightarrow} G. \end{aligned} \tag{4}$$

The full space group Pbnm is generated by (see Fig. 2 in ref. 18) translations $T_{x,y,z}$ (three Bravais primitive vectors correspond to $T_a \equiv T_x T_y$, $T_b \equiv T_x^{-1} T_y$ and $T_c = T_z^2$) and the following three generators $\{\Pi_b, \Pi_n, \Pi_m\}$

$$\begin{aligned} \psi_{(k_a,k_b,k_c)} &\xrightarrow{\Pi_n} e^{i\frac{k_a+k_c}{2}} \frac{i(\sigma_x - \sigma_y)}{\sqrt{2}} \nu_x \tau_x \cdot e^{i\frac{k_c \nu_z + k_a \tau_z}{2}} \psi_{(k_a,-k_b,k_c)}, \\ \psi_{(k_a,k_b,k_c)} &\xrightarrow{\Pi_b} e^{i\frac{k_b}{2}} \frac{i(\sigma_x + \sigma_y)}{\sqrt{2}} \tau_x \cdot e^{i\frac{k_b \tau_z}{2}} \psi_{(-k_a,k_b,k_c)}, \\ \psi_{(k_a,k_b,k_c)} &\xrightarrow{\Pi_m} i\sigma_z \nu_x \psi_{(k_a,k_b,-k_c)}, \\ \psi_{\mathbf{k}} &\xrightarrow{I} \psi_{-\mathbf{k}}. \end{aligned} \tag{5}$$

where Π_b (glide plane $\perp\hat{\mathbf{a}} \equiv \hat{\mathbf{x}} + \hat{\mathbf{y}}$) and Π_n (glide plane $\perp\hat{\mathbf{b}} \equiv \hat{\mathbf{y}} - \hat{\mathbf{x}}$) represent two glide symmetries, while Π_m is the mirror reflection and $\boldsymbol{\sigma}$ denotes the pseudo-spin subspace. $\boldsymbol{I} = \Pi_b \Pi_n \Pi_m$ represents the inversion symmetry.

The tight-binding Hamiltonian of $SrIrO_3$ has the following form[18]

$$\begin{aligned} H_{\mathbf{k}} =& \mathrm{Re}\left(\epsilon_{\mathbf{k}}^{\mathrm{p}}\right)\tau_x + \mathrm{Im}\left(\epsilon_{\mathbf{k}}^{\mathrm{p}}\right)\sigma_z\tau_y + \epsilon_{\mathbf{k}}^{z}\nu_x \\ &+ \mathrm{Re}\left(\epsilon_{\mathbf{k}}^{\mathrm{d}}\right)\nu_x\tau_x + \mathrm{Im}\left(\epsilon_{\mathbf{k}}^{\mathrm{d}}\right)\nu_y\tau_y + \mathrm{Re}\left(\epsilon_{\mathbf{k}}^{\mathrm{po}}\right)\sigma_y\nu_z\tau_y \\ &+ \mathrm{Im}\left(\epsilon_{\mathbf{k}}^{\mathrm{po}}\right)\sigma_x\nu_y\tau_z + \mathrm{Re}\left(\epsilon_{\mathbf{k}}^{\mathrm{do}}\right)\sigma_y\nu_x\tau_y + \mathrm{Im}\left(\epsilon_{\mathbf{k}}^{\mathrm{do}}\right)\sigma_x\nu_x\tau_y \\ &+ \mathrm{Re}\left(\epsilon_{\mathbf{k}}^{\mathrm{d1}}\right)\sigma_y\nu_y\tau_x + \mathrm{Im}\left(\epsilon_{\mathbf{k}}^{\mathrm{d1}}\right)\sigma_x\nu_y\tau_x . \end{aligned} \tag{6}$$

Here various coefficient functions have been defined in ref. 18 with additional term $\epsilon_{\mathbf{k}}^{\mathrm{d1}}$

$$\epsilon_{\mathbf{k}}^{\mathrm{d1}} = t_{\mathrm{d1}}\left(\cos(k_y) + i\cos(k_x)\right)\cos(k_z), \tag{7}$$

where t_{1D} is the inter-layer next nearest-neighbour hopping due to non-vanishing rotation and tilting in local oxygen octahedra. The crystal momentum (k_a, k_b, k_c) relates to (k_x, k_y, k_z) simply by

$$k_a = \frac{k_x + k_y}{a}, \quad k_b = \frac{k_y - k_x}{a}, \quad k_c = \frac{2k_z}{c}. \tag{8}$$

The basis $\psi_{(k_a,k_b,k_c)}$ we use here is different with the basis $\psi_{(k_x,k_y,k_z)}$ in ref. 18 by a unitary rotation $U_{\mathbf{k}}$

$$\psi_{(k_a,k_b,k_c)} \equiv U_{\mathbf{k}}^{\dagger}\psi_{(k_x,k_y,k_z)}, \quad U_{\mathbf{k}} = e^{i\frac{k_z}{2}\nu_z} e^{i\frac{k_x}{2}\tau_z}. \tag{9}$$

In this basis $\psi_{(k_a,k_b,k_c)}$, the tight-binding Hamiltonian is related to $H_\mathbf{k}$ by the unitary transformation $U_\mathbf{k}$ in equation 6:

$$H(\mathbf{k}) \equiv H(k_a,k_b,k_c) = U_\mathbf{k}^\dagger H_\mathbf{k} U_\mathbf{k}. \tag{10}$$

Clearly in $k_c = \frac{\pi}{c}$(or $k_z = \frac{\pi}{2}$) plane the non-vanishing terms in $H_\mathbf{k}$, as in the basis $\psi_{(k_x,k_y,k_z)}$, contain only Pauli matrices $(\tau_x, \sigma_z\tau_y)$, $v_y\tau_y$ and $\sigma_{x,y}v_{x,z}\tau_y$. All these Pauli matrices anti-commute with

$$\mathcal{C} = \sigma_z v_y \tau_z, \tag{11}$$

that is, they all have chiral symmetry $\mathcal{C}$. However, in $\psi_{(k_a,k_b,k_c)}$ representation, the chiral symmetry $\mathcal{C}$ written as

$$\mathcal{C} = \sigma_z v_x \tau_z. \tag{12}$$

K-theory classification procedure and topological invariants. To understand surface states on $[1\bar{1}0]$ surface for instance, let us focus on a 1D system $H_\mathbf{k}$ in momentum space parametrized by fixed momentum $k_a \in \left[-\frac{\pi}{a}, \frac{\pi}{a}\right)$ and $k_c = \frac{\pi}{c}$. Such a 1D system will have mirror reflection $\Pi_m = i\sigma_z v_x$ in (5), as well as chiral symmetry $\mathcal{C} = \sigma_z v_x \tau_z$ in (12). We will classify such a gapped 1D system (since the bulk gap only closes at two points in $k_c = \frac{\pi}{c}$ plane) to see whether it has non-trivial topology, which may protect gapless surface states.

The classification of a gapped system can be understood from classifying possible symmetry-allowed mass matrices for a Dirac Hamiltonian[27,34]. The mathematical framework of K-theory applies to both global symmetry and certain spatial (crystal) symmetry[35,36]. In 1D such a Dirac Hamiltonian can be written as

$$H^{\rm 1d}_{\rm Dirac} = k_a\gamma_1 + m\gamma_0, \tag{13}$$

with chiral symmetry $\mathcal{C}$

$$\{\mathcal{C},\gamma_0\} = \{\mathcal{C},\gamma_1\} = 0. \tag{14}$$

mirror reflection Π_m

$$[\Pi_m,\gamma_0] = [\Pi_m,\gamma_1] = 0. \tag{15}$$

and $U(1)$ charge conservation Q

$$[Q,\gamma_0] = [Q,\gamma_1] = 0. \tag{16}$$

Any two symmetry generators among $(\mathcal{C}, \Pi_m, Q)$ commute with each other. Mathematically, the classification problem corresponds to the following question: given Dirac matrix γ_1 and symmetry matrices $(\mathcal{C}, \Pi_m, Q)$, what is the classifying space $\mathcal{S}$ of mass matrix γ_0? In particular, since each disconnected piece in classifying space $\mathcal{S}$ corresponds to one gapped 1D phase, how many disconnected pieces does $\mathcal{S}$ contain? Hence, the classification of gapped 1D phases with these symmetries is given by the zeroth homotopy $\pi_0(\mathcal{S})$.

In the K-theory classification, if there are generators that commute with all Dirac matrices and other symmetry generators, such as $U(1)$ charge symmetry generator Q here satisfying $Q^2 = (-1)^F$, then we say the gapped system belong to a complex class. Otherwise, it belongs to a real class. Clearly, our case belongs to the complex class because both Q and M commute with any other matrices.

For the $k_c = \pi/c$ plane on a generic $[xy0]$ side surface parallel to $\hat{c}$ axis, the full symmetry group is generated by $(\mathcal{C}, \Pi_m, Q)$ as mentioned earlier. Note that these three symmetry generators all commute with each other while $Q^2 = (-1)^F$. Together with Dirac matrix γ_1, they form a complex Clifford algebra $Cl_2 \times Cl_2$:

$$\{\gamma_1,\mathcal{C}\}\times\Pi_m\times Q, \tag{17}$$

where the generators inside the parenthesis anti-commute with each other and commute with everything outside the parenthesis. The reason we have $Cl_2 \times Cl_2$ is because we can block-diagonalize the $k_c = \pi$ tight-binding Hamiltonian with respect to their Π_m (mirror reflection) eigenvalue ± 1, and in each subspace the Clifford algebra is Cl_2 (two generator in the parenthesis). Now when we add the mass matrix γ_0, the complex Clifford algebra is extended to $Cl_3 \times Cl_3$ generated by

$$\{\gamma_1,\mathcal{C},\gamma_0\}\times\Pi_m\times Q \tag{18}$$

Therefore, the classifying space of mass matrix γ_0 is determined by the extension problem of Clifford algebra $Cl_2 \times Cl_2 \to Cl_3 \times Cl_3$ and we label such a classifying space as $\mathcal{S} = C_2 \times C_2$. The classification of gapped 1D phases with $(\mathcal{C}, \Pi_m, Q)$ symmetries is hence given by

$$\pi_0(C_2\times C_2) = \pi_0(C_2)\times\pi_0(C_2) = \mathbb{Z}\times\mathbb{Z} \tag{19}$$

There are two integer-valued topological invariants (W^+, W^-) that are 1D winding numbers[26] obtained in block-diagonalized subspace with $\Pi_m = \sigma_z v_y = \pm 1$.

Now let us start to break Π_m and $\mathcal{C}$ symmetries. When we break mirror reflection Π_m (but keep $\mathcal{C}$), the Clifford algebra extension problem becomes $Cl_2 \to Cl_3$

$$\{\gamma_1,\mathcal{C}\}\times Q \to \{\gamma_1,\mathcal{C},\gamma_0\}\times Q \tag{20}$$

and hence the classification is $\pi_0(C_2) = \mathbb{Z}$. The topological invariant is total winding number $W = W^+ + W^-$.

If we break $\mathcal{C}$ but keep Π_m, the extension problem is again $(Cl_1)^2 \to (Cl_2)^2$

$$\{\gamma_1\}\times\Pi_m\times Q \to \{\gamma_1,\gamma_0\}\times\Pi_m\times Q \tag{21}$$

and classification is trivial since

$$\pi_0(C_1\times C_1) = \pi_0(C_1)\times\pi_0(C_1) = 0\times 0 = 0. \tag{22}$$

If we break both $\mathcal{C}$ and Π_m symmetries, our extension problem is $Cl_1 \to Cl_1$

$$\{\gamma_1\}\times Q \to \{\gamma_1,\gamma_0\}\times Q, \tag{23}$$

which leads to a trivial classification $\pi_0(C_1) = 0$. Therefore, we obtain the results in Table 1.

Dislocation spectrum. One implication of weak index $\mathbf{M}$ is the existence of pairs of counter-propagating zero modes localized in a dislocation line respecting mirror symmetry Π_m. The number of zero-mode pairs in each dislocation line is determined by its Burgers vector $\mathbf{B}$: number of helical modes $= \mathbf{B}\cdot\mathbf{M}/2\pi$. (ref. 23)

To check such zero modes due to dislocation line, we consider pair-edge dislocations along $\hat{\mathbf{c}}$ axis, with a pair of Burgers vectors: $(\mathbf{B} = \hat{\mathbf{a}},\ \mathbf{B} = -\hat{\mathbf{a}})$ perpendicular to the dislocation line. The type of dislocation core is illustrated in Fig. 4a. We consider now a 3D box with periodic boundary condition in all $\left(\hat{\mathbf{a}}, \hat{\mathbf{b}}, \hat{\mathbf{c}}\right)$ directions (a 3D torus) so that the system does not have an open surface. Consider every unit cell (with four sublattices, two pseudo-spin species per site) as a lattice 'site' in Fig. 4a, then the system has hopping terms between nearest-neighbour 'sites' following the tight-binding Hamiltonian in equation (6). $\hat{\mathbf{c}}$ direction is still translational invariant and k_z remains a good quantum number, but in $\hat{\mathbf{a}}\hat{\mathbf{b}}$ plane translation symmetry is broken by the dislocation. The dislocation spectrum is obtained when there are 39 sites along $\hat{\mathbf{a}}$ axis and 15 sites along $\hat{\mathbf{b}}$ axis. The location of dislocation with Burgers vector $\mathbf{B} = \hat{\mathbf{a}}$ is at (4,6), which means 4th site along $\hat{\mathbf{a}}$ axis and 6th site along $\hat{\mathbf{b}}$ axis. And the position of dislocation with Burgers vector $\mathbf{B} = -\hat{\mathbf{a}}$ is at (24,12). The dislocation spectrum has displayed in Fig. 4b highlighted by red colour. It shows two pairs of gapless helical modes localized on each dislocation cores.

Effective Hamiltonian of Weyl fermion. By projecting the states around the Weyl node, the two-band Hamiltonian with linear Weyl fermion form can be obtained after adding mv_z, which breaks mirror symmetry and magnetic field $h_{[110]}(\sigma_x + \sigma_y)$, which breaks TR symmetry to $H(\mathbf{k})$ in equation (6). The followings are the coefficients for the effective Hamiltonian describes the Weyl fermion at $\mathbf{k}_1$ of

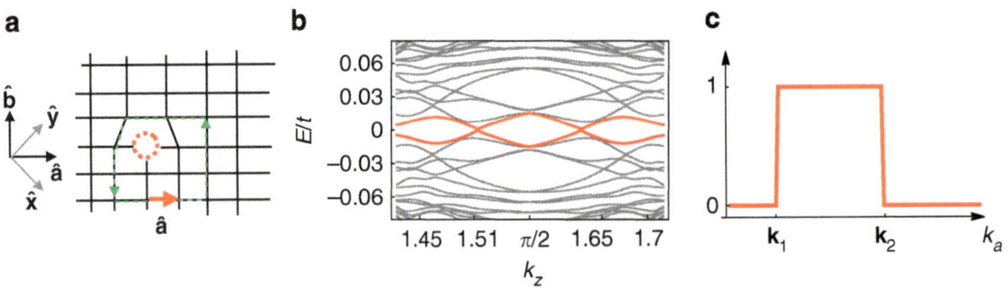

Figure 4 | Disloction spectrum and Chern number. (**a**) Top view of an edge dislocation with Burgers vector $\hat{\mathbf{a}}$ (coloured by red arrow), whose dislocation line is along $\hat{\mathbf{c}}$ axis. Dashed green line denotes a trajectory around the dislocation core (red circle). Clearly when we go around the dislocation once, we need an extra translation by the Burgers vector (the red arrow), equal to Bravais primitive vector $\hat{\mathbf{a}}$ in this case) to return to the starting point. (**b**) The dislocation spectrum for system size 39×15 contains two pairs of gapless helical modes (highlighted with red colour) localized at each dislocation line. (**c**) Chern number as a function of k_a for all occupied bands. Opposite Chern number jump indicates that the Weyl fermion at $\mathbf{k}_1$ and $\mathbf{k}_2$ has different chirality.

equation (2) presented in the main text:

$$\begin{aligned} A_1 &= \frac{2t_p t_d^o \sin^2(k_1)}{\left(t_{2p}^o - t_{1p}^o\right)\cos(k_1)}, \\ A_2 &= \frac{(\sqrt{2}h - m)t_z^o}{\sqrt{2}\left(t_{2p}^o - t_{1p}^o\right)\cos(k_1)}, \\ B_1 &= \frac{t_{d1}(m - \sqrt{2}h)}{\sqrt{2}\left(t_{2p}^o - t_{1p}^o\right)}, \\ D_1 &= \frac{\left[\left(t_{2p}^o - t_{1p}^o\right)^2 - (t_d^o)^2\right]\sin(k_1)}{t_{2p}^o - t_{1p}^o}, \end{aligned} \tag{24}$$

where $t_p, t_{2p}^o, t_{1p}^o, t_d^o, m, t_{d1}, h \equiv h_{[110]}$ are the coefficients in tight-binding Hamiltonian. The Chern number C for all occupied bands as a function of crystal momentum along U→R (or k_a) is shown in Fig. 4c.

References

1. Kane, C. L. & Mele, E. J. Quantum spin hall effect in graphene. *Phys. Rev. Lett.* **95,** 226801 (2005).
2. Kane, C. L. & Mele, E. J. Z_2 topological order and the quantum spin hall effect. *Phys. Rev. Lett.* **95,** 146802 (2005).
3. Bernevig, B. A., Hughes, T. L. & Zhang, S.-C. Quantum spin hall effect and topological phase transition in HgTe quantum wells. *Science* **314,** 1757–1761 (2006).
4. Konig, M. *et al.* Quantum spin hall insulator state in HgTe quantum wells. *Science* **318,** 766–770 (2007).
5. Fu, L., Kane, C. L. & Mele, E. J. Topological insulators in three dimensions. *Phys. Rev. Lett.* **98,** 106803 (2007).
6. Moore, J. E. & Balents, L. Topological invariants of time-reversal-invariant band structures. *Phys. Rev. B* **75,** 121306 (2007).
7. Roy, R. Topological phases and the quantum spin hall effect in three dimensions. *Phys. Rev. B* **79,** 195322 (2009).
8. Hsieh, D. *et al.* A Topological Dirac insulator in a quantum spin hall phase. *Nature* **452,** 970–974 (2008).
9. Hsieh, D. *et al.* Observation of unconventional quantum spin textures in topological insulators. *Science* **323,** 919–922 (2009).
10. Fu, L. Topological crystalline insulators. *Phys. Rev. Lett.* **106,** 106802 (2011).
11. Kargarian, M. & Fiete, G. A. Topological crystalline insulators in transition metal oxides. *Phys. Rev. Lett.* **110,** 156403 (2013).
12. Hsieh, T. H., Liu, J. & Fu, L. Topological crystalline insulators and Dirac octets in antiperovskites. *Phys. Rev. B* **90,** 081112 (2014).
13. Wan, X., Turner, A. M., Vishwanath, A. & Savrasov, S. Y. Topological semimetal and fermi-arc surface states in the electronic structure of pyrochlore iridates. *Phys. Rev. B* **83,** 205101 (2011).
14. Matsuura, S., Chang, P.-Y., Schnyder, A. P. & Ryu, S. Protected boundary states in gapless topological phases. *New J. Phys.* **15,** 065001 (2013).
15. Witczak-Krempa, W., Chen, G., Kim, Y. B. & Balents, L. Correlated quantum phenomena in the strong spin-orbit regime. *Annu. Rev. Condens. Matter Phys.* **5,** 57–82 (2014).
16. Kargarian, M., Wen, J. & Fiete, G. A. Competing exotic topological insulator phases in transition-metal oxides on the pyrochlore lattice with distortion. *Phys. Rev. B* **83,** 165112 (2011).
17. Yang, B.-J. & Nagaosa, N. Emergent Topological Phenomena in Thin Films of Pyrochlore Iridates. *Phys. Rev. Lett.* **112,** 246402 (2014).
18. Carter, J.-M., Shankar, V. V., Zeb, M. A. & Kee, H.-Y. Semimetal and topological insulator in perovskite iridates. *Phys. Rev. B* **85,** 115105 (2012).
19. Zeb, M. A. & Kee, H.-Y. Interplay between spin-orbit coupling and hubbard interaction in $SrIro_3$ and related *Pbnm* perovskite oxides. *Phys. Rev. B* **86,** 085149 (2012).
20. Cao, G. *et al.* Non-Fermi-liquid behavior in nearly ferromagnetic $SrIrO_3$ single crystals. *Phys. Rev. B* **76,** 100402 (2007).
21. Moon, S. J. *et al.* Dimensionality-controlled insulator-metal transition and correlated metallic state in 5d transition metal oxides $Sr_{n+1}Ir_nO3n+1$ (n = 1, 2, and ∞). *Phys. Rev. Lett.* **101,** 226402 (2008).
22. Ran, Y. Weak indices and dislocations in general topological band structures. Preprint at http://arxiv.org/abs/1006.5454 (2010).
23. Ran, Y., Zhang, Y. & Vishwanath, A. One-dimensional topologically protected modes in topological insulators with lattice dislocations. *Nat. Phys* **5,** 298–303 (2009).
24. Schnyder, A. P., Ryu, S., Furusaki, A. & Ludwig, A. W. W. Classification of topological insulators and superconductors in three spatial dimensions. *Phys. Rev. B* **78,** 195125 (2008).
25. Heeger, A. J. *et al.* Solitons in conducting polymers. *Rev. Mod. Phys.* **60,** 781–850 (1988).
26. Schnyder, A. P. & Ryu, S. Topological phases and surface flat bands in superconductors without inversion symmetry. *Phys. Rev. B* **84,** 060504 (2011).
27. Kitaev, A. Periodic table for topological insulators and superconductors. *AIP Conf. Proc.* **1134,** 22–30 (2009).
28. Yang, K.-Y., Lu, Y.-M. & Ran, Y. Quantum hall effects in a weyl semimetal: Possible application in pyrochlore iridates. *Phys. Rev. B* **84,** 075129 (2011).
29. Yi, H. *et al.* Evidence of Topological Surface State in Three-Dimensional Dirac Semimetal Cd_3As_2. *Sci. Rep.* **4,** 6106 (2014).
30. Longo, J., Kafalas, J. & Arnott, R. Structure and properties of the high and low pressure forms of $SrIrO_3$. *J. Solid State Chem.* **3,** 174–179 (1971).
31. Zhao, J. G. *et al.* High-pressure synthesis of orthorhombic $SrIrO_3$ perovskite and its positive magnetoresistance. *J. Appl. Phys.* **103,** 103706 (2008).
32. Matsuno, J. *et al.* Engineering spin-orbital magnetic insulator by tailoring superlattices. Preprint at http://arxiv.org/abs/1401.1066 (2014).
33. Hirai, D., Matsuno, J. & Takagi, H. Fabrication of (111)-oriented $Ca_{0.5}Sr_{0.5}IrO_3$/$SrTiO_3$ superlattices—A designed playground for honeycomb physics. *APL Mat.* **3,** 041508 (2015).
34. Wen, X.-G. Symmetry-protected topological phases in noninteracting fermion systems. *Phys. Rev. B* **85,** 085103 (2012).
35. Morimoto, T. & Furusaki, A. Topological classification with additional symmetries from Clifford algebras. *Phys. Rev. B* **88,** 125129 (2013).
36. Lu, Y.-M. & Lee, D.-H. Inversion symmetry protected topological insulators and superconductors. Preprint at http://arxiv.org/abs/1403.5558 (2014).

Acknowledgements

This work is supported by Natural Science and Engineering Research Council of Canada (NSERC), Center for Quantum Materials at the University of Toronto (Y.C. and H.-Y.K.) and Office of BES, Materials Sciences Division of the U.S. DOE under contract No. DE-AC02-05CH11231 (Y.-M.L.). H.-Y.K. thanks S. Ryu for informing topology of gapless superconductors in Ref. 36. Y.-M.L. and H.-Y.K. acknowledge the hospitality of the Aspen Cetner for Physics supported by National Science Foundation Grant No. PHYS-1066293 where a part of this work was carried out.

Author contributions

All authors performed the theoretical calculations, discussed the results and wrote the manuscript. H.-Y.K. planned and supervised the project.

Additional information

Competing financial interests: The authors declare no competing financial interests.

12

Plexciton Dirac points and topological modes

Joel Yuen-Zhou[1], Semion K. Saikin[2,3], Tony Zhu[4,5], Mehmet C. Onbasli[6], Caroline A. Ross[6], Vladimir Bulovic[5,7] & Marc A. Baldo[5,7]

Plexcitons are polaritonic modes that result from the strong coupling between excitons and plasmons. Here, we consider plexcitons emerging from the interaction of excitons in an organic molecular layer with surface plasmons in a metallic film. We predict the emergence of Dirac cones in the two-dimensional band-structure of plexcitons due to the inherent alignment of the excitonic transitions in the organic layer. An external magnetic field opens a gap between the Dirac cones if the plexciton system is interfaced with a magneto-optical layer. The resulting energy gap becomes populated with topologically protected one-way modes, which travel at the interface of this plexcitonic system. Our theoretical proposal suggests that plexcitons are a convenient and simple platform for the exploration of exotic phases of matter and for the control of energy flow at the nanoscale.

[1] Department of Chemistry and Biochemistry, University of California-San Diego, La Jolla, California 92093, USA. [2] Department of Chemistry and Chemical Biology, Harvard University, Cambridge, Massachusetts 02138, USA. [3] Department of Physics, Kazan Federal University, Kazan 420008, Russian Federation. [4] Department of Physics, Massachusetts Institute of Technology, Cambridge, Massachusetts 02139, USA. [5] Center for Excitonics, Research Laboratory of Electronics, Massachusetts Institute of Technology, Cambridge, Massachusetts 02139, USA. [6] Department of Materials Science and Engineering, Massachusetts Institute of Technology, Cambridge, Massachusetts 02139, USA. [7] Department of Electrical Engineering and Computer Science, Massachusetts Institute of Technology, Cambridge, Massachusetts 02139, USA. Correspondence and requests for materials should be addressed to J.Y.-Z. (email: joelyuen@ucsd.edu).

When ultraviolet–visible light is absorbed by an organic molecular aggregate, it promotes molecules from their ground states to their excited electronic states. The resulting excitations, known as excitons, can migrate between molecules via a mixture of coherent and incoherent processes[1]. Understanding and controlling how this migration of energy occurs is a fundamental problem of chemistry and physics of condensed phases. Furthermore, it is also a technological problem which is relevant to the development of efficient organic solar cells and light-emitting devices as well as all-optical circuitry[2]. Many strategies to control the motion of excitons exist, a particularly interesting one being where they couple to surface plasmons (SPs)[3]. In such strategy, the spatial coherence of plasmons assists the transport of an exciton across length-scales that are orders of magnitude larger than regular exciton diffusion lengths. When the coupling is strong, meaning that the energy exchange between the exciton and plasmon is faster than the respective decay times[4–8], plexcitons (a class of polaritons) emerge[9,10] and energy can migrate ballistically over the coherence length of the plasmon (between 10 and 40 μm)[11]. Besides their importance for energy transport, organic plexcitons promise to be an exciting room-temperature laboratory for the study of light-matter and many-body interactions at the nanoscale[8].

On the other hand, topologically nontrivial states of matter have been a topic of great interest in condensed matter physics owing to the discoveries of the quantum Hall effect[12], and more recently, of topological insulators[13,14]. The systems supporting these states are characterized by topological invariants[15], integer numbers that remain unchanged by weak perturbations. Physically, a nontrivial topological invariant signals the presence of one-way edge modes that are immune against moderate amount of disorder. Even though these phenomena were first conceptualized for fermions in solids, they have been successfully generalized to bosonic systems including photons in waveguides[16–18], ring resonator arrays[19], ultracold atoms in optical lattices[20] and classical electric circuits[21,22]. Furthermore, we have recently proposed an excitonic system consisting of a two-dimensional porphyrin film, which becomes topologically nontrivial in the presence of a magnetic field[23]. A challenging feature of that proposal is the requirement of large magnetic fields ($\gtrsim 10$ T) and cryogenic temperatures to preserve exciton coherence.

In this work, we consider a conceptually different platform, which, by recreating Dirac cones and topologically protected edge modes in plexcitons, avoids the use of large magnetic fields and, under appropriate circumstances, may work at room temperature. In the last year, Dirac and topological polaritons have been proposed in other contexts, such as optomechanical arrays and inorganic materials in optical cavities. All of these works share a common goal to ours, which is the design of exotic modes in strongly coupled light-matter systems. However, there are substantial qualitative and quantitative differences arising from the choices of material (organic exciton versus inorganic exciton[24–26] or mechanical mode[27,28]) and electromagnetic (SP versus microcavity[24–26] or photonic crystal[27,28]) excitations. Hence, the physics involved in our plexciton system contrasts with the other proposals in terms of the energy and length-scales involved in the excitations, the magnitude of the couplings, the generation of nontrivial topology and the experimental conditions for its realization. Organic excitons differ from their inorganic counterparts in that they have large binding energies and are associated with large transition dipole moments. In general, SP electromagnetic fields are strongly confined compared with those in microcavities because of the hybridization of light with charge oscillations in the metal[29]. The combination of all these properties in the organic plexciton context gives rise to strong light-matter interactions even at room temperature and in an open cavity[8]. We believe that, by introducing topological band theory concepts into the realm of plexciton systems, the present article yields a fresh perspective to the degree of control of energy transport which is achievable in the nano- and mesoscales.

Results

Plexciton Dirac points. The setup of interest is depicted in Fig. 1. It consists of three layers: a plasmonic metal modelled with a Drude permittivity ($\epsilon_{\rm m}(\omega)=\epsilon_\infty - \frac{\omega_{\rm P}^2}{\omega^2}$, with constants $\epsilon_\infty = 3.7$, $\omega_{\rm P} = 8.8$ eV, which are representative parameters for Ag), an $a = 80$ nm thick dielectric spacer ($\epsilon_{\rm d}=1$) and an organic layer ($\epsilon_{\rm org}=1$). The spacer is placed to avoid quenching of organic excitons by single-particle excitations in the metal upon close contact[30]. Later on, we will consider the case when this dielectric spacer is also endowed with magneto-optical (MO) properties. For purposes of band-structure engineering, we take the organic layer to be an oblique superlattice of organic nanopillars (Fig. 1a) with unit cell angles β, γ, δ and an angle θ with respect to the x axis of the setup (in the absence of this superlattice, a standard molecular crystal may be used with some trade-offs, as explained below). The nanopillars are rectangular parallelepipeds of densely packed organic chromophores (assuming a van der Waals distance between chromophores of 0.3 nm, $\rho_{\rm np} = 38$ chromophores per nm^3) with volume $V_{\rm np} = W_x W_y W_z$ (Fig. 1b), obtained from growing a J-aggregate film[31,32]. J-aggregation of chromophores results in a collective transition dipole $\boldsymbol{\mu}_{\mathbf{n}}$ for the $\mathbf{n}$-th nanopillar. Dipolar interactions between the collective dipoles couple the various nanopillars. Assuming perfect periodicity of the superlattice and only nearest and next-nearest neighbour (NN and NNN) dipolar interactions, we can write the Hamiltonian for excitations in the plexciton setup as $H = \sum_{\mathbf{k}} H_{\mathbf{k}}$

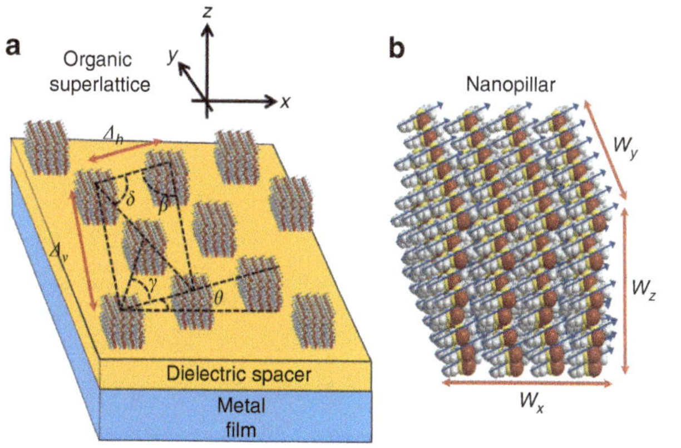

Figure 1 | Plexciton setup. (**a**) It consists of a plasmonic metal film, a dielectric spacer and an organic layer. The latter is taken to be a monoclinic superlattice of organic nanopillars, which makes an angle θ with respect to the x axis and is further characterized by angles β, γ, δ as well as distances between nanopillars Δ_h and Δ_v. When the density of emitters in the organic nanopillars is big enough, the coupling between the excitons in the organic layer and the surface plasmons (SPs) in the metal becomes larger than their linewidths, giving rise to polaritonic eigenmodes that are superpositions of excitons and plasmons, or more succintly, plexcitons. In this article, we shall also consider the case where the dielectric spacer is a magneto-optical (MO) material. The superlattice design is not essential for our purposes, except for guaranteeing a global gap for topological modes spanning all wavevectors. Hence, one may substitute it with a standard molecular crystal with some trade-offs. (**b**) Zoom-in view of a nanopillar. It is a parallelepiped of dimensions W_i along each axis, and consists of closely packed organic emitters (represented by balls and sticks) each featuring a transition dipole (blue arrows).

(see the Methods for details),

$$H_{\mathbf{k}} = \underbrace{\omega_{\mathrm{exc},\mathbf{k}}\sigma_{\mathbf{k}}^{\dagger}\sigma_{\mathbf{k}}}_{\equiv H_{\mathrm{exc},\mathbf{k}}} + \underbrace{\omega(\mathbf{k})a_{\mathbf{k}}^{\dagger}a_{\mathbf{k}}}_{\equiv H_{\mathrm{SP},\mathbf{k}}} + \underbrace{\left[\mathcal{J}(\mathbf{k})a_{\mathbf{k}}\sigma_{\mathbf{k}}^{\dagger} + \mathrm{h.c.}\right]}_{\equiv H_{\mathrm{exc-SP},\mathbf{k}}} \qquad (1)$$

where,

$$\omega_{\mathrm{exc},\mathbf{k}} = \bar{\omega}_{\mathrm{eff}} + 2J_x\cos(k_x\Delta_x) + 2J_y\cos(k_y\Delta_y), \qquad (2)$$

$$\mathcal{J}(\mathbf{k}) = \sqrt{\left(\frac{N_xN_y}{S}\right)\left(\frac{\omega(\mathbf{k})}{2\epsilon_0 L_{\mathbf{k}}}\right)}\, e^{-\alpha_{\mathrm{org}}(\mathbf{k})\bar{z}(\mathbf{k})}\boldsymbol{\mu}\cdot\mathbf{E}(\mathbf{k}). \qquad (3)$$

Here, $\omega_{\mathrm{exc},\mathbf{k}}$ and $\omega(\mathbf{k})$ are dispersions of the exciton and plasmon, respectively. J_i and Δ_i are effective NN hoppings and spacings between nanopillars along the i-th axis, respectively, $\bar{\omega}_{\mathrm{eff}}$ is the effective nanopillar site energy, and we have taken $\boldsymbol{\mu}_{\mathbf{n}} = \boldsymbol{\mu}$ for all $\mathbf{n}$. Equation (3) denotes the $\mathbf{k}$-dependent coupling between a plasmonic mode $(a_{\mathbf{k}}^{\dagger})$ and a collective exciton state $(\sigma_{\mathbf{k}}^{\dagger})$ throughout the organic layer. It is proportional to the square root of the total effective number of nanopillars $\sqrt{N_xN_y}$ coherently coupled to each SP mode[33–35] (N_i is the number of nanopillars along the ith axis) and to the collective nanopillar transition dipole moment $\boldsymbol{\mu}$, which scales as $\sqrt{N_{\mathrm{np}}}$, where N_{np} is the number of chromophores per nanopillar (see the Methods, Modelling of superlattice). Overall, $\mathcal{J}(\mathbf{k})$ scales as $\sqrt{N_xN_yN_{\mathrm{np}}}$, the total number of chromophores in the superlattice, as expected from a plexciton problem[7]. S and $L_{\mathbf{k}}$ denote the SP mode surface area and mode length, respectively, and ε_0 is the permittivity of vacuum.

Figure 2a shows superimposed dispersion curves for $H_{\mathrm{exc},\mathbf{k}}$ and $H_{\mathrm{SP},\mathbf{k}}$ independently (setting $H_{\mathrm{exc-SP}} = 0$ in the calculations, see the Methods for parameters). We assume that $\boldsymbol{\mu}$ is parallel to the xy plane, and design our superlattice to obtain $J_x, J_y > 0$, yielding a dome-like dispersion for $H_{\mathrm{exc},\mathbf{k}}$; that is, it features a maximum at $\mathbf{k} = 0$, behaving as a two-dimensional (2D) H-aggregate[1]. As for $H_{\mathrm{SP},\mathbf{k}}$, its (time-reversal symmetric) dispersion has the shape of a rotationally symmetric fountain, and is nothing more than the 2D rendering of the standard one-dimensional textbook result[36], featuring a linear dependence of the energy at short wavevectors and a plateau at large ones, indicating excitations that are qualitatively closer to light or to charge oscillations in the metal, respectively (Supplementary Notes 1 and 2, as well as Supplementary Figs 1–4). Figure 2b shows the two-plexciton-branch band-structure arising from the diagonalization of equation (1). We notice that anticrossing gaps are opened in the vicinity of where the dispersion curves for $H_{\mathrm{exc},\mathbf{k}}$ and $H_{\mathrm{SP},\mathbf{k}}$ used to cross in Fig. 2a. This is a signature of exciton-SP coupling $H_{\mathrm{exc,SP},\mathbf{k}}$. Given a fixed wavevector direction, whenever these anticrossings occur, the lower-plexciton (LP) branch starts off as being mostly SP at short $\mathbf{k}$ values, but parametrically morphs into mostly exciton at large ones; the opposite happens with the upper-plexciton (UP) branch. However, the most striking feature of Fig. 2b is the appearance of two Dirac cones (see dashed circles) at critical wavevectors $\mathbf{k}^*$ where anticrossings do not happen. Their onset coincides with the directions at which $\mathbf{k}$ is orthogonal to $\boldsymbol{\mu}$. Their physical origin is explained in Fig. 2e, which, in its top panel, shows that the in-plane electric field for the $\mathbf{k}$-th SP mode, $\mathbf{E}_{\perp}(\mathbf{k}) \equiv \mathbf{E}(\mathbf{k}) - \mathbf{E}(\mathbf{k})\cdot\hat{\mathbf{z}}\hat{z}$, is purely parallel to $\mathbf{k}$, $\mathbf{E}_{\perp}(\mathbf{k}) = \mathbf{E}_{\mathbf{k}}(\mathbf{k})$ (blue vector field). If all the dipoles in the organic layer are aligned (in-plane) along $\boldsymbol{\mu}$, their projection onto the SP electric field, which gives rise to the exciton-SP coupling (see equation (3) as well as Supplementary Note 3 and Supplementary Figs 5–8), will wax and wane as a function of the azimuthal angle φ between the fixed dipole and the varying SP wavevector according to $\boldsymbol{\mu}\cdot\mathbf{E}(\mathbf{k}) \propto \cos\varphi$. Clearly, this projection will vanish if $\mathbf{k}$ happens to be orthogonal to $\boldsymbol{\mu}$, that is, at the special angles $\varphi = \frac{\pi}{2}, \frac{3\pi}{2}$, so that any degeneracy between the exciton and the SP modes will remain unlifted along these directions. From this physical picture, we can extract the two essential ingredients for the emergence of the plexciton Dirac cones. First, the dipoles need to be aligned to create an anisotropic $\mathcal{J}(\mathbf{k})$ as a function of

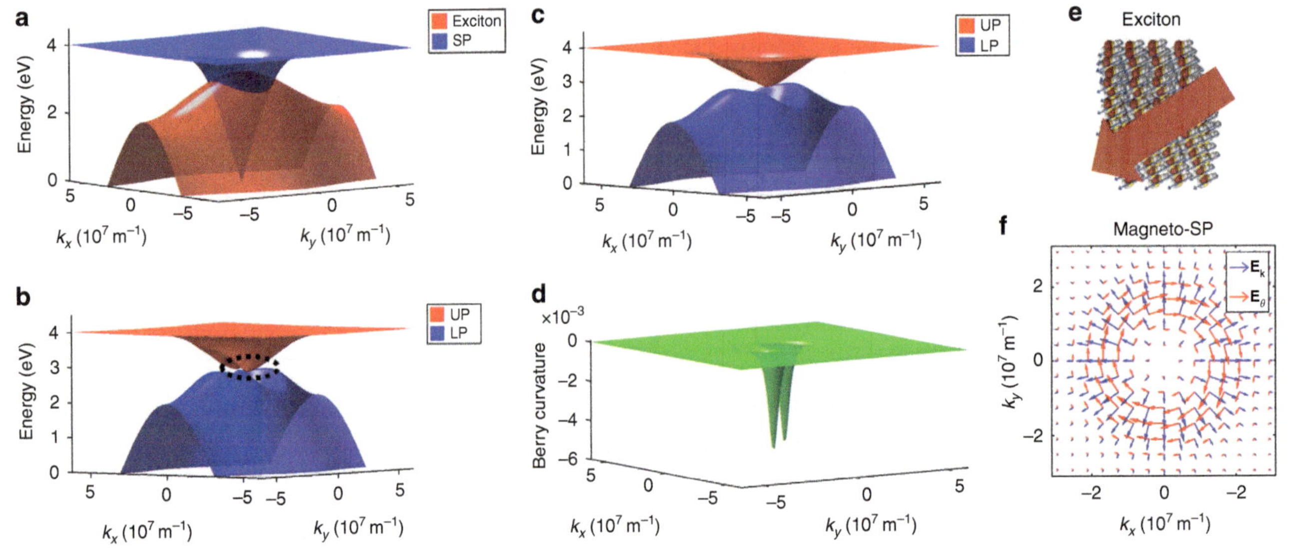

Figure 2 | Bulk plexciton properties. Dispersion relations: (**a**) for SP and exciton (organic layer) modes independently, (**b**) when they couple in the absence of the MO effect, yielding lower (LP) and upper (UP) plexciton branches, which feature two Dirac cones (dashed black circle), and (**c**) when they couple in the presence of the MO effect ($g = 0.3$), lifting the Dirac cones. (**d**) (unnormalized) Berry curvature associated with the LP in **c**. Physical mechanism for the appearance of plexciton Dirac points: (**e**) we specialize to nanopillar transitions, which are parallel to the xy plane and (**f**) plot of magnitude of the electric field of magneto-SP modes as a function of wavevector. In the absence of the MO effect, only the wavevector-parallel components $\mathbf{E}_{\mathbf{k}}$(blue) are present. Thus, the nanopillars experience no coupling with modes whose wavevectors are perpendicular to the transition dipole. Along these directions, degeneracies between the SP and the exciton modes are not lifted, yielding two plexciton Dirac points. Nonzero tangential components $\mathbf{E}_{\theta}$ (red) emerge upon inclusion of the MO effect, lifting these degeneracies.

φ. Second, this alignment needs to be horizontal, as a vertical component of the dipole will couple to the vertical component of the electric field $\mathbf{E}(\mathbf{k}) \cdot \hat{\mathbf{z}}\hat{z}$, and this coupling, unlike its horizontal counterpart, does not vanish for any φ. It is important to note that neither of these properties require the use of the superlattice, which will be exploited for a different purpose (the design of a global gap for topologically protected edge modes, as explained in the next paragraphs). Therefore, a standard organic molecular crystal with aligned transition dipoles lying on the horizontal xy plane will suffice. These plexciton Dirac cones, which, to our knowledge, have not been reported in the past, should be easily detectable by collecting the reflected light spectra upon excitation of the plexcitonic system in a grating, Otto or Kretschmann configurations[36], by systematically scanning across $|\mathbf{k}|$ and φ values. For a general ($|\mathbf{k}|$, φ), the spectrum should consist of two dips as a function of dispersed energy, each associated with the corresponding eigenenergies of the LP and UP. However, the two dips merge at the Dirac cones. In a standard plexciton dispersion measurement, one only scans across $|\mathbf{k}|$. As we are interested in a two-dimensional dispersion, the scan must also be performed across φ. To trace the formation of the Dirac points, Supplementary Note 3 and Supplementary Figs 9–11 show two-dimensional cross-sections of the dispersion curves in Fig. 2.

Plexciton topological modes. Having elucidated the mechanism for the formation of plexciton Dirac cones, we proceed to entertain a more ambitious goal. We aim to engineer topologically protected plexcitons by opening the Dirac cones using a time-reversal symmetry breaking (TRSB) perturbation[15]. To accomplish this, we now assume that the dielectric spacer has MO properties; that is, upon application of a perpendicular magnetic field, its permittivity becomes anisotropic, $\epsilon_d \to \overleftrightarrow{\epsilon}_{MO}$,

$$\overleftrightarrow{\epsilon}_{MO} = \begin{bmatrix} \epsilon_d & ig & 0 \\ -ig & \epsilon_d & 0 \\ 0 & 0 & \epsilon_d \end{bmatrix}. \quad (4)$$

Materials associated with this dielectric tensor exhibit Faraday effect and are of great interest in the fabrication of optical isolators[37]. For the near-infrared and visible spectra, yttrium iron garnets ($Y_3Fe_5O_{12}$, YIG) substituted with Bi (BiYIG), Ce (CeYIG) or other rare earths provide high Faraday rotation with low optical absorption[38,39]. In this article, we are interested in the new SP modes, denoted as magneto-SP modes, arising at the interface of the plasmonic metal and the MO layer. The solution for the latter is highly non-trivial, and we refer the reader to our solution in the Supplementary Notes 1 and 2, which build a perturbation theory on the small MO parameter g, generalizing a simpler calculation of Chiu and Quinn[40]. Figure 2c shows the same calculation as Fig. 2b except for the inclusion of the MO effect. Interestingly, we notice that the Dirac cones have been lifted, yielding a global gap because of the H-aggregate superlattice dispersion. A physical understanding of the latter phenomenon can be obtained by appealing to Fig. 2e again. Within our perturbation theory, the magneto-SP modes differ from their original SP counterparts in that there are additional tangential components ($\mathbf{E}_\theta$, red) to the electric field. The clockwise vortex vector field is a signature of TRSB; it becomes counterclockwise upon change of direction of the magnetic field. This tangential electric field is solely responsible for opening the Dirac cones at the critical angles $\varphi = \frac{\pi}{2}, \frac{3\pi}{2}$, where the original field (blue) ceased to couple to the excitons. Hence, we have concocted a situation where anticrossings occur for all azimuthal angles φ. To characterize the topology of the resulting band-structure, we numerically compute the Berry curvature for each plexciton branch[41]; we show that of the LP in Fig. 2d. Its integral with respect to the Brillouin zone is the so-called Chern number C, an integer which, if nonzero, signals a topologically nontrivial phase. Figure 2d clearly shows that this integral is non-vanishing, and in fact, adds up to $C = -1$ (by the sum rule of Chern numbers, the upper branch necessarily has $C = 1$). Intuitively, it is also clear that most of the nontrivial topology, and hence, Berry curvature, is concentrated in the vicinity of what used to be the Dirac cones. In passing, we note that considerable attention has recently been given to magneto-SPs, where the magnetic field is applied parallel (instead of perpendicular) to the metal film itself, yielding dispersion relations, which are nonreciprocal[40,42]. Curiously, this arrangement does not give us the vortex vector field we are looking for, although it might be intriguing to explore the connection between these magneto-SPs and the ones exploited in our present work, arising from a perpendicular magnetic field.

So far, all the described calculations have been carried out in the bulk. By virtue of the bulk-boundary correspondence[15], we expect topologically protected one-way edge modes associated with this setup. To compute them, it is convenient to keep periodic boundary conditions for the magneto-SP modes, yet consider two domains of excitons on top (Fig. 3a), one with (in-plane) dipoles pointing along $\boldsymbol{\mu}$ (red dipoles) and the other one with vertical dipoles along $\hat{\mathbf{z}}$ (blue dipoles). Pictorially, this setup resembles a donut with two icings, where the donut is the metal with toroidal geometry, and the two icings are the domains of excitons separated by two interfaces located at $y = \pm \frac{L_y}{2}$ and $y = 0$, where L_i is the total width of the simulated sample along i (in our calculations, we take $L_x = 40\,\mu m$, $L_y = 6\,\mu m$). The Chern numbers associated with the bulk LP branch of each domain are $C = -1$ and $C = 0$, respectively. Hence, the plexcitons for the blue domain are topologically trivial. This can be understood by recalling that no plexciton Dirac points occur when dipoles are vertically aligned, regardless of the MO effect. As this system is perfectly periodic along the x-direction, k_x is still a good quantum number, and Fig. 3b shows the corresponding plexciton dispersion. This band-structure is essentially a projection of the gapped 2D bulk band-structures of both domains of plexcitons onto one axis k_x with additional states spanning the gap between the LP and UP branches. Inspection of the nature of these mid-gap states reveals that they have substantial exciton and magneto-SP character, and that they are precisely the edge states we are searching for: one band has positive (negative) dispersion

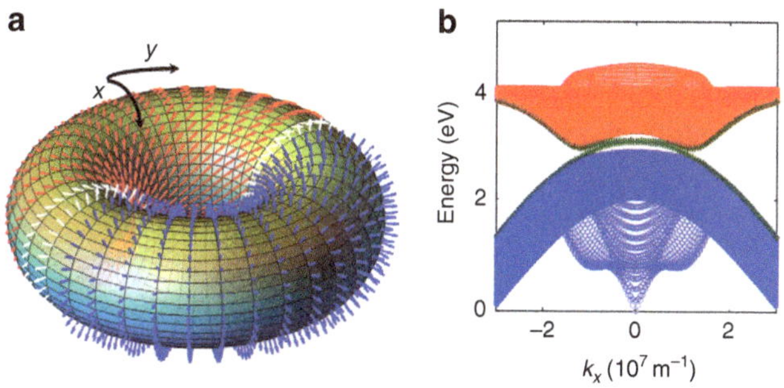

Figure 3 | Topologically protected edge modes. (**a**) Simulation of edge modes where magneto-SPs are computed in the torus geometry. Two domains of organic layers are placed on top of it (just like two icings on a donut). In-plane (red) and out-of-plane (blue) transition dipoles yield topologically nontrivial and trivial plexcitons, respectively. Topologically protected one-way plexcitons appear at the interfaces (white domains). Each interface features a different plexciton direction of motion. (**b**) One-dimensional dispersion relation $\omega(k_x)$ for the setup in **a**. Bulk LPs (blue) and UPs (red) separated by edge modes (green) featuring positive and negative dispersions, respectively, and localized along each interface.

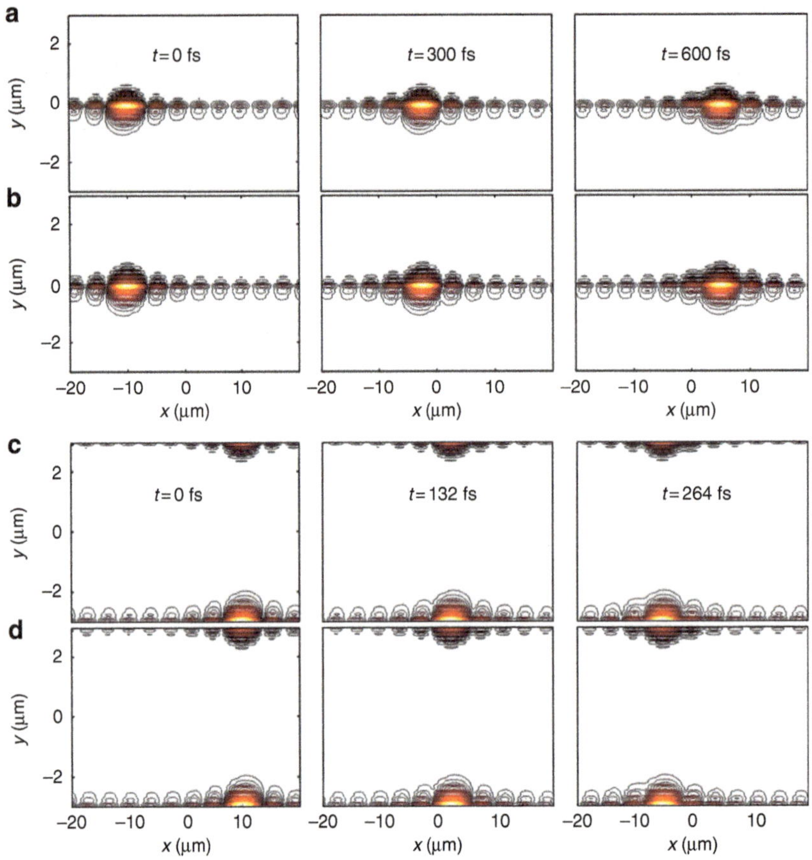

Figure 4 | Snapshots of dynamics of topologically protected edge modes. **(a-d)** The dynamics of an initially localized plexciton along the $y=0$ and $y=\pm\frac{L_y}{2}=\pm3\,\mu$m interfaces of our simulation is shown. **(a-d)** The exciton and magneto-SP components are shown, respectively. These modes, which are robust to disorder, have substantial excitonic and magneto-SP components and travel in opposite directions along different interfaces.

and is localized along $y=0$ ($y=\pm\frac{L_y}{2}$). Thus, by preparing a plexciton wavepacket localized along one of the interfaces, and making sure it is composed of energy states within the global topological gap, one ensures that transport occurs robustly without much probability of backscattering. The reason being that elastic backscattering requires coupling between counterpropagating modes, which are separated by a distance $\frac{L_y}{2}$, which is large compared with the width of the corresponding wavefunctions along $\hat{\mathbf{y}}$.

Figure 3b was generated with the parameters $\epsilon_d=1$, $g=0.3$, yielding a minimum gap between plexciton branches (at the wavevectors $\mathbf{k}^*$ of the original Dirac points) of $2|\mathcal{J}(\mathbf{k}^*)|=0.23$ eV. The crossing of the SP and exciton dispersion curves happens at 2.86 eV. Given typical linewidths associated with the various dissipative mechanisms at room temperature ($\gamma_{\text{exc,rel}}\sim5$ MeV, $\gamma_{\text{exc,deph}}\sim40$ MeV, $\gamma_{\text{SP,rel}}\sim10$ MeV, where rel and deph stand for relaxation and dephasing), our parameters lie within the strong-coupling regime, namely, $2|\mathcal{J}(\mathbf{k}^*)|>\sim10$ meV (ref. 7). Ideally, we would like to have all edge states lying within the global topological gap. For the present simulation, only a fraction of the latter (close to the anticrossings) lies within the global topological gap; the rest are degenerate with bulk plexcitons. This scenario arises as a consequence of the large exciton dispersion anisotropy ($J_x\neq J_y$) compared with the strength of $|\mathcal{J}(\mathbf{k})|$ for all $\mathbf{k}$. The latter is a result of optimization: thicker MO spacers yield larger values of $|\mathcal{J}(\mathbf{k}^*)|$ but, due to the evanescent nature of SP modes, also reduce the overall couplings $|\mathcal{J}(\mathbf{k})|$ for $\mathbf{k}\neq\mathbf{k}^*$ (Supplementary Note 3). Typical MO garnets have similar g but higher ϵ_d values (for an external magnetic field of 0.01 T, $\epsilon_d=6.25$ and $g=0.1$ (ref. 38)), yielding a strong index mismatch in our setup. As explained in Supplementary Note 3, our future aim is to boost $|\mathcal{J}(\mathbf{k}^*)|$ by considering various MO layer thicknesses a as well as novel MO garnet compositions, which maximize the g/ϵ_d ratio, the latter of which is not fundametally limited. Strategies may include the consideration of MO garnet sphere arrays[43], plasmonic/magnetic metal nanostructures[44], Ce substituted YIGs[45] or Eu nanocrystals[46].

Figure 4a,b shows snapshots of the dynamics associated with the edge states lying within the global gap. Figures 4a,b and c,d show an edge plexciton wavepacket that starts localized at $x=-\frac{L_x}{4}$ and $x=\frac{L_x}{4}$, respectively, and track the one-way (to the left or to the right) nature of its micrometre-scale motion within the subpicosecond timescale. The dispersion of the edge plexcitons is such that a subset falls within the light cone ($\omega>ck_x/\sqrt{\epsilon_d}$) so far-field excitation and detection of this fraction is possible via direct interaction with the organic layer. The rest of the plexcitons can be probed using the already mentioned SP measurement techniques, by launching plexcitons exciting the metallic layer itself. Furthermore, the ballistic and one-way nature of these modes can in principle be demonstrated using fluorescence microscopy[47]. As explained, the robustness of one-way transport for these edge modes, even in the presence of disorder, relies on the global topological gap, which is a consequence of our 2D H-aggregate-type superlattice design (see the Methods for details). In terms of experimental feasibility, we note that some compromise on topological protection might be acceptable if using a plain molecular crystal is much easier than constructing the superlattice as long as scattering between k_x states is not strong.

Discussion

To summarize, we have described the design of exotic plexcitons via a judicious choice of material and electromagnetic excitation modes. We showed that Dirac cones and topologically protected edge states emerge from relatively simple hybrid organic/inorganic nanostructures. Even though we have not precisely identified an explicit MO material which fully satisfies our requirements, we believe its design is within reach, and is the subject of our present investigations. It is also worth noting that the physical origin of the described edges states is different from that of edge plasmons in disk geometries[48,49], although the connections are worth exploring. An interesting extension of this work is the consideration of plexcitonic Dirac point opening without TRSB, which might give rise to rich phases in analogy to other Dirac systems[50]. In general, the possibility of directed migration of excitation energy at the nano- and mesoscale offers exciting prospects in light-harvesting and all-optical circuit architectures. Furthermore, given the recent experimental discovery of nonlinear many-body effects such as Bose–Einstein condensation of organic cavity-polaritons[51–53] and plexcitons[54] at room temperature, the introduction of the novel features described in this letter enriches the scope of these materials as a test-bed for novel many-body quantum phenomena.

Methods

Hamiltonian of the plexciton setup. A quantum mechanical description of the plexciton setup is given by a Hamiltonian,

$$H = H_{\mathrm{exc}} + H_{\mathrm{SP}} + H_{\mathrm{exc-SP}}, \tag{5}$$

where each of the terms denotes the energetic contributions from the excitons in the organic layer, the SPs and the coupling between them. More specifically ($\hbar = 1$),

$$\begin{aligned} H_{\mathrm{exc}} &= \sum_{\mathbf{n},s} \omega_{\mathbf{n}} \sigma_{\mathbf{n}}^{\dagger} \sigma_{\mathbf{n}} \\ &+ \sum_{\mathbf{n} \neq \mathbf{n}'} \left(J_{\mathbf{n}\mathbf{n}'} \sigma_{\mathbf{n}}^{\dagger} \sigma_{\mathbf{n}'} + \mathrm{h.c.} \right), \end{aligned} \tag{6}$$

$$H_{\mathrm{SP}} = \sum_{\mathbf{k}} \omega(\mathbf{k}) a_{\mathbf{k}}^{\dagger} a_{\mathbf{k}}, \tag{7}$$

$$H_{\mathrm{excSP}} = \sum_{\mathbf{k},\mathbf{n}} \mathcal{J}_{\mathbf{kn}} a_{\mathbf{k}} \sigma_{\mathbf{n}}^{\dagger} e^{i\mathbf{k}\cdot\mathbf{r}_{\mathbf{n}}} + \mathrm{h.c.} \tag{8}$$

Here, $\sigma_{\mathbf{n}}^{\dagger}(\sigma_{\mathbf{n}})$ and $a_{\mathbf{k}}^{\dagger}(a_{\mathbf{k}})$ label the creation (annihilation) operators for the collective exciton at the **n**-th nanopillar and the **k**-th SP mode, respectively, where **n** and **k** are (two-dimensional) in-plane vectors denoting a position and a wavevector, respectively. J-aggregation of chromophores results in a collective transition dipole $\boldsymbol{\mu}_{\mathbf{n}}$ at an excitation energy $\omega_{\mathbf{n}}$, whereas the dispersion energy of the **k**-th SP mode is denoted $\omega(\mathbf{k})$. Dipolar interactions $J_{\mathbf{nn}'}$ couple the various nanopillars. The coupling between the exciton and the SP depends on the average in-plane location $\mathbf{r}_{\mathbf{n}}$ of the **n**-th nanopillar, and is also dipolar in nature,

$$\mathcal{J}_{\mathbf{kn}} = -\sqrt{\frac{\omega(\mathbf{k})}{2\epsilon_0 S L_{\mathbf{k}}}} e^{-\alpha_{org}(\mathbf{k})\bar{z}(\mathbf{k})} e^{i\mathbf{k}\cdot\mathbf{r}_{\mathbf{n}}} \mu_{\mathbf{n}} \cdot \mathbf{E}(\mathbf{k}). \tag{9}$$

Here, S is the SP mode surface area, $L_{\mathbf{k}}$ denotes a vertical (z-direction) mode-length of the SP, which guarantees that the total energy of a SP prepared at the **k**th mode is quantized at the energy $\omega(\mathbf{k})$, $\mathbf{E}(\mathbf{k})$ is an appropriately scaled electric field of the corresponding mode, and $e^{-\alpha_{\mathrm{org}}(\mathbf{k})\bar{z}(\mathbf{k})}$ yields a mean-field average of the interaction of the evanescent SP field (with decay constant α_{org} in the organic layer) over the chromophores at different vertical positions of the nanopillar; it optimizes the interaction such that one may assume the nanopillar is a point-dipole located at the mean height $\bar{z}(\mathbf{k})$. The latter average renders the originally 3D system into an effectively 2D one. Detailed derivations of equations (6)–(9) are available in the Supplementary Notes 1–3.

Assuming perfect periodicity of the superlattice ($\omega_{\mathbf{n}} = \bar{\omega}$, $\boldsymbol{\mu}_{\mathbf{n}} = \boldsymbol{\mu}$) and only NN and NNN dipolar interactions, we can re-express H_{exc} (equation (6)) in terms of **k** modes. As explained in Supplementary Note 3, it is possible to approximate H_{exc} up to $O(|\mathbf{k}|^2)$ as arising from an *effective* simplified rectangular (rather than monoclinic) lattice aligned along the x, y axes, and with NN interactions only. This is a reasonable thing to do, as the topological effects we are interested arise at relatively long wavelengths. Within this approximation, we may construct Fourier modes $\sigma_{\mathbf{n}}^{\dagger} = \frac{1}{\sqrt{N_x N_y}} \sum_{\mathbf{k}'} \sigma_{\mathbf{k}'}^{\dagger} e^{-i\mathbf{k}'\cdot\mathbf{r}_{\mathbf{n}}}$ (here, N_i is the effective number of nanopillars along the i-th direction) and rewriting equation (5) in reciprocal space, we obtain $H = \sum_{\mathbf{k}} H_{\mathbf{k}}$, where $H_{\mathbf{k}}$ is given by equation (1).

Modelling of superlattice. For our simulations, both of the band-structure and the edge states, we chose the length parameters $\Delta_h = 100$, $\Delta_v = 88$, $W_x = 10$, $W_y = 75$, $W_z = 70$ nm, and take $\beta = 13.1°$ (see Supplementary Note 3 for explanation of choice of parameters). Denoting the transition dipole $\boldsymbol{\mu} = \mu\hat{\boldsymbol{\mu}}$, we estimate $\boldsymbol{\mu} = \sqrt{N_{\mathrm{np}}} \times 12\,D = 1.7 \times 10^4\,D$ (D = Debye), where $N_{\mathrm{np}} = \rho_{\mathrm{np}} V_{\mathrm{np}}$ is the number of chromophores in the nanopillar, $\rho_{\mathrm{np}} = 37$ chromophores per nm^3; $\hat{\boldsymbol{\mu}}$ is the *in-plane* unit vector making an angle of $\alpha = 224°$) with respect to $\hat{\mathbf{x}}$, that is, $\hat{\boldsymbol{\mu}} = \cos\alpha\hat{\mathbf{x}} + \sin\alpha\hat{\mathbf{y}}$. Choosing the simulation values for H_{exc} in equation (2) to be $\Delta_x = \Delta_y = 50$ nm, we obtained the effective parameters $\bar{\omega}_{\mathrm{eff}} = 0.57$ eV, $J_x = 1.04$ meV and $J_y = 0.31$ meV. As explained in Supplementary Note 3, for strong exciton-plasmon coupling, the relevant density of chromophores is the weighed average between the density in the nanopillars and the null one in the void space. The density and vertical thickness of this superlattice are in line with the typical parameters that yield strong coupling in plexciton systems[7].

Experimental considerations. In terms of the fabrication of the plexciton setup, we warn that the creation of BiYIG layers typically need high temperature and oxygen, which is incompatible with deposition on Ag or organic materials. Hence, the MO layer could first be deposited on garnet substrates such as GGG ($Gd_3Ga_5O_{12}$) (111), and subsequently floated off by dissolving or polishing the substrate. One should then transfer the film on a Ag-coated substrate and the organic layer may be deposited and patterned on garnet.

Effects of disorder and global gap. It is important to note that, owing to the topological nature of these states, perfect lattices are not required, so the robustness of the one-way edge states holds as long as orientational and site energy disorder induce perturbations which are smaller than the topological anticrossings. We tested these ideas by simulating lattices with disorder in the site energies ($\bar{\omega}_{\mathrm{eff}} \to \bar{\omega}_{\mathrm{eff}} + \Delta\bar{\omega}_{\mathrm{eff}}$) as well as in the orientations of the dipoles ($\hat{\boldsymbol{\mu}} \to \cos\Delta_\phi[\cos(\alpha + \Delta_\alpha)\hat{\mathbf{x}} + \sin(\alpha + \Delta_\alpha)\hat{\mathbf{y}}] + \sin\Delta_\phi\hat{\mathbf{z}}$), where Δ_j are chosen to be Gaussian random variables centred at 0 and having disorder widths σ_j for each $j = \bar{\omega}_{\mathrm{eff}}$, α, ϕ. By systematically varying these widths independently and keeping track of the presence of the one-way edge states, we noticed that the latter survive under large amounts of disorder, whose thresholds are approximately located at $\sigma_{\bar{\omega}_{\mathrm{eff}}} \sim 0.25$ eV, $\sigma_\alpha \sim 57°$ and $\sigma_\phi \sim 28°$. Furthermore, we suspect that these values are lower bounds, as the disorder in these simulations is exacerbated by the torus configuration of our simulation, where periodicity along the x axis is conserved. In other words, owing to the latter, for a given superlattice coordinate along the y axis, we fixed the same values of disorder Δ_j across all x (that is, the disorder was perfectly correlated along x).

One may check that if $J_x \leq 0$ or $J_y \leq 0$, the global topological gap is not guaranteed anymore, and edge modes may become degenerate with bulk modes. Under certain circumstances, these two types of modes could hybridize because of impurity potentials, yielding channels connecting one edge to the other, opening backscattering channels. Importantly, however, the resulting band-structure would formally remain topologically nontrivial (in terms of the Chern numbers $C \neq 0$ for the plexciton bands) even in the absence of such global gap, so even if perfect one-way transport is not observed in these cases, signatures of the latter may remain. The latter observation applies if it is experimentally more convenient to use of a standard molecular crystal (with negligible exciton dispersion) rather than the proposed superlattice. These issues will be explored in future work. In the mean time, it suffices to note that, as a proof of concept, a global gap that hosts topologically protected edge states can be obtained by using an H-aggregate-type organic superlattice like the one suggested in this article.

References

1. Saikin, S. K., Eisfeld, A., Valleau, S. & Aspuru-Guzik, A. Photonics meets excitonics: natural and artificial molecular aggregates. *Nanophotonics* **2,** 21–38 (2013).
2. Baldo, M. & Stojanović, V. Optical switching: Excitonic interconnects. *Nat. Photon* **3,** 558–560 (2009).
3. Andrew, P. & Barnes, W. L. Energy transfer across a metal film mediated by surface plasmon polaritons. *Science* **306,** 1002–1005 (2004).
4. Bellessa, J., Bonnand, C., Plenet, J. C. & Mugnier, J. Strong coupling between surface plasmons and excitons in an organic semiconductor. *Phys. Rev. Lett.* **93,** 036404 (2004).
5. Govorov, A. O. *et al.* Exciton-plasmon interaction and hybrid excitons in semiconductor-metal nanoparticle assemblies. *Nano Lett.* **6,** 984–994 (2006).
6. Salomon, A., Gordon, R. J., Prior, Y., Seideman, T. & Sukharev, M. Strong coupling between molecular excited states and surface plasmon modes of a slit array in a thin metal film. *Phys. Rev. Lett.* **109,** 073002 (2012).

7. González-Tudela, A., Huidobro, P. A., Martín-Moreno, L., Tejedor, C. & García-Vidal, F. J. Theory of strong coupling between quantum emitters and propagating surface plasmons. *Phys. Rev. Lett.* **110,** 126801 (2013).
8. Törmä, P. & Barnes, W. Strong coupling between surface plasmon polaritons and emitters: a review. *Rep. Prog. Phys.* **78,** 013901 (2015).
9. Fofang, N. T., Grady, N. K., Fan, Z., Govorov, A. O. & Halas, N. J. Plexciton dynamics: Exciton-plasmon coupling in a J-aggregate-Au nanoshell complex provides a mechanism for nonlinearity. *Nano Lett.* **11,** 1556–1560 (2011).
10. Manjavacas, A., Abajo, F. J. G. d. & Nordlander, P. Quantum plexcitonics: Strongly interacting plasmons and excitons. *Nano Lett.* **11,** 2318–2323 (2011).
11. Kooyman, R. P. H. *in Handbook of Surface Plasmon Resonance* (eds Schasfoort, R. B. M. & Tudos, A. J.) 15–34 (Roy. Soc. Chem., 2008).
12. Klitzing, K. v., Dorda, G. & Pepper, M. New method for high-accuracy determination of the fine-structure constant based on quantized hall resistance. *Phys. Rev. Lett.* **45,** 494–497 (1980).
13. Kane, C. L. & Mele, E. J. Quantum spin Hall effect in graphene. *Phys. Rev. Lett.* **95,** 226801 (2005).
14. Bernevig, B. A., Hughes, T. L. & Zhang, S.-C. Quantum spin Hall effect and topological phase transition in HgTe quantum wells. *Science* **314,** 1757–1761 (2006).
15. Bernevig, B. A. *Topological Insulators and Topological Superconductors* (Princeton Univ., 2013).
16. Haldane, F. D. M. & Raghu, S. Possible realization of directional optical waveguides in photonic crystals with broken time-reversal symmetry. *Phys. Rev. Lett.* **100,** 013904 (2008).
17. Khanikaev, A. B. *et al.* Photonic topological insulators. *Nat. Mater.* **12,** 233–239 (2013).
18. Lu, L., Joannopoulos, J. D. & Soljačić, M. Topological Photonics. *Nat. Photon* **8,** 821–829 (2014).
19. Hafezi, M., Mittal, S., Fan, J., Migdall, A. & Taylor, J. M. Imaging topological edge states in silicon photonics. *Nat. Photon* **7,** 1001–1005 (2013).
20. Aidelsburger, M. *et al.* Measuring the Chern number of Hofstadter bands with ultracold bosonic atoms. *Nat. Phys.* **11,** 162–166 (2015).
21. Ningyuan, J., Owens, C., Sommer, A., Schuster, D. & Simon, J. Time- and site-resolved dynamics in a topological circuit. *Phys. Rev. X* **5,** 021031 (2015).
22. Albert, V. V., Glazman, L. I. & Jiang, L. Topological properties of linear circuit lattices. *Phys. Rev. Lett.* **114,** 173902 (2015).
23. Yuen-Zhou, J., Saikin, S. S., Yao, N. Y. & Aspuru-Guzik, A. Topologically protected excitons in porphyrin thin films. *Nat. Mater.* **13,** 1026–1032 (2014).
24. Jacqmin, T. *et al.* Direct observation of Dirac cones and a flatband in a honeycomb lattice for polaritons. *Phys. Rev. Lett.* **112,** 116402 (2014).
25. Karzig, T., Bardyn, C.-E., Lindner, N. H. & Refael, G. Topological polaritons. *Phys. Rev. X* **5,** 031001 (2015).
26. Bardyn, C.-E., Karzig, T., Refael, G. & Liew, T. Topological polaritons and excitons in garden-variety systems. *Phys. Rev. B* **91,** 161413 (2015).
27. Schmidt, M., Peano, V. & Marquardt, F. Optomechanical Dirac physics. *N. J. Phys.* **17,** 023025 (2015).
28. Peano, V., Brendel, C., Schmidt, M. & Marquardt, F. Topological phases of sound and light. *Phys. Rev. X* **5,** 031011 (2015).
29. Kuttge, M., de Abajo, F. J. G. & Polman, A. Ultrasmall mode volume plasmonic nanodisk resonators. *Nano Lett.* **10,** 1537–1541 (2010).
30. Ford, G. W. & Weber, W. H. Electromagnetic interactions of molecules with metal surfaces. *Phys. Rep* **113,** 195–287 (1984).
31. Kuhn, H. & Kuhn, C. *Chromophore Coupling Effects* 1–40 (World Scientific, 1996).
32. Bradley, M. S., Tischler, J. R. & Bulovic, V. Layer-by-layer J-aggregate thin films with a peak absorption constant of 106 cm-1. *Adv. Mater.* **17,** 1881–1886 (2005).
33. Herrera, F., Peropadre, B., Pachon, L. A., Saikin, S. K. & Aspuru-Guzik, A. Quantum nonlinear optics with polar J-aggregates in microcavities. *J. Phys. Chem. Lett.* **5,** 3708–3715 (2014).
34. Coles, D. M. *et al.* Strong coupling between chlorosomes of photosynthetic bacteria and a confined optical cavity mode. *Nat. Commun.* **5,** 5561 (2014).
35. Spano, F. Optical microcavities enhance the exciton coherence length and eliminate vibronic coupling in J-aggregates. *J. Chem. Phys.* **142,** 184707 (2015).
36. Novotny, L. & Hecht, B. *Principles of Nano-Optics* (Cambridge Univ., 2012).
37. Dionne, G. F., Allen, G., Haddad, P., Ross, C. & Lax, B. Circular polarization and nonreciprocal propagation in magnetic media. *Linc. Lab. J* **15,** 323–340 (2005).
38. Kuzmiak, V., Eyderman, S. & Vanwolleghem, M. Controlling surface plasmon polaritons by a static and/or time-dependent external magnetic field. *Phys. Rev. B* **86,** 045403 (2012).
39. Onbasli, M., Goto, T., Sun, X., Huynh, N. & Ross, C. A. Integration of bulk-quality thin film magneto-optical cerium-doped yttrium iron garnet on silicon nitride photonic substrates. *Opt. Express* **22,** 25183–25192 (2014).
40. Chiu, K. & Quinn, J. Magneto-plasma surface waves in solids. *Nuovo Cimento B* **10,** 1–20 (1972).
41. Fukui, T., Hatsugai, Y. & Suzuki, H. Chern numbers in discretized brillouin zone: Efficient method of computing (spin) Hall conductances. *J. Phys. Soc. Jpn.* **74,** 1674–1677 (2005).
42. Yu, Z., Veronis, G., Wang, Z. & Fan, S. One-way electromagnetic waveguide formed at the interface between a plasmonic metal under a static magnetic field and a photonic crystal. *Phys. Rev. Lett.* **100,** 023902 (2008).
43. Christofi, A., Stefanou, N. & Papanikolaou, N. Periodic structures of magnetic garnet particles for strong Faraday rotation enhancement. *Phys. Rev. B* **89,** 214410 (2014).
44. Temnov, V. *et al.* Active magneto-plasmonics in hybrid metal-ferromagnet structures. *Nat. Photon* **4,** 107–111 (2010).
45. Sun, X. Y. *et al.* Single-step deposition of cerium-substituted yttrium iron garnet for monolithic on-chip optical isolation. *ACS Photon* **2,** 856–863 (2015).
46. Hasegawa, Y. Magnetic semiconductor EuO, EuS, and EuSe nanocrystals for future optical devices. *Chem. Lett.* **42,** 2–7 (2013).
47. Akselrod, G. M. *et al.* Visualization of exciton transport in ordered and disordered molecular solids. *Nat. Commun.* **5,** 3646 (2014).
48. Fetter, A. L. Edge magnetoplasmons in a two-dimensional electron fluid confined to a half-plane. *Phys. Rev. B* **33,** 3717–3723 (1986).
49. Song, J. C. W. & Rudner, M. S. Chiral plasmons without magnetic field. Preprint at http://arxiv.org/abs/1506.04743 (2015).
50. Yoshimura, Y., Imura, K.-I., Fukui, T. & Hatsugai, Y. Characterizing weak topological properties: Berry phase point of view. *Phys. Rev. B* **90,** 155443 (2014).
51. Daskalakis, K., Maier, S., Murray, R. & Kéna-Cohen, S. Nonlinear interactions in an organic polariton condensate. *Nat. Mater.* **13,** 271–278 (2014).
52. Bittner, E., Zaster, S. & Silva, C. Thermodynamics of exciton/polaritons in one and two dimensional organic single-crystal microcavities. *Phys. Chem. Chem. Phys.* **14,** 3226–3233 (2012).
53. Bittner, E. R. & Silva, C. Estimating the conditions for polariton condensation in organic thin-film microcavities. *J. Chem. Phys.* **136,** 034510 (2012).
54. Rodriguez, S., Feist, J., Verschuuren, M., Garcia-Vidal, F. & Gomez-Rivas, J. Thermalization and cooling of plasmon-exciton polaritons: towards quantum condensation. *Phys. Rev. Lett.* **111,** 166802 (2013).

Acknowledgements

J.Y.-Z thanks Justin C.W. Song for a discussion on edge magnetoplasmons. J.Y.-Z acknowledges partial support from startup funds at UC San Diego. J.Y-Z., T.Z., V.B. and M.B. were supported by an Energy Frontier Research Center funded by the US Department of Energy, Office of Science, Office of Basic Energy Sciences under Award Number DESC0001088. S.K.S. was supported by the Defense Threat Reduction Agency grant HDTRA1-10-1-0046. S.K.S. is also grateful to the Russian Government Program of Competitive Growth of Kazan Federal University. Finally, C.A.R. and M.C.O. acknowledge support of the Solid-State Solar-Thermal Energy Conversion Center (S3TEC), award DE-SC0001299, and FAME, a STARnet Center of SRC supported by DARPA and MARCO.

Author contributions

J.Y.Z. and S.K.S. conceived the original idea, developed the theoretical formalism in the main text and Supplementary Information and carried out the numerical work. J.Y.Z., S.K.S., T.Z., M.C.O., C.A.R., V.B., and M.A.B. developed and discussed ideas, contributed their technical expertise at every stage of the project, and wrote the final version of the manuscript and Supplementary Information.

Additional information

Competing financial interests: The authors declare no competing financial interests.

Multiple tipping points and optimal repairing in interacting networks

Antonio Majdandzic[1], Lidia A. Braunstein[1,2], Chester Curme[1], Irena Vodenska[1,3], Sary Levy-Carciente[1,4], H. Eugene Stanley[1] & Shlomo Havlin[1,5]

Systems composed of many interacting dynamical networks—such as the human body with its biological networks or the global economic network consisting of regional clusters—often exhibit complicated collective dynamics. Three fundamental processes that are typically present are failure, damage spread and recovery. Here we develop a model for such systems and find a very rich phase diagram that becomes increasingly more complex as the number of interacting networks increases. In the simplest example of two interacting networks we find two critical points, four triple points, ten allowed transitions and two 'forbidden' transitions, as well as complex hysteresis loops. Remarkably, we find that triple points play the dominant role in constructing the optimal repairing strategy in damaged interacting systems. To test our model, we analyse an example of real interacting financial networks and find evidence of rapid dynamical transitions between well-defined states, in agreement with the predictions of our model.

[1] Center for Polymer Studies and Department of Physics, Boston University, 590 Commonwealth Avenue, Boston, Massachusetts 02215, USA. [2] Instituto de Investigaciones Físicas de Mar del Plata (IFIMAR), Physics Department, Universidad Nacional de Mar del Plata-CONICET, Funes 3350, 7600 Mar del Plata, Argentina. [3] Administrative Sciences Department, Metropolitan College, Boston University, Boston, Massachusetts 02215 USA. [4] Economics and Social Sciences Faculty, Central University of Venezuela, 1040 Caracas, Venezuela. [5] Department of Physics, Bar-Ilan University, 52900 Ramat-Gan, Israel. Correspondence and requests for materials should be addressed to A.M. (email: antem@bu.edu).

Most real networks are not isolated structures but interact with other network structures. As a result, much research has been focused recently on the dynamics of interdependent[1–8] and multilayer[9–11] networks. Recent studies on network repair[12–14] have shown the importance of recovery of nodes as a process that leads to reverse transitions, hysteresis effects and such phenomena as spontaneous recovery[12,15].

The cardiovascular and nervous systems in the human body are examples of two dynamically interacting physiological networks[16]. Diseases often result from complex pathological conditions that involve a dynamical interaction with positive or negative feedback between different functional subsystems in the body. Similarly, in the global economy there is a hierarchy of clustered and tightly connected countries, often grouped geographically, which are further interconnected to one large global interacting economic and financial network[17–19]. To understand the behaviour of these systems using network science, we develop a model of interacting networks with nodes that can recover from failure and we examine the resulting phase diagram.

Our model of a generic system consisting of interacting dynamical networks captures the important events found in real-world interacting networks, that is, node failure[20–23], systemic damage propagation[24] and node recovery[12,15,25]. We first analiticaly solve the model in the mean field theory (MFT) approximation and confirm the results in numerical simulations. The phase diagram of this system is characterized by a number of phases, phase transition lines and tipping points[26–29]. We show that triple points play a critical role in devising an optimal strategy for efficient repairing of interconnected network systems. By analysing the network of credit default swaps (CDSs) of sovereign countries, we demonstrate the application of our model to a real system and show that all of its parameters are experimentally accessible.

Results

Model. The structure of our system for the $n=2$ case is modelled as follows. We start with two isolated networks, network A and network B, and for simplicity we assume that both networks have the same number of nodes N and the same degree distribution $f(k)$ (these assumptions can be relaxed but the results stay qualitatively similar). We assume that within each network the nodes are randomly connected. Now, to allow networks A and B to interact, we introduce interdependency links that connect nodes across the two networks[2]. This can be achieved in different ways and we use a simple one-to-one dependency: each node in network A is dependent on exactly one node in network B and vice versa. The pairs of nodes of both networks are chosen randomly.

The dynamic behaviour of our system is governed by two categories of event—failure and recovery—and we assume that every node is in either a failed or an active state. Node failure can result from internal failure or from the spread of damage from neighbour nodes in either the same network or the interdependent network. We thus assume that there are three ways a node can fail. The first way is the internally induced failure, when a node's internal integrity has been compromised, for example, an organ in the body can fail due to a malfunction within the organ or a company can fail due to bad management. The second type of failure is externally induced failure through failure propagation due to connections with failed nodes within the node's own network. Finally, there is a failure induced through the dependency link as a result of being dependent on a failed node from another (opposite) network. Apart of these three types of failures, we assume the existence of associated simple recovery processes for every type of failure. We specify quantitatively each of these processes below.

For internal failures (I), we assume that in both networks any node can fail due to internal problems, independent of other nodes. For each node in network A we assume that there is probability $p_A dt$ that the node will fail internally during any time period dt. The equivalent parameter in network B is p_B.

Every node in network A and network B is connected by links to nearby nodes in its own network. These nodes constitute the node's neighbourhood. The number of links a node has within the network indicates its degree or connectivity, denoted by k. If a large number of nodes in a node's neighbourhood have failed, that is, if the neighbourhood is substantially damaged, we assume that the probability that the node itself will fail is increased. This is modelled by external failures (E). As in refs 12,30, we use a threshold rule to define a substantially damaged neighbourhood, which is a neighbourhood containing $\leq m$ active nodes, where m is a fixed integer threshold. If node j has $>m$ active neighbours during time dt, we consider its neighbourhood to be 'healthy' and there is no risk of external failure. On the other hand, if j has $\leq m$ active neighbours during time dt, there is a probability $r_A\, dt$ (for network A) or $r_B\, dt$ (for network B) that node j will externally fail. For certain systems it is more appropriate to define a fractional threshold $0 \leq m_{frac} \leq 1$ as in ref. 15. That is, the minimum number of active nodes as a requirement for a 'healthy' neighbourhood is replaced by a minimum fraction of active nodes in the neighbourhood. In the example of random regular network that we consider below, both are equivalent and related by $m = km_{frac}$.

In the case of two interdependent networks (A and B), we assume that each node in the first network is dependent on a node in the second network via an interdependent link and vice versa. We assume that if one node in the pair fails, there is a finite (but not 100%) probability, $r_d\, dt$, that during time dt the other node in the pair will fail as well (dependency failure: type D). This represents the probability that the damage will spread through the interdependency link.

We also assume that there is a reversal process, a recovery from each of these three types of failure. A node recovers from an internal failure after a time period $\tau \neq 0$, it recovers from an external failure after time τ' and from a dependency failure after time τ''. In simulations, and without loss of generality, we use $\tau=100$ and for simplicity we set $\tau'=\tau''=1$, to take into account the assumption that real-world systems usually require a longer time period to recover from internal problems (physical faults) then from a lack of environmental support. However, changing the numerical values does not introduce any qualitative difference.

For the node activity notation, we assume that every node is in one of two states: active or failed. A node is considered active in the observed moment, if it is not experiencing internal (I), external (E) or dependency (D) failure.

Mean field theory. We characterize this system by studying the order parameters chosen naturally as the fraction of active nodes in network A and network B, z_A and z_B, respectively. To simplify the calculation, however, we first concentrate on the complementary and equally intuitive fraction of failed nodes a_A and a_B, in networks A and B, respectively ($a_A = 1 - z_A$, $a_B = 1 - z_B$).

Using the MFT presented in Methods, we obtain two coupled equations that connect a_A and a_B, which the system must satisfy in the equilibrium

$$a_A = p_A^* + r_d a_B (1 - p_A^*) + \sum_k f(k) F(k, a_A) \left[r_A - p_A^* r_A - r_A r_d a_B + p_A^* r_A r_d a_B \right] \quad (1)$$

$$a_B = p_B^* + r_d a_A (1 - p_B^*) + \sum_k f(k) F(k, a_B) \left[r_B - p_B^* r_B - r_B r_d a_A + p_B^* r_B r_d a_A \right] \quad (2)$$

Here $F(k, x)=\sum_{j=0}^{m}\binom{k}{j}x^{k-j}(1-x)^{j}$ and we have also introduced simplifying parameters $p_A^* \equiv 1-e^{-p_A\tau}$ and $p_B^* \equiv 1-e^{-p_B\tau}$, to make the equations more elegant and to reduce the number of parameters by replacing p_A, p_B and τ that appear as a product. We find that the parameters p_A^* and p_B^* are very convenient to work with, because they correspond to the fraction of internally failed nodes in network A and network B, respectively.

Despite the seeming complexity of equations (1) and (2), it is noteworthy that there are only two unknown variables, a_A and a_B, and that all other parameters are fixed. These two equations define two curves in the (a_A, a_B) plane.

Figure 1a shows a graphical representation of the curves for a random regular[31] network (in which all the nodes have the same degree) with degree of $k=16$ and threshold $m=8$, for the symmetric parameter values $p_A^*=p_B^*=0.16$, $r_A=r_B=0.60$ and $r_d=0.15$. The size of each network is $N=2\times10^4$. The blue curve is a graphical representation of equation (1) and the brown curve is defined by equation (2). The curves, similar to two 'ropes', create a 'knot' that can have up to nine intersections, representing mathematical solutions of the system of equations. However, not all of these solutions represent observable and stable physical states. To see that, observe one of the curves in Fig. 1a, for example, the blue curve described by equation (1). If we increase damage done to network B (that is, we increase a_B) and keep everything else constant, some damage will undoubtedly spread to network A. Thus, we expect that when a_B is increased, a_A must also increase (it would be very unusual if one network improves its activity as a result of damaging the other network). We conclude that the parts of the blue and brown curve that produce physical solutions are only those where a_A and a_B increase together or decrease together along the curve. This elimination leaves only four states in Fig. 1a that are stable (green circles), whereas the other five states are unstable (red crosses), for this particular choice of parameters. In simulated finite networks, when the network system evolves according to the rules of the model, at $t=0$ we have a freedom to set initial conditions for the activities. Systems initially prepared to have a pair of values (a_A, a_B) corresponding to an unstable solution of equations (1) and (2) will be disturbed by a small fluctuation of a_A or a_B, owing to the system dynamics, and the values of a_A or a_B will rapidly change until one of the stable states is reached. Systems that are initially prepared to have values of a_A or a_B corresponding to a stable solution will fluctuate around these values, until perhaps a large finite fluctuation occurs and the system 'jumps' to another stable state. In general, for any choice of parameters, we have between one and four stable (physical) states. Figure 1b shows the scenario for the same network system when $p_A^*=0.20$, $p_B^*=0.24$, $r_A=r_B=0.60$ and $r_d=0.15$. In this case we have two stable states and one unstable state.

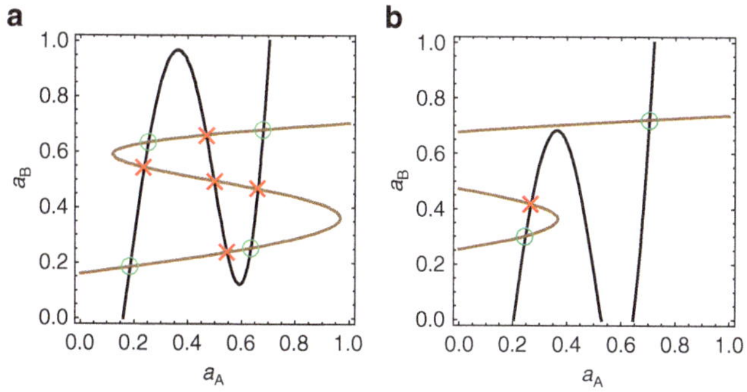

Figure 1 | Graphical representations of the mean field equations. (**a**) The blue and brown curves represent equations (1) and (2), respectively, for $p_A^*=p_B^*=0.16$, $r_A=r_B=0.60$ and $r_d=0.15$, in a system with two interdependent networks ($k=16$, $m=8$). There are nine intersections, representing mathematical solutions for network activities a_A and a_B. Four of them are stable solutions (green circles) representing physical states that we also observe in our simulations and five are unstable solutions (red crosses). (**b**) Example for $p_A^*=0.20$, $p_B^*=0.24$, $r_A=r_B=0.60$ and $r_d=0.15$. Here we obtain two stable solutions and one unstable solution. The two stable solutions correspond to 11 state (both networks are at high activity) and 22 state (both networks are at low activity).

In the Methods, section 'Additional phase diagrams', we show diagrams for $z_A=1-a_A$ for a range of different values of p_A^* and all other parameters fixed. This MFT calculation agrees well with the states that we observe in our simulations, as we will demonstrate below.

The four stable solutions found above correspond to the following four scenarios: '11' or 'up-up', when there is high activity in both network A and network B; '12' or 'up-down' when there is high activity in network A and low activity in network B; '21' or 'down-up' when there is low activity in network A and high activity in network B; and '22' or 'down-down', when there is low activity in both network A and network B.

Depending on the parameters, we obtain between one and four stable states. Each of the states exists in a certain volume of the multi-dimensional space of parameters. Results of the MFT calculation for a particular set of parameters are presented in Fig. 2a–d as a phase diagram with four layers. Figure 2 shows the regions in which each of the four states exist in the (p_A^*, p_B^*) parametric sub-space, when other parameters are fixed at values $r_A=r_B=0.60$ and $r_d=0.15$, with k and m remaining the same as before.

For example, in Fig. 2a the green area indicates the region where the 11 state exists. This state (phase) is bounded with a smooth boundary of three colours. If the boundary is crossed (by increasing p_A^* and p_B^*), the system makes a transition to state 12 (if the orange line is crossed), state 22 (if the blue line is crossed) or state 21 (if the purple line is crossed). The arrows indicate transitions. In Fig. 2a, there are two triple points (black points) that mark the change in the transition type and where three different states can exist. The blue area in Fig. 2b indicates the 22 state. This layer of the phase diagram has two triple points as well and three possible transitions ($22\rightarrow12$, $22\rightarrow11$ and $22\rightarrow21$).

Figure 2c,d show the regions of state 21 (purple) and state 12 (orange), respectively. Each has two different transitions and one critical point. For example, there are two possible ways out of state 21 (Fig. 2c): by a transition to the 11 (green arrow) state or the 22 (blue arrow) state. It is noteworthy that the different state regions (Fig. 2a–d) are not disjoint sets but there is an overlap, resulting in twofold, threefold or even fourfold hysteresis regions.

The state in which the system is found depends on the initial conditions or the system's past. There are a total of 10 different transitions ($11\rightarrow12$, $11\rightarrow22$, $11\rightarrow21$, $12\rightarrow11$, $12\rightarrow22$, $21\rightarrow11$, $21\rightarrow22$, $22\rightarrow12$, $22\rightarrow21$ and $22\rightarrow11$) that connect different layers of the phase diagram (states 11, 12, 21 and 22), much similar to elevators connecting different floors. Transitions $12\rightarrow21$ and $21\rightarrow12$ are the only missing ('forbidden') combinations. Although regions 12 and 21 do overlap, there is no direct transition connecting these two states. The set of all allowed and forbidden transitions is presented in Fig. 2e. The total phase diagram (all four layers on top of each other) is presented in Fig. 3. Here, coloured lines represent the boundaries of four states, with each colour corresponding to the boundary of one state, for example, the green line is a boundary of the 11 state. Rich critical phenomena with discontinuous hybrid phase transitions and second-order transitions have been recently

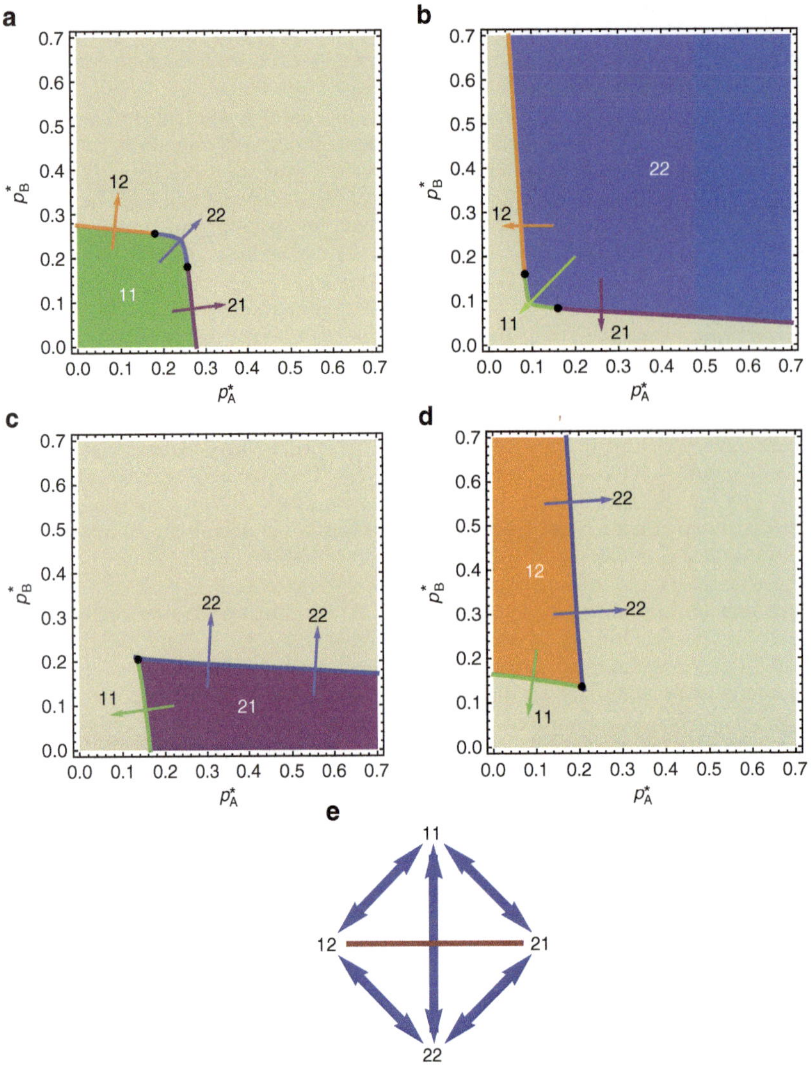

Figure 2 | Four layers of the phase diagram and the transitions connecting them. (**a**) Region of 11 state, in green. Possible transitions are 11→12 (orange line), 11→22 (blue line) and 11→21 (purple line). This layer of the phase diagram has two triple points, marked as black points. (**b**) Region of 22 state (blue), with two triple points and three transitions. (**c**) Region of 21 state (purple), with two transition lines (to 11 and 22 state) that merge in a critical point. (**d**) Region of 12 state (orange), with two transition lines (to 11 and 22 state) that merge in a critical point. (**e**) Illustration showing states (11, 12, 21 and 22) with allowed (blue arrows) and forbidden (red line) transitions.

discovered in multiplex networks. In particular, Baxter *et al.*[32] introduced weak bootstrap percolation and weak pruning percolation in multiplex networks, which have potential applications in infrastructure recovery and information security, and can even provide a way to diagnose missing layers in a multiplex network.

We next can examine the activity profile for various cross-sections in the phase diagram. In Fig. 3 we choose two representative cross-sections (dashed straight lines) to measure activity $z_A = 1 - a_A$ as p_A^* and p_B^* change. The black dashed line is defined by the equation $p_B^*=0.1+4/3p_A^*$ and the red dashed line by $p_B^*=0.4-p_A^*$. Figure 4a shows the activity measured in simulations of network A as we move along the black dashed line, changing both p_A^* and p_B^*, and preserving the relation $p_B^*=0.1+4/3p_A^*$. We perform simulations for various initial conditions and find (Fig. 4a) three different states denoted by green, orange and blue colours, which we identify as 11, 12 and 22 states, respectively. We find four different transitions: 11→12, 12→22, 12→11 and 22→12. The solid lines show the MFT prediction (equations (1) and (2)) for the activity of network A. The good agreement shows that the MFT correctly captures all the properties of the system. We note that qualitative agreement between the MFT and the simulations is better for higher values of k, because for higher k the fluctuations are smaller, which improves the accuracy of the MFT. Figure 4b shows the activity when moving along the red dashed line. Here we obtain four states and six different transitions.

The phase diagram of a system of $n=2$ interacting networks (Fig. 3) is much richer than the phase diagram of a single network with damage and recovery[12]. The analytical results we presented here for $n=2$ can be generalized to n interacting networks in any topological configuration, although as n increases they become increasingly difficult to visualize. In general, a system with n interacting networks can have up to 2^n physical states.

The problem of optimal repairing. Knowing and understanding the phase diagram of interacting networks enable us to answer some fundamental and practical questions. A partially or completely collapsed system of $n \geq 2$ interacting networks in

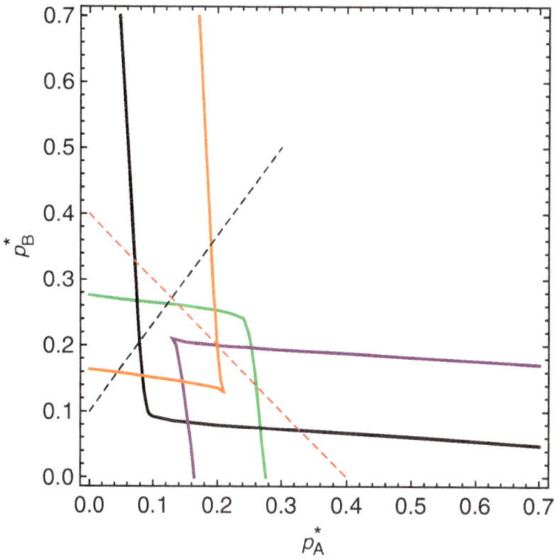

Figure 3 | Total phase diagram with all four layers. Solid lines represent the borders of region 11 (green), 22 (blue), 12 (orange) and 21 (purple). Dashed lines represent cross-sections where we calculate the activity profile, shown in Fig. 4. It is noteworthy that there is a small central 'window' where all four states are possible.

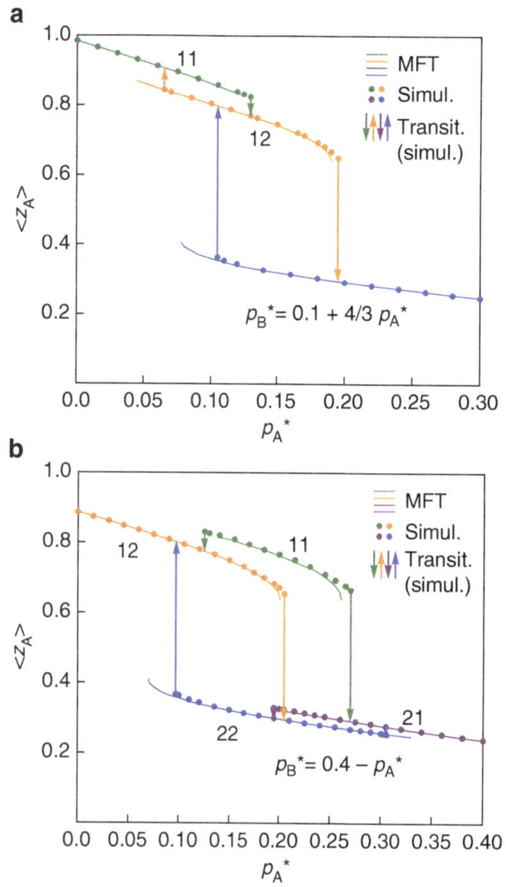

Figure 4 | States with transitions and hysteresis loops for two activity profiles. (**a**) Activity z_A of network A, as measured in simulations (dots) and predicted by MFT (solid lines), along the cross-section defined by the black dashed line in Fig. 3. Parameters p_A^* and p_B^* are changed, preserving the relation $p_B^*=0.1+4/3p_A^*$. Transitions are denoted by arrows. (**b**) Same for the cross-section defined by $p_B^*=0.4-p_A^*$ (red dashed line in Fig. 3). Here we obtain four states and six different transitions, giving rise to more complex hysteresis loops. Network parameters in all cases are ($k=16$, $m=8$).

which some of them are in the low activity state is a scenario common in medicine, for example, when diseases or traumas affect the human body and a few organs are simultaneously damaged and need to be treated, and the interaction between the organs is critical. It is also common in economics, when two or more coupled sectors of the economy[18] experience simultaneous problems, or when a few geographical clusters of countries experience economic difficulties. The practical question that arises is: what is the most efficient strategy to repair such a system? Many approaches are possible if resources are unlimited, but this is usually not the case and we would like to minimize the resources that we spend in the repairing process.

For simplicity, consider two interacting networks, both damaged (low activity). Is repairing both networks simultaneously the more efficient approach, or repairing them one after the other? What is the minimum amount of repair needed to make the system fully functional again? In other words, what is the minimum number of nodes we need to repair, to bring the system to the functional 11 ('up–up') state, and how do we allocate repairs between the two networks? An optimal repairing strategy is essential when resources needed for repairing are limited or very expensive, when the time to repair the system is limited, or when the damage is still progressing through the system, threatening further collapse, and a quick and efficient intervention is needed.

We show below that this problem is equivalent to finding the minimum Manhattan distance between the point in the phase diagram where the damaged system is currently situated and the recovery transition lines to the 11 region. The Manhattan distance between two points is defined as the sum of absolute horizontal and vertical components of the vector connecting the points, with defined vertical and horizontal directions. It is a driving distance between two points in a rectangular grid of streets and avenues. In our phase diagram, it is equal to $|\Delta p_A^*|+|\Delta p_B^*|$. It turns out that two triple points of the phase diagram play a very important role in this fundamental problem. We find that these special points have a direct practical meaning and are not just a topological or thermodynamic curiosity.

To show this, we start by making some simplifying but reasonable assumptions. First, we assume that only internal failures can be repaired by human hands, as these failures are physical faults in nodes (any external and dependency failures and recoveries are 'environmental', and are a spontaneous recognition of the changing neighbourhood of a node). We mentioned above that the parameters p_A^* and p_B^* correspond to fractions of internally failed nodes in networks A and B, respectively. This implies that the number of internally failed nodes repaired in, say, network A, is directly proportional to the change of p_A^*. Hence, repairing nodes in networks A and B means decreasing p_A^* or p_B^*. We also assume that these repairs are done fast enough that there is only a small probability that the newly repaired nodes will internally fail again before the repair process is completed. The total number of repaired nodes is therefore $N_{rep}=N\left(|\Delta p_A^*|+|\Delta p_B^*|\right)$ and it is proportional to the Manhattan distance between the starting and final point in the phase diagram.

To optimize repairing we need to minimize this metric. Figure 5 shows the solution to the minimization problem and a detailed discussion is provided in the Methods section. The different colours in Fig. 5 correspond to the different optimal repair strategies, which depend on the failure state of the system. If the system is initially at point S_1, both networks are in a low

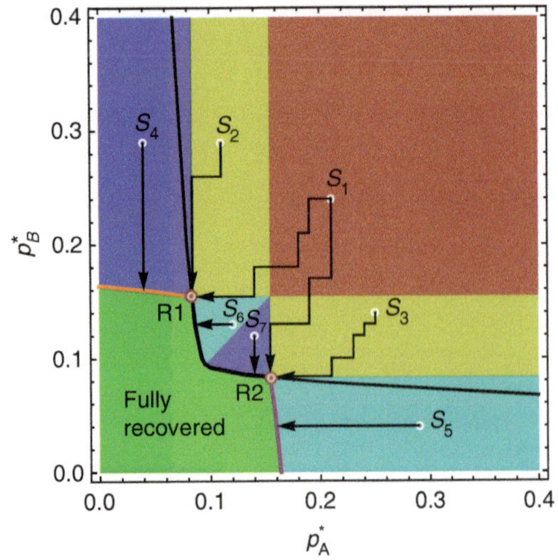

Figure 5 | Optimal repairing strategies. The optimal repairing procedure (least expensive in terms of the number of individual node repairs) depends on the initial condition of the collapsed system. The total cost of repair is $|\Delta p^*_A| + |\Delta p^*_B|$ and the problem of optimal repairing translates into finding the minimal Manhattan distance from the point (in the phase diagram) where the collapsed system is initially situated (S_i) to the nearest border of the green region where it becomes fully functional. For a system having the initial condition within the red section (for example, point S_1), there are two solutions: it is equally optimal to reach any of the two triple points R1 and R2 by decreasing p^*_A and p^*_B. For the systems starting in the yellow regions, it is optimal to reach only one triple point, R1, for the sector containing point S_2, or R2 for the sector containing point S_3. Starting in the dark blue regions it is optimal to decrease p^*_B only, that is, repairing only network B. Similarly, in the light blue regions it is optimal to decrease p^*_A only. Triple points play a crucial role when both networks are initially significantly damaged (red and yellow regions).

activity state, that is, they are non-functional. Our goal is to decrease p^*_A and p^*_B, and arrive to the region where the system is fully recovered (the green region) by performing a minimal number of repairs, that is, minimal N_{rep}. We find that for any point in the red region there are actually two closest points in the green region, at an equal Manhattan distance away from the red region point. These two points are the triple points R1 and R2 shown in Fig. 5, which also correspond to the triple points in Fig. 2b. Although R1 may be closer to point A than R2 by Euclidian distance, the Manhattan distance is the same. Thus, two equally good repairing strategies are available. One involves allocating more node repairs to network A and the other allocating more repairs to network B. For the yellow regions (points S_2 and S_3), the closest points by Manhattan distance are R1 (for point S_2) or R2 (for point S_3). Here, only one triple point represents the optimal solution. It is noteworthy that the path samples in Fig. 5 are 'zig-zag' in shape (to highlight that we are minimizing $|\Delta p^*_A| + |\Delta p^*_B|$); however, even when a diagonal path (direct straight line) to a triple point is used, the Manhattan distance is the same. For the dark blue regions (points S_4 and S_7), the optimal strategy is to decrease p^*_B only, until the system is recovered. Similarly, for the light blue regions (points S_5 and S_6), the optimal strategy is to decrease only p^*_A.

From our optimal repairing strategy analysis we find that the order of repair (the specific path taken between the initial point and final point) does not affect the final result. Minimizing the Manhattan distance only determines the optimal destination point. Therefore, there is actually a set of paths corresponding to equally optimal repairing processes.

States and transitions in real-world networks. In relatively small networks ($N \approx 10$–1,000) fluctuations are very large. Thus, in small network systems exhibiting multistability it is possible to observe phase flipping[12,15,33] between different states. Figure 6a shows the fraction of active nodes for both networks, in time, for

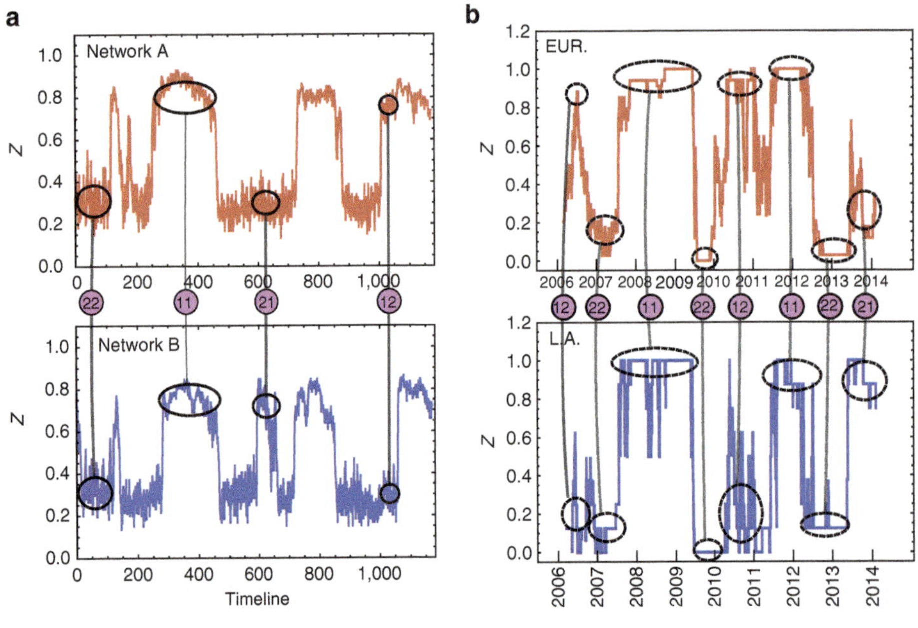

Figure 6 | Collective dynamics in simulated and real interacting networks. (**a**) Simulation of the networks' dynamics, activity versus time, for $N = 100$ and failure parameters $p^*_A = p^*_B = 0.21$, $r_A = r_B = 0.60$ and $r_d = 0.15$ shows the switching of the system between four different states. We can easily identify four collective states: 11, 22, 12 and 21. (**b**) Dynamics of two CDS geographical networks consisting of 17 European and 8 Latin American countries, showing very similar behaviour: individual networks switching between well-defined high-activity and low-activity states, as well as correlated collective behaviour of the two networks in interaction. We identify collective states 11, 22, 12 and 21, and mark them with connected black ovals. It is noteworthy that, as the CDS value grows with risk, a higher activity in a CDS network corresponds to bad economic news.

a symmetric choice of parameters, $p_A^*=p_B^*=0.21$, $r_A=r_B=0.60$ and $r_d=0.15$, when each network has only $N=100$ nodes. Large fluctuations cause the system to jump between the different states allowed for this set of parameters. It is noteworthy that interdependent links cause the two networks to have partially dependent and correlated dynamics. Very often a transition in one network triggers a transition in the other. In Fig. 6a we can identify examples of all four global states: 22, 11, 21 and 12. For example, at time $t\approx 400$ both networks are in the high activity state (11), whereas at $t\approx 620$ network A is in the low activity and network B in the high activity state (21).

As many real-world interacting network systems have a small number of nodes, in those systems we can potentially uncover dynamics similar to what we observe in our model networks. As an example of a real system, we investigate the interacting sovereign 5-year CDS system, consisting of 25 European Union (EU) and Latin American countries (see Methods for the full list of countries) that began to issue the CDSs from 2005. We divide countries into two groups on a geographical basis: 8 countries belong to the Latin American group and 17 belong to the European Union group. Sovereign CDSs are financial instruments, for which the value reflects the probability that the reference country will default on its debts. Each country has a CDS value assigned and this value changes in time reflecting the economic news about this country and the perceived risk of default, which results in a time series that we can observe. CDSs are highly sensitive to important economic news, positive or negative. There is also a significant contagion and influence between the countries, especially between those with strong economic ties, which is reflected in the correlation between their CDSs. These characteristics make the CDS signals a candidate for modelling using our interacting network approach.

We can draw a parallel between the CDS system and our model network if we assume that each country (with its associated CDS signal) can be represented as a node, which has connections (links) to other countries within its own geographical region, as well as ties with countries from another continent. In this case, we might expect that random and independent bad (or good) economic news appearing in any given country have behaviour similar to random internal processes in nodes in our artificial model (random internal failures/recoveries). When economic problems in one country propagate to a neighbouring country within the same geographical region, the process resembles the external failures in our artificial model, whereas interaction between countries from different continents may be modelled by the interdependent links from our network model. For the CDS network system we also suppose that the fractional definition for the threshold (m_{frac}) is somewhat more natural then the absolute definition, as it is less dependent on the country size or its importance, that is, the number of links a country has to other countries.

We study the international CDS system during the period between June 2005, the earliest date when CDSs traded for all countries, and February 2014. We apply the network model to it as follows. We represent each country with one node that can have two states: active or failed. As the raw CDS values are continuous by nature and our model uses binary node states (up or down), we perform a trend mapping procedure to form a binary signal (0 or 1) for each country. In particular, for each time t, we consider the interval $[t-252, t]$ of 252 business days (the usual number of business days in a year). If the CDS value of a country has a net increase during that period, we consider the node of the country to be active at t (state $=1$). If it does not, it is inactive (state $=0$). Having individual binary signals for each country, we can calculate the average activity $0\leq z(t)\leq 1$ for both EU and Latin American networks. The resulting time series for EU and LA activities are shown in Fig. 6b. First, we note that the two geographical networks spend most of time having either a significantly high activity or significantly low activity (that is, there is an indication for two well-defined single-network states). We confirm this by measuring the frequency distribution of network activities (Fig. 7a,b), which exhibit a strong bimodality in z. The CDS network system in Fig. 6b shows rapid transitions between the high and low activity states, much similar to the artificial network system in Fig. 6a. Figure 7c shows the calculated correlations between binary signals of pairs of individual nodes. The correlation matrix reveals two strongly correlated blocks, which we identify as Latin American block (numbers 1–8) and EU block (numbers 9–25).

In Fig. 6b, we also observe that the two networks sometimes make transitions simultaneously, but not always. This behaviour also resembles the behaviour observed in the artificial networks in Fig. 6a.

Finally, we find that it is possible to estimate numerical values for all the model parameters of this real system (internal p_{EU}^*, p_{LA}^*; external $m_{\text{frac,EU}}$, $m_{\text{frac,LA}}$, r_{EU}, r_{LA}; and interdependent r_d) from the data. The basic idea is that for each parameter we identify an observation experiment in which this particular parameter dominates, enabling us to effectively isolate individual parameters from the noise of many others. For example, when both networks (EU and LA) are in the high activity phase, most of the failures are in fact internal failures. This allows us to almost directly estimate p_{EU}^* and p_{LA}^* from Fig. 6b, by observing $p_{EU}^*=1-\langle z_{EU}\rangle$ and $p_{LA}^*=1-\langle z_{LA}\rangle$. External failures are most significant when a network is in a low activity state. Interdependent parameter r_d can be estimated by studying the correlation between $z_{EU}(t)$ and $z_{LA}(t)$, as this is an increasing function of r_d. Threshold parameters $m_{\text{frac,EU}}$ and $m_{\text{frac,LA}}$ can be estimated by exploiting the fact that they most significantly determine the fraction of time that each network spends in the high, or low, activity states. We describe in detail our procedures for numerically estimating these model parameters in the Supplementary Methods. The procedure for estimating some of these parameters is also illustrated in Supplementary Fig. 1. Numerical results for the parameter estimates are presented in Supplementary Table 1. Our dynamical network model also independently predicts that the typical fluctuation size of $z(t)$ is not uniform for all values of z, but has a spike around $z\approx\frac{1}{2}$. We observe this phenomenon in both our simulations and the real network dynamics (Supplementary Fig. 2).

Discussion

Interacting networks appear across many disciplines, from medicine, physics, biology and ecology, finance and economics, to infrastructure (refs 34,35). We propose a generic model that captures some of the most common processes found in real interacting networks—node failure, systemic damage propagation and node recovery. We report several intriguing results. Our solution of the model produces a rich phase diagram with a number of tipping points (critical points, triple points and transition lines). Using the phase diagram, we solve a fundamental problem of the optimal repairing strategy for a damaged system consisting of interconnected networks. Solving this problem enables us to determine the minimum set of node repairs required to repair a failed interconnected network system; thus, it becomes fully functional again. Remarkably, we find that the triple points from our phase diagram play the dominant role in constructing the optimal repairing strategy in damaged interacting systems. This implies that triple points are not only a thermodynamic or topological curiosity, but they have a very direct practical meaning and application, specifically in

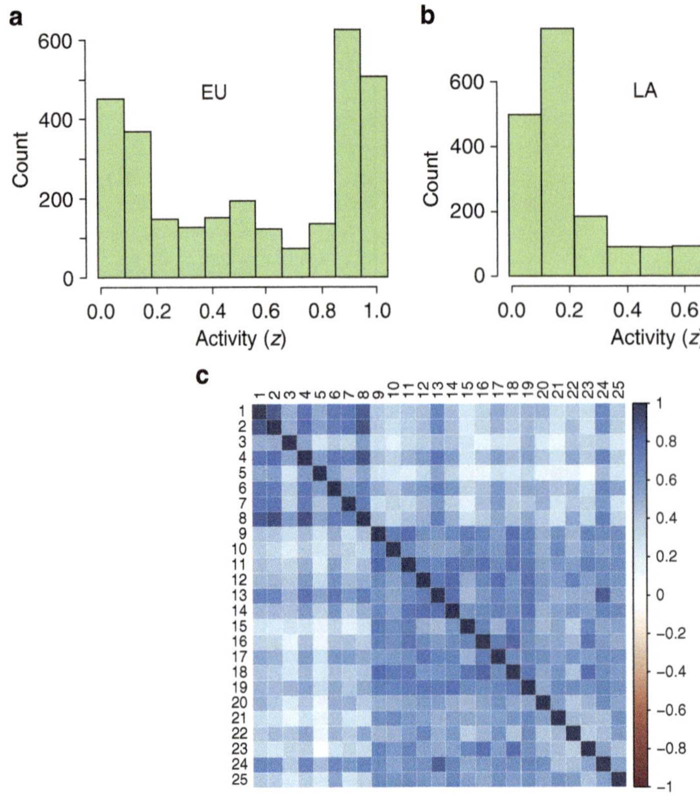

Figure 7 | Real network dynamics. (**a**) Activity density plot for the EU network reveals bimodality, an indication of the existence of two states. (**b**) Same for the LA network. (**c**) Correlation matrix of binary CDS signals with the EU (1-8) and LA (9-25) block. Separation into blocks reinforces our initial decision to sort the countries by geographic location.

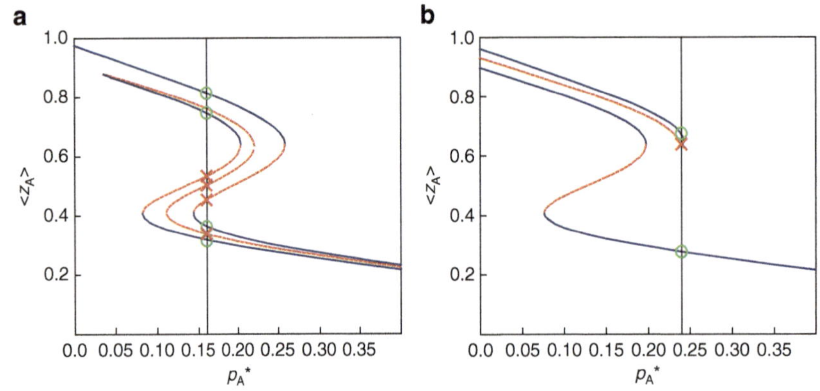

Figure 8 | Activity of network A versus internal failure rate. (**a**) Activity $z_A = 1 - a_A$ obtained by solving equations (1) and (2), for a range of p_A^* values, in a system of two interdependent networks ($k = 16$, $m = 8$). Blue lines correspond to stable physical states and red dotted lines represent unstable solutions. In Fig. 8a, the same parameters as in Fig. 1a are used, except p_A^*, which is not fixed but varied. When $p_A^* = 0.16$ (vertical black line), the corresponding values on the blue solid lines (green circles) match the graphical solutions in Fig. 1a (also green circles). (**b**) An analogous relationship holds between Figs 1b and 8b, in which case two stable states exist.

constructing repairing strategies in interacting systems. The problem of functional repair using minimal intervention is relevant in medicine, economics and other disciplines, when repairs are invasive or expensive. Finally, we apply our model to a selected real system: interconnected networks of CDSs, for two interconnected groups of countries. We propose a methodology to measure all of the model parameters in a real system, by using observational experiments in which a particular parameter dominates the behaviour of the system (see Supplementary Information).

Methods

Damage conductivity parameters. Parameters r_A and r_B are introduced, because they describe how easily the damage is spread through the network. When $r = 0$ there is no damage spread between the nodes, and when $r = 1$ there is perfect damage conduction. Assuming that external failures occur with certainty would mean fixing r to be equal to 1. In the case of a single network with recovery, it has been shown[12] that many important phenomena (for example, spontaneous recovery) are lost when $r = 1$. The most interesting parts of the phase diagram are in fact where r is far from 1.

It is noteworthy that our choice of the value for r_d is quite limited. If r_d is too large, we find that the damage spreads through dependency links extremely efficiently and the only possible stable state is total system collapse. The extreme

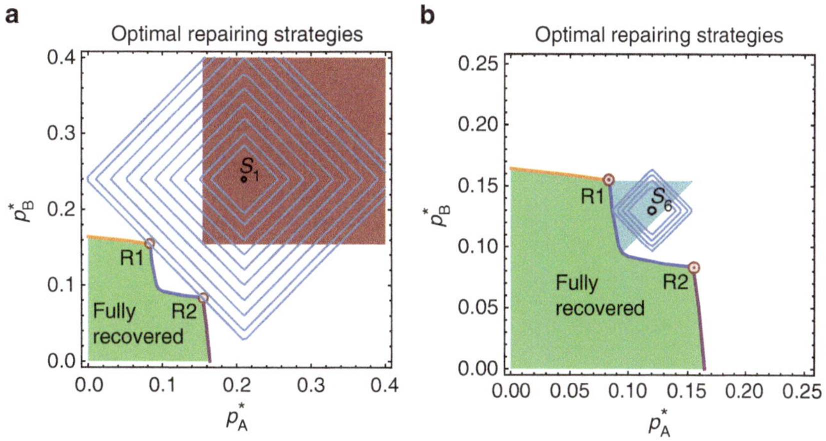

Figure 9 | Minimum Manhattan distance problem in two examples. (a) Finding the minimum Manhattan distance between point S_1 in the red sector and the green region where the system is fully functional. Equidistant curves are plotted in grey and form a 'diamond' shape. The largest 'diamond', barely touching the green region and having its centre at point S_1, suggests there are two equally optimal solutions to the minimization problem: points R1 and R2. **(b)** The same geometrical construction for point S_6 in the light blue region, suggests a unique solution: decreasing p_A^*.

vulnerability of interdependent networks is well known[2,22]. As there is always at least one functional stable state in biological or man-made systems, total system collapse as the only stable state is not realistic. Thus, we need the r_d parameter to 'soften' the dependency links[22] and allow a more realistic behaviour and its value should not be close to 1.

Mean field theory. Fractions a_A and a_B denote the fraction of nodes that are failed due to any of the three types of failures: internal (I), external (E) or dependency failure (D). We denote the probabilities that a node at a time of observation experiences a failure of I, E or D type as $P(I)$, $P(E)$ and $P(D)$, respectively. As a first approximation, we assume that these failures are mutually independent events. Considering network A first, we write an expression for the probability $a_{A,k}$ that a node of degree k in network A is failed. The node can fail due to I, E or D events, or to a combination of them. Using the inclusion–exclusion principle for independent events, we write

$$a_{A,k} = P(I) + P(E) + P(D) - P(I)P(E) - P(I)P(D) - P(E)P(D) + P(I)P(E)P(D). \quad (3)$$

Next, we separately calculate $P(I)$, $P(E)$ and $P(D)$.

$P(I)$ is also the average fraction of internally failed nodes in a network, as internal failures are independent events. This is a Poisson process on individual nodes[12,36] and therefore $P(I)=e^{-p_A\tau}$. As parameters p_A and τ come in this expression as a product, we can replace them with a single parameter, $p_A^* \equiv e^{-p_A\tau}$, which is bounded and also has the property $0 \le p_A^* \le 1$ and thus $P(I)=p_A^*$ for network A.

Next, we calculate $P(E)$, the probability that a randomly chosen node with degree k has externally failed. Focusing once again on network A, without a loss of generality, we let $F(k)$ be the probability that a node of degree k in network A is located in a critically damaged neighbourhood (where fewer than $m+1$ nodes are active). By definition, the time-averaged fraction of failed nodes (for any reason) in network A is $0 \le a_A \le 1$. In a mean-field approximation, this is also the average probability that a randomly chosen node in that network has failed. Using combinatorics, we obtain $F(k, a_A) = \sum_{j=0}^{m} \binom{k}{j} a_A^{k-j} (1-a_A)^j$ (ref. 12). The probability that a node of degree k in network A has externally failed is then $P(E) = r_A F(k, a_A)$. An analogous result is valid for network B.

Finally, we calculate $P(D)$, the probability that a node has failed due to the failure of its dependent counterpart node in the other network. For network A, this probability is equal to the product of parameter r_d and the probability that a counterpart node in B has failed: $P(D) = r_d a_B$. In network B by analogy, this probability is equal to $r_d a_A$.

Writing equation (3) for both networks and inserting the results for $P(I)$, $P(E)$ and $P(D)$ after summing over all k (and noting $a_A = \sum_k f(k) a_{A,k}$ and $a_B = \sum f(k) a_{B,k}$), we get a system of two coupled equations that describes the system of networks,

$$a_A = p_A^* + r_d a_B (1 - p_A^*) + \sum_k f(k) F(a_A) \left[r_A - p_A^* r_A - r_A r_d a_B + p_A^* r_A r_d a_B \right]. \quad (4)$$

$$a_B = p_B^* + r_d a_A (1 - p_B^*) + \sum_k f(k) F(a_B) \left[r_B - p_B^* r_B - r_B r_d a_A + p_B^* r_B r_d a_A \right]. \quad (5)$$

Additional phase diagrams. Figure 8a shows the collection of stable solutions (solid blue lines) and unstable solutions (dashed red lines) for the activity $z_A = 1 - a_A$ of network A, with parameter values as used in Fig. 1a, but for a range of different values of p_A^*. The solid black line indicates $p_A^* = 0.16$, the value of p_A^* used in Fig. 1a. Green circles in this figure correspond to the stable states found in Fig. 1a and red crosses correspond to the unstable solutions for z_A form Fig. 1a. Figure 8b shows an analogous phase diagram for the parameters with values as in Fig. 8a, again for a range of p_A^*.

Forbidden transitions. Transition lines for $12 \rightarrow 21$ and $21 \rightarrow 12$ do not appear in our phase diagram and it is quite easy to understand why. Let us assume that the transition line for $12 \rightarrow 21$ does exist. To obtain that transition, the idea would be to simultaneously increase p_A^* and decrease p_B^* (that is, increase the damage in one part of the system and decrease in another part). Suppose we are in phase 12 and infinitesimally close to the supposed transition line. Considering the local geometry of this line, we may be able to observe its angle with respect to the p_A^* axis. If a transition occurs when increasing p_A^* and decreasing p_B^*, the tangent on the supposed line would have an angle of $\theta \in [0, \frac{\pi}{2}]$. From here, it follows that by increasing p_A^* only, while keeping p_B^* constant, we would also make a transition (cross the transition line). The only other possibility would be that we were moving along the transition line, but this is easy to disprove because it would imply that the transition does not depend on p_A^*. If increasing p_A^* only causes a transition, the transition must end in state 22 and not in 21. This is because if we only increase p_A^*, we increase damage to both network A (directly) and network B (indirectly, through the interdependent links).

Geometry of the Manhattan distance minimization problem. The optimal strategies shown in different colours in Fig. 5 are derived from the geometrical reasoning shown in Fig. 9. Figure 9a shows a plot of a series of curves consisting of points at identical Manhattan distances from point S_1 (equidistant curves). They produce a 'diamond' shape, and the minimal Manhattan distance between point S_1 and the green region translates into the task of 'fitting' the diamond so that it just touches the green region and its centre is at S_1. The diamond in Fig. 9a touches the green region at two points—triple points, which are the solution to the minimization problem. Figure 9b shows the solution for point S_6 in the light blue region. Here the solution suggests a different strategy—decreasing only p_A^*.

Credit default swaps. Figure 6b shows an analysis of 5-year sovereign debt CDSs for a set of European countries: France, Germany, Italy, Spain, Portugal, Belgium, Austria, Denmark, Sweden, Greece, Ukraine, Hungary, Poland, Croatia, Slovenia, Romania, Bulgaria and Slovakia. This is the set of European countries that had a sovereign debt CDS in 2005. The set of Latin American countries we analyse consists of Brazil, Colombia, Argentina, Mexico, Venezuela, Chile, Peru and Panama. A CDS is typically used to transfer the credit exposure of fixed income products from one party to another. The buyer of the CDS is then obligated to make periodic payments to the seller of the CDS until the swap contract matures. In return, the seller of the CDS agrees to compensate (pay off) the seller who holds this third party debt if this (third party) defaults on the issued debt.

A CDS is, in effect, an insurance against non-payment of a debt owed by a third party. The buyer of a CDS does not have to hold the debt of the third party but can speculate on the possibility that the third party will indeed default and the buyer can purchase the CDS for this speculative purpose. CDSs were developed in the 1990s and, given their simple structure and flexible conditions, they are now a major part of the credit derivative activity in the OTC market used to hedge credit risk. One of the most important aspects of a CDS is the definition of the 'credit event' that triggers the CDS. These events include bankruptcy, obligation acceleration, obligation default, failure to pay, repudiation (moratorium) and restructuring. In the case of the sovereign bond market, the last three are typically included in the contracts. CDSs are used by investors to hedge exposure to a fixed income instrument, to speculate on likelihood of a third party (reference asset) default, or to invest in foreign country credit without currency exposure.

References

1. D'Agostino, G. & Scala, A. (Eds.) *Networks of Networks: The Last Frontier of Complexity* (Springer, 2014).
2. Buldyrev, S. V., Parshani, R., Paul, G., Stanley, H. E. & Havlin, S. Catastrophic cascade of failures in interdependent networks. *Nature* **464,** 1025–1028 (2010).
3. Gao, J., Buldyrev, S. V., Stanley, H. E. & Havlin, S. Networks formed from interdependent networks. *Nat. Phys.* **8,** 40–48 (2012).
4. Radicchi, F. & Arenas, A. Abrupt transition in the structural formation of interconnected networks. *Nat. Phys.* **9,** 717–720 (2013).
5. Vespignani, A. Complex networks: the fragility of interdependency. *Nature* **464,** 984–985 (2010).
6. Donges, J. F., Schultz, H. C. H., Marwan, N., Zou, Y. & Kurths, J. Investigating the topology of interacting networks. *Eur. Phys. J. B* **84,** 635–651 (2011).
7. Reis, S. D. S. *et al.* Avoiding catastrophic failure in correlated networks of networks. *Nat. Phys* **10,** 762–767 (2014).
8. Bashan, A. *et al.* The extreme vulnerability of interdependent spatially embedded networks. *Nat. Phys.* **9,** 667–672 (2013).
9. Kivelä, M., Arenas, A., Barthelemy, M., Gleeson, J. P., Moreno, Y. & Porter, M. A. Multilayer networks. *J. Complex Networks* **2,** 203–271 (2014).
10. Bianconi, G. Multilayer networks: dangerous liaisons? *Nat. Phys.* **10,** 712–714 (2014).
11. Boccaletti, S. *et al.* The structure and dynamics of multilayer networks. *Phys. Rep.* **544,** 1–122 (2014).
12. Majdandzic, A. *et al.* Spontaneous recovery in dynamical networks. *Nat. Phys.* **10,** 34–38 (2014).
13. Liu, C. *et al.* Modeling of self-healing against cascading overload failures in complex networks. *Eur. Phys. Lett.* **107,** 68003 (2014).
14. Podobnik, B. *et al.* Predicting lifetime of dynamical networks experiencing persistent random attacks. *Sci. Rep.* **5,** 14286 (2015).
15. Podobnik, B. *et al.* Network risk and forecasting power in phase-flipping dynamical networks. *Phys. Rev. E* **89,** 042807 (2014).
16. Bashan, A. *et al.* Network physiology reveals relations between network topology and physiological function. *Nat. Commun.* **3,** 702 (2012).
17. Caldarelli, G., Chessa, A., Pammolli, F., Gabriell, A. & Puliga, M. Reconstructing a credit network. *Nat. Phys.* **9,** 125–126 (2013).
18. Curme, C. *et al.* Emergence of statistically validated financial intraday lead-lag relationships. *Quant. Finance* **15,** 1375–1386 (2015).
19. Huang, X., Vodenska, I., Havlin, S. & Stanley, H. E. Cascading failures in bi-partite graphs: model for systemic risk propagation. *Sci. Rep.* **3,** 1219 (2013).
20. Albert, R., Jeong, H. & Barabási, A. L. Error and attack tolerance of complex networks. *Nature* **406,** 378–382 (2000).
21. Schneider, C. M., Moreira, A. A., Andrade, Jr J. S., Havlin, S. & J Herrmann, H. Mitigation of malicious attacks on networks. *Proc. Natl Acad. Sci. USA* **108,** 3838–3841 (2011).
22. Parshani, R. *et al.* Interdependent networks: reducing the coupling strength leads to a change from a first to second order percolation transition. *Phys. Rev. Lett.* **105,** 048701 (2010).
23. Simonsen, I., Buzna, L., Peters, K., Bornholdt, S. & Helbing, D. transient dynamics increasing network vulnerability to cascading failures. *Phys. Rev. Lett.* **100,** 218701 (2008).
24. Buzna, L., Peters, K. & Helbing, D. Modelling the dynamics of disaster spreading in networks". *Phys. A Stat. Mech. Appl.* **363,** 132–140.
25. Stippinger, M. & Kertész, J. Enhancing resilience of interdependent networks by healing. *Physica A* **416,** 481–487 (2014).
26. Dorogovtsev, S. N., Goltsev, A. V. & Mendes, J. F. F. Critical phenomena in complex networks. *Rev. Mod. Phys.* **80,** 1275–1335 (2008).
27. Valdez, L. D., Macri, P. A., Stanley, H. E. & Braunstein, L. A. Triple point in correlated interdependent networks. *Phys. Rev. E* **88,** 050803 (R) (2013).
28. Wieland, S. & Nunes, A. Asymmetric coevolutionary voter dynamics. *Phys. Rev. E* **88,** 062809 (2013).
29. Valdez, L. D., Macri, P. A. & Braunstein, L. A. A triple point induced by targeted autonomization on interdependent scale-free networks". *J. Phys. A Math. Theor.* **47,** 055002 (2014).
30. Watts, D. J. A simple model of global cascades on random networks". *Proc. Natl Acad. Sci.* **99,** 5766–5771 (2002).
31. Cohen, R. & Havlin, S. *Complex Networks: Structure, Robustness, and Function* (Cambridge Univ. Press, 2010).
32. Baxter, G. J., Dorogovtsev, S. N., Mendes, J. F. F. & Cellai, D. Weak percolation on multiplex networks. *Phys. Rev. E* **89,** 042801 (2014).
33. Kesselring, T. A., Franzese, G., Buldyrev, S. V., Herrmann, H. J. & Stanley, H. E. Nanoscale dynamics of phase flipping in water near its hypothesized liquid-liquid critical point. *Sci. Rep.* **2,** 474 (2012).
34. Newman, M. E. J. *Networks: An Introduction* (Oxford Univ. Press, 2010).
35. Albert, R. & Barabasi, A. L. Statistical mechanics of complex networks. *Rev. Mod. Phys.* **74,** 47–97 (2002).
36. Saaty, T. L. *Elements of Queueing Theory: with Applications* (McGraw-Hill, 1961).

Acknowledgements

We thank the DTRA, NSF (grants CMMI-1125290, CHE-1213217, PHY-1505000 and SES-1452061), Keck Foundation, European Commission FET Open Project (FOC 255987 and FOC-INCO 297149) and Office of Naval Research for financial support. S.H. acknowledges the European LINC and MULTIPLEX (EU-FET project 317532) projects, the Deutsche Forschungsgemeinschaft (DFG), the Israel Science Foundation, ONR and DTRA for financial support. L.A.B. thanks UNMdP and FONCyT, PICT 0429/13 for financial support. S.L.C. gratefully acknowledges the financial support of the Fulbright Program for visiting scholars. A.M. thanks Bijeli Zeko for useful discussions.

Author contributions

A.M., H.E.S. and S.H. designed the research. A.M., L.A.B., C.C., I.V. and S.L.C. performed the research. I.V. collected CDS data. A.M., L.A.B. and C.C. analysed data. A.M. wrote the main text. I.V. and S.L.C. wrote parts of the Supplementary Information. All the authors participated in the result discussion and revisions of the manuscript.

Additional information

Competing financial interests: The authors declare no competing financial interests.

14

Emergent long-range synchronization of oscillating ecological populations without external forcing described by Ising universality

Andrew E. Noble[1,2], Jonathan Machta[2,3] & Alan Hastings[1]

Understanding the synchronization of oscillations across space is fundamentally important to many scientific disciplines. In ecology, long-range synchronization of oscillations in spatial populations may elevate extinction risk and signal an impending catastrophe. The prevailing assumption is that synchronization on distances longer than the dispersal scale can only be due to environmental correlation (the Moran effect). In contrast, we show how long-range synchronization can emerge over distances much longer than the length scales of either dispersal or environmental correlation. In particular, we demonstrate that the transition from incoherence to long-range synchronization of two-cycle oscillations in noisy spatial population models is described by the Ising universality class of statistical physics. This result shows, in contrast to all previous work, how the Ising critical transition can emerge directly from the dynamics of ecological populations.

[1] Department of Environmental Science and Policy, University of California, Davis, California 95616, USA. [2] Department of Physics, University of Massachusetts, Amherst, Massachusetts 01003, USA. [3] Santa Fe Institute, 1399 Hyde Park Road, Santa Fe, New Mexico 87501, USA. Correspondence and requests for materials should be addressed to A.E.N. (email: andrew.e.noble@gmail.com).

Synchrony of dynamics across space has been a subject of intense study across a diverse array of scientific disciplines[1–5]. The study of synchrony of oscillations in population numbers across space and time has a long history in ecology[6] and has provided great insights into fundamental issues of population dynamics[3,5]. The absence of spatial synchrony is generally considered to be the key to persistence in exploiter–victim systems[5,7–9], and may be vitally important in the conservation of species[3,5,7] and disease eradication[10]. There is long-standing recognition of the need for new models and measures of spatial synchrony relevant to ecology[3,5,10,11] that do not depend upon the simplifying assumptions of the Kuramoto[1,12] and related models[2,4].

Three well-known causes of spatial synchrony are the long-range correlation of environmental variation (known as the Moran effect[6]), entrainment to neighbouring populations via trophic interactions and dispersal of individuals among habitat patches[3,5,13]. In the absence of environmental noise, short-range dispersal is sufficient to phase-lock spatial populations over arbitrarily large distance scales[7]. In the more realistic case of environmental variation, short-range dispersal has been shown to synchronize population oscillations only over distance scales comparable to the correlation length of the environmental noise[5,14–16]. The remaining question is whether and how synchrony over distances much longer than the scale of environmental correlations can arise from short-range dispersal in the presence of strong noise. This paper is the first to demonstrate the emergence of long-range synchronization and model-independent power-law scalings in noisy cyclic populations. Note that our results apply to the onset of spatial synchrony of oscillations in population numbers far from the extinction threshold, and our work is distinct from previous investigations of model independent power laws near the extinction threshold[17–19].

Many of the most prominent studies of spatial synchrony have focused on examples where the underlying dynamics are two cycles, including the dynamics of childhood diseases[10] and small mammals[13]. Because both of these examples have been argued to arise in part from the seasonal forces that impact the vast majority of ecological populations, two-cycle behaviour is truly a ubiquitous phenomenon. That two cycles are so common is expected from the structure of overcompensatory models in population biology, such as the logistic map, because of the large range of growth rates for which the asymptotic dynamics lie in an exact two-cycle. With the addition of noise, approximate two-cycle behaviour can be found over an even larger range of growth rates[20] along the entire period-doubling sequence and through the two-banded chaotic region, as can be seen in the bifurcation diagram for any quadratic map[21].

In statistical physics, the canonical model for the emergence of long-range order from the collective behaviour of local interactions is the Ising model[22–25]. The Ising model describes emergent phenomena at a phase transition where a simple twofold symmetry is broken. Physical systems with unrelated microscopic dynamics, such as magnetic, liquid–vapour and binary alloy systems, exhibit the power-law scaling behaviour of the Ising model at a critical phase transition. The Ising model thus defines a 'universality class'—a set of many disparate systems that are all characterized by identical power-law scaling behaviour over large distances despite wide differences in local interactions and short-range behaviour. Here, for a suite of ecological models, we will show that the power-laws describing asymptotic behaviour at the emergence of long-range synchronization are in the two-dimensional (2D) Ising universality class, providing a powerful tool for understanding the emergence of collective synchronization in spatial populations with local dispersal.

This quantitative correspondence between ecological population models and the Ising model is quite surprising and novel. The Ising universality class describes phase transitions for systems in thermal equilibrium, that is, systems with a well-defined energy and temperature[22–25]. On the other hand, ecological population models are dynamical, dissipative systems lying far from thermal equilibrium. Previous attempts to apply the Ising model to population ecology simply assume thermal equilibrium and lack any clear ecological motivation. In this paper, we show how a quantitative correspondence to the Ising model emerges naturally in spatially extended dynamical models of ecological populations, providing new insights for both fields.

Here, we first introduce a suite of discrete-time spatial population models and define a statistic for discriminating between incoherent and synchronous behaviour. Next, we demonstrate that the asymptotic behaviour of ecological population dynamics at the emergence of long-range synchronization is described by the Ising universality class. Finally, we define new statistics of spatial synchrony, each corresponding to a statistic of the Ising model, that can rapidly detect sudden changes in the spatial synchrony of ecological populations. We anticipate that our methods will be broadly applicable to the study of transitions in the spatial synchrony of cyclic populations, and we underscore the importance of universality for ecological modelling.

Results

Definition of the synchronization order parameter. We investigate the long-range synchronization of two-cycle oscillations in several discrete-time spatial population models with local dispersal and uncorrelated environmental noise: a single-species Ricker model (Methods equations (4) and (5))[26,27] with two different dispersal kernels, a single-species Logistic model (Methods equations (4) and (6))[28–31] and a version of the Nicholson–Bailey Host–Parasitoid model (Methods equations (7) and (8))[32]. The analogue of the local 'spin' variable at a lattice site in the Ising model is a local two-cycle amplitude, $m_{j,t}$ (Methods equation (9)), defined at each habitat patch j and generation t in the spatial population models. A spatiotemporal average of the $m_{j,t}$ is the synchronization order parameter, m (Methods equation (11)), which discriminates between phases of incoherence and long-range synchronization: an order parameter value of zero signals complete incoherence (the disordered phase), while nonzero values measure the degree of long-range synchronization (the ordered phase). On a landscape with one or two spatial dimensions (one-dimensional (1D) or 2D) and L habitat patches along each side (such that the total number of habitat patches is $N=L$ or $N=L^2$, respectively), finite-size estimates of the synchronization order parameter, m_L (Methods equation (13)), can be obtained from numerical simulations.

Ising universality at the critical transition. We present results for 1D and 2D ecological models with various numbers of habitat patches. The synchronization order parameter as a function of uncorrelated environmental noise level is plotted in Fig. 1a,b,d,e and Supplementary Figs 1a,b and 2a,b. On 1D landscapes, for sufficiently large N, the order parameter approaches zero in all the population models, and we find no evidence of long-range synchronization at nonzero noise levels. The same behaviour occurs in the 1D Ising model[22–25]. The behaviour on 2D landscapes is quite different. A continuous transition in the synchronization order parameter sharpens with increasing landscape size. Near this 'critical point', synchronized habitat clusters appear on the landscape and form fluctuating fractal patterns (Fig. 1c,f;

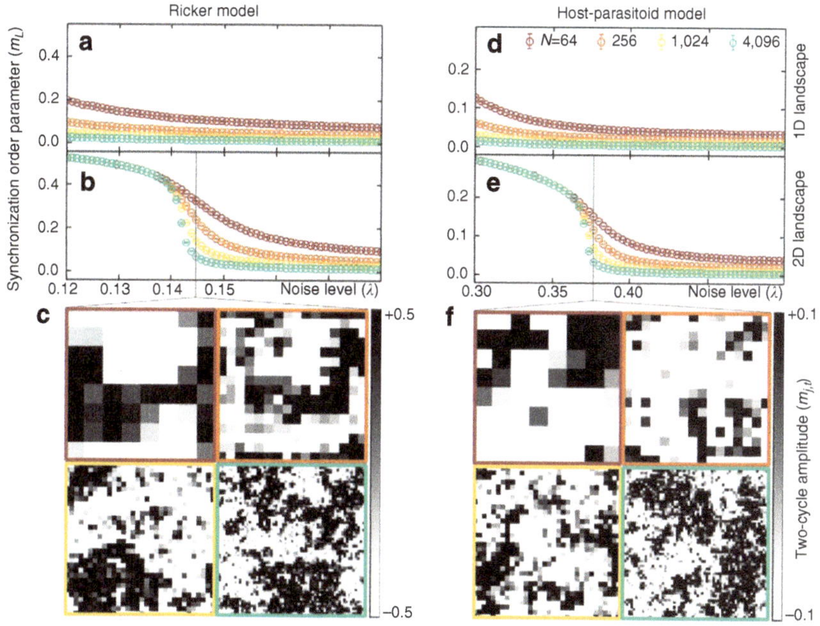

Figure 1 | Critical transitions at the onset of collective synchronization in spatial Ricker and host-parasitoid models on 2D landscapes. (**a**,**d**) On large 1D landscapes (large $N = L$) with short-range dispersal and nonzero uncorrelated environmental noise levels (λ), collective synchronization of spatial populations cannot emerge from local dispersal alone. (**b**,**e**) On 2D landscapes (where $N = L^2$), collective synchronization can emerge from local dispersal at a noise-induced transition. A continuous change in numerical estimates of the synchronization order parameter, m_L, from near zero (the disordered, incoherent phase) to larger values (the ordered, synchronous phase), sharpens as the size of the landscape increases. (**c**,**f**) Near the transition, on 2D landscapes ranging in size from $N = 64$ (upper-left) to 4,096 (lower-right), density plots of local two-cycle amplitudes, the $m_{j,t}$, show emergent long-range order in the fractal coexistence of synchronized habitat patches. In statistical physics, a continuous transition from a disordered phase to a phase ordered by emergent long-range correlations defines a 'critical transition'. Critical transitions at the onset of collective synchronization also occur in the 2D Logistic and Ricker-Moore models. Note that symbols are displayed with s.e.m. error bars, but actual error bars are often much smaller than the symbol size.

Supplementary Figs 1c and 2c) associated with the emergence of long-range order and critical slowing down (Supplementary Fig. 3). These results demonstrate both that long-range synchronization can be an emergent phenomenon on 2D landscapes even in the presence of noise, and also that the existence of a critical transition from incoherence to synchrony is independent of the details of the local ecological dynamics.

We now quantify the strong correspondence between the critical behaviour of the Ising model and the emergence of long-range synchronization on 2D landscapes. In addition to finite-size estimates of the order parameter, m_L (Methods equation (13)), we define two statistics of the local two-cycle amplitudes motivated by the Ising correspondence: the *susceptibility*, χ_L (Methods equation (14)), measures the variance of two-cycle amplitudes, and the fluctuation correlation length, ξ_L (Methods equation (16)), measures the length scale of long-range spatial correlations among two-cycle amplitudes. We find 'finite-size scaling collapses' in which measurements of m_L, χ_L and ξ_L on landscapes of different sizes and noise levels, when multiplied by appropriate powers of L given by the Ising 'critical exponents', lie on universal curves (Methods equation (24); Fig. 2a–c; Supplementary Figs 4–7). We also identify a scaling collapse across models: statistics of the different ecological population models, multiplied by model-dependent pre-factors, collapse onto corresponding Ising statistics near the critical transition (Methods equation (24); Fig. 2d–f; Supplementary Figs 8 and 9).

While it is not the main focus of our work, the slow dynamics near the Ising critical point is also ecologically interesting. The Ising model equipped with the relevant non-conservative dynamics is referred to as 'model A'[33–35]. Critical dynamics described by model A are universal and known to hold across a wide spectrum of physical systems near and far from thermal equilibrium[36–46]. We expect that the critical dynamics of all of the spatial population models studied here fall into the model-A dynamic universality class.

Ising universality explains the robustness and broad applicability of our results and provides a quantitative description of scaling behaviour at the onset of long-range synchronization that is independent of the details of local population dynamics. In particular, universality allows us to conclude that long-range synchronization among ecological oscillators can emerge from short-range dispersal even when a detailed description of the local dispersal behaviour and dynamics is unknown. We have explicitly demonstrated identical scaling behaviour in the spatial Ricker model with either the nearest-neighbour or Moore neighbourhood dispersal kernel (Fig. 2; Supplementary Figs 4,7 and 9), and universality ensures that the same scaling behaviour would be found for any localized dispersal kernel. Ecological populations exhibit even greater variability in their local connectivity but are nonetheless expected to be in the (random) Ising universality class[23] as long as dispersal remains short-ranged.

We can determine the conditions for long-range synchronization as a function of the parameters in each ecological model. For the spatial Ricker model, this information is summarized in Fig. 3 as a 'phase diagram' for incoherence and synchrony (see also Supplementary Fig. 10). We find that long-range synchronization emerging from local dispersal can persist in the presence of strong uncorrelated environmental noise. We conjecture that all critical points on the phase boundary lie in the 2D Ising universality class. The shape of the phase boundary as a function of growth rate, dispersal fraction and noise level will be qualitatively similar in the other population models.

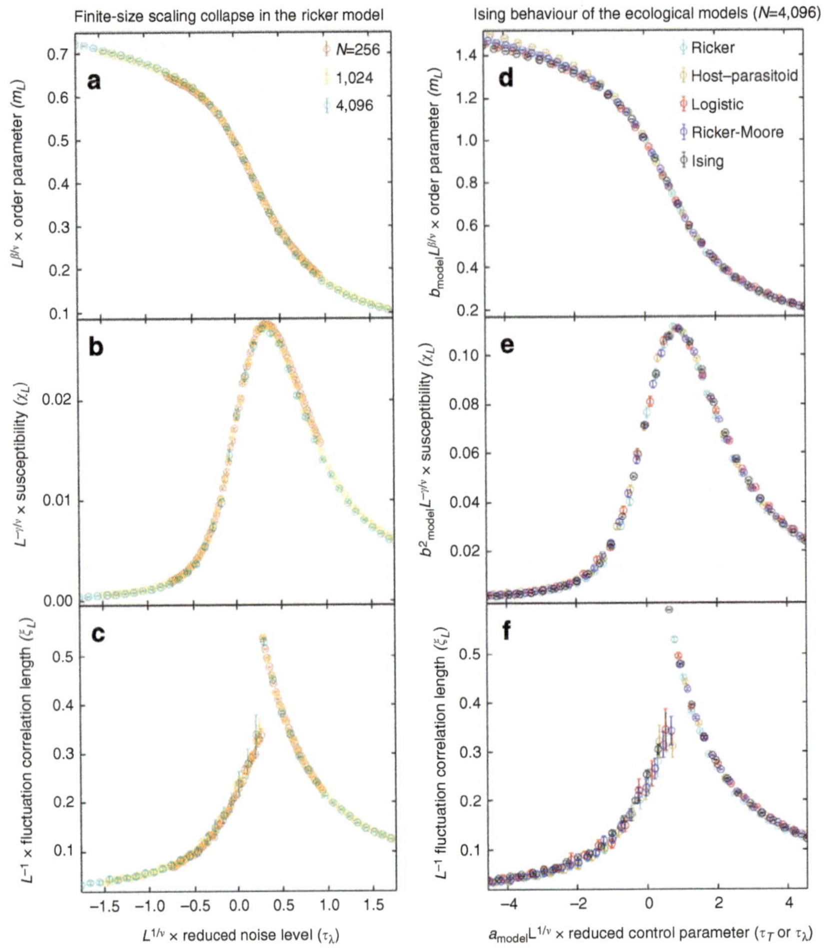

Figure 2 | Ising universality near the onset of collective synchronization in ecological populations on 2D landscapes. Scaled measurements of the order parameter (m_L), susceptibility (χ_L) and fluctuation correlation length (ξ_L) for diverse ecological models and 2D landscape sizes (where $N=L^2$) coincide with Ising model results, showing the existence of universal behaviour. (**a–c**) Finite-size scaling collapse (as a function of L and the Ising critical exponents, $\nu=1$, $\beta=1/8$, $\gamma=7/4$) demonstrates universal Ising behaviour of the Ricker model over a range of ecologically relevant landscape sizes. The same finite-size scaling behaviour is found in all the ecological models that we analyse. (**d–f**) Scaled collapse of ecological statistics onto the corresponding Ising statistics for $N=4{,}096$. Scaling parameters, a_{model} and b_{model}, are chosen separately for each model; without loss of generality, $a_{\text{Ising}}=b_{\text{Ising}}=1$. These results present strong evidence that the onset of collective synchronization is a critical transition in the Ising universality class. Universality explains the robustness and broad applicability of our results: long-range synchronization and power-law scalings emerge from local dispersal independent of the details of local dynamics, including population regulation, dispersal and landscape connectivity. Note that symbols are displayed with s.e.m. error bars, but actual error bars are often much smaller than the symbol size.

Rapid detection of sudden changes in spatial synchrony. Given that large-scale disturbances are not uncommon, the study of transient dynamics is especially important in ecology[47]. The rapid formation of complex patterns of spatial synchrony can be difficult to infer on timescales short enough to allow for an effective response. Using the relatively short time-series data typically available, new measures are needed to rapidly detect sudden changes in spatial synchrony. We introduce three such measures, each corresponding to a statistic of the Ising model: the synchronization index, S_t (Methods equation (17)), measures instantaneous synchronization as a non-spatial, two-generation, simple moving average; the susceptibility index, V_t (Methods equation (19)), measures the magnitude of fluctuations in synchronization windowed in both space and time; the correlation length index, CL_t (Methods equation (23)), measures a distance scale proportional to the size of synchronized clusters and is also windowed in both space and time. The spatial windowing of V_t and CL_t allows statistical power to be gathered across space in the absence of a long time series[48].

We demonstrate the application of these new measures to the rapid detection of sudden changes in spatial synchrony from incoherence to synchronized cluster formation and growth. In Fig. 4, starting with an ensemble of incoherent population configurations at generation zero ($t=0$), we calculate trajectories of S_t, V_t and CL_t emerging from the dynamics of a Ricker model where the growth rate, dispersal fraction and noise level correspond to either the synchronous phase (the signal, red trajectories) or the incoherent phase (the baseline, blue trajectories). The non-spatial index, S_t (Fig. 4a), is outperformed by the spatially windowed indices, V_t and CL_t (Fig. 4b,c), each of which allows for detection of a sudden change in spatial synchrony within two or three generations.

Synchronized cluster formation and growth in ecological models is analogous to a 'quench' in the Ising model with 'non-conserved' dynamics, where the size of clusters increases according to a power-law in time[22–25]. This power-law is universal[22–25], and therefore, independent of the details of local dynamics. The same power-law behaviour can be found in the

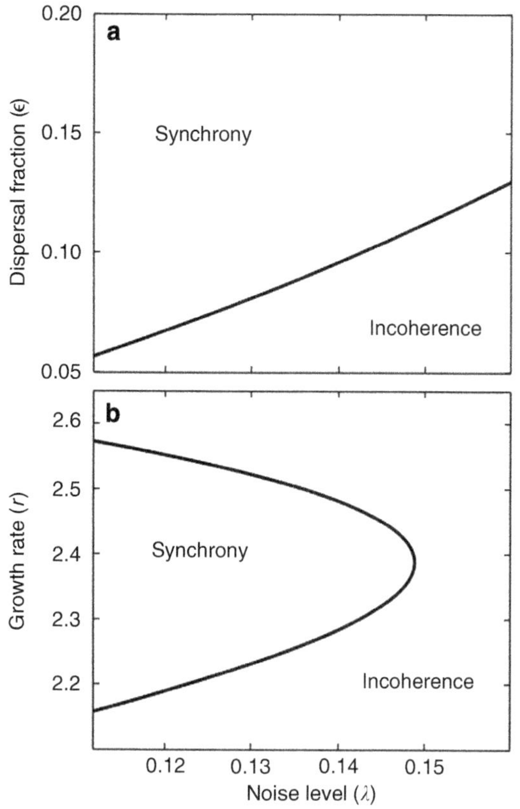

Figure 3 | Phase diagram for incoherence and collective synchronization in the Ricker model on 2D landscapes. The parameters of the Ricker model are dispersal fraction (ϵ), growth rate (r) and noise level (λ). For any given choice of parameters, there is a single asymptotic phase: either incoherence (with synchronization order parameter zero) or synchrony (with synchronization order parameter nonzero). These two phases are separated by a boundary of Ising critical transitions. For (**a**) fixed growth rate and (**b**) fixed dispersal fraction, we plot cross-sections of the critical boundary. Note that long-range synchronization can emerge from local dispersal despite strong uncorrelated environmental noise. A 'reentrant transition' from synchrony to incoherence at high growth rate, as seen in **b**, is associated with the disapperance of two-banded behaviour in the bifurcation diagram of the underlying Ricker map.

Ricker model (Fig. 4c) and in the host–parasitoid model of Rost *et al.*[32] These observations suggest that the sudden, spontaneous emergence of sharp ecological boundaries between clusters of habitat patches with opposite-sign synchronization, as seen in Fig. 4d, might occur in many types of ecological systems with local dispersal and uncorrelated environmental noise.

Discussion

By demonstrating the existence of Ising critical transitions in the spatial synchrony of noisy spatial populations, our work shows, for the first time, how long-range synchronization in ecological systems can emerge in the absence of external forcing (the Moran effect[6]). Beyond the important case of transitions between incoherence and two-cycle synchrony, our results suggest a general correspondence between emergent long-range synchronization in ecology and the universality classes of critical transitions in statistical physics. The methods developed here can be readily generalized and applied to establish those correspondences. This high degree of model independence implies that the possibility of emergent long-range synchronization must be carefully considered when seeking to understand the causes and consequences of spatial synchrony[3,5–16,27,32,49–56], including in the dynamics of childhood diseases, such as measles and pertussis[10], in the biological control of outbreaks of insects, such as bark beetles and gypsy moths[5] and in the mitigation of synchronous masting, which threatens food security in agro-ecosystems[56].

Finding appropriate measures of spatial synchrony in 2D ecological systems has been a serious challenge[5]. We have shown how the Ising correspondence introduces new statistics, both asymptotic measures and transient indices, to the study of spatial synchrony. In contrast to the Kuramoto model[1] and other spatially implicit approaches to quantifying spatial synchrony, these new statistics of spatial synchrony depend on the full spatiotemporal dynamics of a noisy population. Ising-like transient indices should be especially useful in quantifying spatial synchrony over timescales relevant to ecology and in the rapid detection of sudden transitions[57].

This work and others underscore the vital importance of universality for making robust, quantitative predictions about transitions to long-range order in ecological populations where the full complexity of local dynamics could never be modelled explicitly. Connections between spatial population models and the universality classes of statistical physics have previously been demonstrated in phase transitions to flocking[58], to extinction[17,19] and to other ordered phases relevant to ecological systems[25,59]. Universality of transitions in the spatial synchrony of noisy populations promises to remain a fruitful area of investigation.

Methods

Ising model and Ising universality. In ecology, the modelling of spatial populations begins with a dynamical systems approach, while in statistical physics, the modelling of a spin system in thermal equilibrium begins with a probability distribution called the Gibbs distribution[22–25]

$$P(\text{state}) \propto \exp(-E(\text{state})/T), \tag{1}$$

where the spin system is in thermal contact with a thermal reservoir that maintains a fixed temperature, T, and the energy of the state is E(state). (Boltzmann's constant is set to one in equation (1).) The energy function typically models local interactions among nearest-neighbour spins. In the Ising model[22–25], lattice sites are indexed by an integer, i, and the local spin, the s_i, is either up $(+1)$ or down (-1). The Ising energy function can be written as a sum over products of neighbouring spins

$$E = -J\sum_{\text{all } i}\left(s_i \sum_{j\in\mathcal{N}_i} s_j\right), \tag{2}$$

where $\mathcal{N}_i$ is the set of indices for the neighbouring spins of lattice site i and J is a coupling constant. The overall negative sign means that the energy is lower (higher) when neighbouring spins are more (less) aligned. Because the probability of any given configuration of local spin states over the lattice is given by the Gibbs distribution, configurations with a high level of local alignment are the most probable at low temperatures. The expected sum of all the spins

$$m \equiv \left\langle \frac{1}{N}\sum_{j=1}^{N} s_j \right\rangle, \tag{3}$$

defines the order parameter, m. Nonzero values signal an ordered phase in which spins are globally aligned; a zero value signals a disordered phase.

A 1D chain of spins will only order at zero temperature[22–25]. For 2D lattices, a phase transition from disorder to order occurs at a nonzero critical temperature, where the order parameter goes continuously to zero. Near the critical point, the fluctuation correlation length diverges: clusters of aligned spins form on all length scales, and clusters of opposite-sign spins coexist in a fluctuating fractal pattern. These emergent, long-range correlations render many details of the local interactions unimportant to emergent behaviour near the critical point. Indeed, the critical point of the Ising model is universal: the same power-law scaling relationships are common to critical points observed in seemingly unrelated magnetic, liquid–gas and binary alloy systems[22–25], as well as systems far from thermal equilibrium[36–46]. The numerical values of the exponents in those power-laws quantitatively define the Ising universality class[22–25].

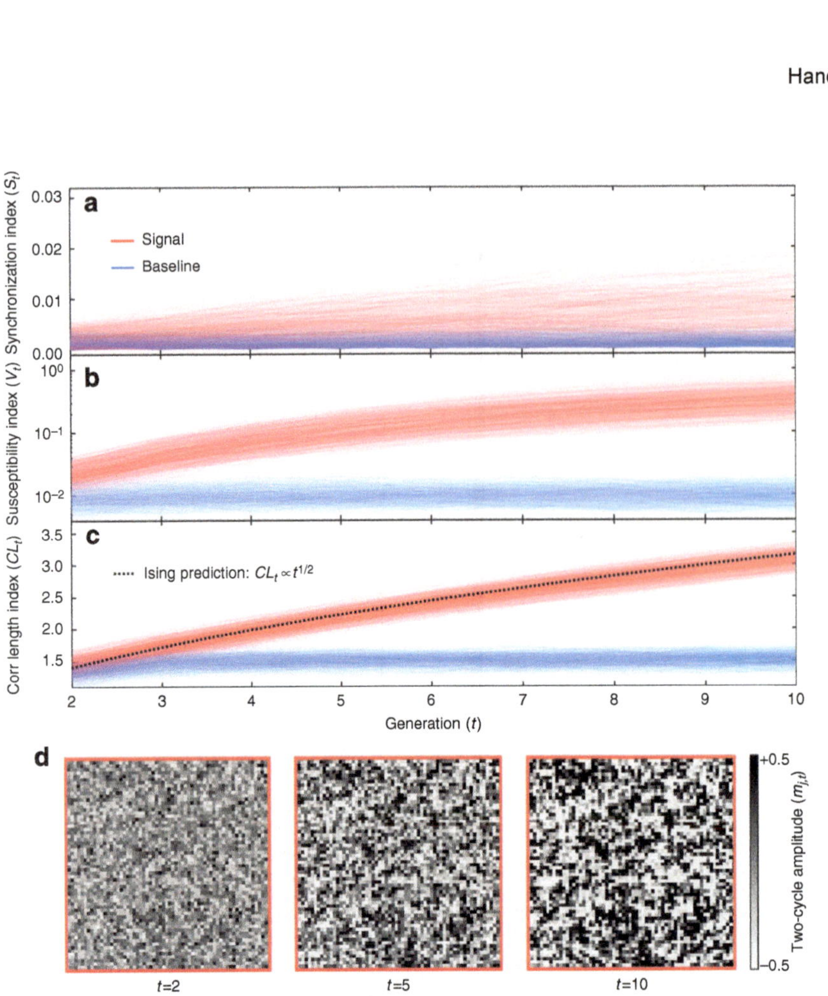

Figure 4 | Rapid detection of sudden changes in spatial synchrony from incoherence to synchronized cluster formation and growth. Synchronized cluster formation and growth in ecological models is analogous to a 'quench' in the dynamical Ising model. Starting with an ensemble of 1,000 incoherent population configurations at generation zero ($t=0$), we calculate trajectories of the (**a**) synchronization index (S_t), (**b**) susceptibility index (V_t) and (**c**) correlation length index (CL_t) emerging from the dynamics of a Ricker model on a 2D landscape of $N=L^2=4{,}096$ habitat patches. The growth rate, dispersal fraction and noise level correspond to either the synchronous phase (the signal, red trajectories) or the incoherent phase (the baseline, blue trajectories). Each index is averaged over two generations and can be calculated starting at $t=2$. The spatially windowed indices, V_t and CL_t, outperform the non-spatial index, S_t, and allow for rapid detection of a sudden change in spatial synchrony within two or three generations. (**d**) Snapshots at $t=2$, 5 and 10 of local two-cycle amplitudes, the $m_{j,t}$, are shown along a representative trajectory of the Ricker model with parameters corresponding to the synchronous phase. Sharp ecological boundaries between clusters of habitat patches with opposite-sign synchronization can emerge spontaneously on 2D landscapes with local dispersal and uncorrelated environmental noise.

Ecological population models. We start by investigating long-range synchronization in three models of single-species populations with local dispersal and uncorrelated environmental noise. The population density of habitat patch j at generation t is a continuous random variable, $X_{j,t}$. The geometry of N habitat patches on a spatial landscape is given by a linear or square lattice with periodic boundary conditions and one or two spatial dimensions. (If the linear dimension is L, then the landscape size is $N=L^d$ where $d=1, 2$.) The spatial dynamics are given by noisy coupled maps

$$X_{j,t+1} = (1-\epsilon)X_{j,t}f(X_{j,t}) + \frac{\epsilon}{z}\sum_{k\in\mathcal{N}_j} X_{k,t}f(X_{k,t}). \tag{4}$$

In each generation, local populations are regulated by the per capita density dependence of a noisy quadratic map, $f(X_{j,t})$, in which random environmental variation in population regulation, uncorrelated in both space and time, is modelled as multiplicative white noise. Afterwards, a fraction of each local population, ϵ, disperses evenly among z neighbouring habitat patches, as indexed by the elements of $\mathcal{N}_j$, the set of indices for the neighbouring spins of lattice site j. Note that changing the number of spatial dimensions does not change the fraction of a local population that disperses.

For any finite-sized landscape and nonzero noise levels, the asymptotic ($t\to\infty$) phase of the system is extinction ($X_j=0$ for all j). Nonetheless, typical extinction times rapidly increase with the number of habitat patches. In numerical simulations for finite N, we gather statistics on the single, long-lived, metastable state that is conjectured to be fully stable in the $N\to\infty$ (or 'thermodynamic') limit. In the following, 'asymptotic' will refer to times much larger than one generation but much less than the extinction time.

The Ricker model is equation (4) with a nearest-neighbour dispersal kernel on a linear or square lattice of habitat patches ($z=2d$) and local populations regulated by a noisy Ricker map[20,26]

$$f(X_{j,t}) = e^{r(1-X_{j,t})}(1+\lambda\xi_{j,t}), \tag{5}$$

where r is the growth rate, $\xi_{j,t}$ are standard normal deviates, uncorrelated in both space and time, and λ is the noise level, that is, the s.d. of the environmental noise distribution. Equation (5) is a well-known extension of the per capita density dependence of the Ricker map[26] to include the effects of multiplicative uncorrelated environmental noise[20]. A similar coupled map has previously been used to understand the causes of intraspecific spatial synchrony in large-scale moth and aphid populations[27].

The Logistic model is equation (4) with a nearest-neighbour dispersal kernel on a linear or square lattice of habitat patches ($z=2d$) and local populations regulated by a noisy logistic map[30]

$$f(X_{j,t}) = r(1-X_{j,t})(1+\lambda\xi_{j,t}). \tag{6}$$

The study of noisy logistic maps has a long history[30]. Pattern formation in coupled logistic maps with local two-cycle oscillations has been investigated previously[31].

The Ricker-Moore model is the same as the Ricker model but with a broader dispersal kernel. On 1D landscapes, the dispersal fraction is evenly distributed among the nearest and next-to-nearest neighbours ($z=4$). On 2D landscapes, the dispersal fraction is evenly distributed among the habitat patches of the Moore neighbourhood (the $z=8$ nearest and next-to-nearest neighbours on a square lattice).

The host-parasitoid model is a mechanistic two-species model in which the population dynamics of two non-interbreeding cohorts of hosts are coupled together by a shared parasitoid. The spatial dynamics of the host (H) and parasitoid (P) is taken from equation (8) of Rost *et al.*[32]

$$\begin{aligned} N_{j,t+2} &= (1-\epsilon)N_{j,t}f_N(N_{j,t},P_{j,t}) + \frac{\epsilon}{z}\sum_{k\in\mathcal{N}_j} N_{k,t}f_N(N_{k,t},P_{k,t}), \\ P_{j,t+1} &= (1-\epsilon)N_{j,t}f_P(P_{j,t}) + \frac{\epsilon}{z}\sum_{k\in\mathcal{N}_j} N_{k,t}f_P(P_{k,t}). \end{aligned} \tag{7}$$

with local populations regulated by a noisy Nicholson–Bailey coupled map

$$\begin{aligned} f_N(N_{j,t},P_{j,t}) &= e^{r(1-N_{j,t})}\frac{1-e^{-aP_{j,t}}}{aP_{j,t}}\left(1+\lambda\xi^N_{j,t}\right), \\ f_P(P_{j,t}) &= \frac{1-e^{-aP_{j,t}}}{aP_{j,t}}\left(1+\lambda\xi^P_{j,t}\right), \end{aligned} \tag{8}$$

where r is the growth rate of the host, a is the searching efficiency of the parasitoid, the $\xi^N_{j,t}$ and the $\xi^P_{j,t}$ are two sets of standard normal deviates, uncorrelated in space and time and with each other, and λ is the noise level. A nearest-neighbour dispersal kernel is assumed, where ϵ is the dispersal fraction for both host and parasitoid. Rost *et al.*[32] demonstrated the existence of coarsening behaviour and an asymptotic phase of spatial synchrony. Phase separation and coarsening in two-variable, discrete-time systems with local interactions have also been studied in the physics literature[60,61]. Our work focuses on the emergence of long-range synchronization from incoherence at a critical transition and the Ising universality of that transition.

Synchronization order parameter. In the Ricker, Logistic and Ricker–Moore models, we define a two-cycle amplitude for each local population given by

$$m_{j,t} \equiv \frac{1}{2}(-1)^{t+1}(X_{j,t+1}-X_{j,t}). \tag{9}$$

The spatial average of the $m_{j,t}$ defines the instantaneous synchronization

$$m_t \equiv \frac{1}{N}\sum_{j=1}^{N} m_{j,t}. \tag{10}$$

In the physics literature, order parameters are defined by an ensemble average denoted by angle brackets. Here, the expectation value of m_t taken over the probability distribution of the $X_{j,t}$ at long times defines the synchronization order parameter, m

$$m \equiv \langle m_t \rangle \quad \text{for } t \text{ large}. \tag{11}$$

An order parameter value of zero signals complete incoherence (the disordered phase); nonzero values measure the degree of partial synchronization (the ordered phase). The synchronization order parameter of the host-parasitoid model is the same as in equations (9)–(11) but with $N_{j,t}$ substituted for $X_{j,t}$ in the definition of the two-cycle amplitude. In our numerical estimates, we obtain the same results if $P_{j,t}$ is substituted instead. The onset of long-range synchrony in both the host and parasitoid populations is conjectured to occur at the same critical point.

To gather some intuition for the emergence of an Ising critical transition in the synchronization order parameter, we first consider a population in which density, x_t, is regulated by the deterministic Ricker map. This is a limiting case of the Ricker model. The limiting value of the two-cycle amplitude, as defined in equation (9), is

$$m_t = \frac{1}{2}(-1)^{t+1}(x_{t+1}-x_t), \tag{12}$$

and the limiting value of the synchronization order parameter, as defined in equation (11), is the average value of m_t in the asymptotic regime. For growth rates in the steady-state range, the asymptotic population density is a constant, $x_t = x^*$, and the order parameter vanishes, $m = 0$. The asymptotic system is clearly invariant under time translations, that is, $t \to t+n$ for any integer n. For growth rates in the two-cycle range, the asymptotic population density oscillates between two distinct values, x_1^* and x_2^*. The order parameter takes on one of two possible nonzero values, $m = \pm(x_1^* - x_2^*) \neq 0$, depending on the temporal phase of the two-cycle oscillation. The asymptotic system remains invariant to even-period, but not odd-period, time translations. Therefore, in the deterministic limit, increasing the growth rate from the steady state to the two-cycle range breaks a twofold, or $\mathcal{Z}_2$, symmetry of the asymptotic steady-state system. In the full Ricker model, high noise levels wash out any cycling in the underlying density dependence, and the asymptotic spatially averaged population density tends to a constant value as the size of the landscape becomes very large. In this phase, the synchronization order parameter vanishes. Only at sufficiently low noise levels can cycling emerge in asymptotic spatially averaged population density. In this phase, the asymptotic system is no longer invariant to odd-period time translations. A discrete twofold symmetry is broken in the transition from incoherence to synchrony, and this is the symmetry breaking pattern of the Ising universality class[22–25].

Asymptotic measures of spatial synchrony. We define new measures of spatial synchrony that complement existing approaches[3,5–16,27,32,49–56]. Each asymptotic measure is identical to a statistic in the Ising model if the $m_{j,t}$ are the Monte Carlo trajectories of a local spin rather than a time series of local two-cycle amplitudes. Each transient measure corresponds to an asymptotic measure but can be estimated at a single generation of the population model and does not require the existence of a long nor stationary time series.

Integrated autocorrelation time. This statistic measures the characteristic timescale over which correlations in trajectories of the instantaneous synchronization decay. Our estimates and error bars, as plotted in Supplementary Fig. 3, are based on p. 433 of Janke[62].

Synchronization order parameter (m_L). Because critical transitions occur only in the limit of infinite system size, numerical estimates of the synchronization order parameter on finite-sized $L \times L$ landscapes are denoted with a subscript, m_L, and are based on a spatial average of the $|m_{j,t,L}|$ (see, for example, equation (3) of Janke[62]). The estimate of the synchronization order parameter, as plotted in Fig. 1 and elsewhere, is

$$m_L \equiv \frac{1}{MN}\sum_{t=1}^{M}\sum_{j=1}^{N}|m_{j,t,L}|, \tag{13}$$

where N is the number of samples in the spatial averaging (the number of habitat patches on the landscape) and M is the number of samples in the time averaging (the length of the time series). Error bars are estimated based on the integrated autocorrelation time[63].

Susceptibility (χ_L). Defined as

$$\chi_L \equiv \frac{1}{M}\sum_{t=1}^{M}\sum_{j=1}^{N} m^2_{j,t,L} - Nm_L^2, \tag{14}$$

the susceptibility measures the strength of fluctuations in the synchronization order parameter. Error bars are estimated in a block bootstrap[63] with 100 blocks.

Fluctuation correlation length (ξ_L). This statistic measures the characteristic length scale over which correlations in local two-cycle amplitudes decay. On a landscape of infinite size, let the displacement between habitat patches i and j be $\mathbf{r}_{i,j}$. The strength of correlations in the $m_{j,t}$, as a function of $\mathbf{r}_{i,j}$, is given by the spatially averaged connected correlation function[63]

$$G(\mathbf{r}) \equiv \frac{1}{MN}\sum_{t=1}^{M}\sum_{i,j\text{s.t.}\mathbf{r}=\mathbf{r}_{i,j}}(m_{i,t}m_{j,t}-m^2). \tag{15}$$

Over large distances, we assume that $G(\mathbf{r})$ decays exponentially. The fluctuation correlation length is the characteristic length scale of that exponential decay[62]

$$\xi \equiv -\lim_{|\mathbf{r}|\to\infty}\frac{|\mathbf{r}|}{\ln G(\mathbf{r})}. \tag{16}$$

Our estimates of mean values and error bars for the fluctuation correlation length on finite-sized landscapes, ξ_L, are based on p. 23 of Janke[64].

Critical noise levels (λ_c). Our estimates of critical noise levels, λ_c, at the onset of long-range synchronization are based on crossings of Binder's cumulant, U[65]. In the 2D Ising model, the critical value, U^*, is universal and lies between -1.830 and -1.835 (ref. 39). We estimate U^* in each of the four population models and find a value consistent with the Ising result.

Transient measures of spatial synchrony. Each transient measure of spatial synchrony corresponds to an asymptotic measure but can be estimated at a single generation of the population model and does not require the existence of a long nor stationary time series. Transient measures are defined on a 2D landscape of $N = L^2$ habitat patches. We omit an explicit subscript 'L' on the symbols below to simplify notation.

Synchronization index (S_t). This statistic is an estimate of the synchronization order parameter windowed in time. We define the index as the absolute value of a two-generation simple moving average of the instantaneous synchronization (equation (10))

$$S_t \equiv \left|\frac{1}{2}\sum_{t'=t-1}^{t} m_{t'}\right|. \tag{17}$$

Susceptibility index (V_t). This statistic is an estimate of susceptibility windowed in both space and time. We define a square spatial window covering $N' = L/2 \times L/2$ habitat patches of the $L \times L$ landscape. With periodic boundary conditions, that window can be placed on the landscape in N possible ways. Let the kth placement of the spatial window cover N' habitat patches with their locations indexed by the elements of $\mathcal{M}_k$. For each placement in each generation, we calculate the average value of the two-cycle amplitudes (equation (10)) that fall within the spatial window

$$m_{k,t} \equiv \frac{1}{N'}\sum_{j\in\mathcal{M}_k} m_{j,t}. \tag{18}$$

Over two generations, we obtain $2N$ estimates of $m_{k,t}$. The absolute value of the mean of the estimates is simply the synchronization index defined above.

The susceptibility index is given by

$$V_t \equiv N'\left(\frac{1}{2N}\sum_{t'=t-1}^{t}\sum_{k=1}^{N} m_{k,t}^2 - S_t^2\right), \quad (19)$$

and provides a measure of variability in local synchronization.

Correlation length index (CL_t). This statistic defines a characteristic length scale for the size of synchronized clusters using the same approach to spatial and temporal windowing as in the definition of the susceptibility index. In the absence of long-range correlations and a long time series, we cannot obtain a good estimate of the long-distance exponential decay of the connected correlation function (equation (15)) that defines the finite-sized fluctuation correlation length, ξ_L (equation (16)). Instead, we define CL_t to be a length scale determined by the short-distance exponential decay of a spatially windowed correlation function. On the kth spatial window covering $N' = L/2 \times L/2$ habitat patches at generation t (see the discussion of spatial windowing in the definition of V_t above), we define the correlation function, $G_{k,t}(\mathbf{r})$, as

$$G_{k,t}(\mathbf{r}) \equiv \frac{1}{N'}\sum_{i,j \text{ s.t. } \mathbf{r}=\mathbf{r}_{i,j}} \left(m_{i,t}m_{j,t} - m^2\right). \quad (20)$$

This function can be averaged over N possible placements of the spatial window in each of two generations, defining $G_t(\mathbf{r})$ as

$$G_t(\mathbf{r}) \equiv \frac{1}{2N}\sum_{t'=t-1}^{t}\sum_{k=1}^{N} G_{k,t}(\mathbf{r}). \quad (21)$$

We hypothesize that $G_t(\mathbf{r})$ decays exponentially even over short distances such that

$$G_t(\mathbf{r}) \propto e^{-4|\mathbf{r}|/CL_t}, \quad (22)$$

and that CL_t can be estimated simply as

$$CL_t = \frac{4}{\log G_t(\mathbf{r}=(0,0)) - \log G_t(\mathbf{r}=(0,1))}. \quad (23)$$

The overall scale factor of 4 is chosen such that CL_t is roughly equal to the typical diameter of a synchronized cluster. In a sudden transition from incoherence to cluster formation and growth, a typical trajectory of CL_t lies on the curve $CL_t \propto t^{1/2}$, as shown in Fig. 4c. This is the expected power-law scaling for coarsening behaviour in the Ising model with non-conserved dynamics[22–25].

Finite-size scaling and evidence of Ising universality. Strictly speaking, the Ising critical transition occurs only on an infinite lattice of spins. Nonetheless, critical behaviour can be studied in numerical Monte Carlo simulations of finite-sized lattices[22–25]. Finite-size scaling theory has previously been used to provide strong numerical evidence of Ising (or in some cases, 'Ising-like') universality in the critical transitions of probabilistic cellular automata[36] and chaotic map lattices[37–39,43]. Here, we apply similar methods to study critical transitions in coupled maps of non-chaotic, noisy two-cycles with applications to ecology. We first define the analogue of reduced temperature in our noisy population models and then introduce our finite-size scaling hypothesis.

Reduced control parameters. In studies of the Ising model, the reduced temperature (τ_T) is defined as $(T-T_c)/T$, where T is the temperature and T_c is the critical temperature. To investigate the Ising universality of noise-induced critical transitions in ecological models, we define the reduced noise level (τ_λ) as $(\lambda-\lambda_c)/\lambda$, where λ is the noise level and λ_c is the critical noise level. If we want to refer to either τ_T or τ_λ, we will simply use the symbol τ and call this the reduced control parameter.

Data collapse for ecological statistics. Near a critical transition in the Ising model and the ecological models, we gather finite-size measurements of the synchronization order parameter, m_L, susceptibility, χ_L, and fluctuation correlation length, ξ_L, for various landscape sizes, L, and reduced control parameter values, τ. For each model, we find that these measurements can be scaled to lie on a function, F_m, F_χ or F_ξ, that is independent of both the model and the landscape size[22–25]

$$\begin{aligned} m_L &= b_{\text{model}}^{-1} L^{-\beta/\nu} F_m\left(a_{\text{model}} L^{1/\nu}\tau\right), \\ \chi_L &= b_{\text{model}}^{-2} L^{\gamma/\nu} F_\chi\left(a_{\text{model}} L^{1/\nu}\tau\right), \\ \xi_L &= L F_\xi\left(a_{\text{model}} L^{1/\nu}\tau\right). \end{aligned} \quad (24)$$

The powers of L are determined by the 2D Ising critical exponents ($\nu = 1$, $\beta = 1/8$, $\gamma = 7/4$) and are model independent. The model-dependent scale factors are a_{model} and b_{model}, where a_{model} is a scaling of the reduced noise level and b_{model} is a scaling of local two-cycle amplitudes

$$m_{j,t,L} \to b_{\text{model}} m_{j,t,L}. \quad (25)$$

The asymptotic measures of spatial synchrony, m_L, χ_L and ξ_L, defined above in terms of the $m_{j,t,L}$, scale with b_{model} as

$$\begin{aligned} m_L &\to b_{\text{model}} m_L, \\ \chi_L &\to b_{\text{model}}^2 \chi_L, \\ \xi_L &\to \xi_L. \end{aligned}$$

We choose $a_{\text{Ising}} = b_{\text{Ising}} = 1$ without loss of generality. For the ecological models, corrections to scaling and quantitative estimates of critical exponents will be published elsewhere.

Monte Carlo simulations. Our numerical methods for estimating asymptotic statistics characterizing noisy ecological populations are based on well-established methods in the statistical physics literature on Monte Carlo simulations[62–65]. In each Monte Carlo simulation, we iterate equation (4) or (7) for a burn-in period of T_{burnin} generations before running for T_{steps} generations and gathering M samples of each statistic at regular intervals. In all simulations, initial population densities are drawn from a normal distribution with mean μ_0 and s.d. σ_0 (negative numbers, if drawn, are thrown out), but results are independent of initial conditions, so long as local population densities are nonzero. Details follow on the Monte Carlo simulations used to estimate the statistics plotted in the main text and Supplementary Information.

Fig. 1: (**a,b**) On the basis of Monte Carlo simulations, with $T_{\text{burnin}} = 2\times10^6$, $T_{\text{steps}} = 2\times10^7$ and $M = 1\times10^5$, of the Ricker model, with $r = 2.3$, $\epsilon = 0.1$ and λ one of 64 evenly spaced values on the interval [0.12,0.18]. Snapshots are shown in **c** for $\lambda = 0.145$. Initial population densities are drawn from a normal distribution with $\mu_0 = 0.50$ and $\sigma_0 = 0.1$. (**d,e**) On the basis of Monte Carlo simulations, with $T_{\text{burnin}} = 2\times10^6$, $T_{\text{steps}} = 2\times10^7$ and $M = 1\times10^5$, of the host–parasitoid model, with $r = 1.5$, $\epsilon = 0.1$, $a = 20$ and λ one of 64 evenly spaced values on the interval [0.3,0.5]. Snapshots are shown in **f** for $\lambda = 0.376$. Initial population densities are drawn from a normal distribution with $\mu_0 = 0.50$ and $\sigma_0 = 0.1$.

Fig. 2: (**a–c**) On the basis of Monte Carlo simulations, with $T_{\text{burnin}} = 2\times10^7$, $T_{\text{steps}} = 2\times10^8$ and $M = 1\times10^5$, of the Ricker model, with $r = 2.3$, $\epsilon = 0.1$. We scanned 72 evenly spaced values of λ on the interval [0.135,0.150] but plotted statistics over a smaller range closer to the critical transition. Initial population densities are drawn from a normal distribution with $\mu_0 = 0.50$ and $\sigma_0 = 0.1$. The critical noise level, $\lambda_c = 0.14131(4)$, is estimated from the critical value of Binder's cumulant and used to calculate the reduced noise level as plotted on the horizontal axis. (**d–f**) The points and error bars are the $N = 4{,}096$ series in panels b, d and f of Supplementary Figs 4–8 (see details below), after a model-dependent scaling of the horizontal and vertical axes by a_{model} and b_{model}. Without loss of generality, we define $a_{\text{Ising}} = b_{\text{Ising}} = 1$. Then we estimate $a_{\text{Ricker}} = 2.60$, $a_{\text{host-parasitoid}} = 3.15$, $a_{\text{Logistic}} = 2.85$, $a_{\text{Ricker-Moore}} = 2.65$, and $b_{\text{Ricker}} = 2.028$, $b_{\text{host-parasitoid}} = 4.382$, $b_{\text{Logistic}} = 7.228$, $b_{\text{Ricker-Moore}} = 2.179$.

Fig. 3: The phase boundary plotted in this figure is an interpolation of the maximal values of fluctuation correlation length in Supplementary Fig. 10 (see details below).

Fig. 4: On the basis of Monte Carlo simulations, with $T_{\text{burnin}} = 0$ and $T_{\text{steps}} = 10$, of the Ricker model, with $\epsilon = 0.1$ and $\lambda = 0.12$. For the signal (red lines), $r = 2.3$; for the baseline behaviour (blue lines), $r = 1.5$. In each scenario, trajectories of the synchronization, susceptibility and correlation length indices are plotted for 1,000 initial population configurations. Initial population densities are drawn from a normal distribution with $\mu_0 = 1.00$ and $\sigma_0 = 0.1$.

Supplementary Fig. 1: same as Fig. 1a–c but for the Logistic model, with $r = 3.3$, $\varepsilon = 0.1$ and λ one of 64 evenly spaced values on the interval [0.04,0.07]. Snapshots are shown in panel (C) for $\lambda = 0.0519$.

Supplementary Fig. 2: (A,B) same as Fig. 1 a,b but for the Ricker–Moore model, with $r = 3.3$, $\varepsilon = 0.1$ and λ one of 64 evenly spaced values on the interval [0.13,0.19]. Snapshots are shown in panel (C) for $\lambda = 0.154$.

Supplementary Fig. 3: integrated autocorrelation times of the synchronization order parameter are estimated from the same Monte Carlo simulations used to generate Fig. 1.

Supplementary Fig. 4: panels D–F are the same as Fig. 2 a–c. Panels A–C in this figure plot the same statistics without finite-size scaling.Supplementary Fig. 5: same as Supplementary Fig. 4 but for the Host–Parasitoid model, with $r = 1.5$, $\varepsilon = 0.1$, $\alpha = 20$ and λ one of 72 evenly spaced values on [0.347,0.397]. Initial population densities, for both host and parasitoid, are drawn from a normal distribution with $\mu_0 = 0.50$ and $\sigma_0 = 0.1$. The critical noise level is $\lambda_c = 0.3696(2)$.

Supplementary Fig. 6: same as Supplementary Fig. 4 but for the Logistic model, with with $r = 3.3$, $\varepsilon = 0.1$ and λ one of 72 evenly spaced values $[0.0475, 0.0550]$. The critical noise level is $\lambda_c = 0.05074(2)$.

Supplementary Fig. 7: same as Supplementary Fig. 4 but for the Ricker–Moore model, with $r = 2.3$, $\varepsilon = 0.1$ and λ one of 72 evenly spaced values on $[0.144, 0.159]$. The critical noise level is $\lambda_c = 0.15016(5)$.

Supplementary Fig. 8: based on simulations of the 2D Ising model using the Wolff algorithm[63], with $T_{burnin} = 1\times10^6$, $T_{steps} = 1\times10^7$ and $M = 1\times10^5$, and $1/T$ (in units where Botzmann's constant is set to unity) one of 64 evenly spaced values on the interval $[0.385, 0.526]$. Initial spin configurations are fully aligned (all spin values are $+1$). The critical temperature and theoretical predictions for an infinite lattice are well known in the physics literature[22–25].

Supplementary Fig. 9: same as Fig. 2 d–f but for the A–C $N = 256$ and D–F $N = 1{,}024$ series.

Supplementary Fig. 10: based on Monte Carlo simulations, with $T_{burnin} = 2\times10^6$, $T_{steps} = 2\times10^7$ and $M = 1\times10^5$, of the Ricker model where $N = 4{,}096$ and λ is one of 26 evenly spaced values on the interval $[0.11, 0.16]$. (A) $r = 2.3$ and ε is one of 30 evenly spaced values on $[0.05, 0.20]$. (B) $\varepsilon = 0.1$ and r is one of 30 evenly spaced values on $[2.10, 2.65]$.

References

1. Kuramoto, Y. in *International Symposium on Mathematical Problems in Theoretical Physics*. (ed. Arakai, H.) 420–422 (Springer, 1975).
2. Acebron, J. A., Bonilla, L. L., Vicente, C. P. J., Ritort, F. & Spigler, R. The Kuramoto model: a simple paradigm for synchronization phenomena. *Rev Mod. Phys.* **77,** 137–185 (2005).
3. Björnstad, O. N., Ims, R. A. & Lambin, X. Spatial population dynamics: Analyzing patterns and processes of population synchrony. *Trends Ecol. Evol.* **14,** 427–432 (1999).
4. Strogatz, S. H. *Sync: The Emerging Science of Spontaneous Order* (Hyperion, 2003).
5. Liebhold, A., Koenig, W. D. & Björnstad, O. N. Spatial synchrony in population dynamics. *Annu. Rev. Ecol. Evol. Syst.* **35,** 467–490 (2004).
6. Moran, P. A. P. The statistical analysis of the Canadian Lynx cycle. 2. Synchronization and meteorology. *Aust. J. Zool.* **1,** 291–298 (1953).
7. Earn, D. J. D., Levin, S. A. & Rohani, P. Coherence and conservation. *Science* **290,** 1360–1364 (2000).
8. Keeling, M. J., Wilson, H. B. & Pacala, S. W. Deterministic limits to stochastic spatial models of natural enemies. *Am. Nat.* **159,** 57–80 (2002).
9. Dey, S. & Joshi, A. Stability via asynchrony in *Drosophila* metapopulations with low migration rates. *Science* **312,** 434–436 (2006).
10. Grenfell, B. T., Björnstad, O. N. & Kappey, J. Travelling waves and spatial hierarchies in measles epidemics. *Nature* **414,** 716–723 (2001).
11. Blasius, B., Huppert, A. & Stone, L. Complex dynamics and phase synchronization in spatially extended ecological systems. *Nature* **399,** 354–359 (1999).
12. Goldwyn, E. E. & Hastings, A. The roles of the Moran effect and dispersal in synchronizing oscillating populations. *J. Theor. Biol.* **289,** 237–246 (2011).
13. Björnstad, O. N., Stenseth, N. C. & Saitoh, T. Synchrony and scaling in dynamics of voles and mice in northern Japan. *Ecology* **80,** 622–637 (1999).
14. Satake, A. & Iwasa, Y. The synchronized and intermittent reproduction of forest trees is mediated by the Moran effect, only in association with pollen coupling. *J. Ecol.* **90,** 830–838 (2002).
15. Satake, A. & Iwasa, Y. Spatially limited pollen exchange and a long-range synchronization of trees. *Ecology* **83,** 993–1005 (2002).
16. Fox, J. W., Vasseur, D. A., Hausch, S. & Roberts, J. Phase locking, the Moran effect and distance decay of synchrony: experimental tests in a model system. *Ecol. Lett.* **14,** 163–168 (2011).
17. Täuber, U. C., Howard, M. & Vollmayr-Lee, B. P. Applications of field-theoretic renormalization group methods to reaction-diffusion problems. *J. Phys.* **38,** R79 (2005).
18. Yaari, G., Ben-Zion, Y., Shnerb, N. M. & Vasseur, D. A. Consistent scaling of persistence time in metapopulations. *Ecology* **93,** 1214–1227 (2012).
19. Bonachela, J. A., Muñoz, M. A. & Levin, S. A. Patchiness and demographic noise in three ecological examples. *J. Statist. Phys.* **148,** 723–739 (2012).
20. Schaffer, W. M., Ellner, S. & Kot, M. Effects of noise on some dynamical models in ecology. *J. Math. Biol.* **24,** 479–523 (1986).
21. Hastings, A., Hom, C. L., Ellner, S., Turchin, P. & Godfray, H. C. J. Chaos in ecology: is mother nature a strange attractor? *Annu. Rev. Ecol. Syst.* **24,** 1–33 (1993).
22. Goldenfeld, N. *Lectures on Phase Transitions and the Renormalization Group* (Westview Press, 1992).
23. Cardy, J. *Scaling and Renormalization in Statistical Physics* (Cambridge Univ. Press, 1996).
24. Sethna, J. *Statistical Mechanics: Entropy, Order Parameters and Complexity* (Oxford Univ. Press, 2006).
25. Solé, R. V. *Phase Transitions* (Princeton Univ. Press, 2011).
26. Ricker, W. E. Stock and recruitment. *J. Fish. Res. Board Can.* **11,** 559–623 (1954).
27. Hanski, I. & Woiwod, I. P. Spatial synchrony in the dynamics of Moth and Aphid populations. *J. Anim. Ecol.* **62,** 656–668 (1993).
28. May, R. M. *Stability and Complexity in Model Ecosystems* 1st edn (Princeton Univ. Press, 1973).
29. May, R. M. Simple mathematical models with very complicated dynamics. *Nature* **261,** 459–467 (1976).
30. Crutchfield, J. P., Farmer, J. D. & Huberman, B. A. Fluctuations and simple chaotic dynamics. *Phys. Rep.* **92,** 45–82 (1982).
31. Oppo, G. L. & Kapral, R. Discrete models for the formation and evolution of spatial structure in dissipative systems. *Phys. Rev. A* **33,** 4219 (1986).
32. Rost, M., Várkonyi, G. & Hanski, I. Patterns of 2-year population cycles in spatially extended host-parasitoid systems. *Theor. Popul. Biol.* **59,** 223–233 (2001).
33. Hohenberg, P.C. & Halperin, B. I. Theory of dynamic critical phenomena. *Rev. Mod. Phys.* **49,** 435–479 (1977).
34. Bassler, K. E. & Schmittmann, B. Critical dynamics of nonconserved Ising-like systems. *Phys. Rev. Lett.* **73,** 3343–3346 (1994).
35. Haake, F., Lewenstein, M. & Wilkens, M. The irrelevance of detailed balance for the dynamical critical exponent. *Z. Phys. B Condens. Matter* **55,** 211–218 (1984).
36. Grinstein, G., Jayaprakash, C. & He, Y. Statistical mechanics of probabilistic cellular automata. *Phys. Rev. Lett.* **55,** 2527–2530 (1985).
37. Miller, J. & Huse, D. Macroscopic equilibrium from microscopic irreversibility in a chaotic coupled-map lattice. *Phys. Rev. E* **48,** 2528–2535 (1993).
38. Marcq, P., Chaté, H. & Manneville, P. Universal critical behavior in two-dimensional coupled map lattices. *Phys. Rev. Lett.* **77,** 4003–4006 (1996).
39. Marcq, P., Chaté, H. & Manneville, P. Universality in Ising-like phase transitions of lattices of coupled chaotic maps. *Phys. Rev. E* **55,** 2606–2626 (1997).
40. Täuber, U. C., Akkineni, V. & Santos, J. Effects of violating detailed balance on critical dynamics. *Phys. Rev. Lett.* **88,** 045702 (2002).
41. Risler, T., Prost, J. & Julicher, F. Universal critical behavior of noisy coupled oscillators. *Phys. Rev. Lett.* **93,** 175702 (2004).
42. Risler, T., Prost, J. & Julicher, F. Universal critical behavior of noisy coupled oscillators: a renormalization group study. *Phys. Rev. E* **72,** 016130 (2005).
43. Marcq, P., Chaté, H. & Manneville, P. Critical properties of phase transitions in lattices of coupled logistic maps. *Prog. Theor. Phys. Suppl.* **161,** 244–250 (2006).
44. Sieberer, L. M., Huber, S. D., Altman, E. & Diehl, S. Dynamical critical phenomena in driven-dissipative systems. *Phys. Rev. Lett.* **110,** 195301 (2013).
45. Sieberer, L. M., Huber, S. D., Altman, E. & Diehl, S. Nonequilibrium functional renormalization for driven-dissipative Bose-Einstein condensation. *Phys. Rev. B* **89,** 134310 (2014).
46. Täuber, U. C. & Diehl, S. Perturbative field-theoretical renormalization group approach to driven-dissipative bose-einstein criticality. *Phys. Rev. X* **4,** 021010 (2014).
47. Hastings, A. Timescales, dynamics, and ecological understanding. *Ecology* **91,** 3471–3480 (2010).
48. Dakos, V., van Nes, E. H., Donangelo, R., Fort, H. & Scheffer, M. Spatial correlation as leading indicator of catastrophic shifts. *Theor. Ecol.* **3,** 163–174 (2010).
49. Grenfell, B. T. *et al.* Noise and determinism in synchronized sheep dynamics. *Nature* **394,** 674–677 (1998).
50. Rohani, P., Earn, D. J. D. & Grenfell, B. T. Opposite patterns of synchrony in sympatric disease metapopulations. *Science* **286,** 968–971 (1999).
51. Ims, R. A. & Andreassen, H. P. Spatial synchronization of vole population dynamics by predatory birds. *Nature* **408,** 194–196 (2000).
52. Post, E. & Forchhammer, M. C. Synchronization of animal population dynamics by large-scale climate. *Nature* **420,** 168–171 (2002).
53. Björnstad, O. N., Peltonen, M., Liebhold, A. M. & Baltensweiler, W. Waves of larch budmoth outbreaks in the European Alps. *Science* **298,** 1020–1023 (2002).
54. Abbott, K. C. A dispersal-induced paradox: synchrony and stability in stochastic metapopulations. *Ecol. Lett.* **14,** 1158–1169 (2011).
55. Vasseur, D. A. & Fox, J. W. Phase-locking and environmental fluctuations generate synchrony in a predator-prey community. *Nature* **460,** 1007–1010 (2009).
56. Rosenstock, T. S., Hastings, A., Koenig, W. D., Lyles, D. J. & Brown, P. H. Testing Moran's theorem in an agroecosystem. *Oikos* **120,** 1434–1440 (2011).
57. Dai, L., Korolev, K. S. & Gore, J. Slower recovery in space before collapse of connected populations. *Nature* **496,** 355–358 (2013).
58. Vicsek, T., Czirók, A., Ben-Jacob, E., Cohen, I. & Shochet, O. Novel type of phase transition in a system of self-driven particles. *Phys. Rev. Lett.* **75,** 1226–1229 (1995).
59. Solé, R. V. & Bascompte, J. *Self-Organization in Complex Ecosystems* (Princeton Univ. Press, 2006).
60. Kapral, R., Oppo, G. L. & Brown, D. B. Phase separation and growth in a two-variable discrete model. *Physica A* **147,** 77–89 (1987).
61. Vilar, J. M. & Solé, R. V. Effects of noise in symmetric two-species competition. *Phys. Rev. Lett.* **80,** 4099 (1998).
62. Janke, W. in *Quantum Simulations of Complex Many-body Systems: From Theory to Algorithms*. (eds Grotendorst, J., Marx, D. & Muramatsu, A.) 423–445 (John von Neumann Institute for Computing, 2002).
63. Newman, M. E. J. & Barkema, G. T. *Monte Carlo Methods in Statistical Physics* (Oxford Univ. Press, 1999).
64. Janke, W. Monte in *Computational Physics*. (eds Hoffmann, K. H. & Schreiber, M.) 10–43 (Springer, 1996).
65. Binder, K. & Heermann, D. W. *Monte Carlo Methods in Statistical Physics: an Introduction* (Springer, 1986).

Acknowledgements

We are grateful to Carl Boettiger, Jeff Gore, Marcel Holyoak and Kirill Korolev for reading and commenting on an early version of this manuscript. We thank Erwin Frey, Nigel Goldenfeld, Ilkka Hanski and Edward Ott for productive conversations. Simulations were performed on a high-performance computing cluster housed within the Computer Science and Engineering department at UC Davis. We are grateful to Bill Broadley and Terri Knight for their excellent administration of that resource. This work is supported by the National Science Foundation under INSPIRE Grant No. 1344187.

Author contributions

This study has its origins in the lab of A.H. The results are the product of a tight-knit interdisciplinary collaboration among the three authors. J.M. and A.H. contributed to the development and analysis of the results, and to the writing and editing of the manuscript. A.E.N. developed analytical results, ran numerical simulations and took primary responsibility for the manuscript.

Additional information

Competing financial interests: The authors declare no competing financial interests.

15

A spin-liquid with pinch-line singularities on the pyrochlore lattice

Owen Benton[1], L.D.C. Jaubert[1], Han Yan[1] & Nic Shannon[1]

The mathematics of gauge theories lies behind many of the most profound advances in physics in the past 200 years, from Maxwell's theory of electromagnetism to Einstein's theory of general relativity. More recently it has become clear that gauge theories also emerge in condensed matter, a prime example being the spin-ice materials which host an emergent electromagnetic gauge field. In spin-ice, the underlying gauge structure is revealed by the presence of pinch-point singularities in neutron-scattering measurements. Here we report the discovery of a spin-liquid where the low-temperature physics is naturally described by the fluctuations of a tensor field with a continuous gauge freedom. This gauge structure underpins an unusual form of spin correlations, giving rise to pinch-line singularities: line-like analogues of the pinch points observed in spin-ice. Remarkably, these features may already have been observed in the pyrochlore material $Tb_2Ti_2O_7$.

[1] Okinawa Institute of Science and Technology Graduate University, Onna-son, Okinawa 904-0495, Japan. Correspondence and requests for materials should be addressed to O.B. (email: owen.benton@oist.jp).

Gauge symmetries are paramount in the understanding of many of the most fundamental theories of physics. Recent decades have seen an increasing appreciation of the role of gauge theories in condensed matter physics, emerging from the long-wavelength description of the collective behaviour of electrons. Emergent gauge theories have proved particularly important in the study of spin-liquids, strongly fluctuating, disordered magnetic states, the description of which lies beyond the familiar territory of Landau theory[1–6].

The use of a gauge theory to describe the fluctuations of a spin-liquid is exemplified by the case of the spin-ice materials $R_2M_2O_7$ (R = Ho, Dy, M = Ti, Sn) (refs 7,8). At low temperatures, the spin configurations in a spin-ice are subject to a constraint directly analogous to Gauss' law for a magnetic field and consequently may be described in terms of a gauge theory. Among the many striking consequences of this is the observation of pinch-point singularities in the magnetic neutron scattering structure factor[9], as observed in $Ho_2Ti_2O_7$ (ref. 10), cf. Fig. 1a. Pinch-point scattering has also been observed in the putative quantum spin-ice $Tb_2Ti_2O_7$ (refs 11–13). However, in this case, the experimental scattering shows pronounced butterfly-like features in the non-spin-flip (NSF) channel and the scattering in the spin-flip (SF) channel shows narrow arm-like features extending along the $\langle 111 \rangle$ directions of reciprocal space, neither of which features are predicted for a spin-ice. This raises the question of whether other types of spin-liquid may be found amongst rare-earth pyrochlore magnets.

Here we introduce a different kind of spin-liquid on the pyrochlore lattice. This spin-liquid arises on the phase diagram of a realistic model for pyrochlore magnets. As with spin-ice, the theory of this spin-liquid contains a gauge symmetry. The nature of this theory is fundamentally different to the Maxwellian theory which describes spin-ice, but just as the emergent gauge structure of spin-ice reveals itself in pinch-point scattering, so the gauge structure of this spin-liquid has striking consequences for scattering experiments. We will show that at low temperatures, this gauge structure leads to line-like singularities along the $\langle 111 \rangle$ directions of reciprocal space, which we dub 'pinch lines' since they are extended versions of the pinch-points exhibited in spin-ice. This is particularly interesting in the light of neutron scattering results on the pyrochlore magnets $Tb_2Ti_2O_7$ and $Yb_2Ti_2O_7$, which show strong, sharpening features along the $\langle 111 \rangle$ directions of reciprocal space. Indeed, our theory is able to account for several features of the diffuse scattering observed in $Tb_2Ti_2O_7$ (refs 11–13), which are unaccounted for by a theory based on a spin-ice model.

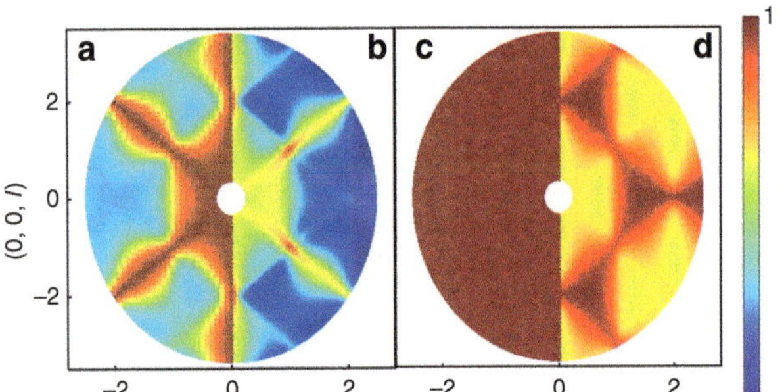

Figure 1 | Comparison of correlations in spin-ice with those of the spin-liquid discussed in this work. Predictions for polarized neutron scattering experiments are shown in the (**a**,**b**) SF and (**c**,**d**) NSF channels, as measured by Fennell *et al.*[10,11]. (**a**) Prediction for spin-ice in the SF channel exhibiting pinch-point singularities. (**b**) Prediction for scattering in the SF channel in the spin-liquid discussed in this work. (**c**) Prediction for spin-ice in the NSF channel. This channel is completely featureless in a nearest neighbour model for spin-ice—as shown here—and develops smooth maxima at the zone boundaries in the presence of long range dipole interactions[10]. (**d**) Prediction for scattering in the NSF channel in the spin-liquid discussed in this work. In contrast to spin-ice, the spin-liquid discussed here exhibits singular features in both SF and NSF scattering. Results are taken from classical Monte Carlo simulation of the nearest neighbour model $\mathcal{H}_{ex}$ (equation (1)), as described in the text.

Results

Spin-liquid regime in a model for pyrochlore magnets. We begin with the most general, symmetry-allowed, Hamiltonian for nearest neighbour anisotropic exchange on the pyrochlore lattice[14–16]:

$$\mathcal{H}_{ex} = \sum_{\langle ij \rangle} \mathbf{S}_i \cdot \bar{\mathcal{J}}_{ij} \cdot \mathbf{S}_j, \quad \bar{\mathcal{J}}_{01} = \begin{pmatrix} J_2 & J_4 & J_4 \\ -J_4 & J_1 & J_3 \\ -J_4 & J_3 & J_1 \end{pmatrix} \tag{1}$$

where the exchange matrix $\bar{\mathcal{J}}_{01}$ couples nearest neighbours along the $\mathbf{r}_{01} = (0,1,1)$ direction and the other exchange matrices can be generated from $\bar{\mathcal{J}}_{01}$ using point group operations. As shown in refs 16,17, it is possible to map out the entire classical ground-state phase diagram of equation (1) by an exact transcription of the Hamiltonian in terms of local fields defined on each tetrahedron[16–18]:

$$\mathcal{H}_{ex} = \frac{1}{2}\sum_{\text{tet}} \left[\Delta_{A_2} m_{A_2}^2 + \Delta_E \mathbf{m}_E^2 + \Delta_{T_2} \mathbf{m}_{T_2}^2 + \Delta_{T_1-} \mathbf{m}_{T_1-}^2 + \Delta_{T_1+} \mathbf{m}_{T_2+}^2 \right] + E_0 \tag{2}$$

where all the coefficients $\Delta_\alpha \geq 0$, E_0 is the ground-state energy and the sum runs over all tetrahedra in the lattice.

The five fields m_{A_2}, $\mathbf{m}_E$, $\mathbf{m}_{T_2}$, $\mathbf{m}_{T_1-}$, $\mathbf{m}_{T_1+}$ appearing in equation (2) are defined in Supplementary Table 1. They transform according to the A_2, E, T_2, T_1 irreducible representations of the point group and have respective dimension 1,2,3,3 and 3. Along a line of points in parameter space the three ordered phases which respectively maximize the fields $\mathbf{m}_{T_1-}$, $\mathbf{m}_{T_2}$, $\mathbf{m}_E$ become degenerate. This line includes the point $J_1 = J_2 = J_4 = 0$, $J_3 < 0$ (cf. Fig. 2a). For parameter sets along this line of points we have $\Delta_{A_2} = \Delta_E = \Delta_{T_1-} = 0$, Δ_{A_2}, $\Delta_{T_1+} > 0$ and the Hamiltonian is given by

$$\mathcal{H}_{ex}^{CL} = \frac{1}{2}\sum_{\text{tet}} [\Delta_{A_2} m_{A_2}^2 + \Delta_{T_1+} \mathbf{m}_{T_1+}^2] + E_0. \tag{3}$$

In a classical ground state of equation (3) it must be the case that

$$m_{A_2} = 0, \quad \mathbf{m}_{T_1+} = 0 \tag{4}$$

for every tetrahedron in the lattice. All of the results derived in this paper flow from the implementation of these constraints. These provide an exact description of the classical ground states along the line in parameter space where the three phases in Fig. 2a are degenerate. Observing the consequences of these constraints does not, however, require precise fine-tuning of the Hamiltonian to equation (3). These constraints will also dominate the physics at finite temperatures for any choice of parameters where Δ_{A_2}, Δ_E, $\Delta_{T_1-} << \Delta_{A_2}$, Δ_{T_1+}, such that energy cost of having a finite value of the fields $\mathbf{m}_{T_1-}$, $\mathbf{m}_{T_2}$, $\mathbf{m}_E$ is much lower than the cost to have a finite value of m_{A_2}, $\mathbf{m}_{T_1+}$.

The constraints in equation (4) are insufficient to select an ordered ground state in themselves. In such circumstances, fluctuations may select a preferred ordered state via the order-by-disorder mechanism, but Monte Carlo simulations indicate that they fail to do so, down to temperatures 3 orders of magnitude below the scale of the bare coupling (see Fig. 2b,c). The system

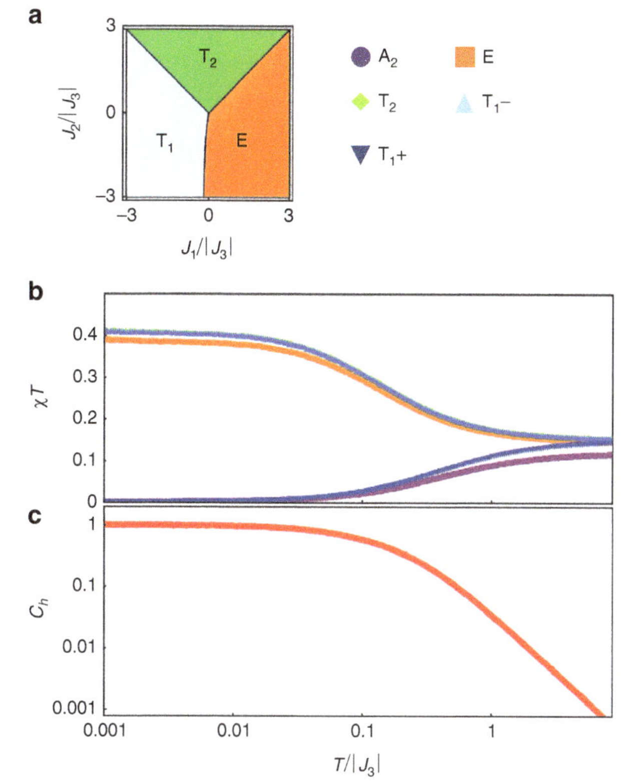

Figure 2 | Evidence of spin-liquid behaviour from Monte Carlo simulation. (**a**) Classical ground-state phase diagram of $\mathcal{H}_{ex}$ (equation (1)) for $J_3 < 0$, in the plane $J_4 = 0$, showing how ordered phases with symmetry T_1, E and T_2 meet at the point $J_1 = J_2 = 0$ (ref. 16). (**b**) Order-parameter susceptibilities and (**c**) heat capacity calculated in classical Monte Carlo simulation of $\mathcal{H}_{ex}$ (equation (1)), for parameters $J_1 = J_2 = J_4 = 0$, $J_3 < 0$. No phase transition is observed down to $T = 0.001\,|J_3|$. Instead, the order-parameter susceptibilities of neighbouring ordered phases exhibit a Curie law crossover, characteristic of a Coulombic spin-liquid[54]. The symbols used for different symmetry channels are shown in the inset.

thus remains in a disordered but highly correlated state down to low temperature.

Theory of the spin-liquid regime. We can understand the correlations of the spin-liquid from equation (4). The demand that the fields m_{A_2} and $\mathbf{m}_{T_1+}$ vanish everywhere leaves the fields $\{\mathbf{m}_E, \mathbf{m}_{T_2}, \mathbf{m}_{T_1-}\}$ with freedom to fluctuate in the ground state. The spatial variation of these fluctuations is constrained by the fact that neighbouring tetrahedra share a spin, therefore a fluctuation of the local fields on one tetrahedron affects the values of the local fields on the neighbouring tetrahedra. The fields $\mathbf{m}_E$, $\mathbf{m}_{T_2}$, $\mathbf{m}_{T_1-}$ must therefore fluctuate in a correlated manner in order to avoid inducing violations of equation (4). In what follows we show how these correlated fluctuations can be understood in terms of the fluctuations of a tensor field with a continuous gauge freedom.

The constraints on the spatial variation of $\mathbf{m}_E$, $\mathbf{m}_{T_2}$, $\mathbf{m}_{T_1-}$ may be obtained from the continuity of fields between A and B sublattice tetrahedra. The ground-state constraints (equation (4)) in fact imply a set of local conservation laws, on the lattice. These conservation laws in turn suggest that a coarse-graining approach can be successful in describing the fluctuations of $\mathbf{m}_E$, $\mathbf{m}_{T_2}$, $\mathbf{m}_{T_1-}$, and, unlike the global conservation laws which underpin hydrodynamic theories, these fully local conservation laws can have consequences even for short wavelength fluctuations, as we shall see. Expanding the local constraints to leading order in a gradient expansion we find

$$\nabla \cdot \mathbf{m}_{T_1-} = 0 \tag{5}$$

$$\begin{pmatrix} \partial_x m_E^1 \\ -\frac{1}{2}\partial_y m_E^1 + \frac{\sqrt{3}}{2}\partial_y m_E^2 \\ -\frac{1}{2}\partial_z m_E^1 - \frac{\sqrt{3}}{2}\partial_z m_E^2 \end{pmatrix} + \frac{\sqrt{3}}{2}\nabla \times \mathbf{m}_{T_2} - \frac{3}{2}\sin\left(\phi'_{T_1}\right)\begin{pmatrix} \partial_y m^z_{T_1-} + \partial_z m^y_{T_1-} \\ \partial_z m^x_{T_1-} + \partial_x m^z_{T_1-} \\ \partial_x m^y_{T_1-} + \partial_y m^x_{T_1-} \end{pmatrix} = 0 \tag{6}$$

where the angle ϕ'_{T_1} is a function of the exchange parameters, defined in the Methods section.

We wish to resolve the constraints (5) and (6) naturally using a gauge-theoretic approach. Note that since $\mathbf{m}_{T_1-}$ appears in both constraints, we cannot simply introduce separate gauge fields to resolve equations (5) and (6). Instead, we incorporate the eight components of $\{\mathbf{m}_E, \mathbf{m}_{T_2}, \mathbf{m}_{T_1-}\}$ into a traceless tensor field $\mathcal{B}$:

$$\mathcal{B} = \begin{pmatrix} m_E^1 & -\frac{\sqrt{3}}{2}m^z_{T_2} - \frac{3\sin(\phi'_{T_1})}{2}m^z_{T_1-} & \frac{\sqrt{3}}{2}m^y_{T_2} - \frac{3\sin(\phi'_{T_1})}{2}m^y_{T_1-} \\ \frac{\sqrt{3}}{2}m^z_{T_2} - \frac{3\sin(\phi'_{T_1})}{2}m^z_{T_1-} & -\frac{1}{2}m_E^1 + \frac{\sqrt{3}}{2}m_E^2 & -\frac{\sqrt{3}}{2}m^x_{T_2} - \frac{3\sin(\phi'_{T_1})}{2}m^x_{T_1-} \\ -\frac{\sqrt{3}}{2}m^y_{T_2} - \frac{3\sin(\phi'_{T_1})}{2}m^y_{T_1-} & \frac{\sqrt{3}}{2}m^x_{T_2} - \frac{3\sin(\phi'_{T_1})}{2}m^x_{T_1-} & -\frac{1}{2}m_E^1 - \frac{\sqrt{3}}{2}m_E^2 \end{pmatrix} \tag{7}$$

Satisfaction of equation (6), along with the condition $\mathrm{Tr}[\mathcal{B}] = 0$ is guaranteed by the introduction of a symmetric, tensor field $\mathcal{Y}$ and writing

$$\mathcal{B} = \mathscr{D} \cdot \mathcal{Y}, \quad \mathscr{D} \equiv \begin{pmatrix} 0 & -\partial_z & \partial_y \\ \partial_z & 0 & -\partial_x \\ -\partial_y & \partial_x & 0. \end{pmatrix} \tag{8}$$

The form of the matrix gauge field $\mathcal{Y}$ is then constrained by equation (5), which is satisfied if we take $\mathcal{Y}$ of the form

$$\mathcal{Y} = \begin{pmatrix} \psi & W_z & W_y \\ W_z & \psi & W_x \\ W_y & W_x & \psi \end{pmatrix}. \tag{9}$$

We can generate alternative forms of $\mathcal{Y}$ by applying Abelian gauge transformations to equation (9) of the form

$$\mathcal{Y}_{\mu\nu} \rightarrow \mathcal{Y}_{\mu\nu} + \partial_\mu \partial_\nu \zeta \tag{10}$$

The transformations of equation (10) leave the flux matrix $\mathcal{B}$, and therefore the physical spin system, unchanged. The form of $\mathcal{Y}$ in equation (9) thus corresponds to a specific choice of gauge. The theory of the spin-liquid is therefore invariant under a group of gauge transformations $\zeta \in \mathbb{R}$. Abelian gauge transformations of a similar form to equation (10), acting on tensor fields also appear in the linearized theory of general relativity[19] and the theory of $S = 2$ gauge fields[20].

At low temperatures, where there are only fluctuations of the local fields $\mathbf{m}_E$, $\mathbf{m}_{T_2}$, $\mathbf{m}_{T_1-}$, the free energy will be controlled by the entropy of these fluctuations. Coarse-graining over some volume much larger than a unit cell but much smaller than the whole system, there will be more states available (and therefore more entropy) with small values of $\mathbf{m}_E$, $\mathbf{m}_{T_2}$ and $\mathbf{m}_{T_1-}$ (ref. 5). The most general symmetry-allowed Gaussian free energy describing small fluctuations of these fields, when written in terms of the tensor field $\mathcal{Y}$, takes the form

$$\mathcal{F}_{SL} = T\int \frac{d^3\mathbf{r}}{V_{u.c.}}\left[\lambda_a \mathrm{Tr}\left[(\mathscr{D}\cdot\mathcal{Y})\cdot(\mathscr{D}\cdot\mathcal{Y})^T\right] + \lambda_b \mathrm{Tr}[(\mathscr{D}\cdot\mathcal{Y})\cdot(\mathscr{D}\cdot\mathcal{Y})] + \lambda_c \sum_\mu (\mathscr{D}\cdot\mathcal{Y})^2_{\mu\mu}\right] \tag{11}$$

which is invariant under the gauge transformations of equation (10).

Consequences for neutron scattering experiments. The distinctive nature of this spin-liquid, and of the theory which describes it (equation (11)), can be revealed by neutron scattering experiments. This can be seen by calculating the correlation functions of the local fields $\mathbf{m}_{\mathrm{E}}$, $\mathbf{m}_{\mathrm{T}_2}$, $\mathbf{m}_{\mathrm{T}_1-}$ in momentum space. In addition to displaying pinch-point singularities at zone centres, these correlation functions are singular approaching any momentum **q** which is along the (h, h, h) directions of reciprocal space, or along any direction related by the lattice symmetry to (h, h, h). This contrasts with the case of the Coulombic spin-liquid which occurs in the case of spin-ice, where the correlation functions are only singular at the Brillouin zone centre. Since the fields $\mathbf{m}_{\mathrm{E}}$, $\mathbf{m}_{\mathrm{T}_2}$, $\mathbf{m}_{\mathrm{T}_1-}$ are simply linear combinations of the spins, this singular behaviour will also show up in the spin structure factor $S(\mathbf{q})$, measurable in neutron scattering experiments.

In the vicinity of one of these singularities, at $T=0$, the scattering is approximated by

$$S\left(\mathbf{K}+\mathbf{q}_{\|}+\mathbf{q}_{\perp}\right) \approx \sum_{\alpha\beta} \gamma_{\alpha\beta}\left(\mathbf{K}, \mathbf{q}_{\|}\right) \frac{q_{\perp}^{\alpha} q_{\perp}^{\beta}}{\mathbf{q}_{\perp}^{2}} \qquad (12)$$

where **K** is a reciprocal lattice vector, $\mathbf{q}_{\|}$ is parallel to a $\langle 111 \rangle$ direction and $\mathbf{q}_{\perp}$ is orthogonal to that direction. The coefficients $\gamma_{\alpha\beta}$ determine the orientation of the singularity in **q**-space. Their dependence on the Brillouin zone **K** may be thought of as a form factor determining the contribution of the fluctuations of each field $\mathbf{m}_{\lambda}$ to the scattering in each Brillouin zone. The dependence on $\mathbf{q}_{\|}$ is smooth and near a zone centre **K** one may write $\gamma_{\alpha\beta}(\mathbf{K}, \mathbf{q}_{\|}) \approx \gamma_{\alpha\beta}(\mathbf{K}, \mathbf{0})$.

For $\gamma_{\alpha\beta} \not\propto \delta_{\alpha\beta}$ the structure factor in the limit $q_{\perp} \to 0$ will depend on the direction of approach and we have a singularity, along the entire $\langle 111 \rangle$ direction. Equation (12) has the form of a pinch-point singularity extended into a line. We therefore will refer to it as a 'pinch-line' singularity.

These pinch lines can be observed by taking planar cuts through the scattering, which intersect these lines away from reciprocal lattice vectors (Fig. 3a). This is illustrated using a $T=0$ calculation of $S(\mathbf{q})$ from the continuum theory (equation (11)) in Fig. 3b,e. For comparison, we show in Fig. 3c,f, the same quantity calculated at finite temperature within classical Monte Carlo simulation.

The simulation results show sharp features in the structure factor approaching the $\langle 111 \rangle$ directions, as predicted by the theory (equation (10)). There is a small broadening of these singularities, coming from the finite temperature thermal fluctuations present in the Monte Carlo simulation. These features are even more clearly visible in the correlation functions of the local fields $\{\mathbf{m}_{\mathrm{E}}, \mathbf{m}_{\mathrm{T}_2}, \mathbf{m}_{\mathrm{T}_1-}\}$; see Supplementary Fig. 1 and Supplementary Note 1. The presence of the pinch lines in the simulation results is a strong validation of our theory of the spin-liquid regime.

The continuum theory (equation (10)) was derived from local constraints, with associated local conservation laws, and the structure of the theory is inherited from the structure of those local constraints. This leads us to expect that the pinch-line singularities will be robust features of the spin-liquid, even at short wavelengths. We have confirmed this expectation using two independent, lattice-based calculations. Firstly, the sharpening of the scattering around the $\langle 111 \rangle$ directions is clearly seen in the Monte Carlo simulations in Fig. 3c,f. Secondly, we have also performed a $1/\mathcal{N}$ calculation of the spin correlations along the lines of that performed for the Heisenberg model in ref. 21. This calculation also predicts pinch-line singularities along the $\langle 111 \rangle$ directions of reciprocal space, as shown in Fig. 3d,g. It is therefore apparent that these singularities are a robust feature of the spin-liquid, arising from the structure of its ground-state constraints, which is captured by the continuum theory derived in this work.

Discussion

Thus far we have uncovered a spin-liquid described by a tensor field carrying a continuous gauge symmetry, arising in a particular limit of a realistic model for magnetism on the pyrochlore lattice (equation (1)). The signal feature of this spin-liquid is sharp line-like singularities along $\langle 111 \rangle$ directions of reciprocal space, which occur in addition to pinch-point singularities at zone centres. These pinch-line singularities are unique to the spin-liquid discussed in this paper and as such provide a very discriminating smoking-gun signature of this magnetic state. In the light of this discovery, it is interesting to consider two known pyrochlore materials, which are often discussed in the context of spin-liquid physics: $Tb_2Ti_2O_7$ and $Yb_2Ti_2O_7$.

$Tb_2Ti_2O_7$ has long been a focal point for discussion of three-dimensional spin-liquid physics[22–24]. While equation (1) alone may not constitute a complete quantitative model for the physics of $Tb_2Ti_2O_7$ it is interesting to compare observations on $Tb_2Ti_2O_7$ with the phenomenology of the spin-liquid. Polarized neutron scattering experiments on $Tb_2Ti_2O_7$ have shown evidence of singular scattering at Brillouin zone centres, but the form of this scattering looks rather different to a typical spin-ice, especially in the non-spin-flip (NSF) channel. At the same time, the data presented in ref. 11, shows bright, narrow features extending along the $\langle 111 \rangle$ directions.

As a point of comparison to these experiments, the behaviour of the structure factor $S(\mathbf{q})$ in the spin-flip (SF) and non-spin-flip (NSF) channels, appropriate to a polarized neutron scattering experiment with initial polarization $\mathbf{n}\|(1, -1, 0)$, is shown in Fig. 1, for the same set of exchange parameters as in Fig. 3. Narrow prominences are visible in the SF channel along the $\langle 111 \rangle$ directions (Fig. 1b). There are also pinch points in both channels at Brillouin zone centres. The distribution, orientation and polarization dependence of the pinch points observed in (ref. 11) is the same as that in Fig. 1b,d. In particular, we are able to reproduce the shape of the features in the NSF channel, something which cannot be done with a spin-ice-based description. The possibility that the theory described in this work could apply to $Tb_2Ti_2O_7$ is lent weight by a recent attempt at parameterizing a pseudo-spin Hamiltonian for $Tb_{2+x}Ti_{2-x}O_{7+y}$ (ref. 25) which places it close to the three-way phase boundary at which this spin-liquid emerges in our classical treatment.

Spin-liquid behaviour at finite temperature does not rule out the possibility of an magnetic order at lower temperature. Indeed, recent experiments have demonstrated the presence of competing ordering phenomena in $Tb_2Ti_2O_7$, with quadrupolar[26,27] and short range ordered antiferromagnetic states[12,28,29] being observed depending on the sample stoichiometry and experimental cooling protocol. This is consistent with the nature of the spin-liquid considered in our manuscript, which sits at the confluence of many competing orders. In particular, we note that the ground-state manifold of the spin-liquid contains states consistent with the $\mathbf{q}^*=(\pm 1/2, \pm 1/2, \pm 1/2)$ order observed under field cooled conditions[12,28]. These states can only be connected to the other states of the spin-liquid by rotation of an $\mathcal{O}(L)$ number of spins, where L is the linear size of the system. This may suggest an explanation for the sensitivity to how the system is cooled—namely that field cooling may drive the system into a state from which it is hard to reach the other parts of the ground-state manifold.

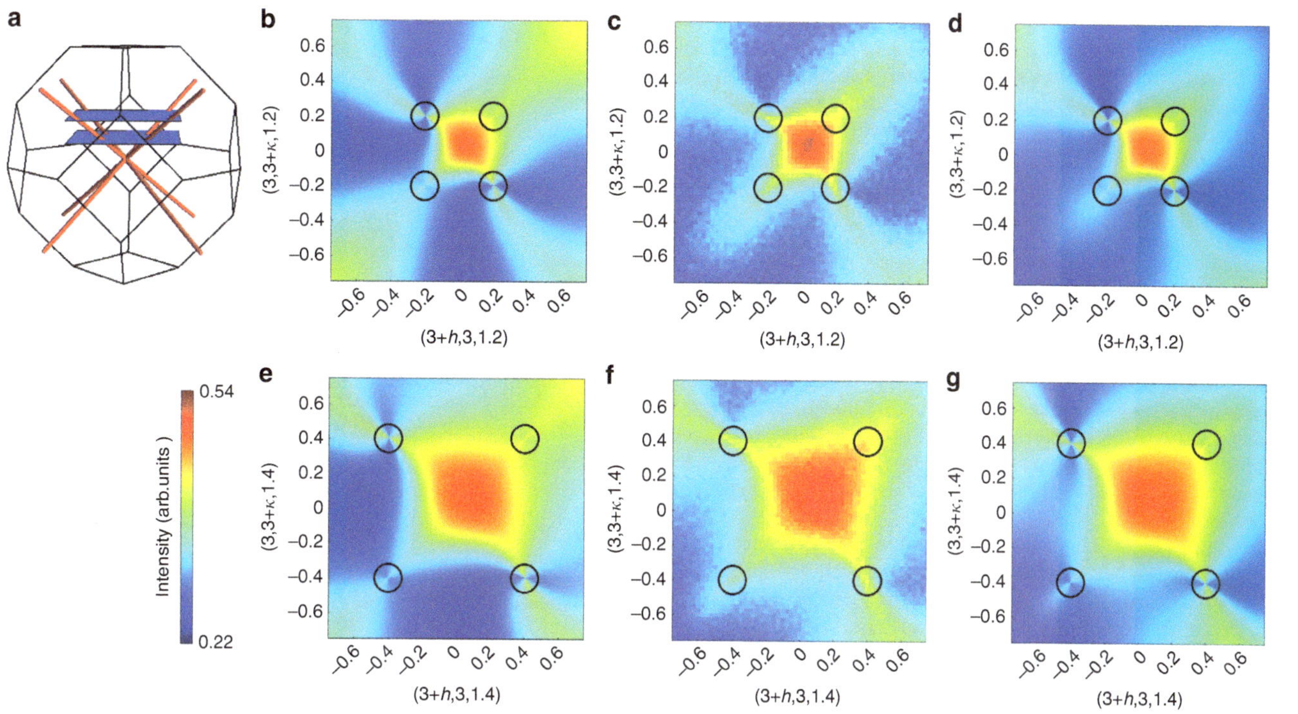

Figure 3 | Gauge structure of the spin-liquid as revealed through pinch-line singularities. (**a**) Location of pinch-line singularities in reciprocal space. (**b**,**e**) Spin structure factor $S(\mathbf{q})$ in parallel planes in reciprocal space in the Brillouin zone centred on $\mathbf{K} = (3,3,1)$, as calculated from the continuum theory (equation (11)). Singular features are visible where these planes intersect ⟨111⟩ directions, as indicated by the black circles in each panel. These pinch-line singularities, equation (12), are characteristic of the gauge structure of the spin-liquid. (**c**,**f**) Spin structure factor calculated in finite temperature Monte Carlo simulation, in the same regions of reciprocal space. The pinch lines appear in the simulation results as sharp features around the point where ⟨111⟩ directions intersect the plane. (**d**,**g**) A calculation of the structure factor made with a lattice-based $1/\mathcal{N}$ theory, also exhibiting pinch-line singularities. The simulations were performed at a temperature $T = 0.001$ K for a cluster of $N = 256{,}000$ sites and dimensions $40a_0 \times 40a_0 \times 10a_0$ where a_0 is the linear size of a cubic unit cell. Results were calculated for parameters $J_1 = 0.042$, $J_2 = 0.122$, $J_3 = -0.118$ and $J_4 = -0.04$ meV, with anisotropic g-tensor $g_\perp / g_\| = 1/3$, in approximate correspondence to $Tb_2Ti_2O_7$ (ref. 55). Since the crystal field ground state in $Tb_2Ti_2O_7$ is a non-Kramers doublet, the finite value of $g_\perp$ should be thought of as coming from mixing with the low-lying crystal field excitation.

The combination of spin-liquid physics and prominent features in the scattering along $\mathbf{q}||(1, 1, 1)$ is also strongly reminiscent of the discussion surrounding another pyrochlore: $Yb_2Ti_2O_7$ (refs 30–34). Indeed, it has recently been argued that the unusual physics of this material springs from competition between the E and T_1 regions of the phase diagram in Fig. 2a (refs 16,35). In this context, it is not unreasonable to imagine that the physics of the paramagnetic phase of $Yb_2Ti_2O_7$ may be influenced by a nearby spin-liquid phase of the form described here. This provides an interesting alternative scenario to quantum spin-ice physics in that material.

One concern which arises, in any comparison with experiment, is the extent to which the validity of this theory depends on detailed, fine-tuning of parameters. At first sight, this might seem like a serious obstacle, since it is unlikely that any real material would exist exactly at the point where three different ordered phase meet. However, in practice, a moderate detuning of parameters is only likely to be important in determining the nature of the competing (classical) ground state. As long as experiments are carried out in the disordered phase, at a temperature such that violations of the constraint, equation (4), are rare, the long-wavelength physics will still be described by equation (11), and pinch lines can be observed, albeit with a finite width coming from thermal fluctuations. The robustness of pinch lines against a finite density of thermally excited violations of equation (4) is evidenced by our Monte Carlo simulations (Fig. 3c,f), which incorporate thermal excitations out of the spin-liquid ground-state manifold. Thus, at finite temperature, the signature features of the spin-liquid, including pinch lines, should remain observable for a finite region of parameter space.

Another important question is the way in which quantum fluctuations will affect the properties of the spin-liquid at low temperartures. In the one case which is fully explored, quantum spin-ice, quantum tunnelling between different spin-ice configurations stabilizes a quantum spin-liquid ground state with the same U(1) gauge structure as the parent, classical spin-liquid[36–41]. Meanwhile, stronger off-diagonal exchange interactions between individual spins drive the system to order at low temperatures[42–44]. However in both cases, classical spin-liquid behaviour is still observed over long-length scales, at finite temperature. Similarly, studies of the quantum $S = 1/2$ Heisenberg model on the pyrochlore lattice find similar spin correlations[45–47] to those predicted by the gauge-theoretic description of the classical problem[21]. In the present case, the nature of the ground state in the presence of quantum fluctuations is an open problem, with both ordered and quantum spin-liquid phases a realistic possibility. However, the simplest estimate of the effect of quantum fluctuations, within linear spin-wave theory, suggests that the ground state is disordered for a finite range of parameters around the classically degenerate point[16]. And, given the large entropy associated with the classical spin-liquid, it seems likely that the system will retain many of its operational features—including the extended pinch lines—at finite temperature, regardless of its quantum ground state.

The theory presented in this work provides a fundamentally different paradigm to the emergent electromagnetism known from spin-ice and possesses a gauge freedom bearing an intriguing similarity to that appearing in the linearized theory of general relativity. This leads to the possibility of a unified theory of classical spin-liquids on the pyrochlore lattice, and a classification of the above based on their associated gauge freedoms and the consequent singularities in their correlation functions. These issues will be explored further elsewhere.

The discovery of a classical spin-liquid is also a promising starting point to search for new quantum spin-liquid ground states. Previous experience suggests that most quantum spin-liquids are found in models close to a point of high classical degeneracy, and in the case of spin-ice the same gauge symmetry underpins both the classical and quantum spin-liquid states. We therefore hope that this work can open the way for even richer physics to be discovered upon the inclusion of quantum fluctuations.

In conclusion, we have demonstrated the existence of an unusual kind of classical spin-liquid phase on the pyrochlore lattice, described by the fluctuations of a tensor field with a continuous gauge freedom. The nature of this spin-liquid is revealed by pinch-line singularities in correlations which could be observed in neutron scattering experiments.

Methods

Decomposition of Hamiltonian in terms of local fields. It was shown in ref. 16 that the generalized model for nearest neighbour exchange on the pyrochlore lattice (equation (1)) may be exactly rewritten in terms of local fields, defined on the pyrchlore tetrahedra

$$\mathcal{H}_{ex} = \frac{1}{2}\sum_{tet}\left(\Delta_{A_2} m_{A_2}^2 + \Delta_E \mathbf{m}_E^2 + \Delta_{T_2}\mathbf{m}_{T_2}^2 + \Delta_{T_1+}\mathbf{m}_{T_1+}^2 + \Delta_{T_1-}\mathbf{m}_{T_1-}^2\right) + \text{constant.} \tag{13}$$

The fields are labelled by the irreducible representations of the point group according to which they transform. These fields are defined in Supplementary Table I.

The angle ϕ'_{T_1} which appears in the definitions of $\mathbf{m}_{T_1+}$ and $\mathbf{m}_{T_1-}$ and equation (6) is chosen such that there is no bilinear coupling between $\mathbf{m}_{T_1+}$ and $\mathbf{m}_{T_1-}$ and such that

$$a_{T_1-} \leq a_{T_1+}. \tag{14}$$

Note that this convention for the definition of the T_1 symmetric fields is different to that chosen in ref. 16.

Monte Carlo simulation. The classical Monte Carlo simulations used to obtain the results in Figs 1–3 are based on the Metropolis algorithm with parallel tempering[48,49] and over-relaxation[50]. The spins are treated as classical vectors of fixed length $|S_i| = 1/2$ with local updates using the Marsaglia method[51].

Following common practice in Monte Carlo simulations, the order-parameter susceptibilities appearing in Fig. 2 are calculated according to the following formula:

$$\chi = \frac{N}{T}\left(\langle \mathbf{m}_\lambda^2\rangle - \langle|\mathbf{m}_\lambda|\rangle^2\right) \tag{15}$$

where N is the number of spins in the system, T is the temperature and m_λ are the local fields appearing in equation (2).

Lattice-based calculation of the structure factor. For the purposes of comparison with the continuum theory developed in the main text, we have also performed some lattice-based calculations of the correlations in the spin-liquid regime.

These calculations follow a method which has been previously been shown successful in understanding the correlations of disordered phases of spin-ice[9] the Heisenberg model on the pyrochlore lattice[21,52], and protons in water ice[53].

In this approach the constraints on the lengths of the spins

$$\mathbf{S}_i^2 = S^2 \tag{16}$$

are only enforced on average

$$\langle \mathbf{S}_i^2\rangle = S^2. \tag{17}$$

Equation (17) is enforced by means of a Lagrange multiplier λ added to the Hamiltonian. We write

$$\beta H \to \beta H_\lambda = \beta H + \lambda\sum_i \mathbf{S}_i^2 \tag{18}$$

where β is the inverse temperature.

Using a Fourier transformation βH_λ may be written as

$$\beta H_\lambda = \frac{1}{2}\tilde{S}(-\mathbf{q})\cdot\mathcal{M}(\mathbf{q})\cdot\tilde{S}(\mathbf{q}) \tag{19}$$

where $\tilde{S}(\mathbf{q})$ is a 12-component vector formed from the Fourier transforms of the three spin components on each of the four sublattices.

The correlations of $\tilde{S}(\mathbf{q})$ are then

$$\langle \tilde{S}_i(-\mathbf{q})\tilde{S}_j(\mathbf{q})\rangle = \left(\mathcal{M}^{-1}(\mathbf{q})\right)_{ij} \tag{20}$$

and λ can be chosen such that equation (17) is obeyed.

Where $\mathcal{M}(\mathbf{q})$ possesses flat bands of eigenvalues at the bottom of its spectrum—as is the case in the spin-liquid regime—the limit $T \to 0$ of the correlation function becomes a projection matrix, projecting into the subspace described by the associated eigenvectors[53]. This projection operator can be thought of as enforcing the local ground-state constraints[9].

It is this, zero-temperature, limit of the correlation function which is plotted in Fig. 3d,g of the main text.

The approach outlined here can be constructed as a perturbative expansion in powers of $1/\mathcal{N}$, where $\mathcal{N}$ is a number of copies of the system and the spin length constraint (equation (17)) becomes

$$\frac{1}{\mathcal{N}}\sum_{\alpha=1}^{\mathcal{N}} \mathbf{S}_{i,\alpha}^2 = S^2 \tag{21}$$

This method is described in more detail in ref. 17.

Data availability:. This is a theoretical work. The authors declare that the data supporting the findings of this study are available within the article and its supplementary information.

References

1. Lee, P. A. & Nagaosa, N. Gauge theory of the normal state of high-T_c superconductors. *Phys. Rev. B* **46,** 5621–5639 (1992).
2. Wen, X. G. Quantum orders and symmetric spin liquids. *Phys. Rev. B* **65,** 165113 (2002).
3. Moessner, R. & Sondhi, S. L. Three-dimensional resonating-valence-bond liquids and their excitations. *Phys. Rev. B* **68,** 064411 (2003).
4. Balents, L. Spin liquids in frustrated magnets. *Nature* **464,** 199–208 (2010).
5. Henley, C. L. The 'Coulomb phase' in frustrated systems. *Annu. Rev. Condens. Matter Phys.* **1,** 179–210 (2010).
6. Gingras, M. J. P. & McClarty, P. A. Quantum spin ice: a search for gapless quantum spin liquids in pyrochlore magnets. *Rep. Prog. Phys.* **77,** 056501 (2014).
7. Castelnovo, C., Moessner, R. & Sondhi, S. L. Magnetic monopoles in spin ice. *Nature* **451,** 42–45 (2008).
8. Castelnovo, C., Moessner, R. & Sondhi, S. L. Spin ice, fractionalization and topological order. *Annu. Rev. Condens. Matter Phys.* **3,** 35–55 (2012).
9. Henley, C. L. Power-law spin correlations in pyrochlore antiferromagnets. *Phys. Rev. B* **71,** 014424 (2005).
10. Fennell, T. *et al.* Magnetic Coulomb phase in the spin ice $Ho_2Ti_2O_7$. *Science* **326,** 415–417 (2009).
11. Fennell, T., Kenzelmann, M., Roessli, B., Haas, M. K. & Cava, R. J. Power-law spin correlations in the pyrochlore antiferromagnet $Tb_2Ti_2O_7$. *Phys. Rev. Lett.* **109,** 017201 (2012).
12. Fritsch, K. *et al.* Antiferromagnetic correlations at (1/2, 1/2, 1/2) in the ground state of the pyrochlore magnet $Tb_2Ti_2O_7$. *Phys. Rev. B* **87,** 094410 (2013).
13. Guitteny, S. *et al.* Anisotropic propagating excitations and quadrupolar effects in $Tb_2Ti_2O_7$. *Phys. Rev. Lett.* **111,** 087201 (2013).
14. Curnoe, S. H. Quantum spin configurations in $Tb_2Ti_2O_7$. *Phys. Rev. B* **75,** 212404 (2007).
15. Ross, K. A., Savary, L., Gaulin, B. D. & Balents, L. Quantum excitations in quantum spin ice. *Phys. Rev. X* **1,** 021002 (2011).
16. Yan, H., Benton, O., Jaubert, L. & Shannon, N. Living on the edge: ground-state selection in quantum spin-ice pyrochlores. Preprint at http://arxiv.org/abs/1311.3501 (2013).
17. Benton, O. *Classical and Quantum Spin Liquids on the Pyrochlore Lattice.* PhD thesis, Univ. Bristol (2015).
18. McClarty, P., Curnoe, S. H. & Gingras, M. J. P. Energetic selection of ordered states in a model of the $Er_2Ti_2O_7$ frustrated pyrochlore XY antiferromagnet. *J. Phys. Conf. Ser.* **145,** 012032 (2009).
19. Misner, C. W., Thorne, K. S. & Wheeler, J. A. in *Gravitation.* (ed. W. H. Freeman) (San Francisco, 1973).

20. Fronsdal, C. Massless fields with integer spin. *Phys. Rev. D* **18,** 3624–3629 (1978).
21. Isakov, S. V., Gregor, K., Moessner, R. & Sondhi, S. L. Dipolar spin correlations in classical pyrochlore magnets. *Phys. Rev. Lett.* **93,** 167204 (2004).
22. Gardner, J. S. *et al.* Cooperative paramagnetism in the geometrically frustrated pyrochlore antiferromagnet in $Tb_2Ti_2O_7$. *Phys. Rev. Lett.* **82,** 1012–1015 (1999).
23. Molavian, H. R., Gingras, M. J. P. & Canals, B. Dynamically induced frustration as a route to a quantum spin ice state in $Tb_2Ti_2O_7$ via virtual crystal field excitations and quantum many-body effects. *Phys. Rev. Lett.* **98,** 157204 (2007).
24. Gardner, J. S., Gingras, M. J. P. & Greedan, J. E. Magnetic pyrochlore oxides. *Rev. Mod. Phys.* **82,** 53–107 (2010).
25. Takatsu, H. *et al.* Quadrupole order in the frustrated pyrochlore $Tb_{2+x}Ti_{2-x}O_{7+y}$. Preprint at http://arxiv.org/abs/1506.04545 (2015).
26. Taniguchi, T. *et al.* Long-range order and spin-liquid states of polycrystalline $Tb_{2+x}Ti_{2-x}O_{7+y}$. *Phys. Rev. B* **87,** 060408 (2013).
27. Kadowaki, H., Takatsu, H., Taniguchi, T., Fåk, B. & Ollivier, J. Composite spin and quadrupole wave in the ordered phase of $Tb_{2+x}Ti_{2-x}O_{7+y}$. *Spin* **5,** 1540003 (2015).
28. Kermarrec, E. *et al.* Gapped and gapless short range ordered magnetic states with (1/2, 1/2, 1/2) wavectors in the pyrochlore magnet $Tb_{2+x}Ti_{2-x}O_{7+\delta}$. *Phys. Rev. B* **92,** 245114 (2015).
29. Guitteny, S. *et al.* Mesoscopic correlations in $Tb_2Ti_2O_7$. *Phys. Rev. B* **92,** 144412 (2015).
30. Bonville, P. *et al.* Transitions and spin dynamics at very low temperature in the pyrochlores $Yb_2Ti_2O_7$ and $Gd_2Sn_2O_7$. *Hyperfine Interact.* **156,** 103–111 (2004).
31. Ross, K. A. *et al.* Two-dimensional kagome correlations and field induced order in the ferromagnetic XY pyrochlore $Yb_2Ti_2O_7$. *Phys. Rev. Lett.* **103,** 227202 (2009).
32. Thompson, J. D. *et al.* Rods of neutron scattering intensity in $Yb_2Ti_2O_7$: compelling evidence for significant anisotropic exchange in a magnetic pyrochlore oxide. *Phys. Rev. Lett.* **106,** 187202 (2011).
33. Chang, L. J. *et al.* Higgs transition from a magnetic Coulomb liquid to a ferromagnet in $Yb_2Ti_2O_7$. *Nat. Commun.* **3,** 992 (2012).
34. Yaouanc, A., Maisuradze, A. & Dalmas de Réotier, P. Influence of short range correlations on the μSR polarization functions in the slow dynamic limit: application to the quantum spin-liquid system $Yb_2Ti_2O_7$. *Phys. Rev. B* **87,** 134405 (2013).
35. Jaubert, L. D. C. *et al.* Are multiphase competition and order-by-disorder the keys to understanding $Yb_2Ti_2O_7$? *Phys. Rev. Lett.* **115,** 267208 (2015).
36. Hermele, M., Fisher, M. P. A. & Balents, L. Pyrochlore photons: the U(1) spin liquid in a $S = 1/2$ three-dimensional frustrated magnet. *Phys. Rev. B* **69,** 064404 (2004).
37. Benton, O., Sikora, O. & Shannon, N. Seeing the light: experimental signatures of emergent electromagnetism in a quantum spin ice. *Phys. Rev. B* **86,** 075154 (2012).
38. Shannon, N., Sikora, O., Pollmann, F., Penc, K. & Fulde, P. Quantum ice: a quantum Monte Carlo study. *Phys. Rev. Lett.* **108,** 067204 (2012).
39. Savary, L. & Balents, L. Coulombic quantum liquids in spin-1/2 pyrochlores. *Phys. Rev. Lett.* **108,** 037202 (2012).
40. Hao, Z., Day, A. G. R. & Gingras, M. J. P. Bosonic many-body theory of quantum spin ice. *Phys. Rev. B* **88,** 144402 (2014).
41. McClarty, P. A. *et al.* Chain-based order and quantum spin liquids in dipolar spin ice. *Phys. Rev. B* **92,** 094418 (2015).
42. Castro-Neto, A. H., Pujol, P. & Fradkin, E. Ice: a strongly correlated proton system. *Phys. Rev. B* **74,** 024302 (2006).
43. Banerjee, A., Isakov, S. V., Damle, K. & Kim, Y.-B. Unusual liquid state of hard-core bosons on the pyrochlore lattice. *Phys. Rev. Lett.* **100,** 047208 (2008).
44. Kato, Y. & Onoda, S. Numerical evidence of quantum melting of spin ice: quantum-to-classical crossover. *Phys. Rev. Lett.* **115,** 077202 (2015).
45. Canals, B. & Lacroix, C. Pyrochlore antiferromagnet: a three-dimensional quantum spin liquid. *Phys. Rev. Lett.* **80,** 2933–2936 (1998).
46. Canals, B. & Lacroix, C. Quantum spin liquid: the Heisenberg antiferromagnet on the three-dimensional pyrochlore lattice. *Phys. Rev. B* **61,** 1149–1159 (2000).
47. Huang, Y., Chen, K., Deng, Y., Prokof'ev, N. & Svistunov, B. Spin-ice state of the quantum heisenberg antiferromagnet on the pyrochlore lattice. Preprint at http://arxiv.org/abs/1511.08285 (2015).
48. Swendsen, R. H. & Wang, J. S. Replica Monte-Carlo simulation of spin glasses. *Phys. Rev. Lett.* **57,** 2607–2609 (1986).
49. Geyer, C. J. in *Proceedings of the 23rd Symposium on the Interface*, 156 (Seattle, Washington, 1991).
50. Creutz, M. Overrelaxation and Monte-Carlo simulation. *Phys. Rev. D* **36,** 515–519 (1987).
51. Marsaglia, G. Choosing a point from the surface of a sphere. *Ann. Math. Stat.* **43,** 645–646 (1972).
52. Conlon, P. & Chalker, J. T. Absent pinch points and emergent clusters: further neighbor interactions in the pyrochlore Heisenberg antiferromagnet. *Phys. Rev. B* **81,** 224413 (2010).
53. Isakov, S. V., Moessner, R., Sondhi, S. L. & Tennant, D. A. Analytical theory for proton correlations in common-water ice I_h. *Phys. Rev. B* **91,** 245152 (2015).
54. Jaubert, L. D. C. *et al.* Topological-sector fluctuations and Curie-law crossover in spin ice. *Phys. Rev. X* **3,** 011014 (2013).
55. Cao, H. *et al.* Ising versus XY anisotropy in frustrated $R_2Ti_2O_7$ compounds as 'seen' by polarized neutrons. *Phys. Rev. Lett.* **103,** 056402 (2009).

Acknowledgements

This work was supported by the Theory of Quantum Matter Unit of the Okinawa Institute of Science and Technology Graduate University.

Author contributions

The analytic calculations were performed by O.B. and H.Y. The Monte Carlo simulations were written and performed by L.D.C.J. The manuscript was written by O.B. with input from all authors. N.S. supervised the project.

Additional information

Competing financial interests: The authors declare no competing financial interests.

16

Structural reducibility of multilayer networks

Manlio De Domenico[1,*], Vincenzo Nicosia[2,*], Alexandre Arenas[1] & Vito Latora[2,3]

Many complex systems can be represented as networks consisting of distinct types of interactions, which can be categorized as links belonging to different layers. For example, a good description of the full protein-protein interactome requires, for some organisms, up to seven distinct network layers, accounting for different genetic and physical interactions, each containing thousands of protein-protein relationships. A fundamental open question is then how many layers are indeed necessary to accurately represent the structure of a multilayered complex system. Here we introduce a method based on quantum theory to reduce the number of layers to a minimum while maximizing the distinguishability between the multi-layer network and the corresponding aggregated graph. We validate our approach on synthetic benchmarks and we show that the number of informative layers in some real multilayer networks of protein-genetic interactions, social, economical and transportation systems can be reduced by up to 75%.

[1] Departament d'Enginyeria Informática i Matemátiques, Universitat Rovira I Virgili, Avda Paisos Catalans 26, Tarragona 43007, Spain. [2] School of Mathematical Sciences, Queen Mary University of London, London E1 4NS, UK. [3] Dipartimento di Fisica ed Astronomia, Università di Catania and INFN, Catania I-95123, Italy. * These authors contributed equally to this work. Correspondence and requests for materials should be addressed to M.D.D. (email: manlio.dedomenico@urv.cat) or to V.N. (email: v.nicosia@qmul.ac.uk).

Network science has shown that characterizing the stucture of a complex system is fundamental when it comes to understanding its dynamical properties[1–3]. In particular, the basic units of most real-world systems are subject to different types of interactions occurring at comparable time scales. For instance, this is the case of social systems, where individuals can have political or financial relationships[4], or can be interacting using different communication channels, including face-to-face interactions, e-mail, Twitter, Facebook, phone calls and so on[5,6]. Similarly, in biological systems basic constituents such as proteins can have physical, co-localization, genetic or many other types of interactions. Recently, it has been shown that retaining such multi-dimensional information[7] and modelling the structure of interdependent and multilayer systems respectively through interdependent[8] and multilayer networks[9–12] reveals new non-trivial structural properties[13–20] and relevant emergent physical phenomena[21–27].

However, some of the interaction layers considered in the multidimensional representation of a system can be redundant or uninformative. Then, a simple question arises about the possibility of reducing the structure of a multilayer network, that is, of considering a smaller number of layers, while retaining as much information as possible about the whole system. This problem has both theoretical and practical implications. From a theoretical point of view, it is always desirable to find the most economical description of a phenomenon, that is, the one which retains all the salient aspects of the system avoiding unnecessary redundancy. From a practical point of view, the computation of even basic structural descriptors for interdependent and multilayer networks, such as clustering coefficient, centrality, motifs abundance and all the measures based on paths and walks, scales superlinearly or even exponentially with the number of layers[17] and can thus result unfeasible already for medium-sized networks. Therefore, finding an optimal configuration consisting of a minimal number of layers becomes a fundamental requirement when dealing with real-world systems.

Inspired by a similar question arising in quantum physics when one needs to quantify the distance between mixed quantum states[28], we propose here a method to aggregate some of the layers of a multilayer system while maximizing its distinguishability from the aggregated network. The method is based on a purely information theoretic perspective, which makes use of the definition of Von Neumann entropy of a graph. We test our procedure on synthetic and real-world multilayer networks, showing that different levels of structural reduction are possible, depending on the overall organization of the network.

Results

Von Neumann entropy of a multilayer network. In quantum mechanics, there are pure states, describing the system by means of a single vector in the Hilbert space, and mixed states, corresponding to statistical ensembles of pure states. The most general quantum system can then be described by the so-called density operator ρ, a semidefinite positive matrix with eigenvalues summing up to 1, which encodes all the information about the statistical ensemble of pure states of the system[29]. The Von Neumann (entanglement) entropy, which is the natural extension of the Shannon information entropy to quantum operators, is a widely adopted descriptor to measure the mixedness of a quantum system, although other measures, satisfying extensivity or non-extensivity, have been lately introduced and studied[30]. The Von Neumann entropy is defined for any density operator ρ. In particular, if the Von Neumann entropy is zero the system is in a pure state, otherwise it is in a mixed state. In general, the larger the Von Neumann entropy, the more mixed the state is.

It has been recently shown that the Von Neumann entropy can also be used to characterize (single layer) graphs[31,32]. Given a graph G represented by the adjacency matrix A, the Von Neumann entropy of G is defined as the Shannon entropy of the spectrum of the rescaled combinatorial Laplacian $\mathcal{L}_G$ associated to G (see Methods). This entropy has been interpreted as the entanglement of the statistical ensemble of pure states where each pure state is one of the edges of the graph[33]. According to this interpretation, a graph is in a pure state if and only if it consists of exactly one edge, corresponding to a Von Neumann entropy $h_A = 0$, and is in a mixed state otherwise, yielding $h_A > 0$.

Here we use a similar formalism to characterize multilayer networks, where we assume that each layer represents one possible state of the system, that is, a network state. We propose to use the Von Neumann entropy to quantify the distinguishability between a multilayer network (or a reduced configuration of its original layers) and the network obtained by aggregating all its layers in a single-layer graph.

Let us consider a multilayer network with the N nodes replicated along the different layers[12]. Such a network can be represented by the set $\mathcal{A} = \{A^{[1]}, A^{[2]}, \ldots, A^{[M]}\}$, whose elements are the $N \times N$ adjacency matrices of the M layers[11,17]. This particular multilayer structure is known as multiplex in the literature[12]. We define the Von Neumann entropy $H(\mathcal{A})$ of a multilayer network as the sum of the Von Neumann entropies of its M layers, that is, $H(\mathcal{A}) = \sum_{\alpha=1}^{M} h_{A^{[\alpha]}}$ where $h_{A^{[\alpha]}} = -\sum_{i=1}^{N} \lambda_i^{[\alpha]} \log_2(\lambda_i^{[\alpha]})$ and $\lambda_i^{[\alpha]}$ are the eigenvalues of the rescaled Laplacian matrix associated to the adjacency matrix $A^{[\alpha]}$ of layer α (see Methods). In the case of more general multilayer networks, where more complicated patterns of interlayer connections are allowed, it is still possible to calculate the Von Neumann entropy by considering the supra-adjacency matrix introduced by Gomez *et al.*[21], obtained as a special flattening of the rank-4 adjacency tensor, an even more general representation of multilayer networks[10].

Quantifying the reducibility of a multilayer network. The Von Neumann entropy of a multilayer network $\mathcal{A} = \{A^{[1]}, A^{[2]}, \ldots, A^{[M]}\}$ explicitly depends on the actual number of layers M and on the structure of each layer, so that in general its value will change if we consider a reduced multilayer network in which some of the layers of the original system have been combined together by means of an appropriate aggregation method. A particular case is represented by the aggregated graph associated to $\mathcal{A}$, which is the one-layer network whose adjacency matrix A is obtained by summing the adjacency matrices of all the M layers of $\mathcal{A}$, that is, $A = A^{[1]} + A^{[2]} + \ldots + A^{[M]}$. The Von Neumann entropy of the aggregated graph is h_A. In general, if we start from an M layer multiplex network $\mathcal{A} = \{A^{[1]}, \ldots, A^{[M]}\}$ and aggregate some of the original layers of $\mathcal{A}$, we obtain a reduced multilayer network $\mathcal{C} = \{C^{[1]}, C^{[2]}, \ldots, C^{[X]}\}$ with $X \leq M$ layers, where the adjacency matrix $C^{[\alpha]}$, where $\alpha = 1, \ldots, X$ is either one of the adjacency matrices of the original layers of $\mathcal{A}$ or the sum of two or more of them. We then consider the entropy per layer of the multilayer network $\mathcal{C}$:

$$\bar{H}(\mathcal{C}) = \frac{H(\mathcal{C})}{X} = \frac{\sum_{\alpha=1}^{X} h_{C^{[\alpha]}}}{X} \tag{1}$$

and we propose to quantify the distinguishability between the multilayer network $\mathcal{C}$ and the corresponding aggregated graph A through the relative entropy:

$$q(\mathcal{C}) = 1 - \frac{\bar{H}(\mathcal{C})}{h_A} \tag{2}$$

The larger $q(\mathcal{C})$, the more distinguishable is the multilayer network $\mathcal{C}$ from the corresponding aggregated graph A. It is worth noting that if all the layers of the multilayer network $\mathcal{C}$ are identical, then $q(\mathcal{C}) = 0$, as $\mathcal{C}$ and the aggregated graph are totally equivalent. Conversely, a value $q(\mathcal{C}) > 0$ indicates that the representation with X layers is distinguishable from the aggregated one; hence the multilayer structure must be preserved. Intuitively, if the aggregation of two layers does not result in a decrease of the relative entropy with respect to the multiplex in which the two layers are kept separated, then one would prefer the reduced configuration, which is more compact. However, it is possible to show (see Methods) that if we consider a multilayer $\mathcal{C}$ with X layers and the reduced configuration $\mathcal{C}'$ with X-1 layers obtained from $\mathcal{C}$ by aggregating two of its layers, then in general $q(\mathcal{C}')$ can be smaller than, equal to, or even larger than $q(\mathcal{C})$. This is due to the fact that the entropy per layer $\bar{H}(\mathcal{C})$ can either increase or decrease as a consequence of the aggregation of two layers (see Supplementary Fig. 1 and Supplementary Table 1). As we show in detail in Methods, our goal is to find argmax $[q(\mathcal{C})]$, that is, the optimally reduced multiplex $\mathcal{C}_{\max}$ yielding the maximum value of distinguishability from the aggregated graph. If we denote by $M_{\rm opt}$ the number of layers corresponding to the maximum value of relative entropy max$[q(\bullet)]$, we can then define the reducibility of a multilayer network $\mathcal{A}$ as:

$$\chi(\mathcal{A}) = \frac{M - M_{\rm opt}}{M - 1}, \quad (3)$$

which is the ratio between the number of reductions (M–$M_{\rm opt}$) and the total possible number of potentially reducible layers (M–1). It is worth noting that $\chi(\mathcal{A}) = 0$ if the system cannot be reduced, that is, when $M_{\rm opt} = M$, while $\chi(\mathcal{A}) = 1$ only if $M_{\rm opt} = 1$, that is, if the M layers can indeed be reduced into a single one (that is, the aggregated network).

The optimal configuration of aggregated layers is the one that maximizes the relative entropy $q(\bullet)$, but finding such a configuration would in general require the enumeration of all the possible partitions of a set of M objects (the layers), which is a well-known NP-hard problem (that is, its solution requires a computational time that scales at least exponentially with M). To overcome this problem, we adopt a greedy agglomerative hierarchical clustering algorithm[34] to explore the space of partitions, based on a concept of distance similar to the one adopted in quantum physics to quantify the distance between mixed quantum states[28]. More specifically, capitalizing on the concept of Von Neumann entropy of a graph, we use the quantum Jensen-Shannon divergence to quantify the (dis-) similarity between all pairs of layers of a multilayer network (see Methods). At each step of the algorithm, we consider the pair of layers having the smallest value of quantum Jensen–Shannon divergence and we aggregate them, obtaining a new multilayer network with one layer less. The rationale behind this choice is that the aggregation of a pair of similar layers is more desirable than the aggregation of two very dissimilar layers, as the latter can introduce artificial structural patterns. The result of this procedure is a dendrogram (see Fig. 1), that is, a hierarchical diagram where each of the M leaves is associated to one of the original layers of the system, each internal node indicates the aggregation of layers (or of clusters of layers) together and the root corresponds to the fully aggregated graph. At the mth step of the algorithm, we obtain a multilayer with M–m layers, for which we can compute the associated value of relative entropy $q(\bullet)$. The cut of the dendrogram corresponding to the maximal value of $q(\bullet)$ identifies the (sub-)optimal configuration of layers in terms of distinguishability with respect to the aggregated graph. The whole procedure proposed is sketched in Fig. 1 and can be summarized as follows: (i) compute the quantum Jensen–

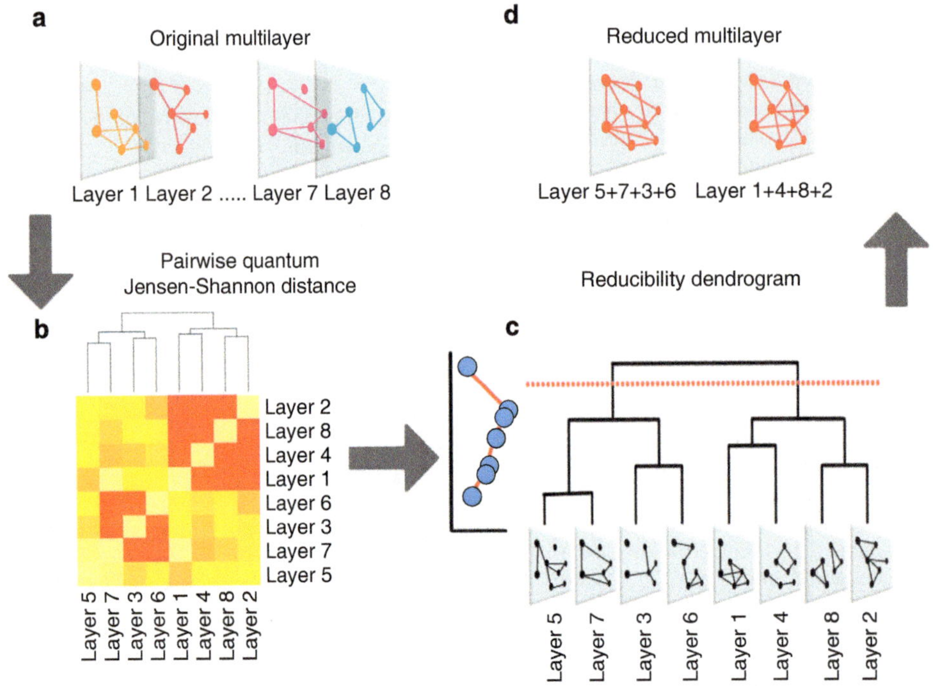

Figure 1 | Layer aggregation and structural reducibility of multilayer networks. Given a multilayer network (**a**), we compute the Jensen-Shannon distance $\mathcal{D}_{JS}$ between each pair of its layers (**b**), which is a proxy for layer redundancy. Such resulting distance matrix allows to perform a hierarchical clustering, whose output is a hierarchical diagram (a dendrogram) whose leaves represent the initial layers and internal nodes denote layer merging (**c**). At each step, the two clustered layers (or group of layers) corresponding to the smallest value of $\mathcal{D}_{JS}$ are aggregated and the quality of the new layer configuration in terms of distinguishability from the aggregated graph is quantified by the global quality function $q(\bullet)$, shown by the curve on the left-hand side of **c**. The best partition is the one for which $q(\bullet)$ is maximal (**d**).

Shannon distance matrix between all pairs of layers; (ii) perform hierarchical clustering of layers using such distance matrix and use the relative entropy $q(\bullet)$ as the quality function for the resulting partition; (iii) finally, choose the partition that maximizes the relative entropy, that is, the distinguishability from the aggregated graph.

Reduction of synthetic multilayer networks. To shed light on the impact of the structural properties of a multilayer network on the results obtained through the proposed layer reduction procedure, we considered different synthetic multilayer benchmarks. Each benchmark consists of several layers characterized by specific features or by a given amount of correlation. In Fig. 2 we report the case of a multilayer network in which the layers are obtained by rewiring different percentages of the edges of the same original layer. The layers of the resulting multilayer network are characterized by an increasing amount of edge overlap (see Methods). As shown in the figure, the hierarchical clustering procedure first aggregates layers characterized by smaller rewiring, which are more similar to each other, and then proceeds to the aggregation of layers obtained for larger values of rewiring. The monotonically decreasing behaviour of the relative entropy $q(\bullet)$, shown in Fig. 2c, confirms that in this case the best representation of the system is the one in which all the layers are kept distinct. In fact, independently of the fraction of edges actually rewired, on average a pair of layers exhibits a relatively small redundancy, as each of the rewired layers carries some information that is not included in the other layers (this multilayer has an overall edge overlap smaller than 5%).

The results obtained on several other synthetic multilayer networks suggest that layers with high edge overlap and similar structure, for example, characterized by highly overlapping communities, tend to be aggregated earlier (see Supplementary Note 1, Supplementary Figs 2,3 and 4).

Reduction of multilayer biological networks. To test the usefulness of our method on real-world systems, we consider here the multilayer networks obtained by taking into account different types of genetic interactions in 13 organisms of the Biological General Repository for Interaction Datasets (BioGRID[35]). This is a public database that stores and disseminates genetic and protein interaction information about simple organisms and humans (http://thebiogrid.org), and currently holds over 720,000 interactions obtained from both high-throughput data sets and individual focused studies, as derived from over 41,000 publications in the primary literature. We use BioGRID 3.2.108 (updated to 1 Jan 2014). In this data set, the networks represent protein–protein interactions and the layers correspond to interactions of different nature, that is, physical (labelled 'Phys' in the following), direct ('Dir'), co-localization ('Col'), association ('Ass') and suppressive ('GSup'), additive ('GAdd') or synthetic genetic ('GSyn') interaction. The number of layers identified for each organism ranges from three to seven.

In Fig. 3 we show the results obtained on three organisms (*Caenorhabditis elegans*, *Mus* and *Candida*). Despite the multilayer networks corresponding to these organisms have a similar number of layers (six for *C. elegans*, seven for *Mus* and *Candida*), each of them is characterized by a peculiar level of structural reducibility. In particular, in the case of *C. elegans* no layer aggregation is possible at all, as the maximum value of $q(\bullet)$ is obtained for the multilayer in which all the six layers are kept distinct. Hence, the reducibility is $\chi = 0$. Conversely, in the case of *Mus* and *Candida* some pairs of layers carry redundant structural information and can be thus aggregated. Remarkably, the reducibility for *Candida* is $\chi = 0.5$, corresponding to three redundant interaction layers out of seven. Here, the layer associated to genetic synthetic interactions is first aggregated with the layer encoding genetic additive interactions, while direct interactions are aggregated with physical ones. For other organisms, the value of reducibility can be as high as $\chi = 0.75$ (see Table 1 for details).

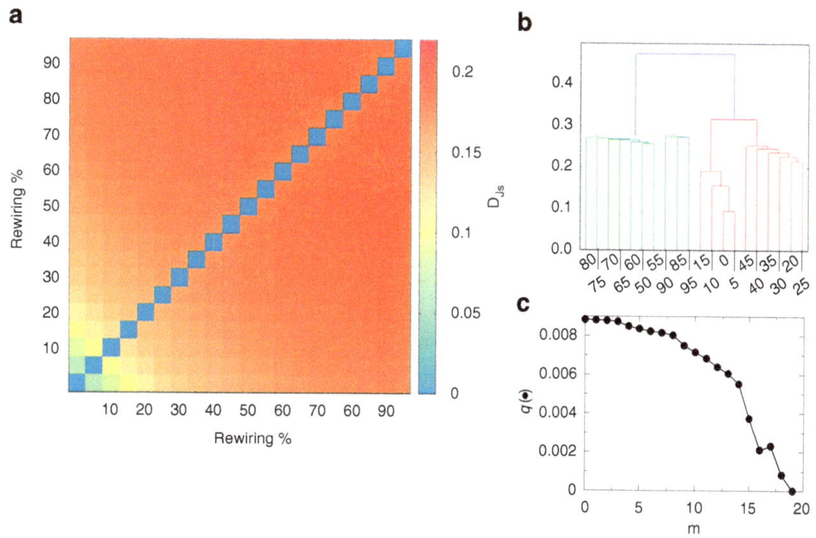

Figure 2 | Multilayer benchmark. We considered a benchmark multilayer network with $N = 5{,}000$ nodes and $M = 20$ layers. The first layer is a scale-free graph with $P(k) \sim k^{-3}$, whereas the other layers are obtained by rewiring an increasing percentage of the edges of the first layer, from 5% up to 95%. By doing so, each pair of layers is characterized by a different amount of edge redundancy (the total overlap of the multilayer is <5%). (**a**) The heat map shows the Jensen-Shannon distance between the 20 layers, where each layer is identified by the corresponding percentage of rewiring. (**b**) The hierarchical clustering procedure successively merges layers with a decreasing percentage of redundant edges. (**c**) In this case $q(\bullet)$ is a decreasing function of m, as each layer has some unique edges that are not present in the others. The best representation of the multilayer is that in which all the layers are kept separated, even if $q(\bullet)$ remains almost constant in the first few aggregation steps (corresponding to the aggregation of pairs of layers with a rewiring smaller than 15%).

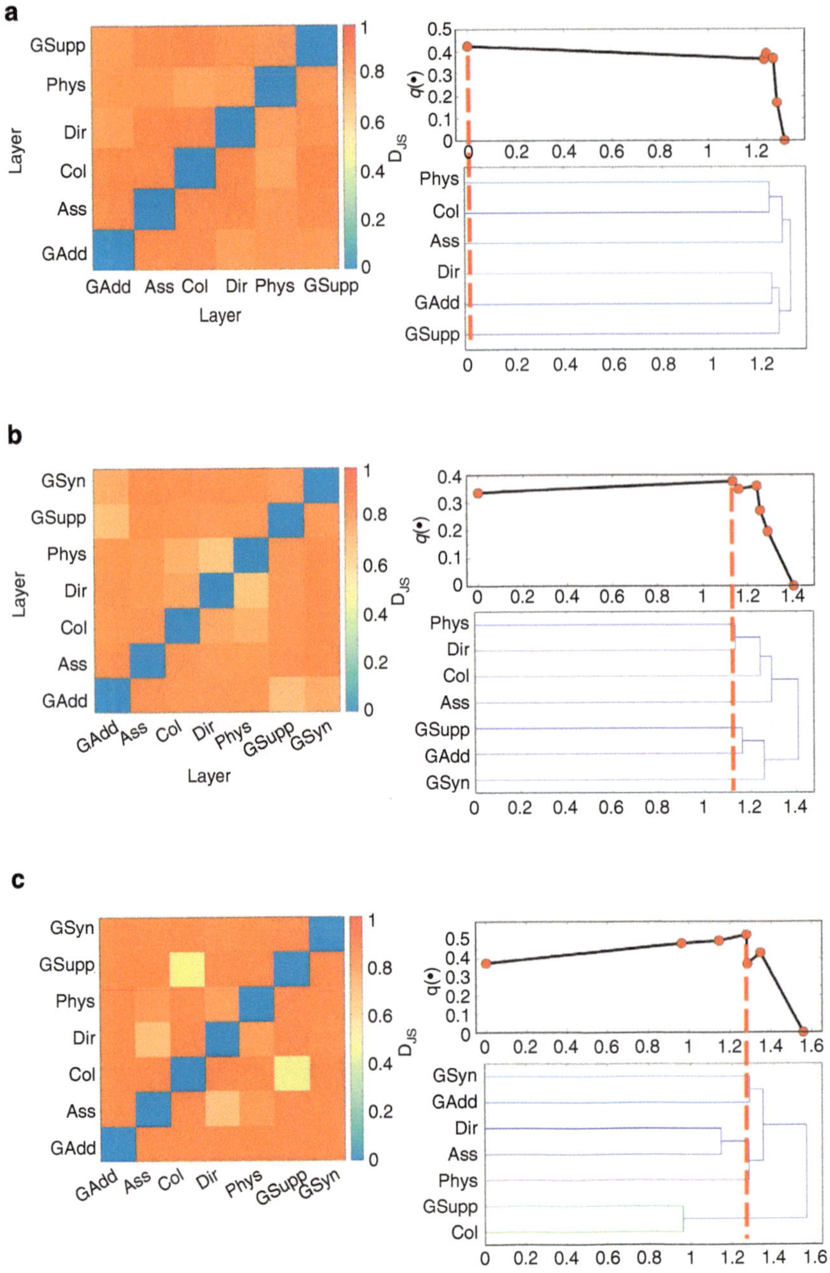

Figure 3 | Layer aggregation of protein-genetic interaction networks. The multilayer protein-genetic networks of different species have different levels of reducibility. We show the heat map of the Jensen-Shannon divergence, together with the dendrogram resulting from hierarchical clustering and the corresponding values of $q(\bullet)$, in 3 of the 13 species considered in this study. The dashed red lines identify the maximum of the global quality function $q(\bullet)$. For some organisms (such as *C. elegans*, reported in **a**), such maximum is obtained by leaving all the layers separate and no aggregation is possible, whereas for some other species a few layers carry redundant information, for example, in (**b**) *Mus* and in (**c**) *Candida*.

In Fig. 4 we summarize the results obtained by applying the proposed layer aggregation procedure to all the 13 multilayer genetic interaction networks of the BioGRID data set. This particular visualization allows to compare the structural reducibility of all organisms simultaneously. Not all multilayer networks can be reduced to a smaller number of layers, suggesting that for some organisms layer aggregation should be avoided. For instance, this is the case of *C. elegans* (nematode), *Arabidopsis thaliana* (cress) and *Bos taurus* (mammal), where no global maximum is present, except for $m=0$, that is, the initial multilayer in which all layers are kept distinct. In other cases, some of the layers are clearly redundant, as happens for instance in *Saccharomyces cerevisiae* (yeast) and *Drosophila melanogaster* (common fruit fly), where a maximum of $q(\bullet)$ is present at $m=2$.

Note that the reducibility values obtained for the above mentioned biological networks are conditioned to the completeness of the corresponding data sets. As a matter of fact, although the protein interactions of some organisms are well known and thoroughly characterized as in the case of *S. cerevisiae* or *D. melanogaster*, for some other organisms the information is only partial or incomplete. Hence, we cannot estimate a priori how the partial information contained in these networks is indeed affecting the values of reducibility that we observe.

Table 1 | Reducibility of empirical multilayer networks.

Network	N	M	M_{opt}	max[$q(\bullet)$]	χ
Arabidopsis	6981	7	5	0.436	0.33
Bos	326	4	3	0.494	0.33
Candida	368	7	4	0.527	0.50
C. elegans	3880	6	4	0.390	0.40
Drosophila	8216	7	5	0.426	0.33
Gallus	314	6	4	0.505	0.40
Human HIV-1	1006	5	2	0.499	0.75
Mus	7748	7	6	0.376	0.17
Plasmodium	1204	3	2	0.500	0.50
Rattus	2641	6	4	0.504	0.40
S. cerevisiae	6571	7	4	0.115	0.50
S. pombe	4093	7	4	0.197	0.50
Xenopus	462	5	3	0.424	0.50
Arxiv coauthorship	14065	13	11	0.231	0.17
Terrorist network	78	4	2	0.239	0.67
FAO Trade network	184	340	182	0.354	0.47
London Tube	369	13	12	0.441	0.08
Airports Europe	1064	175	165	0.667	0.06
Airports Asia	1130	213	202	0.653	0.05
Airports North America	2040	143	136	0.686	0.05

Number of nodes (N), number of layers in the original system (M), number of layers (M_{opt}) corresponding to the maximal value of the quality function (max[$q(\bullet)$]) obtained through the greedy hierarchical clustering procedure, and the value of the reducibility (χ) for several biological, social, economical and technological multilayer networks. Notice that the structure of the three continental air networks and of the London metropolitan transportation system cannot be substantially reduced, in accordance with the fact that in these systems layer redundancy is purposedly avoided. Conversely, social and biological systems exhibit higher levels of redundancy and allow for the merging of up to 75% of the layers.

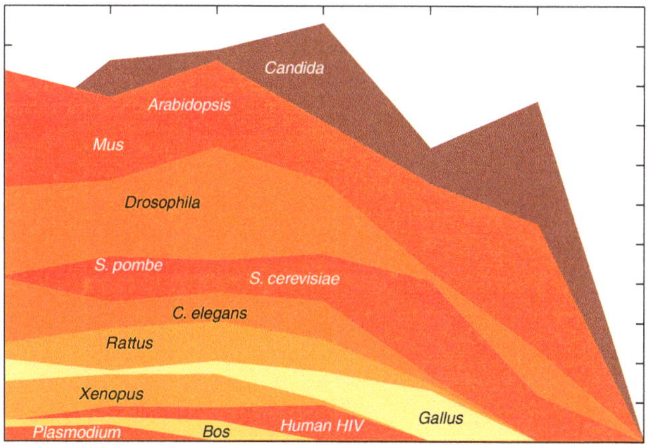

Figure 4 | Structural reducibility of protein-genetic networks in the BioGRID data set. The global quality function $q(\bullet)$ versus the number of merges in the hierarchical clustering procedure for the protein-genetic interaction multilayer networks of all the 13 organisms considered in this study (the plots are vertically rescaled to avoid overlaps). The values of $q(\bullet)$ are not reported in the y axis, because only the existence of a global maximum, and the corresponding value of m in the x axis is meaningful for the analysis. For each organism, $q(\bullet)$ has a maximum corresponding to the partition of the layers which minimizes layer redundancy at the cost of a small loss of information.

Discussion

Nowadays, larger and more detailed data sets describing diverse natural and man-made systems are being produced at an increasingly fast rate. This data deluge has provided an unprecedented amount of information about social, biological and technological phenomena, allowing a better characterization of the structure of different complex systems and a more in-depth understanding of the mechanisms underpinning their functioning. On the one hand, multilayer networks represent a natural framework to properly take into account all the different kinds of relationships connecting the units of a system, in a coherent manner. On the other hand, dealing with multilayer graphs introduces new computational challenges, which might limit the applicability of the multilayer approach to large systems. As a matter of fact, the evaluation of the multilayer version of even the most basic network descriptors, such as average shortest path length, node clustering coefficient, node betweenness and network motifs, tend to scale exponentially with the number of layers of the system and might become too computationally demanding already for medium-sized systems.

A fundamental observation is that not all the available levels of interaction among the constituents of a complex system have the same importance and some of them might be redundant, irrelevant or uninformative, with respect to the overall structure of the system. Hence, comes the idea of providing a consistent way to aggregate some of the layers of a multilayer network according to their similarity, as measured by the quantum Jensen–Shannon divergence, and of looking for configurations of layers that guarantee the maximum possible distinguishability from the fully aggregated graph and still use a minimal number of layers. The proposed approach allows to effectively reduce the redundancy of a multilayer network, as extensively shown in the paper for the case of the protein–genetic interaction networks of several different species.

However, the applicability of this method is not limited to biological systems. As an example, we have applied it also to social[17] and economical systems, coauthorship networks[36], metropolitan transportation networks[24] and continental air transportation systems[20] (see Table 1). A particularly interesting case is that of the FAO (Food and Agriculture Organization of the United Nations) worldwide food import/export network, an economic network in which layers represent products, nodes are countries and edges at each layer represent import/export relationships of a specific food product among countries. We collected the data from http://www.fao.org and built the multilayer network corresponding to trading in 2010. In Fig. 5 we show the distance matrix and the network visualization of three representative layers. The hierarchical clustering procedure reveals that up to 158 out of the 340 available layers can indeed be reduced, yielding a value of χ close to 50%. Intriguingly, the layers that are aggregated in the earlier stages of the clustering procedure correspond to products characterized by similar import/export patterns, as happens for instance for the layers associated to nuts, cocoa, dried and prepared fruits, roasted coffee and coffee-related products, which mainly involve export from Australia, China and Africa to European countries and the United States.

Conversely, the number of layers in the multilayer networks of airline transportation systems cannot be substantially reduced (the few allowed aggregations correspond to layers associated to very small companies, operating on just one or two routes), in agreement with the fact that airline companies tend to minimize the overlap of routes with other operators, to avoid strong competition. This result indicates that the connectivity among airports is practically not redundant for any airline, as expected for a modern large-scale transport infrastructure. Similar results are obtained for the London metropolitan transportation network, in which the overlap among different lines is purposely avoided to guarantee a more efficient coverage of the metropolitan area. In this case, the optimal solution corresponds to the multiplex network in which all the transportation lines are kept

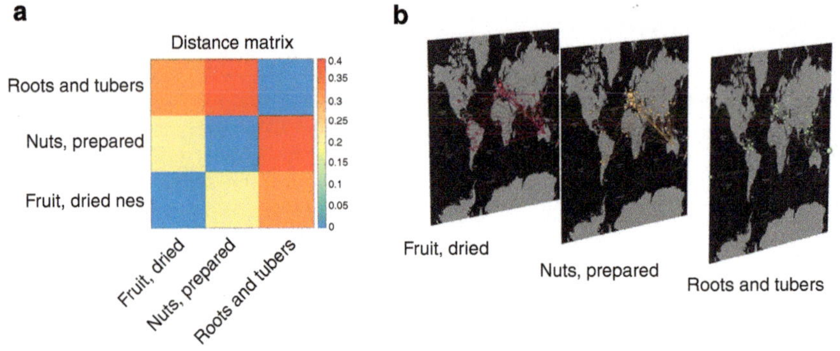

Figure 5 | Structural reducibility of the FAO worldwide food import/export network. The distance matrix of three layers of the FAO worldwide food import/export data set, corresponding to three specific products (that is, 'roots and tubers', 'prepared nuts' and 'dried fruit'), is shown in **a**, whereas the topology of the three layers is reported in **b**. The layers corresponding to 'prepared nuts' and 'dried fruits', which are more similar to each other (that is, closer with respect to the Jensen-Shannon divergence), are indeed aggregated by the algorithm in a single cluster, whereas the 'roots and tubers' layer, which is characterized by a remarkably different topology as evident from **b**, is kept separated. Map tiles By Stamen Design, under CC BY 3.0. Data by OpenStreetMap, under CC BY SA.

separated, with the only exception of the Circle Line and the Hammersmith and City Line, which, as expected, are aggregated together, as they considerably overlap in Zone 1 and Zone 2 (they actually share the same tracks and stations between Hammersmith and Liverpool Street).

We would like to clearly point out that by quantifying the reducibility of a multilayer network one obtains information about the structural redundancy of the different layers of the system. However, in the particular case in which the interaction layers are functionally similar, as in the case of unimodal transportation networks or multidisciplinary collaboration networks (but not for gene–protein interaction networks), the optimal multilayer network resulting from the reduction procedure proposed in the study might be also employed, at least to some extent, to characterize the dynamical behaviour of the system. We are confident that this aspect will be the subject of further research in the field.

It is worth noticing that although the problem of reducing the number of layers of a multilayer network can be tackled from different perspectives and might in principle be solved using different techniques (most of which are still to be explored), the framework provided by the Von Neumann entropy of graphs allows to formulate this problem in a natural way, and to use a standardised set of tools -borrowed from quantum physics- to define similarity relationships among layers (in terms of Jensen–Shannon divergence) and to construct a quality function able to identify optimal configurations of layers in terms of distinguishability from the aggregated graph. We would also like to stress that the problem of obtaining more compact representations of multilayer networks is interesting *per se* and we expect that the present work will trigger the investigation of more sophisticated methods for its solution. Beyond the structural reducibility, the reducibility of a multilayer network, while preserving its dynamics and function, remains an outstanding research problem[37–39].

We find quite remarkable that the formal analogy between quantum systems and multilayer networks allows to formulate the problem of layer reducibility in terms of quantum entropy divergence, and we believe that this analogy should be further exploited, as it might effectively provide a novel perspective on the characterization of the structure of multilayer complex systems.

Methods

Von Neumann entropy of single-layer networks. Given a graph $G(V, E)$ with $N=|V|$ nodes and $K=|E|$ edges, represented by the adjacency matrix $A=\{a_{ij}\}$, where $a_{ij}=1$ if node i and node j are connected through an edge, the Von Neumann entropy of G is defined as:

$$h_A = -\mathrm{Tr}\left[\mathcal{L}_G \log_2 \mathcal{L}_G\right] \quad (4)$$

where $\mathcal{L}_G = c\times(D-A)$ is the combinatorial Laplacian associated to the graph[31] G rescaled by $c = 1/(\sum_{i,j\in V} a_{ij}) = \frac{1}{2K}$ and D is the diagonal matrix of the degrees of the nodes. Formally, $\mathcal{L}_G$ has all the properties of a density matrix (that is, it is positive semi-definite and $\mathrm{Tr}(\mathcal{L}_G)=1$) and it is easy to prove that h_A can be written in terms of the set $\{\lambda_1, \lambda_2, \ldots, \lambda_N\}$ of eigenvalues of $\mathcal{L}_G$:

$$h_A = -\sum_{i=1}^{N} \lambda_i \log_2(\lambda_i), \quad (5)$$

that is, the Von Neumann entropy of a density matrix corresponds to the Shannon entropy of its power spectrum.

In Supplementary Methods and Supplementary Fig. 5 we discuss an efficient procedure to approximate the Von Neumann entropy of a graph that avoids the computation of the whole spectrum of $\mathcal{L}_G$.

Jensen-Shannon distance between graphs. Given two density matrices ρ and σ, it is possible to quantify to which extent ρ is different from σ by means of the Kullback-Liebler divergence:

$$\mathcal{D}_{KL}(\rho||\sigma) = \mathrm{Tr}\left[\rho\left(\log_2(\rho) - \log_2(\sigma)\right)\right] \quad (6)$$

which represents the information gained about σ when the expectation is based only on ρ. However, $\mathcal{D}_{KL}(\cdot||\cdot)$ is not a metric, as it is not symmetric with respect to its arguments (that is, $\mathcal{D}_{KL}(\rho||\sigma) \neq \mathcal{D}_{KL}(\sigma||\rho)$) and it does not satisfy the triangular inequality. A more suitable quantity to measure the dissimilarity between two density operators is the Jensen-Shannon divergence. If we call $\mu = \frac{1}{2}(\rho+\sigma)$ the new density matrix obtained as the mixture of the two operators, the Jensen-Shannon divergence between ρ and σ is defined as:

$$\mathcal{D}_{JS}(\rho||\sigma) = \frac{1}{2}\mathcal{D}_{KL}(\rho||\mu) + \frac{1}{2}\mathcal{D}_{KL}(\sigma||\mu) = h(\mu) - \frac{1}{2}[h(\rho)+h(\sigma)]. \quad (7)$$

By definition, $\mathcal{D}_{JS}$ is a reflexive and symmetric relation. In addition, it is possible to prove that $\sqrt{\mathcal{D}_{JS}}$, usually called Jensen-Shannon distance, takes values in [0,1] and satisfies all the properties of a metric if applied to qbits[40]. Some recent numerical arguments[41] have shown that $\sqrt{\mathcal{D}_{JS}}$ behaves similar to a metric as well, when applied to any pair of mixed quantum states, although a rigorous proof is still lacking. We decided to employ the quantum Jensen-Shannon divergence to quantify the distance, in terms of information gain/loss, between the normalized Laplacian matrices associated to two distinct networks.

The quality function $q(\bullet)$. The relative entropy defined in equation (2) quantifies the distinguishability of a multilayer network from the corresponding aggregated graph. Here we show that $q(\bullet)$ is an appropriate quality function to maximize, to detect the configuration of layers $\mathcal{C}_{max}$ corresponding to the highest possible distinguishability. In general, $q(\bullet)$ can either increase or decrease as a result of the aggregation of two layers, depending on several factors such as the relative density of the two graphs or their actual wiring patterns. In Supplementary Table 1 we report and discuss several illustrative examples.

If we start from the original M-layer multiplex network $\mathcal{A} = \{A^{[1]}, A^{[2]}, \ldots, A^{[M]}\}$ and aggregate some of its layers, we obtain a new multiplex $\mathcal{C} = \{C^{[1]}, C^{[2]}, \ldots, C^{[X]}\}$ with $X \leq M$ layers, where the adjacency matrix of each layer $C^{[\alpha]}$ is either one of the adjacency matrices of the original multiplex or the result of the aggregation of two or more of them. In particular, each of the M original layers of $\mathcal{A}$ will contribute to exactly one of the layers of the reduced multiplex $\mathcal{C}$. If we denote by Γ_α the layer of the reduced multiplex to which the original layer $A^{[\alpha]}$ contributes, then we can express each layer $C^{[\ell]}$ of $\mathcal{C}$ as

$$C^{[\ell]} = A^{[1]}\delta_{\ell,\Gamma_1} + A^{[2]}\delta_{\ell,\Gamma_2} + \ldots + A^{[M]}\delta_{\ell,\Gamma_M} \quad (8)$$

where $\delta_{\ell,\Gamma_\alpha} = 1$ if either the original layer $A^{[\alpha]}$ has been aggregated with other layers to form the new layer $C^{[\ell]}$ or if $C^{[\ell]} \equiv A^{[\alpha]}$.

If we consider the multilayer network $\mathcal{C}$ with X layers and the reduced multilayer network $\mathcal{C}'$ with X–1 layers obtained from $\mathcal{C}$ as a consequence of the aggregation of two layers, we want to find the conditions under which $q(\mathcal{C}')>q(\mathcal{C})$ or, equivalently, $\bar{H}(\mathcal{C}')<\bar{H}(\mathcal{C})$. For the sake of simplicity, and without loss of generality, we assume that the reduced configuration $\mathcal{C}'$ is obtained by aggregating layers $C^{[1]}$ and $C^{[2]}$ into a new layer $C^{[1]}+C^{[2]}$. After some algebra, the inequality $q(\mathcal{C}')>q(\mathcal{C})$ reduces to

$$h_{C^{[1]}+C^{[2]}} < h_{C^{[1]}} + h_{C^{[2]}} - \bar{H}(\mathcal{C}), \quad (9)$$

that is, the quality function $q(\bullet)$ increases as a result of the aggregation, if the entropy of the aggregated layers is smaller than the difference between the sum of the entropies of the layers to be aggregated and the entropy per layer before the aggregation. It is useful to rewrite equation (9) as:

$$\bar{H}(\mathcal{C}) < h_{C^{[1]}} + h_{C^{[2]}} - h_{C^{[1]}+C^{[2]}} = \Delta H \quad (10)$$

where ΔH is the difference of entropy due to the aggregation. This means that $q(\bullet)$ increases if the value of ΔH associated to the two aggregated layers is higher than the entropy per layer of the layer configuration before the aggregation. In general, the Von Neumann entropy is sub-additive, meaning that the entropy of a state obtained as the mixing of two other states is smaller than the sum of the entropies of the two original states, that is, $\Delta H = h_{C^{[1]}} + h_{C^{[2]}} - h_{C^{[1]}+C^{[2]}} \geq 0$. However, as we extensively show in Supplementary Note 2, Supplementary Fig. 1 and Supplementary Table 1, this is not always the case when we aggregate two graphs, so that the Von Neumann entropy of the resulting graph can be either larger or smaller than the sum of the Von Neumann entropies of the two original graphs, that is, the aggregation of two layers can sometimes violate sub-additivity. This happens in at least two cases, that is, when one aggregates layers with very different edge densities or when the aggregation would create structural patterns that did not exist in any of the two original layers, which are both examples of undesirable aggregation (see Supplementary Fig. 1 and Supplementary Table 1). In such cases, equation (10) is automatically not satisfied (remember that $\bar{H}(\mathcal{C}) \geq 0$) and the quality function $q(\bullet)$ decreases.

The condition to have an increase of $q(\bullet)$ expressed by equation (9) can be also written in terms of the Jensen–Shannon divergence of the layers to be aggregated. For the sake of simplicity, let us assume that the two layers $C^{[1]}$ and $C^{[2]}$ aggregated to obtain the new configuration $\mathcal{C}'$ have the same number of links. In this case, the inequality in equation (9) is equivalent to

$$\mathcal{D}_{JS}\left(\rho^{[1]}\|\rho^{[2]}\right) < \frac{h_{C^{[1]}} + h_{C^{[2]}}}{2} - \bar{H}(\mathcal{C}), \quad (11)$$

where $\rho^{[1]}$ and $\rho^{[2]}$ are the density matrices corresponding to layers $C^{[1]}$ and $C^{[2]}$, respectively. The first term on the right-hand side of inequality (11) is the entropy per layer of the multilayer network formed by the two layers that have been aggregated, so that the quality function $q(\bullet)$ increases if the difference between the entropy per layer of this smaller multilayer and the entropy per layer of the full multilayer network is larger than the Jensen–Shannon divergence of the density matrices to be aggregated. In the limiting case in which $C^{[1]}$ and $C^{[2]}$ are identical (that is, $\mathcal{D}_{JS}(\rho^{[1]}\|\rho^{[2]}) = 0$), this leads to an increase of $q(\bullet)$ only if $h_{C^{[1]}} > \bar{H}(\mathcal{C})$, or equivalently if $h_{C^{[1]}} > \frac{1}{X}\sum_{\alpha=1}^{X} h_{C^{[\alpha]}}$, that is, if the entropy of each of the two layers is larger than the entropy per layer of the multilayer network before the merge. In conclusion, an increase of $q(\bullet)$ usually corresponds either to the aggregation of two layers that do not violate sub-additivity or to the merge of layers having very similar structure. Hence, by maximizing $q(\bullet)$ one tends to avoid layer configurations that might contain spurious structural patterns or redundant layers.

Hierarchical clustering. We measure the information lost by merging two layers of a multilayer graph in a single network by comparing the Von Neumann entropy of the compressed multilayer network with the original representation. The main hypothesis is that if the value of the Jensen–Shannon distance between the Laplacian matrices associated to layers α and β is small, then the two layers can be safely merged in a single one without loosing too much information. Conversely, if $\mathcal{D}_{JS}(\mathcal{L}_\alpha, \mathcal{L}_\beta)$ is large, then the two layers provide different information about the relationships among the nodes of the system. In this case, it would be better to leave the two layers separated, as their aggregation will result in a substantial loss of information.

We perform a classical hierarchical clustering of the M layers using the Jensen–Shannon distance to quantify the dissimilarity among (clusters of) layers. At each step of the algorithm, we aggregate the two clusters of layers, which are separated by the smallest value of $\mathcal{D}_{JS}$, and then we update the distances between the newly formed cluster and the remaining ones according to Ward's linkage. By iterating this procedure M–1 times, we obtain a dendrogram, that is, a hierarchical diagram whose M leaves are associated to the original layers of the system, internal nodes indicate merges of (clusters of) layers and the root corresponds to the aggregated graph. The quality of the layer organization obtained after m steps of the hierarchical clustering algorithm is measured by the relative entropy $q(\bullet)$.

To verify whether the proposed greedy clustering procedure is able to find good approximations of the real optimal configuration of layers, we compared the solution corresponding to the optimal cut of the dendrogram with the actual optimal configuration of layers of each of the 13 multilayer networks obtained from the BioGRID data set. For each multilayer network, the optimal configuration of layers was found through exhaustive enumeration of all the possible partitions of the set of layers. The results are reported and discussed in Supplementary Note 3, Supplementary Table 2 and Supplementary Fig. 6, and confirm that the greedy clustering algorithm performs a quite efficient exploration of the quality function landscape, yielding (sub-)optimal solutions associated to values of $q(\bullet)$ that are between 76% and 100% of the actual global optimum. This is quite a remarkable result, especially if we consider that the greedy algorithm performs only M–1 steps (that is, less than seven steps for all the BioGRID multilayer networks), while the exhaustive exploration of all the partitions of a set of M elements requires a number of operations equal to the Mth Bell number, which increases super-exponentially with M.

We notice that the same hierarchical clustering algorithm can be potentially applied with any other measure able to quantify the difference between layers, not just with $\mathcal{D}_{JS}$. The only caveat here is that if the employed measure is not a metric then the classical linkage schemes, including Ward's linkage, cannot be employed directly, so that at each step it is necessary to recompute the distance between the new layer resulting from the last merge and all the remaining layers.

A stand-alone implementation of the algorithm for the reduction of multilayer networks described above is available at https://github.com/KatolaZ/multired. Another implementation of the algorithm is already included in muxViz (https://github.com/manlius/muxViz), a software for the multilayer analysis of networks.

References

1. Albert, R. & Barabási, A.-L. Statistical mechanics of complex networks. *Rev. Mod. Phys.* **74,** 47 (2002).
2. Newman, M. E. The structure and function of complex networks. *SIAM Rev.* **45,** 167–256 (2003).
3. Boccaletti, S., Latora, V., Moreno, Y., Chavez, M. & Hwang, D.-U. Complex networks: structure and dynamics. *Phys. Rep.* **424,** 175–308 (2006).
4. Padgett, J. F. & Ansell, C. K. Robust action and the rise of the Medici, 1400-1434. *Am. J. Sociol.* 1259–1319 (1993).
5. Krackhardt, D. Cognitive social structures. *Soc. Networks* **9,** 109–134 (1987).
6. Wasserman, S. & Faust, K. *Social Network Analysis: Methods and Applications. Structural Analysis in the Social Sciences* (Cambridge University Press, 1994).
7. Cardillo, A. *et al.* Emergence of network features from multiplexity. *Sci. Rep.* **3,** 1344 (2013).
8. Buldyrev, S. V., Parshani, R., Paul, G., Stanley, H. E. & Havlin, S. Catastrophic cascade of failures in interdependent networks. *Nature* **464,** 1025–1028 (2010).
9. Mucha, P. J., Richardson, T., Macon, K., Porter, M. A. & Onnela, J.-P. Community structure in time-dependent, multiscale, and multiplex networks. *Science* **328,** 876–878 (2010).
10. De Domenico, M. *et al.* Mathematical formulation of multilayer networks. *Phys. Rev. X* **3,** 041022 (2013).
11. Nicosia, V., Bianconi, G., Latora, V. & Barthelemy, M. Growing multiplex networks. *Phys. Rev. Lett.* **111,** 058701 (2013).
12. Kivelä, M. *et al.* Multilayer networks. *J. Complex Networks* **2,** 203–271 (2014).
13. Kurant, M. & Thiran, P. Layered complex networks. *Phys. Rev. Lett.* **96,** 138701 (2006).
14. Szell, M., Lambiotte, R. & Thurner, S. Multirelational organization of large-scale social networks in an online world. *Proc. Natl Acad. Sci. USA* **107,** 13636–13641 (2010).
15. Solá, L. *et al.* Eigenvector centrality of nodes in multiplex networks. *Chaos* **23,** 033131 (2013).
16. Bianconi, G. Statistical mechanics of multiplex networks: Entropy and overlap. *Phys. Rev. E* **87,** 062806 (2013).
17. Battiston, F., Nicosia, V. & Latora, V. Structural measures for multiplex networks. *Phys. Rev. E* **89,** 032804 (2014).
18. Sanchez-Garcia, R. J., Cozzo, E. & Moreno, Y. Dimensionality reduction and spectral properties of multilayer networks. *Phys. Rev. E* **89,** 052815 (2014).
19. De Domenico, M., Solé-Ribalta, A., Omodei, E., Gómez, S. & Arenas, A. Ranking nodes in interconnected multilayer networks reveals their versatility. *Nat. Commun.* doi:10.1038/ncomms7868 (2015).
20. Nicosia, V. & Latora, V. Measuring and modelling correlations in multiplex networks. *Preprint at* http://arxiv.org/abs/1403.1546 (2014).
21. Gómez, S. *et al.* Diffusion dynamics on multiplex networks. *Phys. Rev. Lett.* **110,** 028701 (2013).
22. Cellai, D., López, E., Zhou, J., Gleeson, J. P. & Bianconi, G. Percolation in multiplex networks with overlap. *Phys. Rev. E* **88,** 052811 (2013).

23. Estrada, E. & Gómez-Gardeñes, J. Communicability reveals a transition to coordinated behavior in multiplex networks. *Phys. Rev. E* **89,** 042819 (2014).
24. De Domenico, M., Solé-Ribalta, A., Gómez, S. & Arenas, A. Navigability of interconnected networks under random failures. *Proc. Natl Acad. Sci. USA* **111,** 8351–8356 (2014).
25. Granell, C., Gómez, S. & Arenas, A. Dynamical interplay between awareness and epidemic spreading in multiplex networks. *Phys. Rev. Lett.* **111,** 128701 (2013).
26. Radicchi, F. & Arenas, A. Abrupt transition in the structural formation of interconnected networks. *Nat. Phys.* **9,** 717–720 (2013).
27. Nicosia, V., Bianconi, G., Latora, V. & Barthelemy, M. Non-linear growth and condensation in multiplex networks. *Phys. Rev. E* **90,** 042807 (2014).
28. Majtey, A., Lamberti, P. & Prato, D. Jensen-Shannon divergence as a measure of distinguishability between mixed quantum states. *Phys. Rev. A* **72,** 052310 (2005).
29. Dirac, P. A. M. *The Principles of Quantum Mechanics* 27 (Oxford University Press, 1981).
30. Rossignoli, R., Canosa, N. & Ciliberti, L. Generalized entropic measures of quantum correlations. *Phys. Rev. A* **82,** 052342 (2010).
31. Braunstein, S. L., Ghosh, S. & Severini, S. The Laplacian of a graph as a density matrix: a basic combinatorial approach to separability of mixed states. *Ann. Combinatorics* **10,** 291–317 (2006).
32. Passerini, F. & Severini, S. in *Developments in Intelligent Agent Technologies and Multi-Agent Systems: Concepts and Applications* 66–76 (2011).
33. De Beaudrap, N., Giovannetti, V., Severini, S. & Wilson, R. Interpreting the Von Neumann entropy of graph laplacians, and coentropic graphs. *Preprint at* http://arxiv.org/abs/1304.7946 (2013).
34. Kaufman, L. & Rousseeuw, P. J. *Finding Groups in Data: An Introduction to Cluster Analysis* (John Wiley and Sons, New York, 1990).
35. Stark, C. *et al.* Biogrid: a general repository for interaction datasets. *Nucleic Acids Res.* **34,** D535–D539 (2006).
36. De Domenico, M., Lancichinetti, A., Arenas, A. & Rosvall, M. Identifying modular flows on multilayer networks reveals highly overlapping organization in interconnected systems. *Phys. Rev. X.* **5,** 011027 (2015).
37. Solé-Ribalta, A. *et al.* Spectral properties of the Laplacian of multiplex networks. *Phys. Rev. E* **88,** 032807 (2013).
38. Cozzo, E., Banos, R. A., Meloni, S. & Moreno, Y. Contact-based social contagion in multiplex networks. *Phys. Rev. E* **88,** 050801 (2013).
39. Lee, K.-M., Brummitt, C. D. & Goh, K.-I. Threshold cascades with response heterogeneity in multiplex networks. *Phys. Rev. E* **90,** 062816 (2014).
40. Briët, J. & Harremoës, P. Properties of classical and quantum Jensen-Shannon divergence. *Phys. Rev. A* **79,** 052311 (2009).
41. Lamberti, P., Majtey, A., Borras, A., Casas, M. & Plastino, A. Metric character of the quantum Jensen-Shannon divergence. *Phys. Rev. A* **77,** 052311 (2008).

Acknowledgements

A.A. and M.D.D. are supported by MINECO through Grant FIS2012-38266; by the EC FET-Proactive Project PLEXMATH (grant 317614) and the Generalitat de Catalunya 2009-SGR-838. A.A. also acknowledges partial financial support from the ICREA Academia and the James S. McDonnell Foundation. V.L. and V.N. acknowledge support from the EC FET-Proactive Project LASAGNE (grant 318132), funded by the European Commission. V.L. also acknowledge support from the EPSRC project GALE EP/K020633/1. This research used Queen Mary MidPlus computational facilities, supported by QMUL Research-IT and funded by EPSRC grant EP/K000128/1.

Author contributions

M.D.D. and V.N. contributed equally to this work. All authors conceived the study, performed the numerical experiments, collected and analysed the data, and wrote the manuscript. All authors approved the final version of the manuscript

Additional information

Competing financial interests: The authors declare no competing financial interests.

17

The power of a critical heat engine

Michele Campisi[1] & Rosario Fazio[1,2]

Since its inception about two centuries ago thermodynamics has sparkled continuous interest and fundamental questions. According to the second law no heat engine can have an efficiency larger than Carnot's efficiency. The latter can be achieved by the Carnot engine, which however ideally operates in infinite time, hence delivers null power. A currently open question is whether the Carnot efficiency can be achieved at finite power. Most of the previous works addressed this question within the Onsager matrix formalism of linear response theory. Here we pursue a different route based on finite-size-scaling theory. We focus on quantum Otto engines and show that when the working substance is at the verge of a second order phase transition diverging energy fluctuations can enable approaching the Carnot point without sacrificing power. The rate of such approach is dictated by the critical indices, thus showing the universal character of our analysis.

[1] NEST, Scuola Normale Superiore & Istituto Nanoscienze-CNR, Pisa I-56126, Italy. [2] ICTP, Strada Costiera 11, Trieste 34151, Italy. Correspondence and requests for materials should be addressed to M.C. (email: michele.campisi@sns.it).

Increasing the power output of a heat engine has a corresponding cost in terms of reduced efficiency η, or, in equivalent terms, a larger deviation $\Delta\eta = \eta^C - \eta$ from the Carnot efficiency η^C. A question that is currently the object of vigorous research efforts is whether it is possible to devise a heat engine that outputs finite power at Carnot efficiency[1-5]. In the following we address this question by focussing on the quantity

$$\dot{\Pi} \doteq \frac{\mathcal{P}}{\Delta\eta} \tag{1}$$

that is, the ratio of power output $\mathcal{P}$ over $\Delta\eta$. We shall call $\dot{\Pi}$ the performance rate. It is trivially possible to increase the power without affecting the efficiency by scaling the size of the working substance (WS): An array of N identical engines working in parallel provides an N-fold larger power than each of them, at the same efficiency. In this case the output work per cycle and efficiency scale as $\mathcal{P} \sim N$, and $\Delta\eta \sim 1$, consequently $\dot{\Pi} \sim N$ (here the symbol $\sim$ means 'scales as'). Note that this linear increase in performance rate does not represent any real gain as it is achieved at the cost of a corresponding linear increase of resources. The question is therefore 'Can the scaling of the performance rate $\dot{\Pi}$ be improved beyond linear' in order to have a true gain? Note that a positive answer implies (in an asymptotic sense) a positive answer to the fundamental question posed above: Imagine having a WS made up of N constituents (or resources); if $\dot{\Pi} \sim N^{1+a}$, with $a>0$, then one could approach Carnot efficiency as $\Delta\eta \sim \mathcal{P}/\dot{\Pi} \sim N^{-a} \to 0$ by increasing N, while keeping the 'power per resource' fixed, that is, $\mathcal{P} \sim N$.

In the following we make a substantial step towards answering the question above. The main idea that we pursue is that an interaction between the N constituents of the WS could provide the extra scaling power that is needed for a positive resolution, see Fig. 1. In order to address that question quantitatively we consider a special engine cycle that is well studied in the literature (see, for example, refs 6–8 and references therein), namely, a quantum version of the Otto engine[9-14], see Fig. 2. We show that a universal behaviour, with anomalous scaling of the performance rate

$$\dot{\Pi} \sim N^{1+(\alpha - z\nu)/(d\nu)} \tag{2}$$

emerges when the WS is on the verge of a second-order phase transition. Here α,ν,z are the specific heat, correlation length and dynamical critical exponent, and d is the number of dimensions of the WS. Note that the performance rate contains two concurrent contributions, one stemming from the scaling of the heat capacity, that is, the exponent α, and the other stemming from the scaling of the relaxation time, that is, the exponent z. Since $d,\nu>0$, in order to have a more than linear behaviour of $\dot{\Pi}$ one needs a WS with a critical point characterized by the inequality $\alpha - z\nu > 0$. In fact, as we explain below in detail, the stronger condition

$$\alpha - z\nu \geq 1 \tag{3}$$

would ensure the asymptotic approach towards the Carnot point without giving up power per resource. Is that possible? The finiteness of internal energy implies the bound $\alpha \leq 1$. Critical slowing down ($z \geq 0$) would then imply $\alpha - z\nu \leq 1$ (for example, in the three-dimensional Ising model, $\alpha \simeq 0.12$, $\nu \simeq 0.63$ (ref. 15) and $z \simeq 2.35$ (ref. 16); hence $\alpha - z\nu \simeq -0.28$). A number of theoretical and experimental works report on the possibility of the exotic phenomenon of critical speeding up, $z \leq 0$ (refs 17–20). In this case the two terms might add up above unity (for example, recent experimental studies report $\alpha \simeq 0.38$ (ref. 21) and $z\nu \simeq -0.7$ (ref. 19); hence $\alpha - z\nu \simeq 1.08 > 1$, for $Dy_2Ti_2O_7$). We conclude that there currently appears to be no fundamental reason hindering the possibility of approaching Carnot efficiency without giving up power per resource. This is certainly possible up to some threshold size $\bar{N}$ for which the weaker condition $\alpha - z\nu > 0$ suffices. Below we illustrate how such powerful critical engines should be designed. At the heart of our result is the recognition that scaling of the performance, that is,

$$\Pi = \frac{W_{\text{out}}}{\Delta\eta} \tag{4}$$

where W_{out} is the work output per cycle, is dictated by the heat capacity of the WS (which can notably diverge at the critical point), a crucial and simple fact that was never noticed before, see equation (9).

Results

N-body quantum Otto engine. The quantum Otto engine (see Fig. 2) is a four-stroke engine based on a WS with Hamiltonian

$$H^{\text{WS}}(t) = \lambda(t)K. \tag{5}$$

(Following the current literature we call these engines 'quantum Otto engines' although they bear no other quantum feature

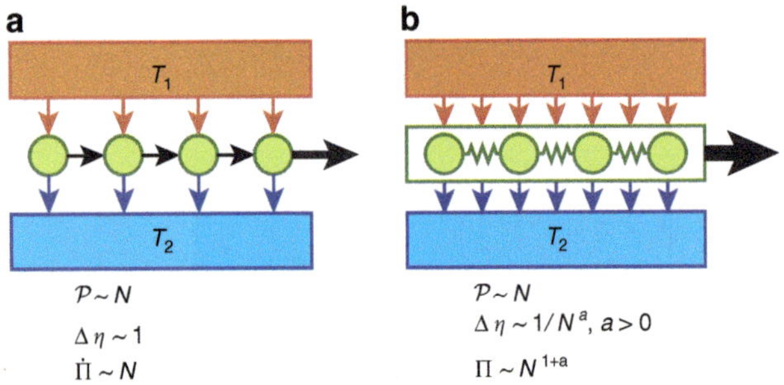

Figure 1 | Main idea at the basis of the results. (**a**) N identical devices operating in parallel provide a power $\mathcal{P}$ that scales like N, at fixed efficiency. (**b**) When an interaction among the N parallel devices is turned on, the approach to Carnot efficiency is enabled.

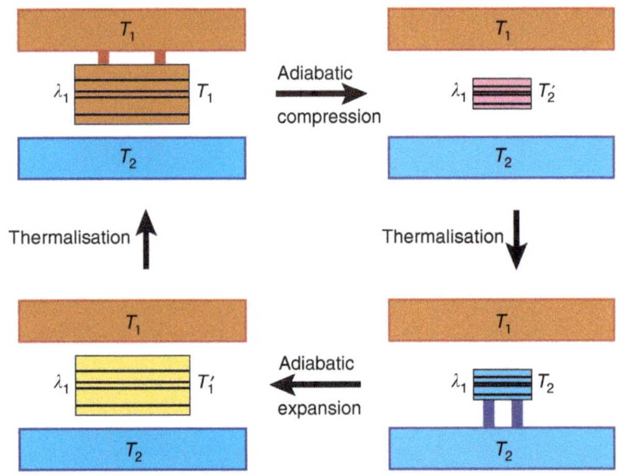

Figure 2 | A quantum Otto engine. At time $t=0$ the WS is in thermal equilibrium with bath 1 and $\lambda = \lambda_1$. During the first stroke the WS undergoes a thermally isolated transformation where the Hamiltonian switches from $H_1^{\text{WS}} = \lambda_1 K$ to $H_2^{\text{WS}} = \lambda_2 K$. The second stroke consists in thermalization with bath 2, with λ being kept fixed at λ_2. During the third stroke the Hamiltonian goes back to $H_1^{\text{WS}} = \lambda_1 K$ while in thermal isolation. The fourth stroke consists in letting the system thermalize with bath 1, with λ being kept fixed at λ_1.

besides the discreteness of the spectrum. Accordingly, there is nothing genuinely quantum in our treatment.) The heats released in the baths during the thermalization strokes are $Q_i = \mp \lambda_i [U_K(\beta_2\lambda_2) - U_K(\beta_1\lambda_1)]$, where $U_K(\theta) = \mathrm{Tr}\,K e^{-\theta K}/\mathrm{Tr}\,e^{-\theta K}$ is the internal energy associated to the base Hamiltonian K at inverse temperature θ, and $-$, $+$ is for $i=1, 2$, respectively. The work performed by the engine $W_{\mathrm{out}} = Q_1 + Q_2$ is thus

$$W_{\mathrm{out}} = (\lambda_2 - \lambda_1)[U_K(\beta_2\lambda_2) - U_K(\beta_1\lambda_1)]. \quad (6)$$

For $\beta_1 < \beta_2$, the condition $W_{\mathrm{out}} \geq 0$, $Q_1 \geq 0$ defines the regime of operation of the engine as a heat engine, implying $\lambda_2/\lambda_1 \geq \beta_1/\beta_2$. The efficiency is

$$\eta = \frac{W_{\mathrm{out}}}{Q_1} = 1 - \frac{\lambda_2}{\lambda_1} \leq 1 - \frac{\beta_1}{\beta_1} = \eta^{\mathrm{C}}. \quad (7)$$

The efficiency is smaller than η^{C}, in accordance with the heat engine fluctuation relation[22,23].

Using equation (7) the output work of the quantum Otto engine can be expressed as a function of $\lambda_1, \beta_1, \beta_2$ and $\Delta\eta = \lambda_2/\lambda_1 - \beta_1/\beta_2$ as

$$W_{\mathrm{out}} = \lambda_1(\Delta\eta - \eta^{\mathrm{C}})[U_K(\beta_1\lambda_1 + \beta_2\lambda_1\Delta\eta) - U_K(\beta_1\lambda_1)]. \quad (8)$$

This expression allows us to study the performance Π. In the region where linear approximation holds, that is, $\Delta\eta \ll 1$ (namely $\beta_1\lambda_1$ is close to $\beta_2\lambda_2$), the scaling of the engine's performance Π with N is given by the scaling of $\partial W_{\mathrm{out}}/\partial\Delta\eta|_{\Delta\eta=0}$, that is, the slope of $W_{\mathrm{out}}(\Delta\eta)$ at the origin. This is illustrated in Fig. 3. In the case of N devices in parallel it is $W_{\mathrm{out}} = N W_{\mathrm{out}}^{\mathrm{single}}$ (here $W_{\mathrm{out}}^{\mathrm{single}}$ denotes the work of each single device); hence trivially $\partial W_{\mathrm{out}}/\partial\Delta\eta\,|_{\Delta\eta=0} \sim \Pi \sim N$, and, accordingly $\dot{\Pi} \sim N$, as we noted before. It follows that in order to boost the scaling of the performance it is necessary to have the slope $\partial W_{\mathrm{out}}/\partial\Delta\eta|_{\Delta\eta=0}$ to scale more than linearly.

By taking the derivative of equation (8) with respect to $\Delta\eta$ we obtain the central relation

$$\Pi \sim \frac{\partial W_{\mathrm{out}}}{\partial\Delta\eta}\,|_{\Delta\eta=0} \sim N c_K(\beta_1\lambda_1) \quad (9)$$

where $c_K(\theta) = -(1/N)\theta^2 \partial U_K/\partial\theta$ is the specific heat of the WS. To understand the physics behind the emergence of equation (9),

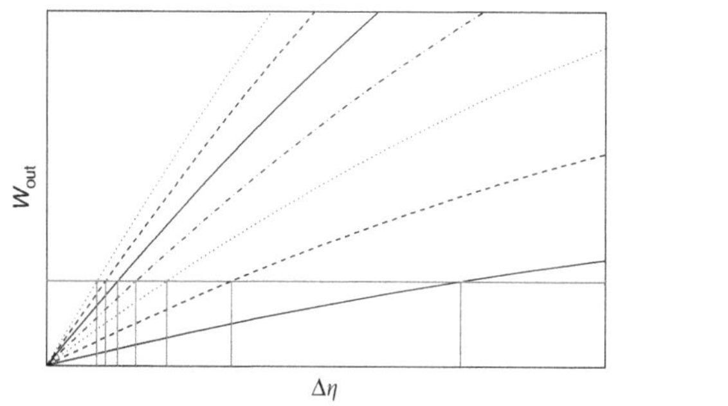

Figure 3 | Graphical demonstration of equation (9). The graph shows various plots of $W_{\mathrm{out}}(\Delta\eta)$ for increasing values of N (from lower to upper curves). At a fixed work output (horizontal red line), the curves are intercepted at decreasing values of $\Delta\eta$. If the slope at the origin increases with a power N^a, $\partial W_{\mathrm{out}}/\partial\Delta\eta\,|_{\Delta\eta=0} \sim N^a$; then, provided for larger and larger N the intercept occurs in the region where the linear approximation holds for the Nth curve, the approach towards $\Delta\eta = 0$ occurs as $1/N^a$; hence $\Pi = W_{\mathrm{out}}/\Delta\eta \sim N^a \sim \partial W_{\mathrm{out}}/\partial\Delta\eta\,|_{\Delta\eta=0}$. Similarly, one might increase (lower) the value of W_{out} with N, for example, as $W_{\mathrm{out}}(\Delta\eta) = wN^b$. Accordingly, provided for larger and larger N the intercept still occurs in the region where the linear approximation holds, the approach towards $\Delta\eta = 0$ occurs as $1/N^{a-b}$. Still, $\Pi = W_{\mathrm{out}}/\Delta\eta \sim N^a \sim \partial W_{\mathrm{out}}/\partial\Delta\eta\,|_{\Delta\eta=0}$.

consider working at some point that is very close to the Carnot point, that is, chose λ_1 and λ_2 so that their ratio λ_1/λ_2 is very close to β_2/β_1. After the adiabatic compression stroke the WS reaches a new temperature $T_2' = 1/(k_B\beta_2')$, which is very close to the cold bath temperature $T_2 = 1/(k_B\beta_2)$: $T_2' = T_1\lambda_2/\lambda_1 = T_2 + \Delta T_2$, with a small ΔT_2. The larger the heat capacity $C_K = Nc_K$ of the WS, at the thermalization point, the larger the heat exchanged $Q_2 = C_K\Delta T_2$ during the subsequent thermalization with the cold bath. Likewise for the subsequent expansion and thermalization. Since $W_{\mathrm{out}} = Q_1 + Q_2$, the larger the heat capacity the larger the work output, and accordingly the larger the performance.

Equation (9) tells us that in order to achieve super-linear scaling of the performance, one needs a WS with an anomalous scaling of the specific heat. Recall that for ordinary substances the heat capacity is extensive, $C_K \sim N$, namely $c_K = C_K/N \sim 1$. What is needed for improved performance is $c_K \sim N^a$, with some $a > 0$. This can happen at the verge of a phase transition. The physical reason is that at a phase transition finite exchanges of heat are accompanied by infinitesimal changes of temperature, that is, the specific heat diverges. This is because at the transition point the energy intake is not employed to heat up but rather to make the change of phase.

For a second-order phase transition, finite size scaling predicts a peak in the specific heat whose height and width scale, respectively, as $\bar{c}_K \sim N^{\alpha/(dv)}$ and $\delta \sim N^{-1/(dv)}$ (ref. 24) where d is the dimensionality of the system, and α and v are the specific heat and the correlation length critical exponents. Accordingly we predict the possibility of a boosted scaling of the perfmormance $\Pi \sim N^{1+\alpha/(dv)}$. Writing the performance rate as $\dot{\Pi} = \mathcal{P}/\Delta\eta = W_{\mathrm{out}}/(\mathcal{T}\Delta\eta) = \Pi/\mathcal{T}$ we see that its scaling is determined by the scaling of Π and of the cycle time $\mathcal{T}$. The latter is dominated by the thermalization time[2], which, according to finite-size scaling theory scales with the dynamical critical exponent z as $\mathcal{T}_{\mathrm{relax}} \sim N^{z/d}$ (ref. 25). This gives equation (2).

Critical engine design. We illustrate how a quantum Otto engine can be designed to achieve the predicted performance rate in equation (2). Let the substance be described by a Hamiltonian K, displaying, in the infinite size limit, a second-order phase transition at the inverse critical temperature θ_{C}. Let the two baths have the inverse temperatures β_1 and β_2. We assume we can stretch/compress the spectrum of the WS and implement accordingly the time-dependent Hamiltonian in equation (5) with $\lambda(t) \in [\lambda_1, \lambda_2]$. When λ takes on the value λ_i, the critical temperature is rescaled to $\beta_{\mathrm{C}}^i = \theta_{\mathrm{C}}/\lambda_i$. We choose, for example, λ_1 so that $\beta_1 = \beta_{\mathrm{C}}^1$ (that is $\lambda_1 = \theta_{\mathrm{C}}/\beta_{\mathrm{C}}^1$). In this way the temperature of bath 1 coincides with the critical temperature of $H_1^{\mathrm{WS}} = \lambda_1 K$. We next choose λ_2 as the solution of $W_{\mathrm{out}} = N^{1+z/d}w$, for some fixed w. That gives the corresponding efficiency $\eta = 1 - \lambda_2/\lambda_1$. Note that since $\mathcal{T} \sim N^{z/d}$, the power per constituent is fixed: $\mathcal{P} = W_{\mathrm{out}}/\mathcal{T} \sim N$. Since the slope of the graph $W(\Delta\eta)$ grows with the heat capacity as $\Pi \sim N^{1+\alpha/(dv)}$, we have $\Delta\eta = W_{\mathrm{out}}/\Pi \sim N^{1+z/d}/N^{1+\alpha/(dv)} \sim N^{-(\alpha - vz)/dv}$. Hence if $\alpha - vz > 0$ the solution λ_2 gets closer and closer to $\lambda_1\beta_1/\beta_2$. Accordingly the efficiency $\eta = 1 - \lambda_2/\lambda_1$ gets closer and closer to $\eta^{\mathrm{C}} = 1 - \beta_1/\beta_2$. As discussed in the caption of Fig. 3, the previous statement is correct as long as λ_2 falls in the region of validity of the linear approximation (that is, in graphical terms, if the corresponding $\Delta\eta$ falls in the region where the curve $W(\Delta\eta)$ is well approximated by a straight line passing through the origin). The extension of that region corresponds to the width of the specific heat peak, which, as mentioned above, universally scales as $\delta \sim N^{-1/(dv)}$. So, in order to keep pace with the shrinking of the linear approximation region, the approach towards $\Delta\eta = 0$ should

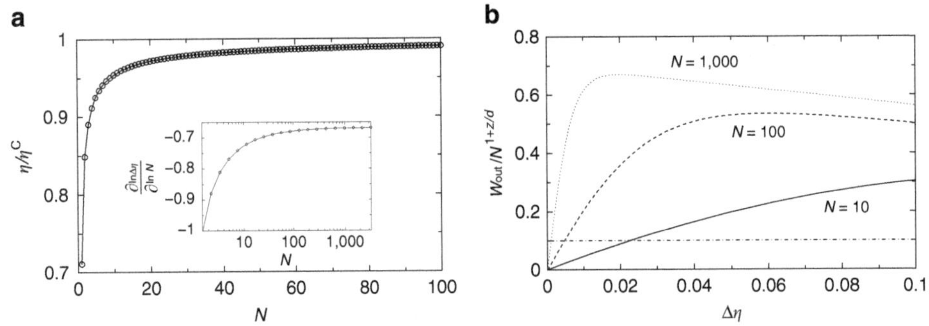

Figure 4 | Efficiency and shrinking of linear region in a critical sped-up engine. (**a**) Approach towards the Carnot point of a critical working substance characterized by equation (3) at a fixed power per constituent. Plots were calculated with critical indices reported recently for $Dy_2Ti_2O_7$, namely $\alpha \simeq 0.38$ (ref. 21) and $z\nu \simeq -0.7$ (ref. 19). The value of ν was obtained from the relation $\nu=(2\text{-}\alpha)/d$ (ref. 27). Carnot point was at $\eta^C=1/2$ and $w=W_{out}/N^{1+z/d}=0.1$. The inset shows the quantity $\partial \ln \Delta\eta/\partial \ln N$ as a function of N, that is, the exponent characterizing the approach to the Carnot point that reaches the expected value $-(\alpha-z\nu)/(d\nu) \simeq -0.67$. (**b**) Rescaled output work $W_{out}/N^{1+z/d}$ as a function of $\Delta\eta$. Note that the linear region shrinks around the origin of axes as N increases, and that the intercept with $W_{out}/N^{1+z/d}=0.1$ occurs within that region for all curves.

be not slower than $N^{-1/(d\nu)}$. Since that would occur at a pace of $N^{-(\alpha-\nu z)/(d\nu)}$, the condition ensuring that it actually asymptotically occurs is given by equation (3).

In Fig. 4a we illustrate how in such a case the engine design that we have described above actually results in an asymptotic approach towards the Carnot point, at a fixed power per constituent. Figure 4b illustrates the shrinking of the linear region, due to the narrowing of the peak width. In making Fig. 4 we have taken full advantage of one of the most striking aspects of our analysis, namely its universality. The details of the specific model are not essential, all that counts are the critical exponents. The specific heat peak reads $c_K(\theta)=N^{\alpha/(d\nu)}\theta^2 U_1\Delta U f(\Delta U(\theta-\theta_C)N^{1/(d\nu)})$, where $f(x)$ is some bell-shaped function (its exact shape is not relevant). The corresponding internal energy of the WS around the critical point reads $U_K(\theta)=U_0-N^{1+(\alpha-1)/(d\nu)}U_1 F(\Delta U(\theta-\theta_C)N^{1/(d\nu)})$, where $F'(x)=f(x)$. The plots are obtained by using the latter formula, with equation (6) used to evaluate the work output, with the choice $f(x)=\text{sech}^2(x)$, $F(x)=\tanh(x)$ and $U_1=\Delta U_1=1$. The plot illustrates the approach towards Carnot efficiency at a fixed power per constituent, with the predicted scaling exponent $-(\alpha-z\nu)/(d\nu)$.

Discussion

The idea that phase transitions could enable the attainment of Carnot efficiency at finite power was previously hinted by Polettini *et al.*[5], but was never pursued before. The present work confirms that intuition in a fully fledged way based on universality and finite-size scaling theory and most importantly by accounting for the first time for the effect of criticality on the time of operation hence of the power.

As is clear from Fig. 3, independent of how the slope $\partial W_{out}/\partial\Delta\eta|_{\Delta\eta=0}$ scales, $W_{out}(\Delta\eta=0)=0$; hence, exactly at the Carnot point all quantum Otto engines deliver null work and hence null power. Our statement should be accordingly understood in a weaker asymptotic sense, namely, that it is possible to get as close as one wants to the Carnot point without giving up the power per constituent. This is the viewpoint that also inspires ref. 2.

We remark that our linear condition $\Delta\eta \ll 1$ substantially differs from the condition of linear response regime (that is, $\beta_2-\beta_1 \ll \beta_2$ or $\eta \ll 1$) that was investigated previously in refs 1, 3–5. In our case the two temperatures need not be close to each other.

We stress that all obstacles hindering the realization of the critical powerful Carnot engines appear to be of technological nature, rather than fundamental. One major difficulty stems from the necessity of implementing the Hamiltonian (5) containing a global coupling $\lambda(t)$, which can be very challenging in practice. Another difficulty stems from the fact that in order to be able to asymptotically approach the Carnot point, one should accordingly have an increasing degree of accuracy with which λ_1 and λ_2 are controlled.

Lastly, it is worth stressing that the result in equation (9) holds in general as a consequence of the linear approximation valid for $\Delta\eta \ll 1$, and is accordingly not restricted to the case of critical phenomena. This tells that one route towards the improvement of the performance of a WS is to increase its specific heat. That could be achieved by increasing its number of constituents, a possibility that we have investigated here, or by manipulating any other parameter entering the Hamiltonian K. Recently reported cases of improved performances[26] can in fact be interpreted in terms of increased heat capacity.

References

1. Benenti, G., Saito, K. & Casati, G. 'Thermodynamic bounds on efficiency for systems with broken time-reversal symmetry'. *Phys. Rev. Lett.* **106,** 230602 (2011).
2. Allahverdyan, A. E., Hovhannisyan, K. V., Melkikh, A. V. & Gevorkian, S. G. 'Carnot cycle at finite power: attainability of maximal efficiency'. *Phys. Rev. Lett.* **111,** 050601 (2013).
3. Proesmans, K. & Van den Broeck, C. 'On-sager coefficients in periodically driven systems'. *Phys. Rev. Lett.* **115,** 090601 (2015).
4. Brandner, K., Saito, K. & Seifert, U. 'Thermodynamics of micro- and nano-systems driven by periodic temperature variations'. *Phys. Rev. X* **5,** 031019 (2015).
5. Polettini, M., Verley, G. & Esposito, M. 'Efficiency statistics at all times: Carnot limit at finite power'. *Phys. Rev. Lett.* **114,** 050601 (2015).
6. Benenti, G., Casati, G., Prosen, T. & Saito, K. Fundamental aspects of steady state heat to work conversion. Preprint at arXiv:1311.4430 (2013).
7. Kosloff, R. & Levy, A. 'Quantum heat engines and refrigerators: continuous devices'. *Annu. Rev. Phys. Chem.* **65,** 365–393 (2014).
8. Gelbwaser-Klimovsky, D., Niedenzu, W. & Kurizki, G. *'Chapter Twelve—Thermodynamics of Quantum Systems under Dynamical Control'* 329–407 (Academic Press, 2015).
9. Scully, M. O. 'Quantum afterburner: Improving the efficiency of an ideal heat engine'. *Phys. Rev. Lett.* **88,** 050602 (2002).
10. Feldmann, T. & Kosloff, R. 'Quantum four-stroke heat engine: Thermodynamic observables in a model with intrinsic friction'. *Phys. Rev. E* **68,** 016101 (2003).
11. Quan, H., Liu, Y.-X., Sun, C. & Nori, F. 'Quantum thermodynamic cycles and quantum heat engines'. *Phys. Rev. E* **76,** 031105 (2007).
12. Niskanen, A. O., Nakamura, Y. & Pekola, J. P. 'Information entropic superconducting microcooler'. *Phys. Rev. B* **76,** 174523 (2007).

13. Allahverdyan, A. E., Johal, R. S. & Mahler, G. 'Work extremum principle: Structure and function of quantum heat engines'. *Phys. Rev. E* **77,** 041118 (2008).
14. Rofinagel, J., Abah, O., Schmidt-Kaler, F., Singer, K. & Lutz, E. 'Nanoscale heat engine beyond the carnot limit'. *Phys. Rev. Lett.* **112,** 030602 (2014).
15. Pelissetto, A. & Vicari, E. 'Critical phenomena and renormalization-group theory'. *Phys. Rep* **368,** 549–727 (2002).
16. Matz, R., Hunter, D. L. & Jan, N. 'The dynamic critical exponent of the three-dimensional Ising model'. *J. Stat. Phys.* **74,** 903–908 (1994).
17. Zappoli, B. *et al.* 'Anomalous heat transport by the piston effect in supercritical fluids under zero gravity'. *Phys. Rev. A* **41,** 2264–2267 (1990).
18. Boukari, H., Briggs, M. E., Shaumeyer, J. N. & Gammon, R. W. 'Critical speeding up observed'. *Phys. Rev. Lett.* **65,** 2654–2657 (1990).
19. Grams, C. P., Valldor, M., Garst, M. & Hemberger, J. 'Critical speeding-up in the magne-toelectric response of spin-ice near its monopole liquid-gas transition'. *Nat. Commun.* **5,** 4853 (2014).
20. Tavora, M., Rosch, A. & Mitra, A. 'Quench dynamics of one-dimensional interacting bosons in a disordered potential: elastic dephasing and critical speeding-up of thermalization'. *Phys. Rev. Lett.* **113,** 010601 (2014).
21. Higashinaka, R., Fukazawa, H., Deguchi, K. & Maeno, Y. 'Low temperature specific heat of dy2ti2o7 in the kagome ice state'. *J. Phys. Soc. Jpn.* **73,** 2845–2850 (2004).
22. Campisi, M. 'Fluctuation relation for quantum heat engines and refrigerators'. *J. Phys. A: Math. Theor.* **47,** 245001 (2014).
23. Campisi, M., Pekola, J. & Fazio, R. 'Nonequilibrium fluctuations in quantum heat engines: theory, example, and possible solid state experiments'. *New J. Phys.* **17,** 035012 (2015).
24. Fisher, M. E. & Barber, M. N. 'Scaling theory for finite-size effects in the critical region'. *Phys. Rev. Lett.* **28,** 1516–1519 (1972).
25. Suzuki, M. 'Static and dynamic finite-size scaling theory based on the renormalization group approach'. *Prog. Theor. Phys.* **58,** 1142–1150 (1977).
26. Cakmak, S., Altintas, F. & Mustecaplioglu, O. E. Lipkin-Meshkov-Glick Model in a quantum Otto Cycle. Preprint at arXiv:1510.04495 (2015).
27. Huang, K. *Statistical Mechanics* 2nd edn (Wiley, 1987).

Acknowledgements

This research was supported by UNICREDIT, the 7th European Community Framework Programme under grant agreements no. 623085 (IEF-NeQuFlux), no. 600645 (IP-SIQS), no. 618074 (STREP-TERMIQ) and by the COST action MP1209 'Thermodynamics in the quantum regime'.

Author contributions

M.C. conceived the idea. M.C. and R.F. conducted the work, analysed the results and drew the conclusions.

Additional information

Competing financial interests: The authors declare no competing financial interests.

18

Arrays of individually controlled ions suitable for two-dimensional quantum simulations

Manuel Mielenz[1], Henning Kalis[1], Matthias Wittemer[1], Frederick Hakelberg[1], Ulrich Warring[1], Roman Schmied[2], Matthew Blain[3], Peter Maunz[3], David L. Moehring[3,†], Dietrich Leibfried[4] & Tobias Schaetz[1,5]

A precisely controlled quantum system may reveal a fundamental understanding of another, less accessible system of interest. A universal quantum computer is currently out of reach, but an analogue quantum simulator that makes relevant observables, interactions and states of a quantum model accessible could permit insight into complex dynamics. Several platforms have been suggested and proof-of-principle experiments have been conducted. Here, we operate two-dimensional arrays of three trapped ions in individually controlled harmonic wells forming equilateral triangles with side lengths 40 and 80 μm. In our approach, which is scalable to arbitrary two-dimensional lattices, we demonstrate individual control of the electronic and motional degrees of freedom, preparation of a fiducial initial state with ion motion close to the ground state, as well as a tuning of couplings between ions within experimental sequences. Our work paves the way towards a quantum simulator of two-dimensional systems designed at will.

[1] Albert-Ludwigs-Universität Freiburg, Physikalisches Institut, Hermann-Herder-Strasse 3, Freiburg 79104, Germany. [2] Department of Physics, University of Basel, Klingelbergstrasse 82, Basel 4056, Switzerland. [3] Sandia National Laboratories, PO Box 5800 Albuquerque, New Mexico 87185-1082, USA. [4] Time and Frequency Division, National Institute of Standards and Technology, 325 Broadway, Boulder, Colorado 80305, USA. [5] Albert-Ludwigs-Universität Freiburg, Freiburg Institute for Advanced Studies, Albertstr. 19, 79104 Freiburg, Germany. † Present address: The Intelligence Advanced Research Projects Activity, College Park, MD, USA. Correspondence and requests for materials should be addressed to U.W. (email: ulrich.warring@physik.uni-freiburg.de).

Richard Feynman was one of the first to recognize that quantum systems of sufficient complexity cannot be simulated on a conventional computer[1]. He proposed to use a quantum mechanical system instead. A universal quantum computer would be suitable, but practical implementations are a decade away at best. However, universality is not required to simulate specific quantum models. It is possible to custom-build an analogue quantum simulator (AQS) that allows for preparation of fiducial input states, faithful implementation of the model-specific dynamics and for access to the crucial observables. Simulations on such AQSs could impact a vast variety of research fields[2], that is, physics[3], chemistry[4] and biology[5], when studying dynamics that is out of reach for numerical simulation on conventional computers.

Many experimental platforms have been suggested to implement AQSs[6–9]. Different experimental systems provide certain advantages in addressing different physics. Results that are not conventionally tractable may be validated by comparing results of different AQSs simulating the same problem[10,11]. Over the last two decades, many promising proof-of-principle demonstrations have been made using photons[6], superconductors[7], atoms[8] and trapped atomic ions[9]. Trapped ions in particular have seen steady progress from demonstrations with one or two ions[12–18] to addressing aspects of quantum magnets[19] with linear strings of 2–16 ions[13,20] and self-ordered two-dimensional crystals containing more than 100 ions[21]. Ions are well suited to further propel the research since they provide long-range interaction and individual, fast controllability with high precision[22].

Two-dimensional trap-arrays may offer advantages over trapping in a common potential, because they are naturally suited to implement tuneable couplings in more than one spatial dimension. Such couplings are, in most cases, at the heart of problems that are currently intractable by conventional numerics[10,23]. Our approach is based on surface-electrode structures[24] originally developed for moving ion qubits through miniaturized and interconnected, linear traps as proposed in refs 25,26. This approach is pursued successfully as a scalable architecture for quantum computer, see, for example, ref. 27. For AQSs, it is beneficial to have the trapped ion ensembles coupled all-to-all so they evolve as a whole. This is enabled by our array architecture with full control over each ion. Individual control allows us to maintain all advantages of single trapped ions while scaling the array in size and dimension[28–30].

Optimized surface electrode geometries can be found for any periodic wallpaper group as well as quasi-periodic arrangements, as, for example, Penrose-tilings[29]. A first step, trapping of ions in two-dimensional arrays of surface traps, has been proposed[15] and demonstrated[31]. Boosting the strength of interaction to a level comparable to current decoherence rates requires inter-ion distances d of a few tens of micrometres. Such distances have been realized in complementary work, where two ions have been trapped in individually controlled sites of a linear surface-electrode trap at d between 30 and 40 μm. The exchange of a single quantum of motion, as well as entangling spin–spin interactions have been demonstrated in this system[32,33]. The increase in coupling strength was achieved with a reduction of the ion-surface separation to order d and the concomitant increase in motional heating due to electrical noise. Recently, methods for reducing this heating by more than two orders of magnitude with either surface treatments[34–36] or cold electrode surfaces[37–39] have been devised.

Here, we demonstrate the precise tuning of all relevant parameters of a two-dimensional array of three ions trapped in individually controlled harmonic wells on the vertices of equilateral triangles with side lengths 80 and 40 μm. In the latter, Coulomb coupling rates[32] approach current rates of decoherence. Dynamic control permits to reconfigure Coulomb and laser couplings at will within single experiments. We initialize fiducial quantum states by optical pumping, Doppler and resolved sideband cooling to near the motional ground state. Our results demonstrate important prerequisites for experimental quantum simulations of engineered two-dimensional systems.

Results

Trap arrays and control potentials. Our surface ion trap chip is fabricated in similar manner to that described in ref. 40 and consists of two equilateral triangular trap arrays with side length of $\simeq 40$ and $\simeq 80$ μm, respectively (Fig. 1a,b), both with a distance of $\simeq 40$ μm between the ions and the nearest electrode surface. The shapes of radio-frequency (RF) electrodes of the arrays are optimized by a linear-programming algorithm that yields electrode shapes with low fragmentation, and requires only a single RF-voltage source for operation[29,30]. To design different and even non-periodic arrays for dedicated trap distances, we can apply the same algorithm to yield globally optimal electrode shapes[29]. Resulting electrode shapes may look significantly different, but will have comparable complexity, spatial extent and the same number of control electrodes per trap site. Therefore, we expect that different arrays will not require different fabrication techniques (Methods). The two arrays are spaced by $\simeq 5$ mm on the chip, and only one of them is operated at a given time. Although we achieve similar results in both arrays, the following discussion is focussed on the 80 μm array.

Three-dimensional confinement of $^{25}Mg^+$ ions is provided by a potential ϕ_{RF} oscillating at Ω_{RF} from a single RF electrode driven at $\Omega_{RF}/(2\pi) = 48.3$ MHz with an approximate peak voltage $U_{RF} = 20$ V. Setting the origin of the coordinate system at the centre of the array and in the surface plane of the chip, the RF potential features three distinct trap sites at $\mathbf{T0} \simeq (-46,0,37)$ μm, $\mathbf{T1} \simeq (23, -23\sqrt{3}, 37)$μm, and $\mathbf{T2} \simeq (23, 23\sqrt{3}, 37)$μm. Owing to the electrode symmetry under rotations of $\pm 2\pi/3$ around the z-axis, it is often sufficient to consider **T0** only, as all our findings apply to **T1** and **T2** after an appropriate rotation. Further, the RF potential exhibits another trap site at $\simeq (0,0,81)$ μm (above the centre of the array); this 'ancillary' trap is used for loading as well as for re-capturing ions that escaped from the other trap sites. We approximate the RF confinement at position $\mathbf{r}$ by a pseudopotential $\phi_{ps}(\mathbf{r}) = Q/(4m\Omega_{RF}^2)E_{RF}^2(\mathbf{r})$, cp. ref. 41, where Q denotes the charge and m the mass of the ion, and $E_{RF}(\mathbf{r})$ is the field amplitude produced by the electrode. Calculations of trapping potentials are based on ref. 42 and utilizing the software package[43]. Equipotential lines of ϕ_{ps} are shown in Fig. 1c–e.

Near **T0** we can approximate ϕ_{ps} up to second order and diagonalize the local curvature matrix to find normal modes of motion described by their mode vectors $\mathbf{u}_1$, $\mathbf{u}_2$ and $\mathbf{u}_3$, which coincide (for the pure pseudopotential) with x, y and z; we use $\mathbf{u}_j$ with $j = \{1,2,3\}$ throughout our manuscript to describe the mode vectors of a single ion near **T0**. We find corresponding potential curvatures of $\kappa_{ps,1} \simeq 3.0 \times 10^8$ V m^{-2}, $\kappa_{ps,2} \simeq 5.9 \times 10^7$ V m^{-2} and $\kappa_{ps,3} \simeq 9.2 \times 10^7$ V m^{-2}, whereas mode frequencies can be inferred from these curvatures as $\omega_j \simeq \sqrt{(Q/m)\kappa_{ps,j}}$, with $j = \{1,2,3\}$: $\omega_1/(2\pi) \simeq 5.4$ MHz, $\omega_2/(2\pi) \simeq 2.4$ MHz and $\omega_3/(2\pi) \simeq 3.0$ MHz. Further, the Mathieu parameters $q_i = 2(Q/m)U_{RF}/\Omega_{RF}^2 \kappa_{RF,i}(\mathbf{r})$, where $\kappa_{RF,i}(\mathbf{r})$ denotes the curvature of ϕ_{RF} along direction $i = \{x,y,z\}$, at **T0** are: $q_x \simeq -0.32$, $q_y \simeq 0.14$ and $q_z \simeq 0.18$.

To gain individual control of the trapping potential at each site, it is required to independently tune local potentials near **T0**, **T1** and **T2** (Methods), that is, to make use of designed local electric

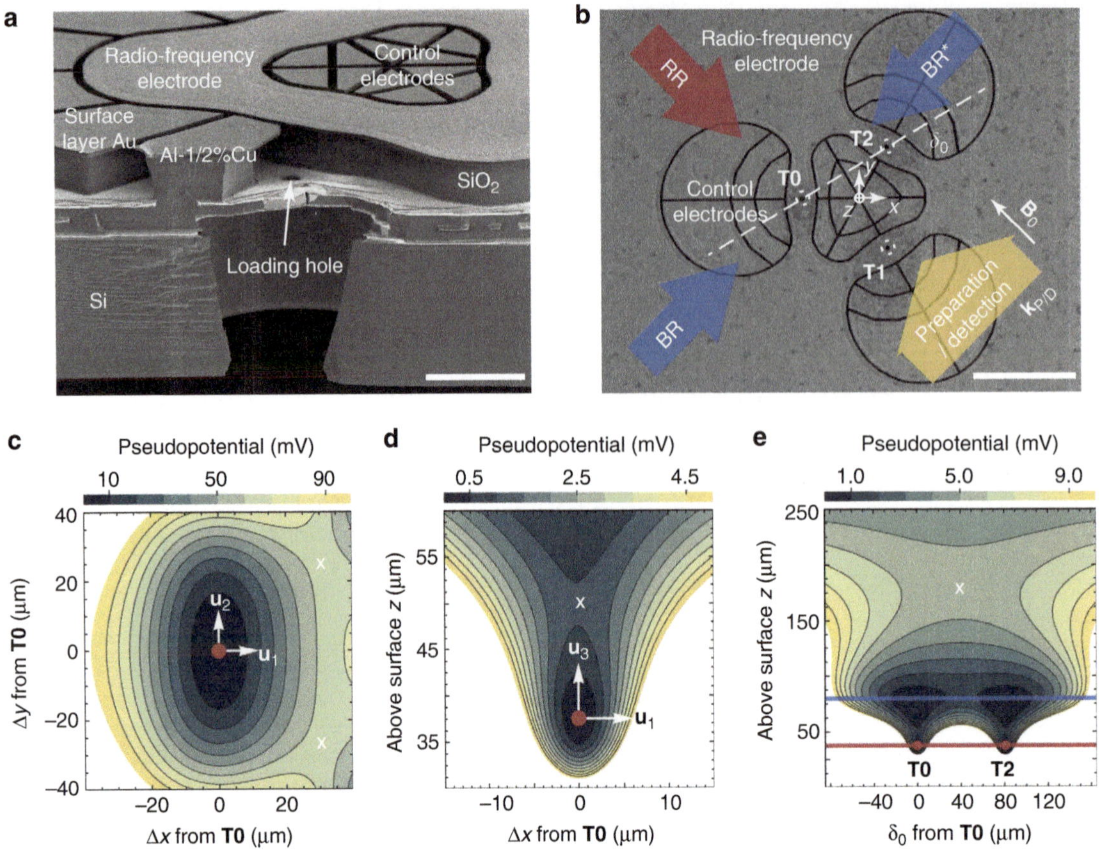

Figure 1 | Surface-electrode ion trap featuring three individual traps. (**a**) Scanning electron microscope (SEM) image of a cleaved copy of our chip; white scale bar, 20 μm. It provides a cross-sectional view vertically through the trap chip (bottom half of image) and a top view of the horizontal, planar trap electrode surface (top half of image) of the 40 μm array. Buried electrode interconnects as well as the overhangs of electrodes that shield trapped ions from insulating surfaces are exposed in this view. A loading channel, vertically traversing the chip, collimates a neutral atom beam from an oven on the backside of the chip. (**b**) SEM top-view of the 80 μm array, dark lines indicate gaps between individual electrodes and dashed circles highlight the three trap sites at **T0**, **T1** and **T2** that lie 40 μm above the electrode plane (white scale bar, 80 μm); corresponding loading holes appear as dark spots. A vertical plane connecting **T0** and **T2** is shown as a dotted line and labelled with δ_0. The single RF electrode extends beyond the image area and encloses 30 control electrodes grouped into four islands depicted in the central part of the image, enabling the control of individual trap sites. Laser beams (coloured arrows) are parallel to the chip surface and wave vectors $\mathbf{k}_{P/D}$ of preparation and detection beams are parallel to the magnetic quantization field $\mathbf{B}_0$ (white arrow). (**c-e**) The pseudopotential $\phi_{ps}(x,y,z)$ of the 80 μm array in different planes is shown, trap sites marked by red dots, motional mode vectors $\mathbf{u}_j$ with $j=\{1,2,3\}$ at **T0** are represented by white arrows, and saddle points are illustrated by white crosses. In **e**, the heights of **T0**, **T2** and the ancillary trap are indicated by red and blue lines, respectively.

fields and curvatures. To achieve this, we apply sets of control voltages to 30 designated control electrodes (see Fig. 1 for details). In the following, a control voltage set is described by a unit vector $\hat{\mathbf{v}}_c \equiv (\hat{v}_{c,1}, \ldots, \hat{v}_{c,30})$, with corresponding dimensionless entries $\hat{v}_{c,n}$ with $n=\{1,\ldots,30\}$, and result in a dimensionless control potential

$$\hat{\phi}_c = \sum_{n=1}^{30} \hat{v}_{c,n}\hat{\phi}_n(\mathbf{r}), \tag{1}$$

where $\hat{\phi}_n(\mathbf{r})$ is the potential resulting when applying 1 V to the n-th electrode following a basis function method[44,45]. We scale $\hat{\phi}_c$ by varying a control voltage U_c and yielding a combined trapping potential

$$\phi(\mathbf{r}) = \phi_{ps}(\mathbf{r}) + U_c\hat{\phi}_c(\mathbf{r}). \tag{2}$$

Bias voltages applied to the control electrodes are, in turn, fully described by $\mathbf{U}_c = U_c\hat{\mathbf{v}}_c$.

To design a specific $\hat{\phi}_c$, we consider the second order Taylor expansion for a point $\mathbf{r}_0$ and small displacements $\Delta\mathbf{r}$:

$$\hat{\phi}_c(\mathbf{r}_0+\Delta\mathbf{r}) \simeq \hat{\phi}_c(\mathbf{r}_0) + [\partial_k]^{\mathrm{T}}\hat{\phi}_c(\mathbf{r})|_{\mathbf{r}=\mathbf{r}_0}\cdot\Delta\mathbf{r} + \frac{1}{2}\Delta\mathbf{r}^{\mathrm{T}}\cdot[\partial_k\partial_l]\hat{\phi}_c(\mathbf{r})|_{\mathbf{r}=\mathbf{r}_0}\cdot\Delta\mathbf{r},$$

where $[\partial_k]^{\mathrm{T}}\hat{\phi}_c(\mathbf{r})|_{\mathbf{r}=\mathbf{r}_0}$ is the local gradient and $[\partial_k\partial_l]\hat{\phi}_c(\mathbf{r})|_{\mathbf{r}=\mathbf{r}_0}$ is the traceless and symmetric matrix with indices k and $l=\{x, y, z\}$ that describes the local curvature; square brackets denote vectors/matrices, ∂ partial derivatives and the superscript $\mathbf{T}$ the transpose of a vector. We constrain local gradients in their three degrees of freedom (DoF) and local curvatures in their five DoF at **T0**, **T1** and **T2**, and solve the corresponding system of 24 linear equations to yield $\hat{\mathbf{v}}_c$. In principle, it would be sufficient to use 24 control electrodes, however, we consider all electrodes and use the extra DoF to minimize the modulus of the voltages we need to apply for a given effect.

In particular, we distinguish two categories of control potentials, denoted by $\hat{\varepsilon}$ and $\hat{\kappa}$, respectively: the first category is designed to provide finite gradients and zero curvatures at **T0**, with zero gradients and curvatures at **T1** and **T2**; for example, $\hat{\phi}_c = \hat{\varepsilon}_x$ provides a gradient along $\hat{\mathbf{x}}$ at **T0**. Control potentials of the second category are designed to provide zero gradients and only curvatures at **T0**, whereas we require related gradients and curvatures to be zero at **T1** and **T2**. For example, we design $\hat{\phi}_c = \hat{\kappa}_{\rm tune}$, with the following non-zero constrains $\partial_y\partial_y\hat{\kappa}_{\rm tune}(\mathbf{r})|_{\mathbf{r}=\mathbf{T0}} = -\partial_z\partial_z\hat{\kappa}_{\rm tune}(\mathbf{r})|_{\mathbf{r}=\mathbf{T0}} = 0.937\times10^7\,\mathrm{m}^{-2}$ with corresponding $U_c = U_{\rm tune}$. Linear combination of multiple control potentials enable us, for example, to locally compensate stray potentials up to second order, to independently control mode frequencies and orientations at each trap site, and, when implementing time-dependent control potentials, to apply directed and phase-controlled mode-frequency modulations or mode excitations.

Optical setup and experimental procedures. We employ eight laser beams at wavelengths near 280 nm, from three distinct laser sources[46], with wave vectors parallel to the xy plane (Fig. 1b) for preparation, manipulation and detection of electronic and motional states of $^{25}\mathrm{Mg}^+$ ions. Five distinct σ^+-polarized beams (two for Doppler cooling, two for optical pumping and one for state detection) are superimposed, with wave vector $\mathbf{k}_{\rm P/D}$ (preparation/detection) aligned with a static homogeneous magnetic quantization field $\mathbf{B}_0 \simeq 4.65\,\mathrm{mT}$ (Fig. 1b). The beam waists (half width at $1/e^2$ intensity) are $\simeq 150\,\mu\mathrm{m}$ in the xy plane and $\simeq 30\,\mu\mathrm{m}$ in z direction, to ensure reasonably even illumination of all three trap sites, while avoiding excessive clipping of the beams on the trap chip. The two Doppler-cooling beams are detuned by $\Delta \simeq -\Gamma/2$ and -10Γ (for initial Doppler cooling and state preparation by optical pumping) with respect to $|\downarrow\rangle \equiv |S_{1/2}, F=3, m_F=+3\rangle \leftrightarrow |P_{3/2}, F=4, m_F=+4\rangle$ with a natural line width $\Gamma/(2\pi) \simeq 42\,\mathrm{MHz}$. The state detection beam is resonant with this cycling transition and discriminates $|\downarrow\rangle$ from $|\uparrow\rangle \equiv |S_{1/2}, F=2, m_F=+2\rangle$, the pseudo-spin states $|\downarrow\rangle$ and $|\uparrow\rangle$ are separated by $\omega_0/(2\pi) \simeq 1{,}681.5\,\mathrm{MHz}$. The resulting fluorescence light is collected with high numerical aperture lens onto either a photomultiplier tube or an electron-multiplying charge-coupled device camera. We prepare (and repump to) $|\downarrow\rangle$ by two optical-pumping beams that couple $|\uparrow\rangle$ and $|S_{1/2}, F=3, m_F=+2\rangle$ to states in $|P_{1/2}\rangle$ from where the electron decays back into the ground state manifold and population is accumulated in $|\downarrow\rangle$. We can couple $|\downarrow\rangle$ to $|\uparrow\rangle$ via two-photon stimulated-Raman transitions[25,47,48], while we can switch between two different beam configurations labelled BR* + RR with $\Delta\mathbf{k}_x \| \hat{\mathbf{x}}$ and BR + RR with $\Delta\mathbf{k}_y \| \hat{\mathbf{y}}$. The beam waists are $\simeq 30\,\mu\mathrm{m}$ in the xy plane and $\simeq 30\,\mu\mathrm{mm}$ in z direction.

We load ions by isotope-selective photoionization from one of three atomic beams collimated by $4\,\mu\mathrm{m}$ loading holes located beneath each trap site (Fig. 1). We can also transfer ions from one site to any neighbouring site via the ancillary trap by applying suitable potentials to control electrodes and a metallic mesh (with high optical transmission) located $\simeq 7\,\mathrm{mm}$ above the surface. Typically, experiments start with 2 ms of Doppler cooling, optionally followed by resolved sideband cooling, and $|\downarrow\rangle$ preparation via optical pumping. We use 30 channels of a 36-channel arbitrary waveform generator with 50 MHz update rate[49] to provide static (persistent over many experiments) and dynamic (variable within single experiments) control potentials. Each experiment is completed by a pulse for pseudo-spin detection of duration $\simeq 150\,\mu\mathrm{s}$ that yields $\simeq 12$ counts on average for an ion in $|\downarrow\rangle$ and $\simeq 0.8$ counts for an ion in $|\uparrow\rangle$. Specific experimental sequences are repeated 100–250 times.

Initially, we calibrate three (static) control potentials $\hat{\varepsilon}_x$, $\hat{\varepsilon}_y$ and $\hat{\varepsilon}_z$ to compensate local stray fields[50] with a single ion near **T0**, whereas we observe negligible effects on the local potentials near **T1** and **T2** (Methods). Rotated versions of these control potentials are used to compensate local stray fields near **T1** and **T2**. Near each site, we achieve residual stray field amplitudes $\leq 3\,\mathrm{V\,m}^{-1}$ in the xy plane and $\leq 900\,\mathrm{V\,m}^{-1}$ along z, currently limited by our methods for detection of micromotion.

With the stray fields approximately compensated, we characterize the trap near **T0** with a single ion (Methods). We find mode frequencies of $\omega_1/(2\pi) \simeq 5.3\,\mathrm{MHz}$, $\omega_2/(2\pi) \simeq 2.6\,\mathrm{MHz}$ and $\omega_3/(2\pi) \simeq 4.1\,\mathrm{MHz}$ with frequency drifts of about $2\pi \times 0.07\,\mathrm{kHz}\,(60\,\mathrm{s})^{-1}$; mode frequencies and orientations are altered by local stray curvatures on our chip, in particular, $\mathbf{u}_1$ and $\mathbf{u}_3$ are rotated in the xz plane, while $\mathbf{u}_2$ remains predominantly aligned along y. We obtain heating rates for the modes $\mathbf{u}_1$ of 0.9(1) quanta ms^{-1}, $\mathbf{u}_2$ of 2.2(1) quanta ms^{-1} and $\mathbf{u}_3$ of 4.0(3) quanta ms^{-1}.

Control of mode configurations at individual trap sites. The ability to control mode frequencies and orientations at each site with minimal effect on local trapping potentials at neighbouring sites is essential for the static and dynamical tuning of inter-ion Coulomb couplings. We experimentally demonstrate individual mode-frequency control using $\hat{\kappa}_{\rm tune}$. To this end we measure local mode frequencies with a single ion near **T0** or **T2** (Methods). Tuning of about $\pm 2\pi \times 80\,\mathrm{kHz}$ of ω_2 near **T0** is shown in Fig. 2 as blue data points, accompanied by residual changes of about $\mp 2\pi \times 1\,\mathrm{kHz}$ in the corresponding mode frequency near the neighbouring site **T2**, depicted by red data points. To infer local control curvatures, we describe the expected detuning $\Delta\omega_2$ due to $\hat{\kappa}_{\rm tune}$ at **T0** (analogously at **T2**) by

$$\Delta\omega_2 = \sqrt{\omega_2^2 + U_{\rm tune}\left(\frac{Q}{m}\partial_y\partial_y\hat{\kappa}_{\rm tune}(\mathbf{r})|_{\mathbf{r}=\mathbf{T0}}\right)} - \omega_2, \qquad (4)$$

where we neglect a small misalignment of $\mathbf{u}_2$ from y. The prediction of equation (4) is shown as a blue/dashed line in Fig. 2. The blue/solid line results from a fit with a function of the form of equation (4) to the data yielding a control curvature of $1.164(3) \times 10^7\,\mathrm{m}^{-2}$. The inset magnifies the residual change in frequency near **T2**. Here, a fit (red/solid line) reveals a curvature of $-0.012(2) \times 10^7\,\mathrm{m}^{-2}$. Residual ion displacements of $\Delta z = -2.95(3)\,\mu\mathrm{m}$ from **T0** and $\Delta z = -2.9(4)\,\mu\mathrm{m}$ from **T2**, respectively, suffice to explain deviations between experimentally determined and designed curvature values and are below our current limit of precision locating the ions in that direction. In future experiments, curvature measurements may be used to further reduce stray fields.

We also implement a dynamic $U_{\rm tune}(t)$, to adiabatically tune ω_2 near **T0** within single experiments: we prepare our initial state by Doppler cooling, followed by resolved sideband cooling of mode $\mathbf{u}_2$ to an average occupation number $\bar{n}_2 \simeq 0.1$ and optical pumping to $|\downarrow\rangle$. In a next step, we apply a first adiabatic ramp from $U_{\rm tune,A} = 0\,\mathrm{V}$ to $U_{\rm tune,B}$ between 0 and 2.3 V (corresponding to a measured frequency difference $\Delta\omega_2/(2\pi) \simeq 430\,\mathrm{kHz}$) within $t_{\rm ramp} = 7.5$ to $120\,\mu\mathrm{s}$ and, subsequently, couple $|\downarrow\rangle$ and $|\uparrow\rangle$ to mode $\mathbf{u}_2$ with pulses of BR + RR tuned to sideband transitions that either add or subtract a single quantum of motion. If the ion is in the motional ground state, no quantum can be subtracted and the spin state remains unchanged when applying the motion subtracting sideband pulse. The motion-adding sideband can always be driven, and comparing the spin-flip probability of the two sidebands allows us to determine the average occupation of the dynamically tuned mode[48]. We find that the average occupation numbers are independent of the duration of the ramp and equal to those obtained by remaining in a static

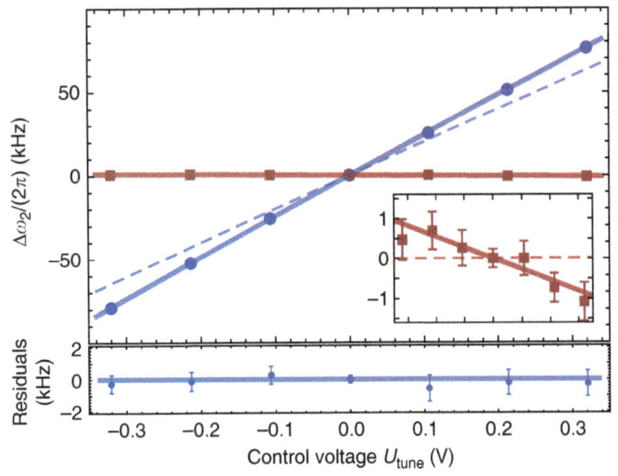

Figure 2 | Individual control of mode frequencies. Frequency change $\Delta\omega_2/(2\pi)$ probed with a single ion near **T0** (blue dots) as a function of U_{tune}. The intended control curvature $0.937 \times 10^7\,m^{-2}$ (along the *y* direction) yields the blue/dashed line (cp. equation (4)), while a fit to the data (blue/solid line) returns a control curvature of $1.164(3) \times 10^7\,m^{-2}$ (corresponding residuals shown in the bottom graph). The remaining change of the corresponding mode frequency of a single ion near **T2** (red squares) is shown in the inset, for the full range of U_{tune}. A fit to these data (red/solid lines) results in a residual control curvature of $-0.012(2) \times 10^7\,m^{-2}$. Ideally, $\hat{\kappa}_{tune}$ would create no additional curvature at **T2** (red/dashed line). We attribute the difference between designed and measured values to residual ion displacements from **T0** and **T2**. Each data point represents the average of 250 experiments with error bars (for some data smaller than symbols) denoting the s.e.m. Residual variations of experimental parameters, for example, changes of stray potentials, can result in day-to-day variations of measurement outcomes that require recalibration to remain within our stated statistical uncertainties.

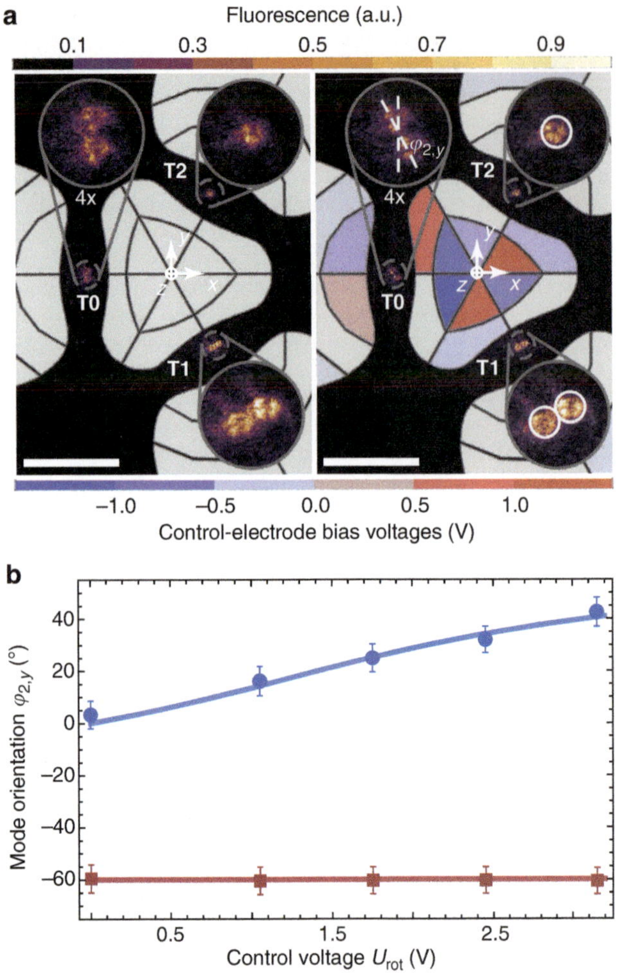

Figure 3 | Individual control of mode orientations. (**a**) Electron-multiplying charge-coupled device images of pairs of ions near **T0** and **T1**, and a single ion near **T2**; white scale bars are 50 µm and blow ups of each site are magnified by four. Schematics of control electrodes are coloured according to their bias voltage $\boldsymbol{U}_{rot}$. Ion pairs align along $\mathbf{u}_2$, the lowest-frequency mode. The left image captures ion positions for $U_{rot} = 0$ V and, here, $\omega_2/(2\pi) = 1.9(2)$ MHz near **T0**. The right image illustrates the rotation effect for $U_{rot} = 2.45$ V: Mode $\mathbf{u}_2$ near **T0** with $\omega_2/(2\pi) = 1.8(2)$ MHz is rotated by $\varphi_{2,y} = 31(5)°$, whereas ion positions near **T1** and **T2** remain unchanged; white circles indicate initial ion positions (for $U_{rot} = 0$). (**b**) Mode $\mathbf{u}_2$ orientation in the *xy* plane for **T0** (blue dots) and **T1** (red squares), described by $\varphi_{2,y}$, derived from a total of 14 images as a function of U_{rot}; error bars denote our systematic uncertainty. The data are in good agreement with our predictions of the effect of $\hat{\kappa}_{rot}$ (solid lines).

potential for t_{ramp}, that is, the motion is unaffected by the dynamic tuning.

We rotate mode orientations near **T0** in the *xy* plane with a control-potential $\hat{\kappa}_{rot}$, while setting additional constraints to keep gradients and curvatures of the local trapping potential constant at **T1** and **T2** (Methods). We determine the rotation of mode orientations from electron-multiplying charge-coupled device images of two ions near **T0** that align along $\mathbf{u}_2$ (axis of weakest confinement). Simultaneously, we trap one or two ions near **T1** and **T2** to monitor residual changes in ion positions and mode orientations (and frequencies) because of unwanted local gradients and curvatures of $\hat{\kappa}_{rot}$. We take 14 images for five different $\hat{\kappa}_{rot}$ values, while constantly Doppler cooling all ions and exciting fluorescence. Figure 3a shows two images for $U_{rot} = 0$ V (left) and $U_{rot} = 2.45$ V (right). Schematics of control electrodes are overlaid to the images and coloured to indicate their bias voltages U_{rot}. Ion positions (in the *xy* plane) are obtained with an uncertainty of ± 0.5 µm, yielding uncertainties for inferred angles $\varphi_{2,y}$ of ± 5°. Here, $\varphi_{2,y}$ denotes the angle between local mode $\mathbf{u}_2$ and *y*. Figure 3b shows measured $\varphi_{2,y}$ for ions near **T0** (blue dots) and **T1** (red squares) and compares them with our theoretical expectation (solid lines), further described in the Methods. We tune $\varphi_{2,y}$ between 0° and 45° near **T0**, enabling us to set arbitrary mode orientations in the *xy* plane, whereas ion positions (mode orientations) near **T1** and **T2** remain constant within ± 0.5 µm (better than ± 5°) in the *xy* plane.

A complementary way of characterizing mode orientations and frequencies, now with respect to $\Delta\mathbf{k}_x$ and/or $\Delta\mathbf{k}_y$ is to analyse the probability of finding $|\downarrow\rangle$ after applying $|\downarrow\rangle \leftrightarrow |\uparrow\rangle$ (carrier) or motional sideband couplings for variable duration. If all modes of a single ion are prepared in their motional ground state, the ratio of Rabi frequencies of carrier and sideband couplings is given by the Lamb-Dicke parameter[48], which is for $\mathbf{u}_1$ and $\Delta\mathbf{k}_x$:

$$\eta_{1,x} = \Delta\mathbf{k}_x \cdot \mathbf{u}_1 \sqrt{\frac{\hbar}{2m\omega_1}} = |\Delta\mathbf{k}_x| \sqrt{\frac{\hbar}{2m\omega_1}} \cos(\varphi_{1,x}), \quad (5)$$

where $\varphi_{1,x}$ is the angle between $\mathbf{u}_1$ and $\Delta\mathbf{k}_x$. The differences of carrier and sideband transition frequencies reveal the mode frequencies, whereas ratios of sideband and carrier Rabi-frequencies determine Lamb-Dicke parameters and allow for finding the orientation of modes.

We use a single ion near **T0** to determine the orientations and frequencies of two modes relative to $\Delta\mathbf{k}_x$. We apply another control potential $\hat{\kappa}_{rot2}$, designed to rotate $\mathbf{u}_1$ and $\mathbf{u}_3$ in the *xz*

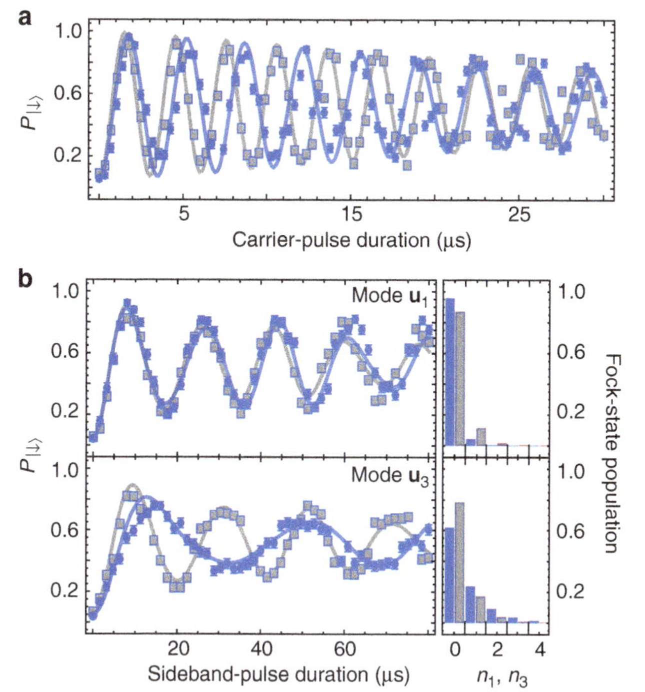

Figure 4 | Measuring mode orientations via laser couplings. Probability of $|\downarrow\rangle$ after applying carrier and sideband couplings via $\Delta\mathbf{k}_x$ with a single ion near **T0**. Mode orientations are set with static $\hat{\kappa}_{rot2}$ (rotations in the xz plane) for $U_{rot2}=1.62$ V (blue dots) and -2.43 V (grey squares). (**a**) Shows the carrier transitions, $|\uparrow, n_1, n_3\rangle \leftrightarrow |\downarrow, n_1, n_3\rangle$, whereas **b** represents the $\mathbf{u}_1$-sideband transitions, $|\uparrow, n_1, n_3\rangle \leftrightarrow |\downarrow, n_1+1, n_3\rangle$, and the $\mathbf{u}_3$-sideband transitions, $|\uparrow, n_1, n_3\rangle \leftrightarrow |\downarrow, n_1, n_3+1\rangle$. From combined model fits to all transitions (for each $\hat{\kappa}_{rot2}$), we find angles $\varphi_{1,x}=24.7(2)°$ for $U_{rot2}=-1.62$ V (blue lines) and 36.1(2)° for $U_{rot2}=-2.43$ V (grey lines) of mode $\mathbf{u}_1$ relative to $\Delta\mathbf{k}_x$. Histograms in **b** display derived Fock-state populations with thermal average occupation numbers between $\simeq 0.05$ and $\simeq 0.6$. Each data point is the average of 250 experiments and error bars (for some data smaller than symbols) denote the s.e.m. Residual variations of experimental parameters, for example, changes of stray potentials, can result in day-to-day variations of measurement outcomes that require recalibration to remain within our stated statistical uncertainties.

plane near **T0**, and implement carrier and sideband couplings to both modes with $\Delta\mathbf{k}_x$ after resolved sideband cooling and initializing $|\uparrow\rangle$. In Fig. 4, the probability of $|\downarrow\rangle$ is shown for different pulse durations of carrier couplings (top) and sideband couplings to mode $\mathbf{u}_1$ (middle) and $\mathbf{u}_3$ (bottom). Data points for $U_{rot2}=-1.62$ V are shown as blue rectangles and for -2.43 V as grey rectangles. We fit each data set to a theoretical model (blue and grey lines) to extract the angles[51] and distributions of Fock-state populations of each mode (shown as histograms): we find $\varphi_{1,x}=24.7(2)°$ for $U_{rot2}=-1.62$ V and $\varphi_{1,x}=36.1(2)°$ for $U_{rot2}=-2.43$ V, whereas average occupation numbers range between $\simeq 0.05$ and $\simeq 0.6$. Adding measurements along $\Delta\mathbf{k}_y$ and taking into account that the normal modes have to be mutually orthogonal would allow to fully reconstruct all mode orientations. With resolved sideband cooling on all three modes, we can prepare a well-defined state of all motional DoF.

Discussion

We characterized two trap arrays that confine ions on the vertices of equilateral triangles with side lengths 80 and 40 μm. We developed systematic approaches to individually tune and calibrate control potentials in the vicinity of each trap site of the 80-μm array, by applying bias potentials to 30 control electrodes. With suitably designed control potentials, we demonstrated precise individual control of mode frequencies and orientations. By utilizing a multi-channel arbitrary waveform generator, we also dynamically changed control potentials within single experimental sequences without adverse effects on spin or motional states. Further, we devised a method to fully determine all mode orientations (and frequencies) based on the analysis of carrier and sideband couplings. Measured heating rates are currently comparable to the expected inter-ion Coulomb coupling rate of $\Omega_{ex}/(2\pi)\simeq 1$ kHz for $^{25}Mg^+$ ions in the 40-μm array at mode frequencies of $\simeq 2\pi\times 2$ MHz (ref. 32). This coupling rate sets a fundamental time scale for effective spin–spin couplings[33]. To observe coherent spin–spin couplings, ambient heating needs to be reduced. Decreases in heating rates of up to two orders of magnitude would leave Ω_{ex} considerably higher than competing decoherence rates and allow for coherent implementation of fairly complex spin–spin couplings. Such heating rate reductions have been achieved in other surface traps by treatments of the electrode structure[34–36] and/or cryogenic cooling of the electrodes[37–39]. The couplings in question have been observed in one dimension in a cryogenic system[32,33].

Currently, we can compensate stray fields, set up normal mode frequencies and directions for all three ions and initialize them for a two-dimensional AQS, that is, prepare a fiducial initial quantum state for ions at each trap site. A complete AQS may use the sequence presented in Fig. 5. A dynamic ramp adiabatically transforms the system between two control sets, labelled as A and B, that realize specific mode frequencies and orientations at each site. Set A may serve to globally initialize spin-motional states of ions, potentially with more than one ion at each site, that could be the ground state of a simple initial Hamiltonian. At all sites, mode frequencies and orientations need to be suitable (bottom left of Fig. 5) to enable global resolved sideband cooling, ideally preparing ground states for all motional modes. A first ramp to set B combined with appropriate laser fields may be used to adiabatically or diabatically realize a different Hamiltonian, for example, by turning on complex spin–spin couplings. Mode frequencies and orientations are tuned such that the Coulomb interactions between ions can mediate effective spin–spin couplings, for example, all mode vectors $\mathbf{u}_1$ are rotated to point to the centre of the triangle (bottom right of Fig. 5). During the application of such interactions, the ground state of the uncoupled system can evolve into the highly entangled ground state of a complex coupled system. In contrast, diabatic ramping to set B will quench the original ground state and the coupled system will evolve into an excited state that is not an eigenstate. After a final adiabatic or diabatic ramp back to set A, we can use global (or local) laser beams to read out the final spin states at each site.

In this way, our arrays may become an arbitrarily configurable and dynamically reprogrammable simulator for complex quantum dynamics. It may enable, for example, the observation of photon-assisted tunnelling, as required for experimental simulations of synthetic gauge fields[52,53] or other interesting properties of finite quantum systems, such as thermalization, when including the motional DoF[54]. Concentrating on spin–spin interactions, the complex entangled ground states of spin frustration can be studied in the versatile testbed provided by arrays of individually trapped and controlled ions[30,55]. Arrays with a larger number of trap sites could realize a level of complexity impossible to simulate on conventional computers[56,57].

Methods

Design of arrays used in the experiments. The design of arrays used in the expeiments is based on the methods described in ref. 29. In particular, we use the Mathematica package for surface atom and ion traps[43] to globally optimize the RF electrode shape for maximal curvature with a given amplitude of the RF drive, whereas producing smooth continuous electrode shapes that require a single RF drive to operate the array. We specify the desired trap site positions as well as the ratio and orientation of normal-mode frequencies as a fixed input to the optimization algorithm for the pseudopotential, that is, we define that the high-frequency mode (for all three sites) lies within the *xy* plane and points towards the virtual centre of the array. Resulting electrode regions held to ground are subdivided into separated control electrodes that provide complete and independent control over the eight DoF at each site.

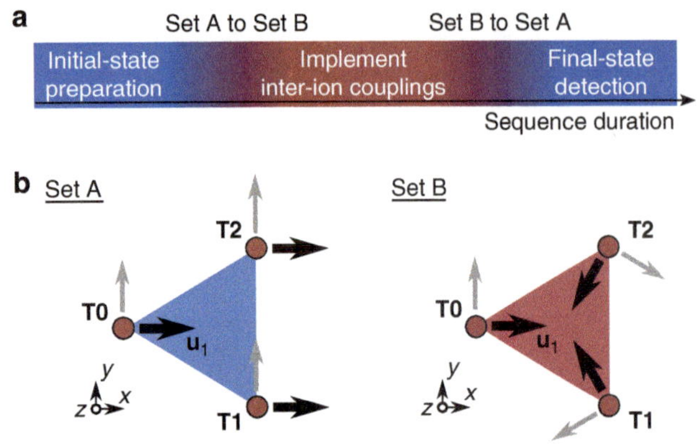

Figure 5 | Generic experimental sequence for tuneable inter-ion couplings. (**a**) Time-line of an experiment starting with the initialization of an fiducial quantum state of all ions in the array, followed by an adiabatic (or diabatic) ramp of control potentials between sets A and B that reconfigure the normal-mode structure from the setup, see **b**. Then, appropriate laser fields implement inter-ion couplings required for the AQS. The simulation completes with ramping the control potentials back to set A, where the individual spin states can be detected. (**b**) Examples for configurations of motional DoF are illustrated by the arrows that show the orientation of $\mathbf{u}_1$ at the three trap sites (red dots). Set A may be applied when globally and/or locally preparing and detecting spin-motional states, whereas set B can establish specific inter-ion Coulomb couplings to mediate, for example, effective spin-spin couplings for AQS, cp. refs 33,55.

Array scaling for future realisations. To ensure that our approach can be scaled to more than three trapping sites, we compare designs of arrays containing different numbers of sites, N_{sites}, that are optimized by the algorithm described in ref. 29. Here, we assume a fixed ratio of $h/d = 1/2$, where h denotes the distance of the sites to the nearest electrode surface and d is the inter-site distance. Further, we specify for all arrays that the high-frequency mode is aligned orthogonally to the *xy* plane at each site, in contrast to our demonstrated arrays (see Fig. 1 for details). This unique mode configuration permits a fair comparison of geometries with increasing N_{sites}. To illustrate the optimal electrode shapes, we present four examples of triangular arrays with $N_{\text{sites}} = \{3,6,18,69\}$ in Fig. 6a–d. To enable the same level of individual control as demonstrated for both of our three-site arrays, we would have to subdivide the optimized ground electrodes into $\geq 8 \times N_{\text{sites}}$ control electrodes. We find that the inner areas converge to fairly regular electrode shapes for larger N_{sites}, whereas electrodes closer to the border are deformed to compensate for edge effects (see Fig. 6d for details). However, the spatial extent and complexity of all electrodes remains comparable to the arrays used in our experiments and, thus, fabrication of these larger arrays can be accomplished by scaling the applied techniques (see below).

To quantify the geometric strength of individual trap sites independently of m, U_{RF}, Ω_{RF} and h, we consider the dimensionless curvature κ of the pseudopotential that we normalize to the highest possible curvature for a single site[29]. We show optimized κ for arrays with N_{sites} between 1 and 102, as well as, the value for $N_{\text{sites}} = \infty$ in Fig. 6e; a fully controlled array with $N_{\text{sites}} = 102$ should be sufficient to study quantum many-body dynamics that are virtually impossible to simulate on a conventional computer. We find that k for $N_{\text{sites}} = 102$ is reduced by about a factor of two compared with $\kappa \simeq 0.87$ for $N_{\text{sites}} = 3$, whereas $\kappa \simeq 0.07$ for $N_{\text{sites}} = \infty$; see ref. 29 for a detailed discussion of infinite arrays. The decrease in trap curvature can be compensated in experiments by adjusting U_{RF} and Ω_{RF} correspondingly, or by reducing h. Further, we estimate that trapping depths remain on the same order of magnitude for increasing N_{sites} compared with our demonstrated arrays (cp. Fig. 1d). For an infinite array it has been shown that depths of a few mV are achievable[30]. Note, that in surface-electrode traps the trapping potential is less deep along z than in the *xy* plane, and ion-escape points (closest and lowest saddle point of the pseudopotential) typically lie above each site. In experiments, we may apply a constant bias potential to the control electrodes, surrounding ground planes, and the mesh (cover plane) to increase the depth along z to a level where trapping is

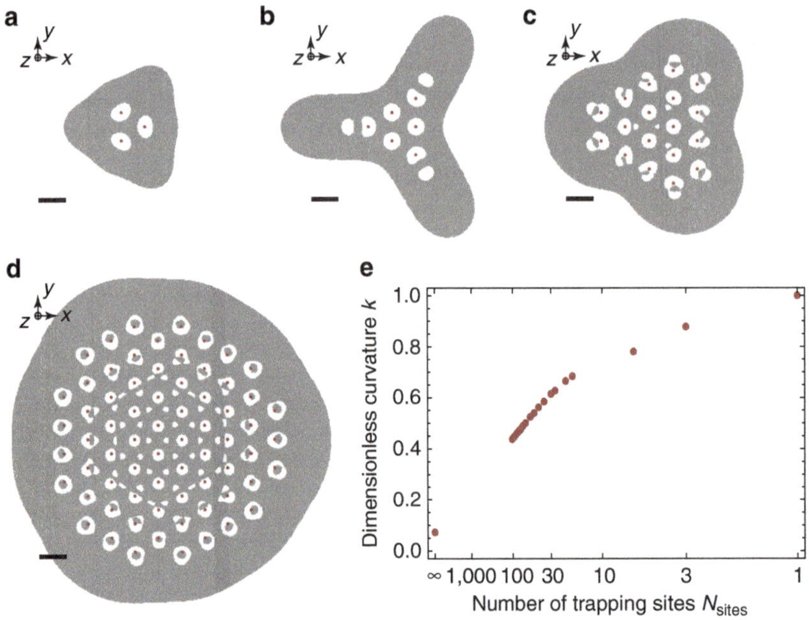

Figure 6 | Scale-up the number of sites in triangular arrays. (**a-d**) Optimized surface-electrode shapes for 3, 6, 18 and 69 trapping sites, where we assume $h/d = 1/2$ (as for our 80 μm array); black scale bars have fixed length d. The high-frequency mode is aligned orthogonally to the *xy* plane in each trapping site, to permit comparability of the different geometries. RF and ground electrodes are coloured in grey and white, respectively, and individual trap sites are marked with red dots. In a functional array, we will have to further subdivide the ground electrodes to enable individual control at each site, as demonstrated in our work with three sites. (**e**) Optimized dimensionless curvature κ as a function of the number of trapping sites in the triangular arrays, quantifying the strength of the individual traps. For a single site $\kappa = 1.0$ and for an infinite triangular lattice $\kappa \simeq 0.07$. In experimental realizations, we can compensate this decrease in trap curvature by adjusting U_{RF} and Ω_{RF}, or by reducing h.

routinely achieved, while reducing the depth in the xy plane[30]. With such measures in place, we are fairly confident that ions created by photoionization from a hot atomic beam can be loaded and cooled into the local minima of larger arrays.

Architecture of our trap chip. The $10 \times 10\,\mathrm{mm}^2$ Si substrate of our trap chip is bonded onto a $33 \times 33\,\mathrm{mm}^2$ ceramic pin grid array (CPGA); the electrodes of the trap arrays are wire-bonded with aluminium wires to the pins of the CPGA, with independent pins for the RF electrodes of the two arrays. The trap chip contains four aluminum-1/2% copper metal layers, that are electrically connected by tungsten vertical interconnects thereby allowing 'islanded' control electrodes in the top electrode layer (Fig. 1). The buried electrical leads are isolated by intermediate SiO_2 layers, nominally 2 µm thick, while the surface layer is spaced by 10 µm from the buried layers. All electrodes are mutually separated by nominally 1.2–1.4 µm gaps and a 50-nm gold layer is evaporated on the top surfaces in a final fabrication step. The trap chip fabrication is substantially the same as that described in the Supplement to ref. 40. Each control electrode is connected to ground by 820 pF capacitors located on the CPGA to minimize potential changes due to capacitive coupling to the RF electrodes.

Compensation of stray potentials at each site. For compensation of local stray fields in the xy plane, we vary the strength of individual control potentials $\hat{\varepsilon}_x$ and $\hat{\varepsilon}_y$ and find corresponding coefficient settings where we obtain a maximal Rabi rate of the detection transition and/or minimal Rabi rates of micromotion-sideband transitions probed with $\Delta\mathbf{k}_x$ and $\Delta\mathbf{k}_y$; resulting in residual stray-field amplitudes of $\lesssim 3\,\mathrm{V\,m}^{-1}$. For compensation along z, we vary the strength of individual $\hat{\varepsilon}_z$ to minimize a change in ion position due to a modulation of U_{RF}. The depth of field of our imaging optics aids to detect changes in z-position via blurring of images of single ions trapped at each site, within an uncertainty of about $\pm 5\,\mu\mathrm{m}$. This corresponds to residual stray-field amplitudes of $\simeq 900\,\mathrm{V\,m}^{-1}$ for typical trapping parameters.

Mode frequency and heating rate measurements. To measure mode frequencies, we Doppler-cool the ion and pump to $|\downarrow\rangle$. Then, we apply a motional excitation pulse with fixed duration $t_{\mathrm{exc}} = 100\,\mu\mathrm{s}$ to a single control electrode. The pulse produces an electric field oscillating at a frequency ω_{exc} that excites the motion, if ω_{exc} is resonant with a mode frequency, and we can detect mode amplitudes of > 100 nm along $\mathbf{k}_{\mathrm{D}}$ via the Doppler effect. In the experiments, we vary ω_{exc} and obtain resonant excitations at ω_j with $j = \{1,2,3\}$. By repeating measurements, we record $\simeq 50$ consecutive frequency values for each mode frequency over the course of $\Delta t \simeq 1$ h with a single ion near **T0**. The results are consistent with linear changes in frequencies, with rates $\Delta\omega_1/\Delta t = -2\pi \times 0.090(3)\,\mathrm{kHz}\,(60\,\mathrm{s})^{-1}$, $\Delta\omega_2/\Delta t = -2\pi \times 0.064(1)\,\mathrm{kHz}\,(60\,\mathrm{s})^{-1}$ and $\Delta\omega_3/\Delta t = -2\pi \times 0.063(5)\,\mathrm{kHz}\,(60\,\mathrm{s})^{-1}$.

For the heating rate measurements, we add multiple resolved-sideband cooling pulses after Doppler cooling to our sequence and determine mode temperatures from the sideband ratios for several different delay times[58]. In our experiments, we either use $\Delta\mathbf{k}_x$ to iteratively address $\mathbf{u}_1$ and $\mathbf{u}_3$ or $\Delta\mathbf{k}_y$ to address only $\mathbf{u}_2$. For this, we prepare similar mode orientations as presented in Fig. 4, find initial mode temperatures after cooling to $\bar{n}_j \lesssim 0.3$, and obtain corresponding heating rates.

Potentials for individual control. As a representative example for designing control potentials, we discuss $\hat{\kappa}_{\mathrm{rot}}$ that serves to rotate the normal modes in the xy plane. At position **T0**, the constraints are:

$$[\partial_k \partial_l]\hat{\kappa}_{\mathrm{rot}}(\mathbf{r})|_{\mathbf{r}=\mathbf{T0}} = \begin{pmatrix} -1.60 & 1.75 & 0 \\ 1.75 & 0.84 & 0 \\ 0 & 0 & 0.76 \end{pmatrix} \times 10^7\,\mathrm{m}^{-2}, \qquad (6)$$

for k and $l = \{x,y,z\}$, while local gradients at all three trap sites and local curvatures at **T1** and **T2** are required to be zero. We add diagonal elements in $[\partial_k\partial_l]\hat{\kappa}_{rot}(\mathbf{r})|_{\mathbf{r}=\mathbf{T0}}$ to reduce changes of the $\mathbf{u}_2$ frequency during variation of $\hat{\kappa}_{\mathrm{rot}}$ around our initial mode configurations. The mode configurations in the real array deviate from those derived from the ϕ_{ps} due to additional curvatures near each trap site generated by stray potentials on our chip. Ideally, we would design control potentials for mode rotations such that all frequencies stay fixed. This is only possible if we explicitly know the initial mode configuration. In addition, we keep mode vectors tilted away from z to sufficiently Doppler cool all modes during state initialization. Similarly, we design $\hat{\kappa}_{\mathrm{rot2}}$ to rotate modes in the xz plane.

Model for varying mode orientations. To model the rotation angle $\varphi_{2,y}$ of $\mathbf{u}_2$ near **T0** as a function of $\hat{\kappa}_{\mathrm{rot}}$, we consider the final trapping curvature at **T0** (analogously for neighbouring sites):

$$[\partial_k\partial_l]\phi_{\mathrm{fin}}(U_{\mathrm{rot}})|_{\mathbf{r}=\mathbf{T}_0} = [\partial_k\partial_l]\phi_{\mathrm{ini}}|_{\mathbf{r}=\mathbf{T}_0}, + U_{\mathrm{rot}}[\partial_k\partial_l]\kappa_{\mathrm{rot}}|_{\mathbf{r}=\mathbf{T}_0}, \qquad (7)$$

where $\phi_{\mathrm{ini}}(\mathbf{r})$ represents the initial potential, that is, the sum of the pseudopotential, stray potential and additional control potentials (used for stray field compensation). The local curvatures (mode frequencies and vectors) of $\phi_{\mathrm{ini}}(\mathbf{r})$ near **T0** are estimated from calibration experiments. For simplicity, we reduce equation (7) to two dimensions (in the xy plane) and find corresponding eigenvectors and eigenvalues for U_{rot} between 0.0 and 3.0 V. We obtain angles $\varphi_{2,y}(U_{\mathrm{rot}})$ of the eigenvector $\mathbf{u}_2$ and we show resulting values as an interpolated solid line in Fig. 3b.

Similarly, we model the effect of $\hat{\kappa}_{\mathrm{tune}}$ on ω_2. We assume that for $U_{\mathrm{tune}} = 0$, the corresponding mode vector $\mathbf{u}_2$ is aligned parallel to y. This is the case for pure RF confinement (cp. Fig. 1c) and sufficiently small stray curvatures. We design $\hat{\kappa}_{\mathrm{tune}}$ to tune the curvature along y, and the curvature as a function of U_{tune} (along this axis) is described by: $\partial_y\partial_y\phi_{\mathrm{fin}}(U_{\mathrm{tune}})|_{\mathbf{r}=\mathbf{T}_0} = \partial_y\partial_y\phi_{\mathrm{ini}}|_{\mathbf{r}=\mathbf{T}_0} + U_{\mathrm{tune}}\partial_y\partial_y\hat{\kappa}_{\mathrm{tune}}|_{\mathbf{r}=\mathbf{T}_0}$. Finally, we insert this into $\omega_2 = \sqrt{Q/m\partial_y\partial_y\phi_{\mathrm{ini}}|_{\mathbf{r}=\mathbf{T}_0}}$ to find equation (4).

References

1. Feynman, R. P. Simulating physics with computers. *Int. J. Theor. Phys.* **21,** 467–488 (1982).
2. Georgescu, I. M., Ashhab, S. & Nori, F. Quantum simulation. *Rev. Mod. Phys.* **86,** 153–185 (2014).
3. Schaetz, T., Monroe, C. R. & Esslinger, T. Focus on quantum simulation. *N. J. Phys.* **15,** 085009 (2013).
4. Lanyon, B. P. *et al.* Towards quantum chemistry on a quantum computer. *Nat. Chem* **2,** 106–111 (2010).
5. Fleming, G. R., Huelga, S. F. & Plenio, M. B. Focus on quantum effects and noise in biomolecules. *N. J. Phys.* **13,** 115002 (2011).
6. Aspuru-Guzik, A. & Walther, P. Photonic quantum simulators. *Nat. Phys* **8,** 285–291 (2012).
7. Houck, A. A., Tureci, H. E. & Koch, J. On-chip quantum simulation with superconducting circuits. *Nat. Phys* **8,** 292–299 (2012).
8. Bloch, I., Dalibard, J. & Nascimbene, S. Quantum simulations with ultracold quantum gases. *Nat. Phys* **8,** 267–276 (2012).
9. Blatt, R. & Roos, C. F. Quantum simulations with trapped ions. *Nat. Phys* **8,** 277–284 (2012).
10. Cirac, J. I. & Zoller, P. Goals and opportunities in quantum simulation. *Nat. Phys* **8,** 264–266 (2012).
11. Leibfried, D. Could a boom in technologies trap Feynman's simulator? *Nature* **463,** 608–608 (2010).
12. Leibfried, D. *et al.* Trapped-ion quantum simulator: experimental application to nonlinear interferometers. *Phys. Rev. Lett.* **89,** 247901 (2002).
13. Friedenauer, A., Schmitz, H., Glueckert, J. T., Porras, D. & Schaetz, T. Simulating a quantum magnet with trapped ions. *Nat. Phys* **4,** 757–761 (2008).
14. Schmitz, H. *et al.* Quantum walk of a trapped ion in phase space. *Phys. Rev. Lett.* **103,** 090504 (2009).
15. Schmitz, H. *et al.* The arch of simulating quantum spin systems with trapped ions. *Appl. Phys. B* **95,** 195–203 (2009).
16. Zahringer, F. *et al.* Realization of a quantum walk with one and two trapped ions. *Phys. Rev. Lett.* **104,** 100503 (2010).
17. Gerritsma, R. *et al.* Quantum simulation of the Dirac equation. *Nature* **463,** 68–71 (2010).
18. Matjeschk, R. *et al.* Experimental simulation and limitations of quantum walks with trapped ions. *N. J. Phys.* **14,** 035012 (2012).
19. Porras, D. & Cirac, J. I. Effective quantum spin systems with trapped ions. *Phys. Rev. Lett.* **92,** 207901 (2004).
20. Islam, R. *et al.* Emergence and frustration of magnetism with variable-range interactions in a quantum simulator. *Science* **340,** 583–587 (2013).
21. Britton, J. W. *et al.* Engineered two-dimensional ising interactions in a trapped-ion quantum simulator with hundreds of spins. *Nature* **484,** 489–492 (2012).
22. Blatt, R. & Wineland, D. Entangled states of trapped atomic ions. *Nature* **453,** 1008–1015 (2008).
23. Hauke, P., Cucchietti, F. M., Tagliacozzo., L., Deutsch, I. & Lewenstein, M. Can one trust quantum simulators? *Rep. Prog. Phys* **75,** 082401 (2012).
24. Seidelin, S. *et al.* Microfabricated surface-electrode ion trap for scalable quantum information processing. *Phys. Rev. Lett.* **96,** 253003 (2006).
25. Wineland, D. J. *et al.* Experimental issues in coherent quantum-state manipulation oftrapped atomic ions. *J. Res. Natl. Inst. Stand. Technol.* **103,** 259 (1998).
26. Kielpinski, D., Monroe, C. & Wineland, D. J. Architecture for a large-scale ion-trap quantum computer. *Nature* **417,** 709–711 (2002).
27. Amini, J. M. *et al.* Toward scalable ion traps for quantum information processing. *N. J. Phys.* **12,** 033031 (2010).
28. Schaetz, T. *et al.* Towards (scalable) quantum simulations in ion traps. *J. Modern Opt.* **54,** 2317–2325 (2007).
29. Schmied, R., Wesenberg, J. H. & Leibfried, D. Optimal surface-electrode trap lattices for quantum simulation with trapped ions. *Phys. Rev. Lett.* **102,** 233002 (2009).
30. Schmied, R., Wesenberg, J. H. & Leibfried, D. Quantum simulation of the hexagonal Kitaev model with trapped ions. *N. J. Phys.* **13,** 115011 (2011).
31. Sterling, R. C. *et al.* Fabrication and operation of a two-dimensional ion-trap lattice on a high-voltage microchip. *Nat. Commun* **5,** 3637 (2014).
32. Brown, K. R. *et al.* Coupled quantized mechanical oscillators. *Nature* **471,** 196–199 (2011).

33. Wilson, A. C. *et al.* Tunable spin-spin interactions and entanglement of ions in separate potential wells. *Nature* **512,** 57–60 (2014).
34. Allcock, D. T. C. *et al.* Reduction of heating rate in a microfabricated ion trap by pulsed-laser cleaning. *N. J. Phys.* **13,** 123023 (2011).
35. Hite, D. A. *et al.* 100-fold reduction of electric-field noise in an ion trap cleaned with *in-situ* argon-ion-beam bombardment. *Phys. Rev. Lett.* **109,** 103001 (2012).
36. Daniilidis, N. *et al.* Surface noise analysis using a single-ion sensor. *Phys. Rev. B* **89,** 245435 (2014).
37. Deslauriers, L. *et al.* Scaling and suppression of anomalous heating in ion traps. *Phys. Rev. Lett.* **97,** 103007 (2006).
38. Labaziewicz, J. *et al.* Suppression of heating rates in cryogenic surface-electrode ion traps. *Phys. Rev. Lett.* **100,** 013001 (2008).
39. Labaziewicz, J. *et al.* Temperature dependence of electric field noise above gold surfaces. *Phys. Rev. Lett.* **101,** 180602 (2008).
40. Tabakov, B. *et al.* Assembling a ring-shaped crystal in a microfabricated surface ion trap. *Phys. Rev. Applied* **4,** 031001 (2015).
41. Ghosh, P. K. *Ion Traps* (Clarendon, 1995).
42. Schmied, R. Electrostatics of gapped and finite surface electrodes. *N. J. Phys.* **12,** 023038 (2010).
43. Schmied, R. SurfacePattern: a Mathematica package for surface atom and ion traps. https://atom.physik.unibas.ch/people/romanschmied/code/SurfacePattern.php (2009-2015).
44. Hucul, D. *et al.* On the transport of atomic ions in linear and multidimensional ion trap arrays. *Quant. Inf. Comput.* **8,** 501–578 (2008).
45. Blakestad, R. B. *et al.* Near-ground-state transport of trapped-ion qubits through a multidimensional array. *Phys. Rev. A* **84,** 032314 (2011).
46. Friedenauer, A. *et al.* High power all solid state laser system near 280 nm. *Appl. Phys. B* **84 (2006).**
47. Monroe, C. *et al.* Resolved-sideband Raman cooling of a bound atom to the 3D zero-point energy. *Phys. Rev. Lett.* **75,** 4011–4014 (1995).
48. Leibfried, D., Blatt, R., Monroe, C. & Wineland, D. Quantum dynamics of single trapped ions. *Rev. Mod. Phys.* **75,** 281–324 (2003).
49. Bowler, R., Warring, U., Britton, J. W., Sawyer, J. & Amini, B. C. Arbitrary waveform generator for quantum information processing with trapped ions. *Rev. Sci. Instrum* **84,** 033108 (2013).
50. Berkeland, D. J., Miller, J. D., Bergquist, J. C., Itano, W. M. & Wineland, D. J. Minimization of ion micromotion in a Paul trap. *J. Appl. Phys.* **83,** 5025–5033 (1998).
51. Kalis, H. *et al.* Motional-Mode Analysis of Trapped Ions. Preprint at http://arxiv.org/abs/1605.01272 (2016).
52. Bermudez, A., Schaetz, T. & Porras, D. Synthetic gauge fields for vibrational excitations of trapped ions. *Phys. Rev. Lett.* **107,** 150501 (2011).
53. Bermudez, A., Schaetz, T. & Porras, D. Photon-assisted-tunneling toolbox for quantum simulations in ion traps. *N. J. Phys.* **14,** 053049 (2012).
54. Clos, G., Porras, D., Warring, U. & Schaetz, T. Time-resolved observation ofthermalization in an isolated quantum system, Preprint at http://arxiv.org/abs/1509.07712 (2015).
55. Schneider, C., Porras, D. & Schaetz, T. Experimental quantum simulations of many-body physics with trapped ions. *Rep. Prog. Phys* **75,** 024401 (2012).
56. Shi, T. & Cirac, J. I. Topological phenomena in trapped-ion systems. *Phys. Rev. A* **87,** 013606 (2013).
57. Nielsen, A. E. B., Sierra, G. & Cirac, J. I. Local models of fractional quantum hall states in lattices and physical implementation. *Nat. Commun* **4,** 2864 (2013).
58. Turchette, Q. A. *et al.* Heating of trapped ions from the quantum ground state. *Phys. Rev. A* **61,** 063418 (2000).

Acknowledgements

This work was supported by DFG (SCHA 972/6-1). Sandia National Laboratories is a multiprogram laboratory managed and operated by Sandia Corporation, a wholly owned subsidiary of Lockheed Martin Corporation, for the US Department of EnergyÕs National Nuclear Security Administration under Contract No. DE-AC04-94AL85000. All statements of fact, opinion or analysis expressed in this paper are those of the authors and do not necessarily reflect the official positions or views of the Office of the Director of National Intelligence (ODNI) or the Intelligence Advanced Research Projects Activity. We thank J. Denter for technical assistance. Further, we are grateful for helpful comments on the manuscript given by S. Todaro, K. McCormick and Y. Minet.

Author contributions

M.M., H.K. and U.W. participated in the design of the experiment and built the experimental apparatus. M.M., H.K., M.W., F.H. and U.W. collected data and analysed results. M.M., U.W., D.L. and T.S. wrote the manuscript. R.S. and D.L. participated in the design of the trap arrays and the experiment. M.B., P.M. and D.L.M. participated in the design and fabricated the trap chips. T.S. participated in the design and analysis of the experiment. M.M. and H.K. contributed equally to this work and all authors discussed the results and the text of the manuscript.

Additional information

Competing financial interests: The authors declare no competing financial interests.

19

Thermodynamics of quantum systems with multiple conserved quantities

Yelena Guryanova[1], Sandu Popescu[1], Anthony J. Short[1], Ralph Silva[1,2] & Paul Skrzypczyk[1]

Recently, there has been much progress in understanding the thermodynamics of quantum systems, even for small individual systems. Most of this work has focused on the standard case where energy is the only conserved quantity. Here we consider a generalization of this work to deal with multiple conserved quantities. Each conserved quantity, which, importantly, need not commute with the rest, can be extracted and stored in its own battery. Unlike the standard case, in which the amount of extractable energy is constrained, here there is no limit on how much of any individual conserved quantity can be extracted. However, other conserved quantities must be supplied, and the second law constrains the combination of extractable quantities and the trade-offs between them. We present explicit protocols that allow us to perform arbitrarily good trade-offs and extract arbitrarily good combinations of conserved quantities from individual quantum systems.

[1] H.H. Wills Physics Laboratory, University of Bristol, Tyndall Avenue, Bristol BS8 1TL, UK. [2] Département de Physique Théorique, Université de Genève, 1211 Genève, Switzerland. Correspondence and requests for materials should be addressed to S.P. (email: s.popescu@bristol.ac.uk).

Thermodynamics is one of the most successful theories of nature that we have. Since its inception almost 200 years ago it has survived the transition from classical mechanics to relativistic and quantum mechanics, with its conceptual basis unchanged. The realm of thermodynamics has also been considerably extended, with recent years witnessing the extension of thermodynamics from dealing with macroscopic systems to individual quantum systems and black holes. From its earliest days, thermodynamics was also generalized to deal not only with energy but also with various conserved quantities, introducing grand canonical ensembles, chemical potentials and so on. Again, its conceptual basis remained unchanged.

In more detail, the idea of the grand canonical ensemble, where not only energy, but also the number of particles, is a conserved quantity, goes all the way back to Gibbs[1]. A milestone was the work in 1957 of Jaynes[2,3], who, coming from a Bayesian perspective, suggested the generalization of thermodynamics to arbitrary conserved quantities through the principle of maximum entropy. The idea of the 'generalized Gibbs ensemble' is by now commonly used in quantum statistical mechanics (see for example, refs 4–8). More recently, refs 9,10 considered Landauer erasure given access to an 'angular momentum bath' instead of a thermal bath, and demonstrated that information can be erased without an energy cost, provided the analogous cost is paid in angular momentum.

Very recently, there has been much renewed interest in the foundations of thermodynamics coming from the field of quantum information. On the one hand, the so-called 'single-shot information theory', which was developed initially to study finite-size effects in quantum cryptography, has proven useful for studying finite-size effects and fluctuations in quantum thermodynamics, which has lead to the development of 'single-shot quantum thermodynamics'[11–17]. On the other hand, inspired by so-called 'resource theories', which have proved to be very powerful for studying quantum information tasks, such as the theory of entanglement[18], purity[19] or asymmetry[20], the 'resource theory of quantum thermodynamics'[21] was developed that, in combination with single-shot framework, has generated a lot of interest and already shown itself to be a fruitful approach to quantum thermodynamics[22–36]. Our work proceeds along the lines of single-shot thermodynamics and resource theories, where a number of initial results concerning the thermodynamics of multiple conserved quantities were presented in refs 37,38.

Here we present a generalized version of thermodynamics that deals with individual quantum systems and multiple—commuting or non-commuting—conserved quantities. What we will show is that unlike standard thermodynamics, where the second law constrains how much of the conserved quantity (energy) can be extracted from a non-equilibrium system, in the form of work, here there is no constraint on the amount of a single conserved quantity that can be extracted. In fact, we can extract as much of any individual conserved quantity as we like, if we supply an appropriate amount of other conserved quantities. What the second law constrains is the combination of conserved quantities that can be extracted—that is, the second law is seen to limit the trade-off of extractable quantities, which for the standard case of a single conserved quantity reduces to constraining how much can be extracted. At the same time, this generalized version of thermodynamics suggests that it may be worthwhile revisiting the basic concepts of the subject. Indeed, to understand the above phenomena we present an alternative viewpoint, which reinterprets some of the standard thermodynamics quantities, and is perhaps more natural and compelling. Closely related, independent work was performed by Lostaglio *et al.*[39] and Halpern *et al.*[40]. In contrast to here, these works focus more on the nature of the thermal state itself when there are multiple conserved quantities.

Results

Overview. Here we consider the standard general framework of thermodynamics that consists of a thermal bath, an external system out of equilibrium with respect to the bath and a number of batteries, in which we will store various conserved quantities, which are extracted from the system and bath. In our case, following Jaynes[2,3], we take the 'thermal bath' to be simply a collection of particles, each described by a generalized thermal state

$$\tau = e^{-(\beta_1 A_1 + \ldots + \beta_k A_k)}/\mathcal{Z} \tag{1}$$

where A_i are various conserved quantities, β_i are the associated inverse temperatures and $\mathcal{Z}$ is the generalized partition function. Two things are important to note: first, the quantities A_i may or may not commute, and even when they commute they may or may not be functionally dependent on one another. Second, and most importantly, energy need not be one of the conserved quantities or indeed play any role. Since energy is the generator of time evolution, such a thermal bath may not arise naturally by thermal equilibration, but have to be created externally (for example, if the Hamiltonian is zero, then no evolution occurs). Yet, as we will see, the thermodynamic flavour of the theory remains.

The batteries are systems that can each store one of the conserved quantities A_i. In our paper we will consider the batteries either explicitly or implicitly, as explained later. The system can be an individual quantum particle. Finally, the actions that we allow to be performed must conserve either exactly or on average each of the quantities A_i, which is the content of the first law.

A central result of standard thermodynamics—the content of the second law—is that if we have access only to a single thermal bath, it is impossible to extract energy, in the ordered form of work, out of it, that is $W := \Delta E^{\text{batt}} \leq 0$, where ΔE^{batt} is the change in the average energy of the battery. We show that, in our case, there is no limit on how much of any single conserved quantity A_i can be extracted, even though we have access only to a single generalized bath. More precisely, there is no limit on $W_{A_i} := \Delta A_i^{\text{batt}}$. There is, however, a global limit.

In particular, to each conserved quantity we can associate an entropic quantity $\beta_i A_i$ (the entropic nature of this quantity will be explained later). We will show that these quantities can be almost perfectly interconverted for one another inside the bath. As a result, because of the first law (conservation of A_i between bath and battery) the only constraint on the W_{A_i} given just a thermal bath is that

$$\sum_i \beta_i W_{A_i} \leq 0. \tag{2}$$

In standard thermodynamics the second law also says that if we have access to a system out of equilibrium with respect to the bath, then we can extract work, but we are limited by the change in free energy of the system, $W \leq -\Delta F_s$. In our case, we define an entropic quantity, the 'free entropy' of the system relative to the generalized bath, $\tilde{F}_s$,

$$\tilde{F}_s := \sum_i \beta_i \langle A_i^s \rangle - S_s, \tag{3}$$

where S_s is the system entropy and show that

$$\sum_i \beta_i W_{A_i} \leq -\Delta \tilde{F}_s. \tag{4}$$

We will show, with a minimal number of assumptions, that we can implement any trade-offs between conserved quantities satisfying equation (2) using the bath, and extract any combination of W_{A_i} satisfying equation (4) from a system, up to an arbitrarily small deficit because of the finite nature of the protocols. In particular, if all the conserved quantities commute, we will give explicit protocols that works for both implicit and explicit batteries, assuming exact conservation of the A_i. For more general non-commuting quantities, we will obtain the same results for implicit batteries, or explicit batteries with average conservation, but leave open the question of how to deal with strict conservation of non-commuting quantities when considering explicit battery systems.

The generalized thermal state. Here we consider in more detail the generalized thermal state given in equation (1)[2,3].

We begin by recalling that there are two ways to define the thermal state—by maximizing the von Neumann entropy, given appropriate constraints, or by minimizing the free energy. We start with the former. Consider a system in state σ with Hamiltonian H and average energy $\langle H\rangle := \mathrm{tr}[H\sigma] = \bar{E}$. There are many states σ that have this particular average energy; the thermal state is the state that maximizes the entropy $S(\sigma) = -\mathrm{tr}(\sigma \ln \sigma)$, subject to the average energy constraint. Solving the maximization problem we get $\tau(\beta) = \frac{1}{\mathcal{Z}} e^{-\beta(\bar{E})H}$, where $\mathcal{Z}$ is the partition function and the inverse temperature β is implicitly determined by the average $\bar{E}$.

In our framework we need the generalization of this idea to the case of multiple conserved quantities. In particular, we consider k quantities $A_i, i \in \{1, \cdots, k\}$ and place no restrictions on the relations between them: they may or may not commute; when they commute they may or may not be functionally dependent on one another. An example of two commuting and functionally dependent quantities are the Hamiltonian H and angular momentum L, where $H = L^2/2I$. In this case the average of one does not uniquely determine the average of the other; however, the range of admissible values is constrained, that is, such that $|\langle L\rangle| \leq \sqrt{2I\langle H\rangle}$. An example of two non-commuting conserved quantities is L_x and L_y. The generalized thermal state $\tau(\beta_1,\ldots,\beta_k)$ is then the state, which maximizes the entropy S subject to the constraint that the conserved quantities A_i have average value $\langle A_i\rangle = \overline{A_i}$. It is found to be

Definition 1. Generalized thermal state

$$\tau(\beta_1, \ldots, \beta_k) = \frac{e^{-\sum_i \beta_i A_i}}{\mathcal{Z}}. \quad (5)$$

where, β_i is the inverse temperature conjugate to A_i, and the generalized partition function is $\mathcal{Z} = \mathrm{tr}(e^{-\sum_i \beta_i A_i})$.

Note that, in general, each β_i is a function of all of the averages $\overline{A_i}$. In the case that the A_i commute, the proof is a simple generalization of the standard proof. For non-commuting observables, the proof is more involved[41].

The second way to define the thermal state (when only energy is conserved) is to fix the inverse temperature $\beta = \frac{1}{T}$ and ask for the density matrix that minimizes the free energy $F(\rho) = \langle H\rangle - TS(\rho)$. The state that solves this optimization has exactly the Gibbs form $\tau(\beta) = e^{-\beta H}/\mathcal{Z}$. Since β is given, the average energy is now implicitly defined, in contrast to the case above, where the average energy was given and the inverse temperature derived.

The idea is to do the same in the case of multiple conserved quantities, and recover the generalized thermal state via a generalized free energy. However, in the standard definition of free energy the temperature is the constant multiplying the entropy. Since we have no notion of multiple entropies, we are not afforded a way of coupling all the inverse temperatures. This is easily overcome if instead of the free energy we define $\tilde{F}(\rho) = \beta\langle H\rangle - S(\rho)$, and it is trivial to generalize this quantity to the case of multiple conserved observables.

Definition 2. Free Entropy. The free entropy of a system ρ is

$$\tilde{F}(\rho) = \sum_i \beta_i \langle A_i\rangle - S(\rho), \quad (6)$$

The free entropy is always defined with respect to a set of inverse temperatures β_i. The generalized thermal state is then the state that minimizes $\tilde{F}$ with fixed β_i. For a complete proof of this fact, see the Supplementary Note 1.

Conceptual viewpoint. As noted in the introduction, the effects presented in this paper suggest that it may be worthwhile to revisit the basic concepts of thermodynamics. A key aspect of this is the conceptual shift from the free energy to the free entropy.

First, we would like to emphasize that the change from the usual free energy to the free entropy is not a simple mathematical manipulation, but marks a fundamental conceptual difference. Indeed, in the standard approach to considering multiple conserved quantities, such as when considering the grand canonical ensemble, one introduces the chemical potential μ such that the free energy becomes

$$F(\rho) = \langle H\rangle + \mu\langle N\rangle - TS(\rho) \quad (7)$$

where N is the particle number operator. In this way, energy is singled out as the privileged quantity, with the chemical potential acting as the 'exchange rate' between particle number and energy (and in the same way temperature acts as the exchange rate between entropy and energy). We argue that there is no reason to single out the energy, or any other quantity for that matter. In fact, it is possible to conceive of situations in which everything is degenerate in energy, and thus where energy plays absolutely no role. We are thus lead to introduce the free entropy, which naturally and uniquely treats all quantities on an equal footing.

A second argument for considering free entropy over the free energy is that the latter might give one incorrect intuition. Indeed, in the standard treatment, the free energy puts bounds on how much of the conserved quantity (energy) can be extracted, and one may be tempted to think that even when we have multiple conserved quantities, thermodynamics is about the bounds that constrain the extraction of individual quantities. However, as we will show, this is not the case, and there are no such bounds. The only limitation is on the trade-off between the conserved quantities, and this is precisely governed by the free entropy. It is only in the standard case of a single conserved quantity that one can choose to consider the free energy, or the free entropy, with both constraining the amount of work that can be extracted.

We also note that the thermal state is the state that minimizes the free energy only when the temperature is positive; if the temperature is negative the thermal state (at negative temperature) is instead the state that maximizes the free energy. On the other hand, for all temperatures (positive or negative), the thermal state always minimizes the free entropy.

Finally, we note that it is the difference in free entropy between ρ_s and τ_s that captures the thermodynamic usefulness of the system. This difference is exactly equal to the relative entropy between these two states, $\tilde{F}(\rho_s) - \tilde{F}(\tau_s) = D(\rho_s \,\|\, \tau_s)$, where $D(\rho \,\|\, \sigma) = \mathrm{tr}(\rho \log \rho - \rho \log \sigma)$. This is demonstrated explicitly in the Methods. This highlights the entropic nature of $\tilde{F}$.

Following on from the introduction of the free entropy, one can go a step further. Since the quantities $\beta_i\langle A_i\rangle$ all appear alongside the entropy in the definition of the free entropy, it suggests that they might be thought of as 'entropic quantities'. Note that this is true even in the standard case, where energy is

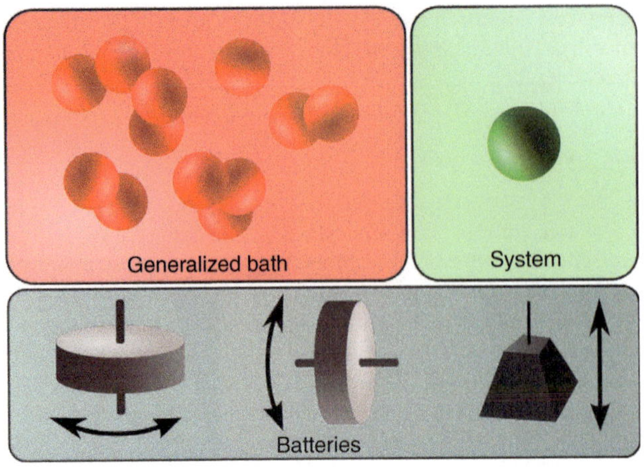

Figure 1 | Schematic diagram of the set-up. We have a generalized bath, a collection of particles each in the generalized thermal state (1). We have a system that is out of equilibrium with respect to the generalized bath. Finally, we have the collection of batteries, one for each conserved quantity. In the example depicted, the three conserved quantities are the angular momentum in the *x* direction, angular momentum in the *z* direction and the energy, with the three corresponding batteries being turntables spinning around the *z* and *x* axes, respectively, and a weight. We can perform any interaction between the components as long as each of the conserved quantities is conserved (the first law).

Table 1 | Summary of the results contained in the paper.

	Commuting	Non-commuting
Implicit batteries		
Second law	✓	✓
Protocol	✓	✓
Explicit batteries (strict conservation)		
Second law	✓	✓
Protocol	✓	?
Explicit batteries (ave. conservation)		
Second law	✓	✓
Protocol	✓	✓*

The second law equations (9) holds in all instances.
*Designates that the result holds only for explicit batteries with continuous spectra.

the only conserved quantity; there one might think of $\beta\langle H\rangle$ as an entropic quantity.

Importantly, the sum of these entropic quantities of the batteries is the object onto which the second law of thermodynamics applies. It says that the increase in this sum is constrained by the decrease in the free entropy of the system relative to the bath equation (4).

The set-up. The set-up is similar in spirit to that of previous lines of work[25,42]. We consider the interaction of generalized thermal baths with quantum systems and batteries (Fig. 1). There are a number of conserved quantities, A_1 to A_k, which may or may not commute or functionally depend on each other. The generalized thermal bath consists of an unbounded collection of systems, each of which is in a generalized thermal state as defined by equation (5). Any given protocol will involve only a finite set of systems in the bath, whose combined thermal state can be written as $\tau_b(\beta_1,\ldots,\beta_k)$. We also want to consider an additional quantum system ρ_s that is both initially uncorrelated from and out of equilibrium with respect to the generalized bath, that is, $\rho_{sb}=\rho_s\otimes\tau_b(\beta_1,\ldots,\beta_k)$ and $\rho_s\neq\tau_s(\beta_1,\ldots,\beta_k)$. The main question we ask is how much of each of the conserved quantities can be extracted from the system (in conjunction with the bath, and stored in an associated battery).

In the interest of being clear, we proceed by concentrating on a scenario with only two conserved quantities, *A* and *B*, since this already captures the majority of the physics contained in the general case of *k* conserved quantities.

In order to talk about the extraction of the conserved quantities, there are two ways in which one can proceed: by either including battery systems implicitly, or explicitly, in the formalism. In the former case, one allows the global amount of each quantity stored in the system and bath to change, and defines the changes as the amount of '*A*-type work' and '*B*-type work' that have been extracted from (or done on) the global system. The idea is that because of global conservation laws, when the A_i of the system and bath changes, this change is compensated by a corresponding change to the external environment (the implicit battery).

In the latter case, one introduces explicit battery systems, which by definition only accept a single type of work (that is, an *A*-type battery and a *B*-type battery). Here by definition the amount of *A* stored in the *A*-type battery is the *A*-type work, and similarly for *B*. We enforce that the global amount of *A* and *B* stored in the system, bath and battery is constant, either strictly (the entire distribution is conserved) or on average.

In the main text we consider the case of implicit batteries. We do this since dealing with implicit batteries simplifies the considerations and allows us to focus on what is arguably the most important part of the protocols, namely the interaction between system and bath. Obviously, it is preferable to have the full protocol including batteries explicitly. In doing so, there are many subtleties, which also arise in the case of standard thermodynamics. In particular, we need to impose 'no cheating' conditions that make sure that we do not make illegitimate use of batteries as sources of free entropy[25,42,43]. The danger stems from the fact that the batteries are systems out of equilibrium with respect to the bath. In Supplementary Note 2 we show how to include explicit batteries for a number of cases, as specified in Table 1.

More concretely, when considering implicit battery systems the class of allowed transformations consists of all global unitary transformations U on the system and bath. After such a transformation, the global state is $\rho'_{sb}=U(\rho_s\otimes\tau_b(\beta_A,\beta_B))U^\dagger$ with the reduced state of the system and bath given by the reductions, $\rho'_s=\mathrm{tr}_b[\rho'_{sb}]$ and $\rho'_b=\mathrm{tr}_s[\rho'_{sb}]$, respectively. We define the *A*-type and *B*-type work to be

$$\begin{aligned}\Delta W_A &= -\Delta A_s-\Delta A_b\\ \Delta W_B &= -\Delta B_s-\Delta B_b\end{aligned}\tag{8}$$

where $\Delta A_s=\mathrm{tr}[(A_s(\rho'_s-\rho_s)]$, $\Delta A_b=\mathrm{tr}[(A_b(\rho'_b-\tau_b)]$ and analogously for ΔB_s and ΔB_b. In addition, note that if our protocol involves multiple bath systems then $A_b=\sum_i A_b^{(i)}$, where $A_b^{(i)}$ acts non-trivially only on bath system *i*, that is, A_b is the sum of the local *A* for each system (and analogously for B_b). In equation (8) we are equating the average change of *A* and *B*, because of the unitary transformation, with the amount of *A*-type and *B*-type work that has been extracted from the system and bath. As such, our framework automatically incorporates the first law of thermodynamics for each of the conserved quantities.

Finally, an additional unrelated problem, but which often plays an important role, concerns the precise structure of the bath. In usual treatments, we may consider particles in the bath that have any energy-level spacing, such that their occupation probabilities can match any probabilities in the external system. This is used to construct efficient protocols. When considering other quantities than energy, we may be faced with quantities whose spectrum is fixed, such as angular momentum. In addition, extra constraints or relationships may exist between the different conserved quantities. This results in additional difficulties. To address these, and to remain as general as possible, we will consider baths with a minimal amount of accessible structure in terms of the eigenvalues of the conserved quantities.

The second law. Of great interest to us is the particular form that the second law of thermodynamics takes in the present setting. In the classical thermodynamic setting, the second law states that if one only has access to a thermal bath, then no work can be extracted, and that the maximal amount of work that can be extracted from a non-equilibrium system interacting with a thermal bath is bounded by the change in its free energy.

In our framework of multiple conserved quantities, we will see that the second law constrains the different combinations of conserved quantities that can be extracted from the system. In particular, we will show below that in the present framework, the amount of A-type work and B-type work that can be extracted is constrained such that

$$\beta_A \Delta W_A + \beta_B \Delta W_B \leq -\Delta \tilde{F}_s, \qquad (9)$$

where $\Delta \tilde{F}_s = \tilde{F}(\rho_s') - \tilde{F}(\rho_s)$. In the case where there is no system, or when the system is left in the same state, $\rho_s' = \rho_s$, then $\Delta \tilde{F}(\rho_s) = 0$, and we obtain as a corollary

$$\beta_A \Delta W_A + \beta_B \Delta W_B \leq 0. \qquad (10)$$

Equations (9) and (10) constitute the second law when one has multiple conserved quantities (with and without a system).

To prove the second law, equation (9), we will need to use two further formulae, as well as the first laws, equation (8). First, since we restrict to unitary transformations, the total entropy of the global system remains unchanged, $S(\rho_{sb}') = S(\rho_{sb})$, and from the fact that the system and bath are initially uncorrelated, along with sub-additivity, we have

$$\Delta S_s + \Delta S_b \geq \Delta S_{sb} = 0 \qquad (11)$$

where $\Delta S_s = S(\rho_s') - S(\rho_s)$, and analogously for ΔS_b and ΔS_{sb}. Second, since the bath starts in the thermal state $\tau(\beta_A, \beta_B)$, which is a minimum of the free entropy (by definition), its free entropy cannot decrease during the protocol; thus,

$$\Delta \tilde{F}_b = \beta_A \Delta A_b + \beta_B \Delta B_b - \Delta S_b \geq 0 \qquad (12)$$

Now, eliminating all quantities on the bath, by substituting from the first laws, equation (8), and from equation (11), we finally arrive at

$$-\beta_A(\Delta A_s + \Delta W_A) - \beta_B(\Delta B_s + \Delta W_B) + \Delta S_s \geq 0 \qquad (13)$$

which, after re-arranging and identifying terms, is straightforwardly seen to be equation (9), as desired. Thus, the first law, in conjunction with the lack of initial correlations (and sub-additivity), and the extremality of the generalized thermal state imply in a direct manner that systems obey a second law of the form given. We note that the proof does not rely on any particular properties of A and B, which need not even commute.

At this point it is worth briefly returning to the issue of implicit versus explicit batteries. If explicit batteries are included, then the unitary operations have to be extended to act on the system, bath and explicit batteries. Crucially, equation (11), and as a consequence the second law, equation (9), can be shown to hold when we are careful to avoid cheating via batteries. Details are provided in Supplementary Note 2.

In the remaining we will study to what extent we can saturate equations (9) and (10), depending upon the properties of the conserved quantities (whether they commute or not), whether we consider implicit or explicit batteries and whether we consider strict or average conservation.

Commuting observables. We will now specialize to the case of commuting observables, where we have access to joint eigenstates, and show how the second law can be saturated, both in terms of trading resources, and when extracting resources from a non-equilibrium system.

In order to remain as general as possible, we want to assume as little as possible about the structure of the generalized thermal bath. What we will require is that there exists a system in the bath (of which we can take arbitrarily many copies) with $d \geq 3$ states, $|a_i, b_i\rangle$, for $i \in \{0, 1, 2, \ldots d-1\}$, which are the joint eigenstates of A_b and B_b such that $A_b|a_i, b_i\rangle = a_i|a_i, b_i\rangle$ and $B_b|a_i, b_i\rangle = b_i|a_i, b_i\rangle$. We then need only three requirements. First, that the three eigenvalues of each observable are distinct, $a_0 \neq a_1 \neq a_2$ and $b_0 \neq b_1 \neq b_2$ (note that one consequence of this is to rule out the case in which either A or B is proportional to the identity, and thus trivially cannot be changed). Second, that the observables should be sufficiently different. In particular, that

$$\frac{a_1 - a_0}{b_1 - b_0} \neq \frac{a_2 - a_0}{b_2 - b_0}, \qquad (14)$$

which amounts to saying that A and B should not be affinely related to each other, in which case they should not be thought of as different quantities. Third, that in the thermal state the joint eigenstates should not have the same populations. In particular, it must be that

$$\begin{aligned} x &:= \beta_A(a_1 - a_0) + \beta_B(b_1 - b_0) \neq 0, \\ y &:= \beta_A(a_2 - a_0) + \beta_B(b_2 - b_0) \neq 0. \end{aligned} \qquad (15)$$

If both x and y simultaneously vanish, then all three states have the same populations, in which case the system looks maximally mixed in this subspace. When trading quantities inside the bath, this will be the only problematic case. However, when we come to processing non-equilibrium systems, we will require simultaneously $x \neq 0$ and $y \neq 0$ in order for the bath to have enough structure to allow us to approach reversibility. We will also see that, depending on how close to reversible we want to be, we will have to exclude a small set (non-dense and of measure zero) of joint values for x and y, which are rationally related, as will be explained later. Below we outline the main ideas, and present all the details in the Methods.

We start by considering the task of trading resources within the generalized bath. That is, we consider the situation where we only have access to a generalized bath (and no external system). We will show that we can perform a unitary transformation such that, first, its free entropy changes by an arbitrarily small amount,

$$\Delta \tilde{F}_b = \beta_A \Delta A_b + \beta_B \Delta B_b \leq \epsilon \qquad (16)$$

where we used that $\Delta S_b = 0$ by definition for unitary transformations. Second, the change ΔA_b or ΔB_b can be made arbitrarily large, that is

$$\Delta A_b \geq \eta \quad \text{or} \quad \Delta B_b \geq \eta \qquad (17)$$

with the other appropriately constrained by equations (16); (generally having large magnitude but opposite sign). If the above two conditions can be satisfied, we say that we can exchange A for B in an essentially reversible manner.

To show that this is possible, we proceed in two steps. We provide an explicit protocol that exchanges a suitably chosen two-dimensional subspace within the bath, and calculate the change in $\Delta \tilde{F}_b$, ΔA_b and ΔB_b that this produces. We then show that by repeating this protocol a sufficient number of times we achieve equations (16) and (17).

The explicit protocol takes the bath as n copies of $\tau(\beta_A, \beta_B)$, and a two dimensional subspace that consists of states that differ in population by Δq, and differ in the number of systems in the state $|a_1, b_1\rangle$ by Δn_1. Then, as we show in the Methods, by interchanging the population of two such states, we can achieve

$$0 < \Delta \tilde{F}_b \leq y \Delta q, \tag{18}$$

$$\frac{|\Delta A_b|}{\Delta \tilde{F}_b} = O(\Delta n_1), \quad \frac{|\Delta B_b|}{\Delta \tilde{F}_b} = O(\Delta n_1), \tag{19}$$

where the sign of ΔA_b or ΔB_b can be chosen arbitrarily, with the other quantity generally having the opposite sign, in accordance with equation (16). Hence, $\Delta \tilde{F}_b$ can be made as small as desired by making Δq arbitrarily small, while independently the relative change $|\Delta A_b|/\Delta \tilde{F}_b$ can be made as large as desired by increasing Δn_1.

Finally, by repeating the above protocol a sufficient number of times, one can trade arbitrary amounts of the conserved quantities from a generalized bath by sacrificing an arbitrarily small amount of free entropy. In particular, to achieve $\Delta A_b^{\text{total}} \geq \eta$ with $\Delta \tilde{F}_b^{\text{total}} \leq \epsilon$, one can perform the protocol above $(\epsilon/\Delta \tilde{F}_b)$ times, with $\Delta A_b/\Delta \tilde{F}_b \geq \eta/\epsilon$.

We now move onto the task of extracting resources from a single quantum system. In the above we showed that arbitrarily good interconversions can be enacted, given access only to a generalized bath. We now move on to the scenario of having a quantum system out of equilibrium with respect to the bath. Our goal is to show that we can saturate the second law given by equation (9) arbitrarily well—that is, that we can extract conserved quantities from a non-equilibrium system such that $\beta_A \Delta W_A + \beta_B \Delta W_B$ is as close as desired to the system's decrease in free entropy.

Let us consider that we have a state ρ_s, which in terms of its eigenstates and eigenvalues is given by $\rho_s = \sum_i p_i |\psi_i\rangle\langle\psi_i|$, and by convention we take the eigenvalues to be ordered, $p_{n+1} \leq p_n$. In general, the eigenbasis of the state will not coincide with the joint eigenbasis of the conserved quantities A and B. The first step is to pre-process the system, to bring it to a diagonal form in this basis. As we show in the Methods, it is always possible to do so, without even utilizing the bath, such that $\Delta \tilde{F}_s = \beta_A \Delta A_s + \beta_B \Delta B_s$, that is, in a fully reversible way that saturates equation (9). Note that in the case of explicit battery systems this pre-processing step is more complicated, nevertheless, the ideal pre-processing can still be arbitrarily well approximated, as shown in full details in Supplementary Note 3.

Now, having bought the system to diagonal form, we want to consider a protocol that moves a small population δp between two eigenstates, which have populations p_0 and p_1, respectively. We can implement such changes by finding two levels in the bath whose ratio of populations are close to p_0/p_1, and then swapping the 2D subspaces of the system and bath. Therefore, to carry out any such transformation requires a finely spaced set of different population ratios in the bath.

In the Methods we show that, except for a non-dense and measure zero set of x and y (as defined in equation (15)), we can find two levels in the bath whose ratio of populations is sufficiently close to p_0/p_1, such that after applying the swap operation between the appropriate two-dimensional subspaces, we achieve

$$\beta_A \Delta W_A + \beta_B \Delta W_B = -\Delta \tilde{F}_s + O(\delta p^2) \tag{20}$$

that is, up to a correction of order $O(\delta p^2)$, the combination of conserved quantities extracted, which themselves are order $O(\delta p)$, matches the change in free entropy of the system. Thus, by composing $O(1/\delta p)$ of such transformations, we can implement a protocol that transforms $\rho_s \to \tau_s(\beta_A, \beta_B)$, whereby in each stage the population changes between two states by order $O(\delta p)$, and such that

$$\beta_A \Delta W_A + \beta_B \Delta W_B = -\Delta \tilde{F}_s + O(\delta p). \tag{21}$$

Therefore, by taking δp sufficiently small we can approach the reversible regime, whereby the change in free entropy of the system matches the dimensionless combination of conserved quantities extracted. Combining this protocol with the protocol from the previous section, involving only the generalized bath and the batteries, we can obtain any combination of extracted conserved quantities.

Finally, note that the same protocol can also be used to perform efficient transformations between any two system states (where the final state is full rank), and not just to the thermal state. Our protocol also immediately gives an asymptotic protocol for the interconversion of states with no average work cost: the rate at which one can transform $\rho^{\otimes n}$ into $\sigma^{\otimes nR}$ is given by

$$R = \frac{\tilde{F}(\rho) - \tilde{F}(\tau(\beta_1, \ldots, \beta_k))}{\tilde{F}(\sigma) - \tilde{F}(\tau(\beta_1, \ldots, \beta_k))} \tag{22}$$

Here one can simply run the protocol 'forward' individually on n copies of ρ in order to obtain in the batteries $n(\beta_1 \Delta W_{A1} + \ldots \beta_k \Delta W_{Ak})$, and create $\tau(\beta_1, \ldots, \beta_k)^{\otimes n}$. Then, on nR copies run the protocol 'backwards' to create $\sigma^{\otimes nR}$, having returned each battery so that finally it contains the same amount its associated quantity that it initially contained (on average).

In Supplementary Note 3 we show how these results extend to the case of explicit batteries with either strict or average conservation. We also show in Supplementary Note 4 how the protocol can be made robust to experimental imperfections—that is, without assuming precise knowledge of β_A or β_B.

Non-commuting observables. In this section we will show that when considering implicit batteries the results obtained in the previous section can easily be modified to also work for non-commuting observables. This also extends to explicit batteries with average conservation laws when the batteries have continuous spectrum. However, the same protocols do not obviously generalize to the case of explicit batteries with strict conservation.

Whereas previously by virtue of the commutativity of the observables we could find a joint eigenbasis $|a_i, b_i\rangle$ that was used in our explicit protocols, that is no longer the case for non-commuting observables. Nevertheless, the generalized thermal state is diagonal in the eigenbasis of $\beta_A A + \beta_B B$,

$$e^{-\beta_A A - \beta_B B}/\mathcal{Z} = \sum_i q_i |i\rangle\langle i| \tag{23}$$

Although to each eigenstate we can no longer associate an eigenvalue for A or B, we can still associate an average value,

$$\begin{aligned} \langle a \rangle_i &:= \langle i|A|i\rangle \\ \langle b \rangle_i &:= \langle i|B|i\rangle \end{aligned} \tag{24}$$

The main point is that all of our previous results hold if instead of joint eigenstates $|a_i, b_i\rangle$, with eigenvalues a_i and b_i, we use the eigenstate $|i\rangle$ with average values $\langle a \rangle_i$ and $\langle b \rangle_i$ throughout.

The only subtlety that arises is the structure we need from the bath. We still only need to use three distinct eigenstates, $|0\rangle$, $|1\rangle$ and $|2\rangle$; however, now the necessary structure relates to the average value of the conserved quantities in the eigenstates, First, it must be that

$$\frac{\langle a\rangle_1 - \langle a\rangle_0}{\langle b\rangle_1 - \langle b\rangle_0} \neq \frac{\langle a\rangle_2 - \langle a\rangle_0}{\langle b\rangle_2 - \langle b\rangle_0} \tag{25}$$

otherwise, at the level of the average values, the observables appear affinely related and therefore cannot be sufficiently distinguished to allow for trade-offs. Furthermore, we still need to be able to find eigenstates in the bath that differ in population, just as before. We can define the analogous quantities $\langle x\rangle$ and $\langle y\rangle$, and if they do not simultaneously vanish then the bath will not be maximally mixed in the subspace. Finally, in order to extract resources from systems out of equilibrium with respect to the generalized bath, there must be sufficient structure such that any ratio of populations can be approximated well enough. Again, in complete analogy to the above, if $(\langle x\rangle/\langle y\rangle)$ is irrational, then we have sufficient structure. If on the other hand $(\langle x\rangle/\langle y\rangle)$ is rational, we will again have to exclude a small set of values of $\langle x\rangle$ and $\langle y\rangle$ (non-dense, of zero measure), for which our results will not hold.

Discussion

In this work we have studied a generalization of thermodynamics where there are multiple conserved quantities, where energy may not even be part of the story. We have been interested in what form the second law takes, and showed that it is no longer about restrictions on individual extractable quantities, but rather about the allowed ways that the conserved quantities can be traded-off for one another. Indeed, we found that we can extract as much of any individual conserved quantity as desired, as long as the other conserved quantities are appropriately consumed in the process, with the second law dictating how much of the others are necessarily consumed. In particular, we were led to introduce a dimensionless generalization of the free energy, which we termed the free entropy, that is the central quantity appearing in the second law and dictating the allowed trade-offs. Moreover, given access to any quantum system out of equilibrium with respect to the generalized bath, we showed that its free entropy change bounds the combination of conserved quantities that can be extracted.

Our results hold both for commuting and non-commuting observables, and with the desire to remain as general as possible we made only very mild assumptions about the bath. Indeed, we assumed very little about the relationship between the conserved quantities or their individual structure. The one case that remains open for future research is the case of non-commuting observables, with explicit batteries and strict conservation of the conserved quantities. Although the protocols presented for saturating the second law do not appear to generalize to this case, we do not know whether entirely different constructions will be able to achieve this goal.

Methods

Relation between free entropy and relative entropy. Here we show that the free entropy difference between any state ρ_s and the generalized thermal state $\tau_s = \exp(-\sum_i \beta_i A_i)/\mathcal{Z}$ is equal to the relative entropy difference between these two states. First, note that the free entropy of the thermal state is

$$\begin{aligned}\tilde{F}(\tau_s) &= \sum_i \beta_i \langle A_i\rangle_\tau + \mathrm{tr}(\tau_s \log \tau_s)\\ &= \sum_i \beta_i \langle A_i\rangle_\tau - \sum_i \beta_i \langle A_i\rangle_\tau - \mathrm{tr}(\tau_s)\log \mathcal{Z}\\ &= -\log \mathcal{Z},\end{aligned} \tag{26}$$

where we recall that $\mathcal{Z} = \mathrm{tr}(\exp(-\sum_i \beta_i A_i))$ is the generalized partition function.

Then, it follows that

$$\begin{aligned}D(\rho_s \,||\, \tau_s) &= -S(\rho_s) - \mathrm{tr}(\rho_s \log \tau_s),\\ &= -S(\rho_s) + \sum_i \beta_i \langle A_i\rangle + \log \mathcal{Z},\end{aligned} \tag{27}$$

which demonstrates the claim. Note that this result is completely analogous to the case of standard thermodynamics[44].

Trading resources. Consider the situation where we only have access to a generalized bath (and no external system). We will show that we can perform a unitary transformation such that: (i) the free entropy $\Delta\tilde{F}_b = \beta_A \Delta A_b + \beta_B \Delta B_b$ (since $\Delta S_b = 0$ by definition) changes by an arbitrarily small amount. (ii) The changes $|\Delta A_b|$ and $|\Delta B_b|$ can be made arbitrarily large.

Consider that we have a n copies of $\tau(\beta_A, \beta_B) = e^{-\beta_A A - \beta_B B}/\mathcal{Z} = \sum_i q_i |a_i, b_i\rangle\langle a_i, b_i|$. It will be convenient to label states by the number of systems found in a given eigenstate, which we shall refer to as the occupation. We denote by $|\mathbf{n}, \alpha\rangle \equiv |n_0, n_1, \ldots, n_{d-1}, \alpha\rangle = P_\alpha |a_0, b_0\rangle^{\otimes n_0} |a_1, b_1\rangle^{\otimes n_1} \cdots |a_{d-1}, b_{d-1}\rangle^{\otimes n_{d-1}}$, where P_α is a permutation operator, permuting the bath systems, labelled by α, and $n_0 + n_1 + \ldots + n_{d-1} = n$. Now, we will consider only two states from the d^n which are available, corresponding to

$$\begin{aligned}|\mathbf{n}, \alpha\rangle &= |n_0, n_1, n_2, n_3, \ldots, n_{d-1}, \alpha\rangle\\ |\mathbf{n}', \alpha'\rangle &= |n'_0, n'_1, n'_2, n_3, \ldots, n_{d-1}, \alpha'\rangle\end{aligned} \tag{28}$$

that is, such that only the occupations of the first three levels differ between these states. As such, we have the constraint that $n_0 + n_1 + n_2 = n'_0 + n'_1 + n'_2$. The key step in our protocol is to apply a swap operation between these two states, while leaving all other states unchanged. We assume the bath is sufficiently large so that after this step any modified systems from the bath can be discarded, and any further operations act on fresh bath systems. A direct calculation shows that the change in the average value of each quantity of interest is

$$\Delta A_b = \Delta q(a_{10}\Delta n_1 + a_{20}\Delta n_2) \tag{29}$$

$$\Delta B_b = \Delta q(b_{10}\Delta n_1 + b_{20}\Delta n_2) \tag{30}$$

$$\Delta F_b = \Delta q(x\Delta n_1 + y\Delta n_2) \tag{31}$$

where $\Delta n_k = (n'_k - n_k)$, $a_{k0} = (a_k - a_0)$, $b_{k0} = (b_k - b_0)$, x and y are as defined in equation (15), and

$$\Delta q = \left(1 - \left(\frac{q_1}{q_0}\right)^{\Delta n_1}\left(\frac{q_2}{q_0}\right)^{\Delta n_2}\right)\prod_{i=0}^{d-1} q_i^{n_i} \tag{32}$$

is the difference in populations between the two states. Given $y \neq 0$, we can rewrite equation (31) as

$$\Delta\tilde{F}_b = y\Delta q\Delta n_1\left(\frac{x}{y} + \frac{\Delta n_2}{\Delta n_1}\right) \tag{33}$$

Now, for arbitrary Δn_1, we can find an integer m such that $m/\Delta n_1 < x/y \leq (m+1)/\Delta n_1$. Setting $\Delta n_2 = -m$ in equation (33), we obtain

$$0 < \Delta\tilde{F}_b \leq y\Delta q. \tag{34}$$

Hence, $\Delta\tilde{F}_b$ can be made as small as desired by making Δq arbitrarily small (which can be achieved by increasing n_0). Note that it is crucial that $\Delta\tilde{F}_b \neq 0$. This is because the thermal state is the unique state that minimizes $\tilde{F}_b$. Thus, $\Delta\tilde{F}_b = 0$ implies that the bath is left completely unchanged, which in turn implies that $\Delta A_b = \Delta B_b = 0$; hence, the desired transformation cannot take place. On the other hand, we find that the relative change in the conserved quantities $\Delta A_b/\Delta\tilde{F}_b$ and $\Delta B_b/\Delta\tilde{F}_b$ are

$$\begin{aligned}\frac{\Delta A_b}{\Delta\tilde{F}_b} &= \frac{a_{20}}{y}\left(1 + \frac{\beta_B b_{20}}{y}\left(\frac{a_{10}}{a_{20}} - \frac{b_{10}}{b_{20}}\right)\left(\frac{x}{y} + \frac{\Delta n_2}{\Delta n_1}\right)^{-1}\right)\\ \frac{\Delta B_b}{\Delta\tilde{F}_b} &= \frac{b_{20}}{y}\left(1 + \frac{\beta_A a_{20}}{y}\left(\frac{b_{10}}{b_{20}} - \frac{a_{10}}{a_{20}}\right)\left(\frac{x}{y} + \frac{\Delta n_2}{\Delta n_1}\right)^{-1}\right)\end{aligned} \tag{35}$$

In both cases, the final term satisfies $(x/y + \Delta n_2/\Delta n_1)^{-1} \geq \Delta n_1$ and can, hence, be made as large as desired by increasing Δn_1. This means that the magnitude of $\Delta A_b/\Delta\tilde{F}_b$ and $\Delta B_b/\Delta\tilde{F}_b$ will become arbitrarily large. The sign of ΔA_b will depend on the other constants, but can be modified if desired by choosing m such that $(m-1)/\Delta n_1 \leq x/y < m/\Delta n_1$ above. Note also that if $y = 0$ but $x \neq 0$ we can construct an equivalent proof with the roles of x and y swapped.

Finally, by repeating the above protocol a sufficient number of times, one can trade arbitrary amounts of the conserved quantities from a generalized bath by sacrificing an arbitrarily small amount of free entropy. In particular, to achieve $\Delta A_b^{\mathrm{total}} \geq \eta$ with $\Delta\tilde{F}_b^{\mathrm{total}} \leq \epsilon$, one can perform the protocol above $(\epsilon/\Delta\tilde{F}_b)$ times, with $\Delta A_b/\Delta\tilde{F}_b \geq \eta/\epsilon$.

There are a number of important aspects of the above protocol: first, it relies on a minimal amount of structure in the observables A and B and the bath: it requires that there exist many copies of a bath system with a three-dimensional subspace where the action of the operators are not trivially related (by a shift and rescaling), and that the state is not maximally mixed in this subspace. Moreover, each bath system is taken to be identical, with no additional parameters necessary (that is, we do not require a family of different A_b's and B_b's, similar to the families of Hamiltonians considered in ref. 25). Second, this protocol is a only a proof-of-principle demonstration that trade-offs can be enacted. No attention was paid to the number of generalized thermal states necessary. If one cared about minimizing the resources utilized, then the above protocol would not be used, and more efficient ones would be sought. Finally, the above analysis generalizes beyond two conserved quantities to the general case of k quantities. In this case, equation (14) must hold pairwise for all quantities.

Extracting resources from a single quantum system. Here we consider the scenario involving a quantum system out of equilibrium with respect to the bath. We will show that the second law given by equation (9) can be saturated arbitrarily well—that is, we can extract conserved quantities from a non-equilibrium system such that $\beta_A \Delta W_A + \beta_B \Delta W_B$ is as close as desired to the system's decrease in free entropy.

Let us consider that we have a state ρ_s, which in terms of its eigenstates and eigenvalues, is given by $\rho_s = \sum_i p_i |\psi_i\rangle\langle\psi_i|$, and by convention we take the eigenvalues to be ordered, $p_{n+1} \leq p_n$. In general, the eigenbasis of the state will not coincide with the joint eigenbasis of the conserved quantities A and B. The first step is to pre-process the system to bring it to a diagonal form in this basis. To do so we will not interact with the bath, but simply apply the unitary

$$U_s = \sum_i |a_i, b_i\rangle\langle\psi_i|, \tag{36}$$

on the system such that $\sigma_s = U_s \rho_s U_s^\dagger = \sum_i p_i |a_i, b_i\rangle\langle a_i, b_i|$. Owing to the first laws, equation (8), we have $\Delta W_A = -\Delta A_s$ and $\Delta W_B = -\Delta B_s$. Moreover, since the entropy of the system did not change, $\Delta \tilde{F}_s = \beta_A \Delta A_s + \beta_B \Delta B_s$, and we have immediately

$$-\Delta \tilde{F}_s = \beta_A \Delta W_A + \beta_B \Delta W_B \tag{37}$$

that is, in this state, unsurprisingly, we have a change in the batteries that coincides with the free entropy change of the system, and saturates equation (9). Note that in the case of explicit battery systems, the transformation in equation (36) cannot be perfectly performed at the level of the system, since $[U_s, A_s] \neq 0$ and $[U_s, B_s] \neq 0$. Nevertheless, one can implement a joint unitary on the system and batteries such that at the level of the system the transformation U_s can be approximated arbitrarily well as long as the battery systems are in appropriate states, similarly to the case of standard quantum thermodynamics[42,43]. Full details can be found in Supplementary Note 3.

Now, having brought the system to a diagonal form, we want to consider a transformation $\sigma_s \rightarrow \sigma_s' = \sum_i p_i' |a_i^s, b_i^s\rangle\langle a_i^s, b_i^s|$ in which only two levels of the system change their populations by a small amount. Note that here we denote the eigenvalues of A_s and B_s by a_i^s and b_i^s to differentiate them from the eigenvalues of A_b and B_b, which we will similarly denote by a_i^b and b_i^b. Namely, we would like to perform $p_1' = p_1 + \delta p, p_0' = p_0 - \delta p$. We can implement such changes by finding two levels in the bath whose ratio of populations is close to p_0/p_1 and then swapping the two-dimensional subspaces of the system and bath. To carry out any such transformation requires a finely spaced set of different population ratios in the bath. Let us now consider whether the simple bath systems we have considered so far can provide such possibilities. As before, we bring in a collection of n generalized thermal states, $\tau(\beta_A,\beta_B)$ and consider only the two states in equation (28), $|\mathbf{n}, \alpha\rangle$ and $|\mathbf{n}', \alpha'\rangle$. We shall denote the populations of these states by $q_\mathbf{n}$ and $q_{\mathbf{n}'}$, respectively. The global unitary transformation we will apply is the swap operator between the two-dimensional subspaces of the system and bath, and the identity everywhere else. That is, the operation that performs

$$|a_0, b_0\rangle|\mathbf{n}', \alpha'\rangle \leftrightarrow |a_1, b_1\rangle|\mathbf{n}, \alpha\rangle, \tag{38}$$

while leaving all other states unchanged. By performing this transformation, the population that is shifted between the states of the system, which coincides with the population that is shifted between the states of the bath, is $\delta p = (p_0 q_{\mathbf{n}'} - p_1 q_\mathbf{n})$. Now, the changes in the conserved quantities of the system and bath are found to be

$$\begin{aligned} \Delta A_s &= \delta p(a_1^s - a_0^s), & \Delta B_s &= \delta p(b_1^s - b_0^s) \\ \Delta A_b &= -\delta p(a_{\mathbf{n}'}^b - a_\mathbf{n}^b), & \Delta B_b &= -\delta p(b_{\mathbf{n}'}^b - b_\mathbf{n}^b). \end{aligned} \tag{39}$$

The change in the entropy of the system and the bath are also found to be

$$\begin{aligned} \Delta S_s &= \delta p \log \frac{p_0'}{p_1'} + O(\delta p^2) \\ \Delta S_b &= -\delta p \log \frac{q_\mathbf{n}}{q_{\mathbf{n}'}} + O(\delta p^2) \end{aligned} \tag{40}$$

where we use $\frac{p_0'}{p_1'}$ rather than $\frac{p_0}{p_1}$ since we can always take $p_1' \neq 0$. We want to achieve $\Delta S_s + \Delta S_b = O(\delta p^2)$, which requires that $\log(q_\mathbf{n}/q_{\mathbf{n}'}) = \log(p_0'/p_1') + O(\delta p)$. To see when this condition is satisfied, we first use the explicit form of the probabilities, which show that $q_\mathbf{n}/q_{\mathbf{n}'} = \exp(-x\Delta n_1 - y\Delta n_2)$. Hence, we require $\log(p_0'/p_1') - x\Delta n_1 - y\Delta n_2 = O(\delta p)$. Since p_0' and p_1' are arbitrary, our requirement is that $x\Delta n_1 + y\Delta n_2$ should be able to come within $O(\delta p)$ of any number.

As we can rescale Δn_1 and Δn_2 by an arbitrary integer, it is sufficient to obtain $0 < x\Delta n_1 + y\Delta n_2 \leq O(\delta p)$. If x and y are not rationally related, this is always possible. However, if x/y is rational, and given in reduced form by u/v (where u and v are co-prime integers), then we need

$$0 < \frac{y}{v}(u\Delta n_1 + v\Delta n_2) \leq O(\delta p). \tag{41}$$

From number theory, one can always find Δn_1 and Δn_2 such that the term in brackets is 1 (or -1); hence, we need $|y/v| \leq O(\delta p)$. For a fixed desired accuracy ϵ (of order $O(\delta p)$), and fixed y, this rules out a finite number of x values for which x/y is a rational with a small denominator. Extending this to the x-y plane, we find that we can achieve the desired accuracy everywhere except for a non-dense set of measure zero.

Returning to the entropies, with the above in place, $\Delta S_s + \Delta S_b = O(\delta p^2)$, that is, the system and bath remain essentially uncorrelated after the transformation. Finally, we use once again the fact that the generalized thermal state is a minimum of the free entropy. This implies that the changes in population, of order $O(\delta p)$, change the free entropy only to second order, $\Delta \tilde{F}_b = O(\delta p^2)$. Putting everything together we have

$$\begin{aligned} \Delta \tilde{F}_b &= \beta_A \Delta A_b + \beta_B \Delta B_b - \Delta S_b, \\ &= -\beta_A(\Delta W_A + \Delta A_s) - \beta_B(\Delta W_B + \Delta B_s) + \Delta S_s + O(\delta p^2), \\ &= -\Delta \tilde{F}_s - (\beta_A \Delta W_A + \beta_B \Delta W_B) + O(\delta p^2). \end{aligned} \tag{42}$$

Thus, since the left-hand side is $O(\delta p^2)$, it must be that

$$\beta_A \Delta W_A + \beta_B \Delta W_B = -\Delta \tilde{F}_s + O(\delta p^2) \tag{43}$$

that is, up to a correction of order $O(\delta p^2)$, the combination of conserved quantities extracted, which themselves are order $O(\delta p)$, matches the change in free entropy of the system. Thus, by composing $O(1/\delta p)$ of such transformations, we can implement a protocol that transforms $\rho_s \rightarrow \tau_s(\beta_A,\beta_B)$, whereby in each stage the population changes between two states by order $O(\delta p)$, and such that

$$\beta_A \Delta W_A + \beta_B \Delta W_B = -\Delta \tilde{F}_s + O(\delta p). \tag{44}$$

Therefore, by taking δp sufficiently small we can approach the reversible regime, whereby the change in free entropy of the system matches the dimensionless combination of conserved quantities extracted. Combining this protocol with the protocol from the previous section, involving only the generalized bath and the batteries, we can obtain any combination of extracted conserved quantities.

It is important to stress that our assumptions only changed by a small amount relative to the previous section. In particular, we need a minimal amount of extra structure in the bath, such that it is useful to process arbitrary individual systems out of equilibrium. Furthermore, as in the previous case the protocol presented generalizes in a straightforward manner to the case of k mutually commuting conserved quantities $A_1,\ldots A_k$.

Finally, note that the same protocol can also be used to perform efficient transformations between any two system states (where the final state is full rank), and not just to the thermal state. Our protocol also immediately gives an asymptotic protocol for the interconversion of states with no average work cost: the rate at which one can transform $\rho^{\otimes n}$ into $\sigma^{\otimes nR}$ is given by

$$R = \frac{\tilde{F}(\rho) - \tilde{F}(\tau(\beta_1, \ldots, \beta_k))}{\tilde{F}(\sigma) - \tilde{F}(\tau(\beta_1, \ldots, \beta_k))} \tag{45}$$

Here one can simply run the protocol 'forward' individually on n copies of ρ in order to obtain in the batteries $n(\beta_1 \Delta W_{A1} + \ldots \beta_k \Delta W_{A_k})$ and create $\tau(\beta_1,\ldots,\beta_k)^{\otimes n}$. Then, on nR copies run the protocol 'backwards' to create $\sigma^{\otimes nR}$, having returned each battery so that finally it contains the same amount its associated quantity that it initially contained (on average).

In Supplementary Note 4 we show how these results extend to the case of explicit batteries with either strict or average conservation. In Supplementary Note 4 we also show how the protocol can be made robust to experimental imperfections—that is, without assuming precise knowledge of β_A or β_B.

References

1. Gibbs., J. W. *Elementary Principles in Statistical Mechanics* (Cambridge University Press, 2010) Cambridge Books Online.
2. Jaynes., E. T. Information theory and statistical mechanics. *Phys. Rev.* **106,** 620-630 (1957).
3. Jaynes., E. T. Information theory and statistical mechanics. II. *Phys. Rev.* **108,** 171-190 (1957).
4. Rigol, M., Dunjko, V., Yurovsky, V. & Olshanii., M. Relaxation in a completely integrable many-body quantum system: an ab initio study of the dynamics of the highly excited states of 1d lattice hard-core bosons. *Phys. Rev. Lett.* **98,** 050405 (2007).

5. Kollar, M., Wolf, F. A. & Eckstein., M. Generalized Gibbs ensemble prediction of prethermalization plateaus and their relation to nonthermal steady states in integrable systems. *Phys. Rev. B* **84,** 054304 (2011).
6. Caux, J.-S. & Konik., R. M. Constructing the generalized Gibbs ensemble after a quantum quench. *Phys. Rev. Lett.* **109,** 175301 (2012).
7. Dóra, B., Bácsi, Á. & Zaránd., G. Generalized Gibbs ensemble and work statistics of a quenched Luttinger liquid. *Phys. Rev. B* **86,** 161109 (2012).
8. Pozsgay., B. The generalized Gibbs ensemble for Heisenberg spin chains. *J. Stat. Mech.* **2013,** P07003 (2013).
9. Vaccaro, J. A. & Barnett., S. M. Information erasure without an energy cost. *Proc. R Soc. A* **467,** 1770–1778 (2011).
10. Barnett, S. M. & Vaccaro., J. A. Beyond Landauer erasure. *Entropy* **15,** 4956–4968 (2013).
11. Dahlsten, O. C. O., Renner, R., Rieper, E. & Vedral., V. Inadequacy of von Neumann entropy for characterizing extractable work. *New J. Phys.* **13,** 53015 (2011).
12. del Rio, L., Åberg, J., Renner, R., Dahlsten, O. C. O. & Vedral., V. The thermodynamic meaning of negative entropy. *Nature* **474,** 61–63 (2011).
13. Egloff, D., Dahlsten, O. C. O., Renner, R. & Vedral., V. A measure of majorization emerging from single-shot statistical mechanics. *New J. Phys.* **17,** 073001 (2015).
14. Faist, P., Dupuis, F., Oppenheim, J. & Renner., R. The minimal work cost of information processing. *Nat. Commun.* **6,** 7669 (2015).
15. Halpern, N. Y., Garner, A. J. P., Dahlsten, O. C. O. & Vedral., V. Introducing one-shot work into fluctuation relations. *New J. Phys.* **17,** 095003 (2015).
16. Woods, M. P., Ng, N. & Wehner., S. The Maximum Efficiency of Nano Heat Engines Depends on More than Temperature. Preprint at http://arxiv.org/abs/1506.02322 (2015).
17. Weilenmann, M., Krämer, L., Faist, P. & Renner., R. Axiomatic Relation Between Thermodynamic and Information-Theoretic Entropies. Preprint at http://arxiv.org/abs/1501.06920 (2015).
18. Bennett, C. H., Bernstein, H., Popescu, S. & Schumacher., B. Concentrating partial entanglement by local operations. *Phys. Rev. A* **53,** 2046–2052 (1996).
19. Horodecki, M., Horodecki, P. & Oppenheim., J. Reversible transformations from pure to mixed states and the unique measure of information. *Phys. Rev. A* **67,** 62104 (2003).
20. Gour, G. & Spekkens., R. W. The resource theory of quantum reference frames: manipulations and monotones. *New J. Phys.* **10,** 033023 (2008).
21. Brandão, F. G. S. L., Horodecki, M., Oppenheim, J., Renes, J. & Spekkens., R. W. Resource theory of quantum states out of thermal equilibrium. *Phys. Rev. Lett.* **111,** 250404 (2013).
22. Janzing, D., Wocjan, P., Zeier, R., Geiss, R. & Beth., T. Thermodynamic cost of reliability and low temperatures: tightening Landauer's principle and the second law. *Int. J. Theor. Phys.* **39,** 2717–2753 (2000).
23. Horodecki, M. & Oppenheim., J. Fundamental limitations for quantum and nanoscale thermodynamics. *Nat. Commun.* **4,** 2059 (2013).
24. Åberg., J. Truly work-like work extraction via a single-shot analysis. *Nat. Commun.* **4,** 1925 (2013).
25. Skrzypczyk, P., Short, A. J. & Popescu., S. Work extraction and thermodynamics for individual quantum systems. *Nat. Commun.* **5,** 4185 (2014).
26. Brandão, F. G. S. L., Horodecki, M., Ng, N., Oppenheim, J. & Wehner., S. The second laws of quantum thermodynamics. *Proc. Natl Acad. Sci. USA* **112,** 3275–3279 (2015).
27. Ng, N. H. Y., Mančinska, L., Cirstoiu, C., Eisert, J. & Wehner, S. Limits to catalysis in quantum thermodynamics. *New J. Phys.* **17,** 085004 (2015).
28. Renes., J. M. Work cost of thermal operations in quantum thermodynamics. *EPJ Plus* **129,** 153 (2014).
29. Faist, P., Oppenheim, J. & Renner., R. Gibbs-preserving maps outperform thermal operations in the quantum regime. *New J. Phys.* **17,** 043003 (2015).
30. Lostaglio, M., Jennings, D. & Rudolph., T. Description of quantum coherence in thermodynamic processes requires constraints beyond free energy. *Nat. Commun.* **6,** 6383 (2015).
31. Lostaglio, M., Korzekwa, K., Jennings, D. & Rudolph., T. Quantum coherence, time-translation symmetry, and thermodynamics. *Phys. Rev. X* **5,** 021001 (2015).
32. Ćwikliski, P., Studziski, M., Horodecki, M. & Oppenheim., J. Limitations on the evolution of quantum coherences: towards fully quantum second laws of thermodynamics. *Phys. Rev. Lett.* **115,** 210403 (2015).
33. Lostaglio, M., Müller, M. P. & Pastena., M. Stochastic independence as a resource in small-scale thermodynamics. *Phys. Rev. Lett.* **115,** 150402 (2015).
34. Korzekwa, K., Lostaglio, M., Oppenheim, J. & Jennings., D. The extraction of work from quantum coherence. *New J. Phys.* **18,** 023045 (2016).
35. Alhambra, Á. M., Oppenheim, J. & Perry, C. What is the Probability of a Thermodynamical Transition? Preprint at http://arxiv.org/abs/1504.00020 (2015).
36. Perry, C., Ćwikliski, P., Anders, J., Horodecki, M. & Oppenheim, J. A Sufficient Set of Experimentally Implementable Thermal Operations. Preprint at http://arxiv.org/abs/1511.06553.
37. Halpern, N. Y. & Renes., J. M. Beyond heat baths: generalized resource theories for small-scale thermodynamics. *Phys. Rev. E* **93,** 022126 (2016).
38. Yunger-Halpern., N. Beyond Heat Baths II: Framework for Generalized Thermodynamic Resource Theories. Preprint at http://arXiv:1409.7845 (2015).
39. Lostaglio, M., Jennings, D. & Rudolph, T. Thermodynamic resource theories, non-commutativity and maximum entropy principles. Preprint at http://arxiv.org/abs/1511.04420 (2015).
40. Halpern, N. Y., Faist, P., Oppenheim, J. & Winter, A. Microcanonical and resource-theoretic derivations of the thermal state of a quantum system with noncommuting charges. *Nat. Commun.* **7,** 12051 (2016).
41. Liu., Y.-K. *The Complexity of the Consistency and N-representability Problems for Quantum States.* PhD thesis, Univ. California, (2007).
42. Malabarba, A. S. L., Short, A. J. & Kammerlander., P. Clock-driven quantum thermal engines. *New J. Phys.* **17,** 045027 (2015).
43. Åberg., J. Catalytic coherence. *Phys. Rev. Lett.* **113,** 150402 (2014).
44. Donald., M. J. Free energy and the relative entropy. *J. Stat. Phys.* **49,** 81–87 (1987).

Acknowledgements

We thank J. Hannay for many discussions while carrying out this research. We also thank S. Massar for useful comments. Y.G., P.S. and S.P. acknowledge support from the European Research Council (ERC AdG NLST). S.P. acknowledges support from a Royal Society Wolfson Merit Award. A.J.S. acknowledges support from the Royal Society and FQXi via the SVCF. R.S. acknowledges support from the Swiss National Science Foundation (Starting grant DIAQ). Y.G., A.J.S., R.S. and P.S. acknowledge support from the COST Action MP1209 Thermodynamics in the quantum regime.

Author contributions

All authors contributed to all aspects of this work.

Additional information

Competing financial interests: The authors declare no competing financial interests.

20

Phase nucleation in curved space

Leopoldo R. Gómez[1], Nicolás A. García[1], Vincenzo Vitelli[2], José Lorenzana[3,4] & Daniel A. Vega[1]

Nucleation and growth is the dominant relaxation mechanism driving first-order phase transitions. In two-dimensional flat systems, nucleation has been applied to a wide range of problems in physics, chemistry and biology. Here we study nucleation and growth of two-dimensional phases lying on curved surfaces and show that curvature modifies both critical sizes of nuclei and paths towards the equilibrium phase. In curved space, nucleation and growth becomes inherently inhomogeneous and critical nuclei form faster on regions of positive Gaussian curvature. Substrates of varying shape display complex energy landscapes with several geometry-induced local minima, where initially propagating nuclei become stabilized and trapped by the underlying curvature.

[1] Department of Physics, Universidad Nacional del Sur—IFISUR—CONICET, 8000 Bahía Blanca, Argentina. [2] Instituut-Lorentz, Universiteit Leiden, 2300 RA Leiden, The Netherlands. [3] Institute for Complex Systems, Consiglio Nazionale delle Ricerche, I-00185 Rome, Italy. [4] Physics Department, University of Rome La Sapienza, I-00185 Rome, Italy. Correspondence and requests for materials should be addressed to L.R.G. (email: lgomez@uns.edu.ar).

In its classical picture, nucleation and growth (NG) starts with local fluctuations of an initial metastable phase leading to the formation of a small nucleus of the equilibrium phase[1–4]. Such nucleus involves local changes in the free energy of the system. On one hand, the formation of a nucleus of the less-energetic phase produces a decrease in the total free-energy scaling with its volume. On the other hand, it raises the energy by an amount that scales with the area of the interface separating the two phases.

In the case of flat two-dimensional (2D) systems, the variation in the free energy due to the formation of a nucleus of radius R takes the expression[1–4]:

$$\Delta F = 2\pi R\sigma - \pi R^2 \Delta f, \qquad (1)$$

where σ represents the surface tension and Δf the difference in the local free energies of initial and final phases driving the phase transition. The competition of surface and volume terms produces the activated dynamics of NG. Only those fluctuating nuclei whose size overpass the critical value $R_c = \sigma/\Delta f$ will propagate. All other nuclei collapse due to surface energy.

On the other hand, phases lying on curved backgrounds[5–10] are not only commonly found in nature in systems like viral capsids, pollen grains, radiolaria and others but also they can be obtained in the laboratory in the form of soft crystal or liquid crystal phases by using colloidal particles[11–14], block copolymers[15–19], liquid crystals[20–23] and other self-assembled systems. Examples of phase nucleation on curved geometries in hard-condensed matter may include, among others, graphene[24,25], epitaxial growth of helium[26] and Wigner crystals on rough surfaces[27].

In equilibrium, these phases display regular structures strongly coupled to the underlying geometry that are not only ideal model systems to study the interplay between order and curvature but also potential platforms for technological applications such as soft lithography[15,19] or defect functionalization[28]. Although experiments and theoretical calculations have contributed to unveil equilibrium configurations and energetics of topological defects in curved space, dynamical processes like crystallization and melting still remain marginally explored[29–32].

In this work, we obtain generic features of NG in 2D systems lying on arbitrary curved backgrounds. Our model is based on a Ginzburg–Landau free energy functional, which is sufficiently general, that can be applied to various physical systems and substrate's geometries. For nearly-flat surfaces and geometries of constant curvature, we obtain all the relevant information related to NG in a closed analytical form. On more complex surfaces of varying curvature, we explore the effects of the underlying geometry on NG by using simulations. We show that in general, the free-energy barrier needed to form a critical nuclei is strongly dependent on the underlying curvature, being smaller (larger) for region of positive (negative) curvature. As a consequence, in geometries of varying curvature NG is an inhomogeneous process, first starting in regions of positive curvature. Also, due to the competing effects of positive and negative curvature, geometries of varying curvature are shown to display complex relaxation in free-energy landscapes, with several barriers and local minima, where initially propagating nuclei become stabilized and trapped by the underlying curvature. Additional effects coming from system's thickness and strain contributions during crystallization are also considered and discussed in view of recent experimental work.

Results

Model. Here we present a model to study how the underlying geometry affects the laws of NG in curved space. In general, the geometry of an oriented surface (that is, separating an 'internal region' from an 'external region' of three-dimensional (3D) space) can be completely characterized by the principal curvature maps $\kappa_i(\mathbf{r}) = \pm 1/R_i$ ($i = 1, 2$), where R_1 (R_2) is the radius of the smallest (largest) circle tangent to the surface at $\mathbf{r}$, and the sign is positive (negative) if the circle is contained in the internal (external) region. Alternatively one can use the Gaussian $K(\mathbf{r}) = \kappa_1(\mathbf{r})\kappa_2(\mathbf{r})$ and mean $H(\mathbf{r}) = (\kappa_1(\mathbf{r}) + \kappa_2(\mathbf{r}))/2$ curvature[33].

The case $K = 0$ and $H \neq 0$ represents a plane bent in only one direction. Since this surface can be flattened preserving the areas (Minding's theorem[33]), nucleation can be mapped to the case of a planar surface and thus results to be trivial. Non-trivial cases have non-zero Gaussian curvature but may have zero mean curvature as in the case $\kappa_1 = -\kappa_2$, which represents a saddle. This clearly cannot be flattened preserving the areas and thus nucleation has to be revisited. In other words, the Gaussian represents an 'intrinsic' curvature and, as we shall see, results to be most important in describing physical processes on the surface. Below we show how the Gaussian curvature influences NG in curved geometries.

The mean curvature is not only related with the geometry of the surface, but also it is tied to the definition of 'internal' and 'external' regions. For example, while crest have positive mean curvature and valleys have negative mean curvature, these values can be inverted by exchanging the definition of 'internal' and 'external' regions[33]. As we will see later, mean curvature will modify NG for 2D systems with thickness.

We study the dynamics of first-order phase transitions through a scalar and real order parameter $\psi(\mathbf{r}, t)$. Points on the surface are specified by a system of curvilinear coordinates $\mathbf{r} = (x_1, x_2)$. In these coordinates, an infinitesimal arc length ds is given in Einstein notation by $ds^2 \equiv |d\mathbf{r}|^2 = g_{\alpha\beta} dx^\alpha dx^\beta$, where $g_{\alpha\beta}$ is the metric tensor. Depending on the physical system considered, the order parameter is or is not a conserved quantity[34]. For example, while conserved order parameters are required to study nucleation of binary mixtures or solid solutions, non-conserved order parameters are commonly used in studies of the liquid-to-crystal transition or in magnetic and spin-related phases.

In general, the free energy of a mixed state in a system lying on a curved surface can be expanded in terms of the order parameter:

$$\frac{F}{k_B T} = \int d^2 r \sqrt{g} \left[f(\psi) + \frac{D}{2} g^{\alpha\beta} \partial_\alpha \psi(r) \partial_\beta \psi(\mathbf{r}) \right], \qquad (2)$$

where k_B is the Boltzman's constant, T the temperature and g is the determinant of the metric tensor[33].

In a curved surface the differential of area is $d^2 r \sqrt{g} \equiv dx_1 dx_2 \sqrt{g}$ so the first term in the free energy represents the contribution from a local homogeneous situation, where $f(\psi)$ is the local free-energy areal density of a phase with an order parameter ψ. For a two-phase system this term takes the typical double-well form[36–38], with two local minima corresponding to the initial ($\psi = 0$) and final ($\psi = 1$) phases, separated by a local free-energy barrier. This can be written phenomenologically as:

$$f(\psi) = \frac{1}{4} \eta \psi^2 (\psi - 1)^2 + \frac{3\varepsilon}{2} \left(\frac{\psi^3}{3} - \frac{\psi^2}{2} \right), \qquad (3)$$

where η and ε are constants. Here the local free-energy difference driving the phase transition, typically controlled by temperature, is given by $\Delta f = f(0) - f(1)$. Note that $\Delta f \equiv \varepsilon/4$ is proportional to the degree of supercooling in the system. The second term of equation 2 is the curved-space generalization of the gradient square term $(|\nabla\psi|^2)$ that penalizes the formation of interfaces[1,2].

In this Ginzburg–Landau approach the dynamics of the phase transition can be studied through a relaxational equation of the form:

$$\tau\frac{\partial\psi}{\partial t} = -(i\Delta_{\rm LB})^{2n}\left\{\frac{\delta F}{\delta\psi}\right\}, \qquad (4)$$

where τ is a characteristic time scale (for simplicity here we take $\tau\equiv1$), $i=\sqrt{-1}$, $n\equiv0{,}1$ for non-conserved and conserved order parameters, respectively, and $\Delta_{\rm LB}=\frac{1}{\sqrt{g}}\frac{\partial}{\partial x^i}\left(g^{ij}\sqrt{g}\frac{\partial}{\partial x^j}\right))$ is the Laplace–Beltrami operator, which is the curved-space generalization of the Laplacian[29,30].

Note that a more complex evolution equation in needed for conserved order parameters. For simplicity we concentrate on the time evolution of non-conserved order parameters. However, most results related to NG on curved geometries can be applied to both[34,35].

Critical nuclei and growth and dissolution laws. We start by analysing the fate of a nucleus resulting from a spontaneous fluctuation, whose time evolution follows equation (4).

In the following we make use of geodesic polar coordinates, which are the curved-space generalization of the polar coordinates[33]. We start by locating an origin O in the centre of the nucleus (Fig. 1). Then, a point P located on the surface is associated to coordinates (r,θ). The coordinate r is defined as the geodesic distance between P and the origin O. Closed lines of constant geodesic distance r_0 are called geodesic circles, and at any point are orthogonal to the geodesics starting at O. The coordinate θ of the point P is the angle that the geodesic connecting O and P makes with a reference geodesic starting in O.

In these coordinates the metric of any surface always take the simple form $ds^2=dr^2+G(r,\theta)d\theta^2$, where the function $G(r,\theta)$ depends on the geometry of the surface through the equation $\frac{\partial^2\sqrt{G}}{\partial r^2}+K\sqrt{G}=0$, with K the Gaussian curvature of the surface, and $G(0,\theta)=0$ (ref. 33).

When using geodesic polar coordinates, the evolution equation (4) can be largely simplified to study steady-state solutions representing the growth or collapse of nuclei during NG. Here the evolution equation of a steadily propagating nucleus is decomposed (see Methods) in two equations for the interface profile (equation 5) and the propagation rate (equation 6).

$$D\frac{d^2\psi}{dX^2}+v\frac{d\psi}{dX}=\frac{\partial f(\psi)}{\partial\psi} \qquad (5)$$

$$\frac{dR}{dt}+D\kappa_g(R)=v \qquad (6)$$

where the comoving coordinate X is given by $X=r-R(t)$, $R(t)$ is the radius of the nucleus at time t, $v\equiv\sqrt{2\eta/D}3\varepsilon$ a propagation constant, and the geodesic curvature $\kappa_g(r,g)=\frac{\partial}{\partial r}\ln\sqrt{G(r,\theta)}$ is the local curvature of the geodesic circles providing information about the underlying geometry.

It is interesting to note that for any geometry, the equation for the interface profile is always the same when written in geodesic polar coordinates. Its solution is exactly the same as found by Chan for NG on a plane[36–38]: $\psi(X)=1/(1+\exp(1/2\sqrt{2\eta/D}X)$.

By contrast, equation (6) shows that the evolution of nuclei's sizes is influenced by the underlying geometry through the geodesic curvature κ_g. Note that for a planar geometry $\kappa_g(R)\equiv1/R$ gives the classical growth rate for NG on a plane. Although in the general case this is a complicated nonlinear ordinary differential equation, it can be solved in closed form for simple geometries, as shown below.

In general, a nucleus will grow only if its initial radius R_0 exceeds the critical value R_c, obtained by setting $\frac{dR}{dt}\equiv0$, that is:

$$\kappa_g(R_c)=\frac{1}{R_c^0}, \qquad (7)$$

where $R_c^0=D/v$ represents the critical size for NG on a planar substrate, under the same external conditions.

Thus, equation (7) shows that on a curved substrate, the critical radius will in general depend on both curvature (through κ_g) and supercooling (through R_c^0). A characteristic length associated to the curvature can be defined by $R_K\sim1/\sqrt{|K|}$. We can expect different regimes of NG depending on how large the ratio R_c^0/R_K is.

In the following sections we use equations (4), (6) and (7) to analyse how the geometry modifies NG on substrates with different distributions of curvature.

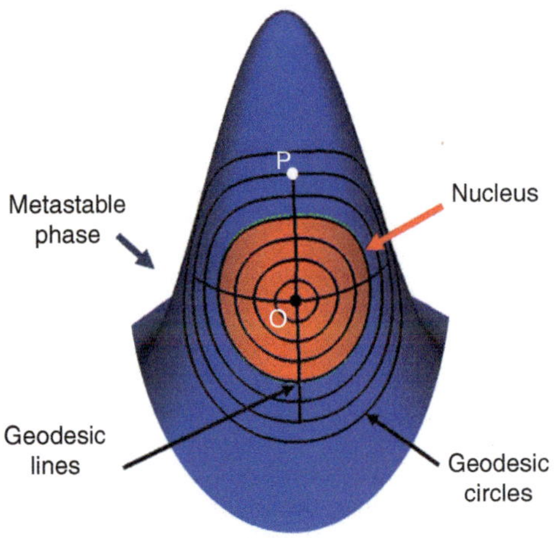

Figure 1 | Geodesic polar coordinates. In this work geodesic polar coordinates are used to describe the growth or collapse of a nucleus (red domain) in the sea of the metastable initial phase (blue). Here a point P on the surface has coordinates (r,θ), being r the geodesic distance to the origin O and θ the angle.

Nucleation on nearly-flat substrates. For cases when the critical nucleus is very small as compared with curvature $(R_c^0/R_K\ll1)$, equally valid for NG far from the coexistence line or for slightly curved geometries, the geodesic curvature can be approximated through a series expansion of the form $\kappa_g(r)=\frac{1}{r}-\frac{1}{3}K_0r+\cdots$, where K_0 is the Gaussian curvature of the substrate evaluated at the centre of the nucleus.

Using this expansion in combination with the equation for the critical nuclei (equation 7), we get a critical radius that can be written in the form:

$$R_c=R_c^0\Gamma_K+\cdots \qquad (8)$$

where Γ_K is a geometric factor given by:

$$\Gamma_K=1-\frac{1}{3}K_0R_c^{0^2} \qquad (9)$$

Thus, under same external conditions, the curvature modifies the process of NG and the critical nucleus will be smaller for regions of positive curvature, like crests and valleys $(K_0>0)$, and larger for negative curved regions like saddles $(K_0<0)$.

The energy needed to form a small nucleus of radius r on the curved substrate can be written as $\Delta F(r)=-A(r)\Delta f+P(r)\sigma$, where $A(r)\sim\pi r^2-\frac{\pi}{12}K_0r^4+\cdots$ and $P(r)\sim2\pi r-\frac{\pi}{3}K_0r^3+\cdots$ represent the area and perimeter of a geodesic circle with a first-order correction due to curvature. The height of the free-energy barrier for nucleation ΔF_c is obtained by evaluating $\Delta F(r)$ at the critical size, and can also be written as:

$$\Delta F_c = \Delta F_c^0 \Gamma_K + \cdots \quad (10)$$

where $\Delta F_c^0 = 2\pi\sigma R_c^0 - \pi\Delta f R_c^{0^2}$ is the energy barrier on a planar geometry.

This last expression shows how (locally) the height of the barrier for nucleation is modified by the underlying geometry, decreasing (increasing) for regions of positive (negative) Gaussian curvature. The reason for this is simply geometrical. Given a small nucleus, the ratio between the perimeter and area enclosed is smaller for surfaces with positive curvature (and larger for negative curvature). Thus, for positive curvature, the contribution of the bulk term of the free energy is larger than the interfacial term, producing a smaller free-energy barrier for nucleation and a smaller critical size.

Note that the last expression for the free-energy barrier for NG in a curved geometry resembles the classical result of heterogeneous nucleation and growth $\Delta F_c = \Delta F_c^0 f(\theta)$, where the free-energy barrier ΔF_c^0 is modified by a function $f(\theta)$ of the wetting contact angle θ of the nuclei with the underlying substrate[1,2]. However, in classical heterogeneous nucleation, the critical size is not modified by the wetting angle θ. This is different from nucleation on curved space, where the underlying curvature affects both free-energy barriers and critical sizes.

These results simply show that the underlying curvature can break the homogeneity of space, such that NG on a substrate of varying curvature becomes inherently inhomogeneous. In the following sections we consider NG for more general cases where the effects of the underlying geometry are much stronger $\left(R_c^0/R_K \sim 1\right)$.

Nucleation on surfaces of constant curvature. For spherical substrates, the Gaussian curvature is constant and positive $K = 1/a^2$, with a the radius of the sphere. Here the geodesic curvature is simply written as $\kappa_g(r) = 1/[a \tan(r/a)]$, and the critical size for NG (equation 7) takes the form:

$$R_c = a \arctan\left(R_c^0/a\right) \quad (11)$$

Thus, the critical size monotonously decreases when decreasing the radius of the spherical substrate, meaning that for positive curvature NG is favoured for higher curvatures. The reason for this is again geometrical: on the surface of spheres, circles have more enclosed area for a given perimeter (as compared with the plane), and the ratio surface-perimeter increases when increasing the curvature.

Given an initial nucleus of size R_0, its temporal evolution is obtained through equation (6), which can be integrated in a closed form giving the implicit relation $R = R(t)$:

$$vt = a\sin(R_c/a)\cos(R_c/a)\ln\left[\frac{\sin((R-R_c)/a)}{\sin((R_0-R_c)/a)}\right] + \cos(R_c/a)^2(R-R_0) \quad (12)$$

This is the growth ($R_0 > R_c$) or dissolution law ($R_0 < R_c$) for NG on spheres, and reduces to the classical result obtained by Chan for NG on a flat plane[36–38], when taking the Euclidean limit $R_c \ll a$.

To test the accuracy of these analytical predictions, here we developed numerical simulations of NG on spherical substrates (see Methods). Figure 2 shows the excellent comparison between the theoretical prediction equation (11) (line) and simulations (symbols) for the critical size R_c as a function of the radius a of the spherical substrate, normalized with the critical size on a plane R_c^0. This plot clearly shows that nucleation is enhanced for high curvatures (smaller spheres). Note also that as the size of the sphere increases, the critical size for NG converges to the result obtained for the flat plane (vanishing curvature).

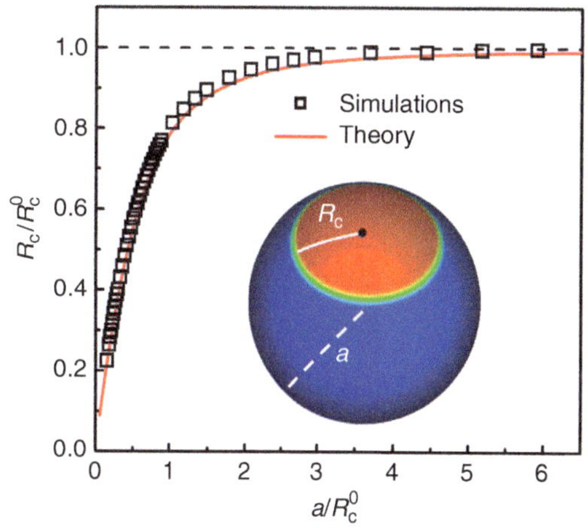

Figure 2 | Critical size for NG on spheres. The plot shows the critical (geodesic) radius R_c as a function of sphere radius a, both quantities normalized with the critical size for nucleation on a plane R_c^0, for simulations (symbols) and theoretical prediction (line). Note how the critical size decreases when increasing the curvature ($a \to 0$). As the underlying curvature decreases ($a \to \infty$) the critical size for NG converges to the value find for the flat plane R_c^0.

We can also consider NG on substrates with constant negative curvature $K = -1/a^2$, surfaces known as pseudospheres[33], which at any point display the same local saddle structure. In this case the geodesic curvature is simply written as $\kappa_g(r) = 1/[a \tanh(r/a)]$, and the expression for the critical radius takes the form:

$$R_c = a \operatorname{arctanh}\left(R_c^0/a\right) \quad (13)$$

Note that in general, all the expression for NG on pseudospheres are identical to the results obtained for spheres, when replacing the harmonic by hyperbolic functions. Contrary to spheres, here the critical size for NG monotonously increase when decreasing the radius a of the pseudosphere.

Figure 3a shows a comparison for the critical size for nucleation on spheres and pseudospheres, where the critical radius R_c is written as a function of the Gaussian curvature K of the substrate $R_c = 1/\sqrt{K} \arctan\left(R_c^0\sqrt{K}\right)$ (this expression is valid for both positive and negative curvature). This plot represents a universal law of NG on substrates of constant curvature, generalizing the well-established results for NG on a plane. Here it is clear that the critical size for nucleation is bigger for pseudospheres as compared with spheres of the same curvature. Planar substrates, indicated with a black dot, have critical sizes between spheres and pseudospheres. Note also that Fig. 3a shows that the critical size for nucleation R_c deviates from the result on a planar geometry R_c^0 for curvatures of the order $R_c^0/R_K \sim 1$, where for this substrates $R_K \equiv a$, as expected.

Figure 3b shows the time evolution of the size $R(t)$ of growing nuclei on different substrates of positive (curves in the red region), null (dashed curve) and negative (curves in the blue region) constant curvature. This plot clearly shows that the growth of nuclei is faster (slower) for substrates of positive (negative) curvature, as compared with the plane. From the plot it is also clear how the evolution law converges to the law of NG on a plane, as curvature of spheres or pseudospheres decreases (dashed line).

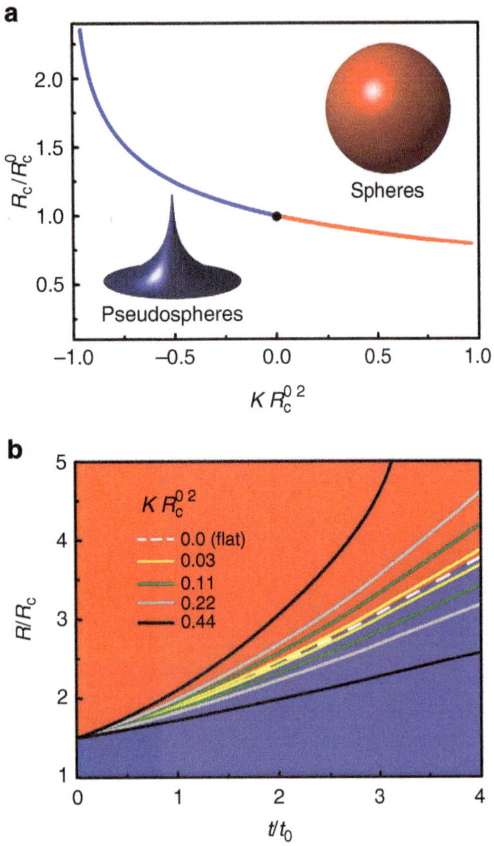

Figure 3 | NG in surfaces of constant curvature. (**a**) Critical size for nucleation R_c as a function of Gaussian curvature K for substrates of constant Gaussian curvature, normalized with the critical size on a plane R_c^0. The plot shows that the critical size for NG is larger for pseudospheres ($K<0$, blue line) as compared with spheres ($K>0$, red line) of the same curvature. Planar substrates ($K=0$), indicated with the black dot, have critical sizes between spheres and pseudospheres. (**b**) Time evolution of growing nuclei $R(t)$ on substrates of positive (curves in the red region) and negative (curves in the blue region) constant curvature. Here R and t are normalized with the critical size R_c and $t_0=R_c/v$. Lines in the red (blue) zone represents the growth of nuclei in spheres (pseudospheres). Continuous lines of the same colour represent substrates on the same curvature (in absolute value), and the dashed line represents the growth law on a planar geometry. The growth of nuclei is faster (slower) for substrates of positive (negative) curvature. Note how the evolution converges to the growth law of NG on a plane as the curvature of spheres or pseudospheres decreases.

In what follows we consider NG on more complex geometries, where the varying curvature will produce richer free-energy landscapes and intricate relaxation paths towards equilibrium.

Nucleation on surfaces of varying curvature. In general, for substrates of non-constant curvature K, it can be difficult to obtain closed analytical expressions for the polar coordinates, needed to describe NG. In addition, for complicated geometries, we can expect the geodesic curvature to be not only function of the geodesic radius r but also of the polar angle θ, $\kappa_g=\kappa_g(r,\theta)$. In such cases the breaking of the radial symmetry makes the simplification of the evolution equation (4) much harder, and in general, other parameters related to shape in addition to size would be needed to describe the nuclei's evolution.

Given that the critical radius is strongly dependent on curvature (Fig. 3a), complicated geometries may induce

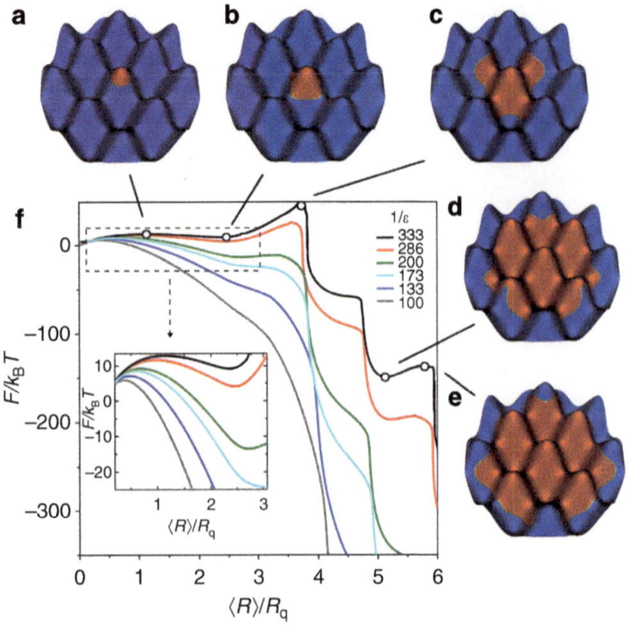

Figure 4 | NG in a sinusoidal geometry. (**a**-**e**) show snapshots of the simulation of growing nuclei seeded at a bump, which display characteristic features in the free-energy landscape for NG. (**f**) Free-energy barriers for NG on a fixed geometry and different degrees of supercoolings ($1/\epsilon$). R_q represents a measure of the roughness of the substrate. For high supercoolings ($1/\epsilon<150$) the free energy display a single barrier. The inset shows the evolution of this barrier with the different supercoolings. The presence of negative curvature induces the formation of local minima an extra barrier for smaller supercoolings ($1/\epsilon \sim 150-200$). Closer to coexistence ($1/\epsilon>250$) a new minima and barrier appear as a consequence of the negative curvature of the next nearest saddles.

complexities in the free-energy landscape for NG. Also, for substrates of varying curvature, we can expect that the opposite effects of positive and negative curvature will induce the acceleration or deceleration of interphases travelling through different regions (Fig. 3b), adding complexity to the relaxation paths towards the equilibrium phase.

To unveil the dynamics of NG on more complex geometries, we developed numerical simulations (see Methods) for geometries in which the surface height at position (x,y) is given by $Z=A\cos(2\pi x/L)\cos(2\pi y/L)$ (see the the surface in Fig. 4a–e), where the curvature is symmetrically distributed into positive and negative curved regions.

Figure 4f shows the free-energy landscape for nuclei initially seeded on the top of a bump (local positive curvature) for different degrees of supercooling $1/\epsilon$. Here $\langle R\rangle$ acts as a collective coordinate and is defined as the geodesic distance between the interface and the centre of the seed averaged along the perimeter, and R_q is the root mean squared distributions of heights, being a parameter representing the substrate's roughness. Figure 4a–e shows shape and sizes of nuclei inducing distinctive points in the free-energy landscapes for NG.

Figure 4f shows that for high supercoolings ($1/\epsilon \sim 100-150$) the system displays a single barrier for NG, as found for systems of constant curvature (plane, spheres or pseudospheres). The inset in Fig. 4f shows the expected enlargement of this barrier as the supercooling decrease.

Remarkably, for low degrees of supercoolings, multiple barriers appear with local minima in between. Starting from low supercooling, a change of slope occurs around $\langle R\rangle/R_q\sim 2.5$, which gives a local minima for $1/\epsilon>170$. This $\langle R\rangle$ corresponds to the region in which the interface of the nuclei starts to approach the first neighbouring saddles (see Fig. 4b for a nucleus of this size). Clearly, this is a geometrical effect coming from an increase of surface energy of nuclei due to the approach to a negative curved region. Thus, initially propagating nuclei become equilibrated and trapped by the negative curvature of the saddles. The escape from these regions requires the overcoming of an extra barrier associated with a much larger critical size $\langle R\rangle/R_q \sim 3.5$ (Fig. 4c).

Here we note that for small supercooling on a periodically curved substrate it is possible to obtain a microemulsion state as a metastable thermodynamic phase. These geometries are such that practically all nucleation events will take place in regions of high positive curvature and the nuclei will be unable to invade regions of negative curvature. Thus, at low supercooling, the system will remain arrested in a metastable microemulsion state that share characteristics of both phases with interesting perspective for applications. The formation of emulsion phases are typical of systems with a conserved order parameter, like the familiar patterns obtained in oil–water mixtures[39]. These are non-equilibrium states that evolve in time according to a self-similar mechanism. However, on flat geometries, emulsions cannot be obtained for non-conserved order parameters, like crystallization, because once a nucleus overcomes the critical size it freely propagates throughout the system.

As the coexistence line is approached, the critical nuclei grow in size, feeling the presence of other curved regions of the substrate. For $1/\epsilon \sim 250-350$ a new local minimum arises for nuclei of sizes around $\langle R\rangle/R_q \sim 5$. This minimum corresponds to the trapping of nuclei by the negative curvature of the next nearest saddle points (Fig. 4d), producing the formation of an extra barrier associated to a larger critical size $\langle R\rangle/R_q \sim 6$.

It is interesting to note here that this second geometry-induced barrier has a much lower height as compared with the previous. This is just a consequence of how the curvature is distributed on these sinusoidal substrates. For the second barrier, the negative curvature of the next nearest saddles is in part compensated by the positive curvature of the nearest positive-curved bumps, such that a lower activation is required to escape from these regions. Thus, close to coexistence, the energy landscape displays a 'ratchet-like' form, where the system relaxes by exploring subsequent metastable states of bigger size and smaller activation energy.

Figure 5 shows the evolution of the critical size R_c for NG as a function of the degree of supercooling $(1/\epsilon)$. As expected, the critical size decreases with supercooling. However, the formation of new minima and extra barriers as supercooling is varied, originate abrupt jumps in the critical size for nucleation.

Note that in complex geometries, it is also possible that the spatial location where the critical nuclei reside, could change as supercooling is varied, originating thus not only jumps in critical sizes but also sudden changes in the surface locations where critical nuclei develop.

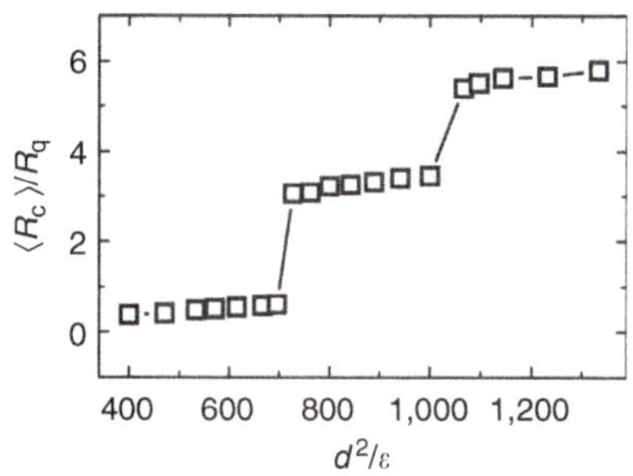

Figure 5 | Critical size for NG in a geometry of varying curvature. As a consequence the varying curvature in degree and sign, the critical size R_c in sinusoidal geometries displays discontinuous jumps when plotting as a function of the supercooling $1/\epsilon$. The x axis is normalized with $d=\sqrt{(D/(2^*\eta))}$.

Effect of the mean curvature. Up to now we have considered the effects of the Gaussian curvature K on NG. This is the only curvature that modifies the laws of NG in 2D systems of negligible thickness. Here we consider physical systems where the thickness δ of nuclei cannot be neglected, but where the NG process can still be considered as fundamentally 2D ($\delta \ll R_c$). In such cases, a system with thickness can be modelled by two parallel surfaces displaced by $\vec{\delta}=\delta\vec{n}$[40,41], where $\vec{n}$ is the normal to the surface (see an scheme of the system in Fig. 6a). As shown below, in such cases not only the Gaussian curvature K but also the mean curvature H plays an important role in NG.

The effects of finite thickness can be studied by considering the energy of formation of a nucleus:

$$\Delta F(r) = -V\Delta f + A\sigma, \tag{14}$$

where the volume V and surface area A can be calculated by considering the nucleus as composed of different layers. On each of these layers the section of the nucleus is (approximately) a

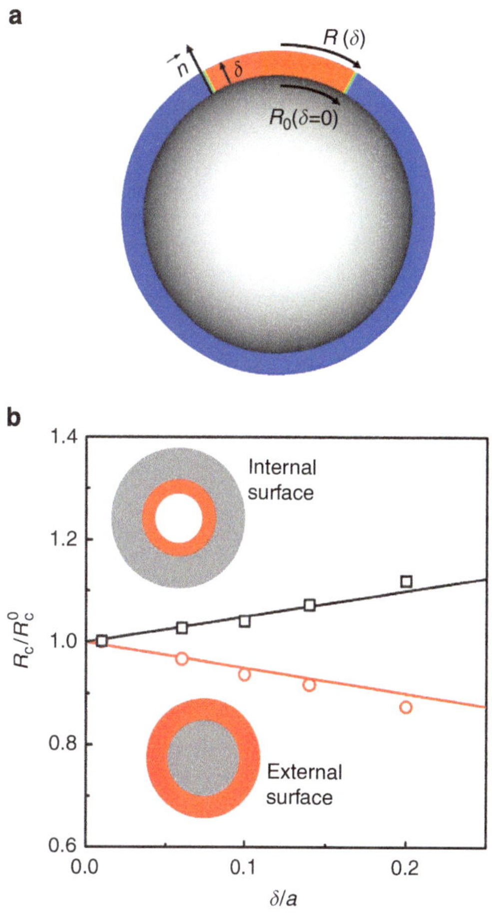

Figure 6 | NG in 2D systems with thickness. (**a**) Cartoon showing a scheme of a nucleus in a surface with normal $\vec{n}$ and thickness δ. Here while in the internal layer the nucleus has a radius $R_0 \equiv R(\delta=0)$, on the external layer the nucleus has a bigger size $R(\delta) > R_0$. (**b**) Nucleation in spherical geometries of radius a for systems with thickness δ. The plot compares simulations (symbols) and theory (lines) for critical size R_c (normalized with the critical size in a planar geometry R_c^0) as a function of thickness δ. The data on the bottom (top) corresponds to NG on external (internal) surfaces, where R_c decreases (grows) with δ.

geodesic circle, but due to the presence of curvature, the radius of these circles will be in general, a function of the thickness. Then, the volume and area can obtained by integrating the area and perimeter of the geodesic circles in the thickness of the system (see Methods).

In the limit of small thickness ($\delta \ll R$) and small curvatures ($R_c^0/R_K \ll 1$), the energy of formation of nuclei can be expanded leading to a simple expression for the critical size for NG (see Methods):

$$R_c = R_c^0 \Gamma_{KH} + \cdots, \tag{15}$$

where again Γ_{KH} represents a geometric factor, but now this factor is a function of both curvatures and thickness:

$$\Gamma_{KH} = 1 - \frac{1}{3} K_0 R_c^{0^2} - \frac{1}{2} H_0 \delta \tag{16}$$

Thus, as previously found for the Gaussian curvature K_0, in general, the presence of non-zero mean curvature H_0 and finite thickness δ will contribute to the formation of critical nuclei during NG. Note however that as the mean curvature is related to the way the surface is embedded in 3D Euclidean space, the critical size for NG will be very different if the process develops on the external or internal surface of the substrate. For example, for the case of cylindrical substrates, it will be much easier to nucleate in the external surface of the cylinder ($\Gamma_{KH} \equiv 1 - \frac{1}{2} H_0 \delta$) than on its internal face ($\Gamma_{KH} \equiv 1 + \frac{1}{2} H_0 \delta$). A similar behaviour was previously observed by using Monte Carlo simulations of NG dynamics in colloidal suspensions[42].

Figure 6b shows the good agreement between the theoretical expression equation (15) and numerical simulations for NG on spheres for nuclei of different thickness δ. Note how the critical size decrease with δ for NG on the external surface of the sphere, but increase with δ on an internal surface.

Effect of elastic frustration. In general, when a crystal is forced to reside on a curved surface, some lattice bonds necessarily compress or stretch, giving rise to an increase of the strain energy, geometrically frustrating the formation of crystal bonds[5–9]. In some cases, such elastic distortions can be released by introducing topological defects in the lattice on particular regions of the substrate.

In a recent experiment, the effects of this elastic frustration on NG has been explored by following the dynamics of 2D growing colloidal crystals on spherical substrates[32]. Here the colloidal particles were constrained to reside on the curved interface of water droplets dispersed in oil. This study revealed the presence of an elastic instability as a main consequence of the high geometric frustration in the crystallites. Due to the rapid increase of the strain energy with size, the nuclei tend to grow in ramified fractal-like form on spheres rather than in the typical compact structure observed in planar geometries.

The effects of curvature and frustration on crystallization has also been recently studied in experiments of 3D heterogeneous nucleation on the surface of curved substrates[43,44]. Here foreign particles are located inside colloidal suspensions, acting as seeds for crystallization, giving rise to a heterogeneous nucleation process, which starts on the surface of a curved hard wall (the surface of the foreign particles)[45]. Such studies have also shown how curvature and geometric frustration can modify NG. It was found that although critical nuclei still form on the surface of the substrates, they tend to detach when growing, mainly because the curvature produce strains in the crystal lattices (some crystal's bonds would need to stretch or compress in order to adjust to the surface geometry), which largely increases the elastic free energy of growing nuclei. Although such NG processes are mainly 3D, and then in principle different to the 2D processes considered here, some similarities are found. For example, due to the effects of mean curvature H_0 and thickness δ the NG is found to be different if the process develops on the outer or inner surface of the substrates.

The effects of elastic frustration can be incorporated in our model by adding the corresponding elastic free energy to the functional in equation 2. In general, the elastic energy of a curved crystalline monolayer on a curved substrate can be written as:

$$F_{\text{strain}} = \frac{1}{2Y} \int d^2 r \sqrt{g} (\Delta_{LB} \chi)^2, \tag{17}$$

where Y is the Young modulus of the crystal and χ is the Airy stress function, which is dependent of both underlying curvature K and presence and location of topological defects in the crystal.

In the case of spherical substrates of radius a, and for small nuclei of size R containing no defects, the elastic free energy results in $F_{\text{strain}} \approx \frac{Y\pi}{384} \frac{R^6}{a^4}$ (ref. 32). Figure 7 shows free-energy barriers for NG on spherical substrates, for defect-free crystal nuclei of different Young modulus. Here in addition to the thermal barrier (red dot), the fast increase in the strain of the nuclei with R produces an absolute minimum in the energy landscape (black dots), such that a growing nuclei would reach an equilibrium size, being unable to propagate throughout the system. One way to escape from this energy minimum is by distorting the nuclei shape in less-compact structures, as observed in the experiments of Meng *et al.*[32]

Note that for cases where the lattice structure is very soft (where topological defects have low core energies), it is conceivable that the system would rapidly relax the strain energy through the formation of topological defects. In such cases, we can expect that the strain-free energy contribution do not play a major role in the nucleation dynamics and therefore we can expect similar results for NG on curved geometries as in previous sections.

Discussion

In this work we have studied NG on curved geometries in systems where the imbalance of volume and surface free energies of nuclei directs the phase transition. It is precisely the change of this imbalance with supercooling and local curvature that makes NG a rich and complex phenomena in curved space.

In general, the critical size and propagation rate of nuclei are both strongly dependent on the underlying geometry. For large

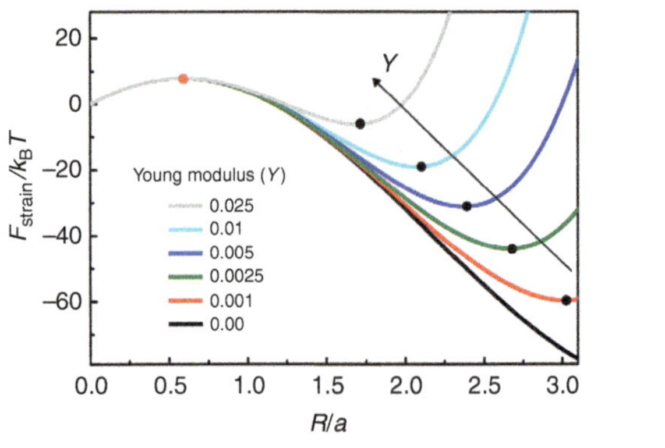

Figure 7 | NG in a spherical substrate under elastic frustration. The plot shows the energy landscape F for NG for different Young modulus Y, where R is the nuclei's geodesic radius and a the sphere radius. For crystal patches having no defects, the free energy displays an absolute minima related to the formation of nuclei reaching equilibrium sizes (black dots) in addition to the conventional thermal barrier (red dot).

curvatures, the critical sizes can be severely modified (Fig. 3a). For example, for cases where the radius of curvature of the substrate is of the order of the critical size for nucleation on the plane $\left(R_c^0/R_K \sim 1\right)$, the critical size for nucleation is reduced in 35% for positive underlying curvature. Note, however, that for the same thermodynamic conditions the critical size becomes about 250% larger for negative underlying curvature.

Because of this effect, substrates of varying curvature impose complex relaxation paths for NG, with several geometry-induced local minima, with interesting applications for phases deposited on substrates of controlled roughness. In general, the characteristic time τ_m that a system spends in a given metastable state roughly scales with its activation energy: $\tau_m \sim \exp(\Delta F/k_B T)$ (refs 1,2). Contrary to NG of two-phase systems in planar geometries, here the different local free-energy minima originated by the varying curvature would give rise to a full spectrum of relaxation times. Such spectrum, having the main information of how the relaxation process is developed, is a fingerprint of the way the curvature is distributed throughout the substrate.

Due to the generality and simplicity of our model, we expect that the results obtained here could be applied to a variety of systems in condensed matter, and other fields concerned with the development of a new phase on a curved surface. Experiments of colloidal crystallization on curved surfaces could be carried out on capillary bridges of negative Gaussian curvature and null mean curvature, used in the studies of 2D curved crystalline structures and defects in refs 13,14. The combined effects of Gaussian K and mean H curvatures and thickness δ on NG could also be experimentally addressed by confining block copolymers, liquid crystals or similar systems into different geometries[16], like corrugated substrates[15,19], spherical shells[20] or toroidal droplets[23]. In all these systems we expect that the interplay between geometry and phase nucleation will lead novel pattern formation phenomena that can be controlled to design useful supramolecular and soft materials.

Methods

Equations describing the steady propagation of nuclei. After the functional derivation in equation (4), the evolution equation for a non-conserved order parameter takes the form:

$$\frac{\partial \psi}{\partial t} = D\Delta_{LB}\psi - \frac{\partial f(\psi)}{\partial \psi} \tag{18}$$

Upon using geodesic polar coordinates, the Laplace–Beltrami operator can be written as[46]:

$$\Delta_{LB}\psi = \frac{\partial^2 \psi}{\partial r^2} + \kappa_g(r,\theta)\frac{\partial \psi}{\partial r} + \frac{1}{r^2}\frac{\partial^2 \psi}{\partial \theta^2}\exp\left[2\int_0^r \kappa_g(u)\mathrm{d}u\right], \tag{19}$$

where $\kappa_g(r,\theta) = \frac{\partial}{\partial r}\ln\sqrt{G(r,\theta)}$ is the local curvature of the geodesic circles. For general geometries the Laplace–Beltrami operator takes a complex form, being dependent on both coordinates (r,θ). However, for surfaces with radial symmetry respect to the origin O, the geodesic curvature is only a function of the geodesic distance $\kappa_g \equiv \kappa_g(r)$. In such cases we expect the nuclei to have radial symmetry, such that the Laplace–Beltrami operator takes the much simplified form $\Delta_{LB}\psi \equiv \frac{\partial^2 \psi}{\partial r^2} + \kappa_g(r,\theta)\frac{\partial \psi}{\partial r}$.

We can now describe steady-state solutions representing the growth or collapse of nuclei during NG. Here the motion of a steadily propagating interface can be described through a single combined coordinate $X = r - R(t)$, where $R(t)$ is the radius of the nuclei at time t. We assume that the interface is narrow respect to the nuclei radius R. Then we can take $\kappa_g(r) \equiv \kappa_g(R)$ and rewrite the evolution equation in the comoving coordinates X, as the system of coupled equations (6) and (7) in the Results section.

Simulations. The simulations of NG in curved space performed in this work were developed by seeding an initial (geodesic) circular nucleus of a prescribed initial radius R_0 on a particular region of the substrate, and studying the temporal evolution through the numerical resolution of equation (18).

The initial circular nuclei were obtained, for any underlying geometry, through a fast marching algorithm[47]. Equation (18) can be accurately solved through a finite difference scheme, forward in time and centred in space and periodic boundary conditions[29,30]. In the case of systems with finite thickness, the simulations were performed by numerically solving equation (18) through finite elements, with periodic and null flux boundary conditions.

For different geometries and degrees of supercoolings ε, the critical sizes R_c were obtained by finding the smallest nuclei, which is able to grow in time. The free-energy landscapes for NG were obtained by inserting the order parameter distribution $\psi(\mathbf{r})$, of growing or collapsing nuclei of size $R(t)$, in the free-energy functional equation (2).

Critical size in systems with thickness. Given a nucleus in a curved system with thickness, some general conclusions can be drawn in the limit of small width δ and small critical size R_c.

To get the volume V and area A associated to the formation during nucleation (equation 14), the nucleus is considered as having different layers, where in each layer the nucleus has a circular shape. Thus, the volume and area can be obtained by integrating the area and perimeter of the geodesic circles in the thickness of the system. Here the perimeter P_δ and area A_δ of these circles can be written in the approximate form[33]:

$$P_\delta \approx 2\pi R(\delta) - \frac{\pi}{3}K(\delta)R(\delta)^3 \tag{20}$$

$$A_\delta = \pi R(\delta)^2 - \frac{\pi}{12}K(\delta)R(\delta)^4 \tag{21}$$

where $R(\delta)$ and $K(\delta)$ are the radius and Gaussian curvature in the δ-layer, given by:

$$R(\delta) \sim (1 + H_0\delta)R_0 \tag{22}$$

$$K(\delta) = \frac{K_0}{1 + 2H_0\delta + K_0\delta^2} \sim \frac{K_0}{1 + 2H_0\delta} \tag{23}$$

Here R_0, K_0 and H_0 represent the size of nuclei and the Gaussian and mean curvatures evaluated on the surface of the substrate (for $\delta \equiv 0$). The above expression for the Gaussian curvature is a well-known result that relates the Gaussian curvature of a surface K with the Gaussian K_0 and mean H_0 curvatures of a parallel surface displaced by $\vec{\delta} = \delta\vec{n}$, with $\vec{n}$ the normal to the surface[40,41].

In this approximation, the area and volume of the nucleus take the simple forms:

$$A \approx \left(2\pi R_0 - \frac{\pi}{3}K_0R_0^3\right)\left(1 + \frac{1}{2}H_0\delta\right)\delta \tag{24}$$

$$V \approx \left(\pi R_0^2 - \frac{\pi}{12}K_0R_0^4\right)(1 + H_0\delta)\delta \tag{25}$$

The replacement of these expressions in the energy of formation of a nucleus with thickness (equation 14), straightforwardly leads to the expression of the critical size for NG (equation 15).

References

1. Kashchiev, D. *Nucleation: Basic Theory with Applications* (Butterworth-Heinemann, 2000).
2. Kelton, K. F. & Greer, A. L. *Nucleation in Condensed Matter: Applications in Materials and Biology* (Pergamon, 2010).
3. Cavagna, A. Supercooled liquids for pedestrians. *Phys. Rep.* **476,** 51–124 (2009).
4. Prestipino, S., Laio, A. & Tosatti, E. Systematic improvement of classical nucleation theory. *Phys. Rev. Lett.* **108,** 225701 (2012).
5. Nelson, D. R. *Defects and Geometry in Condensed Matter Physics* (Cambridge University Press, 2002).
6. Vitelli, V. & Turner, A. M. Anomalous coupling between topological defects and curvature. *Phys. Rev. Lett.* **93,** 215301 (2004).
7. Turner, A. M., Vitelli, V. & Nelson, D. R. Vortices on curved surfaces. *Rev. Mod. Phys.* **82,** 1301 (2010).
8. Vitelli, V. & Nelson, D. R. Defect generation and deconfinement on corrugated topographies. *Phys. Rev. E* **70,** 051105 (2004).
9. Vitelli, V., Lucks, J. B. & Nelson, D. R. Crystallography on curved surfaces. *Proc. Natl Acad. Sci. USA* **103,** 12323–12328 (2006).
10. Bowick, M. J. & Giomi, L. Two-dimensional matter: order, curvature and defects. *Adv. Phys.* **58,** 449–563 (2009).
11. Bausch, A. R. *et al.* Grain boundary scars and spherical crystallography. *Science* **299,** 1716–1718 (2003).
12. Lipowsky, P., Bowick, M. J., Meinke, J. H., Nelson, D. R. & Bausch, A. R. Direct visualization of dislocation dynamics in grain boundary scars. *Nat. Mater.* **4,** 407–411 (2005).
13. Irvine, W. T. M., Vitelli, V. & Chaikin, P. M. Pleats in crystals on curved surfaces. *Nature* **468,** 947–951 (2010).
14. Irvine, W. T. M., Bowick, M. & Chaikin, P. M. Fractionalization of interstitials in curved colloidal crystals. *Nat. Mater.* **11,** 948–951 (2012).
15. Hexemer, A. *Order and Disorder of Block Copolymers and Particles on Surfaces with Topology* (PhD thesis Univ. California Santa Barbara, 2006).
16. Yu, B. *et al.* Confinement-induced novel morphologies of block copolymers. *Phys. Rev. Lett.* **96,** 138306 (2006).

17. Chantawansri, T. L. *et al.* Self-consistent field theory simulations of block copolymer assembly on a sphere. *Phys. Rev. E* **75,** 031802 (2007).
18. Hexemer, A., Vitelli, V., Kramer, E. J. & Fredrickson, G. H. A Monte Carlo study of order and defects on weakly curved surfaces. *Phys. Rev. E* **76,** 051604 (2007).
19. Vega, D. A. *et al.* Coupling between mean curvature and textures in block copolymer thin films deposited on curved substrates. *Soft Matter* **9,** 9385–9391 (2013).
20. Lopez-Leon, T., Koning, V., Devaiah, K. B. S., Vitelli, V. & Fernandez-Nieves, A. Frustrated nematic order in spherical geometries. *Nat. Phys.* **7,** 391–394 (2011).
21. Napoli, G. & Vergori, L. Extrinsic curvature effects on nematic shells. *Phys. Rev. Lett.* **108,** 207803 (2012).
22. Kamien, R. D., Nelson, D. R., Santangelo, C. D. & Vitelli, V. Extrinsic curvature, geometric optics, and lamellar order on curved substrates. *Phys. Rev. E* **80,** 051703 (2009).
23. Pairam, E. *et al.* Stable nematic droplets with handles. *Proc. Natl Acad. Sci. USA* **110,** 9295–9300 (2013).
24. Lyuksyutov, I., Naumovets, A. G. & Pokrovsky, V. *Two-Dimensional Crystals* (Academic, 1992).
25. Xue, J. *et al.* Scanning tunnelling microscopy and spectroscopy of ultra-flat graphene on hexagonal boron nitride. *Nat. Mat.* **10,** 282–285 (2011).
26. Balibar, S., Alles, H. & Parshin, A. Y. The surface of helium crystals. *Rev. Mod. Phys.* **77,** 317–370 (2005).
27. Jiang, H.-W. & Dahm, A. J. Conduction properties of a new two-dimensional sliding charge-density wave. *Phys. Rev. Lett.* **62,** 1396–1399 (1989).
28. DeVries, G. A. *et al.* Divalent metal nanoparticles. *Science* **315,** 358–361 (2007).
29. Gómez, L. R. & Vega, D. A. Relaxational dynamics of smectic phases on a curved substrate. *Phys. Rev. E* **79,** 031701 (2009).
30. García, N. A., Register, R. A., Vega, D. A. & Gómez, L. R. Crystallization dynamics on curved surfaces. *Phys. Rev. E* **88,** 012306 (2013).
31. García, N. A., Pezzutti, A. D., Register, R. A., Vega, D. A. & Gómez, L. R. Defect formation and coarsening in hexagonal 2D curved crystals. *Soft Matter* **11,** 898–907 (2015).
32. Meng, G., Paulose, J., Nelson, D. R. & Manoharan, V. N. Elastic instability of a crystal growing on a curved surface. *Science* **343,** 634–637 (2014).
33. Struik, D. J. *Lectures on Classical Differential Geometry* (Dover, 1961).
34. Chaikin, P. M. & Lubensky, T. C. *Principles of Condensed Matter Physics* (Cambridge University Press, 1995).
35. Cahn, J. W. & Hilliard, J. E. Free energy of a nonuniform system. III. Nucleation in a two-component incompressible fluid. *J. Chem. Phys.* **31,** 688–699 (1959).
36. Chan, S. Steady-state kinetics of diffusionless first order phase transformations. *J. Chem. Phys.* **67,** 5755–5762 (1977).
37. Jou, H. & Lusk, M. T. Comparison of Johnson-Mehl-Avrami-Kologoromov kinetics with a phase-field model for microstructural evolution driven by substructure energy. *Phys. Rev. B* **55,** 8114–8121 (1997).
38. Iwamatsu, M. Minimum free-energy path of homogenous nucleation from the phase-field equation. *J. Chem. Phys.* **130,** 244507 (2009).
39. Bray, A. J. Theory of phase-ordering kinetics. *Adv. Phys.* **43,** 357–459 (1994).
40. Safran, S. A. Curvature elasticity of thin films. *Adv. Phys.* **48,** 395–448 (1999).
41. Hyde, S. T. *et al. The Language of Shape: The Role of Curvature in Condensed Matter: Physics, Chemistry and Biology* (Elsevier, 1997).
42. Cacciuto, A., Auer, S. & Frenkel, D. Onset of heterogeneous crystal nucleation in colloidal suspensions. *Nature* **428,** 404–406 (2004).
43. Ziese, F., Maret, G. & Gasser, U. Heterogeneous nucleation and crystal growth on curved surfaces observed by real-space imaging. *J. Phys. Condens. Matter* **25,** 375105 (2013).
44. Sandomirski, K. *et al.* Heterogeneous crystallization of hard and soft spheres near flat and curved walls. *Eur. Phys. J. Special Topics* **223,** 439–454 (2014).
45. de Villeneuve, V. W. A. *et al.* Colloidal hard-sphere crystal growth frustrated by large spherical impurities. *Science* **309,** 1231–1233 (2005).
46. Faraudo, J. Diffusion equation on curved surfaces. I. Theory and application to biological membranes. *J. Chem. Phys.* **116,** 5831–5841 (2002).
47. Sethian, J. A. & Smereka, P. Level set methods for fluid interfaces. *Annu. Rev. Fluid Mech.* **35,** 341–372 (2003).

Acknowledgements

This work was supported by the National Research Council of Argentina, CONICET, ANPCyT and Universidad Nacional del Sur.

Author contributions

L.R.G. and D.A.V. conceived the work; L.R.G. developed the model and carried out analytical calculations; NAG and LRG performed the simulations; all authors discussed and analysed the results; L.R.G., V.V., J.L. and D.A.V. contributed to the manuscript preparation.

Additional information

Competing financial interests: The authors declare no competing financial interests.

21

Fast escape of a quantum walker from an integrated photonic maze

Filippo Caruso[1], Andrea Crespi[2,3], Anna Gabriella Ciriolo[3], Fabio Sciarrino[4] & Roberto Osellame[2,3]

Escaping from a complex maze, by exploring different paths with several decision-making branches in order to reach the exit, has always been a very challenging and fascinating task. Wave field and quantum objects may explore a complex structure in parallel by interference effects, but without necessarily leading to more efficient transport. Here, inspired by recent observations in biological energy transport phenomena, we demonstrate how a quantum walker can efficiently reach the output of a maze by partially suppressing the presence of interference. In particular, we show theoretically an unprecedented improvement in transport efficiency for increasing maze size with respect to purely quantum and classical approaches. In addition, we investigate experimentally these hybrid transport phenomena, by mapping the maze problem in an integrated waveguide array, probed by coherent light, hence successfully testing our theoretical results. These achievements may lead towards future bio-inspired photonics technologies for more efficient transport and computation.

[1] LENS, QSTAR & Dipartimento di Fisica e Astronomia, Università di Firenze, via Nello Carrara 1, I-50019 Sesto Fiorentino, Italy. [2] Istituto di Fotonica e Nanotecnologie, Consiglio Nazionale delle Ricerche (IFN-CNR), Piazza Leonardo da Vinci 32, I-20133 Milano, Italy. [3] Dipartimento di Fisica, Politecnico di Milano, Piazza Leonardo da Vinci 32, I-20133 Milano, Italy. [4] Dipartimento di Fisica, Sapienza Università di Roma, Piazzale Aldo Moro 5, I-00185 Roma, Italy. Correspondence and requests for materials should be addressed to F.C. (email: filippo.caruso@lens.unifi.it) or to R.O. (email: roberto.osellame@polimi.it).

Transport problems are very popular in several fields of science, as biology, chemistry, sociology, information science, physics and even in everyday life. One of the most challenging transport problems is represented by efficiently traversing a maze, that is, finding the exit in the shortest possible time of a topologically complex network of interconnected sites (Fig. 1). The efficiency in reaching the exit of a maze dramatically decreases with the number of sites in the structure, rapidly making this problem intractable[1].

The problem of solving mazes has fascinated mankind since the ancient times. One famous maze is the Cretan one, designed by the architect Daedalus, build to hold the mythological creature Minotaur that was eventually killed by the hero Theseus. To find the Minotaur he used the most typical maze-solving strategy: exploring several possible alternatives, while marking the visited paths (by a ball of thread). Around 60 years ago, Shannon realized the first ever experiment on maze-solving that was based on physical means, in particular an electromagnetic mouse Theseus[2]. Nowadays, the availability of new physical, chemical and biological systems has opened up the way for traversing a maze with a parallel exploration of all possible transport channels at the same time. For instance, in ref. 3 a maze is experimentally solved by filling it with a Belousov–Zhabotinsky reaction mixture and then exploiting the superposition effect of travelling chemical wavefronts. More recently, it was shown that this parallel addressing can be indirectly obtained by the chemo-attractant waves emitted by the oat flake placed at the destination site, while a plasmodium slime walks directly to the exit[4]. This demonstrates the crucial role of interference to find the maze's exit in a more efficient way.

In the framework of quantum mechanics, even a single particle, represented by a wavefunction, shows interference effects. Exploiting this property, a quantum walker is able to propagate in the fastest way inside perfectly ordered lattices[5,6]; however, localization phenomena may occur when disorder is present[7–10]. Quantum walks find applications to energy transport[11] and quantum information[12–15] with polynomial as well as exponential speedup[16], for example, Grover search algorithm[17], universal models for quantum computation[18], state transfer in spin and harmonic networks[19–21] and recent proposals on web page ranking[22]. Recently, the maze problem has been converted into a quantum search problem to get a quadratic speedup[23]. Interestingly enough, the interplay of interference and noise effects can further enhance quantum transport over complex networks, as recently observed for energy transport phenomena in light-harvesting proteins[24–28] and proposed for noise-assisted quantum communication[29]. In particular, it is extremely difficult to study quantum transport phenomena in biological systems, as well as to change in a controlled way the problem parameters to fully understand their role. For this reason, it is very important to develop a perfectly controlled artificial platform that can be used to simulate, understand and engineer these phenomena.

In the last years, several technological platforms have been employed to investigate quantum transport phenomena, such as NMR[30,31], trapped ions[32,33], neutral atoms[34] and several photonic schemes as bulk optics[35,36], fibre loop configurations[37,38] and miniaturized integrated waveguide circuits[39–43]. Among these, a very interesting experimental platform is represented by three-dimensional waveguide arrays, fabricated by femtosecond laser micromachining[41,44–46]. Femtosecond laser waveguide writing[47] enables to fabricate high-quality optical waveguides, directly buried in the bulk of a transparent substrate. Ultrashort laser pulses are focused at the desired depth in the substrate and nonlinear absorption processes induce localized and permanent refractive index increase; translation of the sample at uniform speed allows to draw guiding paths in the substrate with unique three-dimensional design freedom. Many diverse quantum phenomena[48,49] can be observed and simulated by means of such structures: in particular, a powerful analogy can be exploited between the Schrödinger equation, describing the evolution of a wavepacket in a two-dimension potential, and the equations describing the paraxial evolution of light into a dielectric structure, such as a waveguide array. In particular, an array of coupled waveguides is equivalent to a two-dimensional array of quantum wells. The temporal evolution of a single quantum particle, placed initially in a certain well, can be mapped to the spatial evolution along the propagation direction of a single photon, injected initially in a certain waveguide.

Here, we investigate the role of a partial suppression of interference effects in the transport dynamics through maze-like graphs. In particular, we theoretically demonstrate that an optimal mixing of classical and quantum dynamics leads to a remarkably efficient transmission of energy/information from the input to the exit door of a generic maze. In addition, we show that it is possible to reproduce experimentally these dynamics in a photonic simulator, unfolding the maze onto a femtosecond-laser-written three-dimensional waveguide array, where noise is implemented by modulating the propagation constants of the waveguides during the writing process. The results provide a clear demonstration that a controlled amount of decoherence in the walker can produce an enhanced transport efficiency in escaping the maze and that these phenomena can be investigated in an experimentally accessible platform and not only in abstract models.

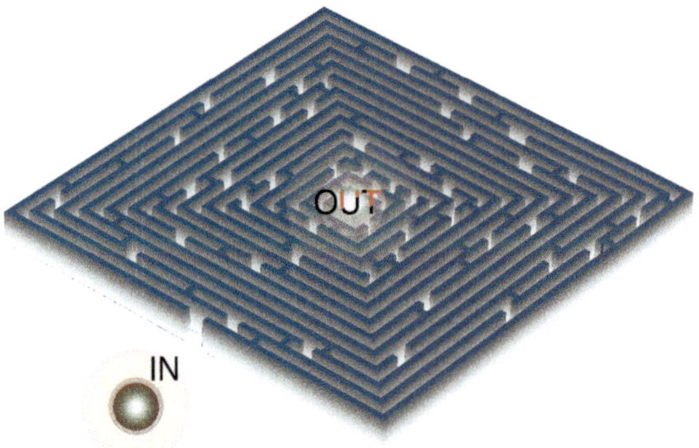

Figure 1 | Maze problem. Pictorial view of a maze with single input (IN) and output (OUT) ports. An ideal walker has to travel from IN to OUT in the shortest possible time.

Results

Theory. The maze structure is created here by the so-called random Depth-First Search algorithm applied on a square lattice of N nodes[50] (see the Methods section: Maze construction, together with Supplementary Fig. 1). The transport model is represented by a walker entering the maze in some initial (IN) site or input door and moving over the structure until reaching a final (OUT) site or exit door (maze's solution).

Following the framework of quantum stochastic walks[28,51], the density matrix ρ describing the state of the system evolves according to the Lindblad master equation:

$$\frac{d\rho}{dt} = -(1-p)i[H,\rho] + p\sum_{ij}\left(L_{ij}\rho L_{ij}^{\dagger} - \frac{1}{2}\left\{L_{ij}^{\dagger}L_{ij},\rho\right\}\right). \quad (1)$$

A purely unitary evolution, given by the hermitian Hamiltonian H, which implements the quantum walk dynamics, is mixed with an incoherent evolution describing a classical random walk, given by the operators $L_{i,j}$. The balance between the two parts of the Lindblad superoperator is given by the value of the parameter p. In particular, for $p=0$ a fully coherent (pure interference) dynamics is observed, whereas $p=1$ corresponds to the case of classical random walk, that is, classical random hopping with no interference; for intermediate values, a mixing of the two types of behaviour is obtained. An irreversible transfer process from the exit site to an external sink is added and the walker's probability in getting the exit at time t is quantified by transfer efficiency function to the sink $\mathcal{E}(p,t)$, whose values are in the range (0, 1). Further technical details are given in Supplementary Note 1.

As shown in the left side of Fig. 2, the transfer efficiency for a maze of about one thousand sites, for a given time (linearly increasing with the maze size), is more than five order of magnitudes larger when one partially suppresses interference effects ($p\simeq 0.1$, that is, $\sim 10\%$ of mixing), with respect to the limiting cases of purely coherent and fully classical dynamics. Such transport enhancement is based on an intricate interplay between coherence and noise and shows peculiar features that makes it a fascinating field to investigate. In fact, an analogous optimal mixing has been very recently demonstrated over a large family of complex networks for $p\simeq 0.1$ (ref. 28) and experimentally observed in ref. 52 (for the robustness of this mixing value see Supplementary Note 1 and Supplementary Fig. 2). In addition, noise-enhanced transport dynamics was observed even for totally regular and ordered graphs[53] (where an intuitive picture of this optimality can be given in terms of a 'momentum rejuvenation'), thus evidencing how this phenomenon cannot be explained as just a cross-over from disorder-induced coherent localization towards classic diffusive regime. As in ref. 53, we can analyse the transport inefficiency in terms of the average dwelling time in the network, which we define as $\bar{t}=\int_0^\infty P(t)dt$, with $P(t)$ being the population remaining on the network, that is the probability that at time t the energy quantum has failed to exit the network—see the right side of Fig. 2. This further supports the behaviour observed above for the transfer efficiency at long time scales (Fig. 2 left), showing that our particular choice of the time t for the plotted $\mathcal{E}(p,t)$ does not affect our conclusions.

One can consider how in the noiseless case the particle undergoes discrete diffraction in the structure: the strong interference effects given by full coherence generate bright and dark zones, even if the wavefunction does not strictly localize, and this may limit the transfer efficiency between two distant sites of the graph. Adding an optimal quantity of noise may help in suppressing the fine-grained interference pattern while keeping the wavefunction spread almost as in the ballistic case, without reaching the diffusive limit where the transport dynamics is much slower. Although the Lindblad model introduces decoherence only through direct classical transitions (T_1-like processes), a similar behaviour would be obtained by considering a pure dephasing process (T_2-like)—see (refs 27–29,53).

Experimental realization. Taking advantage of the unique three-dimensional fabrication capabilities of femtosecond laser waveguide writing, we implement a simulator of quantum stochastic walks by engineering an integrated photonic device probed by laser light. In fact, the probability distribution at the output for a single photon is perfectly reproduced by the intensity distribution of coherent light in the waveguide array. The maze structure is mapped onto a three-dimensional waveguide array, in which each waveguide represents a site of the maze. In particular, our experimental study is focused on the maze configuration shown in Fig. 3a, composed of 18 sites, taken as a significant example for observing the dynamics predicted by our theoretical model.

A first problem that has to be addressed is how to map in a waveguide system the topology of the links between the sites of our maze. Whereas in an arbitrary maze structure transfer between adjacent sites can be inhibited by walls, in waveguide arrays the coupling between two waveguides is solely determined by their relative distance. Thus, the geometry of the array needs to be engineered to keep far enough from each other waveguides that must not couple. This might not be possible if the maze graph is too complex. In our case, however, it was possible to unfold the maze graph in Fig. 3a, by considering chains with side tails, onto the partially linear and more feasible structure in Fig. 3b. Note that this unfolded geometry is not unique, other configurations being conceivable in principle with the same distances between equally coupled sites.

Another experimental issue is the realization of the exit door (that is, OUT site). In the theoretical model, this site should behave like a sink that absorbs energy irreversibly. In our

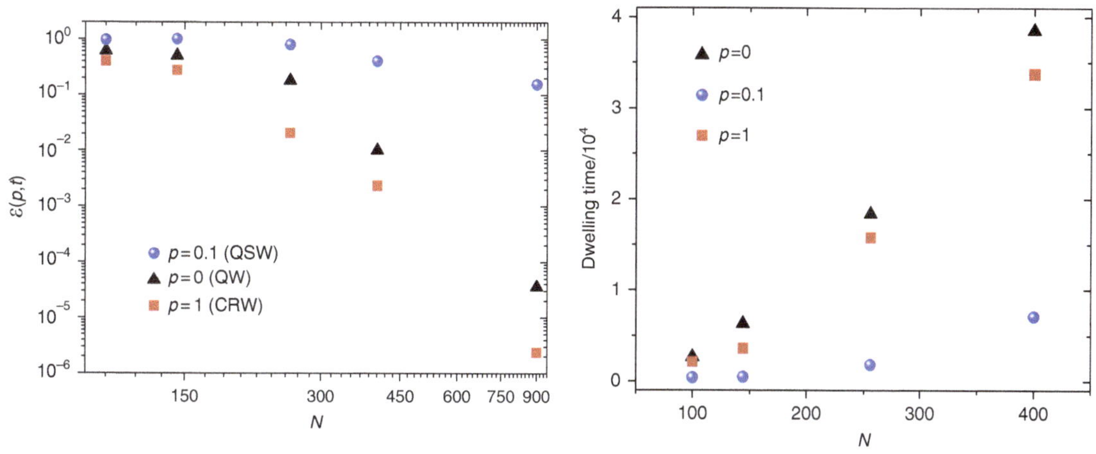

Figure 2 | Transport efficiency for different sizes. Left: Transfer efficiency $\mathcal{E}(p,t)$ as a function of the size N of the maze, for a time scale t linearly increasing with N, that is, $t=10\,N$. For a maze with $N=900$ nodes, the optimal mixing $p\simeq 0.1$ provides a transfer efficiency that is about five orders of magnitude larger than the perfectly coherent (quantum, that is, $p=0$) and fully noisy (classical, that is, $p=1$) regimes. The trend of the curves with the maze complexity N, for the different values of p, indicates that even higher speedup can be achieved for increasingly larger mazes. Right: Dwelling time $\bar{t}$ as a function of the size N of the maze, for $p=0$, 0.1, 1.

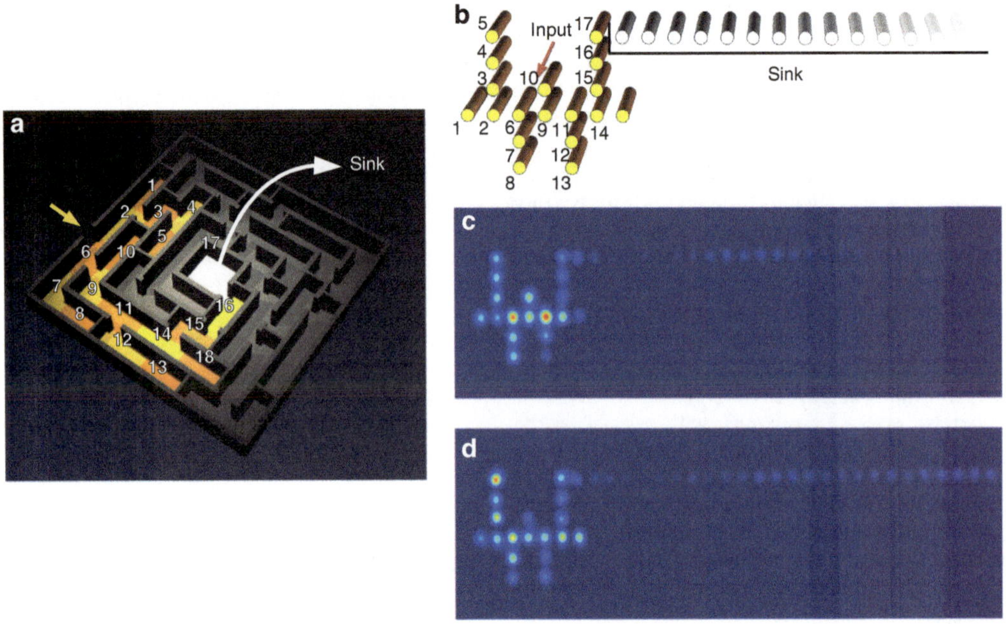

Figure 3 | Implementing the maze. (**a**) Maze structure that is experimentally implemented; (**b**) unfolding of the maze into an almost linear graph, where each node is represented by a wave guide; (**c,d**) snapshots of the light diffusion for uniform (**c**) and noisy structure (**d**). The latter pictures correspond both to a propagation length of 60 mm. The noisy configuration is *noise 3* in Fig. 4.

photonic implementation, the sink is implemented by a long chain of waveguides (62 waveguides), which approximates well a one-way energy transfer process, with negligible probability for the light to be coupled back to the system.

Structures composed of uniform waveguides correspond to the purely coherent case (QW). Fully coherent transport dynamics in such maze can be studied straightforwardly by fabricating arrays with different lengths and characterizing the output distribution when coherent light is injected in the desired initial site (IN). It is worth noting that in this realization, the evolution parameter *t*, considered in the theoretical model, is mapped onto the propagation length, which we still label as *t*.

A controlled amount of noise is introduced in the structure by segmenting the waveguides corresponding to the sites of the maze. This is achieved by modulating the writing speed in the fabrication process, which induce a proportional variation of the propagation constant, while keeping the coupling coefficient unvaried[46]. The value of the propagation constant variation in each segment is randomly picked from a uniform distribution with a given amplitude; the same distribution is used for every waveguide within the same array. The random variation of the propagation constants is equivalent to a random variation of the site energy due to the interaction with an incoherent environment[25], hence effectively adding also pure dephasing in the dynamics. This approach has been extensively tested by numerical simulations of this specific implementation as compared with the theoretical Lindblad model discussed in the previous section (for further details, see Supplementary Notes 2 and 3, together with Supplementary Figs 3 and 4).

The waveguide array implementing the sink is in all cases composed by uniform, not-segmented, waveguides. To characterize the transfer efficiency to the sink, the output facet of each fabricated structure is imaged onto a CMOS camera (examples of snapshots are shown in Fig. 3c,d), the light intensity on the maze and sink regions of the array are numerically integrated and the fraction of light in the sink is calculated. Technical details of the characterization procedure are given in the Methods section (Characterization measurements: experimental details).

Twenty-four structures were fabricated with the transverse layout as in Fig. 3b, implementing six different propagation lengths for both the noiseless, fully coherent, situation and three different noise configurations with the same strength (that is, same amplitude of propagation constant distribution). Waveguide arrays were inscribed in EAGLE2000 (Corning) glass substrates, by femtosecond laser writing. A Yb-based amplified laser system (FemtoREGEN, HighQLaser) was used, providing laser pulses with 400 fs duration and 300 nJ energy at 1 MHz repetition rate. The laser was focused in the substrate by a 0.45 numerical aperture, $\times 20$ microscope objective, compensated for spherical aberrations at 170 μm below the glass surface, which is the average depth of the fabricated structures. The waveguides yield single-mode operation at the wavelength of 850 nm and the coupling coefficient between nearest-neighbouring waveguides is $\kappa = 0.40\,\mathrm{mm}^{-1}$. The amplitude of the random distribution of the propagation constants, adopted in the noise implementation, is $\Delta\beta_{max} = 0.40\,\mathrm{mm}^{-1}$. The modulation of the propagation constant is achieved by proportionally varying the waveguide writing speed in the 10–40 $\mathrm{mm\,s}^{-1}$ range (see also Supplementary Fig. 5 for details). In fact, varying the writing speed means changing the amount of deposited energy in the material, which, in the above range, causes a proportional variation of refractive index change and thus of $\Delta\beta$. The value of the propagation constant is modified every 3 mm of waveguide length.

Transfer efficiency. The transfer efficiency to the sink, calculated theoretically with the method reported in ref. 28, is shown in Fig. 4 for a maze with the layout presented in Fig. 3. Such layout represents the actual structure that we have experimentally implemented and is considered both for the case of fully coherent quantum transport and for the case of partially incoherent transport with $p = 0.1$. Figure 4 shows the experimentally retrieved transfer efficiencies, each point corresponding to a physically different structure fabricated to implement a certain noise map and a given propagation length. The average between the points corresponding to the three different noise implementations is also shown (right panel).

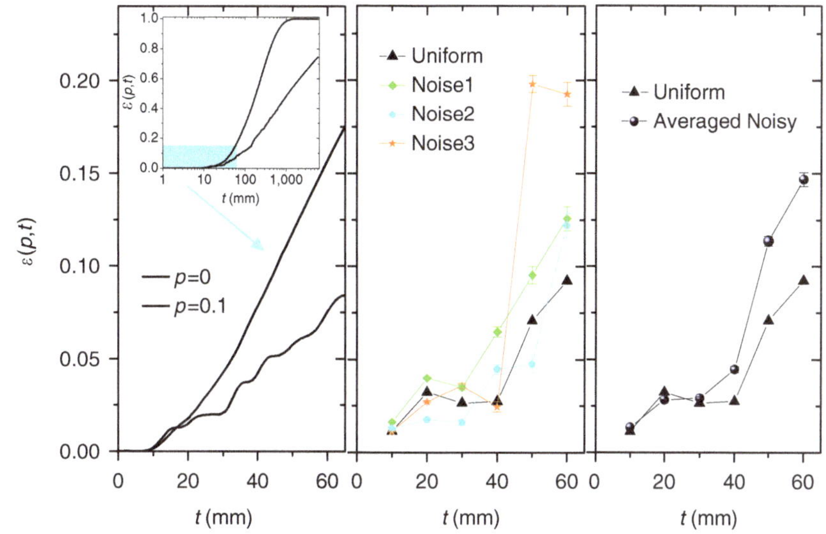

Figure 4 | Role of noise in the transfer efficiency time evolution. Left: Theoretical behaviour of the transfer efficiency $\mathcal{E}(p, t)$ as a function of the evolution parameter t for two values of the mixing parameter p, corresponding to QW ($p=0$) and QSW ($p=0.1$), for the maze in Fig. 3. Inset: a larger time scale is shown in order to point out the remarkable efficiency enhancement when there is a partial suppression of interference. Centre: Experimental results for both uniform (triangles) and three noisy realizations of the structure reported in Fig. 3. Where not shown the error bars are smaller than the mark size. Right: as in the middle panel, but only with the uniform case (triangles) and with the average efficiency over the noise realizations (circles). In all three panels, the time t is in units of mm because it is experimentally mapped into the propagation length of the three-dimensional waveguide array.

The physical quantity that is measured experimentally is the fraction of light present in the sink after a certain propagation, and not the fraction of light that is transferred to the sink. In case propagation losses are the same both in the maze waveguides and in the sink waveguides, the two quantities indeed correspond. As a matter of fact, the modulation of the writing speed produces additional losses in the waveguides of the maze with respect to the waveguides of the sink and this causes in general an overestimation of the transfer efficiency. However, we characterized accurately such additional losses in our structures and simulated their impact on the estimation of the transfer efficiency. The consequent systematic error contribution has been directly taken into account in Fig. 4, whereas the random error contribution is reported with the error bars. Details on the treatment of noise and error contributions in the efficiency estimation are given in the Methods section (Characterization measurements: experimental details).

A very good agreement between theoretical and experimental curves is observed both for the noiseless, fully coherent case, and for the partially coherent transport when considering the average of our 'noisy' waveguide implementations. Therefore, this experimental evidence is consistent with our claim that the interplay of noise and interference effects leads to higher efficiency in finding the way out from the maze.

To further assess our experimental observation of a noise-induced enhancement in transfer efficiency in this platform, we fabricated other photonic structures implementing the maze of Fig. 3 at an evolution parameter $t=60$ mm, but employing a different femtosecond laser writing setup for the fabrication (see Supplementary Note 4 for details). We fabricated one uniform structure and 16 different random 'noisy' implementations, 8 with a noise strength $\Delta\beta_{max}=0.12\,\mathrm{mm}^{-1}$ and 8 implementing a noise strength $\Delta\beta_{\mathrm{max}}=0.40\,\mathrm{mm}^{-1}$. As shown in Fig. 5, we can experimentally observe a transfer efficiency of 12.5% for the uniform case, to be compared with an average transfer efficiency of 14.1% and 22.2% for the $\Delta\beta_{\mathrm{max}}=0.12\,\mathrm{mm}^{-1}$ and $\Delta\beta_{\mathrm{max}}=0.40\,\mathrm{mm}^{-1}$ cases, respectively. The experimental data are in good agreement with the simulations, which take into account small differences in the waveguide properties with respect to those fabricated with the previous system. It can be noticed that the distribution of the measurements (open circles in Fig. 5) around the mean value spreads with increasing disorder: however, the increase in the average transfer efficiency is clear when the amount of noise approaches the optimum value.

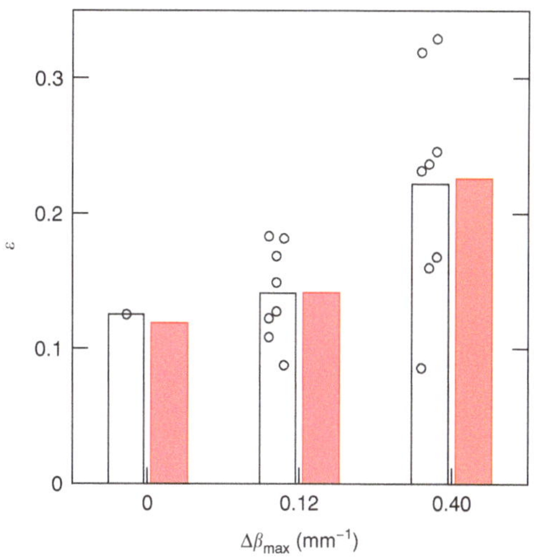

Figure 5 | Averaging over different noise realizations. Experimentally measured transfer efficiency in photonic maze structures with the topology of Fig. 3 and propagation length $t=60$ mm. Both uniform and different noise maps with $\Delta\beta_{\mathrm{max}}=0.12\,\mathrm{mm}^{-1}$ and $\Delta\beta_{\mathrm{max}}=0.40\,\mathrm{mm}^{-1}$ have been implemented. The light grey (left) bars indicate the average over several experimental results obtained for different noise maps with the same amplitude $\Delta\beta_{\mathrm{max}}$ (single measurements are reported as open circles). The red (right) bars represent the expected average transfer efficiency, numerically simulated for 1,000 different noise maps with the given $\Delta\beta_{\mathrm{max}}$. Error bars are smaller than the mark size.

Discussion

To summarize, here we have studied both theoretically and experimentally the dynamics of a walker travelling in a maze, having a single path from the input door (starting point) to the exit (solution). By considering a model that mixes the behaviour of a classical walker and a quantum one, we have found an optimal condition leading to extremely efficient and fast transmission. For large enough maze size, this leads to a remarkably high enhancement of more than five order of magnitudes in the transfer efficiency with respect to both the classical and purely quantum limits. This result is a clear example that decoherence is not always a detrimental phenomenon that should be avoided in quantum processes and it may provide some insight on the reason why nature evolution has made the observation of purely coherent phenomena so difficult.

By exploiting the unique capabilities of the femtosecond laser writing technology, we have unfolded the maze and implemented it in a three-dimensional waveguide array, where a suitable modulation of the waveguide properties allowed us to mimic a partial decoherence of the walker. Our measurements have faithfully confirmed the theoretical predictions and, in particular, the remarkable role of a partial suppression of interference in enhancing transport dynamics in mazes. It is also worth noting that our technological platform has enabled an experimental simulation of a noise-assisted problem in well-controlled conditions and over complex topologies, and can thus represent a very powerful tool for further studies in this direction. These results, together with future full circuit reconfigurability, will pave the way to much more complex integrated photonics devices exploiting interference, quantum features and noise effects for improved problem-solving efficiency, and remarkably fast transmission of information in ICT applications and of energy in novel solar technologies.

Another experimental demonstration of enhanced quantum transport by controlled decoherence has been reported during the preparation of this manuscript[54].

Methods

Maze construction. Depth-First Search algorithm is the simplest maze generation algorithm and is based on the following iterative procedure that is applied to a regular square grid of N nodes, where all neighbour sites are separated by a wall[50]. One starts from a random node and then search for a random neighbour that has not considered yet. If so, the wall between these two sites is knocked down; otherwise, one backs up to the previous node. This procedure is repeated until all sites of the grid have been visited. By doing so, the final structure is a maze where we have only one path connecting the IN to the OUT node, that is, a so-called two-dimensional perfect maze with no closed loops. Applying this procedure to larger and larger square lattices, one obtains increasing large maze graphs (see Supplementary Fig. 1 for details).

Characterization measurements: experimental details. The fabricated structures are probed by coherent light. Laser light at 850 nm wavelength is injected into the input waveguide. The output distribution is imaged onto a CMOS camera by means of a 0.12 numerical aperture objective. Numerical integration on different parts of the acquired image allows to retrieve the fraction of light present in the sink waveguide array. The advantages of this method are, on one hand, insensitivity to coupling losses of the input beam and, on the other hand, the possibility of a fast acquisition of the output of many waveguides.

A careful analysis of the measurement error has been performed. A possible source of error is the quantization of the intensity levels of the CMOS sensor, as well as its finite spatial resolution. To analyse this error contribution, we simulated the numerical integration of gaussian modes, with the same size as the measured ones, but random intensity and peak position, discretized both in the (256) intensity levels and in the pixels of the spatial profile. Supplementary Fig. 6 shows the (normalized) difference between the numerically calculated integral and the analytic integral of the gaussian profile, for 1,000 randomly distributed modes, as a function of the peak intensity. Note that, given a certain peak intensity, the numerical integral may give different values depending on the position of the peak, because of pixel discretization. Systematic and random errors are almost independent on the peak intensity and they have been taken into account in data elaboration assuming that the acquired image contains n modes of random uniformly distributed intensity. As a matter of fact, the contribution of such errors on the experimentally measured efficiencies is relevant only for the shortest lengths, where the light intensity in the sink is particularly low.

Furthermore, cascading waveguide segments with different fabrication speed (to mimic noise) causes small additional losses, at each interface between different waveguide segments. Importantly, these losses are present only in the waveguides representing the maze and not in the sink waveguides. Because of those losses, measuring the fraction of power *present* in the sink after a certain propagation distance, with respect to the overall output power, as is done in our characterization process, may not correspond precisely to the fraction of input power *transferred* to the sink. In fact, the transfer efficiency is slightly overestimated.

Overall additional losses can be measured quite accurately, by simply measuring and comparing the insertion losses of the fabricated devices (ration of the overall output power over the input power). However, the precise contribution of these losses on each measured transfer efficiency can be hardly retrieved. In fact, the transfer process is not uniform during the propagation and depends on the *random* noise map implemented.

Thus, to statistically quantify such overestimation, we numerically simulated the light propagation in waveguide structures analogous to the fabricated ones, for 100 random different noise distributions (with always the same amplitude, as the one adopted in the experiment), both in the case of waveguides with no losses (which corresponds to the ideal situation) and in the case of waveguides yielding uniform additional losses with respect to the waveguides of the sink, in such a way that the overall losses of the structure correspond to the experimentally measured ones (~ 2 dB additional losses for the longest arrays). We evaluated in each case the estimation error of the transfer efficiency and calculated the error statistical distribution. Supplementary Fig. 7 shows the average error, together with its standard deviation, as a function of the propagation length. The effect of these losses reveals to be small (the systematic component is $<3\%$ for 60 mm, correspondent to our longest fabricated devices) and does not significantly influence our experimental observation of an increase in transfer efficiency in the cases in which noise is added.

The data that support the findings of this study are available from the corresponding authors upon request.

References

1. Hyafil, L. & Rivest, R. L. Constructing optimal binary decision trees is NP-complete. *Inform. Proc. Lett.* **5,** 15–17 (1976).
2. Shannon, C. Presentation of a maze solving machine. *Trans. 8th Conf. Cybernetics: Circular, Causal and Feedback Mechanisms in Biological and Social Systems* 169–181 (New York, USA, 1951).
3. Steinbock, O., Toth, A. & Showalter, K. Navigating complex labyrinths: optimal paths from chemical waves. *Science* **267,** 868–871 (1995).
4. Adamatzky, A. Slime mold solves maze in one pass, assisted by gradient of chemo-attractants. *IEEE Trans. NanoBiosci.* **11,** 131–134 (2012).
5. Aharonov, Y., Davidovich, L. & Zagury, N. Quantum random walks. *Phys. Rev. A* **48,** 1687 (1993).
6. Kempe, J. Quantum random walks: an introductory overview. *Contemp. Phys.* **44,** 307 (2003).
7. Keating, J. P., Linden, N., Matthews, J. C. F. & Winter, A. Localization and its consequences for quantum walk algorithms and quantum communication. *Phys. Rev. A* **76,** 012315 (2007).
8. Schwartz, T., Bartal, G., Fishman, S. & Segev, M. Transport and Anderson localization in disordered two-dimensional photonic lattices. *Nature (London)* **446,** 52 (2007).
9. Lahini, Y. *et al.* Anderson localization and nonlinearity in one-dimensional disordered photonic lattices. *Phys. Rev. Lett.* **100,** 013906 (2008).
10. Moonseok, K. *et al.* Maximal energy transport through disordered media with the implementation of transmission eigenchannels. *Nat. Photonics* **6,** 581585 (2012).
11. Mülken, O. & Blumen, A. Continuous-time quantum walks: Models for coherent transport on complex networks. *Phys. Rep.* **502,** 37–87 (2011).
12. Santha, M. in *Theory and Applications of Models of Computation. Lecture Notes in Computer Science* (eds Agrawal, M., Du, D. Z., Duan, Z. H. & Li, A. S.) Vol. 4978, 31 (Springer, 2008).
13. Farhi, E. & Gutmann, S. Quantum computation and decision trees. *Phys. Rev. A* **58,** 915–928 (1998).
14. Childs, A., Farhi, E. & Gutmann, S. An example of the difference between quantum and classical random walks. *J. Quant. Inf. Proc* **1,** 35 (2002).
15. Childs, A. Universal computation by quantum walk. *Phys. Rev. Lett.* **102,** 180501 (2009).
16. Ambainis, A. Quantum walks and their algorithmic applications. *Int. J. Quantum Inform.* **1,** 507 (2003).
17. Grover, L. K. Quantum mechanics helps in searching for a needle in a haystack. *Phys. Rev. Lett.* **79,** 325 (1997).
18. Childs, A. M., Gosset, D. & Webb, Z. Universal computation by multiparticle quantum walk. *Science* **339,** 791 (2013).
19. Bose, S. Quantum communication through an unmodulated spin chain. *Phys. Rev. Lett.* **91,** 207901 (2003).
20. Christandl, M., Datta, N., Ekert, A. & Landahl, A. J. Perfect state transfer in quantum spin networks. *Phys. Rev. Lett.* **92,** 187902 (2004).

21. Plenio, M. B., Hartley, J. & Eisert, J. Dynamics and manipulation of entanglement in coupled harmonic systems with many degrees of freedom. *New J. Phys.* **6,** 36 (2004).
22. Sánchez-Burillo, E., Duch, J., Gómez-Gardeñes, J. & Zueco, D. Quantum navigation and ranking in complex networks. *Sci. Rep.* **2,** 605 (2012).
23. Kumar, N. & Goswami, D. Quantum algorithm to solve a maze: converting the maze problem into a search problem. Preprint at http://arxiv.org/abs/1312.4116 (2013).
24. Engel, G. S. *et al.* Evidence for wavelike energy transfer through quantum coherence in photosynthetic systems. *Nature* **446,** 782 (2007).
25. Mohseni, M., Rebentrost, P., Lloyd, S. & Aspuru-Guzik, A. Environment-assisted quantum walks in photosynthetic energy transfer. *J. Chem. Phys.* **129,** 174106 (2008).
26. Plenio, M. B. & Huelga, S. F. Dephasing-assisted transport: quantum networks and biomolecules. *New J. Phys.* **10,** 113019 (2008).
27. Caruso, F., Chin, A. W., Datta, A., Huelga, S. F. & Plenio, M. B. Highly efficient energy excitation transfer in light-harvesting complexes: The fundamental role of noise-assisted transport. *J. Chem. Phys.* **131,** 105106 (2009).
28. Caruso, F. Universally optimal noisy quantum walks on complex networks. *New J. Phys.* **16,** 055015 (2014).
29. Caruso, F., Huelga, S. F. & Plenio, M. B. Noise-enhanced classical and quantum capacities in communication networks. *Phys. Rev. Lett.* **105,** 190501 (2010).
30. Du, J. *et al.* Experimental implementation of the quantum random-walk algorithm. *Phys. Rev. A* **67,** 042316 (2003).
31. Ryan, C. A., Laforest, M., Boileau, J. C. & Laflamme, R. Experimental implementation of a discrete-time quantum random walk on an NMR quantum-information processor. *Phys. Rev. A* **72,** 062317 (2005).
32. Schmitz, H. *et al.* Quantum walk of a trapped ion in phase space. *Phys. Rev. Lett.* **103,** 090504 (2009).
33. Zähringer, F. *et al.* Realization of a quantum walk with one and two trapped ions. *Phys. Rev. Lett.* **104,** 100503 (2010).
34. Karski, M. *et al.* Quantum walk in position space with single optically trapped atoms. *Science* **325,** 174 (2009).
35. Broome, M. A. *et al.* Discrete single-photon quantum walks with tunable decoherence. *Phys. Rev. Lett.* **104,** 153602 (2010).
36. Kitagawa, T. *et al.* Observation of topologically protected bound states in photonic quantum walks. *Nat. Commun.* **3,** 882 (2012).
37. Schreiber, A. *et al.* A 2D quantum walk simulation of two-particle dynamics. *Science* **336,** 55–58 (2012).
38. Jeong, Y.-C., Di Franco, C., Lim, H.-T., Kim, M. & Kim, Y.-H. Experimental demonstration of delayed-choice decoherence suppression. *Nat. Commun.* **5,** 4522 (2013).
39. Peruzzo, A. *et al.* Quantum walks of correlated photons. *Science* **329,** 1500–1503 (2010).
40. Perets, H. B. *et al.* Realization of quantum walks with negligible decoherence in waveguide lattices. *Phys. Rev. Lett.* **100,** 170506 (2008).
41. Owens, J. O. *et al.* Two-photon quantum walks in an elliptical direct-write waveguide array. *New J. Phys.* **13,** 075003 (2011).
42. Sansoni, L. *et al.* Two-particle bosonic-fermionic quantum walk via integrated photonics. *Phys. Rev. Lett.* **108,** 010502 (2012).
43. Crespi, A. *et al.* Anderson localization of entangled photons in an integrated quantum walk. *Nat. Photonics* **7,** 322 (2013).
44. Crespi, A., Corrielli, G., Della Valle, G., Osellame, R. & Longhi, S. Dynamic band collapse in photonic graphene. *New J. Phys.* **15,** 013012 (2013).
45. Rechtsman, M. C. *et al.* Photonic Floquet topological insulators. *Nature* **496,** 196–200 (2013).
46. Corrielli, G., Crespi, A., Della Valle, G., Longhi, S. & Osellame, R. Fractional Bloch oscillations in photonic lattices. *Nat. Commun.* **4,** 1555 (2013).
47. Della Valle, G., Osellame, R. & Laporta, P. Micromachining of photonic devices by femtosecond laser pulses. *J. Opt. A Pure Appl. Opt.* **11,** 013001 (2009).
48. Longhi, S. Quantum-optical analogies using photonic structures. *Laser Photonics Rev.* **3,** 243–261 (2009).
49. Szameit, A. & Nolte, S. S. Discrete optics in femtosecond-laser-written photonic structures. *J. Phys. B At. Mol. Opt. Phys.* **43,** 163001 (2010).
50. Shimon, E. *Graph Algorithms* 2nd ed (Cambridge Univ., 2011).
51. Whitfield, J. D., Rodríguez-Rosario, C. A. & Aspuru-Guzik, A. Quantum stochastic walks: a generalization of classical random walks and quantum walks. *Phys. Rev. E* **81,** 022323 (2010).
52. Viciani, S., Lima, M., Bellini, M. & Caruso, F. Observation of noise-assisted transport in an all-optical cavity-based network. *Phys. Rev. Lett.* **115,** 083601 (2015).
53. Li, Y., Caruso, F., Gauger, E. & Benjamin, S. C. Momentum rejuvenation underlies the phenomenon of noise-assisted quantum energy flow. *New J. Phys.* **17,** 013057 (2015).
54. Biggerstaff, D. N. *et al.* Enhancing quantum transport in a photonic network using controllable decoherence. Preprint at http://arxiv.org/abs/1504.06152 (2015).

Acknowledgements

We acknowledge M.B. Plenio for carefully reading the manuscript and very useful and stimulating discussions. This work was supported by the European Union through the project FP7-ICT-2011-9-600838 (QWAD Quantum Waveguides Application and Development), the European Research Council (ERC-Starting Grant 3D-QUEST, 3D-Quantum Integrated Optical Simulation, grant agreement no. 307783, http://www.3dquest.eu/), EU FP7 Marie-Curie Programme (Career Integration Grant, project no. 293449) and by national grants as PRIN (Programmi di ricerca di rilevante interesse nazionale) project AQUASIM (Advanced Quantum Simulation and Metrology) and MIUR-FIRB project (RBFR10M3SB). We acknowledge QSTAR for computational resources, partially based also on GPU programming by NVIDIA Tesla C2075 GPU computing processors.

Authors contributions

F.C., F.S. and R.O. conceived the whole project. A.C., A.G.C. and R.O. carried out the experiment; F.C. led and carried out the theoretical work. A.C. and A.G.C. exploited ultra-fast laser writing process to realize the maze structure. F.C. and A.C. performed the numerical simulations of the transport dynamics. F.C., F.S. and R.O. supervised the project. All authors contributed to the discussion, analysis of the results and the writing of the manuscript.

Additional information

Competing financial interests: The authors declare no competing financial interests.

Loss-tolerant state engineering for quantum-enhanced metrology via the reverse Hong-Ou-Mandel effect

Alexander E. Ulanov[1,2], Ilya A. Fedorov[1,3], Demid Sychev[1], Philippe Grangier[4] & A.I. Lvovsky[1,3,5]

Highly entangled quantum states, shared by remote parties, are vital for quantum communications and metrology. Particularly promising are the NOON states—entangled N-photon wavepackets delocalized between two different locations—which outperform coherent states in measurement sensitivity. However, these states are notoriously vulnerable to losses, making them difficult to both share them between remote locations and recombine in order to exploit interference effects. Here we address this challenge by utilizing the reverse Hong-Ou-Mandel effect to prepare a high-fidelity two-photon NOON state shared between two parties connected by a lossy optical medium. We measure the prepared state by two-mode homodyne tomography, thereby demonstrating that the enhanced phase sensitivity can be exploited without recombining the two parts of the NOON state. Finally, we demonstrate the application of our method to remotely prepare superpositions of coherent states, known as Schrödinger's cat states.

[1] Russian Quantum Center, 100 Novaya Street, Skolkovo, Moscow 143025, Russia. [2] Moscow Institute of Physics and Technology, Dolgoprudny 141700, Russia. [3] P.N. Lebedev Physics Institute, Leninskiy prospect 53, Moscow 119991, Russia. [4] Laboratoire Charles Fabry, Institut d'Optique Graduate School, CNRS, Université Paris-Saclay, Palaiseau 91127, France. [5] Institute for Quantum Science and Technology, University of Calgary, Calgary, Alberta, Canada T2N 1N4. Correspondence and requests for materials should be addressed to A.I.L. (email: lvov@ucalgary.ca).

In the current race towards the practical implementation of quantum techniques for information processing and communications, a strong trend is to design loss-tolerant quantum protocols, such as the preparation of non-local superpositions of quasi-classical light states[1], discord-assisted remote state preparation[2], quantum illumination[3], undoing the effect of losses on continuous-variable entanglement[4] and the preparation of single-qubit entangled states over a long distance[5–8].

In this article we are interested in N00N states $|N :: 0\rangle = (|N,0\rangle + |0,N\rangle)/\sqrt{2}$, which are useful in linear-optical quantum computation[9,10], quantum-optical state engineering[11,12] and the preparation of photon-number path entanglement[13,14]. But the most important potential application of these states is as a resource for quantum enhanced metrology[14–19]. Interference measurements with N00N states exhibit super-resolving properties: the number of fringes per wavelength equals *N*, in contrast to a single fringe in the case of coherent states. This property can be exploited for precise measurement of diverse physical quantities. Widespread application of N00N states for metrology is however precluded by their extreme sensitivity to losses. When exposed to even moderate losses, the degree of entanglement and hence the super-resolution potential of the N00N states dramatically degrade to an extent that eliminates any advantage[20,21].

In the present work, we address this challenge by developing a technique to losslessly produce N00N states between parties that are separated by a lossy quantum channel. In addition, using N00N states usually requires bringing back together the two entangled parts, introducing more propagation losses. But we will show that this second step is not required, and that super-resolution for the optical phase can be obtained remotely, by using homodyne detection[11].

Results

Concept. For N00N state production, we exploit some peculiar properties of the Hong-Ou-Mandel (HOM) effect[22], a well-known quantum interference phenomenon in which two indistinguishable photons that are overlapped on a symmetric beam splitter (BS) always emerge in the same output mode, preparing the N00N state

$$|11\rangle \rightarrow |2 :: 0\rangle = \frac{|2,0\rangle + |0,2\rangle}{\sqrt{2}} \qquad (1)$$

in the beam splitter output. Our experiment relies upon the reverse version of the HOM effect[23], in which the measurements in the two output modes of the BS project them onto single-photon states. Because of the time-reversible nature of quantum mechanics, such projection is equivalent to projecting the state of the input onto the two-photon N00N state (1). If each of the beam splitter inputs is, in turn, entangled with other modes, these modes become entangled with each other, thanks to entanglement swapping[24]. Whereas the original HOM setting creates a two-photon N00N state, extension to any even *N* is straightforward (see Supplementary Section 'Preparation of high-order N00N states').

Specifically, consider a set-up in which two pairs of modes (A,C) and (B,D) are prepared in two-mode squeezed states:

$$|\Psi_{AC/BD}\rangle \propto |0,0\rangle + \gamma|1,1\rangle + \gamma^2|2,2\rangle + \cdots, \qquad (2)$$

by means of non-degenerate parametric down-conversion. Modes C and D are then mixed on a symmetric BS, the outputs of which are subjected to measurement in the photon-number basis via single-photon counting modules (SPCMs), as shown in Fig. 1.

In the weak-squeezing limit $|\gamma|^2 \ll 1$, every SPCM click is likely to be caused by no more than one photon. Then, a coincidence click in both SPCMs correspond to projection on state $|1,1\rangle_{CD}$. Due to the unitary nature of the BS operation, this event assures that modes C and D were initially in the N00N state (1). This corresponds to two pairs of photons having been produced in either of two crystals; therefore, now the remaining modes A and B also share a two-photon N00N state. The application of the reverse HOM effect to N00N state preparation has been proposed by Kok *et al.*[13], albeit not in a remote fashion as we do here.

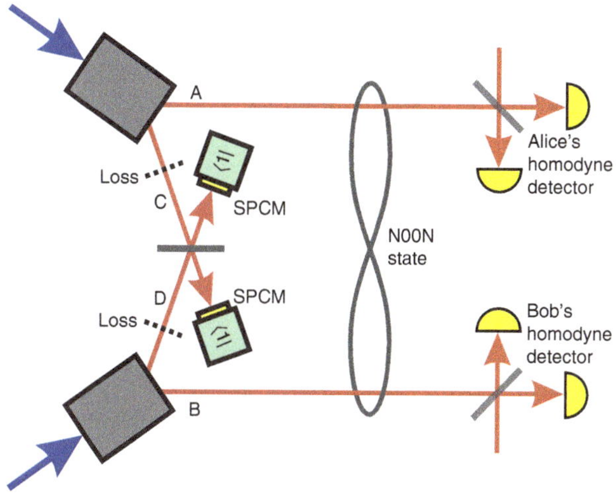

Figure 1 | Conceptual scheme of the experiment. A coincidence click of the SPCMs projects modes C and D onto the two-photon N00N state due to the reverse HOM effect. This, in turn, prepares modes A and B in the same state, thanks to entanglement swapping.

A remarkable feature of this scheme is its robustness to losses in channels C and D. Such losses only reduce the probability of the two SPCMs to click, but, if the clicks do occur and the down-conversion amplitude γ is sufficiently small, the leading term in the state of channels A and B is still the two-photon N00N state. We show in Supplementary equation (8) and Supplementary Fig. 4 that the fidelity of this state is bounded from below by $F \gtrsim 1 - 4\gamma^2$ for all loss values. Therefore, by pumping the down-conversion crystals sufficiently weakly, one can prepare a N00N state of arbitrarily high purity, no matter how high the losses. Our method is equally robust with respect to low quantum efficiencies of the SPCMs, because these are effectively equivalent to additional losses in modes C and D[25]. Note, however, that our technique does not help eliminate losses in modes A and B, as well as the effect of detector imperfection in these modes.

N00N state tomography. Conditioned on coincidence clicks of the two SPCMs, we characterize the state of modes A and B by means of homodyne tomography (Methods section). The measured states in both cases are very close to the ideal state (1) that has suffered a $1-\eta = 45\%$ loss due to imperfect detection efficiency[26,27] (Fig. 2a).

To illustrate the reverse HOM effect and the critical significance of the matching between modes C and D for proper N00N state preparation, we measured the behaviour of the state in modes A and B as a function of increasing temporal mismatch between modes C and D. As evidenced by Fig. 3, the fraction of the biphoton component $|1,1\rangle$ in that state exhibits a dip that is characteristic as a signature of the HOM effect. The tomographic state reconstruction for the case of complete mismatch is shown in Fig. 2b. In addition to a macroscopic biphoton component, no off-diagonal terms are present in this case because no coherence between modes A and B can emerge in the absence of interference.

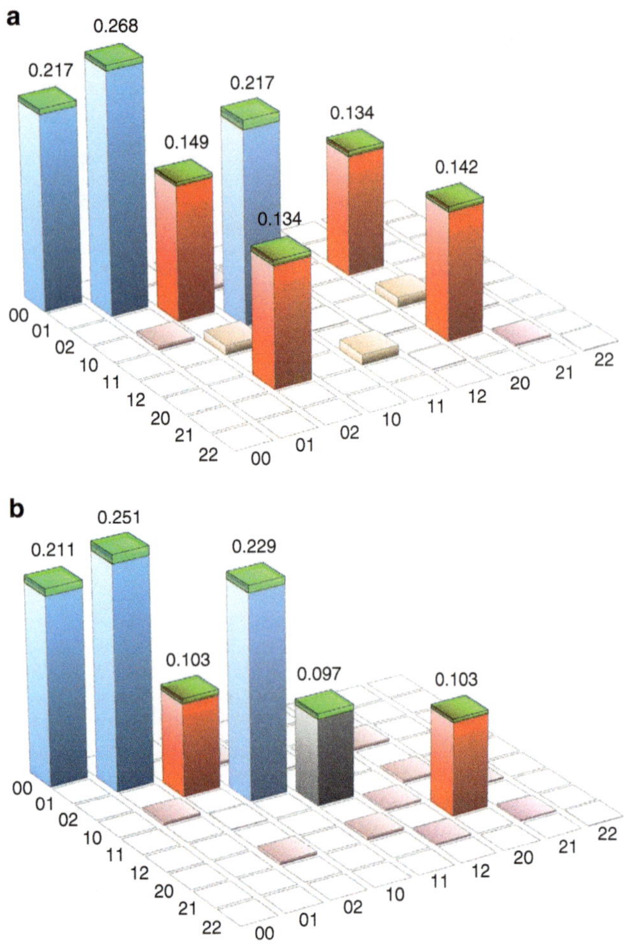

Figure 2 | Density matrix of the conditionally prepared state of modes A and B. The data are displayed with a 10-dB total loss in modes C and D before the BS, reconstructed in the Fock basis. No correction is made for losses in any of the channels. The numbers along the horizontal axes are the indices of the bra and ket elements of that matrix. The components shown in red correspond to the ideal NOON state, others appear due to losses. (**a**) When modes C and D are matched to each other, a two-photon NOON state is prepared. The $|1, 1\rangle\langle 1, 1|$ component amounts to 0.01. (**b**) In the case of a 5-ps mismatch between modes C and D, the HOM interference and any coherence between modes A and B are eliminated. The emerged $|1, 1\rangle\langle 1, 1|$ component is shown in grey. Green hats show the statistical error of the reconstruction.

Phase sensitivity enhancement. The enhanced phase sensitivity is manifested by the mean values of observables $X_A X_B$ and $X_A^2 X_B^2$, where $X_A = Q_A \cos\theta_A + P_A \sin\theta_A$ and $X_B = Q_B \cos\theta_B + P_B \sin\theta_B$, are quadrature operators of modes A and B. For the one-photon $|1 :: 0\rangle = (|1, 0\rangle + |0, 1\rangle)/\sqrt{2}$ and two-photon NOON states one has, respectively,

$$\begin{aligned}\langle 1 :: 0|X_A X_B|1 :: 0\rangle &= -\frac{\eta}{2}\sin\Delta\theta,\\ \langle 1 :: 0|X_A^2 X_B^2|1 :: 0\rangle &= \frac{1+2\eta}{4}\end{aligned} \tag{3}$$

and

$$\begin{aligned}\langle 2 :: 0|X_A X_B|2 :: 0\rangle &= 0,\\ \langle 2 :: 0|X_A^2 X_B^2|2 :: 0\rangle &= \frac{1}{4} + \eta + \frac{\eta^2}{2}\cos(2\Delta\theta)\end{aligned} \tag{4}$$

where $\Delta\theta = \theta_A - \theta_B$ and we assumed the phases in modes C and D to be constant.

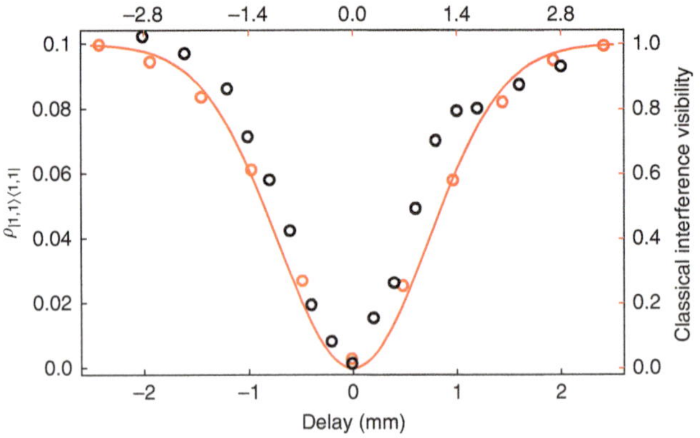

Figure 3 | The reverse HOM dip. Quantum and classical interference as a function of the delay between modes C and D. The blue points represent the weight of the $|1, 1\rangle\langle 1, 1|$ component in the reconstructed density matrix (obtained from homodyne tomography, not from photon counting). The red points and red curve correspond to the visibility of the classical interference between the master laser pulses that underwent the same spectral filtering as quantum modes C and D. The horizontal axes for the classical and quantum cases are scaled by a factor of $\sqrt{2}$ with respect to each other to match the theoretically expected widths of the interference patterns.

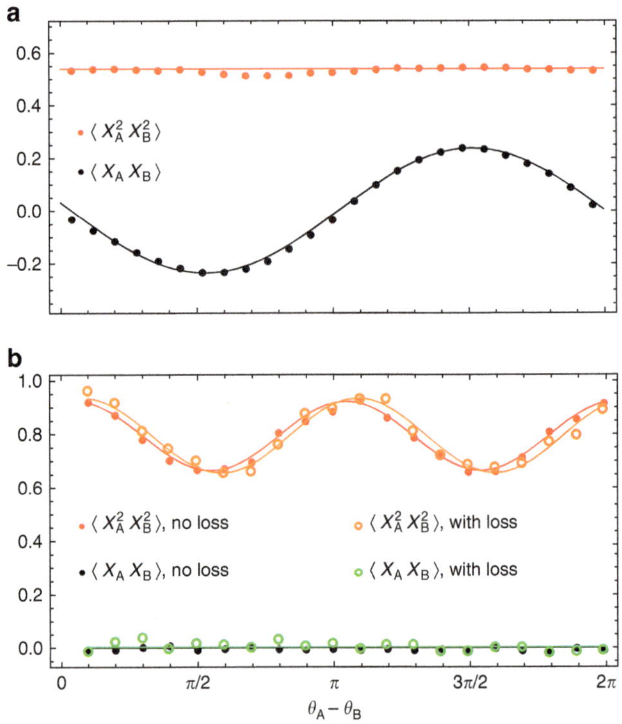

Figure 4 | Periodic phase-dependent behaviour of NOON states. Dependence of the mean and mean-square product of Alice's and Bob's quadratures on the difference $\theta_A - \theta_B$ of Alice's and Bob's phases for states $|1{::}0\rangle$ (**a**) and $|2{::}0\rangle$ (**b**) is shown. Enhanced phase sensitivity is evident for the two-photon NOON state. Solid lines are theoretical predictions.

We see that in order to compare between the one- and two-photon NOON states using quadrature measurements, we need to use different observables. This notwithstanding, the phase dependence of the appropriate observable in each state is as expected, that is, with period $2\pi/N$ for each N-photon NOON state.

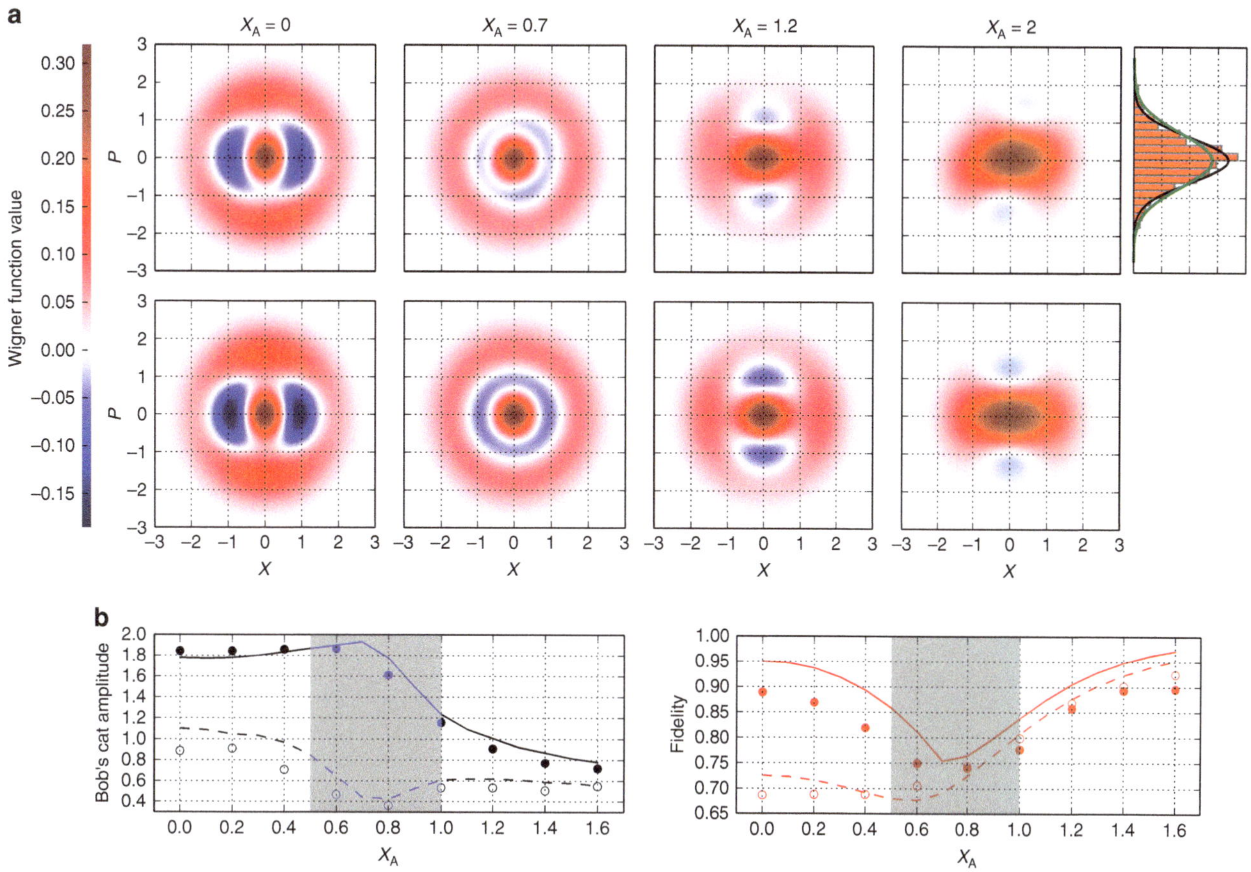

Figure 5 | Schrödinger's cat states. (**a**) Bob's Wigner functions after conditioning on Alice's quadrature measurement, with 55% efficiency correction in Bob's mode. Top row: experimental results. Bottom row: theoretical simulation. Fidelities between experiment and theory are 0.98, 0.98, 0.97, 0.99 from left to right. The top-right panel shows the marginal distribution of Bob's position quadrature in the case of $X_A = 2$, which exhibits squeezing. The black line fits the data, the green line corresponds to the vacuum state. (**b**) Amplitude of the CSS that best approximates the state of Bob's mode versus Alice's quadrature, and the corresponding fidelity. Lines show theoretical predictions. Filled (empty) markers and solid (dashed) lines stand for the data with (without) the efficiency correction. The approximation of Bob's state with a CSS state near $X = 0.7$ (second column) is unreliable because of the unproportionally large two-photon fraction and low phase dependence; this instability causes opposite behaviour of the blue lines in the shaded region.

An experimental check of this behaviour is demonstrated in Fig. 4. In agreement with the results of state reconstruction, the second-order interference of the $|2 :: 0\rangle$ state, generated in the presence of the loss, exhibits the same visibility as without loss.

Loss-tolerant preparation of Schrödinger's cat states. Being path-entangled, the N00N state can also be used for single-mode quantum-state engineering. Associating modes A and B with fictitious observers Alice and Bob, we consider a setting in which Alice, by performing quadrature measurements on her mode, remotely prepares a state in Bob's mode. Neglecting inefficiencies, Alice's quadrature outcome X measured at phase angle θ brings the Bob's mode to the state

$$ {}_{\mathrm{A}}\langle X_\theta|2 :: 0\rangle_{\mathrm{AB}} = {}_{\mathrm{A}}\langle X_\theta|2\rangle_{\mathrm{A}}|0\rangle_{\mathrm{B}} + {}_{\mathrm{A}}\langle X_\theta|0\rangle_{\mathrm{A}}|2\rangle_{\mathrm{B}} \quad (5)$$

where

$$\langle X_\theta|m\rangle_{\mathrm{A}} = e^{im\theta} e^{-X^2} \frac{H_m(X)}{\sqrt{2^{m-1/2} m!}}, \quad (6)$$

are Fock state wavefunctions, with $H_m(X)$ being Hermite polynomials. In this way, by postselecting specific values of Alice's observed quadrature, one can generate arbitrary superpositions of the 0- and 2-photon states, which approximate the even coherent-state superpositions (CSS) $|\alpha\rangle + |-\alpha\rangle$, sometimes viewed as a quantum-optical implementation of the Schrödinger's cat paradox[11,12,28,29].

The states of Bob's mode for different outcomes of Alice's homodyne measurement are displayed in Fig. 5. Projection on $X = 0$ (Fig. 5, first column) results in superposition $0.052|0\rangle - 0.85|2\rangle$, partially mixed due to losses in Alice's channel. After correcting for Bob's homodyne detection inefficiency, this state has a fidelity of 0.88 with the even CSS state of amplitude $\alpha = 1.84$, squeezed with parameter $z = 0.48$. This value compares favourably with state-of-the-art results[12,29], with the added advantage that our protocol is tolerant to the losses in the optical channel between Alice and Bob.

Effective CSS amplitudes and approximation fidelities for other values of X are shown in Fig. 5b. With increasing X, the two-photon fraction initially increases relative to the vacuum because the two-photon state wavefunction $\langle X|2\rangle$ decreases faster than the vacuum wavefunction $\langle X|0\rangle$. Wavefunction $\langle X|2\rangle$ changes sign near the value $X = 0.7$, where ideally a pure two-photon state in Bob's mode should be observed. In practice, due to the losses and finite width of Alice's post-selection window, a phase-insensitive mixture of $0.4|0\rangle\langle 0| + 0.6|2\rangle\langle 2|$ is produced (second column). In this region, the remotely prepared state approximates the CSS poorly because of the high two-photon component. For higher values of X, this two-photon component in Bob's state reduces again, resulting in increasingly faithful approximation of even CSSs with decreasing amplitudes (third column). For very low amplitudes, this state approximates a weakly squeezed vacuum state. This is the case for $X = 2$ (fourth column): Bob's quadrature spectrum exhibits squeezing by 0.65 ± 0.24 dB (without efficiency correction). The corresponding quadrature histogram is shown in the upper right panel of Fig. 5.

Discussion

The protocol developed here addresses the primary challenge in the way of employing nonclassical states of light, particularly the N00N state, for quantum metrology: optical losses. With conventional optical fibres, it allows establishing nearly ideal N00N entanglement over large distances, with a considerably increased tolerance to the loss in the channel connecting the parties. Due to its extreme sensitivity to the overall phase between Alice and Bob (see Supplementary Section 'Phase behavior' for details), our scheme could be used for ultra-precise measurements of the distance between Alice's and Bob's homodyne detectors. Its further advantage is that there is no need to recombine the two parts of the N00N states, thanks to the homodyne detections.

In many cases, however, it is likely that a simple laser interferometric measurement would do better than our scheme due to the much larger number of available photons. Our scheme becomes advantageous in settings where the amount of light transmitted between the two parties is limited. Suppose, for instance, that a lossy biological medium, very sensitive to the light intensity, is inserted in channels C and/or D. Then it will be possible to get high-precision interferometric measurements through this medium, with almost negligible light intensity, as long as channels A, B and the associated homodyne detections have a high quality.

At this stage our experiment is a proof-of-principle only, and its application in practical settings would involve a number of challenges—in particular, the synchronization of the photons in channels C and D within the inverse down-conversion bandwidth. Furthermore, the production of higher-order N00N states would be required to achieve a significant practical gain over interferometry with coherent states, which are not limited in photon number. Our scheme can be generalized to enable producing N00N states with an arbitrary even N. To this end, we can use a sequence of $N/2$ reverse HOM measurements to subject modes C and D to operator $\hat{c}^N + \hat{d}^N$ using the recipe of Kok *et al.*[13]. An event in which the application of this operator has been successful (all N SPCMs have clicked) implies that either C or D initially contained at least N photons, but is unable to distinguish which one. This means, in turn, that modes A and B are in the $|N :: 0\rangle$ state. Similarly to the two-photon case, the down-conversion amplitude γ must be sufficiently low in order to reduce the contribution of higher-order terms in the output state. We present a detailed theoretical and numerical analysis of this scheme in the Supplementary Section 'Preparation of high-order N00N states'. In particular, we find that the fidelity of the prepared state for a given γ does not strongly depend on the losses in modes C and D. This enables preparation of high N00N states at reasonable rates, even with quite lossy channels.

In addition to interferometry, the generalization of our scheme to higher-order N00N states may have other applications, based on ultraremote preparation of quantum states of light, as illustrated by the 'cat state' generation presented above. In that case, fibre transmission over long distances may be used, with possible quantum cryptography applications. The present scheme is therefore a valuable addition to the quantum-state engineering toolbox, for both discrete- and continuous-variable degrees of freedom of optical modes[11,30,31].

Methods

Experimental details. We employ a pulsed Ti:Sapphire laser (Coherent Mira 900D) with a wavelength of 780 nm, mean power 1.3 W, repetition rate of 76 MHz and a pulse width of ~ 1.6 ps. Most of the laser output is directed into an lithium triborate crystal for frequency doubling. We obtain up to 300 mW second harmonic; after subsequent spectral cleaning, about 100 mW remain. Then we implement parametric down-conversion in two periodically poled potassium titanyl phosphate crystals in a type II spectrally and spatially degenerate, but polarization non-degenerate configuration. The single-photon detection is implemented using SPCMs by Excelitas. The modes entering the SPCMs are spectrally filtered and delivered to the detectors by means of single-mode fibres, which ensures proper preparation of the heralded photon mode[26,27]. Including the filters and fibres, the quantum efficiency of these detectors is estimated as $\eta_{SPCM} \sim 0.15$.

Without losses, the experimental double-coincidence event rate is ~ 100 Hz, which corresponds to a probability of $\sim 10^{-6}$ per pulse and the down-conversion amplitude $\gamma^2 \sim 0.007$. With a total of 10 dB loss equally distributed between modes C and D, the coincidence rate decreases by a factor of 10 due to the reduction of the equivalent SPCM efficiency down to ~ 0.1.

A total of 500,000 quadrature samples have been acquired in each setting for the tomographic reconstruction of the N00N state. The density matrix was obtained from these data using the iterative maximum likelihood algorithm[32–34], which in case of N00N states requires the knowledge of the phase difference between two local oscillators at each moment in time. This information was collected from the quadrature correlations exhibited by $|1 :: 0\rangle$ state (Fig. 4, top), which was produced upon a single click of one of the SPCMs. The rate of these events was about 100 kHz, which allowed us to evaluate the quadrature correlation and hence the phase at any moment in time with a sufficient precision.

References

1. Ourjoumtsev, A., Ferreyrol, F., Tualle-Brouri, R. & Grangier, P. Preparation of non-local superpositions of quasi-classical light states. *Nat. Phys.* **5,** 189–192 (2009).
2. Dakić, B. *et al.* Quantum discord as resource for remote state preparation. *Nat. Phys.* **8,** 666–670 (2012).
3. Lopaeva, E. D. *et al.* Experimental realization of quantum illumination. *Phys. Rev. Lett.* **110,** 153603 (2013).
4. Ulanov, A. E. *et al.* Undoing the effect of loss on quantum entanglement. *Nat. Photon.* **9,** 764–768 (2015).
5. Simon, C. & Irvine, W. T. M. Robust long-distance entanglement and a loophole-free Bell test with ions and photons. *Phys. Rev. Lett.* **91,** 110405 (2003).
6. Moehring, D. L. *et al.* Entanglement of single atom quantum bits at a distance. *Nature* **449,** 68–71 (2007).
7. Hofmann, J. *et al.* Heralded entanglement between widely separated atoms. *Science* **337,** 72–75 (2012).
8. Hensen, B. *et al.* Experimental loophole-free violation of a Bell inequality using entangled electron spins separated by 1.3 km. *Nature* **526,** 682–686 (2015).
9. Dowling, J. P., Franson, J. D., Lee, H. & Milburn, G. J. Towards scalable linear-optical quantum computers. *Expt. Asp. Quantum Comput.* **3,** 205–213 (2005).
10. Lim, Y. L., Beige, A. & Kwek, L. C. Repeat-until-success linear optics distributed quantum computing. *Phys. Rev. Lett.* **95,** 030505 (2005).
11. Andersen, U. L., Neergaard-Nielsen, J. S., van Loock, P. & Furusawa, A. Hybrid discrete- and continuous-variable quantum information. *Nat. Phys.* **11,** 713–719 (2015).
12. Etesse, J., Bouillard, M., Kanseri, B. & Tualle-Brouri, R. Experimental generation of squeezed cat states with an operation allowing iterative growth. *Phys. Rev. Lett.* **114,** 193602 (2015).
13. Kok, P., Lee, H. & Dowling, J. P. Creation of large-photon-number path entanglement conditioned on photodetection. *Phys. Rev. A* **65,** 052104 (2002).
14. Nagata, T., Okamoto, R., O'Brien, J. L., Sasaki, K. & Takeuchi, S. Beating the standard quantum limit with four-entangled photons. *Science* **316,** 726–729 (2007).
15. Edamatsu, K., Shimizu, R. & Itoh, T. Measurement of the photonic de Broglie wavelength of entangled photon pairs generated by spontaneous parametric down-conversion. *Phys. Rev. Lett.* **89,** 213601 (2002).
16. Dowling, J. P. Quantum optical metrology—the lowdown on high-N00N states. *Contemp. Phys.* **49,** 125–143 (2013).
17. Chen, P., Shu, C., Guo, X., Loy, M. M. T. & Du, S. Measuring the biphoton temporal wave function with polarization-dependent and time-resolved two-photon interference. *Phys. Rev. Lett.* **114,** 010401 (2015).
18. Walther, P. *et al.* De Broglie wavelength of a non-local four-photon state. *Nature* **429,** 158–161 (2004).
19. Mitchell, M. W., Lundeen, J. S. & Steinberg, A. M. Super-resolving phase measurements with a multiphoton entangled state. *Nature* **429,** 161–164 (2004).

20. Kacprowicz, M., Demkowicz-Dobrzański, R., Wasilewski, R., Banaszek, K. & Walmsley, I. A. Experimental quantum-enhanced estimation of a lossy phase shift. *Nat. Photon.* **4,** 357–360 (2010).
21. Rubin, M. A. & Kaushik, S. Loss-induced limits to phase measurement precision with maximally entangled states. *Phys. Rev. A* **75,** 053805 (2007).
22. Hong, C. K., Ou, Z. Y. & Mandel, L. Measurement of subpicosecond time intervals between two photons by interference. *Phys. Rev. Lett.* **59,** 2044–2046 (1987).
23. Chen, J., Lee, K. F. & Kumar, P. Deterministic quantum splitter based on time-reversed Hong-Ou-Mandel interference. *Phys. Rev. A* **76,** 031804R (2007).
24. Pan, J.-W., Bouwmeester, D., Weinfurter, H. & Zeilinger, A. Experimental entanglement swapping: entangling photons that never interacted. *Phys. Rev. Lett.* **80,** 3891–3894 (1998).
25. Berry, D. W. & Lvovsky, A. I. Linear-optical processing cannot increase photon efficiency. *Phys. Rev. Lett.* **105,** 203601 (2010).
26. Lvovsky, A. I. *et al.* Quantum state reconstruction of the single-photon Fock state. *Phys. Rev. Lett.* **87,** 050402 (2001).
27. Huisman, S. R. *et al.* Instant single-photon Fock state tomography. *Opt. Lett.* **34,** 2739–2741 (2009).
28. Ourjoumtsev, A., Jeong, H., Tualle-Brouri, R. & Grangier, P. Generation of optical 'Schrödinger cats' from photon number states. *Nature* **448,** 784–786 (2007).
29. Huang, K. *et al.* Optical synthesis of large-amplitude squeezed coherent-state superpositions with minimal resources. *Phys. Rev. Lett.* **115,** 023602 (2015).
30. Jeong, H. *et al.* Generation of hybrid entanglement of light. *Nat. Photon.* **8,** 564–569 (2014).
31. Morin, O. *et al.* Remote creation of hybrid entanglement between particle-like and wave-like optical qubits. *Nat. Photon.* **8,** 570–574 (2014).
32. Lvovsky, A. I. Iterative maximum-likelihood reconstruction in quantum homodyne tomography. *J. Opt. B: Quantum Semiclass. Opt.* **6,** S556–S559 (2004).
33. Řeháček, J., Hradil, Z., Knill, E. & Lvovsky, A. I. Diluted maximum-likelihood algorithm for quantum tomography. *Phys. Rev. A* **75,** 042108 (2007).
34. Lvovsky, A. I. & Raymer, M. G. Continuous-variable optical quantum-state tomography. *Rev. Mod. Phys.* **81,** 299 (2009).

Acknowledgements

We thank J. Dowling, Y. Kurochkin, A. Pushkina and A. Fedorov for helpful discussions, and RQC, NSERC and CIFAR for support. The support from the Ministry of Education and Science of the Russian Federation in the framework of the Federal Program (Agreement 14.582.21.0009) is greatly acknowledged.

Author contributions

All authors participated in the conception and planning of the project, theoretical analysis and writing of the paper. The experiment was performed by A.E.U., I.A.F. and D.S. The data were analysed by A.E.U., I.A.F., D.S. and A.I.L.

Additional information

Competing financial interests: The authors declare no competing financial interests.

23

Microcanonical and resource-theoretic derivations of the thermal state of a quantum system with noncommuting charges

Nicole Yunger Halpern[1], Philippe Faist[2], Jonathan Oppenheim[3] & Andreas Winter[4,5]

The grand canonical ensemble lies at the core of quantum and classical statistical mechanics. A small system thermalizes to this ensemble while exchanging heat and particles with a bath. A quantum system may exchange quantities represented by operators that fail to commute. Whether such a system thermalizes and what form the thermal state has are questions about truly quantum thermodynamics. Here we investigate this thermal state from three perspectives. First, we introduce an approximate microcanonical ensemble. If this ensemble characterizes the system-and-bath composite, tracing out the bath yields the system's thermal state. This state is expected to be the equilibrium point, we argue, of typical dynamics. Finally, we define a resource-theory model for thermodynamic exchanges of noncommuting observables. Complete passivity—the inability to extract work from equilibrium states—implies the thermal state's form, too. Our work opens new avenues into equilibrium in the presence of quantum noncommutation.

[1]Institute for Quantum Information and Matter, Division of Physics, Mathematics, and Astronomy, California Institute of Technology, Pasadena, California 91125, USA. [2]Institute for Theoretical Physics, Department of Physics, ETH Zürich, 8093 Zürich, Switzerland. [3]Department of Physics and Astronomy, University College London, Gower Street, London WC1E 6BT, UK. [4]Departament de Física, Grup d'Informació Quántica, Universitat Autònoma de Barcelona, ES-08193 Barcelona, Spain. [5]ICREA, Pg. Lluís Companys 23, 08010 Barcelona, Spain. Correspondence and requests for materials should be addressed to N.Y.H. (email: nicoleyh@caltech.edu) or to P.F. (email: pfaist@phys.ethz.ch) or to A.W. (email: andreas.winter@uab.cat).

Recently reignited interest in quantum thermodynamics has prompted information-theoretic approaches to fundamental questions[1-4]. The role of entanglement, for example, has been clarified with canonical typicality[5-8]. Equilibrium-like behaviours have been predicted[9-12] and experimentally observed in integrable quantum gases[13,14].

Thermodynamic resource theories offer a powerful tool for analysing fundamental properties of the thermodynamics of quantum systems. Heat exchanges with a bath are modelled with 'free states' and 'free operations'[15-18]. These resource theories have been extended to model exchanges of additional physical quantities, such as particles and angular momentum[18-22].

A central concept in thermodynamics and statistical mechanics is the thermal state. The thermal state has several important properties. First, typical dynamics evolve the system towards the thermal state. The thermal state is the equilibrium state. Second, consider casting statistical mechanics as an inference problem. The thermal state is the state that maximizes the entropy under constraints on physical quantities[23,24]. Third, consider the system as interacting with a large bath. The system-and-bath composite occupies a microcanonical state. Physical observables of the composite, such as the total energy and total particle number, have sharply defined values. The system's reduced state is the thermal state. Finally, in a resource theory, the thermal state is the only completely passive state. No work can be extracted from any number of copies of the thermal state[25,26].

If a small system exchanges heat and particles with a large environment, the system's thermal state is a grand canonical ensemble: $e^{-\beta(H-\mu N)}/Z$. The system's Hamiltonian and particle number are represented by H and N. β and μ denote the environment's inverse temperature and chemical potential. The partition function Z normalizes the state. The system-and-bath dynamics conserves the total energy and total particle number. More generally, subsystems exchange conserved quantities, or 'charges,' Q_j, $j=1, 2,\ldots c$. To these charges correspond generalized chemical potentials μ_j. The μ_j's characterize the bath.

We address the following question. Suppose that the charges fail to commute with each other: $[Q_j, Q_k]\neq 0$. What form does the thermal state have? We call this state 'the non-Abelian thermal state' (NATS). Jaynes applied the principle of maximum entropy to this question[24]. He associated fixed values v_j with the charges' expectation values. He calculated the state that, on satisfying these constraints, maximizes an entropy. This thermal state has a generalized Gibbs form:

$$\gamma_{\mathbf{v}} = \frac{1}{Z} e^{-\sum_{j=0}^{c} \mu_j Q_j} \tag{1}$$

wherein the v_j's determine the μ_j's.

Our contribution is a mathematical, physically justified derivation of the thermal state's form for systems whose dynamics conserve noncommuting observables. We recover the state (1) via several approaches, demonstrating its physical importance. We address puzzles raised in refs 21,27 about how to formulate a resource theory in which thermodynamic charges fail to commute. Closely related, independent work was performed by Guryanova *et al.*[28] We focus primarily on the nature of passive states. Guryanova *et al.*, meanwhile, focus more on the resource theory for multiple charges and on trade-offs amongst types of charge extractions.

In this paper, we derive the NATS's form from a microcanonical argument. A simultaneous eigenspace of all the noncommuting physical charges might not exist. Hence, we introduce the notion of an approximate microcanonical subspace. This subspace consists of the states in which the charges have sharply defined values. We derive conditions under which this subspace exists. We show that a small subsystem's reduced state lies, on average, close to $\gamma_{\mathbf{v}}$. Second, we invoke canonical typicality[7,8]. If the system-and-bath composite occupies a random state in the approximate microcanonical subspace, we argue, a small subsystem's state likely lies close to the NATS. Typical dynamics are therefore expected to evolve a well-behaved system's state towards the NATS. Third, we define a resource theory for thermodynamic exchanges of noncommuting conserved charges. We extend existing resource theories to model the exchange of noncommuting quantities. We show that the NATS is the only possible free state that renders the theory nontrivial: Work cannot be extracted from any number of copies of $\gamma_{\mathbf{v}}$. We show also that the NATS is the only state preserved by free operations. From this preservation, we derive 'second laws' that govern state transformations. This work provides a well-rounded, and novelly physical, perspective on equilibrium in the presence of quantum noncommutation. This perspective opens truly quantum avenues in thermodynamics.

Results

Overview. We derive the NATS's form via three routes: from a microcanonical argument; from a dynamical argument built on canonical typicality; and from complete passivity in a resource theory. Details appear in Supplementary Notes 1–3.

Microcanonical derivation. In statistical mechanics, the form $e^{-\beta(H-\mu N)}/Z$ of the grand canonical ensemble is well known to be derivable as follows. The system of interest is assumed to be part of a larger system. Observables of the composite have fixed values v_j. For example, the energy equals E_0, and the particle number equals N_0. The microcanonical ensemble is the whole-system state spread uniformly across these observables' simultaneous eigenspace. Tracing out the environmental degrees of freedom yields the state $e^{-\beta(H-\mu N)}/Z$.

We derive the NATS's form similarly. Crucially, however, we adapt the above strategy to allow for noncommuting observables. Observables might not have well-defined values v_j simultaneously. Hence a microcanonical ensemble as discussed above, suitable for commuting observables, may not exist. We overcome this obstacle by introducing an approximate microcanonical ensemble Ω. We show that, for every state satisfying the conditions of an approximate microcanonical ensemble, tracing out most of the larger system yields, on average, a state close to the NATS. We exhibit conditions under which an approximate microcanonical ensemble exists. The conditions can be satisfied when the larger system consists of many noninteracting replicas of the system. An important step in the proof consists of reducing the noncommuting case to the commuting one. This reduction relies on a result by Ogata (ref. 29, Theorem 1.1). A summary appears in Fig. 1.

Set-up. Let $\mathcal{S}$ denote a system associated with a Hilbert space $\mathcal{H}$, with a Hamiltonian $H \equiv Q_0$, and with observables (which we call 'charges') $Q_1, Q_2, \ldots, Q_c$. The charges do not necessarily commute with each other: $[Q_j, Q_k]\neq 0$.

Consider N replicas of $\mathcal{S}$, associated with the composite-system Hilbert space $\mathcal{H}^{\otimes N}$. We average each charge Q_j over the N copies:

$$\bar{Q}_j := \frac{1}{N}\sum_{\ell=0}^{N-1} \mathbb{I}^{\otimes \ell} \otimes Q_j \otimes \mathbb{I}^{\otimes (N-1-\ell)} \tag{2}$$

The basic idea is that, as N grows, the averaged operators $\bar{Q}_j$ come increasingly to commute. Indeed, there exist operators $\bar{Y}_j$ that commute with each other and that approximate the averages (ref. 29, Theorem 1.1). An illustration appears in Fig. 2.

Derivation. Since the $\bar{Y}_j$'s commute mutually, they can be measured simultaneously. More importantly, the joint Hilbert space $\mathcal{H}^{\otimes n}$ contains a subspace on which each $\bar{Q}_j$ has prescribed values close to v_j. Let $\mathcal{M}$ denote the subspace. Perhaps

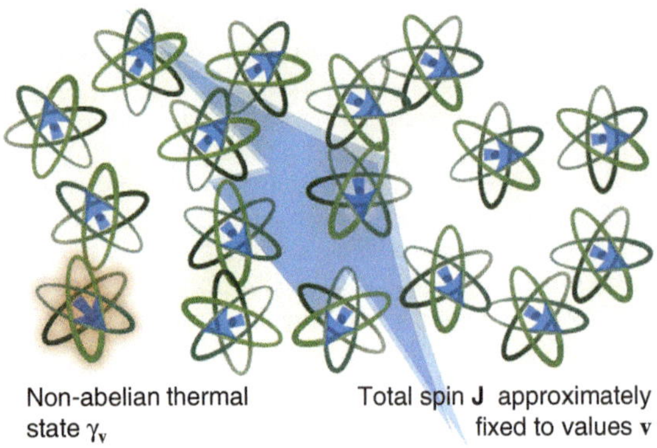

Figure 1 | Non-Abelian thermal state. We derive the form of the thermal state of a system that has charges that might not commute with each other. Example charges include the components J_i of the spin **J**. We derive the thermal state's form by introducing an approximate microcanonical state. An ordinary microcanonical ensemble could lead to the thermal state's form if the charges commuted: suppose, for example, that the charges were a Hamiltonian H and a particle number N that satisfied $[H, N]=0$. Consider many copies of the system. The composite system could have a well-defined energy E_{tot} and particle number N_{tot} simultaneously. E_{tot} and N_{tot} would correspond to some eigensubspace $\mathcal{H}_{E_{tot},N_{tot}}$ shared by the total Hamiltonian and the total-particle-number operator. The (normalized) projector onto $\mathcal{H}_{E_{tot},N_{tot}}$ would represent the composite system's microcanonical state. Tracing out the bath would yield the system's thermal state. But the charges J_i under consideration might not commute. The charges might share no eigensubspace. Quantum noncommutation demands a modification of the ordinary microcanonical argument. We define an approximate microcanonical subspace $\mathcal{M}$. Each state in $\mathcal{M}$ simultaneously has almost-well-defined values of noncommuting whole-system charges: measuring any such whole-system charge has a high probability of outputting a value close to an 'expected' value analogous to E_{tot} and N_{tot}. We derive conditions under which the approximate microcanonical subspace $\mathcal{M}$ exists. The (normalized) projector onto $\mathcal{M}$ represents the whole-system state. Tracing out most of the composite system yields the reduced state of the system of interest. We show that the reduced state is, on average, close to the NATS. This microcanonical derivation of the NATS's form strengthens the link between Jaynes's information-theoretic derivation and physics.

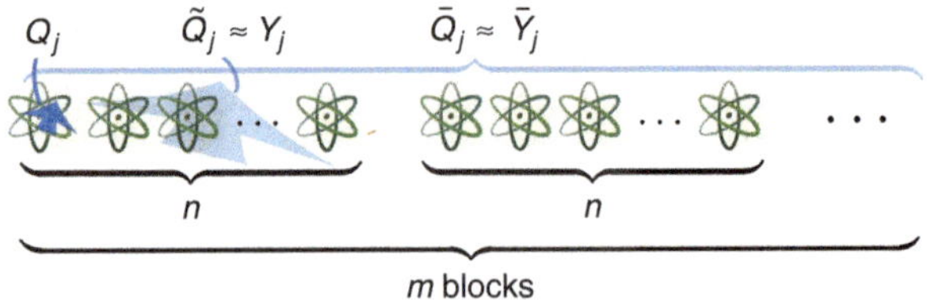

Figure 2 | Noncommuting charges. We consider a thermodynamic system $\mathcal{S}$ that has conserved charges Q_j. These Q_j's might not commute with each other. The system occupies a thermal state whose form we derive. The derivation involves an approximate microcanonical state of a large system that contains the system of interest. Consider a block of n copies of $\mathcal{S}$. Most copies act, jointly, similarly to a bath for the copy of interest. We define $\tilde{Q}_j$ as the average of the Q_j's of the copies in the block. Applying results from Ogata[29], we find operators $\tilde{Y}_j$ that are close to the $\tilde{Q}_j$'s and that commute with each other. Next, we consider m such blocks. This set of m blocks contains $N=mn$ copies of $\mathcal{S}$. Averaging the $\tilde{Q}_j$'s over the blocks, for a fixed j-value, yields a global observable $\bar{Q}_j$. The $\bar{Q}_j$'s are approximated by $\bar{Y}_j$'s. The $\bar{Y}_j$'s are the corresponding averages of the $\tilde{Y}_j$'s. The approximate global charges $\bar{Y}_j$ commute with each other. The commuting $\bar{Y}_j$'s enable us to extend the concept of a microcanonical ensemble from the well-known context in which all charges commute to truly quantum systems whose charges do not necessarily commute.

unsurprisingly, because the $\bar{Y}_j$'s approximate the $\bar{Q}_j$'s, each state in $\mathcal{M}$ has a nearly well-defined value of $\bar{Q}_j$ near v_j. If $\bar{Q}_j$ is measured, the distribution is sharply peaked around v_j. We can also show the opposite: every state with nearly well-defined values v_j of all $\bar{Q}_j$'s has most of its probability weight in $\mathcal{M}$.

These two properties show that $\mathcal{M}$ is an approximate microcanonical subspace for the $\bar{Q}_j$'s with values v_j. The notion of the approximate microcanonical subspace is the first major contribution of our work. It captures the idea that for large N we can approximately fix the values of the noncommuting charges Q_j. An approximate microcanonical subspace $\mathcal{M}$ is any subspace consisting of the whole-system states whose average observables $\bar{Q}_j$ have nearly well-defined values v_j. More precisely, a measurement of any $\bar{Q}_j$ has a high probability of yielding a value near v_j if and only if most of the state's probability weight lies in $\mathcal{M}$.

Normalizing the projector onto $\mathcal{M}$ yields an approximate microcanonical ensemble, Ω. Tracing out every copy of $\mathcal{S}$ but the ℓth yields the reduced state Ω_ℓ. The distance between Ω_ℓ and the NATS $\gamma_{\mathbf{v}}$ can be quantified by the relative entropy

$$D(\Omega_\ell \| \gamma_{\mathbf{v}}) := -S(\Omega_\ell) - \mathrm{Tr}(\Omega_\ell \log \gamma_{\mathbf{v}}). \tag{3}$$

Here $S(\Omega_\ell) := -\mathrm{Tr}(\Omega_\ell \log \Omega_\ell)$ is the von Neumann entropy. The relative entropy D is bounded by the trace norm $\|.\|_1$, which quantifies the distinguishability of Ω_ℓ and $\gamma_{\mathbf{v}}$ (ref. 30):

$$D(\Omega_\ell \| \gamma_{\mathbf{v}}) \geq \frac{1}{2} \|\Omega_\ell - \gamma_{\mathbf{v}}\|_1^2. \tag{4}$$

Our second main result is that if Ω is an approximate microcanonical ensemble, then the average, over systems ℓ, of the relative entropy D between Ω_ℓ and $\gamma_{\mathbf{v}}$ is small:

$$\frac{1}{N} \sum_{\ell=0}^{N-1} D(\Omega_\ell \| \gamma_{\mathbf{v}}) \leq \theta + \theta'. \tag{5}$$

The parameter $\theta = (\text{const.})/\sqrt{N}$ vanishes in the many-copy limit. θ' depends on the number c of charges, on the approximate expectation values v_j, on the eigenvalues of the charges Q_j and on the (small) parameters in terms of which $\mathcal{M}$ approximates a microcanonical subspace.

Inequality (5) capstones the derivation. The inequality follows from bounding each term in equation (3), the definition of the relative entropy D. The entropy $S(\Omega_\ell)$ is bounded with θ. This bound relies on Schumacher's Theorem, which quantifies the size of a high-probability subspace like $\mathcal{M}$ with an entropy $S(\gamma_{\mathbf{v}})$[31]. We bound the second term in the D definition with θ'. This bound relies on the definition of $\mathcal{M}$: outcomes of measurements of the $\bar{Q}_j$'s are predictable up to parameters on which θ' depends.

Finally, we present conditions under which the approximate microcanonical subspace $\mathcal{M}$ exists. Several parameters quantify the approximation. The parameters are shown to be interrelated and to approach zero simultaneously as N grows. In particular, the approximate microcanonical subspace $\mathcal{M}$ exists if N is great enough.

This microcanonical derivation offers a physical counterpoint to Jaynes's maximum-entropy derivation of the NATS's form. We relate the NATS to the physical picture of a small subsystem in a vast universe that occupies an approximate microcanonical state. This vast universe allows the correspondence principle to underpin our argument. In the many-copy limit as $N \to \infty$, the principle implies that quantum behaviours should vanish, as the averages of the noncommuting charges Q_j come to be approximated by commuting $\bar{Y}_j$'s. Drawing on Ogata's Theorem 1.1 (ref. 29), we link thermality in the presence of noncommutation to the physical correspondence principle.

Dynamical considerations. The microcanonical and maximum-entropy arguments rely on kinematics and information theory. But we wish to associate the NATS with the fixed point of dynamics. The microcanonical argument, combined with canonical typicality, suggests that the NATS is the equilibrium state of typical dynamics. Canonical typicality enables us to model the universe's state with a pure state in the approximate microcanonical subspace $\mathcal{M}$. If a large system occupies a randomly chosen pure state, the reduced state of a small subsystem is close to thermal[5–8].

Consider, as in the previous section, N copies of the system $\mathcal{S}$. By Ω, we denoted the composite system's approximately microcanonical state. We denoted by Ω_ℓ the reduced state of the ℓth copy, formed by tracing out most copies from Ω. Imagine that the whole system occupies a pure state $|\psi\rangle \in \mathcal{M}$. Denote by ρ_ℓ the reduced state of the ℓth copy. ρ_ℓ is close to Ω_ℓ, on average, by canonical typicality[7]:

$$\langle \|\rho_\ell - \Omega_\ell\|_1 \rangle \leq \frac{d}{\sqrt{D_M}}. \tag{6}$$

The average $\langle . \rangle$ is over pure states $|\psi\rangle \in \mathcal{M}$. The trace norm is denoted by $\|.\|_1$; $d := \dim(\mathcal{H})$ denotes the dimensionality of the Hilbert space $\mathcal{H}$ of one copy of $\mathcal{S}$; and $D_M := \dim(\mathcal{M})$ denotes the dimensionality of the approximate microcanonical subspace $\mathcal{M}$.

We have bounded, using canonical typicality, the average trace norm between ρ_ℓ and Ω_ℓ. We can bound the average trace norm between Ω_ℓ and the NATS $\gamma_{\mathbf{v}}$, using our microcanonical argument. (Supplementary equation (10) bounds the average relative entropy D between Ω_ℓ and $\gamma_{\mathbf{v}}$. Pinsker's inequality, Ineq. (4), lower bounds D in terms of the trace norm.) Combining these two trace-norm bounds via the triangle inequality, we bound the average distance between ρ_ℓ and $\gamma_{\mathbf{v}}$:

$$\left\langle \frac{1}{N} \sum_{\ell=0}^{N-1} \|\rho_\ell - \gamma_{\mathbf{v}}\|_1 \right\rangle \leq \frac{d}{\sqrt{D}} + \sqrt{2(\theta + \theta')}. \tag{7}$$

If the whole system occupies a random pure state $|\psi\rangle$ in $\mathcal{M}$, the reduced state ρ_ℓ of a subsystem is, on average, close to the NATS $\gamma_{\mathbf{v}}$.

Sufficiently ergodic dynamics is expected to evolve the whole-system state to a $|\psi\rangle$ that satisfies Ineq. (7): suppose that the whole system begins in a pure state $|\psi(t=0)\rangle \in \mathcal{M}$. Suppose that the system's Hamiltonian commutes with the charges: $[H, Q_j] = 0$ for all $j = 1, \ldots, c$. The dynamics conserves the charges. Hence, most of the amplitude of $|\psi(t)\rangle$ remains in $\mathcal{M}$ for appreciable times. Over sufficient times, ergodic dynamics yields a state $|\psi(t)\rangle$ that can be regarded as random. Hence the reduced state is expected be close to $\Omega_\ell \approx \gamma_{\mathbf{v}}$ for most long-enough times t.

Exploring how the dynamics depends on the number of copies of the system offers promise for interesting future research.

Resource theory. A thermodynamic resource theory is an explicit characterization of a thermodynamic system's resources, free states and free operations with a rigorous mathematical model. The resource theory specifies what an experimenter considers valuable (for example, work) and what is considered plentiful, or free (for example, thermal states). To define a resource theory, we specify allowed operations and which states can be accessed for free. We use this framework to quantify the resources needed to transform one state into another.

The first resource theory was entanglement theory[32]. The theory's free operations are local operations and classical communication. The free states are the states that can be easily prepared with local operations and classical communication, the separable states. Entangled states constitute valuable resources. One can quantify entanglement using this resource theory.

We present a resource theory for thermodynamic systems that have noncommuting conserved charges Q_j. The theory is defined by its set of free operations, which we call 'non-Abelian thermal operations' (NATO). NATO generalize thermal operations[15,18]. How to extend thermodynamic resource theories to conserved quantities other than energy was noted in refs 18,20,21. The NATO theory is related to the resource theory in ref. 27.

We supplement these earlier approaches with two additions. First, a battery has a work payoff function dependent on chemical potentials. We use this payoff function to define chemical work. Second, we consider a reference system for a non-Abelian group. The reference system is needed to resolve the difficulty encountered in refs 21,27: There might be no nontrivial operations which respect all the conservation laws. The laws of physics require that any operation performed by an experimenter commutes with all the charges. If the charges fail to commute with each other, there might be no non-trivial unitaries that commute with all of them. In practice, one is not limited by such a stringent constraint. The reason is that an experimenter has access to a reference frame[33–35].

A reference frame is a system W prepared in a state such that for any unitary on a system S which does not commute with the charges of S, some global unitary on WS conserves the total charges and approximates the unitary on S to arbitrary precision. The reference frame relaxes the strong constraint on the unitaries. The reference frame can be merged with the battery, in which the agent stores the ability to perform work. We refer to the composite as 'the battery'. We denote its state by ρ_{W}. The battery has a Hamiltonian H_{W} and charges $Q_{j_{\mathrm{W}}}$, described below.

Within this resource theory, the NATS emerges in two ways:

1. The NATS is the unique state from which work cannot be extracted, even if arbitrarily many copies are available. That is, the NATS is completely passive.
2. The NATS is the only state of S that remains invariant under the free operations during which no work is performed on S.

On proving the latter condition, we prove second laws for thermodynamics with noncommuting charges. These laws provide necessary conditions for a transition to be possible. In some cases, we show, the laws are sufficient. These second laws govern state transitions of a system ρ_{S}, governed by a Hamiltonian H_{S}, whose charges $Q_{j_{\mathrm{S}}}$ can be exchanged with the surroundings. We allow the experimenter to couple ρ_{S} to free states ρ_{R}. The form of ρ_{R} is determined by the Hamiltonian H_{R} and the charges $Q_{j_{\mathrm{R}}}$ attributable to the free system. We will show that these free states have the form of the NATS. As noted above, no other state could be free. If other states were free, an arbitrarily large amount of work could be extracted from them.

Before presenting the second laws, we must define 'work'. In textbook examples about gases, one defines work as $\delta W = p\, dV$, because a change in volume at a fixed pressure can be translated into the ordinary notion of mechanical work. If a polymer is stretched, then $\delta W = F\, dx$, wherein x denotes the polymer's linear displacement and F denotes the restoring force. If B denotes a magnetic field and M denotes a medium's magnetization, $\delta W = B\, dM$. The definition of 'work' can depend on one's ability to transform changes in thermodynamic variables into a standard notion of 'work,' such as mechanical or electrical work.

Our approach is to define a notion of chemical work. We could do so by modelling explicitly how the change in some quantity Q_j can be used to extract $\mu_j\ \delta Q_j$ work. Explicit modelling would involve adding a term to the battery Hamiltonian H_W. Rather than considering a specific work Hamiltonian or model of chemical work, however, we consider a work payoff function,

$$\mathcal{W} = \sum_{j=0}^{c} \mu_j Q_{jw} \tag{8}$$

The physical situation could determine the form of this $\mathcal{W}$. For example, the μ_j's could denote the battery's chemical potentials. In such a case, $\mathcal{W}$ would denote the battery's total Hamiltonian, which would depend on those potentials.

We choose a route conceptually simpler than considering an explicit Hamiltonian and battery system, however. We consider equation (8) as a payoff function that defines the linear combination of charges that interests us. We define the (chemical) work expended or distilled during a transformation as the change in the quantum expectation value $\langle \mathcal{W} \rangle$.

The form of $\mathcal{W}$ is implicitly determined by the battery's structure and by how charges can be converted into work. For our purposes, however, the origin of the form of $\mathcal{W}$ need not be known. $\mathcal{W}$ will uniquely determine the μ_j's in the NATS. Alternatively, we could first imagine that the agent could access, for free, a particular NATS. This NATS's form would determine the work function's form. If the charges commute, the corresponding Gibbs state is known to be the unique state that is completely passive with respect to the observable (8).

In Supplementary Note 3, we specify the resource theory for noncommuting charges in more detail. We show how to construct allowable operations, using the reference frame and battery. From the allowable operations, we derive a zeroth law of thermodynamics.

Complete passivity and zeroth law. This zeroth law relates to the principle of complete passivity, discussed in refs 25,26. A state is complete passive if an agent cannot extract work from arbitrarily many copies of the state. In the resource theory for heat exchanges, completely passive states can be free. They do not render the theory trivial because no work can be drawn from them[17].

In the NATO resource theory, we show, the only reasonable free states have the NATS's form. The free states' chemical potentials equal the μ_j's in the payoff function $\mathcal{W}$, at some common fixed temperature. Any other state would render the resource theory trivial: from copies of any other state, arbitrarily much work could be extracted for free. Then, we show that the NATS is preserved by NATO, the operations that perform no work on the system.

The free states form an equivalence class. They lead to notions of temperature and chemical potentials μ_j. This derivation of the free state's form extends complete passivity and the zeroth law from ref. 17 to noncommuting conserved charges. The derivation further solidifies the role of the NATS in thermodynamics.

Second laws. The free operations preserve the NATS. We therefore focus on contractive measures of states' distances from the NATS. Contractive functions decrease monotonically under the free operations. Monotones feature in 'second laws' that signal whether NATO can implement a state transformation. For example, the α-Rényi relative entropies between a state and the NATS cannot increase.

Monotonicity allows us to define generalized free energies as

$$F_\alpha(\rho_S, \gamma_S) := k_B T D_\alpha(\rho_S \| \gamma_S) - k_B T \log Z \tag{9}$$

wherein $\beta \equiv 1/(k_B T)$ and k_B denotes Boltzmann's constant. γ_S denotes the NATS with respect to the Hamiltonian H_S and the charges Q_{js} of the system S. The partition function is denoted by Z. Various classical and quantum definitions of the Rényi relative entropies D_α are known to be contractive[17,36–39]. The free energies F_α decrease monotonically if no work is performed on the system. Hence the F_α's characterize natural second laws that govern achievable transitions.

For example, the classical Rényi divergences $D_\alpha(\rho_S \| \gamma_S)$ are defined as

$$D_\alpha(\rho_S \| \gamma_S) := \frac{\operatorname{sgn}(\alpha)}{\alpha - 1} \log\left(\sum_k p_k^\alpha q_k^{1-\alpha} \right), \tag{10}$$

wherein p_k and q_k denote the probabilities of ρ_S and of γ_S in the $\mathcal{W}$ basis. The D_α's lead to second laws that hold even in the absence of a reference frame and even outside the context of the average work.

The F_α's reduce to the standard free energy when averages are taken over large numbers. Consider the asymptotic ('thermodynamic') limit in which many copies $(\rho_S)^{\otimes n}$ of ρ_S are transformed. Suppose that the agent has some arbitrarily small probability ε of failing to implement the desired transition. ε can be incorporated into the free energies via a technique called 'smoothing'[17]. The average, over copies of the state, of every smoothed F_α^ε approaches F_1 (ref. 17):

$$\lim_{n\to\infty} \frac{1}{n} F_\alpha^\varepsilon\left((\rho_S)^{\otimes n}, (\gamma_S)^{\otimes n}\right) = F_1 \tag{11}$$

$$= k_B T D(\rho_S \| \gamma_S) - k_B T \log(Z) \tag{12}$$

$$= \langle H_S \rangle_{\rho_S} - k_B T S(\rho_S) + \sum_{j=1}^{c} \mu_j \langle Q_{js} \rangle. \tag{13}$$

We have invoked the relative entropy's definition

$$D(\rho_S \| \gamma_S) := \operatorname{Tr}(\rho_S \log(\rho_S)) - \operatorname{Tr}(\rho_S \log(\gamma_S)). \tag{14}$$

Note the similarity between the many-copy average F_1 in equation (13) and the ordinary free energy, $F = E - TS + \sum_j \mu_j N_j$. The monotonic decrease of F_1 constitutes a necessary and sufficient condition for a state transition to be possible in the presence of a reference system in the asymptotic limit.

In terms of the generalized free energies, we formulate second laws.

Proposition 1: in the presence of a heat bath of inverse temperature β and chemical potentials μ_j, the free energies $F_\alpha(\rho_S, \gamma_S)$ decrease monotonically:

$$F_\alpha(\rho_S, \gamma_S) \geq F_\alpha(\rho_S', \gamma_S') \ \forall \alpha \geq 0 \tag{15}$$

wherein ρ_S and ρ_S' denote the system's initial and final states. The system's Hamiltonian and charges may transform from H_S and Q_{js} to H_S' and Q_{js}'. The NATSs associated with the same Hamiltonians and charges are denoted by γ_S and γ_S'. If

$$\begin{aligned} &[\mathcal{W}, \rho_S'] = 0 \quad \text{and} \\ &F_\alpha(\rho_S, \gamma_S) \geq F_\alpha(\rho_S', \gamma_S') \ \forall \alpha \geq 0 \end{aligned} \tag{16}$$

some NATO maps ρ_S to ρ_S'.

As in ref. 17, additional laws can be defined in terms of quantum Rényi divergences[36–39]. This amounts to choosing, in Proposition 1, a definition of the Rényi divergence which accounts for the possibility that ρ_S and ρ_S' have coherences relative to the $\mathcal{W}_S$ eigenbasis. Several measures are known to be contractive[36–39]. They, too, provide a new set of second laws.

Extractable work: in terms of the free energies F_α, we can bound the work extractable from a resource state via NATO. We consider the battery W separately from the system S of interest. We assume that W and S occupy a product state.

(This assumption is unnecessary if we focus on average work.) Let ρ_W and ρ'_W denote the battery's initial and final states.

For all α,

$$F_\alpha(\rho_S \otimes \rho_W, \gamma_{SW}) \geq F_\alpha(\rho'_S \otimes \rho'_W, \gamma_{SW}). \tag{17}$$

Since $F_\alpha(\rho_S \otimes \rho_W, \gamma_{SW}) = F_\alpha(\rho_S, \gamma_S) + F_\alpha(\rho_W, \gamma_W)$,

$$F_\alpha(\rho'_W, \gamma_W) - F_\alpha(\rho_W, \gamma_W) \leq F_\alpha(\rho_S, \gamma_S) - F_\alpha(\rho'_S, \gamma_S). \tag{18}$$

The left-hand side of Ineq. (18) represents the work extractable during one implementation of $\rho_S \to \rho'_S$. Hence the right-hand side bounds the work extractable during the transition.

Consider extracting work from many copies of ρ_S (that is, extracting work from $\rho_S^{\otimes n}$) in each of many trials. Consider the average-over-trials extracted work, defined as $\mathrm{Tr}(\mathcal{W}[\rho'_W - \rho_W])$. The average-over-trials work extracted per copy of ρ_S is $\frac{1}{n}\mathrm{Tr}(\mathcal{W}[\rho'_W - \rho_W])$. This average work per copy has a high probability of lying close to the change in the expectation value of the system's work function, $\frac{1}{n}\mathrm{Tr}(\mathcal{W}[\rho'_W - \rho_W]) \approx \mathrm{Tr}(W[\rho'_S - \rho_S])$, if n is large.

Averaging over the left-hand side of Ineq. (18) yields the average work $\delta\langle W\rangle$ extracted per instance of the transformation. The average over the right-hand side approaches the change in F_1 (equation (13)):

$$\delta\langle W\rangle \leq \delta\langle H_S\rangle_{\rho_S} - T\,\delta S(\rho_S) + \sum_{j=1}^{c} \mu_j \langle \delta_{js}\rangle \tag{19}$$

This bound is achievable with a reference system, as shown in refs 40,41.

We have focused on the extraction of work defined by $\mathcal{W}$. One can extract, instead, an individual charge Q_j. The second laws do not restrict single-charge extraction. But extracting much of one charge Q_j precludes the extraction of much of another charge, Q_k. In Supplementary Note 3, we discuss the trade-offs amongst extractions of different charges Q_j.

Discussion

We have derived, via multiple routes, the form of the thermal state of a system that has noncommuting conserved charges. First, we regarded the system as part of a vast composite that occupied an approximate microcanonical state. Tracing out the environment yields a reduced state that lies, on average, close to a thermal state of the expected form. This microcanonical argument, with canonical typicality, suggests that the NATS is the fixed point of typical dynamics. Defining a resource theory, we showed that the NATS is the only completely passive state and is the only state preserved by free operations. These physical derivations buttress Jaynes's information-theoretic derivation from the principle of maximum entropy.

Our derivations also establish tools applicable to quantum noncommutation in thermodynamics. In the microcanonical argument, we introduced an approximate microcanonical state Ω. This Ω resembles the microcanonical ensemble associated with a fixed energy, a fixed particle number and so on but accommodates noncommuting charges. Our complete-passivity argument relies on a little-explored resource theory for thermodynamics, in which free unitaries conserve noncommuting charges.

We expect that the equilibrium behaviours predicted here may be observed in experiments. Quantum gases have recently demonstrated equilibrium-like predictions about integrable quantum systems[11,13].

From a conceptual perspective, our work shows that notions previously considered relevant only to commuting charges—for example, the microcanonical subspace—extend to noncommuting charges. This work opens fully quantum thermodynamics to analysis with familiar, but suitably adapted, technical tools.

References

1. Gemmer, J., Michel, M., Michel, M. & Mahler, G. *Quantum Thermodynamics: Emergence of Thermodynamic Behavior within Composite Quantum Systems* (Springer, 2009).
2. Gogolin, C. & Eisert, J. Equilibration, thermalisation, and the emergence of statistical mechanics in closed quantum systems. *Rep. Prog. Phys.* **79**, 056001 (2016).
3. Goold, J., Huber, M., Riera, A., del Rio, L. & Skrzypczyk, P. The role of quantum information in thermodynamics—a topical review. *J. Phys. A: Math. Theor.* **49**, 143001 (2016).
4. Vinjanampathy, S. & Anders, J. Quantum thermodynamics. Preprint at http://arxiv.org/abs/1508.06099 (2015).
5. Goldstein, S., Lebowitz, J. L., Tumulka, R. & Zanghí, N. Canonical typicality. *Phys. Rev. Lett.* **96**, 050403 (2006).
6. Gemmer, J., Michel, M. & Mahler, G. *18 Equilibrium Properties of Model Systems* (Springer, 2004).
7. Popescu, S., Short, A. J. & Winter, A. The foundations of statistical mechanics from entanglement: individual states versus averages. *Nat. Phys.* **2**, 754–758 (2006).
8. Linden, N., Popescu, S., Short, A. J. & Winter, A. Quantum mechanical evolution towards thermal equilibrium. *Phys. Rev. E* **79**, 061103 (2009).
9. Fermi, E., Pasta, J. & Ulam, S. *Studies of Nonlinear Problems I.* Los Alamos Report LA-1940 (Los Alamos Scientific Laboratory, 1955).
10. Kinoshita, T., Wenger, T. & Weiss, D. S. A quantum Newton's cradle. *Nature* **440**, 900–903 (2006).
11. Rigol, M., Dunjko, V., Yurovsky, V. & Olshanii, M. Relaxation in a completely integrable many-body quantum system: an *Ab Initio* study of the dynamics of the highly excited states of 1d lattice hard-core bosons. *Phys. Rev. Lett.* **98**, 050405 (2007).
12. Polkovnikov, A., Sengupta, K., Silva, A. & Vengalattore, M. *Colloquium*: nonequilibrium dynamics of closed interacting quantum systems. *Rev. Mod. Phys.* **83**, 863–883 (2011).
13. Langen, T. *et al.* Experimental observation of a generalized Gibbs ensemble. *Science* **348**, 207–211 (2015).
14. Langen, T., Geiger, R. & Schmiedmayer, J. Ultracold atoms out of equilibrium. *Annu. Rev. Condens. Matter Phys.* **6**, 201–217 (2015).
15. Janzing, D., Wocjan, P., Zeier, R., Geiss, R. & Beth, T. Thermodynamic cost of reliability and low temperatures: tightening landauer's principle and the second law. *Int. J. Theor. Phys.* **39**, 2717–2753 (2000).
16. Brandão, F. G. S. L., Horodecki, M., Oppenheim, J., Renes, J. M. & Spekkens, R. W. Resource theory of quantum states out of thermal equilibrium. *Phys. Rev. Lett.* **111**, 250404 (2013).
17. Brandao, F. G. S. L., Horodecki, M., Ng, N. H. Y., Oppenheim, J. & Wehner, S. The second laws of quantum thermodynamics. *Proc. Natl Acad. Sci. USA* **112**, 3275–3279 (2015).
18. Horodecki, M. & Oppenheim, J. Fundamental limitations for quantum and nano thermodynamics. *Nat. Commun.* **4**, 2059 (2013).
19. Vaccaro, J. A. & Barnett, S. M. Information erasure without an energy cost. *Proc. R. Soc. A* **467**, 1770–1778 (2011).
20. Yunger Halpern, N. & Renes, J. M. Beyond heat baths: generalized resource theories for small-scale thermodynamics. *Phys. Rev. E* **93**, 022126 (2016).
21. Yunger Halpern, N. Beyond heat baths II: Framework for generalized thermodynamic resource theories. Preprint at http://arxiv.org/abs/1409.7845 (2014).
22. Weilenmann, M., Krämer, L., Faist, P. & Renner, R. Axiomatic relation between thermodynamic and information-theoretic entropies. Preprint at http://arxiv.org/abs/1501.06920 (2015).
23. Jaynes, E. T. Information theory and statistical mechanics I. *Phys. Rev.* **106**, 620–630 (1957).
24. Jaynes, E. T. Information theory and statistical mechanics II. *Phys. Rev.* **108**, 171–190 (1957).
25. Pusz, W. & Woronowicz, S. L. Passive states and KMS states for general quantum systems. *Comm. Math. Phys.* **58**, 273–290 (1978).
26. Lenard, A. Thermodynamical proof of the Gibbs formula for elementary quantum systems. *J. Stat. Phys.* **19**, 575 (1978).
27. Lostaglio, M., Jennings, D. & Rudolph, T. Thermodynamic resource theories, non-commutativity and maximum entropy principles. Preprint at http://arxiv.org/abs/1511.04420 (2015).

28. Guryanova, Y., Popescu, S., Short, A. J., Silva, R. & Skrzypczyk, P. Thermodynamics of quantum systems with multiple conserved quantities. *Nat. Commun.* **7,** 12049 (2016).
29. Ogata, Y. Approximating macroscopic observables in quantum spin systems with commuting matrices. *J. Funct. Anal.* **264,** 2005–2033 (2013).
30. Hiai, F., Ohya, M. & Tsukada, M. Sufficiency, KMS condition and relative entropy in von Neumann algebras. *Pacific J. Math.* **96,** 99–109 (1981).
31. Schumacher, B. Quantum coding. *Phys. Rev. A* **51,** 2738–2747 (1995).
32. Horodecki, R., Horodecki, P., Horodecki, M. & Horodecki, K. Quantum entanglement. *Rev. Mod. Phys.* **81,** 865–942 (2009).
33. Aharonov, Y. & Susskind, L. Charge superselection rule. *Phys. Rev.* **155,** 1428–1431 (1967).
34. Kitaev, A., Mayers, D. & Preskill, J. Superselection rules and quantum protocols. *Phys. Rev. A* **69,** 052326 (2004).
35. Bartlett, S. D., Rudolph, T. & Spekkens, R. W. Reference frames, superselection rules, and quantum information. *Rev. Mod. Phys.* **79,** 555–609 (2007).
36. Hiai, F., Mosonyi, M., Petz, D. & Bény, C. Quantum *f*-divergences and error correction. *Rev. Math. Phys.* **23,** 691–747 (2011).
37. Müller-Lennert, M., Dupuis, F., Szehr, O., Fehr, S. & Tomamichel, M. On quantum Rényi entropies: a new generalization and some properties. *J. Math. Phys.* **54,** 122203 (2013).
38. Wilde, M. M., Winter, A. & Yang, D. Strong converse for the classical capacity of entanglement-breaking and Hadamard channels via a sandwiched Rényi relative entropy. *Commun. Math. Phys.* **331,** 593–622 (2014).
39. Jaksic, V., Ogata, Y., Pautrat, Y. & Pillet, C.-A. in *Quantum Theory from Small to Large Scales: Lecture Notes of the Les Houches Summer School.* Vol. **95** (eds Fröhlich, J., Manfred, S., Vieri, M., De Roeck, W. & Cugliandolo, L. F.) (Oxford Univ. Press, 2012).
40. Åberg, J. Catalytic coherence. *Phys. Rev. Lett.* **113,** 150402 (2014).
41. Korzekwa, K., Lostaglio, M., Oppenheim, J. & Jennings, D. The extraction of work from quantum coherence. *N. J. Phys.* **18,** 023045 (2016).

Acknowledgements

We thank David Jennings, Tim Langen, Elliott Lieb (who pointed us to ref. 29), Matteo Lostaglio, Shelly Moram, Joseph M. Renes and Terry Rudolph for interesting conversations. We thank the authors of refs 27,28 for their community spirit. Much of this paper was developed at 'Beyond i.i.d. in Information Theory 2015,' hosted by BIRS. A.W.'s work was supported by the EU (STREP 'RAQUEL'), the ERC (AdG 'IRQUAT'), the Spanish MINECO (grant FIS2013-40627-P) with the support of FEDER funds, as well as by the Generalitat de Catalunya CIRIT, project 2014-SGR-966. J.O. is supported by an EPSRC Established Career Fellowship, the Royal Society and FQXi. N.Y.H. was supported by an IQIM Fellowship and NSF grant PHY-0803371. The Institute for Quantum Information and Matter (IQIM) is an NSF Physics Frontiers Center supported by the Gordon and Betty Moore Foundation. P.F. acknowledges support from the European Research Council (ERC) via grant No. 258932, from the Swiss National Science Foundation through the National Centre of Competence in Research 'Quantum Science and Technology' (QSIT), and by the European Commission via the project 'RAQUEL.' This work was partially supported by the COST Action MP1209.

Author contributions

All authors contributed to all aspects of this work.

Additional information

Competing financial interests: The authors declare no competing financial interests.

24

Statistical moments of quantum-walk dynamics reveal topological quantum transitions

Filippo Cardano[1], Maria Maffei[1], Francesco Massa[1,†], Bruno Piccirillo[1], Corrado de Lisio[1,2], Giulio De Filippis[1,2], Vittorio Cataudella[1,2], Enrico Santamato[1] & Lorenzo Marrucci[1,3]

Many phenomena in solid-state physics can be understood in terms of their topological properties. Recently, controlled protocols of quantum walk (QW) are proving to be effective simulators of such phenomena. Here we report the realization of a photonic QW showing both the trivial and the non-trivial topologies associated with chiral symmetry in one-dimensional (1D) periodic systems. We find that the probability distribution moments of the walker position after many steps can be used as direct indicators of the topological quantum transition: while varying a control parameter that defines the system phase, these moments exhibit a slope discontinuity at the transition point. Numerical simulations strongly support the conjecture that these features are general of 1D topological systems. Extending this approach to higher dimensions, different topological classes, and other typologies of quantum phases may offer general instruments for investigating and experimentally detecting quantum transitions in such complex systems.

[1] Dipartimento di Fisica, Università di Napoli Federico II, Complesso Universitario di Monte Sant'Angelo, via Cintia, Napoli 80126, Italy. [2] CNR-SPIN, Complesso Universitario di Monte Sant'Angelo, Via Cintia, Napoli 80126, Italy. [3] CNR-ISASI, Via Campi Flegrei 34, Pozzuoli (NA) 80078, Italy. † Present address: Faculty of Physics, University of Vienna, Boltzmanngasse 5, Vienna 1090, Austria. Correspondence and requests for materials should be addressed to L.M. (email: lorenzo.marrucci@unina.it).

The presence of topological order in matter is responsible for fundamental phenomena, such as the fractional and integer quantum Hall effects[1,2] or the protected surface states observed in topological insulators[3,4]. Non-trivial topological phases are related to symmetries and can be characterized by specific topological invariants, defined in terms of the energy band eigenstates. Besides physical systems that naturally exhibit such features, the same topological scenarios have been reproduced in easy-to-access artificial systems (quantum simulators), which represent convenient platforms for the investigation of the intriguing properties associated with these states of matter[5–9]. In particular, a variety of simulators based on photonic architectures have been realized[10–16].

In the field of quantum simulation of topological phenomena, quantum walks (QWs) are emerging as a versatile tool[17–19]. In its simplest version, a QW is the discrete-time evolution of a particle (the walker) on a one-dimensional (1D) lattice[20]. At each step, the walker can move to either one of the two nearest-neighbour sites of the lattice, as determined by the configuration of an internal two-state quantum system (the coin). Between consecutive steps, the coin state undergoes a rotation that determines the relative probability amplitudes of the subsequent walker move, thus providing the quantum version of the random choice process (the coin toss) characterizing the familiar classical random walk. Among simulators of topological physics, QWs are attracting a wide attention since this simple quantum dynamics can realize all topological phases occurring in one- and two-dimensional systems of non interacting particles[18]; as a first application of this concept, the formation of topologically protected bound states was observed in a photonic QW[10].

In this paper, we report the study of a QW process that exhibits two distinct topological phases, associated with chiral symmetry in a 1D bipartite lattice. Remarkably, identical symmetries and topological features characterize the Su–Schrieffer–Heeger model (SSH), describing for example the mobile electron dynamics in the poly-acetylene chain[21], and the effective theory for spin-less 1D superconductors showing p-wave pairing[22]. By studying the QW protocol introduced here, we found that the probability distribution moments for the particle position, in the limit of an infinitely long temporal evolution, show a different asymptotic behaviour for the two topological phases: varying an external control parameter, whose value determines the phase of the system, they exhibit a slope discontinuity at the quantum transition. These effects are found to occur also in other 1D topological systems, including models belonging to different topological classes; hence, we conjecture here that they are fully general in 1D. We simulated experimentally such phenomena and proposed a theoretical interpretation based on the dispersion relations and the geometric features of the system eigenstates. Remarkably, our analysis takes into account bulk dynamics only, and not the physical effects manifesting at the edges of the system. With a similar approach topological phases have been recently detected in a non-Hermitian QW,[17,11] while all the models we investigate here are Hermitian.

Results

Quantum-walk protocol based on light spin-orbit interaction. In recent years, QWs have been implemented in a variety of physical architectures[23]. In our photonic platform[24], the walker is encoded in the orbital angular momentum (OAM) of light[25]: discrete positions on the lattice are associated with states $|m\rangle$, where m is an integer, describing a photon carrying $m\hbar$ of OAM along its propagation axis. The coin is encoded in the polarization degree of freedom, that is, in the spin angular momentum (SAM) of photons: vectors $|L\rangle$ and $|R\rangle$, representing left and right circular polarizations (SAM of $\pm\hbar$ per photon), respectively, are the two internal states that determine opposite shift directions in the lattice. The quantum state of the photon after N steps of QW is denoted as $|\psi_N\rangle = \hat{U}_0^N|\psi_0\rangle$, where $|\psi_0\rangle$ is the initial (input) state and $\hat{U}_0$ is the unitary evolution operator of a single QW step. In our protocol, this is realized by cascading two optical elements: a quarter-wave plate (QWP) and a q-plate (QP). A q-plate is essentially a liquid-crystal birefringent cell having the optic axis arranged in a singular pattern[26], with topological charge q (in our case $q = 1/2$). This patterned birefringence gives rise to an engineered optical spin–orbit coupling that induces the polarization-controlled shift of OAM. As specified below (see equation 2), besides the charge q, the action of this device is determined by the value of the optical retardation δ, which can be tuned with an applied electric field[27]; this allows in turn for a fine control of the light spin–orbit interaction mediated by the plate. In our protocol, the QWP is oriented at 90° with respect to the horizontal direction; its action is described by the operator $\hat{W}_{\mathrm{qwp}}$, that transforms the polarization states as follows:

$$\begin{aligned}\hat{W}_{\mathrm{qwp}}|L,m\rangle &= \frac{1}{\sqrt{2}}(|L,m\rangle - i|R,m\rangle),\\ \hat{W}_{\mathrm{qwp}}|R,m\rangle &= \frac{-i}{\sqrt{2}}(|L,m\rangle + i|R,m\rangle).\end{aligned} \tag{1}$$

The q-plate action is described by the operator $\hat{Q}_\delta$, defined as

$$\begin{aligned}\hat{Q}_\delta|L,m\rangle &= \cos(\delta/2)|L,m\rangle + i\sin(\delta/2)|R,m+2q\rangle,\\ Q_\delta|R,m\rangle &= \cos(\delta/2)|R,m\rangle + i\sin(\delta/2)|L,m-2q\rangle.\end{aligned} \tag{2}$$

Here we can observe that the spin–orbit interaction introduced by the plate, proportional to $\sin(\delta/2)$, results in an exchange of angular momenta, that is a conversion from SAM to OAM. The combination of $\hat{Q}_\delta$ and $\hat{W}_{\mathrm{qwp}}$ realizes the operator for a single QW step, that is

$$\hat{U}_0 = \hat{Q}_\delta \hat{W}_{\mathrm{qwp}}. \tag{3}$$

A topological characterization of QWs with chiral symmetry may be defined in terms of the eigenstates of the operator $\hat{U}_0$ which, due to the lattice translation symmetry, are associated with two quasi-energy bands parameterized by a quasi-momentum k, as shown in Fig. 1a. As for momentum k, a periodic quasi-energy variable E replaces here the usual energy, being the time (= step number) a discrete variable. Accordingly, the dispersion relation for our QW is given by

$$E_\delta(k) = \pm\cos^{-1}\left[\frac{\cos(\delta/2) + \sin(\delta/2)\cos(k)}{\sqrt{2}}\right], \tag{4}$$

where the two signs correspond to the two bands. As a consequence, the expression of the associated group velocity is

$$V_\delta(k) = \frac{\mathrm{d}E_\delta(k)}{\mathrm{d}k} = \mp\frac{\sin(\delta/2)\sin(k)}{\sqrt{2 - [\cos(\delta/2) + \sin(\delta/2)\cos(k)]^2}}. \tag{5}$$

The band eigenstates are given by the product of a walker part, that is plane waves with quasi-momentum k in the walker space, and a coin part, denoted as $|\phi_{s,\delta}(k)\rangle$, where $s \in \{1, 2\}$ is the band index. Conveniently, these states can be represented as points on the Poincaré sphere for light polarization, individuated by a three-dimensional unit vector that we refer to as $\mathbf{n}_\delta(k)$; here we

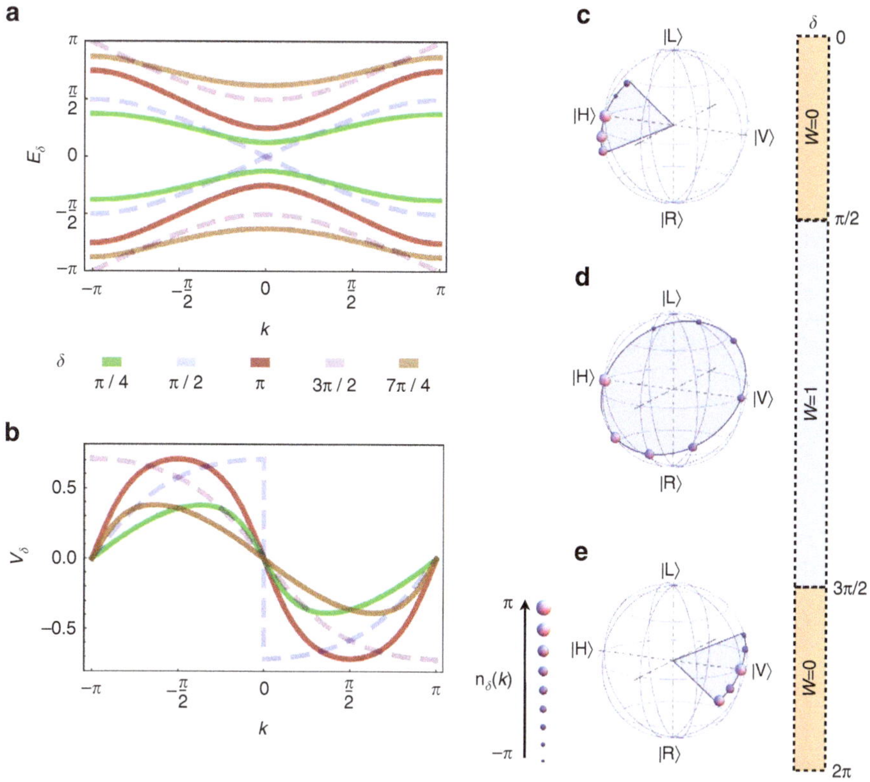

Figure 1 | Topological characterization of the QW process. (**a**,**b**) Dispersion relation for the quasi-energy $E_\delta(k)$ and the group velocity $V_\delta(k)$ of the Bloch bands in the QW; for the latter quantity, we plot the lower band only. The dispersion curves depend on the external parameter δ; the temporal coordinate is a discrete variable, hence the quasi-energy is defined in a Brillouin zone $\{-\pi, \pi\}$. Here we report few examples; colours of the dispersion curves are associated with specific values of δ, as shown in the panel legend. Generally, a finite gap separates the energies of the two bands; only at the two points $\delta=\delta_1=\pi/2$ and $\delta=\delta_2=3\pi/2$ (blue and purple dashed lines, respectively) the gap vanishes for $k=0$ and $k=\pi$, respectively, causing the presence of a discontinuity in the group velocity (see panel **b**). As shown in **c** points δ_1 and δ_2 represent the boundaries between two topological phases. (**c**-**e**) Topology of the Bloch eigenstates. States $|\phi_s(k)\rangle$, representing the polarization (coin) part of the eigenstates of the system, can be represented as points on the surface of a Poincaré sphere, where they are individuated by $\pm\mathbf{n}_\delta(k)$. Here we represent these states as solid spheres, whose radius is related to the value of k, as shown in the legend. As the quasi-momentum k spans the Brillouin zone $\{-\pi, \pi\}$, chiral symmetry forces these states to lie on a great circle of the sphere. The associated winding number W depends on δ, and determines the existence of a non-trivial and a trivial topological phases; the former ($W=1$) occurs when $\delta_1<\delta<\delta_2$, while the latter ($W=0$) is obtained in the remaining part of the interval $\{0, 2\pi\}$. As an example, in **c**-**e** we represent the coin eigenstates of a QW with $\delta=\pi/4$, π, $7\pi/4$, respectively. For $\delta=\pi$ these form a closed loop along the great circle, while in the other cases they go back and forth along a finite arch, whose length depends on the value of δ (when $\delta=0$ or 2π, the arch reduces to a single point).

report the explicit expressions of its components, that is

$$\begin{aligned} n_x(k) &= \frac{\cos(\delta/2)-\sin(\delta/2)\cos k}{\sqrt{2\{1-\cos^2[E_\delta(k)]\}}}, \\ n_y(k) &= -\frac{\sin(\delta/2)\sin k}{\sqrt{2\{1-\cos^2[E_\delta(k)]\}}}, \\ n_z(k) &= -n_y(k). \end{aligned} \tag{6}$$

Chiral symmetry constrains vectors $\mathbf{n}_\delta(k)$ to lie on a great circle of the sphere. When k varies throughout the Brillouin zone $\{-\pi, \pi\}$, the number of closed loops (winding number) described by $\mathbf{n}_\delta(k)$ is a topological invariant, labelled here as W. As shown in Fig. 1c the value of δ in the range $\{0, 2\pi\}$, that is the strength of the spin–orbit interaction, determines the existence of two phases with a different winding number: a phase with $W=2q=1$ occurs when $\delta_1<\delta<\delta_2$, while $W=0$ in the remaining regions. As pointed out in refs 19,28, for time-periodic systems with chiral symmetry a complete topological classification would require introducing two topological invariants, so W alone is not enough. Nevertheless, this is not necessary for the purposes of this work, as we focus our attention on the phase transitions only, while we do not need to identify completely the nature of the topological phases, and it can be shown that all topological phase transitions are associated with a change of W[19,28]. Therefore, to keep the discussion simpler we will use only the winding number W in our discussion. In this sense, we will refer to the phases $W=0$ and 1 as trivial and non-trivial phases, respectively. In Fig. 1a it can be noted that when $\delta=\delta_1$ and $\delta=\delta_2$ the two bands are touching and the dispersion is locally linear at $k=0$ and $k=\pi$, respectively, as typically occurs for quantum transitions between topological phases[3]. Interestingly, comparing the expressions of group velocity $V_\delta=\mathrm{d}E_\delta/\mathrm{d}k$ (equation 5) and vector $\mathbf{n}_\delta(k)$ (equation 6), we can observe that $n_z=-n_y=|V_\delta|$. Thus in the non-trivial phase, where the trajectory of $\mathbf{n}_\delta(k)$ is fixed, the maximum of the group velocity is independent of δ (see the Supplementary Note 1 and the Supplementary Fig.1 for further details). A recently introduced QW protocol, that is the split-step QW, has very similar properties[10,18].

Dynamical moments and topological phases. The features of the energy bands and associated eigenstates have profound consequences on the system dynamics, which shows marked differences in the two topological phases[11,17]. Here we characterize such different behaviour through the analysis of the moments of the probability distribution $P(m)$ associated with the walker position, defined as $M_j = \sum_m m^j P(m)$. In particular, we consider a photon starting its walk in the position $m = 0$ with an arbitrary polarization, that is $|\psi_0\rangle = |0\rangle \otimes |\phi_0\rangle$, where $|\phi_0\rangle$ is a generic state in the coin Hilbert space. Simulations of the QW evolution in the large step-number limit show that the moments M_j assume a constant value (independent of δ) in the non-trivial phase, and undergo abrupt slope variations at the phase transitions, that is at $\delta = \{\delta_1, \delta_2\}$. In the infinite-steps-limit, we proved that these moments have simple asymptotic expressions in terms of the energy band dispersion relations; in particular, those relative to the first and second moments M_1 and M_2 are the following (see Methods for a proof):

$$M_1/N = (s_y - s_z)L(\delta) + O(1/N), \tag{7}$$

$$M_2/N^2 = L(\delta) + O(1/N^2), \tag{8}$$

where $s_i = \langle\phi_0|\hat{\sigma}_i|\phi_0\rangle$, with $i \in \{x, y, z\}$, are the reduced Stokes parameters, calculated as the expectation values of the Pauli operators $\hat{\sigma}_i$ for the coin initial state $|\phi_0\rangle$. Interestingly, we observe that M_2 is independent of the initial coin state. The quantity $L(\delta)$ appearing in equations 7 and 8 is equal to the square of the group velocity V_δ, and hence to the square of n_y, averaged over the Brillouin zone:

$$L(\delta) = \int_{-\pi}^{\pi} \frac{dk}{2\pi} [V_\delta(k)]^2 = \int_{-\pi}^{\pi} \frac{dk}{2\pi} [n_y(k, \delta)]^2. \tag{9}$$

For our QW process, the integral in equation 9 admits a closed form, that is

$$L(\delta) = \begin{cases} 2\sin^2(\delta/4) & \text{for} \quad 0 \le \delta \le \pi/2 \\ 1 - \frac{1}{\sqrt{2}} & \text{for} \quad \pi/2 \le \delta \le 3\pi/2 \\ 2\cos^2(\delta/4) & \text{for} \quad 3\pi/2 \le \delta \le 2\pi \end{cases} \tag{10}$$

In Fig. 2a–c, this expression is plotted as a function of δ; as a consequence of the discontinuity present in the group velocity, $L(\delta)$ is a piecewise function, and manifests abrupt slope variations at $\delta_1 = \pi/2$ and $\delta_2 = 3\pi/2$. As shown in Fig. 3 and thoroughly discussed in the Supplementary Note 2 and in the Supplementary Figs 2–6, by applying a similar approach to other 1D topological systems we were able to observe very similar features. For our QW, in the non-trivial phase $L(\delta)$ is constant; this result can be qualitatively understood by looking at how the dispersion law of the group velocity is modified by a change of δ (Supplementary Fig. 1). On the other hand, in the trivial phase $L(\delta)$ increases or decreases in the two regions $\{0, \pi/2\}$ and $\{3\pi/2, 2\pi\}$, respectively. Relying on a numerical analysis, we find that such features are robust to the presence of disorder arising from possible experimental imperfections in our setup (details are provided in Methods and in Supplementary Fig. 7). In a QW with a finite number of steps, statistical moments M_1 and M_2 have a continuous behaviour, converging to that given by equations 7 and 8 asymptotically as $N \to \infty$. For M_2, this convergence is rapid and visible for values of N that are small enough to be achieved in an experimental simulation, and it is independent of the coin initial state, whereas for M_1 such process is much slower (see Fig. 2). Thus, as an indicator for identifying the quantum transitions, we chose to use $\sqrt{M_2}/N = \frac{\sqrt{\langle \hat{m}^2 \rangle}}{N}$, corresponding to the width of the OAM statistical distribution, normalized to the number of steps. In particular, we implemented a six-step QW and measured the OAM distribution width when tuning the spin–orbit interaction in each q-plate, that is δ, so as to realize the topological quantum transition.

Experimental results. The layout of the apparatus is shown in Fig. 4. A standard heralded single-photon source is realized at the input of the QW system, as discussed in Methods. Before undergoing the QW evolution, a single photon is prepared in localized initial state ($m = 0$), with its polarization (the coin part) described by $|\phi_0\rangle = \alpha|L\rangle + \beta|R\rangle$. The value of the two complex coefficients α and β (with $|\alpha|^2 + |\beta|^2 = 1$) can be tuned by suitably orienting two waveplates that are positioned at input of the setup (see the caption of Fig. 4 for more details). After the QW dynamics, the photon state is analysed in both polarization and OAM so as to determine the output probabilities and the

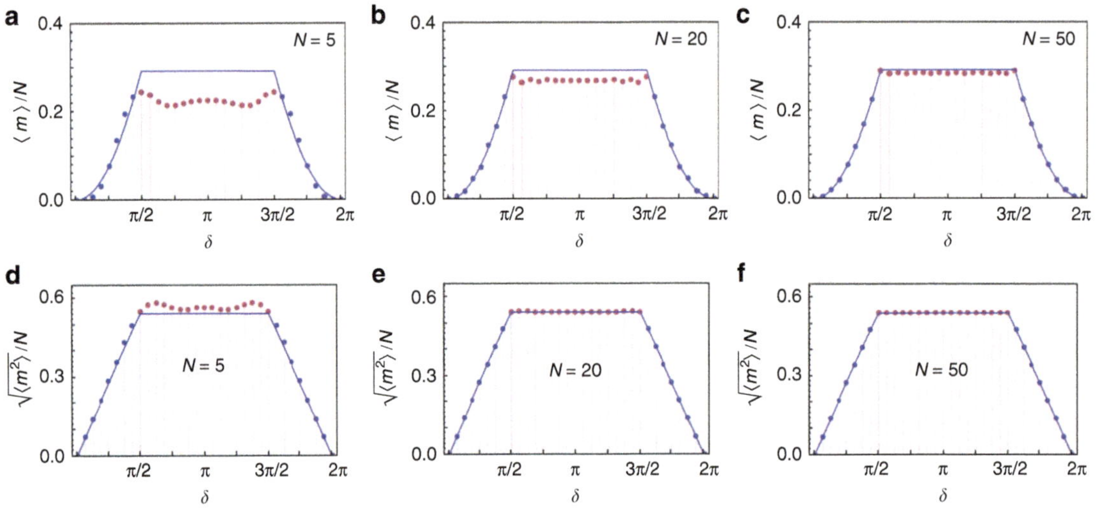

Figure 2 | Analysis of statistical moments in a QW. The system, initially localized at $x = 0$ with a generic coin state $|\phi_0\rangle$, undergoes a QW described by the step operator (3). For every plot, purple points are obtained from a numerical simulation, when varying δ with steps of $\pi/16$ in the range $\{0, 2\pi\}$; continuous blue lines represent the quantity $L(\delta)$ **(a-c)**, or $\sqrt{L(\delta)}$ **(d-f)** (see equation 10). For the simulation, we prepared the coin in the state $\{\alpha, \beta\} = \{1, 0\}$, corresponding to $\{s_1, s_2, s_3\} = \{0, 0, 1\}$. **(a-c)** First-order moment M_1, divided by the number of steps of the walk, as a function of the parameter δ, for a walk of 5, 20 and 50 steps, respectively (as specified in each panel). As N increases, simulated data converge to the values predicted by equation 7. **(d-f)** Square root of the second-order moment, divided by the number of steps N. The figures are organized as in **a-c**. In this case, we can observe that simulated data converge much faster to the asymptotic values reported in equation 8, with a discontinuity emerging even for a walk of few steps.

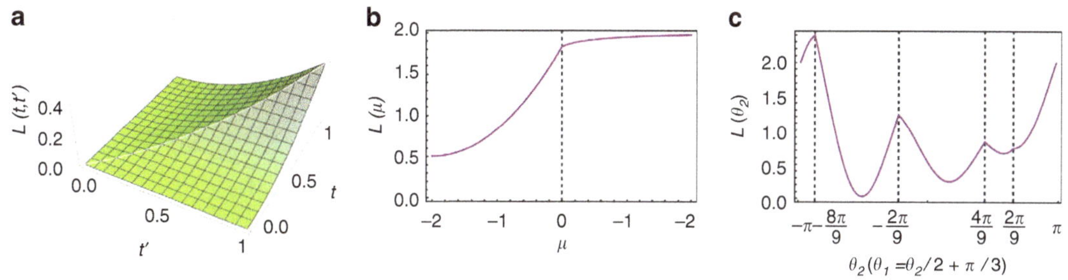

Figure 3 | Dynamical moments reveal topological phase transitions in other 1D topological systems. (**a**,**b**) We plot the expected asymptotic values of the second-order moment for a single particle whose evolution is described by the SSH model and the *p*-wave theory for superconductors, respectively, which have the same symmetries as our QW. In the SSH model (**a**), the value of the topological invariant is determined by the parameters (t, t'). In agreement with our conjecture, non-analyticities of the second-order moment can be observed at the phase change, that is for $t=t'$. In the *p*-wave theory, topological features are defined in terms of the chemical potential μ. Again, an abrupt slope variation can be appreciated at the transition point $\mu=0$. (**c**) Here we consider a QW model having particle-hole symmetry only, in order to show that our results are independent of the specific symmetries characterizing the system. In this model, topological phases are determined in terms of two angles θ_1 and θ_2. These two angles are varied along the trajectory given by the equation $\theta_1=\theta_2/3+\pi/3$; similarly to **a**,**b**, the second-order moment after a long temporal evolution shows non-analyticities at the phase changes, identified by vertical dashed lines in the plot. In all simulations, we considered an initial state localized at a specific lattice site. Properties of these models, along with the analysis of other theories, are discussed in detail in the Supplementary Note 2.

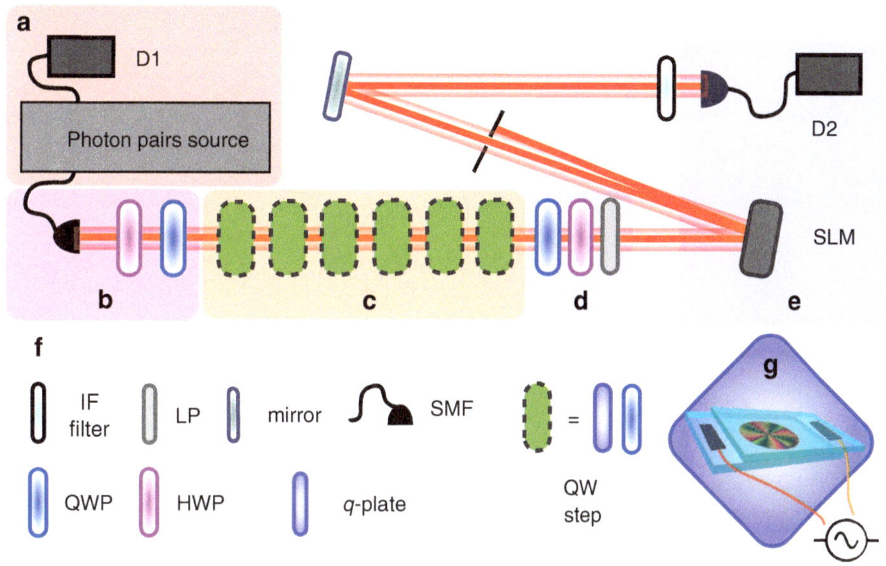

Figure 4 | Layout of the experimental setup. (**a**) As described in Methods section, pairs of indistinguishable and temporally correlated photons are generated at the input of the setup. For each pair, they are split by exploiting their orthogonal polarization: one photon is directly sent to an APD D1 and acts as a trigger for the detection of its correlated partner. The latter indeed, after coupling into a single-mode optical fibre (SMF), is sent to the QW setup and finally detected at D2, in coincidence with detection signals from D1. (**b**) The photons undergoing the QW dynamics exits a SMF in the OAM state $m=0$, and then its polarization is prepared in the state $|\phi_0\rangle=\alpha|L\rangle+\beta|R\rangle$; the values of the two complex coefficients α and β (with $|\alpha|^2+|\beta|^2=1$) are set by using a half-wave plate (HWP) and a QWP (apart from an unimportant global phase). (**c**) After the initial state preparation, the photon goes through the six-step QW, with the single step consisting of a QWP oriented at 90° and a *q*-plate. For each *q*-plate, the value of the optical retardation δ is controlled by the amplitude of an alternating electric field, induced by an external generator[27]. (**d**) At the exit of the QW, a polarization projection is realized using a second HWP-QWP set followed by an LP. (**e**) The OAM state is then analysed by diffraction on a SLM, followed by coupling into a SMF that is directly connected to the detector D2, consisting in a APD. Before photons are detected, IF centred at 800 nm and with a bandwidth of 3.6 nm are used for spectral cleaning. The latter was required since the photons wavelength strongly affects the action of the devices implementing the QW (*q*-plate and QWP). (**f**) List of all optical elements used in our setup. (**g**) Illustrative picture of a *q*-plate, whose optical retardation is controlled by means of an external oscillating electric field. IF, interference filters; LP, linear polarizer; SLM, spatial light modulator.

associated moments. Letting δ vary in the range $\{0, \pi\}$ with steps of $\pi/16$, we determined the corresponding probability distribution of the walker after a six-step QW; a typical example is reported in Fig. 5a. In Fig. 5b,c we plot the measured values for M_1/N and $\sqrt{M_2}/N$, respectively, as a function of δ. Experimental data match well the quantum predictions and, as expected, they are close to the asymptotic limit reported in equations 7 and 8 only in the case of $\sqrt{M_2}/N$. For the latter, the emergence of an abrupt slope variation at $\delta=\pi/2$ can be appreciated; the observed non-analyticity is a signature of the underlying quantum transition. In addition, M_2 is observed to be locked to a constant value in the non-trivial phase. On one hand, this result reflects the features of the group velocity and the strong link between the latter and the geometry and topology of the system eigenstates.

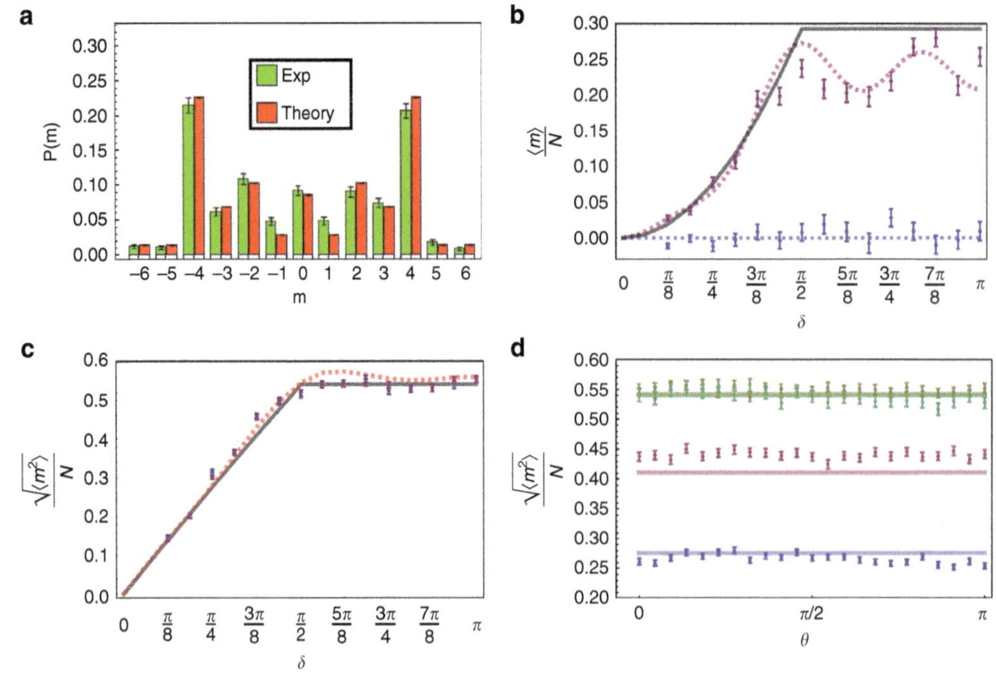

Figure 5 | Theoretical predictions and experimental results. (**a**) Example of a probability distribution for the walker: measured (green, left) and expected (red, right) probability distributions after a six-step QW of a photon initially prepared in the state $m=0$ and $\{\alpha,\beta\}=1/\sqrt{2}\{1,1\}$, for $\delta=2.95$. (**b**,**c**) Measured values of statistical moments M_1/N and $\sqrt{M_2}/N$, respectively, when varying δ in the range $\{\pi/8,\pi\}$ with steps of $\pi/16$, for initial states $\{\alpha,\beta\}=1/\sqrt{2}\{1,1\}$ (blue points), and $\{\alpha,\beta\}=\{0,1\}$ (purple points). Dotted lines represent the corresponding results of numerical simulations. The continuous lines are the asymptotic limits given by equation 8. (**d**) Measured values of $\sqrt{M_2}/N$ when the initial polarization corresponds to the state $\cos(\theta/2)|L\rangle+\sin(\theta/2)|R\rangle$, varying the polar angle $\theta\in[0,\pi]$ with steps of $\pi/22$; data are collected in correspondence of $\delta=\pi/4$ (blue points), $\delta=3\pi/8$ (purple points), $\delta=3\pi/4$ (yellow points) and $\delta=\pi$ (green points). For each of these configurations, continuous lines give the corresponding asymptotic limit obtained from equation 8. When $\delta=3\pi/8$, experimental data are slightly higher then the associated continuous line; nevertheless, the discrepancy is compatible with the results of numerical simulations for a finite number of steps (see **c**). These data confirm the prediction that M_2 is asymptotically independent of the input coin/polarization state. In all plots (**a**-**d**), the error bars represent statistical errors at 1 s.d., calculated assuming Poissonian fluctuations on single counts.

On the other hand, this is not a general feature of 1D topological systems; in the Supplementary Note 2 we show that for other models, such as the SSH, this plateau can be observed in either of the two phases, depending on the specific trajectory followed in the parameters space. Yet, it is possible to show that for our QW (equation 9) and the SSH model (Supplementary Note 3), the value of M_2 is proportional to the integral of n_y^2 over the Brillouin zone and that the latter always shows a constant value, that is a plateau, when $W=1$. Preparing the initial state of the coin ($m=0$ for the walker) in two non-orthogonal polarizations we also verified that M_2 (unlike M_1) is independent of the coin initial conditions, as expected from equation 8 (see Fig. 5b,c). This aspect was further investigated by measuring $\sqrt{M_2}/N$ for several initial polarizations, corresponding to specific points along a meridian of the Poincaré sphere, and repeating the experiment for two values of δ in each topological sector. The final data, reported in Fig. 5d, match well the predicted results and clearly show that the value of the second-order moment is independent of the coin initial state; this guarantees the robustness of our results with respect to imperfections in the initial state preparation.

Discussion

The main result of this work is the idea that a signature of quantum transitions between distinct topological phases is present in the behaviour of suitable dynamical observables that are easily accessed experimentally. In the context of 1D topological systems, we propose that the statistical moments of the particle probability distribution in space (in the large step-number limit) provide a convenient choice of such observables. We experimentally validated this proposal by simulating this topological environment within a specific photonic QW architecture. The measured asymptotic moments as a function of a control parameter show slope discontinuities at the phase change, and hence can be used as direct indicators of the phase change occurrence. It is remarkable that these observables reflect bulk properties only, in contrast with the common strategy that relies on investigating topological physics exclusively through edge effects. Numerical simulations show that these features are robust to experimental imperfections and confirm their presence in several other 1D topological systems, thus strongly supporting the conjecture of their general validity in 1D. Moreover, we are currently investigating the possibility to apply similar concepts to 2D systems. Localization induced by spatial disorder is of course expected to affect the final asymptotic behaviour of the dynamics. However, if the disorder is not very strong, it is plausible that an intermediate time regime will still exist, in which the quasi-nonanalytic behaviour of the moments can be detected. In prospect, our approach based on dynamical moments could be applicable to the investigation and to the detection of other classes of quantum transitions, and to topological phases associated with more complex symmetries or with a higher dimensionality, thus helping to shed new light on the physics of topological phenomena and on their simulation in suitable experimental architectures.

Methods

Asymptotic expressions of first and second moments. The large step-number limit for the walker distribution moments (equations 7 and 8) can be derived conveniently by evaluating M_1 and M_2 in momentum representation, where they are defined as follows:

$$M_1 = \int_{-\pi}^{\pi} \frac{dk}{2\pi} \langle \phi_0 | \left(\hat{U}_0^{\dagger}\right)^N (-i)\frac{d}{dk} \hat{U}_0^N |\phi_0\rangle,$$
$$M_2 = \int_{-\pi}^{\pi} \frac{dk}{2\pi} \langle \phi_0 | \left(\hat{U}_0^{\dagger}\right)^N (-i)^2\frac{d^2}{dk^2} \hat{U}_0^N |\phi_0\rangle. \quad (11)$$

Here N is the step-number, $\hat{U}_0$ is the single-step evolution operator (3), and $|\phi_0\rangle$ is the coin initial state. Expanding the evolution operator as

$$\begin{aligned}\hat{U}_0^N &= \mathrm{Exp}\{-iNE_\delta(k)[\mathbf{n}_\delta(k)\cdot\hat{\sigma}]\} \\ &= \cos[NE_\delta(k)]\hat{I}_2 - i\sin[NE_\delta(k)](\mathbf{n}_\delta(k)\cdot\hat{\sigma})\end{aligned} \quad (12)$$

where $\hat{I}_2$ is the identity matrix in 2D, $\hat{\sigma}=\{\hat{\sigma}_x,\hat{\sigma}_y,\hat{\sigma}_z\}$ and the components of $\mathbf{n}_\delta(k)$ are those reported in equation 6, it is straightforward to obtain the following equations:

$$M_1/N = \int_{-\pi}^{\pi} \frac{dk}{2\pi} V_\delta(k)\langle\phi_0|(\mathbf{n}_\delta(k)\cdot\hat{\sigma})|\phi_0\rangle + O(1/N), \quad (13)$$

$$M_2/N^2 = \int_{-\pi}^{\pi} \frac{dk}{2\pi} [V_\delta(k)]^2 + O(1/N^2), \quad (14)$$

It is important to observe that equations 13 and 14 are valid for any quantum system whose Hilbert space has the same structure as our QW, that is the direct product of an infinite discrete lattice and a two-state quantum degree of freedom. Equation 13, reporting the expression for the first-order moment, can be evaluated considering that $V_\delta = n_z = -n_y$. Applying this substitution, the same equations reads

$$M_1/N = (s_y - s_z)\int_{-\pi}^{\pi} \frac{dk}{2\pi} [V_\delta(k)]^2 + O(1/N). \quad (15)$$

First- and second-order moments reported in equations 14 and 15, respectively, coincide with the equations 7 and 8 of the main text. In the specific case of our QW model, the integral appearing in equations 14 and 15 can be solved analytically. Using the expression for the group velocity reported in equations 5 and 9 yields

$$L(\delta) = \frac{1}{2\pi}\int_0^{2\pi} V_\delta^2 dk = \frac{1}{2\pi}\int_0^{2\pi} \frac{\sin^2(\delta/2)\cos^2(k)}{2-[\cos(\delta/2)+\sin(\delta/2)\sin(k)]^2} dk. \quad (16)$$

This integral can be calculated by the residue theorem passing to the complex variable $z=e^{ik}$. Then we obtain $L(\delta)=\oint f_\delta(z)dz$, where the integral is along the unit circle in the complex z-plane and $f_\delta(z)$ is given by

$$f_\delta(z) = \frac{i(1+z^2)^2\sin^2(\delta/2)}{\pi z\left[(1+z^2)^2\cos(\delta) - z^4 - 10z^2 - 1 - 4iz(z^2-1)\sin(\delta)\right]}. \quad (17)$$

The poles of $f_\delta(z)$ are located on the imaginary axis at $z_k=\{0, i(\sqrt{2}-1)\cot(\delta/4), i(\sqrt{2}+1)\tan(\delta/4), -i(\sqrt{2}+1)\cot(\delta/4)$ and $-i(\sqrt{2}-1)\tan(\delta/4)\}$ $(k=1,\ldots 5)$ and the residues of $f_\delta(z)$ at the poles are given by $2\pi i r_k=\{1, \frac{1}{4}[-\sqrt{2}+2\cos(\delta/2)], \frac{1}{4}[\sqrt{2}-2\cos(\delta/2)], \frac{1}{4}[\sqrt{2}+2\cos(\delta/2)]$ and $\frac{1}{4}[-\sqrt{2}-2\cos(\delta/2)]\}$, respectively. Apart from the pole at $z=0$, when δ varies, the locations of the poles move in the complex plane entering and exiting the unit circle, but only the residues of the poles inside the unit circle contribute to $L(\delta)$. Thus, we find

$$L(\delta) = \begin{cases} 2\pi i(r_1+r_3+r_5) = 2\sin^2(\delta/4) & \text{for } 0\le\delta\le\pi/2 \\ 2\pi i(r_1+r_2+r_5) = 1-\frac{1}{\sqrt{2}} & \text{for } \pi/2\le\delta\le3\pi/2 \\ 2\pi i(r_1+r_2+r_4) = 2\cos^2(\delta/4) & \text{for } 3\pi/2\le\delta\le2\pi \end{cases}. \quad (18)$$

In Supplementary Note 3 and in Supplementary Fig. 6 we show how with the same approach the analytical expression for the function L can be found for the SSH model.

Heralded single-photon source. In the heralded single-photon source (the optical components used to realize this source are not shown in the setup scheme of Fig. 4, for details see ref. 24), laser pulses (100 fs) at 800 nm generated by a titanium–sapphire source (Ti:Sa) with repetition rate 82 MHz shine on a type-I β-barium borate crystal for second harmonic generation; frequency-doubled pulses at 400 nm, with 110 mW average power and linear-horizontal polarization, pump a type-II β-barium borate crystal, cut for collinear and degenerate spontaneous parametric down-conversion. Signal and idler photons, generated in horizontal and vertical linear polarizations, respectively, are spatially separated by means of a polarizing beam-splitter and then coupled into single-mode fibres; the idler photon is directly sent to an avalanche photodiode (APD D1), while the signal one is sent to the QW system. After the QW evolution, such photon is analysed in polarization and OAM and finally detected by APD D2, in coincidence with D1. In this way, the idler photon is used as a trigger, which heralds the presence of the correlated particle passing through the QW setup.

Experimental imperfections and role of disorder. Experimental realizations of quantum protocols and quantum dynamics exploit specific devices whose behaviour may present deviations with respect to their ideal features. Here we individuate possible imperfections that may arise in our system, discussing in turn how they would affect our QW protocol and, as a consequence, the dynamical moments associated with the walker wavefunction. Interestingly, we find that abrupt variations at the phase change are robust to these kinds of disorder.

A first type of imperfection can result from a bad orientation of the QWPs performing the coin rotation. We considered two different coin input, namely $|H\rangle$ and $|R\rangle$ polarizations already used in the experiment, and simulated a 80 steps QW, where at each step a QWP is oriented at $\pm 2°$ with respect to the correct angle; here we assumed that at 0° a QWP implements the operator reported in equation 1. The $\pm$ sign is randomly selected. In Supplementary Fig. 7, we report the results of our numerical analysis. Data clearly show that the second-order moment have no significant differences with respect to the ideal QW; accordingly the phase change can be appreciated still, thus proving the robustness of such indicator. For $|R\rangle$ as initial state, the mean value has a similar behaviour, whereas something different happens for the $|H\rangle$ polarization; in the latter case, ideal QW leads to a vanishing mean value. When QWPs have random bad orientations, this is not true anymore, even though deviations from ideal case are smaller than typical experimental errors affecting our data. A second source of errors can arise from the bad alignment of q-plates. In particular, their centre could be displaced with respect to the propagation axis of the beam. As a first approximation, the q-plate operator reported in equation 2 would be transformed as follows:

$$\begin{aligned}\hat{Q}'_{\delta=\pi}|L,m\rangle &= i(a_m|R,m+1\rangle + b_m|R,m\rangle + c_m|R,m+2\rangle), \\ \hat{Q}'_{\delta=\pi}|R,m\rangle &= i(a_m|L,m-1\rangle + b_m|L,m\rangle + c_m|L,m-2\rangle),\end{aligned} \quad (19)$$

where $\hat{Q}\prime$ is the operator describing the q-plate action when a displacement of its centre is taken into account; here it has been considered the simple case $\delta=\pi$ (the term proportional to $\cos(\delta/2)$ is the identity operator, which is not affected by such displacement). Moreover, we are explicitly neglecting the coupling to spatial modes of light with different radial distributions, as for instance Laguerre–Gauss modes with radial index $p\ge1$. We can note that in equation 19 non-vanishing values for c_m are responsible for next-nearest-neighbour couplings in the OAM lattice, corresponding to an 'off-diagonal' disorder[29]. Coefficients $\{a_m, b_m, c_m\}$ can be computed numerically; interestingly, the effects of the q-plate displacement are not equal for all OAM states; in particular, deviations from the ideal case becomes negligible for OAM modes with high $|m|$; for such states, $a_m\simeq1$ and $b_m\simeq c_m\simeq0$. Similarly to the previous analysis we implemented a numerical simulation of a 80 steps QW for $|H\rangle$ and $|R\rangle$ coin input; at each step the q-plate is displaced randomly along the x or the y axis (the z axis corresponds to the propagation direction of the light beam). We numerically computed coefficients $\{a_m, b_m, c_m\}$, assuming a displacement equal to 5% with respect to the beam transverse dimensions (that is the radius at the waist of the input Gaussian beam). Data reported in Supplementary Fig. 7 show that the effect of such imperfections is negligible. Importantly, the abrupt variation at the phase change can be still appreciated, thus proving the robustness of such features. Even for this case the mean value for $|H\rangle$ has no significant differences when compared with the ideal case, being it very close to zero for all values of δ (within our experimental uncertainties). In conclusion, we want to compare such imperfections with the presence of static disorder, which is typically responsible of localization phenomena. An imperfect orientation of the waveplates leads to a modification of the coin operator that is equal for all OAM states, thus it cannot produce any localization. Effects associated with the q-plate displacement are not uniform in the OAM space, but still they are quite different with respect to the typical static disorder, that consists in random site-dependent phase shifts, equal at each step. Moreover, in both cases alterations of the step operator are different at every step of the walk, thus they represent a temporal disorder. Importantly, our simulations show the robustness of moments non-analyticities as indicators of quantum phase changes in such disordered scenarios. Although the spatial static disorder is certainly not significant in our experimental implementation, it can be obviously important for the generalization of our results to other systems. Localization induced by this kind of disorder is expected to affect the final asymptotic behaviour of the dynamics. However, if the disorder is not very strong, it is plausible that an intermediate time regime will still exist in which a quasi-nonanalytic behaviour of the moments can be detected. If 'localization time' is sufficiently larger than the measurement time, the disorder will not affect the kinks (slightly rounded in real experiments because of the finite measurement time) that signal the occurrence of the topological transitions. Hence, we believe that our main results on the use of moments for detecting topological transitions can be useful also in the context of disordered systems, as long as the disorder is not too strong.

References

1. Thouless, D. J., Kohmoto, M., Nightingale, M. P. & den Nijs, M. Quantized hall conductance in a two-dimensional periodic potential. *Phys. Rev. Lett.* **49,** 405–408 (1982).
2. Zhang, Y., Tan, Y. W., Stormer, H. L. & Kim, P. Experimental observation of the quantum hall effect and berry's phase in graphene. *Nature* **438,** 201–204 (2005).

3. Qi, X. & Zhang, S. Topological insulators and superconductors. *Rev. Mod. Phys.* **83,** 1057–1110 (2011).
4. Hasan, M. Z. & Kane, C. L. *Colloquium* : topological insulators. *Rev. Mod. Phys.* **82,** 3045–3067 (2010).
5. Gomes, K. K., Mar, W., Ko, W., Guinea, F. & Manoharan, H. C. Designer dirac fermions and topological phases in molecular graphene. *Nature* **483,** 306–310 (2012).
6. Atala, M. *et al.* Direct measurement of the zak phase in topological bloch bands. *Nat. Phys.* **9,** 795–800 (2013).
7. Genske, M. *et al.* Electric quantum walks with individual atoms. *Phys. Rev. Lett.* **110,** 190601 (2013).
8. Hauke, P. *et al.* Non-abelian gauge fields and topological insulators in shaken optical lattices. *Phys. Rev. Lett.* **109,** 145301 (2012).
9. Cayssol, J., Dóra, B., Simon, F. & Moessner, R. Floquet topological insulators. *Phys. Status Solidi RRL* **7,** 101–108 (2013).
10. Kitagawa, T. *et al.* Observation of topologically protected bound states in photonic quantum walks. *Nat. Commun.* **3,** 882 (2012).
11. Zeuner, J. M. *et al.* Probing topological invariants in the bulk of a non-hermitian optical system. *Phys. Rev. Lett.* **115,** 040402 (2015).
12. Lu, L., Joannopoulos, J. D. & Soljačić, M. Topological photonics. *Nat. Photon.* **8,** 821–829 (2014).
13. Hafezi, M. Measuring topological invariants in photonic systems. *Phys. Rev. Lett.* **112,** 210405 (2014).
14. Ozawa, T. & Carusotto, I. Anomalous and quantum Hall effects in lossy photonic lattices. *Phys. Rev. Lett.* **112,** 133902 (2014).
15. Hu, W. *et al.* Measurement of a topological edge invariant in a microwave network. *Phys. Rev. X* **5,** 011012 (2015).
16. Mittal, S., Ganeshan, S., Fan, J., Vaezi, A. & Hafezi, M. Observation of the Chern-Simons gauge anomaly. *Nat. Photon.* **10,** 180–183 (2016).
17. Rudner, M. & Levitov, L. Topological transition in a non-hermitian quantum walk. *Phys. Rev. Lett.* **102,** 065703 (2009).
18. Kitagawa, T., Rudner, M. S., Berg, E. & Demler, E. Exploring topological phases with quantum walks. *Phys. Rev. A* **82,** 033429 (2010).
19. Asbóth, J. K. Symmetries, topological phases, and bound states in the one-dimensional quantum walk. *Phys. Rev. B* **86,** 195414 (2012).
20. Venegas-Andraca, S. E. Quantum walks: a comprehensive review. *Quantum Inf. Process.* **11,** 1015–1106 (2012).
21. Su, W., Schrieffer, J. & Heeger, A. Solitons in polyacetylene. *Phys. Rev. Lett.* **42,** 1698–1701 (1979).
22. Read, N. & Green, D. Paired states of fermions in two dimensions with breaking of parity and time-reversal symmetries and the fractional quantum Hall effect. *Phys. Rev. B* **61,** 10267 (2000).
23. Wang, J. & Manouchehri, K. *Physical Implementation of Quantum Walks* (Springer, 2013).
24. Cardano, F. *et al.* Quantum walks and wavepacket dynamics on a lattice with twisted photons. *Sci. Adv.* **1,** e1500087 (2015).
25. Yao, A. M. & Padgett, M. J. Orbital angular momentum: origins, behavior and applications. *Adv. Opt. Photon* **3,** 161–204 (2011).
26. Marrucci, L., Manzo, C. & Paparo, D. Optical spin-to-orbital angular momentum conversion in inhomogeneous anisotropic media. *Phys. Rev .Lett.* **97,** 163905 (2006).
27. Piccirillo, B., D'Ambrosio, V., Slussarenko, S., Marrucci, L. & Santamato, E. Photon spin-to-orbital angular momentum conversion via an electrically tunable q-plate. *Appl. Phys. Lett.* **97,** 241104 (2010).
28. Asbóth, J. K. & Obuse, H. Bulk-boundary correspondence for chiral symmetric quantum walks. *Phys. Rev. B* **88,** 121406 (2013).
29. Zhao, Q. & Gong, J. From disordered quantum walk to physics of off-diagonal disorder. *Phys. Rev. B* **92,** 214205 (2015).

Acknowledgements

We thank Maciej Lewenstein, Pietro Massignan and Alessio Celi for reading a preliminary version of this manuscript and for providing useful comments, Takuya Machida for alerting us to some mistakes appearing in a preprint early version of this paper, and Domenico Paparo and Antonio Ramaglia for lending us some equipment. This work was partly supported by the Future Emerging Technologies FET-Open Program, within the 7th Framework Programme of the European Commission, under Grant No. 255914, PHORBITECH.

Author contributions

F.C., M.M., F.M., E.S. and L.M. devised various aspects of the project and designed the experimental methodology. F.C. and M.M., with contributions from C.d.L., carried out the experiment and analysed the data. B.P. prepared the *q*-plates. F.C., M.M, G.D.F., V.C., E.S. and L.M. developed the theoretical aspects. F.C., M.M. and L.M. wrote the manuscript, with contributions from E.S., G.D.F and V.C. All authors discussed the results and contributed to refining the manuscript.

Additional information

Competing financial interests: The authors declare no competing financial interests.

25

Numerical approach for unstructured quantum key distribution

Patrick J. Coles[1], Eric M. Metodiev[1] & Norbert Lütkenhaus[1]

Quantum key distribution (QKD) allows for communication with security guaranteed by quantum theory. The main theoretical problem in QKD is to calculate the secret key rate for a given protocol. Analytical formulas are known for protocols with symmetries, since symmetry simplifies the analysis. However, experimental imperfections break symmetries, hence the effect of imperfections on key rates is difficult to estimate. Furthermore, it is an interesting question whether (intentionally) asymmetric protocols could outperform symmetric ones. Here we develop a robust numerical approach for calculating the key rate for arbitrary discrete-variable QKD protocols. Ultimately this will allow researchers to study 'unstructured' protocols, that is, those that lack symmetry. Our approach relies on transforming the key rate calculation to the dual optimization problem, which markedly reduces the number of parameters and hence the calculation time. We illustrate our method by investigating some unstructured protocols for which the key rate was previously unknown.

[1]Department of Physics and Astronomy, Institute for Quantum Computing, University of Waterloo, Waterloo, Ontario, Canada N2L3G1. Correspondence and requests for materials should be addressed to P.J.C. (email: pcoles@uwaterloo.ca).

Quantum key distribution (QKD) will play an important role in quantum-safe cryptography, that is, cryptography that addresses the emerging threat of quantum computers[1]. Since its original proposal[2,3], QKD has developed significantly over the past three decades[4,5], both in theory and implementation. Indeed, QKD is now a commercial technology, with the prospect of global QKD networks on the horizon[6,7].

The main theoretical problem in QKD is to calculate how much secret key can be distributed by a given protocol. A crucial practical issue is that the QKD protocols that are easiest to implement with existing optical technology do not necessarily coincide with the protocols that are easiest to analyse theoretically[4]. Currently, calculating the secret key output of a protocol is typically extremely technical, and hence only performed by skilled experts. Furthermore, each new protocol idea requires a new calculation, tailored to that protocol. Ultimately, the technical nature of these calculations combined with the lack of universal tools limits the pace at which new QKD protocols can be discovered and analysed. Here we address this problem by developing a robust, user-friendly framework for calculating the secret key output, with the hope of bringing such calculations 'to the masses'.

The secret key output is typically quantified by the key rate, which refers to the number of bits of secret key established divided by the number of distributed quantum systems. Operationally, this corresponds to the question of how much privacy amplification Alice and Bob must apply to transform their raw key into the final secure key. Analytical simplifications of the key rate calculation can be made for some special protocols that have a high degree of symmetry[8]. Examples of such symmetric protocols, where the signal states have a group-theoretic structure, include the BB84 (ref. 3) and six-state protocols[9]. Indeed the key rate is known for these protocols. However, in practice, lack of symmetry is the rule rather than the exception. That is, even if experimentalists try to implement a symmetric protocol, experimental imperfections tend to break symmetries[10]. Furthermore, it is sometimes desirable due to optical hardware issues to implement asymmetric protocols, for example, as in ref. 11.

We refer to general QKD protocols involving signal states or measurement choices that lack symmetry as 'unstructured' protocols. Some recent work has made progress in bounding the key rate for special kinds of unstructured protocols, such as four-state protocols in refs 12,13 and qubit protocols in ref. 14. Still, there is no general method for computing tight bounds on the key rate for arbitrary unstructured protocols. Yet, these are the protocols that are most relevant to experimental implementations.

This motivates our present work, in which we develop an efficient, numerical approach to calculating key rates. Our ultimate aim is to develop a computer program, where Alice and Bob input a description of their protocol (for example, their signal states, measurement devices, sifting procedure and key map) and their experimental observations, and the computer outputs the key rate for their protocol. This program would allow for any protocol, including those that lack structure.

At the technical level, the key rate problem is an optimization problem, since one must minimize the well-known entropic formula for the key rate[15], over all states ρ_{AB} that satisfy Alice's and Bob's experimental data. The main challenge here is that this optimization problem is unreliable and inefficient. In this work, we give a novel insight that transforming to the dual problem (for example, see ref. 16) resolves these issues, hence paving the way for automated key rate calculations.

Specifically, the unreliable (or unphysical) aspect of the primal problem is that it is a minimization, hence the output will in general be an upper bound on the key rate. But one is typically more interested in reliable lower bounds, that is, physically achievable key rates. Transforming to the dual problem allows one to formulate the problem as a maximization, and hence approach the key rate from below. Therefore, every number outputted from our computer program represents an achievable asymptotic key rate, even if the computer did not reach the global maximum.

The inefficient aspect of the primal problem is that the number of parameters grows as $d_A^2 d_B^2$ for a state ρ_{AB} with $d_A = \dim(\mathcal{H}_A)$ and $d_B = \dim(\mathcal{H}_B)$. For example, if $d_A = d_B = 10$, the number of parameters that one would have to optimize over is 10,000. In contrast, in the dual problem, the number of parameters is equal to the number of experimental constraints that Alice and Bob choose to impose. For example, in the generalization of the BB84 protocol to arbitrary dimensions[17,18], Alice and Bob typically consider two constraints, their error rates in the two mutually-unbiased bases (MUBs). So, for this protocol, we have reduced the number of parameters to something that is constant in dimension. We, therefore, believe that our approach (of solving the dual problem) is ideally suited to efficiently calculate key rates in high dimensions.

We have written a MATLAB program to implement our key rate calculations. To illustrate the validity of our program, we show (Fig. 1) that it exactly reproduces the known theoretical dependence of the key rate on error rate, for both the BB84 and six-state protocols.

But ultimately the strength of our approach is its ability to handle unstructured protocols. We demonstrate this by investigating some unstructured protocols for which the key rates were not previously known. For example, we study a general class of protocols where Alice and Bob measure n MUBs, with $2 \leq n \leq d+1$, in dimension d. Also, we investigate the B92 protocol[19], which involves two signal states whose inner product is arbitrary. Our key rates are higher than known analytical lower bounds[20,21] for B92. Finally, we argue that our approach typically gives markedly higher key rates than those obtained from an analytical approach based on the entropic uncertainty relation[22,23].

We focus on asymptotic key rates in this work. Nevertheless, the optimization problem that we solve is also at the heart of

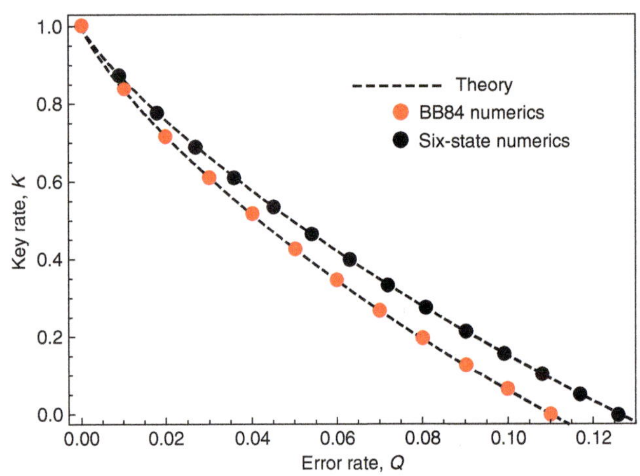

Figure 1 | Key rate for two well-known QKD protocols. Here we compare our numerics (from Theorem 1) with the theoretical curves. The results of our numerical optimization for the BB84 and six-state protocols are respectively shown as red and blue dots. The known theoretical curves for these protocols are also shown as black dashed lines. The dots should be viewed as reliable lower bounds on the key rate, but in this case they happen to be perfectly tight, coinciding with the theoretical curves.

finite-key analysis, for example, see refs 24,25. We, therefore, hope to extend our approach to the finite-key scenario in future efforts. We remark that finite-size effects generally reduce the key rate below its asymptotic value.

In what follows, we first present our main result: a reformulation of the key rate optimization problem in such a way that it is easily computable. We then outline our general framework for treating a broad range of protocols. Finally, we illustrate our approach with various examples.

Results

Setup of the problem. Consider a general entanglement-based QKD protocol involving finite-dimensional quantum systems A and B that are respectively received by Alice and Bob. Note that prepare-and-measure QKD protocols can be recast as entanglement-based protocols, as discussed below. For simplicity of presentation, we consider protocols where Alice's raw key is derived from a measurement on her system, possibly after some post-selection corresponding to a public announcement with a binary outcome, 'pass' or 'fail'. However, our approach can easily be extended to more general protocols.

Let Z_A (Z_B) denote the measurement that Alice (Bob) performs on system A (B) to derive the raw key. Suppose they use one-way direct reconciliation for the classical post-processing and that their error correction is optimal (that is, leaks out the minimum number of bits), then the asymptotic key rate is given by the Devetak–Winter formula[15]:

$$K = H(Z_A|E) - H(Z_A|Z_B). \qquad (1)$$

In equation (1), $H(X|Y) := H(\rho_{XY}) - H(\rho_Y)$ is the conditional von Neumann entropy, with $H(\sigma) := -\mathrm{Tr}(\sigma \log_2 \sigma)$, and

$$\rho_{Z_A Z_B} = \sum_{j,k} \mathrm{Tr}\left[\left(Z_A^j \otimes Z_B^k\right)\rho_{AB}\right] |j\rangle\langle j| \otimes |k\rangle\langle k|, \qquad (2)$$

$$\rho_{Z_A E} = \sum_j |j\rangle\langle j| \otimes \mathrm{Tr}_A\left[\left(Z_A^j \otimes \mathbb{1}\right)\rho_{AE}\right]. \qquad (3)$$

Here ρ_{ABE} is the tripartite density operator shared by Alice, Bob and Eve (and it may be the state after some post-selection procedure, see our general framework below). Also, $\{Z_A^j\}$ and $\{Z_B^k\}$ are the sets of positive operator valued measure (POVM) elements associated with Alice's and Bob's key-generating measurements. In what follows we refer to $\{Z_A^j\}$ as the key-map POVM.

In the previous paragraph and in what follows, we assume that the state shared by Alice, Bob and Eve has an i.i.d. (independent, identically distributed) structure, and hence it makes sense to discuss the state ρ_{ABE} associated with a single round of quantum communication. To avoid confusion, we emphasize that our approach is 'unstructured' in the sense of lacking structure for a given round of quantum communication, but we do impose the i.i.d. structure that relates one round to other rounds. This i.i.d. structure corresponds to Eve doing a so-called collective attack. However, the security of our derived asymptotic key rate also holds against the most general attacks (coherent attacks) if one imposes that the protocol involves a random permutation of the rounds (a symmetrization step) such that the de Finetti theorem[26,27] or the post-selection technique[28] applies.

Typically, Alice's and Bob's shared density operator ρ_{AB} is unknown to them. A standard part of QKD protocols is for Alice and Bob to gather data through local measurements, and in a procedure known as parameter estimation, they use this data to constrain the form of ρ_{AB}. The measurements used for this purpose can, in general, be described by bounded Hermitian operators Γ_i, with the set of such operators denoted by $\vec{\Gamma} = \{\Gamma_i\}$.

From their data, Alice and Bob determine the average value of each of these measurements:

$$\vec{\gamma} = \{\gamma_i\}, \quad \text{with } \gamma_i := \langle \Gamma_i \rangle = \mathrm{Tr}(\rho_{AB}\Gamma_i), \qquad (4)$$

and this gives a set of experimental constraints:

$$C = \{\mathrm{Tr}(\rho_{AB}\Gamma_i) = \gamma_i\}. \qquad (5)$$

We denote the set of density operators that are consistent with these constraints as:

$$\mathcal{C} = \{\rho_{AB} \in \mathcal{P}_{AB} : C\,\text{holds}\} \qquad (6)$$

where $\mathcal{P}_{AB}$ denotes the set of positive semidefinite operators on $\mathcal{H}_{AB}$, and an additional constraint $\langle \mathbb{1} \rangle = 1$ is assumed to be added to the set C to enforce normalization.

Because Alice and Bob typically do not perform full tomography on the state, $\mathcal{C}$ includes many density operators, and hence the term $H(Z_A|E)$ in equation (1) is unknown. To evaluate the key rate, Alice and Bob must consider the most pessimistic of scenarios where $H(Z_A|E)$ takes on its smallest possible value that is consistent with their data. This is a constrained optimization problem, given by

$$K = \min_{\rho_{AB} \in \mathcal{C}} [H(Z_A|E) - H(Z_A|Z_B)] \qquad (7)$$

where Eve's system E can be assumed to purify ρ_{AB} since it gives Eve the most information. Here the number of parameters in the optimization is $(d_A d_B)^2$, corresponding to the number of parameters in a positive semidefinite operator on $\mathcal{H}_{AB}$. We refer to equation (7) as the primal problem.

Main result. Our main result is a reformulation of the optimization problem in equation (7).

Theorem 1: The solution of the minimization problem in equation (7) is lower bounded by the following maximization problem:

$$K \geq \frac{\Theta}{\ln 2} - H(Z_A|Z_B) \qquad (8)$$

where

$$\Theta := \max_{\vec{\lambda}} \left(- \left\| \sum_j Z_A^j R\left(\vec{\lambda}\right) Z_A^j \right\| - \vec{\lambda} \cdot \vec{\gamma} \right), \qquad (9)$$

and

$$R\left(\vec{\lambda}\right) := \exp\left(-\mathbb{1} - \vec{\lambda} \cdot \vec{\Gamma}\right). \qquad (10)$$

In equation (9), the optimization is over all vectors $\vec{\lambda} = \{\lambda_i\}$, where the λ_i are arbitrary real numbers and the cardinality of $\vec{\lambda}$ is equal to that of $\vec{\Gamma}$. Also, $\|M\|$ denotes the supremum norm of M, which is the maximum eigenvalue of M when M is positive semidefinite, as in equation (9).

The proof of Theorem 1 is given in the Methods section. It relies on the duality of convex optimization problems, as well as some entropic identities that allow us to simplify the dual problem. Note that the term $H(Z_A|Z_B)$ in equation (8) is pulled outside of the optimization since Alice and Bob can compute it directly from their data.

The cardinalities of the sets $\vec{\lambda}$ and $\vec{\Gamma}$ are the same. This means that the number of parameters λ_i that one must optimize over, to solve equation (9), is equal to the number of experimental constraints that Alice and Bob have. (More precisely this is the number of independent constraints, since one can eliminate constraints that carry redundant information.) This has the potential to be significantly less than the number of parameters in the primal problem. Indeed, we demonstrate below that equation (9) can be easily solved using MATLAB on a personal computer for a variety of interesting QKD protocols.

Formulating constraints. For a given protocol, how does one decide which constraints to include in the set C? Consider the following remarks. First, adding in more constraints will never decrease the key rate obtained from our optimization. This follows since adding a new constraint gives an additional λ_i to maximize over, while setting this new λ_i to zero recovers the old problem. Second, coarse-graining constraints, that is, merging two constraints $\langle\Gamma_i\rangle=\gamma_i$ and $\langle\Gamma_j\rangle=\gamma_j$ into a single constraint $\langle\Gamma_i+\Gamma_j\rangle=\gamma_i+\gamma_j$, will never increase the key rate obtained from our optimization. This follows since merging two constraints means that two λ_i's are merged into a single λ_i, thus restricting the optimization. Hence, to obtain the highest key rates, one should input all of one's refined knowledge that is available into our optimization. On the other hand, coarse graining reduces the number of constraints and thus may help to simplify the optimization problem, possibly at the cost of a reduced key rate.

One's refined knowledge is captured as follows. In a general entanglement-based protocol, Alice measures a POVM (whose elements may be non-commuting, for example, if she randomly measures one of two MUBs), which we denote as $\Gamma_A=\{\Gamma_{A,i}\}$. Likewise Bob's corresponding POVM is $\Gamma_B=\{\Gamma_{B,i}\}$. Hence, through public discussion, Alice and Bob obtain knowledge of expectation values of the form

$$\mathrm{Tr}\left[\rho_{AB}\left(\Gamma_{A,i}\otimes\Gamma_{B,j}\right)\right]=\gamma_{ij},\quad \text{for each } i,j. \tag{11}$$

These constraints form the set C in equation (5). We remark that it is common in the QKD field to express correlations in terms of average error rates rather than in terms of the joint probability distribution in equation (11). This is an example of the coarse graining that we mentioned above. For simplicity of presentation, we will do this sort of coarse graining for some protocols that we investigate below, although equation (11) represents our general framework for constructing C.

Framework for prepare and measure. While our approach is stated in the entanglement-based scenario, let us note how it applies to prepare-and-measure protocols. Consider a prepare-and-measure protocol involving a set of N signal states $\{|\phi_j\rangle\}$, which Alice sends with probabilities $\{p_j\}$. It is well-known that prepare-and-measure protocols can be recast as entanglement-based protocols using the source-replacement scheme (see, for example, refs 4,8,29). Namely, one forms the entangled state:

$$|\psi_{AA'}\rangle=\sum_j\sqrt{p_j}|j\rangle|\phi_j\rangle. \tag{12}$$

One imagines that Alice keeps system A, while system A' is sent over an insecure quantum channel $\mathcal{E}$ to Bob, resulting in

$$\rho_{AB}=(\mathcal{I}\otimes\mathcal{E})(|\psi_{AA'}\rangle\langle\psi_{AA'}|). \tag{13}$$

The numerical optimization approach described above can then be applied to the state ρ_{AB} in equation (13). However, in addition to the constraints obtained from Alice's and Bob's measurement results, we must add in further constraints to account for the special form of ρ_{AB}. In particular, note that the partial trace over B gives

$$\rho_A=\sum_{j,k}\sqrt{p_jp_k}\langle\phi_k|\phi_j\rangle|j\rangle\langle k|. \tag{14}$$

The form of ρ_A, which is closely related to Gram matrix, depends on the inner products between the signal states, which (we assume) Alice knows. Suppose $\{\Omega_i\}$ is a set of tomographically complete observables on system A, then one can add in the calculated expectation values $\{\omega_i\}$ of these observables into the set of constraints. That is, add

$$\langle\Omega_i\otimes\mathbb{1}\rangle=\omega_i,\quad\text{for each } i \tag{15}$$

to the set C in equation (5). This will capture Alice's knowledge of her reduced density operator.

Framework for decoy states. In decoy-state QKD[30], which aims to combat photon-number splitting attacks, Alice prepares coherent states of various intensities and then randomizes their phases before sending them to Bob. Our framework can handle decoy states simply by allowing for additional signal states to be added to the set $\{|\phi_j\rangle\}$ in equation (12). For example, to treat decoy protocols with partial phase randomization[31], one can consider signal states that are bipartite (on the signal mode S and the reference mode R) of the form

$$\left|\phi_{jkl}\right\rangle=\left|\alpha_je^{i(\theta_k+\phi_l)}\right\rangle_S\otimes\left|\alpha_je^{i\theta_k}\right\rangle_R \tag{16}$$

where α_j is the amplitude of the coherent state associated with the jth intensity setting, θ_k is the kth phase used in phase randomization, and ϕ_l is the phase Alice uses to encode her information (for example, for generating key). Decoy protocols with complete phase randomization are also treatable in our framework, namely, by adding in a signal state for each photon-number basis state (up to a cutoff), and treating multi-photon signals as orthogonal states (so-called 'tagged states') since Eve can perfectly distinguish them.

Framework for MDI QKD. A special kind of prepare-and-measure protocol is measurement-device-independent (MDI) QKD[32]. In MDI QKD, Alice prepares states $\{|\phi_j\rangle\}$ with probabilities $\{p_j\}$ and sends them to Charlie, and Bob does the same procedure as Alice (Fig. 2). Charlie typically does a Bell-basis measurement, however the security proof does not assume this particular form of measurement. Charlie announces the outcome of his measurement, which we denote by the classical register M. Our framework for treating MDI QKD considers the tripartite state ρ_{ABM}, where A and B, respectively, are Alice's and Bob's systems in the source-replacement scheme, playing the same role as system A in equation (12) (see Supplementary Note 1 for elaboration). For our numerics, we impose the constraint that the marginal $\rho_{AB}=\rho_A\otimes\rho_B$ is fixed (since Eve cannot access A and B), with ρ_A and ρ_B given by the form in equation (14). We enforce this constraint using the same approach as used in

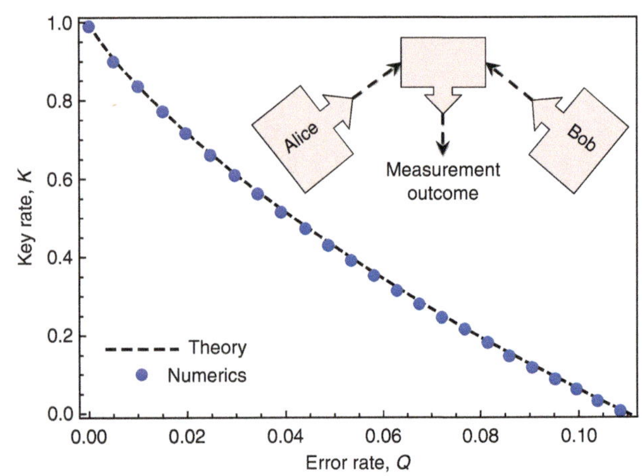

Figure 2 | Key rate for MDI QKD with the BB84 signal states. The inset shows the basic idea of MDI QKD: Alice and Bob each prepare a signal state and send it to an untrusted node, which performs an (untrusted) Bell-basis measurement and announces the outcome. Our numerics (circular dots) essentially reproduce the known theoretical dependence of the key rate on the error rate (dashed curve), which is the same expression as that given in (20). See Supplementary Note 1 for elaboration.

equation (15) to fix ρ_A for prepare-and-measure protocols. The only other constraints we impose are the usual correlation constraints, that is, a description of the joint probability distribution for the standard bases on A, B and M, of the form

$$\mathrm{Tr}[\rho_{ABM}(|j\rangle\langle j| \otimes |k\rangle\langle k| \otimes |m\rangle\langle m|)] = \gamma_{jkm}. \quad (17)$$

Framework for post-selection and announcements. In general, a QKD protocol may involve post-selection. As an example, if Alice sends photons to Bob over a lossy channel, then they may post-select on rounds in which Bob detects a photon. As noted above, for simplicity we consider protocols where the post-selection involves a binary announcement, and Alice and Bob keep (discard) rounds when 'pass' ('fail') is announced. Let $\mathcal{G}$ be the completely positive linear map corresponding to the post-selection. The action of $\mathcal{G}(\cdot)=G(\cdot)G^{\dagger}$ is given by a single Kraus operator G, corresponding to the 'pass' announcement.

The key rate formula (1) should be applied to the post-selected state:

$$\tilde{\rho}_{AB} = \mathcal{G}(\rho_{AB})/p_{\text{pass}} \quad (18)$$

where $p_{\text{pass}}= \mathrm{Tr}(\mathcal{G}(\rho_{AB}))$ is the probability for passing the post-selection filter. We remark that since $\mathcal{G}$ is given by a single Kraus operator, it maps pure states to pure states, and hence taking Eve's system to purify the post-selected state $\tilde{\rho}_{AB}$ is equivalent to taking it to purify ρ_{AB}. Hence applying the key rate formula to $\tilde{\rho}_{AB}$ does not give Eve access to any more than she already has, and hence does not introduce any looseness into our bound. Future extension of our work to more general maps $\mathcal{G}$ will need to carefully account for how Eve's system is affected by $\mathcal{G}$, so as not to lose key rate from this proof technique.

The only issue is that Alice's and Bob's experimental constraints C in equation (5) are still in terms of state ρ_{AB}. To solve for the key rate, one must transform these constraints into constraints on $\tilde{\rho}_{AB}$. For the special case where $\mathcal{G}$ has an inverse $\mathcal{G}^{-1}$ that is also completely positive, one can simply insert the identity channel $\mathcal{I}=\mathcal{G}^{-1}\mathcal{G}$ into the expression $\mathrm{Tr}(\rho_{AB}\Gamma_i)= \mathrm{Tr}(\mathcal{G}^{-1}\mathcal{G}(\rho_{AB})\Gamma_i)$. Using cyclic permutation under the trace, we transform equation (5) into a set of constraints on $\tilde{\rho}_{AB}$,

$$\tilde{C} = \{\mathrm{Tr}(\tilde{\rho}_{AB}\tilde{\Gamma}_i) = \tilde{\gamma}_i\}. \quad (19)$$

where the $\tilde{\Gamma}_i = (\mathcal{G}^{-1})^{\dagger}(\Gamma_i)$ are Hermitian operators, with $(\mathcal{G}^{-1})^{\dagger}$ being the adjoint of $\mathcal{G}^{-1}$, and $\tilde{\gamma}_i=\gamma_i/p_{\text{pass}}$. Note that p_{pass} is determined experimentally and hence the $\tilde{\gamma}_i$ are known to Alice and Bob. More generally, we provide a method for obtaining $\tilde{C}$ for arbitrary $\mathcal{G}$, as described in Supplementary Note 2.

We remark that public announcements can be treated with a simple extension of our post-selection framework. While our framework directly applies to announcements with only two outcomes corresponding to 'pass' or 'fail' (as discussed above), more general announcements can be treated by adding classical registers that store the announcement outcomes. Our treatment of MDI QKD is an example of this approach (Fig. 2 and Supplementary Note 1). Additional examples that could be treated in this way are protocols with two-way classical communication[33] such as advantage distillation.

Outline of examples. We now illustrate our numerical approach for lower bounding the key rate by considering some well-known protocols. First, we consider the BB84 and six-state protocols (Fig. 1), MDI QKD with BB84 states (Fig. 2), and the generalized BB84 protocol involving two MUBs in any dimension (Fig. 3). In each case, the dependence of the key rate on error rate is known, and we show that our numerical approach exactly reproduces these theoretical dependences. After considering these structured protocols, we move on to using our numerical optimization for its intended purpose: studying unstructured protocols. The fact that our bound is tight for the structured protocols mentioned above gives reason to suspect that we will get strong bounds in the unstructured case. We investigate below a protocol involving n MUBs, a protocol involving bases with arbitrary angle between them, and the B92 protocol.

BB84 example. Consider an entanglement-based version of the BB84 protocol[3], where Alice and Bob each receive a qubit and measure either in the $Z=\{|0\rangle, |1\rangle\}$ or $X=\{|+\rangle, |-\rangle\}$ basis, where $|\pm\rangle=(|0\rangle \pm |1\rangle)/\sqrt{2}$. For all protocols that we discuss, we assume perfect sifting efficiency, which can be accomplished asymptotically via asymmetric basis choice[34]. Let us suppose that Alice and Bob each use their Z basis to generate key. For simplicity, suppose they observe that their error rates in the Z and X bases are identical and equal to Q, then it is known (see, for example, ref. 4) that the key rate is given by

$$K = 1 - 2h(Q) \quad (20)$$

where $h(p) := -p\log_2 p - (1-p)\log_2(1-p)$ is the binary entropy.

To reproduce this result using our numerics, we write the optimization problem as follows:

$$\text{Key-map POVM}: Z_A = \{|0\rangle\langle 0|, |1\rangle\langle 1|\} \quad (21)$$

$$\text{Constraints}: \langle \mathbb{1} \rangle = 1 \quad (22)$$

$$\langle E_X \rangle = Q \quad (23)$$

$$\langle E_Z \rangle = Q \quad (24)$$

where the error operators are defined as

$$E_Z := |0\rangle\langle 0| \otimes |1\rangle\langle 1| + |1\rangle\langle 1| \otimes |0\rangle\langle 0| \quad (25)$$

$$E_X := |+\rangle\langle +| \otimes |-\rangle\langle -| + |-\rangle\langle -| \otimes |+\rangle\langle +|. \quad (26)$$

Equations (21)–(24) highlight the fact that, as far as the optimization in equation (9) is concerned, a QKD protocol is defined by the POVM elements used for generating the key and the experimental constraints used for 'parameter estimation' (and also the post-selection map $\mathcal{G}$, but this is trivial for the ideal BB84 protocol). Once these things are specified, the protocol is

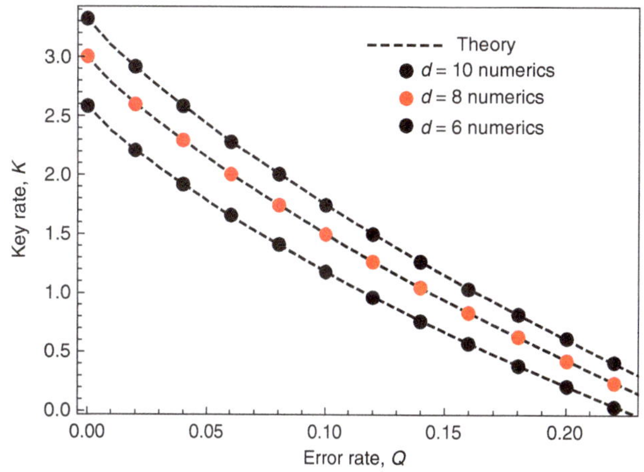

Figure 3 | Higher dimensional analog of BB84, using two MUBs. This plot shows the theoretical key rate as solid curves, and the result of our numerical optimization as circular dots, for $d_A = d_B = d$, with $d = 6$ (blue), $d = 8$ (red), and $d = 10$ (black). Again, the dots should be viewed as reliable lower bounds, but in this case they are perfectly tight.

defined and the key rate is determined. Numerically solving the problem defined in equations (21)–(24) for several values of Q leads to the red dots in Fig. 1, which agree perfectly with the theory curve.

Six state example. Now consider an entanglement-based version of the six-state protocol, where Alice and Bob each measure one of three MUBs (X, Y or Z) on their qubit. Suppose that Alice and Bob observe that their error rates in all three bases are identical, $\langle E_X\rangle = \langle E_Y\rangle = \langle E_Z\rangle = Q$, where

$$E_Y := |y_+\rangle\langle y_+| \otimes |y_+\rangle\langle y_+| + |y_-\rangle\langle y_-| \otimes |y_-\rangle\langle y_-|, \quad (27)$$

with $|y_\pm\rangle = (|0\rangle \pm i|1\rangle)/\sqrt{2}$. (Our definition of E_Y reflects the fact that the standard Bell state is correlated in Z and X but anti-correlated in Y.) To reproduce the known key rate[9,21], we write the optimization problem as:

$$\text{Key-map POVM}: Z_A = \{|0\rangle\langle 0|, |1\rangle\langle 1|\} \quad (28)$$

$$\text{Constraints}: \langle \mathbb{1}\rangle = 1 \quad (29)$$

$$\langle E_{XY}\rangle = Q \quad (30)$$

$$\langle E_Z\rangle = Q, \quad (31)$$

where $E_{XY} := (E_X + E_Y)/2$ quantifies the average error for X and Y. Note that the constraint $\langle E_{XY}\rangle = Q$ is obtained by coarse graining the individual error rates. In theory, one can get a stronger bound on the key rate by splitting up this constraint into $\langle E_X\rangle = Q$ and $\langle E_Y\rangle = Q$. However, our numerics show that this does not improve the key rate, and the constraints in equation (29)–(31) are enough to reproduce the theory curve. Indeed, numerically solving the problem in equation (28)–(31) leads to the blue dots in Fig. 1, which agree with the theory curve.

Two MUBs in higher dimensions example. A distinct advantage of our approach of solving equation (9) instead of the primal problem equation (7) is that we can easily perform the optimization in higher dimensions, where the number of parameters in equation (7) would be quite large. To illustrate this, we consider a generalization of BB84 to arbitrary dimension, where Alice and Bob measure generalized versions of the X and Z bases. This protocol has been implemented, for example, in ref. 18 using orbital angular momentum. Taking Z as the standard basis $\{|j\rangle\}$, Alice's X basis can be taken as the Fourier transform $\{F|j\rangle\}$, where

$$F = \sum_{j,k} \frac{\omega^{-jk}}{\sqrt{d}} |j\rangle\langle k| \quad (32)$$

is the Fourier matrix, with $\omega = e^{2\pi i/d}$, and for simplicity we choose Alice's and Bob's dimension to be equal: $d_A = d_B = d$. Bob's X basis is set to $\{F^*|j\rangle\}$, where F^* denotes the conjugate of F in the standard basis.

Suppose that Alice and Bob observe that their error rates in Z and X are identical. The theoretical key rates[8,17] for the cases $d = 6$, 8, 10 are shown as dashed curves in Fig. 3, while our numerics are shown as circular dots. Clearly there is perfect agreement with the theory.

For our numerics we employ the same constraints as used for BB84 in equation (21)–(24), but generalized to higher d. We again emphasize that the calculation of Θ here is very efficient and can easily handle higher dimension. This is because the number of parameters one is optimizing over is independent of dimension—equal to the number of constraints, which in this case is 3. This is in sharp contrast to the primal problem in equation (7), where the number of parameters is d^4, which would be 10,000 for $d = 10$.

***n* MUBs example**. A simple generalization of the above protocols is to consider a set of n MUBs in dimension d. For example, in prime power dimensions there exist explicit constructions for sets of n MUBs with $2 \leqslant n \leqslant d+1$ (ref. 35). Consisder a protocol where we fix the set of n MUBs, and in each round, Alice and Bob each measure their d dimensional system in one basis chosen from this set. For general n the symmetry group is not known for this protocol[8], so one can consider it an unstructured protocol. Indeed, only for the special cases $n = 2$ and $n = d+1$ do we have analytical formulas for the key rate[8]. Nevertheless it is straightforward to apply our numerics to this protocol for any n. Our results are shown in Fig. 4 for $d = 5$. To obtain these curves we only need three constraints, which are analogous to equation (29)–(31), but generalized such that $\langle E_{XY}\rangle$ is replaced by the average error rate in all $n-1$ bases, excluding the basis used for key generation (the Z basis).

Interestingly, Fig. 4 shows that just adding one basis, going from $n = 2$ to $n = 3$, gives a large jump in the key rate, whereas there are diminishing returns as one adds more bases. This can be seen in the inset of Fig. 4, which plots the error tolerance (that is, the value of Q for which the key rate goes to zero) as a function of n. We have seen similar behaviour for other d besides $d = 5$. After completion of this work, an analytical formula for $n = 3$ was discovered[36], and we have verified that it agrees perfectly with our numerics.

In Supplementary Note 3, we analytically prove the following.

Proposition 2: Our numerical results are perfectly tight for the protocols discussed in Figs 1,3 and 4. That is, for these protocols, our optimization exactly reproduces the primal optimization (equation (7)).

Note that this observation implies that key rate for protocols involving n MUBs (as in Fig. 4) is now known; namely it is given by our numerical optimization.

Arbitrary angle between bases example. While MUBs are a special case, our approach can handle arbitrary angles between the different measurements or signal states. For example, we consider a simple qubit protocol[37] where Alice and Bob each measure either the Z or W basis, where W is rotated by an angle θ away from the ideal X basis. This protocol provides the

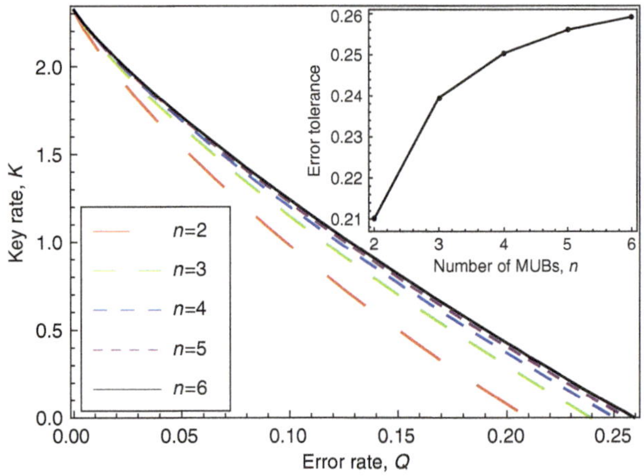

Figure 4 | Protocol where Alice and Bob each use *n* MUBs. The key rate is plotted for various $n \in \{2, 3, 4, 5, 6\}$ and for $d_A = d_B = 5$. This is an unstructured protocol, since for intermediate values of n the symmetry group and hence the key rate is unknown. However, our numerics provides the dependence of key rate on error rate for any n, as shown. The inset shows the error tolerance—the smallest error rate that makes the key rate vanish—as a function of n. Note that the largest jump in the error tolerance occurs from $n = 2$ to $n = 3$.

opportunity to compare our numerical approach to an analytical approach based on the entropic uncertainty principle, introduced in refs 22,23. This is the state-of-the-art method for lower bounding the key rate. So for comparison, Fig. 5 plots the bound obtained from the entropic uncertainty principle for bases Z and W.

We apply our numerical approach with the constraints:

$$\text{Constraints}: \langle \mathbb{1} \rangle = 1 \tag{33}$$

$$\langle \sigma_W \otimes \sigma_W \rangle = 1 - 2Q \tag{34}$$

$$\langle \sigma_Z \otimes \sigma_Z \rangle = 1 - 2Q \tag{35}$$

$$\langle \sigma_Z \otimes \sigma_W \rangle = \langle \sin\theta \rangle (1 - 2Q) \tag{36}$$

$$\langle \sigma_W \otimes \sigma_Z \rangle = (\sin\theta)(1 - 2Q), \tag{37}$$

where σ_Z and σ_W are the Pauli operators associated with the Z and W bases. Figure 5 plots a hierarchy of lower bounds obtained from gradually adding in more of the constraints in equation (33)-(37). As the plot shows, we already beat the entropic uncertainty principle with only the first two constraints. Furthermore, adding in all these constraints gives a markedly higher bound, showing the uncertainty principle gives highly pessimistic key rates for this protocol. From an experimental perspective, Fig. 5 is reassuring, in that small variations in θ away from the ideal BB84 protocol ($\theta = 0$) have essentially no effect on the key rate. Figure 5 also highlights the fact that our approach allows us to systematically study the effect on the key rate of Alice and Bob using more or less of their data. In this example, we see that it is useful to keep data that one will usually discard in the sifting step of the protocol.

B92 example. Next we consider the B92 protocol[19], which is a simple, practical and unstructured protocol. It nicely illustrates our framework because it is inherently a prepare-and-measure protocol and it involves post-selection. In the protocol, Alice sends one of two non-orthogonal states $\{|\phi_0\rangle, |\phi_1\rangle\}$ to Bob. Since the Bloch-sphere angle θ between the two states is arbitrary, with $\langle\phi_0|\phi_1\rangle = \cos(\theta/2)$, the protocol is unstructured. Bob randomly (with equal probability) measures either in basis $B_0 = \{|\phi_0\rangle, |\overline{\phi_0}\rangle\}$ or basis $B_1 = \{|\phi_1\rangle, |\overline{\phi_1}\rangle\}$, where $\langle\phi_0|\overline{\phi_0}\rangle = \langle\phi_1|\overline{\phi_1}\rangle = 0$. If Bob gets outcome $|\overline{\phi_0}\rangle$ or $|\overline{\phi_1}\rangle$, then he publicly announces 'pass', and he assigns a bit value of 1 or 0, respectively, to his key. Otherwise, Bob announces 'fail' and they discard the round.

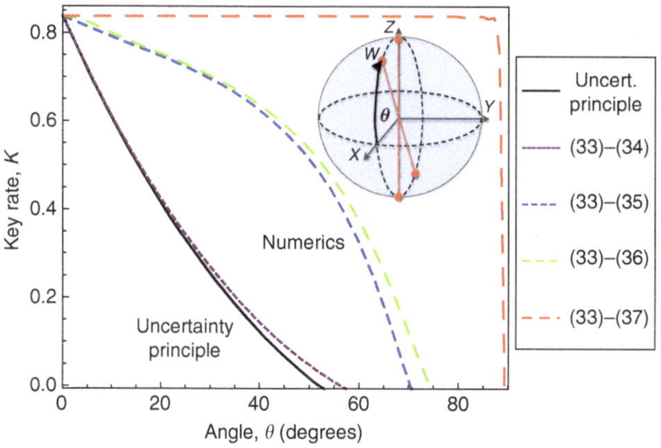

Figure 5 | Protocol where Alice and Bob each measure *Z* or *W*. Here Z is the standard basis and the W basis is rotated by an angle θ away from the X basis. The key rate versus θ is shown with the error rate set to $Q = 0.01$. Our numerics give a hierarchy of four lower bounds on the key rate, corresponding to adding in additional constraints from (33)-(37). All of our bounds are tighter than the bound obtained from the entropic uncertainty principle. The plot indicates that the uncertainty principle gives a dramatically pessimistic key rate, much lower than the true key rate of the protocol.

A detailed description of the constraints we employed for B92 can be found in Supplementary Note 4. Our numerical results are shown in Fig. 6. Figure 6 shows that the optimal angle for maximizing key rate depends on the depolarizing noise p, although small deviations $\pm 5°$ from the optimal angle do not affect the key rate much.

Our results give higher key rates for B92 than refs 20 and 21, which respectively predicted positive key rates for $p \leqslant 0.034$ and $p \leqslant 0.048$, while we predict it for $p \leqslant 0.053$. On the other hand, ref. 38 directly solved the primal problem equation (7) for B92 by brute-force numerics, and achieves positive key rate for $p \leqslant 0.065$. We have verified that the gap between our results and those of ref. 38 is due to the looseness of our usage of the Golden-Thompson inequality (equation (60) in the Methods section). However, ref. 38 only showed a plot for $p \geqslant 0.046$, noting that the numerical optimization for the primal problem did not converge for smaller p values. This highlights a benefit of going to the dual problem, in that we have no trouble with obtaining the full dependence on p.

Discussion

In conclusion, we address one of the main outstanding problems in QKD theory: how to calculate key rates for arbitrary protocols. Our main result is a numerical method for lower-bounding key rates that is both efficient and reliable. It is reliable in the sense that, by reformulating the problem as a maximization, every solution that one's computer outputs is an achievable key rate. It is efficient in the sense that we have reduced the number of parameters in the optimization problem from $d_A^2 d_B^2$ down to the number of experimental constraints, which in some cases is independent of dimension.

The motivation for our work is twofold. First, experimental imperfections tend to break symmetries, so theoretical techniques that exploit symmetries do not apply. Hence, there is no general method currently available for calculating the effect of imperfections on the key rate. Second, it is interesting to ask whether protocols that are intentionally designed to lack symmetry might outperform the well-known symmetric protocols. Such a question cannot be posed without a method for calculating key rates for unstructured protocols. Just to give an example where the key rate is currently unknown, we plan to apply our approach to protocols

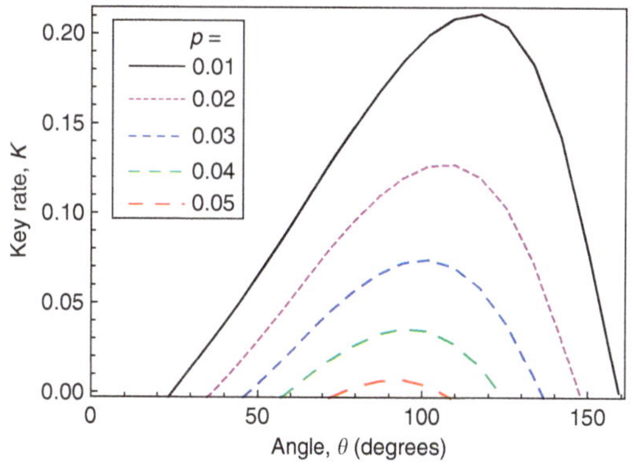

Figure 6 | The B92 protocol. The key rate (in bits per photon sent by Ailce) is plotted versus the Bloch-sphere angle between the two signal states. Curves are shown for various values of the depolarizing probability p.

where a small, discrete set of coherent states are the signal states and information is encoded in the phase[39].

We envision that our method could be a standard tool for QKD researchers. In future work we hope to extend our approach to the finite-key scenario. Indeed, the optimization problem we solve is closely related to one appearing in finite-key analysis[24].

Methods

Outline. Here we prove our main result, Theorem 1. Our proof relies on several technical tools. First is the notion of the duality of optimization, that is, transforming the primal problem to its dual problem. Second, we employ several entropic identities to simplify the dual problem. Third, we use a recent, important result from ref. 40 that solves a relative entropy optimization problem.

For readability, we prove Theorem 1 here for the special case where the key map POVM $Z_A = \{Z_A^j\}$ is a projective measurement, that is, where the Z_A^j are projectors (of arbitrary rank). We postpone the proof for arbitrary POVMs to Supplementary Note 5.

The primal problem. First we rewrite equation (7) as:

$$K = \left[\min_{\rho_{AB}\in\mathcal{C}} H(Z_A|E)\right] - H(Z_A|Z_B), \tag{38}$$

noting that the second term in equation (38), $H(Z_A|Z_B)$, will be determined experimentally and hence can be pulled out of the optimization. We remark that, simply for illustration purposes we used Fano's inequality to upper-bound $H(Z_A|Z_B)$ in our figures; however, in practice $H(Z_A|Z_B)$ would be directly calculated from the data.

Since we only need to optimize the first term, we redefine the primal problem as

$$\alpha := \min_{\rho_{AB}\in\mathcal{C}} H(Z_A|E), \tag{39}$$

and note that we can take E to be a purifying system of ρ_{AB}, since that gives Eve the most information. Next we use a result for tripartite pure states $\rho_{ABE}=|\psi\rangle\langle\psi|_{ABE}$ from refs 41,42 that relates the conditional entropy to the relative entropy:

$$H(Z_A|E) = D\left(\rho_{AB}\middle\|\sum_j Z_A^j \rho_{AB} Z_A^j\right) \tag{40}$$

where the relative entropy is defined by

$$D(\sigma\|\tau) := \mathrm{Tr}(\sigma\log_2\sigma) - \mathrm{Tr}(\sigma\log_2\tau). \tag{41}$$

We remark that the joint convexity of the relative entropy implies that the right-hand side of equation (40) is a convex function of ρ_{AB}. (See ref. 43 for an alternative proof of convexity.) Because of this, and the fact that the constraints in equation (5) are linear functions of ρ_{AB}, equation (39) is a convex optimization problem[16].

It is interesting to point out the connection to coherence[44]. For some set of orthogonal projectors $\Pi=\{\Pi^j\}$ that decompose the identity, $\sum_j \Pi^j=\mathbb{1}$, the coherence (sometimes called relative entropy of coherence) of state ρ is defined as[44]:

$$\Phi(\rho,\Pi) = D\left(\rho\middle\|\sum_j \Pi^j\rho\Pi^j\right). \tag{42}$$

Rewriting the primal problem in terms of coherence gives

$$\alpha = \min_{\rho_{AB}\in\mathcal{C}} \Phi(\rho_{AB}, Z_A). \tag{43}$$

Hence, we make the connection that calculating the secret key rate is related to optimizing the coherence.

This observation is important since the coherence is a continuous function of ρ (Supplementary Note 6). This allows us to argue in Supplementary Note 6 that our optimization problem satisfies the strong duality criterion[16], which means that the solution of the dual problem is precisely equal to that of primal problem.

The dual problem. Now we transform to the dual problem. Due to a pesky factor of ln(2), it is useful to rescale the primal problem as follows:

$$\hat{\alpha} := \alpha\ln(2) = \min_{\rho_{AB}\in\mathcal{C}} \hat{\Phi}(\rho_{AB}, Z_A) \tag{44}$$

where, henceforth, we generally use the notation $\hat{M} := M\ln(2)$, for any quantity M. The dual problem[16] of equation (44) is given by the following unconstrained optimization:

$$\hat{\beta} = \max_{\vec{\lambda}} \min_{\rho_{AB}\in\mathcal{C}} \mathcal{L}\left(\rho_{AB},\vec{\lambda}\right) \tag{45}$$

where $\mathcal{P}$ is the set of positive semidefinite operators:

$$\mathcal{P} = \{\rho_{AB}\in\mathcal{H}_{d_A d_B} : \rho_{AB}\geqslant 0\}. \tag{46}$$

Here the Lagrangian is given by

$$\mathcal{L}\left(\rho_{AB},\vec{\lambda}\right) := \hat{\Phi}(\rho_{AB}, Z_A) + \sum_i \lambda_i[\mathrm{Tr}(\rho_{AB}\Gamma_i) - \gamma_i], \tag{47}$$

where the $\vec{\lambda}=\{\lambda_i\}$ are Lagrange multipliers. Strong duality implies that

$$\hat{\beta} = \hat{\alpha}. \tag{48}$$

In what follows, we go through several steps to simplify the dual problem. It helps to first state the following lemma from refs 42,45.

Lemma 3: For any ρ and $\Pi=\{\Pi^j\}$, the coherence can be rewritten as

$$\Phi(\rho,\Pi) = \min_{\omega\in\mathcal{D}} D\left(\rho\middle\|\sum_j \Pi^j\omega\Pi^j\right) \tag{49}$$

where $\mathcal{D}$ is the set of density operators.

Hence, we have

$$\hat{\Phi}(\rho_{AB}, Z_A) = \min_{\sigma_{AB}\in\mathcal{D}} \hat{D}(\rho_{AB}\|\mathcal{Z}_A(\sigma_{AB})), \tag{50}$$

where we define the quantum channel $\mathcal{Z}_A$ whose action on an operator O is given by

$$\mathcal{Z}_A(O) := \sum_j Z_A^j O Z_A^j. \tag{51}$$

Next, we interchange the two minimizations in (45)

$$\min_{\rho_{AB}\in\mathcal{P}}\min_{\sigma_{AB}\in\mathcal{D}} f\left(\rho_{AB},\sigma_{AB},\vec{\lambda}\right) = \min_{\sigma_{AB}\in\mathcal{D}}\min_{\rho_{AB}\in\mathcal{P}} f\left(\rho_{AB},\sigma_{AB},\vec{\lambda}\right) \tag{52}$$

where

$$f\left(\rho_{AB},\sigma_{AB},\vec{\lambda}\right) := \hat{D}(\rho_{AB}\|\mathcal{Z}_A(\sigma_{AB})) + \sum_i \lambda_i(\langle\Gamma_i\rangle - \gamma_i). \tag{53}$$

Ref. 40 solved a relative entropy optimization problem, a special case of which is our problem:

$$\min_{\rho_{AB}\in\mathcal{P}} f\left(\rho_{AB},\sigma_{AB},\vec{\lambda}\right). \tag{54}$$

From ref. 40, the unique solution of equation (54) is

$$\rho_{AB}^* = \exp\left(-\mathbb{1} - \vec{\lambda}\cdot\vec{\Gamma} + \ln(\mathcal{Z}_A(\sigma_{AB}))\right). \tag{55}$$

Inserting equation (55) into equation (53) gives the optimal value:

$$f\left(\rho_{AB}^*,\sigma_{AB},\vec{\lambda}\right) = -\mathrm{Tr}(\rho_{AB}^*) - \sum_i \lambda_i\gamma_i. \tag{56}$$

In summary, the dual problem becomes

$$\hat{\beta} = \max_{\vec{\lambda}} \eta\left(\vec{\lambda}\right), \tag{57}$$

with

$$\eta\left(\vec{\lambda}\right) := -\max_{\sigma_{AB}\in\mathcal{D}}\left[\mathrm{Tr}(\rho_{AB}^*) + \vec{\lambda}\cdot\vec{\gamma}\right]. \tag{58}$$

A lower bound. We can obtain a simple lower bound on $\eta(\vec{\lambda})$ as follows. The Golden–Thompson inequality states that

$$\mathrm{Tr}(exp(A+B)) \leq \mathrm{Tr}(\exp(A)\exp(B)). \tag{59}$$

Applying this inequality gives:

$$\mathrm{Tr}(\rho_{AB}^*) \leq \mathrm{Tr}\left(R\left(\vec{\lambda}\right)\exp(\ln\mathcal{Z}_A(\sigma_{AB}))\right) \tag{60}$$

$$= \mathrm{Tr}\left(R\left(\vec{\lambda}\right)\mathcal{Z}_A(\sigma_{AB})\right) \tag{61}$$

$$= \mathrm{Tr}\left(\mathcal{Z}_A\left(R\left(\vec{\lambda}\right)\right)\sigma_{AB}\right), \tag{62}$$

where $R(\vec{\lambda}) = \exp(-\mathbb{1} - \vec{\lambda}\cdot\vec{\Gamma})$ was defined in equation (10). Next, note that

$$\max_{\sigma_{AB}\in\mathcal{D}} \mathrm{Tr}\left(\mathcal{Z}_A\left(R\left(\vec{\lambda}\right)\right)\sigma_{AB}\right) = \left\|\mathcal{Z}_A\left(R\left(\vec{\lambda}\right)\right)\right\|. \tag{63}$$

Hence, we arrive at our final result

$$\hat{\beta} \geq \max_{\vec{\lambda}}\left[-\left\|\mathcal{Z}_A\left(R\left(\vec{\lambda}\right)\right)\right\| - \vec{\lambda}\cdot\vec{\gamma}\right], \tag{64}$$

where the right-hand side is denoted as Θ in Theorem 1.

References

1. Campagna, M. *et al.* *Quantum Safe Cryptography and Security* (European Telecommunications Standards Institute, 2015).
2. Wiesner, S. Conjugate coding. *ACM SIGACT News* **15**, 78–88 (1983).
3. Bennett, C. H. & Brassard, G. in *International Conference on Computers, Systems & Signal Processing* 175–179 (Bangalore, India, 1984).
4. Scarani, V. *et al.* The security of practical quantum key distribution. *Rev. Mod. Phys.* **81**, 1301–1350 (2009).
5. Lo, H.-K., Curty, M. & Tamaki, K. Secure quantum key distribution. *Nat. Photon.* **8**, 595–604 (2014).
6. Sasaki, M. *et al.* Field test of quantum key distribution in the Tokyo QKD Network. *Opt. Express* **19**, 10387–10409 (2011).
7. Wang, S. *et al.* Field and long-term demonstration of a wide area quantum key distribution network. *Opt. Express* **22**, 329–342 (2014).
8. Ferenczi, A. & Lütkenhaus, N. Symmetries in quantum key distribution and the connection between optimal attacks and optimal cloning. *Phys. Rev. A* **85**, 052310 (2012).
9. Bruss, D. Optimal eavesdropping in quantum cryptography with six states. *Phys. Rev. Lett.* **81**, 3018–3021 (1998).
10. Gottesman, D., Lo, H.-K., Lütkenhaus, N. & Preskill, J. Security of quantum key distribution with imperfect devices. *Quant. Inf. Comput.* **4**, 325–360 (2004).
11. Fung, C.-H. F. & Lo, H.-K. Security proof of a three-state quantum-key-distribution protocol without rotational symmetry. *Phys. Rev. A* **74**, 042342 (2006).
12. Maroy, O., Lydersen, L. & Skaar, J. Security of quantum key distribution with arbitrary individual imperfections. *Phys. Rev. A* **82**, 032337 (2010).
13. Woodhead, E. Quantum cloning bound and application to quantum key distribution. *Phys. Rev. A* **88**, 012331 (2013).
14. Tamaki, K., Curty, M., Kato, G., Lo, H.-K. & Azuma, K. Loss-tolerant quantum cryptography with imperfect sources. *Phys. Rev. A* **90**, 052314 (2014).
15. Devetak, I. & Winter, A. Distillation of secret key and entanglement from quantum states. *Proc. R. Soc. A* **461**, 207–235 (2005).
16. Boyd, S. & Vandenberghe, L. *Convex Optimization* (Cambridge University Press, 2004).
17. Sheridan, L. & Scarani, V. Security proof for quantum key distribution using qudit systems. *Phys. Rev. A* **82**, 030301 (2010).
18. Mafu, M. *et al.* Higher-dimensional orbital-angular-momentum-based quantum key distribution with mutually unbiased bases. *Phys. Rev. A* **88**, 032305 (2013).
19. Bennett, C. H. Quantum cryptography using any two nonorthogonal states. *Phys. Rev. Lett.* **68**, 3121–3124 (1992).
20. Tamaki, K. & Lütkenhaus, N. Unconditional security of the Bennett 1992 quantum key-distribution protocol over a lossy and noisy channel. *Phys. Rev. A* **69**, 032316 (2004).
21. Renner, R., Gisin, N. & Kraus, B. Information-theoretic security proof for quantum-key-distribution protocols. *Phys. Rev. A* **72**, 012332 (2005).
22. Berta, M., Christandl, M., Colbeck, R., Renes, J. M. & Renner, R. The uncertainty principle in the presence of quantum memory. *Nat. Phys.* **6**, 659–662 (2010).
23. Tomamichel, M., Ci, C., Lim, W., Gisin, N. & Renner, R. Tight finite-key analysis for quantum cryptography. *Nat. Commun.* **3**, 634 (2012).
24. Scarani, V. & Renner, R. Quantum cryptography with finite resources: Unconditional security bound for discrete-variable protocols with one-way postprocessing. *Phys. Rev. Lett.* **100**, 200501 (2008).
25. Sano, Y., Matsumoto, R. & Uyematsu, T. Secure key rate of the BB84 protocol using finite sample bits. *J. Phys. A* **43**, 2677–2681 (2010).
26. Renner, R. *Security of Quantum Key Distribution* (PhD Thesis, ETH Zurich, 2005).
27. Renner, R. Symmetry of large physical systems implies independence of subsystems. *Nat. Phys.* **3**, 645–649 (2007).
28. Christandl, M., König, R. & Renner, R. Postselection technique for quantum channels with applications to quantum cryptography. *Phys. Rev. Lett.* **102**, 020504 (2009).
29. Bennett, C., Brassard, G. & Mermin, N. Quantum cryptography without Bell's theorem. *Phys. Rev. Lett.* **68**, 557–559 (1992).
30. Lo, H.-K., Ma, X. & Chen, K. Decoy state quantum key distribution. *Phys. Rev. Lett.* **94**, 230504 (2005).
31. Cao, Z., Zhang, Z., Lo, H.-K. & Ma, X. Discrete-phase-randomized coherent state source and its application in quantum key distribution. *New J. Phys.* **17**, 053014 (2015).
32. Lo, H.-K., Curty, M. & Qi, B. Measurement-device-independent quantum key distribution. *Phys. Rev. Lett.* **108**, 130503 (2012).
33. Gottesman, D. & Lo, H. K. Proof of security of quantum key distribution with two-way classical communications. *IEEE Trans. Inf. Theory* **49**, 457–475 (2003).
34. Lo, H.-K., Chau, H. F. & Ardehali, M. Efficient quantum key distribution scheme and a proof of its unconditional security. *J. Cryptol.* **18**, 133–165 (2004).
35. Bandyopadhyay, S., Boykin, P. O., Roychowdhury, V. & Vatan, F. A new proof for the existence of mutually unbiased bases. *Algorithmica* **34**, 512–528 (2001).
36. Bradler, K., Mirhosseini, M., Fickler, R., Broadbent, A. & Boyd, R. Finite-key security analysis for multilevel quantum key distribution. Preprint at http://arxiv.org/abs/1512.05447 (2015).
37. Matsumoto, R. & Watanabe, S. Narrow basis angle doubles secret key in the BB84 protocol. *J. Phys. A* **43**, 145302 (2010).
38. Matsumoto, R. in *Proceedings of IEEE International Symposium on Information Theory,* 351-353 (2013).
39. Lo, H. & Preskill, J. Security of quantum key distribution using weak coherent states with nonrandom phases. *Quant. Inf. Comput.* **7**, 431–458 (2007).
40. Zorzi, M., Ticozzi, F. & Ferrante, A. Minimum relative entropy for quantum estimation: Feasibility and general solution. *IEEE Trans. Inf. Theory* **60**, 357–367 (2014).
41. Coles, P. J., Yu, L., Gheorghiu, V. & Griffiths, R. B. Information-theoretic treatment of tripartite systems and quantum channels. *Phys. Rev. A* **83**, 062338 (2011).
42. Coles, P. J. Unification of different views of decoherence and discord. *Phys. Rev. A* **85**, 042103 (2012).
43. Watanabe, S., Matsumoto, R. & Uyematsu, T. Tomography increases key rates of quantum-key-distribution protocols. *Phys. Rev. A* **78**, 042316 (2008).
44. Baumgratz, T., Cramer, M. & Plenio, M. B. Quantifying coherence. *Phys. Rev. Lett.* **113**, 140401 (2014).
45. Modi, K., Paterek, T., Son, W., Vedral, V. & Williamson, M. Unified view of quantum and classical correlations. *Phys. Rev. Lett.* **104**, 080501 (2010).

Acknowledgements

We thank Jie Lin, Adam Winick, and Bailey Gu for technical help with the numerics and for obtaining the data in Fig. 2. We thank Yanbao Zhang and Saikat Guha for helpful discussions. We acknowledge support from Industry Canada, Sandia National Laboratories, Office of Naval Research (ONR), NSERC Discovery Grant, and Ontario Research Fund (ORF).

Author contributions

P.J.C. obtained the conceptional main results. E.M.M. contributed the approach to incorporate post-selection and announcements. N.L. conceived and supervised the project. P.J.C. wrote the manuscript with input from E.M.M. and N.L.

Additional information

Competing financial interests: The authors declare no competing financial interests.

26

Ultrafast and reversible control of the exchange interaction in Mott insulators

J.H. Mentink[1,†], K. Balzer[1] & M. Eckstein[1]

The strongest interaction between microscopic spins in magnetic materials is the exchange interaction J_{ex}. Therefore, ultrafast control of J_{ex} holds the promise to control spins on ultimately fast timescales. We demonstrate that time-periodic modulation of the electronic structure by electric fields can be used to reversibly control J_{ex} on ultrafast timescales in extended antiferromagnetic Mott insulators. In the regime of weak driving strength, we find that J_{ex} can be enhanced and reduced for frequencies below and above the Mott gap, respectively. Moreover, for strong driving strength, even the sign of J_{ex} can be reversed and we show that this causes time reversal of the associated quantum spin dynamics. These results suggest wide applications, not only to control magnetism in condensed matter systems, for example, via the excitation of spin resonances, but also to assess fundamental questions concerning the reversibility of the quantum many-body dynamics in cold atom systems.

[1] Max Planck Research Department for Structural Dynamics, University of Hamburg-CFEL, 22761 Hamburg, Germany. † Present address: Radboud University, Institute for Molecules and Materials, Heyendaalseweg 135, 6525 AJ Nijmegen, The Netherlands. Correspondence and requests for materials should be addressed to J.H.M. (email: j.mentink@science.ru.nl).

Controlling magnetically ordered systems on sub-picosecond timescales is currently a widely studied research area owing to the joint fundamental interest and technological demand for faster and more energy-efficient magnetic storage[1]. The fastest pathways to reverse magnetic order utilize the exchange interaction J_{ex} between microscopic magnetic moments[2–6], which can exceed external magnetic fields by orders of magnitude. Because J_{ex} relies on the electrostatic Coulomb repulsion and the Pauli principle rather than on magnetic dipole forces, it may be modified directly by the action of a laser pulse on the electronic state. This implies an appealing and largely unexplored scenario to control magnetism on the fastest possible timescale. Recently, several experimental studies on magnetic materials have discussed an ultrafast modification of J_{ex} or change of the type of magnetic order by creating a nonequilibrium electron distribution (by photodoping or laser heating)[7–15]. In these cases the spin dynamics after the excitation strongly depends on the relaxation of the electrons, thereby hindering a direct and reversible control of the spin degrees of freedom alone. On the other hand, reversible electrical control of J_{ex} was recently demonstrated in a multiferroic solid state system, where the bond alignment can be changed by a static electric field[16]. Clearly, a natural goal is to achieve control of J_{ex}, which is both reversible and ultrafast, that is, it is active while a laser pulse is on, but leaves the electronic state unexcited after the pulse is switched off.

A versatile approach to reversibly control the dynamics of quantum systems is given by the rectification of time-periodic perturbations. The use of periodic driving to control the dynamics of a quantum system is known in many areas of physics, for example, through effective conservative forces resulting from the AC Stark effect or through the coherent destruction of tunnelling[17,18]. For extended solid state systems, it is well known that particles in a tight-binding band subject to periodic driving evolve under an effective Hamiltonian that has a different band structure[17–22]. On the other hand, the control of the exchange interaction requires to understand how the driving influences both the band structure and the electronic correlations, which determine J_{ex} in equilibrium. This is a highly challenging problem in general, since it implies the solution of a strongly time-dependent many-body problem of an extended system.

In recent years, important insights into the control of J_{ex} have been obtained by studying the effect of periodic driving for one and two spin systems, leading to light-induced Kondo effects[23], the design of an effective low-energy spin Hamiltonian in bosonic double-well systems[24], as well as an optically induced RKKY interaction between localized spins in semiconductor quantum dots by virtual excitation of delocalized excitons[25,26]. The latter has been shown to be effective in extended systems as well, in particular for the dilute ferromagnetic (FM) semiconductors[27,28]. However, since the spin dynamics in ferromagnets requires a change of the total angular momentum, it is difficult to induce fast dynamics by modifying J_{ex}. To the contrary, extended antiferromagnetic (AFM) systems do not suffer from the angular momentum bottleneck, and hence can provide novel opportunities for the ultrafast control of the spin dynamics by modifying J_{ex}.

In this paper, we demonstrate that it is possible to reversibly control J_{ex} in extended AFM Mott insulators by periodically modulating the electronic structure with a frequency ω higher than $J_{ex}/\hbar$, but not resonant to electronic excitations. We investigate a simple driving scheme using time-periodic electric fields, which can be realized both in solid state systems and for cold atoms, and hence suggests wide applications: besides the possibility of manipulating magnetism in solids, for example, via the excitation of spin resonances, we find that in the extreme limit of strong driving one may even achieve a sign reversal of J_{ex}, which is equivalent of letting the system evolve backwards in time and may allow for addressing fundamental questions concerning the reversibility of quantum many-body dynamics[29–31] in cold atom experiments. Furthermore, we show that considerable insight can be obtained from analytical Floquet theory for a few-site cluster under continuous driving, which predicts reversible enhancement, reduction and even complete sign-change of the exchange interaction. The relevance of these results for extended many-body systems may not be clear *a priori*, since in this case a true quasi-steady driven state may always become infinitely excited[32]. For the relatively short-term dynamics of interest here, the predictions of Floquet theory are nevertheless correct, as we demonstrate using numerical calculations for both the classical spin dynamics in high-dimensional Mott insulators and the quantum spin dynamics in low-dimensional Mott insulators.

Results

Floquet theory for a two-site cluster. In this work, we study the repulsive Hubbard model as a model for strongly interacting electrons on a lattice. The Hamiltonian is given by

$$H = -t_0 \sum_{\langle ij \rangle \sigma} c_{i\sigma}^{\dagger} c_{j\sigma} + U \sum_j n_{j\uparrow} n_{j\downarrow}, \tag{1}$$

where $c_{i\sigma}^{\dagger}$ creates an electron at site i with spin $\sigma = \uparrow, \downarrow$, t_0 is the hopping between nearest-neighbour sites, and U the repulsive on-site interaction. Arbitrary time-dependent electric fields $\mathbf{E}(t)$ are incorporated by adding a Peierls phase to the hopping matrix elements (see Methods). Below we set $\hbar = 1$ and measure energy and time in units of the hopping t_0 and the inverse hopping, respectively. Electric fields are measured in units of t_0/ea, where a is the lattice spacing and e the electron charge.

For half-filling and $U/t_0 \gg 1$, the Hubbard model describes a Mott insulator with one electron per site, in which the remaining spin degrees of freedom are coupled by an AFM exchange interaction $J_{ex} = 2t_0^2/U$. The simplest analytical understanding for this result is obtained already for two electrons on two Hubbard sites: for total $S_z = 0$ we have four states. In the atomic limit, two of them ($|\uparrow, \downarrow\rangle$ and $|\downarrow, \uparrow\rangle$) are singly occupied sites at $E_1 = 0$, while the other two states involve a doubly occupied and empty site at energy $E_2 = U$ ($|\uparrow\downarrow, 0\rangle$ and $|0, \uparrow\downarrow\rangle$). In the presence of hopping, the degeneracy is lifted and the lowest states become singlet and triplet states at energies $E_S = -4t_0^2/U$ and $E_T = 0$, respectively. The low-energy spectrum is thus described by a spin Hamiltonian $2J_{ex}\mathbf{S}_1\mathbf{S}_2$ with $J_{ex} = (E_T - E_S)/2 = 2t_0^2/U$. This analytical understanding from the cluster is useful since lattice effects beyond the lowest-order perturbative result only appear in the order t_0^4/U^3. In the same spirit, to gain theoretical insight into the modification of J_{ex} by periodic driving, we first consider the same two-site Hubbard cluster and employ Floquet's theorem[33,34], the analogue of Bloch's theorem in time. When the Hamiltonian is periodic in time with a period $T = 2\pi/\omega$, solutions of the time-dependent Schrödinger equation are given in the form $|\psi(t)\rangle = e^{-i\epsilon_\alpha t}|\psi_\alpha(t)\rangle$, where $|\psi_\alpha(t+T)\rangle = |\psi_\alpha(t)\rangle$ is time periodic and ϵ_α is a quasi-energy defined up to multiples of ω. The Floquet picture describes a system that undergoes virtual absorption and emission of an arbitrary number of photons, as depicted in Fig. 1c for the Mott–Hubbard systems. The unperturbed Floquet sectors are described by the time-averaged Hamiltonian shifted by $n\omega$, and mixing between these Floquet sectors results in a renormalization of quasi-energy levels. A natural procedure is then to adopt an 'adiabatic' principle in which the driving amplitude varies slowly as compared with the driving frequency, $|\dot{E}/E| \ll \omega$, and identify the

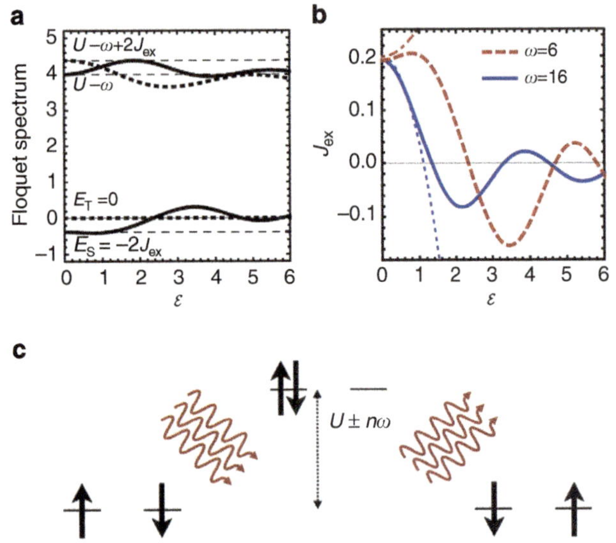

Figure 1 | Floquet spectrum of a two-site Hubbard model. (**a**) Floquet spectrum for $U=10$ and $\omega=6$ as a function of Floquet amplitude $\mathcal{E}$. Thin dashed lines indicate the spectrum without driving. The exchange interaction (J_{ex}) is extracted from the difference between the lowest singlet (E_S) and triplet (E_T) levels. The result is shown in **b** for the same $U=10$ and two different frequencies above ($\omega=16$, blue solid line) and below ($\omega=6$, red dashed line) the Mott gap. For large driving strength $\mathcal{E}\gtrsim 1$, a reversal of J_{ex} is possible. In addition, thin blue dotted ($\omega=16$) and thin red dash dotted ($\omega=6$) lines indicate the modification of J_{ex} as obtained within second-order perturbation theory. (**c**) Illustration of the modification of J_{ex} in the Floquet picture. Red wavy lines indicate the virtual absorption and emission of an arbitrary number n of photons with frequency ω. This induces a coupling to excited states in different Floquet sectors displaced by an energy $U\pm n\omega$.

amplitude-dependent singlet-triplet splitting $\epsilon_T-\epsilon_S$ in the quasi-energy spectrum with the (time dependent) exchange interaction that describes the spin dynamics in the laser driven system on timescales much slower than the driving period T.

The numerical solution of the Floquet spectrum and various analytically tractable limiting cases for the two-site Hubbard model are detailed in the methods section. For a tight-binding model driven by an electric field $E(t)=E_0\cos(\omega t)$, the coupling between Floquet sectors is controlled by the dimensionless Floquet amplitude

$$\mathcal{E}=\frac{eaE_0}{\hbar\omega}, \tag{2}$$

and the time averaging of H corresponds to a coherent reduction of the tunnelling amplitude[17,18] by a factor $J_0(\mathcal{E})$, where J_0 is the Bessel function (see Methods). In Fig. 1, we display the Floquet spectrum and the exchange splitting $J_{ex}(\mathcal{E})$ for a half-filled two-site Hubbard model. In contrast to the limit $\omega\gg U$, t_0, where the only effect would be a renormalization of the hopping by $J_0(\mathcal{E})$ and a corresponding reduction of the exchange splitting at large U by a factor $J_0(\mathcal{E})^2$, one can see that J_{ex} can be both increased and decreased for finite ω, depending on the driving. This is clear already in the perturbative limit for $\mathcal{E}\ll 1$ and $t_0/U\ll 1$, which is given by $J_{ex}=2t_0^2/U+\Delta J_{ex}$ with (see Methods)

$$\Delta J_{ex}=\frac{\mathcal{E}^2t_0^2}{4}\left(\frac{1}{U+\omega}+\frac{1}{U-\omega}-\frac{2}{U}\right), \tag{3}$$

and indicated with thin (dash-)dotted lines in the right panel of Fig. 1. The last term of equation (3) is the reduction of the exchange due to coherent reduction of the tunnelling, while the first two terms derive from the coupling to the $m=\pm 1$ Floquet sector with effectively shifted charge-transfer energies $U\pm\omega$. The net effect is an enhancement (reduction) of J_{ex} for driving frequencies below (above) the Coulomb energy U. For sufficiently strong driving one can even reverse the sign of J_{ex}, thus leading to the remarkable finding of a FM exchange coupling in the half-filled Hubbard model. This happens when $\mathcal{E}$ is of order one, such that coupling to higher Floquet sectors with effective Coulomb energy U-$m\omega$ becomes strongly enhanced, while the direct exchange path is reduced by coherent destruction of tunnelling $\sim J_0(\mathcal{E})$. For larger Floquet amplitudes, the direct exchange path again increases due to the oscillating behaviour of the Bessel function. In the remainder of this paper, we verify that these predictions from the two-site Floquet picture can remain valid for extended condensed matter systems at off-resonant driving with finite pulse duration, in spite of the possible importance of higher-order processes such as multi-photon absorption in the strongly driven regime, and the limited number of cycles in the pulse.

Mean-field spin dynamics in a high-dimensional lattice. A direct prediction of the Floquet theory (equation (3)) is the enhancement (reduction) of the exchange interaction for driving below (above) the Mott gap with weak amplitudes ($\mathcal{E}\ll 1$). A large class of materials for which this might be relevant are three-dimensional bulk systems, such as transition metal oxides, whose low-energy spin dynamics contain uniform spin resonances that can be conveniently described in mean-field theory. For large dimensions, also a numerical solution of the nonequilibrium electron dynamics in the Hubbard model is possible using the dynamical mean-field theory (DMFT, see Methods).

Within DMFT, the equilibrium solution of the Hubbard model at half-filling and low temperature is the Néel state. To asses the exchange interaction in this state, we study the excitation of resonances in the AFM phase in a transverse magnetic field B_x, a setup that was pioneered in ref. 15 for resonant photoexcitation. In equilibrium, the balance of B_x and J_{ex} gives rise to a canting of the magnetic sublattices out of the y-z plane. If J_{ex} is modified under periodic driving, the sub-lattice magnetizations are no longer aligned with the effective field B_{eff} given by external and exchange fields, as illustrated in the inset of Fig. 2. This implies a rotation of the spins in the plane perpendicular to B_x (leaving the total angular momentum S_x conserved), from which the time-dependent modification of the exchange interaction is calculated (see Methods). Hence, in our calculations, J_{ex} is defined as the parameter that describes best the observed transverse spin dynamics as is obtained from the solution of the full electron problem. We stress that this allows us to quantify changes of the exchange interaction independent of the exchange energy stored in the system. This is particularly important for the regime in which absorption is not negligible, where the laser excites mobile carriers that transfer their energy to the spin background on an ultrafast timescale and thus reduce the ordered moment.

We have implemented the DMFT solution of the Hubbard model in a time-dependent external electric field for the infinite-dimensional hyper-cubic lattice with density of states $D(\epsilon)=\exp(-\epsilon^2)/\sqrt{\pi}$ (refs 35,36). The electric field is pointing along the body diagonal of the lattice and represents a laser pulse with frequency ω and a Gaussian envelope that contains 15 cycles per pulse, that is, $E(t)=E_0\sin(\omega t)\exp(-(t-3t_c)^2/t_c^2)$ with $t_c=15\pi/(2.1\omega)$. Figure 2a,b show time traces of the induced change of the exchange interaction $\Delta J_{ex}^c(t)$ for one driving frequency below (a) and above (b) gap, as extracted from the time evolution of the spin degrees of freedom during the pulse. In accordance with the prediction of the Floquet theory, we observe an enhancement (reduction) of the exchange interaction during

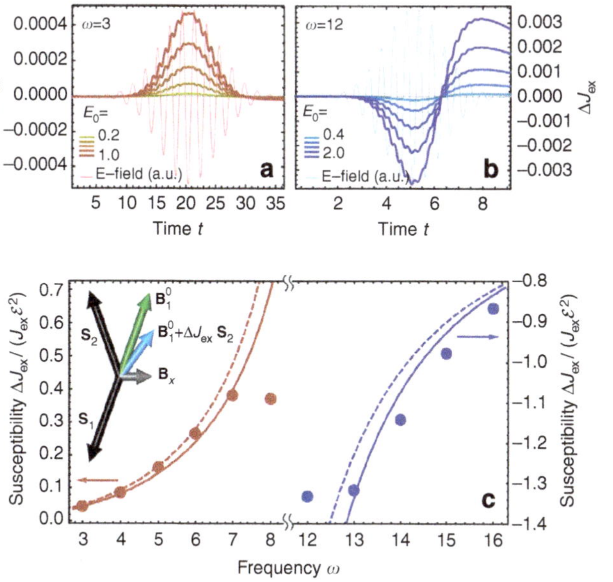

Figure 2 | Laser-induced modification of the exchange interaction. (a,b) Time-dependent change of the exchange interaction (ΔJ_{ex}, thick lines) during the action of a laser pulse, for driving frequencies $\omega=3$ (**a**) and $\omega=12$ (**b**) below and above the Mott gap, respectively. Different colours correspond to results obtained with different amplitude E_0 of the electric field, increasing from light to dark. Thin lines show the time dependence of the electric field (E-field). Numerical results were obtained using dynamical mean-field theory (DMFT) for the hyper-cubic lattice at $U=10$ and initial temperature $T=0.025$. (**c**) The driving susceptibility $\Delta J_{ex}/(J_{ex}\mathcal{E}^2)$ for $\mathcal{E}\to 0$ for frequencies above (blue, right vertical axis) and below gap (red, left vertical axis), obtained from DMFT for the hyper-cubic lattice (disks), from the numerical Floquet spectrum of a two-site Hubbard cluster (solid lines), and from the perturbative result equation (3) (dashed lines). The inset illustrates the canted geometry of the two sub-lattice magnetizations $\mathbf{S}_{1,2}$ (black arrows) induced by a static transverse magnetic field $\mathbf{B}_x$ (grey arrow). In equilibrium, the effective magnetic field $\mathbf{B}_1^0$ (green arrow) is collinear with $\mathbf{S}_1$. A modification of the exchange interaction (ΔJ_{ex}) would rotate the effective field (light blue arrow) with respect to $\mathbf{S}_1$ causing the excitation of a spin resonance. In the DMFT calculations, ΔJ_{ex} is computed from the observed spin precession in this canted geometry.

the application of the field with a frequency below (above) gap. The frequency $\omega=3$ in Fig. 2a is far from the resonance $\omega\approx U$, and we observe that $\Delta J_{ex}^c\approx 0$ after the pulse, demonstrating the reversibility of the effect. Conversely, the driving frequency $\omega=12$ in Fig. 2b is chosen close to the edge of the upper Hubbard band where we observe significant absorption and transient behaviour after the pulse. Hence, although the exchange interaction is modified in this case as well, the effect is not reversible. Note also that the time reached in the present simulations is too short for the photoexcited carriers to relax, hence a description in terms of a quasi-stationary photo-doped state discussed earlier[15] is not yet valid.

A quantitative comparison with the Floquet theory is shown in the bottom panel of Fig. 2, where the 'driving susceptibility' $\Delta J_{ex}/(J_{ex}\mathcal{E}^2)$ for $\mathcal{E}\to 0$ is plotted as a function of the driving frequency. Solid discs show the DMFT results as obtained by a linear fit through the dependence of the ratio $\Delta J_{ex}(t)/J_{ex}$ on $\mathcal{E}^2$ at its maximum. Dashed and solid lines show the results based on the perturbative Floquet formula (equation (3)) and the full Floquet spectrum (non-perturbative in t_0/U), evaluated from the derivative $dJ_{ex}/d\mathcal{E}^2$ at $\mathcal{E}=0$. As expected, in the vicinity of the band edge ($|\omega-U|\sim 2$), strong deviation is found since band absorption is not captured in a cluster picture. Away from the band edge, however, the frequency dependence matches very well, being even in quantitative agreement for the lowest frequencies below gap. This demonstrates the usefulness of the Floquet theory for understanding how off-resonant periodic driving modifies the exchange interaction in extended condensed matter systems by photo-assisted hopping.

One-dimensional quantum spin dynamics. An intriguing prediction of the Floquet analysis is the existence of amplitude and frequency ranges in which the exchange coupling becomes FM. Such a sign-change of J_{ex} cannot cause a transition to a FM state since the Hubbard model equation (1) conserves the total spin. However, even if the system remains AFM, a change of sign of J_{ex} by periodic driving allows for a very non-trivial and unique way to control the spin dynamics, namely, to reverse the time evolution of the undriven system. Such time reversal can be anticipated by considering a pure Heisenberg spin Hamiltonian $H_{ex}=J_{ex}\sum_{\langle ij\rangle}\mathbf{S}_i\mathbf{S}_j$, which gives an accurate description of the low-energy spin dynamics in the half-filled Hubbard model at $U\gg t_0$ if the system is not electronically excited. In the absence of driving, the propagation over a time interval t is given by the evolution operator $\mathcal{U}_{AFM}=\exp(-iH_{ex}t)$. Such evolution can exactly be reversed by the propagation with an exchange interaction J'_{ex} of opposite sign over a time interval $t'=|J_{ex}/J'_{ex}|t$, since for the FM time evolution operator we have $\mathcal{U}_{FM}=\exp(-iH'_{ex}t')=\exp(+iH_{ex}t)=\mathcal{U}_{AFM}^{-1}$, that is, the two time evolution operators are exactly inverse to each other.

To demonstrate that periodic driving of the Hubbard model at large U indeed yields the anticipated time reversal of the spin degrees of freedom, we consider a chain of $L=10$ sites and compute the dynamics using exact diagonalization techniques (see Methods). The system is initially prepared in a classical Néel state $c_{1\uparrow}^{\dagger}c_{2\downarrow}^{\dagger}c_{3\uparrow}^{\dagger}\ldots|0\rangle$ and is evolved under the unperturbed Hamiltonian (1). In a quantum Heisenberg model, the classical Néel state is a highly excited state the energy of which exceeds the thermal energy at the Néel temperature, such that no remanent long-range order is expected at long times, apart from finite size effects. In one dimension, not even the ground state displays long-range order. As a consequence of the spin-flip terms $J_{ex}(S_i^+S_{i+1}^-+S_i^-S_{i+1}^+)$ in the effective AFM Heisenberg model, we thus observe a rapid decay of the total staggered magnetization $M=\frac{1}{L}\sum_{i=1}^{L}(-1)^{i+1}\langle n_{i\uparrow}-n_{i\downarrow}\rangle$ (Fig. 3b). After this initial free evolution to a state where long-range order is suppressed, we ramp on a time-periodic electric field (Fig. 3a), with Floquet amplitude $\mathcal{E}=3.4$ and frequency $\omega/U=0.6$ such that the Floquet theory for a two-site model predicts a reversal of the exchange coupling. Under the periodic driving, one indeed observes a near perfect reversal of the dynamics of $M(t)$ in Fig. 3b, which almost completely recovers to the initial value $M(t=0)$ around $t\approx 100$. Subsequently, $M(t)$ is reduced again by further evolution in the reverse direction, as a consequence of the spin-flip terms in the FM model. This continues until the field is ramped off, after which one observes that the free evolution brings the system again back to the initial state, from which the same rapid decay of $M(t)$ is observed as for the initial free evolution. Hence, we conclude that the periodic driving allows for a reversible control of the spin dynamics for the timescale considered in our simulations. This is further confirmed by the time evolution of the total double occupation $d=\sum_{i=1}^{L}\langle n_{i\uparrow}n_{i\downarrow}\rangle$, which has the same mean value before and after the driving, demonstrating that electronic excitations due to the driving are negligible. The weak oscillations in $d(t)$ are caused by switching on the hopping at

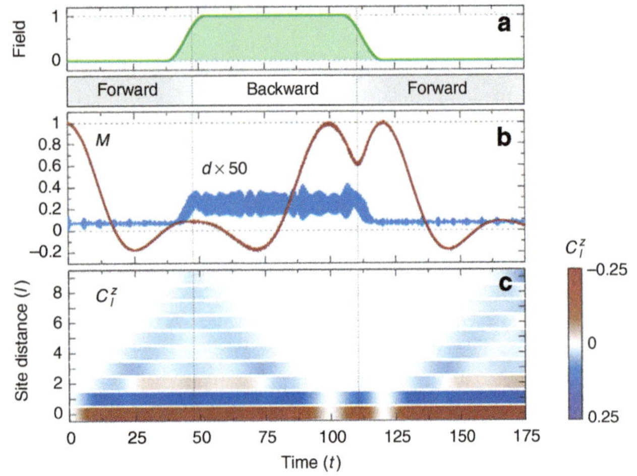

Figure 3 | Time reversal of the spin dynamics by periodic driving. (**a**) Field envelope with cosine-shaped ramps of length $\Delta t = 15$ around $t_f = 45$ and $t = 112.5$. The bar below the field envelope indicates forward (grey) and backward (white) time evolution when the field is off and on, respectively. (**b**) Time evolution of staggered magnetization M (red) and total double occupation d (blue), scaled by a factor of 50, of a 10-site Hubbard chain, showing free evolution for times $t < 37.5$ and $t > 120$ and evolution under an additional periodic driving at frequency $\omega/U = 0.6$ and Floquet amplitude $\mathcal{E} = 3.4$ in between. (**c**) Build up and diminishing of the spin-spin correlation function C_l^z. Numerical results were obtained by exact diagonalization for $U = 50$ and open boundary conditions, starting from a classical Néel state. The colour bar indicates the value of C_l^z, ranging from -0.25 (dark red) through 0 (white) to 0.25 (dark blue).

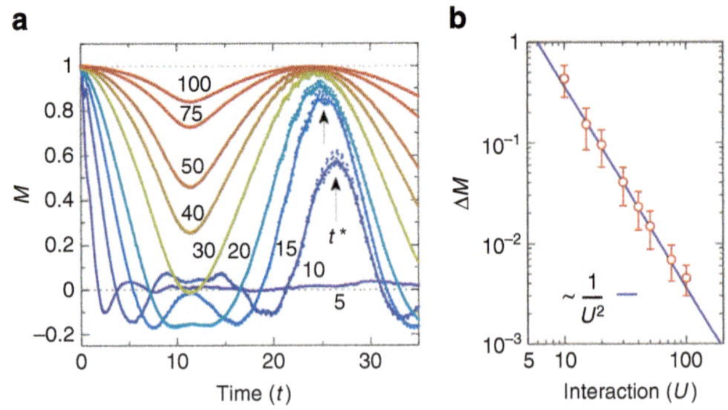

Figure 4 | Quality of the time reversal. (**a**) Time evolution of the staggered magnetization M in a 10-site Hubbard chain, where after fixed forward propagation time $t_f = 10$ the field is ramped up within a time interval $\Delta t = 10$. Different colours correspond to different values of U, as indicated by the corresponding coloured labels. (**b**) Difference ΔM between the initial staggered magnetization and the staggered magnetization at the revival time t^* as a function of U. The error bars represent the magnitude of the short-time fluctuations of M close to t^*. In all calculations, the driving frequency $\omega/U = 0.6$ and Floquet amplitude $\mathcal{E} = 3.4$ were used.

$t = 0$, while the increased mean value of $d(t)$ during driving is due to photo-assisted hopping processes.

Time reversal can be demonstrated not only on the level of local observables. Figure 3c displays the evolution of the spin–spin correlation function $C_l^z = \sum_{|i-j|=l} \frac{1}{N_l} (\langle S_i^z S_j^z \rangle - \langle S_i^z \rangle \langle S_j^z \rangle)$ as a function of distance l and time (N_l is the number of site pairs with distance l). Starting from the initial uncorrelated product state, correlations build up under the evolution of H. Even though the system is small, the spreading of correlations resembles the light-cone effect that has been observed in quantum many-body systems after a quench[37], that is, correlations stay zero outside the light cone $|l| \leq 2\upsilon t$, where υ is a maximal mode velocity[38], while short-range AFM correlations emerge inside the light cone. Further, under the action of the periodic driving, the spin–spin correlations diminish with the same speed, restoring the initially uncorrelated state.

To determine quantitatively how well the time evolution is reversed in our simulations, we computed the difference $\Delta M = 1 - M(t^*)$ between the initial magnetization in the Néel state and the magnetization $M(t^*)$ at the revival time t^* for different values of U (Fig. 4). In all simulations, the system is evolved forward in time for a given time $t_f = 10$, after which the field is ramped on for a period $\Delta t = 10$. As before, we choose $\mathcal{E} = 3.4$ and $\omega = 0.6U$, which gives the same relative change of J_{ex} for all sufficiently large values $U \gg t_0$. The observed scaling $\Delta M \sim 1/U^2$ indicates that in the current setup the deviation from perfect reversal originates from small electronic excitations above the gap, which arises from switching on the hopping in the beginning of the simulation and the ramping on of the field. Since the dynamics of the electronically excited states is not captured by the spin Hamiltonian, it is not time-reversed. While the electronic excitation can be further reduced by slow ramping, the ultimate limit is given by non-Heisenbergian terms in the effective spin Hamiltonian, for which perfect reversal under periodic driving is not expected. For the half-filled Hubbard model in equilibrium, the leading-order correction to the Heisenberg model appears in the order t_0^4/U^3 in the strong-coupling expansion. At least for small times t, the contribution of a perturbation δH proportional to t_0^4/U^3 to the time-reversed Hamiltonian would lead to a scaling $\Delta M \sim 1/U^3$, which is smaller than the electronic excitation in the present case. Note, however, that the times reached in the current simulations are nevertheless long enough to observe near perfect reversal even from a state without magnetic order, in which spin correlations have spread throughout the full chain (Fig. 3).

Discussion

Our results demonstrate ultrafast and reversible electrical control of the exchange interaction in extended fermionic many-body systems by modulation with time-periodic electric fields. These numerical results have recently been corroborated analytically in the high-frequency limit, by showing that the effective spin model for the two-site cluster remains valid for the driven lattice[39]. We emphasize that Floquet amplitudes $\mathcal{E} \sim 0.1$ are well accessible for condensed matter systems, which would lead to relative changes of J_{ex} up to 1%. For example, for a frequency $\hbar\omega = 1$ eV and a lattice spacing of 2 Å, a laser fluence of 1 mJ cm^{-2} in a 100-fs pulse corresponds to a Floquet parameter $\mathcal{E} \sim 0.05$. While realistic condensed matter systems usually involve several correlated bands, we think that our current results are already quite robust for the AFM oxides governed by superexchange interactions. Similar as in the single-band model studied here, superexchange interactions are governed by virtual charge excitations, which will be reversibly modified through the mechanism of photo-assisted hopping between different Floquet sectors. This is further supported by recently presented experiments on canted AFM oxides[40] using terahertz emission spectroscopy[41], which show the first experimental evidence of reversibly controlling exchange interactions by off-resonant pumping below the charge-transfer gap. Furthermore, equation (3) implies a strong enhancement of off-resonant effects at low frequencies. As an extreme limit of this one can anticipate control of spin dynamics by few-cycle terahertz pulses, using both the coupling between the spins to the magnetic field of the light[42], and the modifications of J_{ex} predicted by our work. At the same time, extensions to multi-band models are very interesting and relevant to perform as they usually involve multiple possibly competing exchange paths, which generally yield quantitative differences as well as different dependencies on the driving frequency due to the presence of additional resonances, potentially enabling to achieve even stronger effects. In this connection, we also mention that extensions to multi-band models may be relevant too for the description of metallic ferromagnets. Here static electronic structure calculations already indicate that exchange integrals involving electrons excited to higher bands considerably differ from those in the ground state[43]. Furthermore, a more accurate quantitative description of photo-excited states would also include the effect of dynamic screening. However, we anticipate that this will be important only if mobile carriers are injected, that is, for photo-doping excitations, while it will be a secondary effect for the off-resonant driving investigated here, which leaves the electron distribution unchanged.

While the single-band model is clearly a minimal model for application to condensed matter systems, fermionic cold atom systems resemble the single-band model very accurately. Moreover, such systems may realize the large amplitudes $\mathcal{E}$ needed to achieve the FM exchange at frequencies sufficiently far from resonances $n\omega = U$, which are difficult to realize for most condensed matter systems. It will, therefore, be of fundamental interest to investigate the reversal of the exchange interaction and the associated time reversal of the quantum spin dynamics in cold atom experiments. In these systems recently, great progress has been made to prepare and measure systems with single-site spatial resolution[44,45]. Furthermore, various single-particle Floquet Hamiltonians could be realized in the limit of strong periodic driving without substantial heating by inter-band absorption (see, for example, ref. 46). Cold atoms in optical lattices have been successfully used as a quantum simulator for the dynamics of a quantum quenches in the Bose–Hubbard model, starting from an artificially prepared charge-ordered phase[47], which suggest similar techniques to probe the behaviour of spin systems under time reversal. In the methods section (*cf.* equation (7)), we show that analogous time reversal can be achieved by modulating the amplitude of the hopping instead of its phase, which is easier to control in cold atoms. An intriguing problem to study both theoretically and experimentally is the fundamental question how well the time evolution can be reversed after (much) longer forward evolution time and investigate systematically the influence of small deviations from perfect time reversal. Furthermore, a study of the Loschmidt echo and dynamical phase transitions[48] in cold atoms might be possible by including additional perturbations to the back propagation.

Methods

Floquet theory. The Floquet quasi-energy spectrum can be obtained from the ansatz $|\psi(t)\rangle = e^{-i\epsilon_\alpha t}|\psi_\alpha(t)\rangle$ by expanding $|\psi_\alpha(t)\rangle$ in a Fourier series $|\psi_\alpha(t)\rangle = \sum_m e^{i\omega m t}|\psi_{\alpha,m}\rangle$, where $|\psi_{\alpha,m}\rangle$ is referred to as the component of the wave function in the mth Floquet sector. The Schrödinger equation then achieves a block-matrix structure

$$(\epsilon_\alpha + m\omega)|\psi_{\alpha,m}\rangle = \sum_{m'} H_{m-m'}|\psi_{\alpha,m'}\rangle, \tag{4}$$

where $H_m = (1/T)\int_0^T dt\, e^{i\omega m}H(t)$ are the Fourier components of the Hamiltonian. Different from the usual discussions of Floquet theory for single-particle Hamiltonians, here we focus on the effect of periodic driving on an electronic spectrum that is controlled by electronic correlations.

Time-dependent electric fields are incorporated into the Hubbard Hamiltonian (1) by adding a time-dependent Peierls phase to the hopping matrix elements, $t_{ij}(t) = t_0 \exp[ieaA_{ij}(t)]$, where A_{ij} is the projection of the vector potential along the direction from site i to j (choosing a gauge with zero scalar potential and $E(t) = -\partial_t A(t)$). For the one-dimensional chain with electric field $E_0 \sin(\omega t)$ along the chain, this implies $A_{ij}(t) = -\frac{1}{\omega}(i-j)E_0\cos(\omega t)$. The Fourier components of the Hamiltonian are thus given by

$$H_m = -t_0 \sum_{\langle ij\rangle\sigma}(-1)^m J_m((i-j)\mathcal{E})c^\dagger_{i\sigma}c_{j\sigma}, \tag{5}$$

plus the additional (time-independent) interaction part in the $m=0$ component, where $J_m(x)$ is the mth Bessel function, and the dimensionless parameter $\mathcal{E}=eaE_0/(\hbar\omega)$ measures the strength of the perturbation (cf. equation (2)). For the numerical determination of the Floquet energies, one truncates the number of Floquet sectors in equation (4) to $|n| \le N$, and increases N to reach convergence. The determination of a many-body Floquet spectrum thus requires the diagonalization of a matrix of dimension $N \times D$, where D is the dimension of the Hilbert space. The results presented in Fig. 1 are converged with $N=8$.

In the limit of large frequency, $\omega \gg U$, t_0, Floquet sectors in equation (4) are separated in energy, and one can restrict oneself to the lowest sector $m=0$. This is equivalent to replacing the Hamiltonian with its time average, which leads to the renormalization of the hopping by $J_0(\mathcal{E})$, and a corresponding reduction of the exchange by a factor $J_0(\mathcal{E})^2$. In the perturbative limit where both $t_0/U \ll 1$ and $\mathcal{E} \ll 1$, we expand the Bessel functions $J_n(x) \sim x^n$ for $x \to 0$. To lowest order only states of the $m=0$ and $m=\pm 1$ Floquet sectors have to be taken into account, and the result given by equation (3) follows from standard second-order perturbation theory. Furthermore, an interesting limit for the Mott regime is given by $t_0/U \ll 1$, but allowing for fields of arbitrary amplitude. Because all terms H_m for $m \neq 0$ are proportional to t_0, the perturbative shift of the spin states in the $m=0$ sector is given by a sum over all second-order processes containing precisely one virtual hopping to a higher Floquet sector and back, yielding

$$\frac{J_{ex}(\mathcal{E},\omega)}{J_{ex}(\mathcal{E}=0)} = \sum_{n=-\infty}^{\infty}\frac{J_{|n|}(\mathcal{E})^2}{1+n\omega/U}. \tag{6}$$

The unperturbed exchange is modified by a factor dependent only on ω/U and $\mathcal{E}$. For the parameters $\mathcal{E}= 3.4$ and $\omega/U = 0.6$ chosen for the time reversal, for example, the factor is given by -0.95 and indicates a near perfect sign reversal.

Finally, we note that a similar analysis is possible for the case of periodic modulation of the hopping amplitude, taking $t_0(t) = t_0(1 + A\cos(\omega t))$. As above, we obtain for $U \ll t$

$$\frac{J_{ex}(A,\omega)}{J_{ex}(A=0)} = 1 + \sum_{n=\pm 1}\frac{A^2}{1+n\omega/U}, \tag{7}$$

yielding, for example, a perfect sign reversal for driving above gap $\omega/U = 1.2$ at $A \approx 0.67$. Other than for the field driven case, the perturbation is purely harmonic, and only Floquet sectors $m = \pm 1$ enter this expression.

Dynamical mean-field theory. To solve the electron dynamics in the Hubbard model, we use nonequilibriumDMFT[35,49]. The electric field of the laser is incorporated by the Peierls substitution (see Methods, Floquet theory), so that the light matter interaction within the single-band model is treated to all orders. Within DMFT[50], which becomes exact in the limit of infinite dimensions[51], local correlation functions are obtained from an effective impurity model in which one site of the lattice is coupled to a non-interacting, self-consistently determined bath. The impurity model is solved within the perturbative hybridization expansion (non-crossing approximation[52]). The accuracy of this approach has been tested in equilibrium and for the short-time dynamics by comparison with higher-order hybridization expansions, as well as with numerically exact Quantum Monte Carlo[52–54], which revealed good agreement both in the AFM and paramagnetic Mott insulator regime $t_0 \ll U$. A detailed description of the formalism and of our numerical implementation is given in ref. 15, which studies the same setup (that is, a hyper-cubic lattice with electric field along the body diagonal), yet for different electric fields.

Determination of J_{ex} in DMFT. To asses the exchange interaction from mean-field spin dynamics, we investigate the AFM phase of the Hubbard model supplemented with a term $B_x\sum_i S_{ix}$, which couples the spin $S_{i\alpha} = \frac{1}{2}\sum_{\sigma\sigma'} c^\dagger_{i\sigma}(\hat{\sigma}_\alpha)_{\sigma\sigma'}c_{i\sigma'}$ to a homogeneous magnetic field B_x along the x axis ($\hat{\sigma}_\alpha$ denotes the Pauli matrices; $\alpha = x, y, z$). DMFT allows us to compute the time-dependent expectation value of the electron spin $\langle \mathbf{S}_{1,2}\rangle$ on the two magnetic sublattices. Assuming a rigid macrospin model, the time-dependent exchange interaction can be inferred from these results by inverting the Landau–Lifshitz equation for the dynamics of spins on the mean field $\mathbf{B}^{1,2}_{\rm eff} = B_x\mathbf{e}_x - 2J_{ex}\langle \mathbf{S}_{2,1}\rangle$. It was shown[15] that this approach compares well with the definition of exchange interactions from a time-dependent response formalism[55], as well as to the analytical perturbative result $J_{ex} = 2t_0^2/U$ in equilibrium at large U. In the transverse field, the equilibrium exchange interaction can be determined from the canting induced by B_x, yielding $J^c_{ex} = -B_x/(4\langle S_x\rangle)$. Out of equilibrium, we obtain $J^c_{ex}(t) = -B_x/(4\langle S_x\rangle) + \Delta J^c_{ex}(t)$,

$$\Delta J^c_{ex}(t) = -\frac{1}{4T\langle S_{1x}\rangle}\int_{t-T/2}^{t+T/2}\frac{\langle \dot{S}_{1y}(s)\rangle}{\langle S_{1z}(s)\rangle}ds, \tag{8}$$

where the time averaging is done to extract only the low-frequency component, similar as in the Floquet theory. Note that by calculating the exchange interaction in this way, we have automatically projected out dynamical changes in the (time averaged) local magnetization $|\langle \mathbf{S}_{1,2} \rangle|$, and slight changes of the local moments (double occupation) as a result of the virtual absorption and emission of photons.

Exact diagonalization. To compute the time evolution of the (driven) one-dimensional Hubbard model from the Schrödinger equation with a time-dependent Hamiltonian $H(t)$ and a given initial state $|\psi_0\rangle = c^\dagger_{1\uparrow} c^\dagger_{2\downarrow} c^\dagger_{3\uparrow} \ldots |0\rangle$, we use the Krylov technique[56] in combination with a commutator-free exponential time-propagation scheme[57]. While the Krylov method provides efficient approximations to the time propagator, which are important to treat large Hilbert spaces, the commutator-free exponential time-propagation scheme is related to the Magnus expansion, and, preserving unitarity, allows for a high-order accurate integration of the Schrödinger equation in time.

References

1. Kirilyuk, A., Kimel, A. V. & Rasing, Th. Ultrafast optical manipulation of magnetic order. *Rev. Mod. Phys.* **82,** 2731–2784 (2010).
2. Radu, I. *et al.* Transient ferromagnetic-like state mediating ultrafast reversal of antiferromagnetically coupledspins. *Nature* **472,** 205–208 (2011).
3. Mentink, J. H. *et al.* Ultrafast spin dynamics in multisublattice magnets. *Phys. Rev. Lett.* **108,** 057202 (2012).
4. Wienholdt, S., Hinzke, D., Carva, K., Oppeneer, P. M. & Nowak, U. Orbital-resolved spin model for thermal mag-netization switching in rare-earth-based ferrimagnets. *Phys. Rev. B* **88,** 020406 (2013).
5. Baryakhtar, V. G., Butrim, V. I. & Ivanov, B. A. Exchange relaxation as the mechanism of ultrafast spin re-orientation in two-sublattice ferrimagnets. *JETP Lett.* **98,** 289–293 (2013).
6. Evans, R. F. L., Ostler, T. A., Chantrell, R. W., Radu, I. & Rasing, Th. Ultrafast thermally induced magnetic switching in synthetic ferrimagnets. *Appl. Phys. Lett.* **104,** 082410 (2014).
7. Rhie, H.-S., Dürr, H.A. & Eberhardt, W. Femtosecond electron and spin dynamics in ni/w(110) films. *Phys. Rev. Lett.* **90,** 247201 (2003).
8. Ju, G. *et al.* Ultrafast generation of ferromagnetic or-der via a laser-induced phase transformation in ferh thin films. *Phys. Rev. Lett.* **93,** 197403 (2004).
9. Thiele, J., Buess, M. & Back, C. H. Spin dynamics of the antiferromagnetic-to-ferromagnetic phase transition in ferh on a sub-picosecond time scale. *Appl. Phys. Lett.* **85,** 2857 (2004).
10. Wall, S., Prabhakaran, D., Boothroyd, A. T. & Cavalleri, A. Ultrafast coupling between light, coherent lattice vibrations, and the magnetic structure of semicovalent LaMnO3. *Phys. Rev. Lett.* **103,** 097402 (2009).
11. Först, M. *et al.* Driving magnetic order in a manganite by ultrafast lattice excitation. *Phys. Rev. B* **84,** 241104 (2011).
12. Carley, R. *et al.* Femtosecond laser excitation drives ferromagnetic gadolinium out of magnetic equilibrium. *Phys. Rev. Lett.* **109,** 057401 (2012).
13. Li, T. *et al.* Femtosecond switching of magnetism via strongly correlated spin-charge quantum excitations. *Nature* **496,** 69–73 (2013).
14. Matsubara, M. *et al.* Tuning the ultrafast spin dynamics in carrier-density-controlled ferromagnets. Preprint at http://arxiv.org/abs/1304.2509 (2013).
15. Mentink, J. H. & Eckstein, M. Ultrafast quenching of the exchange interaction in a Mott insulator. *Phys. Rev. Lett.* **113,** 057201 (2014).
16. Ryan, P. J. *et al.* Reversible control of magnetic inter-actions by electric field in a single-phase material. *Nat. Commun.* **4,** 1334 (2013).
17. Dunlap, D. H. & Kenkre, V. M. Dynamical localization of a particle moving under inuence of an electric field. *Phys. Rev. B* **34,** 3625–3633 (1986).
18. Grossmann, F., Dittrich, T., Jung, P. & Hänggi, P. Coherent destruction of tunneling. *Phys. Rev. Lett.* **67,** 516–519 (1991).
19. Lignier, H. *et al.* Dynamical control of matter-wave tunneling in periodic potentials. *Phys. Rev. Lett.* **99,** 220403 (2007).
20. Oka, T. & Aoki, H. Photovoltaic Hall effect in graphene. *Phys. Rev. B* **79,** 081406 (2009).
21. Tsuji, N., Oka, T., Werner, P. & Aoki, H. Dynamical band ipping in fermionic lattice systems: an ac-field-driven change of the interaction from repulsive to attractive. *Phys. Rev. Lett.* **106,** 236401 (2011).
22. Wang, Y. H., Steinberg, H., Jarillo-Herrero, P. & Gedik, N. Observation of Floquet-Bloch states on the surface of a topological insulator. *Science* **342,** 453–457 (2013).
23. Shahbazyan, T., Perakis, I. & Raikh, M. Spin correlations in nonlinear optical response: Light-induced kondo effect. *Phys. Rev. Lett.* **84,** 5896–5899 (2000).
24. Chen, Y.-A., Nascimbeñe, S., Aidelsburger, M., Atala, M., Trotzky, S. & Bloch, I. Controlling correlated tunnel-ing and superexchange interactions with ac-driven optical lattices. *Phys. Rev. Lett.* **107,** 210405 (2011).
25. Piermarocchi, C., Pochung Chen, L., Sham & Steel, D. Optical RKKY interaction between charged semicon-ductor quantum dots. *Phys. Rev. Lett.* **89,** 167402 (2002).
26. Piermarocchi, C. & Quinteiro, G. F. Coherent optical control of spin-spin interaction in doped semiconductors. *Phys. Rev. B* **70,** 235210 (2004).
27. Fernández-Rossier, J., Piermarocchi, C., Chen, P., Mac-Donald, A. & Sham, L. Coherently photoinduced ferromagnetism in diluted magnetic semiconductors. *Phys. Rev. Lett.* **93,** 127201 (2004).
28. Wang, J., Cotoros, I., Dani, K., Liu, X., Furdyna, J. & Chemla, D. Ultrafast enhancement of ferromagnetism via photoexcited holes in GaMnAs. *Phys. Rev. Lett.* **98,** 217401 (2007).
29. Lebowitz, J. L. Boltzmann's entropy and time's arrow. *Phys. Today* **46,** 32–38 (1993).
30. Polkovnikov, A. Microscopic diagonal entropy and its connection to basic thermodynamic relations. *Ann. Phys.* **326,** 486–499 (2011).
31. Caneva, T., Silva, A., Fazio, R., Lloyd, S., Calarco, T. & Montangero, S. Complexity of controlling quantum many-body dynamics. *Phys. Rev. A* **89,** 042322 (2014).
32. D'Alessio, L. & Rigol, M. Long-time behavior of isolated periodically driven interacting lattice systems. *Phys. Rev. X* **4,** 041048 (2014).
33. Floquet, G. Sur les équations différentielles linéaires à coefficients périodiques. *Ann. Sci. Ec. Normale Super* **12,** 47–88 (1883).
34. Grifoni, M. & Hänggi, P. Driven quantum tunnelling. *Phys. Rep.* **304,** 229–354 (1998).
35. Aoki, H., Tsuji, N., Eckstein, M., Kollar, M., Oka, T. & Werner, P. Nonequilibrium dynamical mean-field theory and its applications. *Rev. Mod. Phys.* **86,** 779–837 (2014).
36. Turkowski, V. & Freericks, J. K. Nonlinear response of bloch electrons in infinite dimensions. *Phys. Rev. B* **71,** 085104 (2005).
37. Cheneau, M. *et al.* Light-cone-like spreading of correlations in a quantum many-body system. *Nature* **481,** 484–487 (2012).
38. Calabrese, P. & Cardy, J. Time dependence of correlation functions following a quantum quench. *Phys. Rev. Lett.* **96,** 136801 (2006).
39. Itin, A. P. & Katsnelson, M. I. Effective Hamiltonians for fastly driven many-body lattice systems. Preprint at http://arxiv.org/abs/1401.0402v2 (2015).
40. Mikhaylovskiy, R. V. *et al.* Inverse magneto-refraction as a mechanism for laser modification of spin-spin exchange parameters and subsequent Terahertz emission from iron oxides. Preprint at http://arxiv.org/abs/1412.7094 (2014).
41. Mikhaylovskiy, R. V., Hendry, E., Kruglyak, V. V., Pisarev, R. V. & Kimel, Th. Rasing, and A. V. Terahertz emission spectroscopy of laser-induced spin dynamics in TmFeO3 and ErFeO3 orthoferrites. *Phys. Rev. B* **90,** 184405 (2014).
42. Kampfrath, T. *et al.* Coherent Terahertz control of antiferromagnetic spin waves. *Nat. Photon* **5,** 31–34 (2011).
43. Zhang, G. P., Gu, M. & Wu, X. S. Ultrafast reduction in exchange interaction by a laser pulse: alternative path to femtomagnetism. *J. Phys. Condens. Matter* **26,** 376001 (2014).
44. Bakr, W. S. *et al.* Probing the superuid-to-Mott insulator transition at the single-atom level. *Science* **329,** 547–550 (2010).
45. Endres, M. *et al.* Single-site- and single-atom-resolved measurement of correlation functions. *Appl. Phys. B* **113,** 27–39 (2013).
46. Struck, J. *et al.* Tunable gauge potential for neutral and spinless particles in driven optical lattices. *Phys. Rev. Lett.* **108,** 225304 (2012).
47. Trotzky, S. *et al.* Probing the relaxation towards equilibrium in an isolated strongly correlated one-dimensional bose gas. *Nat. Phys.* **8,** 325–330 (2012).
48. Heyl, M., Polkovnikov, A. & Kehrein, S. Dynamical quantum phase transitions in the transverse-field Ising model. *Phys. Rev. Lett.* **110,** 135704 (2013).
49. Freericks, J. K., Turkowski, V. M. & Zlatić, V. Nonequilibrium dynamical mean-field theory. *Phys. Rev. Lett.* **97,** 266408 (2006).
50. Georges, A., Kotliar, G., Krauth, W. & Rozenberg, M. J. Dynamical mean-field theory of strongly correlated fermion systems and the limit of infinite dimensions. *Rev. Mod. Phys.* **68,** 13–125 (1996).
51. Metzner, W. & Vollhardt, D. Correlated lattice fermions in $d = \infty$ dimensions. *Phys. Rev. Lett.* **62,** 324–327 (1989).
52. Eckstein, M. & Werner, P. Nonequilibrium dynamical mean-field calculations based on the noncrossing approximation and its generalizations. *Phys. Rev. B* **82,** 115115 (2010).
53. Eckstein, M. & Werner, P. Photoinduced states in a Mott insulator. *Phys. Rev. Lett.* **110,** 126401 (2013).
54. Werner, P., Tsuji, N. & Eckstein, M. Nonthermal symmetry-broken states in the strongly interacting hubbard model. *Phys. Rev. B* **86,** 205101 (2012).
55. Secchi, A., Brener, S., Lichtenstein, A. I. & Katsnelson, M. I. Non-equilibrium magnetic interactions in strongly correlated systems. *Ann. Phys.* **333,** 221–271 (2013).
56. Hochbruck, M. & Lubich, C. On Krylov subspace ap-proximations to the matrix exponential operator. *SIAM J. Numer. Anal.* **34,** 1911–1925 (1997).
57. Alvermann, A. & Fehske, H. High-order commutator-free exponential time-propagation of driven quantum sys-tems. *J. Comp. Phys.* **230,** 5930–5956 (2011).

Acknowledgements

We thank U. Bovensiepen, M.I. Katsnelson, A.V. Kimel, A. Lichtenstein, R.V. Mikhaylovskiy, T. Oka, A. Secchi, N. Tsuji and Ph. Werner for fruitful discussions. The exact diagonalization calculations were run in part on the supercomputer HLRN-II of the North-German Supercomputing Alliance. J.H.M. acknowledges funding from the Nederlandse Organisatie voor Wetenschappelijk onderzoek (NWO) by a Rubicon Grant.

Author contributions

The Floquet theory was developed by M.E. and J.H.M. DMFT simulations were run by J.H.M., while K.B. obtained the exact diagonalization results. All authors contributed to the discussion and interpretation of the results and the writing of the manuscript.

Additional information

Competing financial interests: The authors declare no competing financial interests.

27

Coupling of Higgs and Leggett modes in non-equilibrium superconductors

H. Krull[1], N. Bittner[2], G.S. Uhrig[1], D. Manske[2] & A.P. Schnyder[2]

In equilibrium systems amplitude and phase collective modes are decoupled, as they are mutually orthogonal excitations. The direct detection of these Higgs and Leggett collective modes by linear-response measurements is not possible, because they do not couple directly to the electromagnetic field. In this work, using numerical exact simulations we show for the case of two-gap superconductors, that optical pump-probe experiments excite both Higgs and Leggett modes out of equilibrium. We find that this non-adiabatic excitation process introduces a strong interaction between the collective modes, which is absent in equilibrium. Moreover, we propose a type of pump-probe experiment, which allows to probe and coherently control the Higgs and Leggett modes, and thus the order parameter directly. These findings go beyond two-band superconductors and apply to general collective modes in quantum materials.

[1] Lehrstuhl für Theoretische Physik I, Technische Univerität Dortmund, Otto-Hahn Strasse 4, D-44221 Dortmund, Germany. [2] Max-Planck-Institut für Festkörperforschung, Heisenbergstrasse 1, D-70569 Stuttgart, Germany. Correspondence and requests for materials should be addressed to A.S. (email: a.schnyder@fkf.mpg.de).

Collective excitation modes are a characteristic feature of symmetry-broken phases of matter. These collective excitations are due to amplitude and phase fluctuations of the order parameters, which are decoupled in equilibrium systems, as they represent mutually orthogonal excitations. The properties of collective modes are of fundamental interest, as they are a distinguishing feature of any symmetry-broken phase, such as superconductors, charge-density waves or antiferromagnets. For example, superconductors exhibit an amplitude Higgs mode and a phase mode, which are the radial and angular excitations in the Mexican-hat potential of the free energy. In two-band superconductors there exists in addition a Leggett phase mode[1], which corresponds to collective fluctuations of the interband phase difference.

As we show in this work, out-of-equilibrium excitations of symmetry-broken phases lead to a direct coupling between phase and amplitude modes, an effect which is absent in equilibrium systems. Furthermore, we demonstrate that ultrafast pump–probe measurements allow to directly probe and coherently control collective excitation modes. Pump–probe measurements have recently become a key tool to probe the temporal dynamics and relaxation of quantum materials[2–13]. In particular, this technique has been used to measure the oscillations of the amplitude Higgs mode of the one-gap superconductor NbN. It has been shown, both theoretically[14–28] and experimentally[2–4], that a short intense laser pulse of length τ much shorter than the dynamical time scale of the superconductor $\tau_\Delta \simeq h/(2|\Delta|)$ induces oscillations in the order parameter amplitude at the frequency $\omega_{\mathrm{H}} = 2\Delta_\infty/\hbar$, with Δ_∞ the asymptotic gap value.

Although non-equilibrium collective modes in conventional single-gap superconductors are well understood, the investigation of collective excitations in unconventional non-equilibrium superconductors with multiple gaps, such as MgB_2 or iron pnictides, is still in its infancy[29–31]. These multicomponent superconductors have a particularly rich spectrum of collective excitations[1,32,33]. In this study, we simulate the pump–probe process in a two-gap superconductor using a semi-numerical approach based on the density-matrix formalism. This method is exact for mean-field Hamiltonians[15,21], captures the coupling between the superconductor and the electromagnetic field of the pump laser at a microscopic level and allows for the calculation of the pump–probe conductivity, as measured in recent experiments[2,3]. Two-gap superconductors exhibit besides the amplitude Higgs[34] and the phase modes[35,36], also a Leggett mode[1], which results from fluctuations of the relative phase of the two coupled gaps, that is, equal but opposite phase shifts of the two-order parameters, see Fig. 1b. In equilibrium superconductors, the Higgs and Leggett modes are decoupled, as they correspond to mutually orthogonal fluctuations. In contrast to the phase mode, both Higgs and Leggett modes are charge neutral and therefore do not couple to the electromagnetic field[37]. As a consequence, these excitations cannot be detected directly with standard linear-response-type measurements. Observation of these modes has only been possible in special circumstance, for example, when they couple to another order parameter, such as in charge density wave systems[38–41].

Here we show that in a pump–probe experiment both Leggett and Higgs modes can be excited out of equilibrium, and directly observed as oscillations in the absorption spectra at their respective frequencies. We find that the non-adiabatic excitation process of the pump pulse induces an intricate coupling between the two charge-neutral modes, which pushes the frequency of the Leggett mode below the continuum of two-particle excitations. Moreover, the frequencies of the Leggett and Higgs modes and the coupling between them can be controlled by the fluence of the pump pulse. Hence, by adjusting the laser intensity the two modes can be brought into resonance, which greatly enhances their oscillatory signal in the pump–probe absorption spectra.

Results

Pump-excitation process. In pump–probe measurements, the pump laser pulse excites a high density of quasiparticles above the gap of the order parameter, thereby modifying the Mexican-hat potential of the free energy $\mathcal{F}$ (Fig. 1). As a result, the amplitude of the order parameter decreases, reducing the minimum of the free energy. If the pump-pulse-induced changes in $\mathcal{F}$ occur on a faster time scale than the intrinsic response time of the symmetry-broken state, the collective modes start to oscillate at their characteristic frequencies about the new free-energy minimum (see Fig. 1a). In this work we study this non-adiabatic excitation mechanism for two-band superconductors perturbed by a short and intense pump pulse.

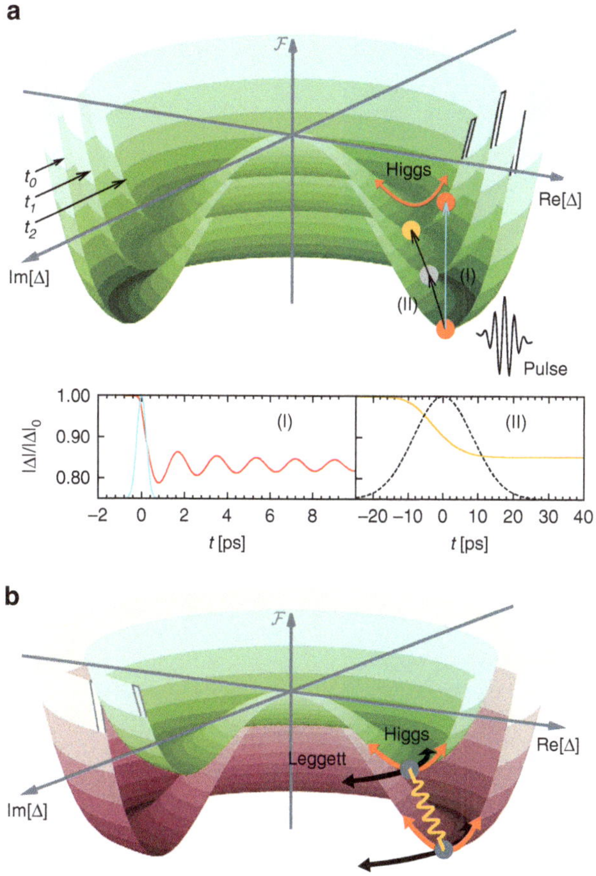

Figure 1 | Illustration of Leggett and Higgs modes. (**a**) Illustration of the excitation process for a one band superconductor. The pump laser pulse modifies the free energy $\mathcal{F}$ on different time scales depending on the pulse duration τ. For $\tau \gg h/(2|\Delta|)$ the superconductor can follow the change in $\mathcal{F}$ adiabatically, resulting in a monotonic lowering of the order parameter $|\Delta|$ (orange trace in inset (II)). For short pulses with $\tau \lesssim h/(2|\Delta|)$, on the other hand, the superconductor is excited in a non-adiabatic manner, which results in oscillations of $|\Delta|$ about the new minimum of $\mathcal{F}$ (red trace in inset (I)). The blue and cyan lines in the two insets represent the Gaussian profiles of the pump pulses. (**b**) Effective free-energy landscape $\mathcal{F}$ for a two-gap superconductor, with green and red representing the Mexican-hat potentials of the smaller and larger gaps, respectively. The amplitude Higgs modes and the phase modes are indicated by red and blue/black arrows, respectively. The Leggett mode corresponds to out-of-phase fluctuations of the phase difference between the two gaps.

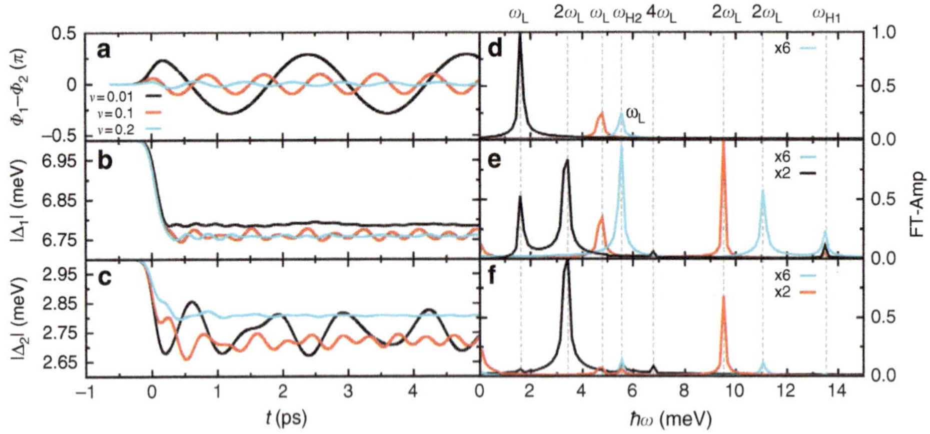

Figure 2 | Leggett phase mode and amplitude Higgs mode oscillations. Numerical simulation of the gap dynamics of a two-gap superconductor after a non-adiabatic excitation by a short intense laser pulse of width $\tau = 0.4$ ps, pump energy $\hbar\omega_0 = 8$ meV and light-field amplitude $|\mathbf{A}_0| = 10 \times 10^{-8}$ Js (Cm)$^{-1}$. (**a**) Phase difference Φ_1-Φ_2 between the two gaps as a function of time t for various interband coupling strengths υ. (**d**) Fourier spectrum of the oscillations in **a**. The frequency of the non-equilibrium Leggett mode oscillation is indicated by ω_L. (**b**,**c**) Gap amplitudes $|\Delta_1|$ and $|\Delta_2|$ as a function of time t for different interband couplings υ. (**e**,**f**) Fourier spectra of the amplitude mode oscillations in **b**,**c**, which display the following frequencies: ω_{H1} and ω_{H2} are the frequencies of the non-equilibrium Higgs modes of gap Δ_1 and Δ_2, respectively; ω_L is the frequency of the non-equilibrium Leggett mode; and higher harmonics of the non-equilibrium Leggett mode are denoted by $2\omega_L$ and $4\omega_L$.

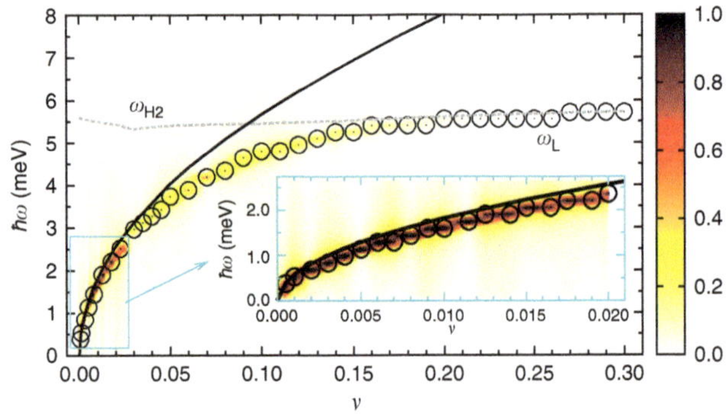

Figure 3 | Leggett phase mode oscillations versus relative interband coupling. Fourier spectrum of the phase mode Φ_1-Φ_2 as a function of relative interband coupling ν for a two-band superconductor perturbed by the same laser pulse as in Fig. 2. The amplitude of the phase fluctuations is indicated by the colour scale with dark red and light yellow representing the highest and lowest amplitudes, respectively. The blue open circles mark the frequency of the non-equilibrium Leggett mode ω_L. The blue solid line represents the frequency of the equilibrium Leggett mode described by equation (2). The dashed grey line indicates the frequency of the Higgs mode ω_{H2}, which coincides with the boundary to the continuum of Bogoliubov quasiparticle excitations, given by twice the asymptotic gap value of the second band $2\Delta_2^\infty$. The inset shows a zoom-in of the blue frame in the main panel.

The Hamiltonian describing the two-band superconductor coupled to the pump laser field is given by $H = H_{BCS} + H_{laser}$, with the two-band Bardeen-Cooper-Schrieffer (BCS) mean-field Hamiltonian

$$H_{BCS} = H_0 + \sum_{\mathbf{k}\in\mathcal{W}} \sum_{l=1}^{2} (\Delta_l c_{\mathbf{k}l\uparrow}^\dagger c_{-\mathbf{k}l\downarrow}^\dagger + \Delta_l^* c_{-\mathbf{k}l\downarrow} c_{\mathbf{k}l\uparrow}), \qquad (1)$$

where $H_0 = \sum_{\mathbf{k}l\sigma} \varepsilon_{\mathbf{k}l} c_{\mathbf{k}l\sigma}^\dagger c_{\mathbf{k}l\sigma}$ denotes the normal state Hamiltonian and $c_{\mathbf{k}l\sigma}^\dagger$ creates electrons with momentum $\mathbf{k}$, band index l and spin σ. The first sum in equation (1) is taken over the set $\mathcal{W}$ of momentum vectors with $|\varepsilon_{\mathbf{k}l}| \leq \hbar\omega_c = 50$ meV, ω_c being the cutoff frequency. The gaps Δ_1 and Δ_2 in the two bands are determined at each temporal integration step from the BCS gap equations with the attractive intraband pairing interactions V_1 and V_2, and the interband coupling $V_{12} = \upsilon V_1$. Motivated by the numbers for MgB_2 (ref. 42), we fix V_1 and V_2 such that the gaps in the initial state take on the values $\Delta_1(t_i) = 7$ meV and $\Delta_2(t_i) = 3$ meV, and study the dynamics of the two-gap superconductor as a function of the relative interband coupling υ. H_{laser} represents the interaction of the pump laser with the superconductor and contains terms linear and quadratic in the vector potential of the laser field, which is of Gaussian shape with central frequency $\hbar\omega_0 = 8$ meV, pulse width $\tau = 0.4$ ps and light-field amplitude $|\mathbf{A}_0|$. We determine the dynamics of Hamiltonian (1) by means of the density matrix approach and solve the resulting equations of motion using Runge–Kutta integration (see Methods).

Pump response. Pumping the two-band superconductor with a short laser pulse of length $\tau \ll \tau_\Delta$ excites a non-thermal distribution of Bogoliubov quasiparticles above the gaps Δ_i, which in turn leads to a rapid, non-adiabatic change in the free-energy landscape $\mathcal{F}$ (Fig. 1). As a result, the collective modes of the superconductor start to oscillate about the new minima of $\mathcal{F}$. This is clearly visible in Fig. 2, which shows the temporal evolution of the gap amplitudes $|\Delta_i|$ and of the phase difference Φ_1-Φ_2 between the two gaps. From the Fourier spectra in Figs 2d–f we can see that three different modes (and their higher harmonics) are excited at the frequencies ω_{H1}, ω_{H2} and ω_L. The two modes at ω_{H1} and ω_{H2} only exist in the dynamics of $\Delta_1(t)$ and $\Delta_2(t)$, respectively, and their peaks are located at the energy of the superconducting gaps $\omega_{Hi} = 2|\Delta_i^\infty|/\hbar$, where Δ_i^∞ denotes the asymptotic gap value[14–20]. This holds for all parameter regimes, even as the laser fluence is increased far beyond the linear absorption region (see Fig. 5). We therefore assign the peaks at ω_{H1} and ω_{H2} to the Higgs amplitude modes of the two gaps. The higher Higgs mode ω_{H1} is strongly damped, because it lies within the continuum of Bogoliubov quasiparticle excitations, which is lower bounded by $2\Delta_2^\infty$. For the lower mode ω_{H2}, on the other hand, the decay channel to quasiparticles is small, as ω_{H2} is at the continuum threshold. This is similar to the non-equilibrium Higgs mode of the single-gap superconductor NbN, whose oscillations have recently been observed over a time period of about 10 ps by pump–probe measurements[2,3].

Interestingly, two-band superconductors exhibit a third collective mode besides the two Higgs modes at a frequency ω_L below the quasiparticle continuum. This mode is most clearly visible in the dynamics of the phase difference Φ_1–Φ_2 (Fig. 2a) and displays a striking dependence on interband coupling strength υ. With decreasing υ its frequency rapidly decreases, whereas its intensity grows. In the limit of vanishing υ, however, the third mode ω_L is completely absent. We thus identify ω_L as the Leggett phase mode, that is, as equal but opposite oscillatory phase shifts of the two coupled gaps. Remarkably, the Leggett phase mode is also observable in the time dependence of the gap amplitudes $\Delta_1(t)$ and $\Delta_2(t)$ (Figs 2b,c), which indicates that Higgs and Leggett modes are coupled in non-equilibrium superconductors.

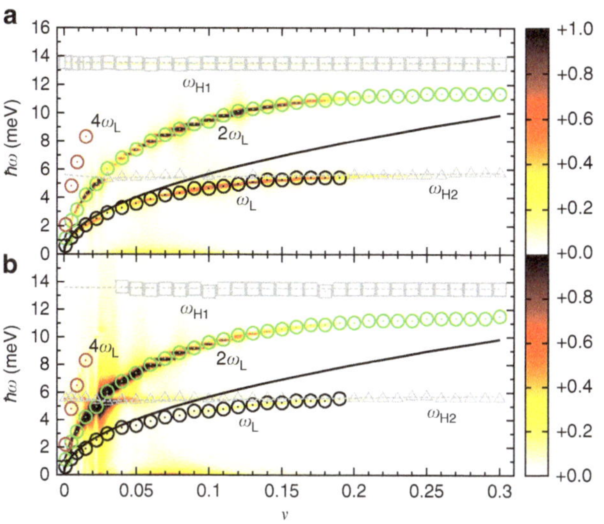

Figure 4 | Amplitude mode oscillations versus relative interband coupling. Fourier spectrum of the amplitude mode oscillations as a function of relative interband coupling υ for (**a**) the superconducting gap Δ_1 in the first band and (**b**) the superconducting gap Δ_2 in the second band. The parameters of the laser pump pulse are the same as in Fig. 2. The amplitude of the oscillations is indicated by the colour scale with dark red and light yellow representing the highest and lowest amplitudes, respectively. The open circles represent the frequencies of the non-equilibrium Leggett mode ω_L and its higher harmonics denoted by $2\omega_L$ and $4\omega_L$. The frequencies of the non-equilibrium Higgs mode of the first and second band, ω_{H1} and ω_{H2}, are indicated by the grey open squares and triangles, respectively. The blue solid line is the frequency of the equilibrium Leggett mode given by equation (2).

To obtain a more detailed picture, we plot in Figs 3 and 4 the energies of the amplitude and phase mode oscillations against the relative interband coupling υ. This reveals that for small υ the non-equilibrium Leggett mode ω_L shows a square root increase, which is in good agreement with the equilibrium Leggett frequency[1,43]

$$\omega_L^0 = 2\sqrt{\Delta_1^\infty \Delta_2^\infty \frac{\upsilon}{V_2 - \upsilon^2 V_1}\left(\frac{1}{\rho_1} + \frac{1}{\rho_2}\right)}, \qquad (2)$$

where ρ_1 and ρ_2 denote the density of states on the two bands. Indeed, as displayed by the inset of Fig. 3, equation (2) represents an excellent parameter-free fit to the numerical data at low υ. For larger υ, on the other hand, the non-equilibrium Leggett mode deviates from the square root behaviour of equation (2). That is, as ω_L approaches the Bogoliubov quasiparticle continuum, it is repelled by the lower Higgs mode ω_{H2}, evidencing a strong coupling between them. As a result, the non-equilibrium Leggett mode is pushed below the continuum and remains nearly undamped for a wide range of υ, which is considerably broader than in equilibrium. Moreover, owing to the dynamical coupling among the collective modes, ω_L and its higher harmonics are observable not only in the phase difference Φ_1–Φ_2 but also in the dynamics of the gap amplitudes $\Delta_i(t)$ (blue and green circles in Fig. 4).

A key advantage of measuring collective modes by pump–probe experiments is that the frequencies of the Higgs modes can be adjusted by the pump fluence. This is demonstrated in Fig. 5, which plots the dynamics of $\Delta_i(t)$ and Φ_1–Φ_2 as a function of integrated pump pulse intensity $|\mathbf{A}_0|^2\tau$. With increasing pump fluence, more Cooper pairs are broken up and superconductivity is more and more suppressed, as reflected in the reduction of the gap amplitudes. At the same time, the frequency of the Higgs oscillations decreases, as it is controlled by the superconducting gaps after pumping. Hence, it is possible to tune the lower Higgs mode ω_{H2} to resonance with ω_L, which strongly enhances the magnitude of the collective-mode oscillations (Fig. 5a,c,e). A

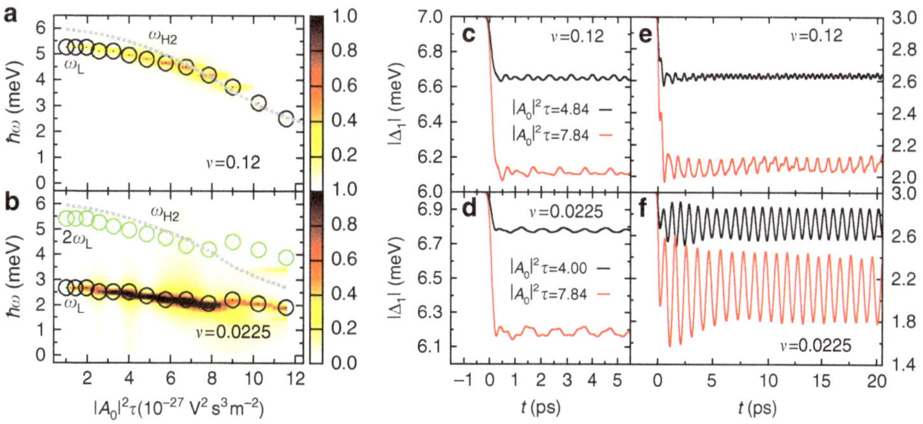

Figure 5 | Fluence dependence of gap dynamics. (**a**,**b**) Fourier spectrum of the phase mode Φ_1-Φ_2 as a function of laser fluence (integrated pulse intensity) $|\mathbf{A}_0|^2\tau$ for two different interband couplings υ with pulse energy $\hbar\omega_0 = 8$ meV and pulse width $\tau = 0.4$ ps. The amplitude of the phase fluctuations is represented by the colour scale with dark red and light yellow indicating high and low amplitudes, respectively. The open circles mark the frequencies of the non-equilibrium Leggett mode ω_L and its higher harmonic $2\omega_L$. The grey dotted lines display the Higgs mode ω_{H2}, which coincides with the boundary to the continuum of Bogoliubov quasiparticle excitations. (**c**-**f**) Gap amplitudes $|\Delta_1|$ and $|\Delta_2|$ as a function of time t for two different interband couplings ν and integrated pulse intensities $|\mathbf{A}_0|^2\tau$.

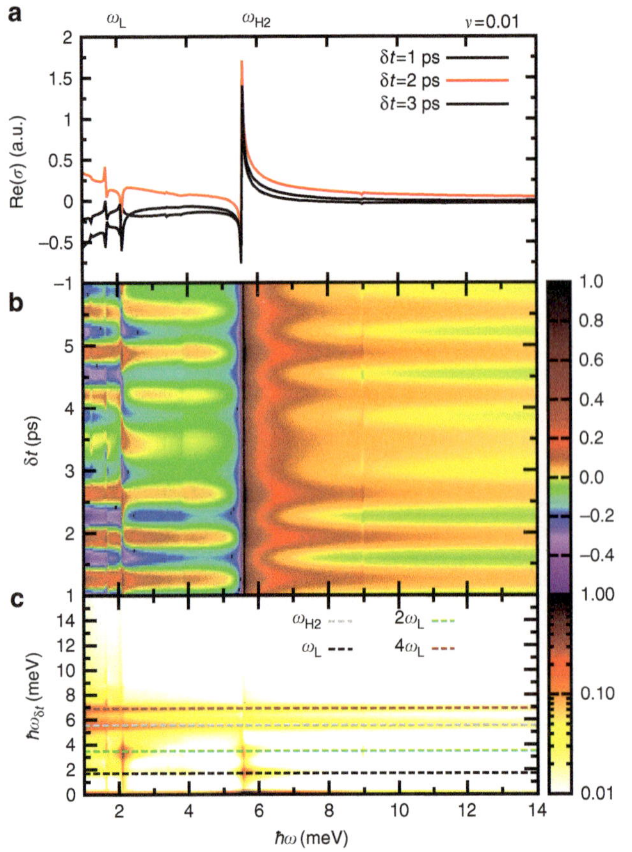

Figure 6 | Pump-probe spectrum. (**a**) Real part of the pump-probe response Re[$\sigma(\delta t,\omega)$] as a function of frequency ω for a two-band superconductor excited by the same pump pulse as in Fig. 2. Both the energy of the non-equilibrium Leggett mode ω_L and the Higgs mode ω_{H2} are visible in the pump-probe signal as sharp peaks. (**b**) Temporal evolution of Re[$\sigma(\delta t,\omega)$]. The intensity of the pump-probe signal is represented by the colour scale with dark red and dark violet, indicating high and low intensities, respectively. (**c**) Fourier spectrum of the pump-probe signal. The pump-probe spectrum oscillates as a function of pump-probe delay time δt with the frequencies of the non-equilibrium Higgs mode ω_{H2} (dashed gray), the frequency of the non-equilibirum Leggett mode ω_L (dashed blue) and the frequencies of higher harmonics of the Leggett mode $2\omega_L$ (dashed green) and $4\omega_L$ (dashed red).

similar enhancement is obtained when ω_{H2} is brought into resonance with twice the frequency of the Leggett mode (Fig. 5b,d,f).

Pump-probe signal. Finally, let us discuss how the Higgs and Leggett modes can be observed in pump-probe spectroscopy. In view of the recent THz pump-THz probe experiments of refs 2–4, we focus on the dynamics of the optical pump-probe conductivity $\sigma(\delta t, \omega) = j(\delta t, \omega)/[i\omega A(\delta t,\omega)]$, where δt is the delay time between pump and probe pulses, $j(\delta t, \omega)$ denotes the current density and $A(\delta t, \omega)$ represents the vector potential of the probe pulse. As the probe pulse has a weak intensity, we neglect terms of second order and higher in the probe field $A(\delta t, \omega)$. Similar to recent experiments[2–7], we take the probe pulse to be very short with width $\tau_{pr} = 0.15$ ps and centre frequency $\hbar\omega_{pr} = 5.5$ meV (see Methods). With this choice, the probe pulse has a broad spectral bandwidth such that the dynamics of the superconductor is probed over a very wide frequency range.

In Fig. 6a we plot the real part of the pump-probe response Re[$\sigma(\delta t, \omega)$] as a function of frequency for fixed δt. The non-equilibrium Leggett mode ω_L and the Higgs-mode ω_{H2} reveal themselves in the pump-probe signal as sharp peaks. Figure 6b shows Re[$\sigma(\delta t, \omega)$] versus delay time δt and frequency ω. Clear oscillations are seen as a function of delay time δt. These are most prominent at the frequencies of the lower Higgs and the Leggett modes, ω_{H2} and ω_L, where $\sigma(\delta t, \omega)$ displays sharp edges as a function of ω (Fig. 6a). Fourier transforming with respect to δt shows that the dominant oscillations are ω_{H2} and ω_L (and its higher harmonics) (Fig. 6b). We therefore predict that both the lower Higgs mode ω_{H2} and the Leggett mode ω_L can be observed in THz pump-THz probe experiments as oscillations of the pump-probe conductivity, in particular at the gap edge $2\Delta_2^\infty/\hbar$ and the Leggett mode frequency ω_L. The higher Higgs mode ω_{H1}, on the other hand, is not visible in the pump-probe signal, as it is strongly damped by the two-particle continuum.

Discussion

Using a semi-numerical method based on the density matrix approach, we have studied the non-equilibrium excitation of Higgs and Leggett modes in two-band superconductors. Although the amplitude Higgs and the Leggett phase mode are decoupled in equilibrium, we find that the out-of-equilibrium excitation process leads to a strong coupling between these two collective modes. As a result, the Leggett phase mode ω_L is pushed below the Bogoliubov quasiparticle continuum and remains undamped for a wide range of interband couplings (Figs 3 and 4). Likewise, the lower Higgs mode ω_{H2} is only weakly damped, as its frequency is at the threshold to the quasiparticle continuum. To maximize the oscillatory signal of these collective modes in the pump-probe spectra, it is necessary to choose the experimental parameters as follows: (i) the pump-pulse duration τ should be smaller than the intrinsic response time of the superconductor $h/(2|\Delta_i|)$, such that the collective modes are excited in a non-adiabatic manner; (ii) the pump-pulse energy needs to be of the order of the superconducting gap (that is, in the terahertz regime), so that Bogoliubov quasiparticles are excited across the gap, but modes at higher energies $\hbar\omega \gg |\Delta_i|$ are not populated; and (iii) the pump-pulse intensity must not exceed a few nJ cm^{-2}, to ensure that the superconducting condensate is only partially broken up, but not completely destroyed. We have predicted that under these conditions both Higgs and Leggett modes can be observed as clear oscillations in the time-resolved pump-probe absorption spectra (Fig. 6). Similarly, we expect that collective mode oscillations are visible in other pump-probe-type experiments, for example, in time-resolved photoemission spectroscopy or time-resolved Raman scattering.

Our findings apply beyond the scope of two-band superconductors to general collective modes in quantum materials. That is, we expect that out-of-equilibrium excitations lead to the coupling of collective modes in any symmetry-broken phase. It would be particularly intriguing to study this in more detail for the case of unconventional exotic superconductors, where several competing orders are present, such as heavy fermion superconductors or high-temperature cuprate and pnictide superconductors. In these systems the pump pulse could be used to induce a transition from one competing order to another. Furthermore, the unconventional pairing symmetries of these superconductors, such as the $d_{x^2-y^2}$-wave pairing of the cuprates, give rise to a multitude of new Higgs modes[44]. Our work indicates that pump-probe experiments will allow to coherently excite and control these novel Higgs modes, which await to be further explored both theoretically and experimentally.

Methods

Model definition. The gap equations for the BCS Hamiltonian $H_{\rm BCS}$ (see equation (1) in the main text) are given by[45]

$$\Delta_1 = \frac{1}{N}\sum_{\mathbf{k}'\in\mathcal{W}}(V_1\langle c_{-\mathbf{k}',\downarrow,1}c_{\mathbf{k}',\uparrow,1}\rangle + V_{12}\langle c_{-\mathbf{k}',\downarrow,2}c_{\mathbf{k}',\uparrow,2}\rangle),$$
$$\Delta_2 = \frac{1}{N}\sum_{\mathbf{k}'\in\mathcal{W}}(V_2\langle c_{-\mathbf{k}',\downarrow,2}c_{\mathbf{k}',\uparrow,2}\rangle + V_{12}\langle c_{-\mathbf{k}',\downarrow,1}c_{\mathbf{k}',\uparrow,1}\rangle), \quad (3)$$

where N is the number of lattice points, V_1 and V_2 denote the intraband interactions and $V_{12}=vV_1$ is the interband coupling. The two-band superconductor is brought out of equilibrium via the coupling to a pump pulse, which is modelled by

$$H_{\rm Laser} = \frac{e\hbar}{2}\sum_{\mathbf{k},\mathbf{q},\sigma,l}\frac{(2\mathbf{k}+\mathbf{q})\mathbf{A}_{\mathbf{q}}(t)}{m_l}c^{\dagger}_{\mathbf{k}+\mathbf{q},\sigma,l}c_{\mathbf{k},\sigma,l} \quad (4)$$

$$+\frac{e^2}{2}\sum_{\mathbf{k},\mathbf{q},\sigma,l}\frac{\left(\sum_{\mathbf{q}'}\mathbf{A}_{\mathbf{q}-\mathbf{q}'}(t)\mathbf{A}_{\mathbf{q}'}(t)\right)}{m_l}c^{\dagger}_{\mathbf{k}+\mathbf{q},\sigma,l}c_{\mathbf{k},\sigma,l},$$

where m_l is the effective electron mass of the lth band and $\mathbf{A}_{\mathbf{q}}(t)$ represents the transverse vector potential of the pump laser. We consider a Gaussian pump pulse described by

$$\mathbf{A}_{\mathbf{q}}(t) = \mathbf{A}_0 e^{-\left(\frac{2\sqrt{\ln 2}t}{\tau}\right)^2}(\delta_{\mathbf{q},\mathbf{q}_0}e^{-i\omega_0 t} + \delta_{\mathbf{q},-\mathbf{q}_0}e^{i\omega_0 t}), \quad (5)$$

with central frequency ω_0, pulse width τ, light-field amplitude $\mathbf{A}_0=|\mathbf{A}_0|\hat{\mathbf{e}}_y$ and photon wave vector $\mathbf{q}_0=q_0\hat{\mathbf{e}}_x$.

Density matrix formalism. To simulate the non-equilibrium dynamics of the two-band superconductor (1), we use a semi-numerical method based on the density matrix formalism. This approach involves the analytical derivation of equations of motions for the Bogoliubov quasiparticle densities $\langle\alpha^{\dagger}_{\mathbf{k},l}\alpha_{\mathbf{k}',l}\rangle$, $\langle\beta^{\dagger}_{\mathbf{k},l}\beta_{\mathbf{k}',l}\rangle$, $\langle\alpha^{\dagger}_{\mathbf{k},l}\beta^{\dagger}_{\mathbf{k}',l}\rangle$ and $\langle\alpha_{\mathbf{k},l}\beta_{\mathbf{k}',l}\rangle$, which are then integrated up numerically using a Runge Kutta algorithm. The Bogoliubov quasiparticle densities are defined in terms of the fermionic operators $\alpha_{\mathbf{k},l}$ and $\beta_{\mathbf{k},l}$, with

$$\alpha_{\mathbf{k},l} = u_{\mathbf{k},l}c_{\mathbf{k},l,\uparrow} - v_{\mathbf{k},l}c^{\dagger}_{-\mathbf{k},l,\downarrow}, \quad (6)$$

$$\beta_{\mathbf{k},l} = v_{\mathbf{k},l}c^{\dagger}_{\mathbf{k},l,\uparrow} + u_{\mathbf{k},l}c_{-\mathbf{k},l,\downarrow}, \quad (7)$$

where $v_{\mathbf{k},l} = \Delta_l(t_i)/|\Delta_l(t_i)|\sqrt{(1-\epsilon_{\mathbf{k},l}/E_{\mathbf{k},l})/2}$, $u_{\mathbf{k},l} = \sqrt{(1+\epsilon_{\mathbf{k},l}/E_{\mathbf{k},l})/2}$ and $E_{\mathbf{k},l} = \sqrt{\epsilon^2_{\mathbf{k},l}+|\Delta_l(t_i)|^2}$. We emphasize that the coefficients $u_{\mathbf{k},l}$ and $v_{\mathbf{k},l}$ do not depend on time, that is, the temporal evolution of the quasiparticle densities is computed with respect to a fixed time-independent Bogoliubov-de Gennes basis in which the initial state is diagonal. The equations of motion for the quasiparticle densities are readily obtained from Heisenberg's equation of motion. Since equation (1) represents a mean-field Hamiltonian, this yields a closed system of differential equations and hence no truncation is needed (for details, see refs 17,29,23,25).

Pump-probe response. All physical observables, such as the current density $\mathbf{j}_{\mathbf{q}_{\rm pr}}(\delta t, t)$, can be expressed in terms of the quasiparticle densities. For the current density we find that

$$\mathbf{j}_{\mathbf{q}_{\rm pr}}(\delta t, t) = -e\hbar\sum_{\mathbf{k},l,\sigma}\frac{2\mathbf{k}+\mathbf{q}_{\rm pr}}{2m_l V}\langle c^{\dagger}_{\mathbf{k},l,\sigma}c_{\mathbf{k}+\mathbf{q}_{\rm pr},l,\sigma}\rangle(\delta t, t)$$
$$-e^2\sum_{\mathbf{k},l,\mathbf{q},\sigma}\frac{\mathbf{A}_{\mathbf{q}_{\rm pr}-\mathbf{q}}}{m_l V}\langle c^{\dagger}_{\mathbf{k},l,\sigma}c_{\mathbf{k}+\mathbf{q},l,\sigma}\rangle(\delta t, t),$$

where $\mathbf{A}_{\mathbf{q}_{\rm pr}}(\delta t, t)$ and $\mathbf{q}_{\rm pr}=|\mathbf{q}_{\rm pr}|\hat{\mathbf{e}}_x$ are the vector potential and the wave vector of the probe pulse, respectively. With this, we obtain the pump–probe conductivity via[23,46]

$$\sigma(\delta t, \omega) = \frac{j(\delta t, \omega)}{i\omega A(\delta t, \omega)}, \quad (9)$$

where $j(\delta t,\omega)$ and $A(\delta t,\omega)$ denote the Fourier transformed y components of the current density $\mathbf{j}_{\mathbf{q}_{\rm pr}}(\delta t, t)$ and the vector potential $\mathbf{A}_{\mathbf{q}_{\rm pr}}(\delta t, t)$, respectively. To compute the effects of the probe pulse, we neglect terms of second order and higher in the probe field $A_{\rm pr}(t)$, as the probe pulse has a very weak intensity.

Numerical discretization and integration. To keep the number of equations of motion manageable, we have to restrict the number of considered points in momentum space. The first restriction is that we only take expectation values with indices $\mathbf{k}$ and $\mathbf{k}+\mathbf{q}\in\mathcal{W}$ into account. This means that we concentrate on the $\mathbf{k}$-values where the attractive pairing interaction takes place. Furthermore, as the external electromagnetic field may add or subtract only momentum $n\mathbf{q}_0$, it is sufficient to consider expectation values with indices $(\mathbf{k}, \mathbf{k}+n\mathbf{q}_0)$, where $n\in\mathbb{Z}$. For small amplitudes $|\mathbf{A}_{\mathbf{q}_0}|$ the off-diagonal elements of the quasiparticle densities decrease rapidly as n increases, as $(\mathbf{k}, \mathbf{k}+n\mathbf{q}_0) = O(|\mathbf{A}_{\mathbf{q}_0}|^{|n|})$. Thus, we set all entries with $n>4$ to 0. With this momentum-space discretization, we obtain of the order of 10^5 equations, which we are able to solve numerically using high-efficiency parallelization. Further, technical details can be found in refs 23,25.

References

1. Leggett, A. J. Number-phase fluctuations in two-band superconductors. *Prog. Theor. Phys.* **36,** 901–930 (1966).
2. Matsunaga, R. & Shimano, R. Nonequilibrium BCS state dynamics induced by intense terahertz pulses in a superconducting NbN film. *Phys. Rev. Lett.* **109,** 187002 (2012).
3. Matsunaga, R. *et al.* Higgs amplitude mode in the BCS superconductors $Nb_{1-x}Ti_xN$ induced by terahertz pulse excitation. *Phys. Rev. Lett.* **111,** 057002 (2013).
4. Matsunaga, R. *et al.* Light-induced collective pseudospin precession resonating with Higgs mode in a superconductor. *Science* **345,** 1145–1149 (2014).
5. Mansart, B. *et al.* Coupling of a high-energy excitation to superconducting quasiparticles in a cuprate from coherent charge fluctuation spectroscopy. *Proc. Natl Acad. Sci. USA* **110,** 4539–4544 (2013).
6. Pashkin, A. & Leitenstorfer, A. Particle physics in a superconductor. *Science* **345,** 1121–1122 (2014).
7. Dal Conte, S. *et al.* Disentangling the electronic and phononic glue in a high-T_c superconductor. *Science* **335,** 1600–1603 (2012).
8. Beck, M. *et al.* Transient increase of the energy gap of superconducting nbn thin films excited by resonant narrow-band terahertz pulses. *Phys. Rev. Lett.* **110,** 267003 (2013).
9. Pashkin, A. *et al.* Femtosecond response of quasiparticles and phonons in superconducting $YBa_2Cu_3O_{7-x}$ studied by wideband terahertz spectroscopy. *Phys. Rev. Lett.* **105,** 067001 (2010).
10. Beck, M. *et al.* Energy-gap dynamics of superconducting NbN thin films studied by time-resolved terahertz spectroscopy. *Phys. Rev. Lett.* **107,** 177007 (2011).
11. Hellmann, S. *et al.* Time-domain classification of charge-density-wave insulators. *Nat. Commun.* **3,** 1069 (2012).
12. Torchinsky, D. H., Mahmood, F., Bollinger, A. T., Božovic', I. & Gedik, N. Fluctuating charge-density waves in a cuprate superconductor. *Nat. Mater.* **12,** 387–391 (2013).
13. Kimel, A. V., Kirilyuk, A., Tsvetkov, A., Pisarev, R. V. & Rasing, T Laser-induced ultrafast spin reorientation in the antiferromagnet $TmFeO_3$. *Nature* **429,** 850–853 (2004).
14. Amin, M. H. S., Bezuglyi, E. V., Kijko, A. S. & Omelyanchouk, A. N. Wigner distribution function formalism for superconductors and collisionless dynamics of the superconducting order parameter. *Low Temp. Phys.* **30,** 661–666 (2004).
15. Volkov, A. F. & Kogan, S. M. Collisionless relaxation of the energy gap in superconductors. *Sov. Phys. JETP* **38,** 1018 (1974).
16. Schnyder, A. P., Manske, D. & Avella, A. Resonant generation of coherent phonons in a superconductor by ultrafast optical pump pulses. *Phys. Rev. B* **84,** 214513 (2011).
17. Papenkort, T., Axt, V. M. & Kuhn, T. Coherent dynamics and pump-probe spectra of BCS superconductors. *Phys. Rev. B* **76,** 224522 (2007).
18. Papenkort, T., Kuhn, T. & Axt, V. M. Coherent control of the gap dynamics of BCS superconductors in the nonadiabatic regime. *Phys. Rev. B* **78,** 132505 (2008).
19. Yuzbashyan, E. A., Tsyplyatyev, O. & Altshuler, B. L. Relaxation and persistent oscillations of the order parameter in fermionic condensates. *Phys. Rev. Lett.* **96,** 097005 (2006).
20. Yuzbashyan, E. A., Altshuler, B. L., Kuznetsov, V. B. & Enolskii, V. Z. Nonequilibrium cooper pairing in the nonadiabatic regime. *Phys. Rev. B* **72,** 220503 (2005).
21. Yuzbashyan, E. A., Altshuler, B. L., Kuznetsov, V. B. & Enolskii, V. Z. Solution for the dynamics of the BCS and central spin problems. *J. Phys. A Math. Gen.* **38,** 7831 (2005).
22. Zachmann, M. *et al.* Ultrafast terahertz-field-induced dynamics of superconducting bulk and quasi-1d samples. *New J. Phys.* **15,** 055016 (2013).
23. Krull, H., Manske, D., Uhrig, G. S. & Schnyder, A. P. Signatures of nonadiabatic BCS state dynamics in pump-probe conductivity. *Phys. Rev. B* **90,** 014515 (2014).
24. Yuzbashyan, E. A., Dzero, M., Gurarie, V. & Foster, M. S. Quantum quench phase diagrams of an *s*-wave BCS -BEC condensate. *Phys. Rev. A* **91,** 033628 (2015).
25. Krull, H. *Conductivity of Strongly Pumped Superconductors.* PhD thesis, TU Dortmund (2014).

26. Kemper, A. F., Sentef, M. A., Moritz, B., Freericks, J. K. & Devereaux, T. P. Direct observation of Higgs mode oscillations in the pump-probe photoemission spectra of electron-phonon mediated superconductors. *Phys. Rev. B* **92,** 224517 (2015).
27. Peronaci, F., Schiró, M. & Capone, M. Transient dynamics of *d*-wave superconductors after a sudden excitation. *Phys. Rev. Lett.* **115,** 257001 (2015).
28. Murakami, Y., Werner, P., Tsuji, N. & Aoki, H. Multiple amplitude modes in strongly coupled phonon-mediated superconductors. *Phys. Rev. B* **93,** 094509 (2016).
29. Akbari, A., Schnyder, A. P., Manske, D. & Eremin, I. Theory of nonequilibrium dynamics of multiband superconductors. *Europhys. Lett.* **101,** 17002 (2013).
30. Dzero, M., Khodas, M. & Levchenko, A. Amplitude modes and dynamic coexistence of competing orders in multicomponent superconductors. *Phys. Rev. B* **91,** 214505 (2015).
31. Demsar, J. *et al.* Pair-breaking and superconducting state recovery dynamics inMgB_2. *Phys. Rev. Lett.* **91,** 267002 (2003).
32. Burnell, F. J., Hu, J., Parish, M. M. & Bernevig, B. A. Leggett mode in a strong-coupling model of iron arsenide superconductors. *Phys. Rev. B* **82,** 144506 (2010).
33. Anishchanka, A., Volkov, A. F. & Efetov, K. B. Collective modes in two-band superconductors in the dirty limit. *Phys. Rev. B* **76,** 104504 (2007).
34. Pekker, D. & Varma, C. Amplitude/Higgs modes in condensed matter physics. *Annu. Rev. Condensed Matter Phys.* **6,** 269–297 (2015).
35. Anderson, P. W. Random-phase approximation in the theory of superconductivity. *Phys. Rev.* **112,** 1900–1916 (1958).
36. Bogoliubov, N. N., Tolmatschev, V. V. & Shirkov, D. V. *New Method in the Theory of Superconductivity* (Publisher Consultants Bureau, New York, 1959).
37. Podolsky, D., Auerbach, A. & Arovas, D. P. Visibility of the amplitude (Higgs) mode in condensed matter. *Phys. Rev. B* **84,** 174522 (2011).
38. Sooryakumar, R. & Klein, M. V. Raman scattering by superconducting-gap excitations and their coupling to charge-density waves. *Phys. Rev. Lett.* **45,** 660–662 (1980).
39. Littlewood, P. B. & Varma, C. M. Gauge-invariant theory of the dynamical interaction of charge density waves and superconductivity. *Phys. Rev. Lett.* **47,** 811–814 (1981).
40. Blumberg, G. *et al.* Observation of Leggett's collective mode in a multiband MgB_2 superconductor. *Phys. Rev. Lett.* **99,** 227002 (2007).
41. Méasson, M.-A. *et al.* Amplitude Higgs mode in the 2*H*-$NbSe_2$ superconductor. *Phys. Rev. B* **89,** 060503 (2014).
42. Mazin, I. & Antropov, V. Electronic structure, electron–phonon coupling, and multiband effects in MgB_2. *Phys. C Superconductivity* **385,** 49–65 (2003).
43. Sharapov, S. G., Gusynin, V. P. & Beck, H. Effective action approach to the Leggett's mode in two-band superconductors. *Eur. Phys. J. B* **30,** 45–51 (2002).
44. Barlas, Y. & Varma, C. M. Amplitude or Higgs modes in *d*-wave superconductors. *Phys. Rev. B* **87,** 054503 (2013).
45. Suhl, H., Matthias, B. T. & Walker, L. R. Bardeen-Cooper-Schrieffer theory of superconductivity in the case of overlapping bands. *Phys. Rev. Lett.* **3,** 552–554 (1959).
46. Shao, C., Tohyama, T., Luo, H.-G. & Lu, H. A numerical method to compute optical conductivity based on the pump-probe simulations. Preprint at http://arXiv.org/abs/1507.01200 (2015).

Acknowledgements

We gratefully acknowledge many useful discussions with A. Avella, S. Kaiser and R. Shimano. G.S.U. and H.K. acknowledge financial support by the DFG in TRR 160 and by the Stiftung Mercator. H.K. thanks the Max-Planck-Institut FKF Stuttgart for its hospitality.

Author contributions

The density matrix simulations were developed and run by H.K., N.B. and A.P.S. All authors contributed to the discussion and interpretation of the results, and to the writing of the paper.

Additional information

Competing financial interests: The authors declare no competing financial interests.

28

Efficient plasmonic emission by the quantum Čerenkov effect from hot carriers in graphene

Ido Kaminer[1], Yaniv Tenenbaum Katan[2], Hrvoje Buljan[3], Yichen Shen[1], Ognjen Ilic[1], Josué J. López[1], Liang Jie Wong[4], John D. Joannopoulos[1] & Marin Soljačić[1]

Graphene plasmons have been found to be an exciting plasmonic platform, thanks to their high field confinement and low phase velocity, motivating contemporary research to revisit established concepts in light-matter interaction. In a conceptual breakthrough over 80 years old, Čerenkov showed how charged particles emit shockwaves of light when moving faster than the phase velocity of light in a medium. To modern eyes, the Čerenkov effect offers a direct and ultrafast energy conversion scheme from charge particles to photons. The requirement for relativistic particles, however, makes Čerenkov emission inaccessible to most nanoscale electronic and photonic devices. Here we show that graphene plasmons provide the means to overcome this limitation through their low phase velocity and high field confinement. The interaction between the charge carriers flowing inside graphene and the plasmons enables a highly efficient two-dimensional Čerenkov emission, giving a versatile, tunable and ultrafast conversion mechanism from electrical signal to plasmonic excitation.

[1]Department of Physics, Massachusetts Institute of Technology, 77 Massachusetts Avenue, Cambridge, Massachusetts 02139, USA. [2]Physics Department and Solid State Institute, Technion, Haifa 32000, Israel. [3]Department of Physics, University of Zagreb, Bijenička c. 32, 10000 Zagreb, Croatia. [4]Singapore Institute of Manufacturing Technology, 2 Fusionopolis Way, Innovis, Singapore 138634, Singapore. Correspondence and requests for materials should be addressed to I.K. (email: kaminer@mit.edu).

Achieving ultrafast conversion of electrical to optical signals at the nanoscale using plasmonics[1,2] is a long-standing goal, owing to its potential to revolutionize electronics and allow ultrafast communication and signal processing. Given their strong field confinement, plasmonic systems combine the benefits of high frequencies (10^{14}–10^{15} Hz) with those of small spatial scales, thus avoiding the limitation of conventional photonic systems. However, the realization of plasmonic sources that are electrically pumped, power efficient and compatible with current device fabrication processes (for example, complementary metal-oxide semiconductors (CMOS)), is a formidable challenge. In recent years, several groups have demonstrated the potential of surface plasmons as a platform for strong and ultrafast light-matter interaction[3–6]. Graphene's tunability and strong field confinement[7–10] have motivated proposals for the use of graphene plasmons (GPs)[7,11–13] in electrically pumped plasmonic sources[14] and in the conversion of electrical energy into luminescence[15–17].

Here we show that under proper conditions charge carriers propagating inside graphene can efficiently excite GPs, through a two-dimensional (2D) Čerenkov emission process. Graphene provides a platform in which the flow of charge alone is sufficient for Čerenkov radiation, eliminating the need for accelerated charge particles in vacuum chambers and opening up an opportunity for the study of the Čerenkov effect (ČE) and its applications, especially as a novel plasmonic source. Unlike other types of plasmon excitations, the 2D ČE manifests as a plasmonic shock wave, analogous to the conventional ČE that creates shockwaves in a three-dimensional (3D) medium. On a quantum mechanical level, this shockwave is reflected in the wavefunction of a single GP emitted from a single hot carrier.

Results

The mechanism of the graphene ČE. The mechanism of 2D ČE is enabled by two fundamental characteristics of graphene. On one hand, hot charge carriers moving with high velocities (up to the Fermi velocity $v_F \approx 10^6\,\mathrm{m\,s^{-1}}$) are considered possible, even in relatively large sheets of graphene (10 μm and more[18]). On the other hand, plasmons in graphene can have an exceptionally slow phase velocity, down to a few hundred times slower than the speed of light[7,9,19]. Consequently, velocity matching between charge carriers and plasmons is possible, enabling the emission of GPs from electrical excitations (hot carriers) at very high rates. This paves the way to devices using the ČE on the nanoscale, a prospect made even more attractive by the dynamic tunability of the Fermi level of graphene. For a wide range of parameters, the emission rate of GPs is significantly higher than the rates previously found for photons or phonons[7,20,21], suggesting that taking advantage of the ČE enables near-perfect energy conversion from electrical energy to plasmons. Of course, it is impossible to reach perfect conversion due to other competing decay processes as discussed below. We further show that, contrary to expectations, plasmons can be created at energies above $2E_F$—thus exceeding energies attainable by photon emission—resulting in a plasmon spectrum that extends from terahertz to near-infrared frequencies and possibly into the visible range. Furthermore, we show that tuning the Fermi energy by external voltage can control the parameters (direction and frequency) of enhanced emission. This tunability also reveals regimes of backward GP emission and regimes of forward GP emission with low angular spread, emphasizing the uniqueness of ČE from hot carriers flowing in graphene. Surprisingly, we find that GP emission can also result from intraband transitions that are made possible by plasmonic losses. These kinds of transitions can become significant and might help explain several phenomena observed in graphene devices, such as current saturation[22], high frequency radiation spectrum from graphene[17,23] and the black body radiation spectrum that seems to relate to extraordinary high electron temperatures[24].

Recent studies[25–27], which focus on cases of classical free charge particles moving outside graphene, have revealed strong Čerenkov-related GP emission resulting from the charge particle–plasmon coupling. In contrast, in this work we focus on the study of charge carriers inside graphene, as illustrated in Fig. 1a. For this purpose, we develop a quantum theory of ČE in graphene. As we shall see, our analysis of this system gives rise to a variety of novel Čerenkov-induced plasmonic phenomena.

The conventional threshold of the ČE in either 2D or 3D ($v > v_p$) may seem unattainable for charge carriers in graphene,

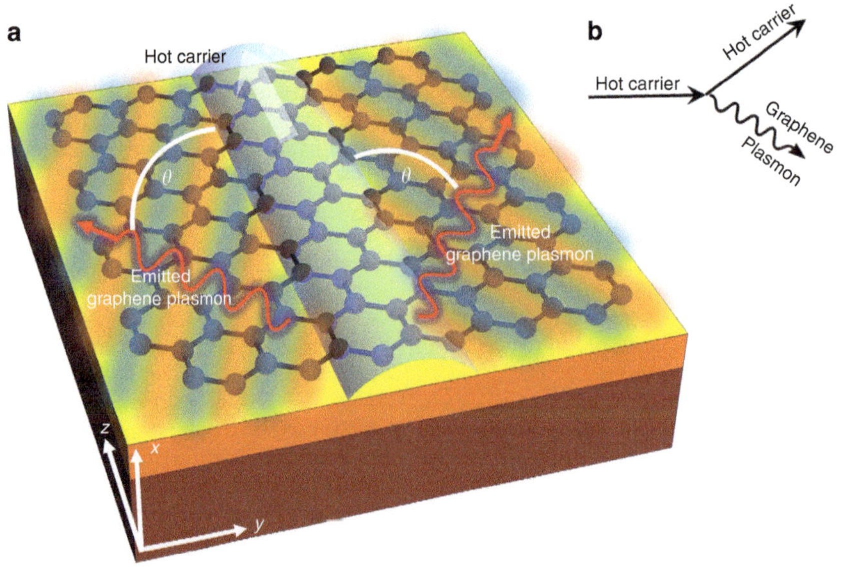

Figure 1 | Illustration of the plasmon emission from charge carriers in graphene via a 2D Čerenkov process. (**a**) GP emission in graphene from a hot carrier flowing inside it. The hot carrier (white arrow marking a transparent-blue arch shape) excites GPs that propagate sideways (glowing red-blue bars) along the graphene surface (plotted on the yellow-orange-red substrate). The Čerenkov angle into which the GPs are emitted is denoted by θ (defined between the wiggling red arrows and the z axis, which is the direction of motion of the hot carrier). (**b**) A diagram describing the GP emission process from a hot carrier in graphene.

because they are limited by the Fermi velocity $v \le v_F$, which is smaller than the GP phase velocity $v_F < v_p$, as shown by the random phase approximation calculations[19,28]. However, we show that quantum effects come into play, to enable these charge carriers to exceed the actual ČE threshold. Specifically, the actual ČE threshold for free electrons is shifted from its classically predicted value by the quantum recoil of electrons on photon emission[29,30]. Because of this shift, the actual ČE velocity threshold can in fact lie below the velocity of charge carriers in graphene, contrary to the conventional predictions. At the core of the modification of the quantum ČE is the linearity of the charge carrier energy–momentum relation (Dirac cone). Consequently, a careful choice of parameters (for example, Fermi energy or hot carrier energy) allows the ČE threshold to be attained—resulting in significant enhancements and high efficiencies of energy conversion from electrical to plasmonic excitation.

The quantum theory of the ČE from hot carriers. The quantum ČE can be described as a spontaneous emission process of a charge carrier emitting into GPs, calculated by Fermi's golden rule[29,30]. In our case, the matrix elements must be obtained from the light–matter interaction term in the graphene Hamiltonian, illustrated by a diagram such as Fig. 1b. To model the GPs, we use the random phase approximation[19,28,31,32], combined with a frequency-dependent phenomenological lifetime[19] that takes into account additional loss mechanisms such as optical phonons and scattering from impurities in the sample (assuming graphene mobility of $\mu = 2{,}000\,\mathrm{cm^2\,V^{-1}\,s^{-1}}$). This approach has been shown to give good agreement with experimental results[8,12,13,33,34]. The graphene sheet is in the yz plane and the charge carrier is moving in the z direction (Fig. 1a). For the case of low-loss GPs, the calculation reduces to the following integral (Lossy GPs are described later in this work—equation (6)).

$$\Gamma = \frac{2\pi}{\hbar}\int_{-\infty}^{\infty} \left|M_{\mathbf{k}_i\to\mathbf{k}_f+\mathbf{q}}\right|^2 \delta\left(E_{\mathbf{k}_i} - \hbar\omega(\mathbf{q}) - E_{\mathbf{k}_f}\right)\frac{d^2\mathbf{q}}{(2\pi)^2/S}\frac{d^2\mathbf{k}_f}{(2\pi)^2/S} \quad (1)$$

$$M_{\mathbf{k}_i\to\mathbf{k}_f+\mathbf{q}} = q_e(2\pi)^2\delta\left(q_y + k_{fy}\right)\delta\left(k_{iz} - q_z - k_{fz}\right)v_F\sqrt{\frac{\hbar q}{\varepsilon_0\tilde{\omega}(\mathbf{q})S^3}}\cdot[\mathrm{SP}] \quad (2)$$

where $M_{\mathbf{k}_i\to\mathbf{k}_f+\mathbf{q}}$ is the matrix element, S is the surface area used for normalization, q_e is the electric charge, ε_0 is the vacuum permittivity, [SP] is the spinor–polarization coupling term and $\tilde{\omega}(\mathbf{q})$ is the GP dispersion-based energy normalization term[35] ($\tilde{\omega}(\mathbf{q}) = \bar{\epsilon}_r\omega\cdot v_p/v_g$, using the group velocity $v_g = \partial\omega/\partial q$). Supplementary Notes 1 and 2 elaborate on these definitions and describe the normalization (respectively). The GP momentum $\mathbf{q} = (q_y, q_z)$ satisfies $\omega^2/v_p^2 = q_y^2 + q_z^2$, with the phase velocity $v_p = v_p(\omega)$ or $v_p(\mathbf{q})$ obtained from the plasmon dispersion relation as $v_p = \omega/q$. The momenta of the incoming (outgoing) charge carrier $\mathbf{k}_i = (k_{iy}, k_{iz})$ ($\mathbf{k}_f = (k_{fy}, k_{fz})$) correspond to energies $E_{\mathbf{k}_i}$ ($E_{\mathbf{k}_f}$) according to the conical momentum-energy relation $E_{\mathbf{k}}^2 = \hbar^2 v_F^2\left(k_y^2 + k_z^2\right)$. The charge velocity is $v = E_{\mathbf{k}}/|\hbar\mathbf{k}|$, which equals a constant (v_F). The only approximation in equation (2), henceforth, comes from the standard assumption of high GP confinement (free space wavelength/GP wavelength much larger than 1)[19]. Substituting equation (1) into equation (2) we obtain (denoting $E_i = E_{\mathbf{k}_i}$):

$$\Gamma = \int_{-\infty}^{\infty} \frac{\alpha c\hbar v_g(\mathbf{q})}{\bar{\epsilon}_r v_p^2(\mathbf{q})/v_F^2}\delta\left(q_y + k_{fy}\right)\delta\left(k_{iz} - q_z - k_{fz}\right) \times \delta\left(E_i - \hbar\omega(\mathbf{q}) - E_{\mathbf{k}_f}\right)|\mathrm{SP}|^2 d^2\mathbf{q}\, d^2\mathbf{k}_f \quad (3)$$

where $\alpha(\approx \frac{1}{137})$ is the fine-structure constant, c is the speed of light and $\bar{\epsilon}_r$ is the relative substrate permittivity obtained by averaging the permittivity on both sides of the graphene. We assume $\bar{\epsilon}_r = 2.5$ for all the figures. As material dispersion is neglected, all spectral features are uniquely attributed to the GP dispersion and its interaction with charge carriers, and not to any frequency dependence of the dielectrics. We further define the angle φ for the outgoing charge and θ for the GP, both relative to the z axis, which is the direction of the incoming charge. This notation allows us to simplify the [SP] for charge carriers inside graphene to $|\mathrm{SP}|^2 = \cos^2(\theta - \varphi/2)$ or $|\mathrm{SP}|^2 = \sin^2(\theta - \varphi/2)$ for intraband or interband transitions, respectively. The emission is restricted to two angles $\theta = \pm\theta_{\check{C}}$, as reflected in the delta function in equation (3) (a clear signature of the ČE) and so we simplify the rate of emission to (see Supplementary Notes 3 and 4 for the complete derivation):

$$\cos(\theta_C) = \frac{v_p}{v_F}\left[1 - \frac{\hbar\omega}{2E_i}\left(1 - \frac{v_F^2}{v_p^2}\right)\right] \quad (4)$$

$$\Gamma_\omega = \frac{2\alpha c}{v_F\bar{\epsilon}_r}\frac{\left|1 - \frac{\hbar\omega}{2E_i}\left(1 + \frac{v_F}{v_p}\cos(\theta_C)\right)\right|}{|\sin(\theta_{\check{C}})|} = \frac{2\alpha c}{v_F\bar{\epsilon}_r}\left|\frac{\sin(\theta_{\check{C}})}{1 - v_p^2/v_F^2}\right| \quad (5)$$

We note in passing that by setting $\hbar \to 0$ in the above expressions we recover the classical Čerenkov angle $\cos(\theta_{\check{C}}) = v_p/v$ and the classical 2D ČE, which can also be obtained from a purely classical electromagnetic calculation. However, although charge particles outside of graphene satisfy $\hbar\omega \ll E_i$, making the classical approximation almost always exact[29], the charges flowing inside graphene can have much lower energies, because they are massless. Consequently, the introduced $\hbar$ terms in equations (4) and (5) modifies the conventional velocity threshold significantly, allowing ČE to occur for lower charge velocities. For example, although the conventional ČE requires charge velocity above the GP phase velocity ($v > v_p$), equation (4) allows ČE below it and specifically requires the velocity of charge carriers in graphene ($v = v_F$) to reside between $v_p > v_F > v_p\left|1 - \frac{2E_i}{\hbar\omega}\right|$. Physically, the latter case involves interband transitions made possible when graphene is properly doped: when the charge carriers are hot electrons (holes), interband ČE requires negatively (positively) doped graphene. Figures 2 and 3 demonstrate this interband ČE that indeed occurs for charge velocities below the conventional velocity threshold. More generally, the inequalities can be satisfied in two spectral windows simultaneously for the same charge carrier, owing to the frequency dependence of the GP phase velocity (shown by the intersection of the red curve with the blue regime in Fig. 2d). Moreover, part of the radiation (or even most of it, as in Fig. 2) can be emitted backward, which is considered impossible for ČE in conventional materials[36,37]. Several spectral cutoffs appear in Figs 2c, 3c and 4c, as seen by the range of non-vanishing blue spectrum. These can be found by substituting $\theta_{\check{C}} = 0$ in equation (4), leading to $\hbar\omega_{\mathrm{cutoff}} = 2E_i/(1 \pm v_F/v_p)$, exactly matching the points where the red curves in Figs 2d, 3d and 4d cross the border of the blue regime. The upper most frequency cutoff marked by the thick orange line in Figs 2–4 occurs at $\hbar\omega = E_i + E_F$ due to the interband transition being limited by the Fermi sea of excited states. Since E_i can be larger than E_F, this implies that GP emission from electrical excitation can be more energetic than photon emission from a similar process (that is limited already by $\hbar\omega \lesssim 2E_F$). Finite temperature will broaden all cutoffs by the expected Fermi–Dirac distribution. However, for most frequencies, the GP losses are a more significant source of broadening.

Including plasmonic losses into the graphene ČE. To incorporate the GP losses (as we do in all the figures), we modify the

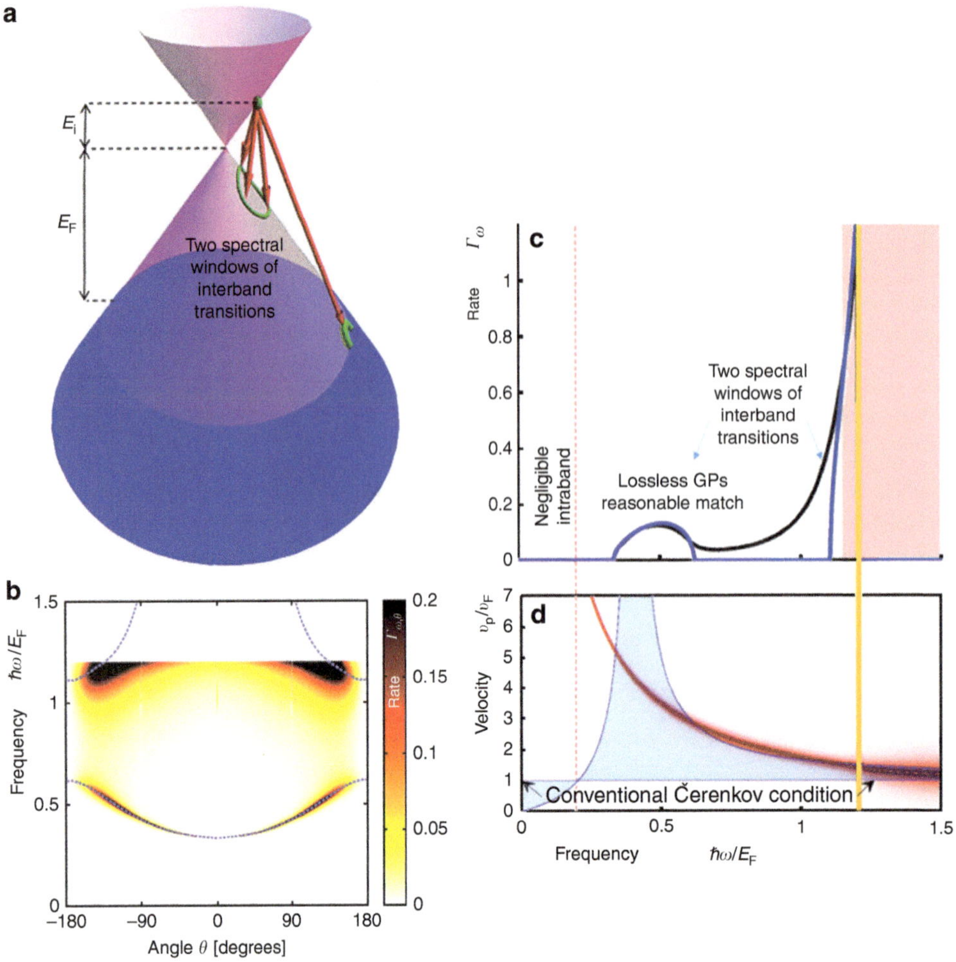

Figure 2 | GP emission from hot carriers. (**a**) Illustration of the possible transitions. The hot carrier (green dot) has a range of potential transitions (red arrows) with distinct final states (green curves and circles), emitting plasmons that satisfy conservation of momentum and energy (corresponding to the height and angle of the red arrows). This way the cone geometry correlates the GP frequency and angle. The projection of these arrows to a 2D plane predicts the in-plane angle θ of the plasmonic emission, matching the (**b**) map of GP emission rate as a function of frequency and angle, equation (6). We find most of the GP emission around the dashed blue curves that are exactly found by the Čerenkov angle equation (4). (**c**) Spectrum of the ČE GP emission process, with the red regime marking the area of high losses (as in ref. 19). Black, emission spectrum with GP losses, equation (6). Blue, lossless emission approximation, equation (5). The thick orange line marks the spectral cutoff due to the Fermi sea, beyond which all states are occupied (therefore, $\hbar\omega < E_i + E_F$). (**d**) Explaining the GP emission with the quantum ČE. The GP phase velocity is plotted as a red curve, with its thickness presenting the GP loss. The blue-shaded regime shows the range of allowed velocities according to the quantum ČE. We find enhanced GP emission in the frequencies for which the red curve crosses the blue regime, either directly or due to the curve thickness. The vertical dotted red line that crosses both **c** and **d** divides between interband to intraband transitions (exactly at $\hbar\omega = E_i$). At the parameters presented in this figure, there is only negligible intraband transitions (zero spectrum on the left of the dotted line). All figures are presented in normalized units, except for the angle shown in degrees. The hot carrier energy $E_i = 0.2E_F$ and $n_s = 3 \times 10^{13}\,\text{cm}^{-2}$ (corresponding to $E_F = 0.639$ eV).

matrix elements calculation by including the imaginary part of the GP wavevector $q_I = q_I(\omega)$, derived independently for each point of the GP dispersion curve[19]. This is equivalent to replacing the delta functions in equation (3) by Lorentzians with $1/\gamma$ width, defining $\gamma(\omega) = q_R(\omega)/q_I(\omega)$. The calculation can be done partly analytically (Supplementary Note 3) yielding:

$$\Gamma_{\omega,\theta} = \frac{\alpha c}{\pi^2 \bar{\epsilon}_r v_p(\omega)} \left| \frac{E_i}{\hbar\omega} - 1 \right| \int_0^{2\pi} d\varphi \begin{cases} \cos^2(\theta - \varphi/2) & \text{intraband transition} \\ \sin^2(\theta - \varphi/2) & \text{interband transition} \end{cases}$$
$$\cdot \frac{\left|\frac{\sin(\theta)}{\gamma(\omega)}\right|}{\left(\frac{v_p(\omega)}{v_F}\left|\frac{E_i}{\hbar\omega} - 1\right| \sin(\varphi) + \sin(\theta)\right)^2 + \left|\frac{\sin(\theta)}{\gamma(\omega)}\right|^2}$$
$$\cdot \frac{|\cos(\theta)/\gamma(\omega)|}{\left(\frac{v_p(\omega)}{v_F}\left|\frac{E_i}{\hbar\omega} - 1\right| \cos(\varphi) + \cos(\theta) - \frac{v_p(\omega)}{v_F}\frac{E_i}{\hbar\omega}\right)^2 + |\cos(\theta)/\gamma(\omega)|^2} \quad (6)$$

The immediate effect of the GP losses is the broadening of the spectral features as shown in Figs 2c, 3c and 4c. Still, the complete analytic theory of equations (4) and (5) matches very well with the exact graphene ČE. For example, regimes of enhanced emission in Figs 2b and 3b match with the blue dashed curves marked according to the Čerenkov angle formula of equation (4) (see Supplementary Note 4 for its derivation in the lossless limit). The presence of GP loss also opens up a regime of quasi-ČE that takes place when the charge velocity is very close to the Čerenkov threshold but does not exceed it. The addition of Lorentzian broadening then closes the gap, creating significant non-zero matrix elements that can lead to intraband GP emission (Fig. 4). This GP emission occurs even for hot electrons (holes) in positively (negatively) doped graphene, with the only change in Fig. 4 being that the upper frequency cutoff is instead shifted to $\hbar\omega \le E_i - E_F$ (eliminating all interband transitions). The dip in

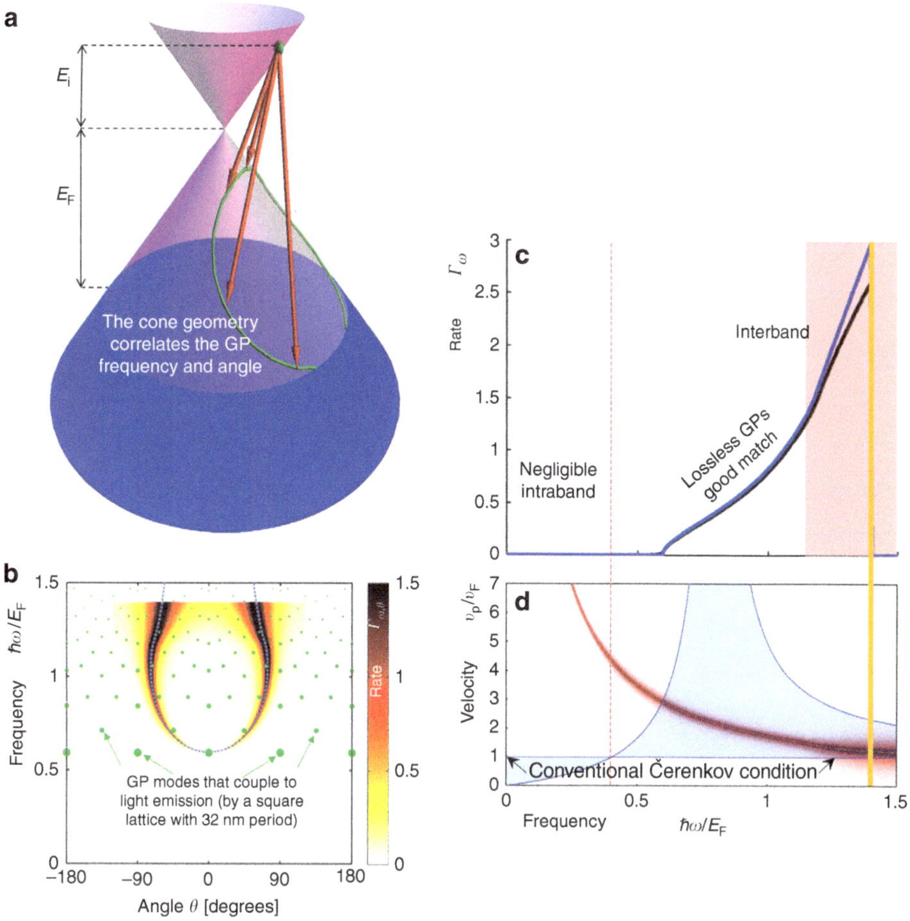

Figure 3 | GP emission from hot carriers. Caption and notations same as in Fig. 2. The green dots in **b** show the GPs can be coupled out, as light, with each dot's size illustrating the strength of the coupling (in this case, the dots correspond to coupling out through a square lattice with a period of 32 nm in both dimensions). The hot carrier energy $E_i = 0.4E_F$. E_F as in Fig. 2.

the spectrum at the boundary between interband and intraband transitions (Fig. 4c) follows from the charge carriers' density of states being zero at the tip of the Dirac cone.

The interband ČE in Fig. 4 shows the possibility of emission of relatively high-frequency GPs, even reaching near-infrared and visible frequencies. These are interband transitions as in Figs 2 and 3, thus limited by $\hbar\omega \leq E_i + E_F$. Such a limit can get to a few eVs, because E_i is controlled externally by the mechanism creating the hot carriers (for example, *p–n* junction, tunnelling current in a heterostructure, scanning tunneling microscope (STM) tip, ballistic transport in graphene with high drain–source voltage and photoexcitation). GPs at near-infrared frequencies have already been shown to exist through direct and indirect experimental evidence[38–41]. The only fundamental limitation is the energy at which the graphene dispersion ceases to be conical ($\sim$1 eV from the Dirac point[42]). Even then, our equations are only modified by changing the dispersion relations of the charge carrier and the GP, and therefore the graphene ČE should appear for E_i as high as (approximately) 3 eV[43]. The equations we presented are still valid, as they are written for a general dispersion relation, with $v_p(\omega)$ and $\gamma(\omega)$ as parameters; thus, the basic predictions of the equations and the ČE features we describe will continue to hold regardless of the precise plasmon dispersion. For example, a recent paper[44] suggests an alternative way of calculating GP dispersion, giving smaller GP phase velocities at high frequencies—this will lead to more efficient GP emission and another intraband regime that can occur without being mediated by the GP loss.

Opportunities for experimental observation of the ČE in graphene. There exist several possible avenues for observing the quantum ČE in GPs, having to do with schemes for exciting hot carriers. For example, apart from photoexcitation, hot carriers have been excited from tunnelling current in a heterostructure[45] and by a biased STM tip[38]. These methods can be used to generate hot carriers with narrow energy distributions but broad angular spreads. To show that many of the ČE features discussed in this study can be observed experimentally even with a broad angular distribution of hot carriers, Figs 2c and 3c plot the spectral distribution of the ČE averaged over the angular degree of freedom. Notably, the spectra contain clear signatures of the ČE (for example, the two distinct spectral peaks in Fig. 2c). This confirms that even with non-directional hot carriers plasmons with the distinctive attributes of Čerenkov radiation can be created and observed with established experimental means.

To generate hot carriers that are also directional, a possible method would be to apply a strong drain–source voltage across a graphene *p–n* junction[46] or in other graphene devices with ballistic transport. This could be achieved by using concepts of graphene electron optics[47], where recent experimental demonstrations have shown controlled and directional ballistic motion of hot carriers by specially designed gates and/or an external magnetic field[47–49]. The generation of directional hot carriers would facilitate the measurement of even more features of the ČE, such as the GP Čerenkov angle (as shown in Figs 2b, 3b and 4b). Yet, another intriguing approach for observing

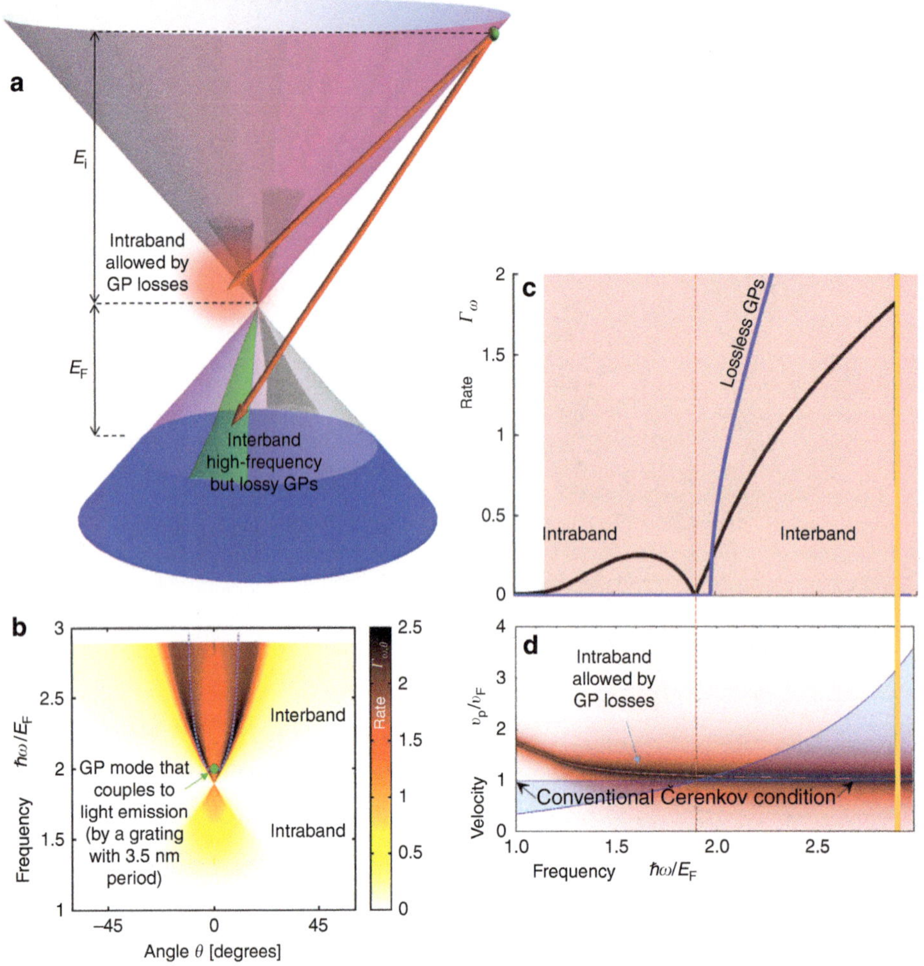

Figure 4 | GP emission from hot carriers. Caption and notations same as in Fig. 2. Unlike conventional ČE, most of the emission occurs in the forward direction with a relatively low angular spread as is shown by **b**. The green dot shows that GPs a particular frequency can be coupled out as light (we assume a grating with period of 3.5 nm). For the parameters used here, ČE emission occurs due to intraband transitions that are becoming allowed by the GP losses, whereas high-frequency (such as $\hbar\omega > 2E_F$) emission occurs due to interband transitions at areas of high GP losses. The hot carrier energy $E_i = 1.9E_F$. E_F as in Fig. 2.

additional features of the graphene ČE could be exciting the hot carriers and measuring the generated ČE with the photon-induced near-field electron microscopy (PINEM)[50,51], which might allow the visualization of the temporal dynamics of the Čerenkov emission. This approach is especially exciting, as the temporal dynamics of the ČE is expected to appear in the form of a plasmonic shockwave (as the conventional ČE appears as a shockwave of light).

If the hot carriers have a broad distribution in both energy and direction, the spectral features will be partially smoothened but still present a clear signature of the graphene ČE. Hot carriers with a wide energy distribution (instead of a single E_i), for instance, may be generated from a tunnelling current or *p*–*n* junction. The ČE spectrum corresponding to an arbitrary hot carrier excitation energy distribution is readily computed by integrating over a weighted distribution of ČE spectra for monoenergetic hot carriers. An example is presented in Fig. 5, where we assume the hot carriers have an exponential energy distribution (multiplied by the appropriate electron energy of states). This scenario can be readily created in a heterostructure[45] having a sheet of graphene isolated from another conductor (possibly graphene as well), or by biasing an STM[38] brought near the graphene surface. In both cases, the hot carriers tunnel through a potential barrier; thus, their distribution can be controlled by modifying the barrier thickness or the bias voltage. Figure 5a,b show that the spectral features are only partially smoothened and still present a clear signature of the graphene ČE. To give a rough estimate for energy conversion efficiency, it is noteworthy that the average GP energies in Fig. 5a,b are $0.93E_F$ and $1.1E_F$, respectively. Dividing by $\max(E_i) + E_F$ (upper limit on the energy invested into a hot carrier), we get 77% and 78% (respectively) energy conversion efficiency. Here we assume that not more than one plasmon is emitted from each hot carrier—the efficiency is higher when GP emission from intraband trasitions is significant, as then multiple emissions from a single hot carrier are probable (as in Fig. 4).

Future studies could directly generalize our approach to regimes of higher rates of excitation of hot carriers, by solving for the steady-state occupation probability. Here we assume the hot carriers are sparse enough so that they relax to the Fermi sea after their GP emission without creating a significant occupation that would exclude the transitions of other hot carriers. Altogether, higher excitation rates are therefore expected to cause gradual saturation of the GP emission and thus reduce the conversion efficiency. Importantly, the conversion efficiency remains high even when the carriers' energy distribution is broad, as implied by the high ČE rate of emission for the representative values of E_i studied here (Figs 2–4 and 5 all show rates on the order of $\Gamma_\omega \sim 1$). Emission rates on the order of $\Gamma_\omega \sim 1$ are two to three orders of magnitude higher than those

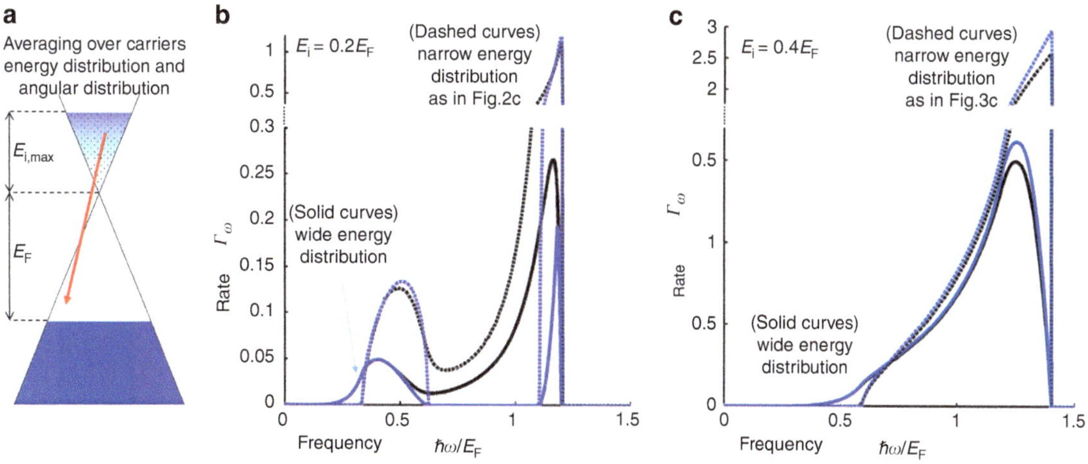

Figure 5 | The spectrum of the Graphene ČE: hot carrier excitation energy having a wide versus narrow distribution. (**a**) Illustration of the distribution of hot carriers, which is taken to be an exponential multiplied by the linear electron density of states in graphene. The exponential decay is (**b**) $e^{E_i/0.1eV}$ with maximum hot carrier energy of $E_{i,max} = 0.2$ eV corresponding to Fig. 2, or (**c**) $e^{E_i/0.2eV}$ with maximum hot carrier energy of $E_{i,max} = 0.4$ eV corresponding to Fig. 3. In both **b** and **c** we plot the exact spectrum (integrating equation (6) over the energy distribution) in solid black and the lossless approximation (integrating equation (5) over the energy distribution) in solid blue. The dashed curves are for the respective cases of narrow energy distribution (matching Figs 2c and 3c). The Fermi energy E_F is as in Figs 2-4.

found in the conventional ČE in a 3D homogeneous dielectric, $\Gamma_\omega = \alpha \frac{v}{c}(1 - v_p^2/v^2)$. For example, in a bulk silicon nitride having $\epsilon_r = 4$ ($v_p = c/\sqrt{\epsilon_r}$), we find values ranging from $\Gamma_\omega = 0.0013$ for $v = 0.6c$ to $\Gamma_\omega = 0.0055$ for $v \to c$. In graphene, the high ČE emission rate over a broad range of E_i owes itself to the low phase velocity and high confinement of GPs over a wide frequency range, which are properties unique to 2D plasmons.

Notably, the ČE emission of GPs can be coupled out as free-space photons by creating a grating or nanoribbons—fabricated in the graphene, in the substrate, or in a layer above it (see, for example, refs 39,52–54)—with two arbitrarily chosen examples marked by the green dots in Figs 3b and 4b. Careful design of the coupling mechanism can restrict the emission to pre-defined frequencies and angles, with further optimization needed for efficient coupling. This clearly indicates that the GP emission, although usually considered as merely a virtual process, can be, in fact, completely real in some regimes, with the very tangible consequences of light emission in terahertz, infrared or possibly visible frequencies. Such novel sources of light could have promising applications due to graphene's dynamic tunability and small footprint (due to the small scale of GPs). Moreover, near-perfect conversion efficiency of electrical energy into photonic energy might be achievable due to the ČE emission rate dominating all other scattering processes. Lastly, unlike plasmonic materials such as silver and gold, graphene is especially exciting in this context, as it is CMOS compatible. Still, further research is needed in the design of gratings and/or cavities to minimize losses in the GP-to-photon conversion.

Discussion

The high efficiency of GP emission through electrical excitation makes it very promising as a plasmonic source. The efficiency of coupling to the plasmons from free space, in contrast, is very low due to momentum mismatch (the GP momentum is two orders of magnitude higher than the momentum of a photon of the same frequency in free space). For regular plasmons, this mismatch is typically overcome using a grating (nanoribbons are also a viable option in graphene); however, this results in relatively low coupling efficiency. In general, the efficiency of out-coupling is similar to the efficiency of in-coupling, for example, 2% has been shown in the best cases of ref. 39. Yet, another approach to solve the momentum mismatch challenge is by using the nonlinear optical response of graphene[55], for which coupling efficiencies approaching 10^{-5} have been reported. The prospect of attaining energy conversion efficiencies on the order of 77–78%, as in the example above, thus makes the quantum ČE from electrical excitation of hot carriers an exciting candidate for GP generation.

Furthermore, electrical excitation is interesting in itself due to its direct generation of radiation, without the need to use another radiation source (as in coupling lasers from free space). The unavoidable disadvantage of this scheme lies in the broad spectral range of the generated plasmons, compared with the highly monochromatic spectrum possible when using an external laser.

The hot carrier lifetime due to GP emission is defined by the inverse of the total rate of GP emission (integrating equation (5) or equation (6)) and can therefore be exceptionally short (down to a few femtosecond). Such short lifetimes are in general agreement with previous research on the subject (for example, see refs 20,21,56) that have shown electron–electron scattering as the dominant cooling process of hot carriers, unless hot carriers of relatively high energies ($E_i \approx 2E_F$ and above) are involved. In this latter case, one expects single-particle excitations to prevail over the contribution of the plasmonic resonances[57]. This is also in agreement with the fact that plasmons with high energies and momenta (in the electron-hole continuum, pink areas in Figs 2–4) are very lossy[7,19]. Additional factors that keep the ČE from attaining near-perfect conversion efficiency include other scattering processes such as acoustic and optical phonon scattering. Owing to the relatively long lifetime from acoustic phonon scattering (hundreds of femtoseconds to several picoseconds[7]), any deterioration due to this effect is not likely to be significant. Scattering by optical phonons is more significant for hot carriers above 0.2 eV, but its contribution is still about an order of magnitude smaller in our regime of interest[20]. Further research is necessary to optimize the emission process, including considerations of multi-plasmon emission and other higher-order effects.

Interestingly, the high rates of GP emission also conform to research of the reverse process—of plasmons enhancing and controlling the emission of hot carriers—that is also found to be particularly strong in graphene[58,59]. This might reveal unexplored relations between ČE and other novel ideas of graphene-based radiation sources that are based on different physical principles[60–64].

It is also worth noting that Čerenkov-like plasmon excitations from hot carriers can be found in other condensed matter systems such as a 2D electron gas at the interface of semiconductors. Long before the discovery of graphene, such systems have demonstrated very high Fermi velocities (even higher than graphene's), while also supporting meV plasmons that can have slow phase velocities, partly due to the higher refractive indices possible in such low frequencies[65]. The ČE coupling, therefore, should not be unique to graphene. In many cases[66,67], the coupling of hot carriers to bulk plasmons is even considered as part of the self-energy of the carriers, although the plasmons are then considered as virtual particles in the process. Nonetheless, graphene offers a unique opportunity where the Čerenkov velocity matching can occur at relatively high frequencies, with plasmons that have relatively low losses. Crucially, these differences are what makes the efficiency of the graphene ČE so high. Continued research into other 2D materials (for example, 2D silver and Beryllium[7,68–70]) may lead to materials with higher frequency, lower loss and higher confinement (lower phase velocity) than GPs. The prospect of higher frequency plasmons is especially exciting, as the ČE radiation intensity increases with frequency (explaining the bluish colour of conventional ČE).

We conclude with some very intriguing yet at this stage admittedly very speculative comments. Effects associated with the highly efficient emission of interband GPs and the unexpected emission of intraband GPs predicted by our quantum ČE theory may have already manifested themselves in current graphene experiments, even in ones that do not involve any optical measurement, such as transistor-based graphene devices[22,45]. For example, such GP emission could be a contributing factor to the effect of current saturation observed in graphene devices[22], as large source–drain voltages can take graphene out of equilibrium and create hot carriers. When these hot carriers cross the energy threshold for significant GPs emission they lose energy abruptly, causing a sudden increase in resistivity. As another example, our graphene ČE might play a role in explaining the surprisingly high frequency of emitted light from graphene[17,23,24], as GPs can couple out as light emission by surface roughness, impurities and so on. This hypothesis is encouraged by reports[17] in which the measurements show characteristics typical of ČE variants, such as threshold values and power scaling behaviour that do not fit simple black body models. If our theory is indeed applicable here, then the extremely high-temperature estimates of the electron gas would need to be modified, to account for the contribution of GP emission in the high-frequency range of the observed spectrum. This would imply a lower black body radiation spectrum and thus lower graphene temperatures than otherwise expected[24]. Finally, as the GP energy can be higher than both E_i and E_F, the ČE could form part of the explanation for the observed frequency up-conversion[23], especially given that multi-plasmon effects are expected due to the high rate of the emission process. Of course, future detailed studies of the systems will be needed to verify the ČE connections proposed here.

References

1. Fang, Y. & Sun, M. Nanoplasmonic waveguides: towards applications in integrated nanophotonic circuits. *Light Sci. Appl.* **4,** 294 (2015).
2. Atwater, H. A. The promise of plasmonics. *Sci. Am.* **296,** 56–62 (2007).
3. MacDonald, K. F., Sámson, Z. L., Stockman, M. I. & Zheludev, N. I. Ultrafast active plasmonics. *Nat. Photonics* **3,** 55–58 (2009).
4. Bharadwaj, P., Bouhelier, A. & Novotny, L. Electrical excitation of surface plasmons. *Phys. Rev. Lett.* **106,** 226802 (2011).
5. Koller, D. M. *et al.* Organic plasmon-emitting diode. *Nat. Photonics* **2,** 684–687 (2008).
6. Huang, K. C. *et al.* Electrically driven subwavelength optical nanocircuits. *Nat. Photonics* **8,** 244–249 (2014).
7. Jablan, M., Soljacic, M. & Buljan, H. Plasmons in graphene: fundamental properties and potential applications. *Proc. IEEE* **101,** 1689 (2013).
8. Alonso-González, P. *et al.* Controlling graphene plasmons with resonant metal antennas and spatial conductivity patterns. *Science* **344,** 1369–1373 (2014).
9. García de Abajo, F. J. Graphene plasmonics: challenges and opportunities. *ACS Photonics* **1,** 135–152 (2014).
10. Vakil, A. & Engheta, N. Transformation optics using graphene. *Science* **332,** 1291–1294 (2011).
11. Grigorenko, A. N., Polini, M. & Novoselov, K. S. Graphene plasmonics. *Nat. Photonics* **6,** 749–758 (2012).
12. Fei, Z. *et al.* Gate-tuning of graphene plasmons revealed by infrared nano-imaging. *Nature* **487,** 82–85 (2012).
13. Chen, J. *et al.* Optical nano-imaging of gate-tunable graphene plasmons. *Nature* **487,** 77–81 (2012).
14. Apalkov, V. & Stockman, M. I. Proposed graphene nanospaser. *Light Sci. Appl.* **3,** 191 (2014).
15. Beams, R., Bharadwaj, P. & Novotny, L. Electroluminescence from graphene excited by electron tunneling. *Nanotechnology* **25,** 055206 (2014).
16. Withers, F. *et al.* Light-emitting diodes by band-structure engineering in van der Waals heterostructures. *Nat. Mater.* **14,** 301–306 (2015).
17. Kim, Y. D. *et al.* Bright visible light emission from graphene. *Nat. Nanotechnol.* **10,** 676–681 (2015).
18. Wang, L. *et al.* One-dimensional electrical contact to a two-dimensional material. *Science* **342,** 614–617 (2013).
19. Jablan, M., Buljan, H. & Soljacic, M. Plasmonics in graphene at infrared frequencies. *Phys. Rev. B* **80,** 245435 (2009).
20. Tse, W. K., Hwang, E. H. & Sarma, S. D. Ballistic hot electron transport in graphene. *Appl. Phys. Lett.* **93,** 023128 (2008).
21. Rana, F., Strait, J. H., Wang, H. & Manolatou, C. Ultrafast carrier recombination and generation rates for plasmon emission and absorption in graphene. *Phys. Rev. B* **84,** 045437 (2011).
22. Meric, I. *et al.* Current saturation in zero-bandgap, top-gated graphene field-effect transistors. *Nat. Nanotechnol.* **3,** 654–659 (2008).
23. Lui, C. H., Mak, K. F., Shan, J. & Heinz, T. F. Ultrafast photoluminescence from graphene. *Phys. Rev. Lett.* **105,** 127404 (2010).
24. Berciaud, S. *et al.* Electron and optical phonon temperatures in electrically biased graphene. *Phys. Rev. Lett.* **104,** 227401 (2010).
25. García de Abajo, F. J. Multiple excitation of confined graphene plasmons by single free electrons. *ACS Nano* **7,** 11409–11419 (2013).
26. Liu, S. *et al.* Coherent and tunable terahertz radiation from graphene surface plasmon polaritons excited by an electron beam. *Appl. Phys. Lett.* **104,** 201104 (2014).
27. Shi, X. *et al.* Caustic graphene plasmons with Kelvin angle. *Phys. Rev. B* **92,** 081404 (2015).
28. Hwang, E. H. & Das Sarma, S. Dielectric function, screening, and plasmons in two-dimensional graphene. *Phys. Rev. B* **75,** 205418 (2007).
29. Ginzburg, V. L. Quantum theory of radiation of electron uniformly moving in medium. *Zh. Eksp. Teor. Fiz.* **10,** 589–600 (1940).
30. Kaminer *et al.* Quantum Čerenkov radiation: spectral cutoffs and the role of spin and orbital angular momentum. *Phys. Rev. X* **6,** 011006 (2016).
31. Pines, D. & Schrieffer, J. R. Approach to equilibrium of electrons, plasmons, and phonons in quantum and classical plasmas. *Phys. Rev.* **125,** 804 (1962).
32. Wunsch, B., Stauber, T., Sols, F. & Guinea, F. Dynamical polarization of graphene at finite doping. *N. J. Phys.* **8,** 318 (2006).
33. Koppens, F. H., Chang, D. E. & Garcia de Abajo, F. J. Graphene plasmonics: a platform for strong light–matter interactions. *Nano Lett.* **11,** 3370–3377 (2011).
34. Ju, L. *et al.* Graphene plasmonics for tunable terahertz metamaterials. *Nat. Nanotechnol.* **6,** 630–634 (2011).
35. Harris, E. G. *A Pedestrian Approach to Quantum Field Theory* (John Willey and Sons, Inc. (1972).
36. Luo, C., Ibanescu, M., Johnson, S. G. & Joannopoulos, J. D. Cerenkov radiation in photonic crystals. *Science* **299,** 368–371 (2003).
37. Xi, S. *et al.* Experimental verification of reversed Cherenkov radiation in left-handed metamaterial. *Phys. Rev. Lett.* **103,** 194801 (2009).
38. Brar, V. W. *et al.* Observation of carrier-density-dependent many-body effects in graphene via tunneling spectroscopy. *Phys. Rev. Lett.* **104,** 036805 (2010).
39. Brar, V. W., Jang, M. S., Sherrott, M., Lopez, J. J. & Atwater, H. A. Highly confined tunable mid-infrared plasmonics in graphene nanoresonators. *Nano Lett.* **13,** 2541–2547 (2013).

40. Zhang, Q. *et al.* Graphene surface plasmons at the near-infrared optical regime. *Sci. Rep.* **4,** 6559 (2014).
41. Tielrooij *et al.* Electrical control of optical emitter relaxation pathways enabled by graphene. *Nat. Phys.* **11,** 281–287 (2015).
42. Geim, A. K. & Novoselov, K. S. The rise of graphene. *Nat. Mater.* **6,** 183–191 (2007).
43. Zhou, S. Y. *et al.* First direct observation of Dirac fermions in graphite. *Nat. Phys.* **2,** 595–599 (2006).
44. Page, A. F., Ballout, F., Hess, O. & Hamm, J. M. Nonequilibrium plasmons with gain in graphene. *Phys. Rev. B* **91,** 075404 (2015).
45. Britnell, L. *et al.* Field-effect tunneling transistor based on vertical Graphene heterostructures. *Science* **335,** 947–950 (2012).
46. Gabor, N. M. *et al.* Hot carrier–assisted intrinsic photoresponse in graphene. *Science* **334,** 648–652 (2011).
47. Rickhaus, P. *et al.* Ballistic interferences in suspended graphene. *Nat. Commun.* **4,** 2342 (2013).
48. Rickhaus, P. *et al.* Guiding of electrons in a few-mode ballistic graphene channel. *Nano Lett.* **15,** 5819–5825 (2015).
49. Rickhaus, P. *et al.* Gate tuneable beamsplitter in ballistic graphene. *Appl. Phys. Lett.* **107,** 251901 (2015).
50. Piazza, L. *et al.* Simultaneous observation of the quantization and the interference pattern of a plasmonic near-field. *Nat. Commun.* **6,** 6407 (2015).
51. Barwick, B., Flannigan, D. J. & Zewail, A. H. Photon-induced near-field electron microscopy. *Nature* **462,** 902–906 (2009).
52. Liu, G. *et al.* Epitaxial graphene nanoribbon array fabrication using BCP-assisted nanolithography. *ACS Nano* **6,** 6786–6792 (2012).
53. Rodrigo, D. *et al.* Mid-infrared plasmonic biosensing with graphene. *Science* **349,** 165–168 (2015).
54. Abbas, A. N. *et al.* Patterning, characterization, and chemical sensing applications of graphene nanoribbon arrays down to 5 nm using helium ion beam lithography. *ACS Nano* **8,** 1538–1546 (2014).
55. Constant, T. J., Hornett, S. M., Chang, D. E. & Hendry, E. All-optical generation of surface plasmons in graphene. *Nat. Phys.* **12,** 124 (2016).
56. Song, J. C. & Levitov, L. S. Energy flows in graphene: hot carrier dynamics and cooling. *J. Phys. Condensed Matter* **27,** 164201 (2015).
57. Hwang, E. H., Hu, B. Y.-K. & Sarma, S. D. Inelastic carrier lifetime in graphene. *Phys. Rev. B* **76,** 115434 (2007).
58. Sundararaman, R., Narang, P., Jermyn, A. S., Goddard, III W. A. & Atwater, H. A. Theoretical predictions for hot-carrier generation from surface plasmon decay. *Nat. Commun.* **5,** 5788 (2014).
59. Brongersma, M. L., Halas, N. J. & Nordlander, P. Plasmon-induced hot carrier science and technology. *Nat. Nanotechnol.* **10,** 25–34 (2015).
60. Rana, F. Graphene terahertz plasmon oscillators. *IEEE Transact. Nanotechnol.* **7,** 91–99 (2008).
61. Mikhailov, S. A. Graphene-based voltage-tunable coherent terahertz emitter. *Phys. Rev. B* **87,** 115405 (2013).
62. Gumbs, G., Iurov, A., Huang, D. & Pan, W. Tunable surface plasmon instability leading to emission of radiation. *J. Appl. Phys.* **118,** 054303 (2015).
63. Ooi, K. J., Chu, H. S., Hsieh, C. Y., Tan, D. T. & Ang, L. K. Highly efficient midinfrared on-chip electrical generation of graphene plasmons by inelastic electron tunneling excitation. *Phys. Rev. Appl.* **3,** 054001 (2015).
64. Wong, L. J., Kaminer, I., Ilic, O., Joannopoulos, J. D. & Soljačić, M. Towards graphene plasmon-based free-electron infrared to X-ray sources. *Nat. Photonics* **10,** 46 (2016).
65. Stern, F. Polarizability of a two-dimensional electron gas. *Phys. Rev. Lett.* **18,** 546 (1967).
66. Chaplik, A. V. Energy spectrum and electron scattering processes in inversion layers. *Sov. Phys. JETP* **33,** 997–1000 (1971).
67. Müller, F. *et al.* Electron-electron interaction in ballistic electron beams. *Phys. Rev. B* **51,** 5099 (1995).
68. Diaconescu, B. *et al.* Low-energy acoustic plasmons at metal surfaces. *Nature* **448,** 57–59 (2007).
69. Nagao, T. in *Proceedings of the SPIE, Ultrafast Phenomena in Semiconductors and Nanostructure Materials XIV* **7600,** 1Q (2010).
70. Rugeramigabo, E. P., Nagao, T. & Pfnür, H. Experimental investigation of two-dimensional plasmons in a DySi 2 monolayer on Si (111). *Phys. Rev. B* **78,** 155402 (2008).

Acknowledgements

We thank Professor L. Levitov and Professor Y.D. Chong for helpful discussions, and Dr Marinko Jablan for useful advice. The work was supported by the US Army Research Laboratory and the US Army Research Office through the Institute for Soldier Nanotechnologies (contract number W911NF-13-D-0001). H.B. acknowledges support from the QuantiXLie and G2D Center of Excellence. The research of L.J.W. was supported by the Science and Engineering Research Council (SERC; grant number 1426500054) of the Agency for Science, Technology and Research (A*STAR), Singapore. J.J.L. was supported in part by the MRSEC Program of the National Science Foundation under award number DMR—1419807. The research of I.K. was partially supported by the Seventh Framework Programme of the European Research Council (FP7–Marie Curie IOF) under grant agreement number 328853—MC–BSiCS.

Author contributions

All authors discussed the results and made critical contributions to the work.

Additional information

Competing financial interests: The authors declare no competing financial interests.

29

Mode engineering for realistic quantum-enhanced interferometry

Michał Jachura[1], Radosław Chrapkiewicz[1], Rafał Demkowicz-Dobrzański[1], Wojciech Wasilewski[1] & Konrad Banaszek[1]

Quantum metrology overcomes standard precision limits by exploiting collective quantum superpositions of physical systems used for sensing, with the prominent example of non-classical multiphoton states improving interferometric techniques. Practical quantum-enhanced interferometry is, however, vulnerable to imperfections such as partial distinguishability of interfering photons. Here we introduce a method where appropriate design of the modal structure of input photons can alleviate deleterious effects caused by another, experimentally inaccessible degree of freedom. This result is accompanied by a laboratory demonstration that a suitable choice of spatial modes combined with position-resolved coincidence detection restores entanglement-enhanced precision in the full operating range of a realistic two-photon Mach-Zehnder interferometer, specifically around a point which otherwise does not even attain the shot-noise limit due to the presence of residual distinguishing information in the spectral degree of freedom. Our method highlights the potential of engineering multimode physical systems in metrologic applications.

[1] Faculty of Physics, University of Warsaw, Pasteura 5, 02-093 Warsaw, Poland. Correspondence and requests for materials should be addressed to R.C. (email: radekch@fuw.edu.pl).

Quantum phenomena can facilitate and boost the performance of imaging techniques[1–4], sensitive measurements in delicate materials[5,6], as well as detection schemes probing subtle physical effects such as gravitational waves[7]. These strategies rely on preparing collective superposition states of multiple probes (for example, photons, atoms) to achieve precision enhancement beyond standard limits[8–17]. In the optical domain, a common strategy for collective state preparation is to realize multiphoton interference in linear circuits, for example, free space or integrated interferometers[18–21], fed with non-classical states of light. Attainable precision can be, however, markedly vulnerable to residual distinguishing information between interfering photons. Standard methods to improve indistinguishability based on filtering are often inadequate, in particular, introducing attenuation that may easily diminish the overall benefit of collective state preparation.

The purpose of this paper is to analyse the interplay between degrees of freedom with different experimental accessibility in two-photon interferometry, which is a canonical example of a quantum-enhanced measurement[22]. We demonstrate that detrimental effects caused by distinguishing information present in one degree of freedom that is beyond experimental control or lacks technical means to improve indistinguishability, can be alleviated by mode engineering in another degree of freedom, even though these two remain completely uncorrelated. This feature is investigated in the case of local phase estimation, whose precision becomes strongly dependent on the operating point if the two photons feeding the interferometer exhibit residual distinguishability. It is shown that a carefully designed preparation and detection scheme for a degree of freedom other than the one causing distinguishability allows one to restore quantum-enhanced precision in the entire operating range of the interferometer. We attribute this effect to non-trivial combination of one- and two-photon interference that turns out to augment phase sensitivity beyond the shot-noise limit. We also present an example indicating that similar enhancement occurs also at higher photon numbers. Prospectively, the results reported here may provide another class of strategies to mitigate effects of imperfections and environmental noise in quantum-enhanced metrology[23–29]. The theoretical analysis is complemented with an experiment investigating sensitivity of a balanced Mach–Zehnder interferometer fed with photon pairs. We determine the precision of local phase estimation around the operating point when the photons coalesce pairwise at the interferometer output ports. In this regime, the residual spectral distinguishability within pairs has a markedly deleterious effect on the attainable precision. Building on recent advances in spatially resolved single-photon detection[1,30–39], we demonstrate that by controlling the input spatial structure of interfering photons and extracting complete spatial information at the detection stage it is nevertheless possible to recover the sub-shot-noise precision. This confirms in proof-of-principle settings the feasibility of mode engineering techniques for quantum-enhanced interferometry.

Results

Realistic two-photon interferometer. A generic two-photon Mach–Zehnder interferometer constructed with a pair of balanced 50/50 beam splitters and fed with photon pairs is shown schematically in Fig. 1a. The phase shift θ between the interferometer arms modulates probabilities of detection events at the output ports that can be grouped into two types: either the photons exit through different paths, producing a coincidence event between the detectors monitoring the ports, or both are found in the same output port leading to a double event. If the two photons are indistinguishable at the input, the first beam splitter generates a coherent superposition of both the photons in one or another arm of the interferometer, which is the simplest case of a N00N state providing sensitivity that approaches the Heisenberg limit[10–12].

In the multimode description of the setup one introduces two sets of annihilation operators for upper path modes $\hat{a}_\mu$ and lower path modes $\hat{b}_\mu$ that are matched pairwise. Individual modes in each arm are mutually orthogonal, that is, $[\hat{a}_\mu, \hat{a}_\nu^\dagger] = [\hat{b}_\mu, \hat{b}_\nu^\dagger] = \delta_{\mu\nu}$. The unitary map $\hat{U}(\theta)$ implemented by the interferometer between the input and the output ports transforms pairs of field operators labelled with the same index μ as

$$\hat{U}^\dagger(\theta)\begin{pmatrix}\hat{a}_\mu\\ \hat{b}_\mu\end{pmatrix}\hat{U}(\theta) = \begin{pmatrix}\cos\frac{\theta}{2} & \sin\frac{\theta}{2}\\ \sin\frac{\theta}{2} & -\cos\frac{\theta}{2}\end{pmatrix}\begin{pmatrix}\hat{a}_\mu\\ \hat{b}_\mu\end{pmatrix}. \quad (1)$$

Partial distinguishability of the interfering photons can be modelled by assuming that at the input the photon in the upper path occupies a certain mode $\hat{a}_1$, while the lower path photon is prepared in a combination of a matching mode $\hat{b}_1$ and another orthogonal mode $\hat{b}_2$ with relative weights $\mathcal{V}$ and $1-\mathcal{V}$, where $\mathcal{V}$ is the visibility parameter specifying the fraction of indistinguishable pairs. To keep the notation concise, we will write the complete two-photon state as pure

$$|\psi\rangle = \hat{a}_1^\dagger\left(\sqrt{\mathcal{V}}\hat{b}_1^\dagger + \sqrt{1-\mathcal{V}}\hat{b}_2^\dagger\right)|\text{vac}\rangle, \quad (2)$$

and trace the final formulas over the index $\mu = 1, 2$ for the initial modes $\hat{b}_\mu$. This is equivalent to taking from the start a reduced density matrix $\mathcal{V}\hat{a}_1^\dagger\hat{b}_1^\dagger|\text{vac}\rangle\langle\text{vac}|\hat{a}_1\hat{b}_1 + (1-\mathcal{V})\hat{a}_1^\dagger\hat{b}_2^\dagger|\text{vac}\rangle\langle\text{vac}|\hat{a}_1\hat{b}_2$. Here $|\text{vac}\rangle$ is the vacuum state of the entire multimode electromagnetic field satisfying $\hat{a}_\mu|\text{vac}\rangle = \hat{b}_\mu|\text{vac}\rangle = 0$ for any μ. Note that the above model includes the case of partly overlapping wavepackets constructed from a continuum of modes, wherein $\hat{b}_1$ and $\hat{b}_2$ can be identified through the standard algebraic technique of Gram–Schmidt orthogonalization.

The general transformation from equation (1) taken with $\mu = 1, 2$ implies the following expression for the probability of a coincidence event

$$p_c(\theta) = 1 - \frac{1}{2}(1+\mathcal{V})\sin^2\theta, \quad (3)$$

while the probability of a double event is $p_d(\theta) = 1 - p_c(\theta)$. These formulas combine expressions for fully indistinguishable and completely distinguishable pairs with respective probabilities $\mathcal{V}$ and $1-\mathcal{V}$. The resulting fringes are depicted in Fig. 1b for $\mathcal{V} = 1$ and 0.93.

As a consequence of the Cramér-Rao bound[40], the minimum uncertainty of any unbiased phase estimate obtained from a measurement using N photon pairs around an operating point θ is given by

$$\Delta^{\text{pair}} = \frac{1}{\sqrt{NF^{\text{pair}}(\theta)}}, \quad (4)$$

where for standard photon counting at output ports of the interferometer the Fisher information $F^{\text{pair}}(\theta)$ is given by a sum of two terms contributed by coincidence and double events[17],

$$F^{\text{pair}}(\theta) = \frac{1}{p_c(\theta)}\left(\frac{dp_c}{d\theta}\right)^2 + \frac{1}{p_d(\theta)}\left(\frac{dp_d}{d\theta}\right)^2. \quad (5)$$

As a reference, we will take the uncertainty of ideal, shot-noise-limited phase measurement $\Delta^{\text{shot}} = 1/\sqrt{2N}$, when $2N$ photons are sent individually to the interferometer. Our figure of merit will be the ratio $\varepsilon = \Delta^{\text{pair}}/\Delta^{\text{shot}}$ of these two uncertainties, with $\varepsilon < 1$ implying that sub-shot-noise precision has been achieved.

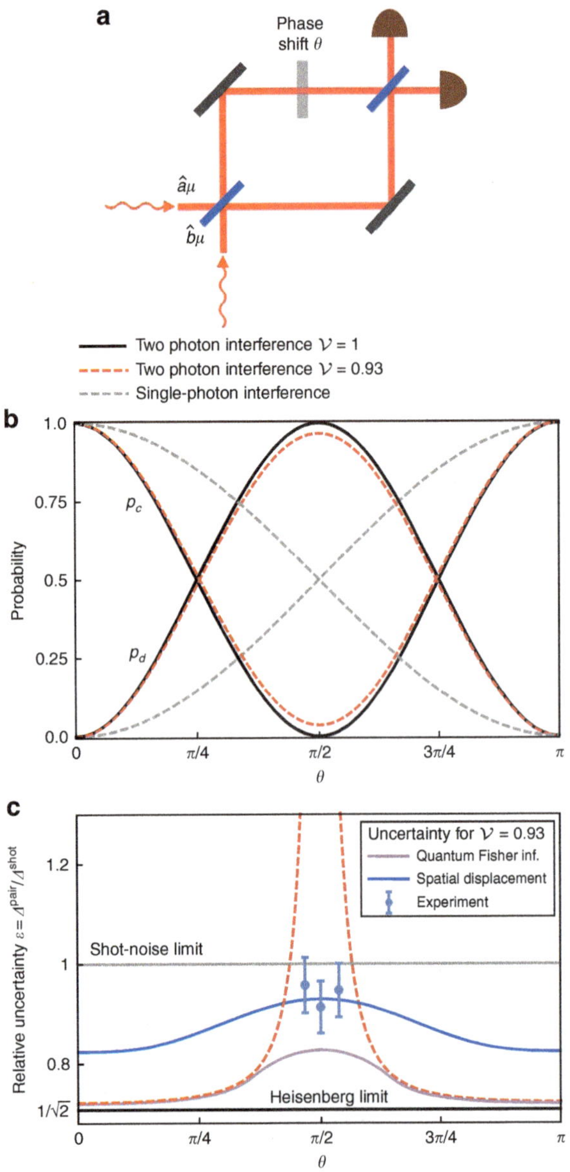

Figure 1 | Two-photon interferometer. (**a**) In the multimode description, two paths are described by families of annihilation operators $\hat a_\mu$ and $\hat b_\mu$ subjected pairwise to a unitary map dependent on the phase shift θ. (**b**) Probabilities of coincidence p_c and double count events p_d at the interferometer output are noticeably affected around $\theta \approx \pi/2$ by imperfect indistinguishability (red dashed lines for the visibility $\mathcal{V}$=0.93) when compared with the ideal case (black solid lines). Grey dashed lines depict standard single-photon interference fringes, when only one input port is illuminated. (**c**) Residual distinguishability has a dramatic effect on the phase estimation uncertainty, shown with correspondingly coded lines, which diverges at $\theta = \pi/2$ for non-unit visibility $\mathcal{V}$. Engineering overlap in an additional degree of freedom of the interfering photons and implementing optimal measurement allows one to restore the sub-shot-noise precision over the entire operating range, shown with the purple solid line when the fraction of overlapping pairs is optimized individually for each operating point. The solid grey line depicts the shot noise limit of relative uncertainty. The blue solid line depicts an explicit strategy based on introducing a fixed transverse displacement between interferometer inputs and performing spatially resolved detection. These predictions are confirmed by the experimentally determined estimation precision, depicted as solid circles with two s.d.s. error bars.

When the interferometer is fed with pairs of perfectly indistinguishable photons characterized by $\mathcal{V}$= 1, we have $\varepsilon = 1/\sqrt{2}$ independently of the operating point of the interferometer. It can be verified that around $\theta = 0$ (equivalent to $\theta = \pi$) and $\theta = \pi/2$ the main contribution to Fisher information defined in equation (5) comes, respectively, from double or coincidence events that occur with vanishing probabilities when approaching these phase values. This is because in the ideal scenario even a small number of rare events provides a sound basis to infer the phase shift. Such a regime corresponds in standard interferometry to dark-fringe operation used, for example, in gravitational wave detectors[7,41].

As seen in Fig. 1c, the precision of phase estimation is affected markedly by the non-ideal indistinguishability of photon pairs. In particular, statistical noise generated by non-vanishing background of coincidence events effectively suppresses information about the phase shift that could be retrieved around $\theta = \pi/2$. We will refer to this operating point as the coincidence dark fringe. An analogous effect would be observed also at $\theta = 0$ if any mechanism generating spurious double events was incorporated into calculations.

Restoring quantum enhancement. The analysis presented above assumed that we have no access to the degree of freedom introducing partial distinguishability. For concreteness, we will consider this degree of freedom to be the spectral one, which means that all measurements performed on the photons are integrated in the frequency domain. Suppose now that we can fully control and measure another, uncorrelated degree of freedom of the photon pairs sent to the interferometer. For the clarity of the argument, it will be convenient to use in this role the transverse spatial characteristics of the photons. Let us consider a scenario when in addition to spectral distinguishability characterized by $\mathcal{V}$ we reduce the spatial overlap of the photons by preparing them in nonorthogonal spatial modes. As a result, even in the regime of perfect spectral indistinguishability only a fraction $\mathcal{D}$ of photon pairs would effectively overlap in space. To account for this scenario we will take the mode index to have two components $\mu = i\chi$, where $i = 1, 2$ refers to the spectral degree of freedom, while $\chi = R, L$ denotes two mutually orthogonal spatial modes. Using this notation, the input state $|\psi_{\mathcal{D}}\rangle$ is described by an expression analogous to equation (2) with the following substitution of creation operators:

$$\begin{aligned} \hat a_1^\dagger &\to \hat a_{1R}^\dagger \\ \hat b_i^\dagger &\to \sqrt{\mathcal{D}}\hat b_{iR}^\dagger + \sqrt{1-\mathcal{D}}\hat b_{iL}^\dagger, \quad i = 1, 2 \end{aligned} \tag{6}$$

and it explicitly reads

$$\begin{aligned} |\psi_{\mathcal{D}}\rangle = \hat a_{1R}^\dagger \Big[&\sqrt{\mathcal{V}}\Big(\sqrt{\mathcal{D}}\hat b_{1R}^\dagger + \sqrt{1-\mathcal{D}}\hat b_{1L}^\dagger\Big) \\ &+ \sqrt{1-\mathcal{V}}\Big(\sqrt{\mathcal{D}}\hat b_{2R}^\dagger + \sqrt{1-\mathcal{D}}\hat b_{2L}^\dagger\Big)\Big]|\mathrm{vac}\rangle \end{aligned} \tag{7}$$

Note that both the spectral components $\hat b_1$ and $\hat b_2$ have been subjected to the same spatial transformation. This is in accordance with our assumption that photon manipulations cannot depend on the inaccessible spectral degree of freedom.

Although the spectral and the spatial degrees of freedom are treated on equal footing in equation (7), the crucial difference is the ability to measure the latter at the interferometer output. Therefore, the phase shift θ can be read out from the paths taken by the photons as well as their transverse spatial properties. To find the optimal strategy, we will resort to the concept of quantum Fisher information $F_Q(\theta)$ (refs 42,43), which defines through an expression analogous to equation (4), the minimum uncertainty of a phase estimate inferred from the entire available characteristics of the physical system used for sensing. When the spatial degree of freedom for coincidence events is taken into account, the explicit expression for quantum Fisher information at the coincidence dark fringe $\theta = \pi/2$, where the effects of spectral distinguishability are most severe, reads

$$F_Q(\pi/2) = 2\frac{1-\mathcal{D}^2}{1-\mathcal{D}\mathcal{V}}. \qquad (8)$$

Detailed derivation of this result is presented in the Methods section. For a given $\mathcal{V}$, the maximum value of the above expression is obtained for $\mathcal{D}^{\text{opt}} = \left(1-\sqrt{1-\mathcal{V}^2}\right)/\mathcal{V}$ and reads $F_Q^{\text{opt}}(\pi/2) = 4\left(1-\sqrt{1-\mathcal{V}^2}\right)/\mathcal{V}^2$. Remarkably, this value gives sub-shot-noise precision for any $\mathcal{V} > 0$. In Fig. 2, we compare the precision implied by numerically computed $F_Q(\theta)$ in the range $0 \leq \theta \leq \pi$ with $\mathcal{D}$ optimized for an individual operating point to a scenario when no mode engineering has been attempted. A cross-section of these plots for $\mathcal{V}$=0.93 has also been shown in Fig. 1c. It is seen that the singularity in precision around the coincidence dark fringe is removed and that the sub-shot-noise operation is ensured across the entire range of θ. Note that this result is achieved without any post-selection of two-photon detection events and no filtering or any other manipulation in the spectral domain has been applied.

The standard strategy to improve spectral indistinguishability in two-photon experiments is to restrict the bandwidth of detected photons using narrowband interference filters. However, for any finite filtering bandwidth the visibility parameter would remain below one and consequently the singularity in the uncertainty of phase estimation around $\theta = \pi/2$ would not be removed. Furthermore, filtering reduces the number of detection events that can be used for estimation. Starting from a simple model for partial distinguishability based on gaussian spectral profiles, we verified that if the number of input photon pairs is taken as a benchmark to calculate the relative uncertainty, spectral filtering does not improve the measurement precision at all.

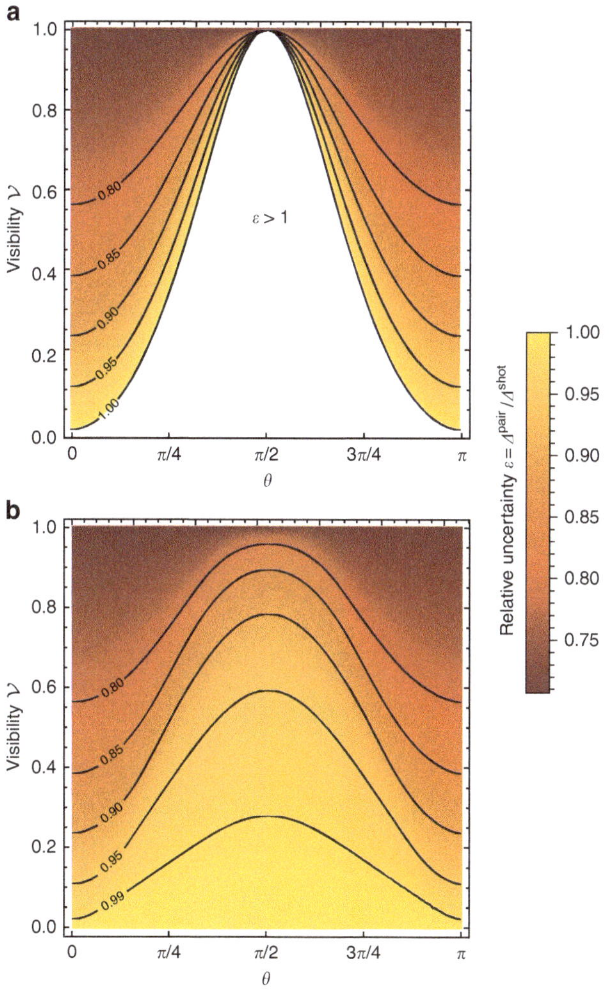

Figure 2 | Enhancement of estimation precision. Relative enhancement $\varepsilon = \Delta^{\text{pair}}/\Delta^{\text{shot}}$ for (**a**) full spatial overlap $\mathcal{D}$= 1 of the two input photons and (**b**) the overlap parameter $\mathcal{D}$ optimized individually for each given spectral visibility $0 \leq \mathcal{V} \leq 1$ and an operating point $0 \leq \theta \leq \pi$. The white area in **a** depicts the region $\varepsilon > 1$ where sub-shot-noise sensitivity is lost. It is seen that spatial mode engineering allows one to restore quantum enhancement across the entire parameter range. The uncertainty of the two-photon scheme Δ^{pair} is given by quantum Fisher information derived in equation (26).

Measurement scheme. To elucidate the origin of the above effect, it is instructive to analyse the operation of the interferometer in the *R*/*L* basis of transverse spatial modes. Detecting both the photons in *RR* modes means that they overlapped spatially at the input and therefore underwent imperfect two-photon interference affected by non-unit spectral visibility $\mathcal{V}$ with fringes shown in Fig. 1b. On the other hand, combinations *RL* and *LR* at the output imply that the upper path photon was initially in the mode *R* and the lower path photon in the mode *L*. Consequently, both the photons propagated through the interferometer as independent particles exhibiting single-photon interference, which for $\theta = \pi/2$ gives the steepest slope of interference fringes as seen in Fig. 1b. If the lower path photon was prepared in a statistical mixture of *R* and *L* modes, only single-photon interference would provide information about the phase shift at $\theta = \pi/2$ and the shot-noise limit could not be surpassed. However, because the lower path photon is sent into the interferometer in a superposition of the modes *R* and *L*, information from two- and one-photon interference can be combined in a coherent way through a suitable choice of the measurement basis at the interferometer output. Strikingly, although neither one- nor two-photon interference used separately beats the shot-noise limit itself, their coherent combination restores quantum enhancement of the measurement.

As derived in the Methods section, the explicit form of the optimal measurement attaining quantum Fisher information at the coincidence dark fringe requires discrimination between double events and two types of coincidence events corresponding to the following projections in the spatial degree of freedom:

$$|\pm\rangle = \sqrt{\frac{\mathcal{D}}{1+\mathcal{D}}}|RR\rangle + \frac{1}{2}\left(\sqrt{\frac{1-\mathcal{D}}{1+\mathcal{D}}} \pm 1\right)|RL\rangle + \frac{1}{2}\left(\sqrt{\frac{1-\mathcal{D}}{1+\mathcal{D}}} \mp 1\right)|LR\rangle \qquad (9)$$

In Fig. 3, we depict the resulting interference fringes for all three types of events when $\mathcal{D}$ is optimized for $\mathcal{V}$=93% at the operating point $\theta = \pi/2$. It is seen that coincidence events resolved in the $\pm$ basis indeed exhibit both one- and two-photon interference providing a rather steep slope at $\theta = \pi/2$, while their overall probability remains relatively low at this operating point. Combination of these features yields sub-shot-noise sensitivity.

In the limit $\mathcal{V} \to 0$ the optimal $\mathcal{D}$ approaches zero, which means that at the input the two photons are nearly fully

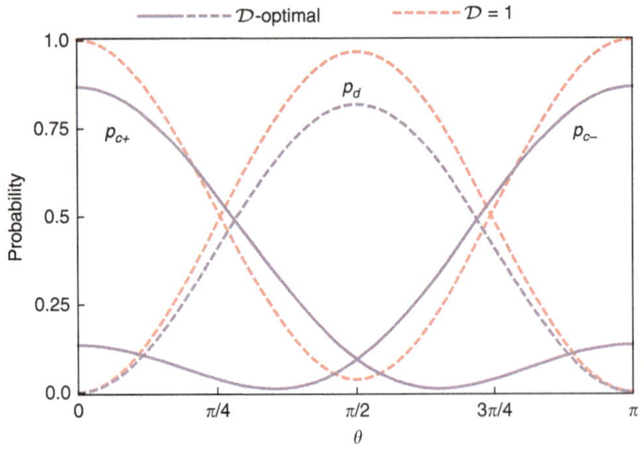

Figure 3 | Optimized interference fringes. The probabilities of double p_d (purple dashed line) and two types of coincidence $p_{c\pm}$ (purple solid lines) events corresponding to projections in the optimal basis $|\pm\rangle$ defined in equation (9) with the overlap parameter $\mathcal{D}$ optimizing quantum Fisher information at $\theta=\pi/2$ for the visibility $\mathcal{V}$=0.93 in the inaccessible degree of freedom. Dashed red lines show standard two-photon interference fringes for the same visibility without engineering an additional degree of freedom.

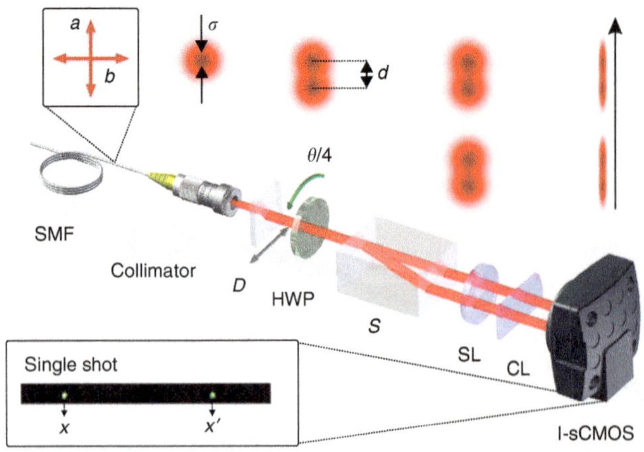

Figure 4 | Experimental setup. Interference between two orthogonally polarized photons a and b delivered by the single-mode fibre (SMF) is realized in the common path configuration with the phase shift θ corresponding to the quadrupled rotation angle of the half-wave plate (HWP). The preceding calcite displacer D introduces transverse displacement between photon modes, which combined with spatially resolved detection restores the sub-shot-noise precision of phase estimation around the coincidence dark fringe. The output ports of the calcite beam separator S are mapped using a spherical lens (SL) and a cylindrical lens (CL) onto the intensified sCMOS camera detecting individual coincidence events with spatial resolution as shown in the inset. The upper right part of the figure depicts the spatial profiles of interferometer modes at the consecutive stages of the setup.

distinguishable in their spatial degree of freedom. In this regime the asymptotic expressions for the optimal measurement basis are $|+\rangle=|RL\rangle$ and $|-\rangle=|LR\rangle$, that is, we need to identify the origin of a photon that has appeared at a given output port. This is equivalent to realizing single-photon interference twice, each time sending a photon into a different input port. The above analysis explains why in the limit of zero visibility $\mathcal{V}=0$ we recover the shot-noise limit as evidenced by Fig. 2b. The described scheme is able to exploit non-classical two-photon interference for any $\mathcal{V}>0$ to achieve sub-shot-noise operation, although unsurprisingly with a diminishing quantum enhancement when $\mathcal{V}\rightarrow 0$.

Spatially resolved detection. The measurement maximizing quantum Fisher information requires implementation of rather exotic projections on two-photon superposition states given in equation (9). This leads to the question about practical realization of the measurement achieving sub-shot-noise precision at a given operating point. Fortunately, as we will now demonstrate, for generic spatial modes sub-shot-noise precision can be restored just by measuring transverse positions of photons emerging from the interferometer. To discuss quantitatively this idea it will be convenient to resort to the paraxial approximation and to introduce two sets of operators $\hat{a}_i(x)$ and $\hat{b}_i(x)$ labelled with a continuous one-dimensional transverse position x. As before, the index $i=1, 2$ labels spectral modes. Suppose now that the photons entering the interferometer along the upper and the lower path are prepared in spatial modes described by respective normalized profiles $u(x)$ and $v(x)$,

$$\begin{aligned} \hat{a}_{iR} &= \int \mathrm{d}x\, u(x)\hat{a}_i(x) \\ \sqrt{\mathcal{D}}\hat{b}_{iR}^{\dagger}+\sqrt{1-\mathcal{D}}\hat{b}_{iL}^{\dagger} &= \int \mathrm{d}x\, v(x)\hat{b}_i(x). \end{aligned} \qquad (10)$$

The overlap parameter now reads $\mathcal{D}=\left|\int \mathrm{d}x\, u^{*}(x)v(x)\right|^2$. The probability of detecting photons in two different output ports at positions x and x' is given by

$$p_c(x,x'|\theta)=\sum_{i,j=1,2}\left|\langle \mathrm{vac}|\hat{a}_i(x)\hat{b}_j(x')\hat{U}(\theta)|\psi_{\mathcal{D}}\rangle\right|^2 \qquad (11)$$

where the summation over $i=1, 2$ stems from tracing over the spectral degree of freedom.

As a concrete example we will take two gaussian modes of width σ displaced by d,

$$u(x)=\frac{1}{\sqrt[4]{2\pi\sigma^2}}e^{-(x+d/2)^2/4\sigma^2}, \quad v(x)=u(x-d) \qquad (12)$$

that can be readily prepared by simple experimental means. In the phase estimation procedure, we will exploit information contained in the relative position $\xi=x-x'$ of photons detected in coincidence events. The probability distribution for this variable is given explicitly by

$$\begin{aligned} p_c(\xi|\theta) &= \int \mathrm{d}x'\, p_c(x'+\xi,x'|\theta) \\ &= \frac{1}{2\sigma\sqrt{\pi}}\left(\cos^4\frac{\theta}{2}e^{-(\xi-d)^2/4\sigma^2}\right. \\ &\left. +\sin^4\frac{\theta}{2}e^{-(\xi+d)^2/4\sigma^2}-\frac{1}{2}\mathcal{V}\sin^2\theta e^{-(\xi^2+d^2)/4\sigma^2}\right). \end{aligned} \qquad (13)$$

When the relative position of the two photons in coincidence events is available, Fisher information defined in equation (5) becomes enhanced by replacing the first term in the sum with the following integral over ξ:

$$\frac{1}{p_c(\theta)}\left(\frac{\mathrm{d}p_c}{\mathrm{d}\theta}\right)^2\rightarrow F_c(\theta)=\int \mathrm{d}\xi\frac{1}{p_c(\xi|\theta)}\left(\frac{\partial p_c(\xi|\theta)}{\partial\theta}\right)^2. \qquad (14)$$

In Fig. 1c, we depict the estimation precision for the ratio $d/\sigma=1.64$, which was used in the experiment described below. It is seen that the precision is brought to the sub-shot-noise regime over the entire operating range.

Experiment. To verify experimentally sub-shot-noise phase sensitivity of the interferometric scheme described above, we constructed an optical setup shown in Fig. 4. The interferometer is fed with 800 nm photon pairs generated via type-II spontaneous parametric down-conversion process, which are synchronized using a delayed line, spatially filtered through a single-mode fibre and delivered to the setup in two mutually orthogonal linear polarizations corresponding to the two input ports of the Mach–Zehnder interferometer. The photons emerging from the fibre were partly separated in space by inserting a 1.9-mm-long calcite displacer D, which results in the displacement of $d = 200\,\mu\text{m}$ between the two ortogonally polarized output paths. The spatial modes can be modelled by gaussian functions defined in equation (12) with $\sigma = 122\,\mu\text{m}$. To ensure temporal stability, the interferometer transformation is implemented in the common path configuration as a half-wave plate (HWP), with the rotation angle equal to quadruple the phase shift θ between the interferometer arms, followed by a calcite beam separator S. The equivalence of this setup with the standard Mach–Zehnder interferometer is evidenced by decomposing the interferometer transformation introduced in equation (1) as

$$\begin{pmatrix} \cos\frac{\theta}{2} & \sin\frac{\theta}{2} \\ \sin\frac{\theta}{2} & -\cos\frac{\theta}{2} \end{pmatrix} = \begin{pmatrix} \cos\frac{\theta}{4} & -\sin\frac{\theta}{4} \\ \sin\frac{\theta}{4} & \cos\frac{\theta}{4} \end{pmatrix} \begin{pmatrix} 1 & 0 \\ 0 & -1 \end{pmatrix} \begin{pmatrix} \cos\frac{\theta}{4} & \sin\frac{\theta}{4} \\ -\sin\frac{\theta}{4} & \cos\frac{\theta}{4} \end{pmatrix}. \tag{15}$$

In the common path configuration the diagonal matrix with entries 1 and -1 describes the HWP in the coordinate system of its principal axes and the two outer matrices correspond to rotation to the laboratory reference frame by an angle $\theta/4$.

We determined through an independent measurement (see Methods section) the residual spectral distinguishability of the photons to yield $\mathcal{V}$=93%. Note that this value of visibility has been used in the example presented in Fig. 1b. The transverse distance between the two spatially separated modes corresponding to the output ports of the interferometer is 3.2 mm. The rear surface of the separator S is imaged by means of a spherical lens onto a single-photon-sensitive camera system (see Methods section) capable of spatially resolved detection of coincidence events[38,39]. The profiles of spatial modes propagating in this case through consecutive stages of the setup are shown in the upper right part of Fig. 4.

We recorded spatially resolved two-photon detection events for three values of the phase shift around the dark fringe, $\theta = \pi/2 - 0.1$, $\pi/2$, $\pi/2 + 0.1$, registering $\sim 6 \times 10^3$ events in each case. Experimentally observed spatial distributions $p_c(x, x'|\theta)$ of coincidence events along with their marginals $p_c(\xi|\theta)$ for $\xi = x - x'$ are presented in Fig. 5. It is seen that the joint position distributions are clearly sensitive to the phase shift θ, in particular, the sign of its deviation from $\pi/2$ can be unambiguously inferred from the asymmetry of the distribution with respect to the diagonal.

To quantify information about the phase shift present in spatial distributions, we performed phase estimation from the actual experimental data and determined the estimation precision. In the preliminary step, we verified the applicability of equation (13) as a statistical model for the collected data assuming independently measured mode parameters σ and d. We used the maximum-likelihood method to fit $\mathcal{V}$ and θ, obtaining for the three cases depicted in Fig. 5b respective phase values $\theta = 1.47(2)$, 1.57(2), 1.69(2) and $\mathcal{V} = 0.93$, which is in agreement with the visibility inferred from the independently measured Hong-Ou-Mandel dip.

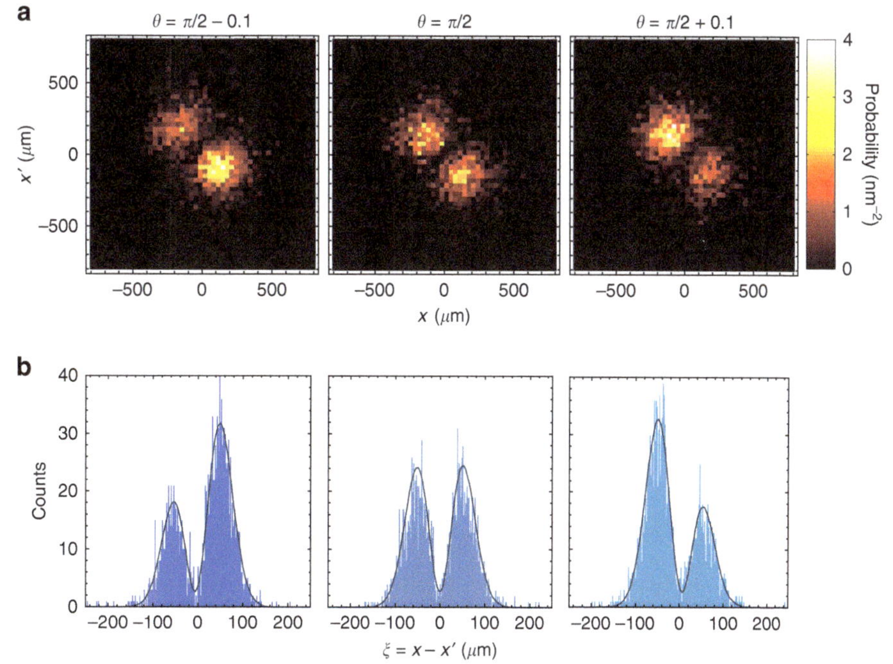

Figure 5 | Spatial distributions of coincidence events. (**a**) Experimentally measured joint distributions $p_c(x, x'|\theta)$ of detecting two photons at positions x and x' in two different output ports of the interferometer for photon pairs characterized by spectral indistinguishability $\mathcal{V}$=0.93 and three phase shifts $\theta = \pi/2 - 0.1$, $\pi/2$, $\pi/2 + 0.1$. (**b**) Useful information about the phase shift can be extracted from the marginal histograms of the relative position between the photons $\xi = x - x'$, shown along with the fitted theoretical model $p_c(\xi|\theta)$ used in the estimation procedure (solid lines).

To determine the actual estimation precision, we divided data obtained for a given HWP setting into ~600 subsets of 10 two-photon detection events and estimated the value of the phase shift separately from each subset. The width of the resulting distribution of individual estimates can be used as a figure for the estimation precision. The choice of the right estimation procedure needs some attention for small sizes of data sets. The maximum-likelihood estimator is known to be asymptotically efficient[40], that is, it saturates the Cramér-Rao bound in the asymptotic limit of infinitely many independent data samples. However, its application to small data sets is not justified owing to potential biasedness. Therefore, we used an estimator that is manifestly unbiased for any data size and yields the precision given by equation (4) at least in the vicinity of a given operation point. Specifically, for an experimentally measured statistical frequency distribution $f(\xi)$ of the relative distance between the two photons, this estimator in the vicinity of $\theta_0 = \pi/2$ is explicitly given by[17]

$$\tilde{\theta}[f] = \theta_0 + \frac{1}{F_c(\theta_0)} \int d\xi \frac{f(\xi)}{p_c(\xi|\theta_0)} \frac{dp_c(\xi|\theta_0)}{d\theta}\bigg|_{\theta=\theta_0}, \quad (16)$$

where $F_c(\theta)$ has been defined in equation (14). In practice, the integral in the above formula is discretized according to the width of the histogram bins. Applying this estimator to individual data subsets yielded a distribution of phase estimates. The estimation precision was determined as the s.d. of this distribution, shown in Fig. 1c for the three phase shifts along with two-sigma error bars. The obtained values, clearly situated below the shot-noise limit, demonstrate quantum-enhanced operation of the interferometer despite partial spectral distinguishability of the input photons.

The analysis presented above was based on processing only the detected sample of photon pairs. This is a routine approach in proof-of-principle experiments employing 'modest-efficiency detectors'[2,44]. The detected events are generated by a random subset of all photon pairs produced by the source and therefore provide fair statistics to characterize the multiphoton interference underlying the observed quantum enhancement. Current advances in source and detector technology should enable soon unconditional demonstrations of quantum-enhanced metrology also in more complex scenarios such as the one described here.

Discussion

We analysed operation of a two-photon Mach–Zehnder interferometer, which is one of the first[22] and technologically most advanced examples of quantum-enhanced metrology, in a scenario that included two degrees of freedom for interfering photons. It was assumed that one degree of freedom was inaccessible experimentally. Residual distinguishability of interfering photons in this degree of freedom has a markedly deleterious effect on the precision of phase estimation around the operating point where coincidences at the interferometer output are suppressed owing to the Hong-Ou-Mandel effect, producing a coincidence dark fringe. We showed that exploiting another, completely uncorrelated degree of freedom over which one has full experimental control can mitigate this effect, restoring quantum-enhanced precision in the entire operating range. This result is based on a rather subtle interplay between one- and two-photon interference. At the coincidence dark-fringe, imperfect two-photon interference alone does not provide any information about the phase shift, while one-photon interference obviously cannot surpass the shot-noise limit on its own. We demonstrated that combining coherently both types of interference through a suitably designed preparation and measurement scheme in the second degree of freedom yields precision below the shot-noise limit. The feasibility of this approach was confirmed in an experiment using the transverse position of the photons as the controllable degree of freedom that could be measured with high resolution using a single-photon sensitive camera system[38,39]. The presented strategy can be also applied to other scenarios involving two independent degrees of freedom with different experimental accessibility, for example, to mitigate effects of residual spatial distinguishability when only area-integrating detection is available, but photons can be suitably manipulated and measured in the spectral domain.

Beyond two-photon interferometry, one can consider a simple scenario involving a higher photon number when the input ports of a Mach-Zenhder interferometer are fed, respectively, with two photons and a single photon, as shown in Fig. 6a. If all three photons are indistinguishable, counting photons at the interferometer outputs yields fringes depicted in Fig. 6b for which Fisher information $F^{(3)} = 7$ irrespectively of the operating point. This is more than twofold improvement over the shot-noise limit for three photons used independently. When the single photon exhibits residual spectral distinguishability with respect to the two photons (assumed to be identical) feeding the other port, minute changes of the shapes of the interference fringes have a marked effect around points $\theta_0 = 2\arctan(1/\sqrt{2})$ and $\pi - \theta_0$ on a relative phase estimation uncertainty $\varepsilon = \Delta^{(3)}/\Delta^{\text{shot}}$ plotted in Fig. 6c, where $\Delta^{(3)}$ is defined analogously to equation (4). According to Fig. 6b, this can be attributed to a non-vanishing background of events when two photons are detected at one output port and the remaining single photon at the other one. Preparing two photons in the same spatial mode $u(x)$ and the single photon sent individually in a partly overlapping mode $v(x)$ allows one to restore quantum enhancement as seen in Fig. 6c. Interestingly, this effect occurs at different operating points compared with the two-photon case.

A worthwhile candidate to analyse the benefits of mode engineering in a scalable multiphoton scenario may be the celebrated Holland–Burnett scheme[45] employing two Fock states with equal photon numbers. It is easy to verify that sub-shot-noise precision at $\theta = \pi/2$ originates from the suppression of odd photon number events at the interferometer outputs, which again is sensitive to residual distinguishability of input photons. A quantitative analysis of this scenario would require developing an efficient approach to deal with multimode multiphoton states. On the other hand, it is known that in the presence of certain common imperfections, such as photon loss, the ultimate precision follows asymptotically the shot-noise type scaling and the quantum enhancement has the form of a multiplicative factor that can be attained via a repetitive use of finite-size multiparticle superposition states[46]. Therefore, results obtained for a fixed number of photons may also prove useful in the asymptotic limit for realistic scenarios.

The presented results are based on the multimode description of a two-photon interferometer, which goes beyond the simplest models typically used to conceive quantum-enhanced measurement schemes. In practice, the applicability of quantum-enhanced techniques depends crucially on the ability to reduce decoherence effects caused by noise and experimental imperfections. Results presented in this paper suggest that in addition to obvious attempts to suppress decoherence effects in interferometry by improving transmission of optical elements, stabilizing phase reference and so on, exploiting the multimode structure of quantum fields can help to achieve the non-classical regime of operation. If the modal structure of the probes is carefully engineered at the input and suitably detected, such a strategy can offer a noticeable improvement in precision even though

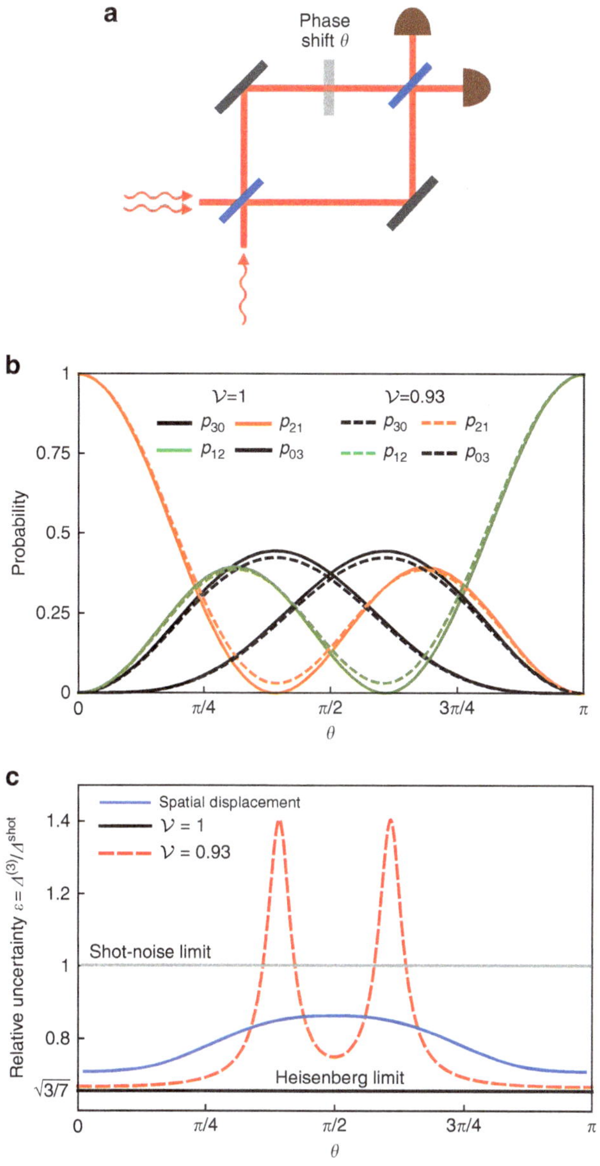

Figure 6 | Three-photon scheme. (**a**) A Mach-Zehnder interferometer fed with a combination of two photons at one input port and a single photon at the other one. (**b**) Probabilities $p_{nn'}(\theta)$ of detecting, respectively, n and n' photons at the interferometer outputs for perfectly indistinguishable photons $\mathcal{V}=1$ and the single photon exhibiting residual distinguishability with respect to the other two mutually identical photons, characterized by $\mathcal{V}=0.93$. (**c**) In the presence of residual distinguishability, the precision of phase estimation, shown with a dashed red line, deteriorates around operating points where dominant contribution to Fisher information comes from the suppression of events $nn'=21$ or 12. This effect can be mitigated by introducing spatial displacement $d=1.45\sigma$ between interferometer inputs, resulting in quantum enhancement across the entire operating range (solid blue line). The solid grey line depicts the shot-noise limit of relative uncertainty.

decoherence effects due to another degree of freedom that we do not have control over remain at the same level. An interesting question is whether analogous strategies can be generalized to quantum-enhanced interferometry with squeezed states of light[7,41,47], boson sampling with linear multiport devices[18–21] or perhaps benefit other quantum technologies such as optical quantum computing[48,49].

Methods

Output two-photon state. For two photons prepared initially in the state $|\psi_{\mathcal{D}}\rangle$ defined in equation (7) it will be convenient to write the output state as a sum of two components,

$$\hat{U}(\theta)|\psi_{\mathcal{D}}\rangle = |\psi_c(\theta)\rangle + |\psi_d(\theta)\rangle \tag{17}$$

Here $|\psi_c(\theta)\rangle$ is the conditional state describing coincidence events, when the two photons leave the interferometer through different ports

$$\begin{aligned}|\psi_c(\theta)\rangle = &-\Big[\sqrt{\mathcal{D}\mathcal{V}}\hat{a}^\dagger_{1R}\hat{b}^\dagger_{1R}\cos\theta \\ &+\sqrt{\mathcal{D}(1-\mathcal{V})}\left(\hat{a}^\dagger_{1R}\hat{b}^\dagger_{2R}\cos^2\frac{\theta}{2}-\hat{a}^\dagger_{2R}\hat{b}^\dagger_{1R}\sin^2\frac{\theta}{2}\right)\\ &+\sqrt{(1-\mathcal{D})\mathcal{V}}\left(\hat{a}^\dagger_{1R}\hat{b}^\dagger_{1L}\cos^2\frac{\theta}{2}-\hat{a}^\dagger_{1L}\hat{b}^\dagger_{1R}\sin^2\frac{\theta}{2}\right)\\ &+\sqrt{(1-\mathcal{D})(1-\mathcal{V})}\left(\hat{a}^\dagger_{1R}\hat{b}^\dagger_{2L}\cos^2\frac{\theta}{2}-\hat{a}^\dagger_{2L}\hat{b}^\dagger_{1R}\sin^2\frac{\theta}{2}\right)\Big]|\mathrm{vac}\rangle,\end{aligned} \tag{18}$$

while $|\psi_d(\theta)\rangle$ corresponds to double events, when both the photons emerge at the same output port of the interferometer:

$$\begin{aligned}|\psi_d(\theta)\rangle = &\tfrac{1}{2}\sin\theta\Big(\sqrt{\mathcal{D}\mathcal{V}}\left[\left(\hat{a}^\dagger_{1R}\right)^2-\left(\hat{b}^\dagger_{1R}\right)^2\right]\\ &+\sqrt{\mathcal{D}(1-\mathcal{V})}\left(\hat{a}^\dagger_{1R}\hat{a}^\dagger_{2R}-\hat{b}^\dagger_{1R}\hat{b}^\dagger_{2R}\right)\\ &+\sqrt{(1-\mathcal{D})\mathcal{V}}\left(\hat{a}^\dagger_{1R}\hat{a}^\dagger_{1L}-\hat{b}^\dagger_{1R}\hat{b}^\dagger_{1L}\right)\\ &+\sqrt{(1-\mathcal{D})(1-\mathcal{V})}\left(\hat{a}^\dagger_{1R}\hat{a}^\dagger_{2L}-\hat{b}^\dagger_{1R}\hat{b}^\dagger_{2L}\right)\Big)|\mathrm{vac}\rangle.\end{aligned} \tag{19}$$

The overall probabilities for double and coincidence events are

$$p_d(\theta) = \langle\psi_d(\theta)|\psi_d(\theta)\rangle = \frac{1}{2}(1+\mathcal{D}\mathcal{V})\sin^2\theta \tag{20}$$

and $p_c(\theta)=1-p_d(\theta)$. Inserting $\mathcal{D}=1$ gives as a special case the result presented in equation (3).

Let us note that all the terms in $|\psi_d(\theta)\rangle$ exhibit identical dependence on θ. Consequently, resolving double events with respect to the spatial degree of freedom cannot yield more information about the phase shift. Therefore, we will focus our attention on coincidence events described by $|\psi_c(\theta)\rangle$. To analyse information about θ when the spectral degree of freedom cannot be accessed, we will treat it formally as another subsystem Ω, writing $\hat{a}^\dagger_{i\chi}\hat{b}^\dagger_{i'\chi'}|\mathrm{vac}\rangle=|\chi\chi'\rangle\otimes|ii'\rangle_\Omega$, where i, $i'=1, 2$ and $\chi, \chi'=R, L$. Tracing the two-photon state over the spectral subsystem yields the reduced density matrix $\hat{\varrho}_c(\theta)=\mathrm{Tr}_\Omega[|\psi_c(\theta)\rangle\langle\psi_c(\theta)|]$, which written in the basis of spatial modes $|RR\rangle, |RL\rangle, |LR\rangle, |LL\rangle$ reads:

$$\hat{\varrho}_c(\theta)=\begin{pmatrix}\mathcal{D}[1-\frac{1}{2}(1+\mathcal{V})\sin^2\theta] & \sqrt{\mathcal{D}(1-\mathcal{D})}(\cos^4\frac{\theta}{2}-\frac{1}{4}\mathcal{V}\sin^2\theta) & \sqrt{\mathcal{D}(1-\mathcal{D})}(\sin^4\frac{\theta}{2}-\frac{1}{4}\mathcal{V}\sin^2\theta) & 0\\ \sqrt{\mathcal{D}(1-\mathcal{D})}(\cos^4\frac{\theta}{2}-\frac{1}{4}\mathcal{V}\sin^2\theta) & (1-\mathcal{D})\cos^4\frac{\theta}{2} & -\frac{1}{4}(1-\mathcal{D})\mathcal{V}\sin^2\theta & 0\\ \sqrt{\mathcal{D}(1-\mathcal{D})}(\sin^4\frac{\theta}{2}-\frac{1}{4}\mathcal{V}\sin^2\theta) & -\frac{1}{4}(1-\mathcal{D})\mathcal{V}\sin^2\theta & (1-\mathcal{D})\sin^4\frac{\theta}{2} & 0\\ 0 & 0 & 0 & 0\end{pmatrix}. \tag{21}$$

Fisher information. If the spatial degree of freedom of the photons at the interferometer output is projected onto the R/L basis, the complete statistics of measurement results is described by the diagonal elements of the density matrix $\hat{\varrho}_c(\theta)$ for coincidence events and the collective probability $p_d(\theta)$ for all double events. An easy calculation yields the corresponding Fisher information

$$F_{R/L}(\theta) = \frac{2(1+\mathcal{D}\mathcal{V})-(\mathcal{D}+1)(\mathcal{V}+1)\sin^2\theta}{1-\frac{1}{2}(\mathcal{V}+1)\sin^2\theta}. \tag{22}$$

At the dark fringe we have $F_{R/L}(\pi/2)=2(1-\mathcal{D})$. This expression, which does not even reach the shot-noise level, can be understood intuitively: information about the phase shift is obtained only from mode-mismatched pairs, when the spatial mode R or L at the output identifies unambiguously the input port of a given photon. The factor $1-\mathcal{D}$ in $F_{R/L}(\pi/2)$ is the overall fraction of these events, while the constant 2 is contributed by single-photon interference exhibited by such pairs.

Clearly, a measurement in the R/L basis neglects information contained in the off-diagonal elements of the density matrix $\hat{\varrho}_c(\theta)$. A measurement strategy that exploits optimally $\hat{\varrho}_c(\theta)$ is described by quantum Fisher information involving the symmetric logarithmic derivative. Its calculation is simplified by switching to a new basis for the spatial degree of freedom,

$$\begin{aligned}|\alpha\rangle &= \sqrt{\frac{2\mathcal{D}}{1+\mathcal{D}}}|RR\rangle+\sqrt{\frac{1-\mathcal{D}}{2(1+\mathcal{D})}}(|RL\rangle+|LR\rangle),\\ |\beta\rangle &= \frac{1}{\sqrt{2}}(|RL\rangle-|LR\rangle),\\ |\gamma\rangle &= \sqrt{\frac{1-\mathcal{D}}{1+\mathcal{D}}}|RR\rangle-\sqrt{\frac{\mathcal{D}}{1+\mathcal{D}}}(|RL\rangle+|LR\rangle).\end{aligned} \tag{23}$$

The conditional density matrix $\hat{\varrho}_c(\theta)$ takes the form

$$\begin{aligned}\langle\alpha|\hat{\varrho}_c(\theta)|\alpha\rangle &= \frac{1+\mathcal{D}}{2}\left(1-\frac{1+\mathcal{V}}{2}\sin^2\theta\right)\\ \langle\alpha|\hat{\varrho}_c(\theta)|\beta\rangle &= \langle\beta|\hat{\varrho}_c(\theta)|\alpha\rangle = \frac{1}{2}\sqrt{1-\mathcal{D}^2}\cos\theta \\ \langle\beta|\hat{\varrho}_c(\theta)|\beta\rangle &= \frac{1-\mathcal{D}}{2}\left(1-\frac{1-\mathcal{V}}{2}\sin^2\theta\right)\end{aligned} \quad (24)$$

while all other elements involving $|\gamma\rangle$ or $\langle\gamma|$ vanish. The symmetric logarithmic derivative $\hat{L}_c(\theta)$, given in general by the implicit formula

$$\frac{d\hat{\varrho}_c}{d\theta} = \frac{1}{2}\left[\hat{L}_c(\theta)\hat{\varrho}_c(\theta)+\hat{\varrho}_c(\theta)\hat{L}_c(\theta)\right] \quad (25)$$

can now be easily found in the two-dimensional subspace spanned by $|\alpha\rangle$ and $|\beta\rangle$. If double events are not resolved in the spatial degree of freedom, quantum Fisher information can be written as

$$\begin{aligned}F_Q(\theta) &= \mathrm{Tr}\left\{\hat{\varrho}_c(\theta)\left[\hat{L}_c(\theta)\right]^2\right\} + \frac{1}{p_d(\theta)}\left(\frac{dp_d}{d\theta}\right)^2\\ &= 2\frac{1-\mathcal{D}^2+(1+\mathcal{D}\mathcal{V})^2\cos^2\theta}{1-\mathcal{D}\mathcal{V}+(1+\mathcal{D}\mathcal{V})\cos^2\theta}\sin^2\theta + 2(1+\mathcal{D}\mathcal{V})\cos^2\theta,\end{aligned} \quad (26)$$

where $p_d(\theta)$ is given by equation (20). Specializing the above expression to $\theta=\pi/2$ yields equation (8), whereas its value for $\mathcal{D}$ optimized individually for a given spectral visibility $0\leq\mathcal{V}\leq1$ and an operating point $0\leq\theta\leq\pi$ is presented in Fig. 2b. The symmetric logarithmic derivative has a simple off-diagonal form at $\theta=\pi/2$

$$\hat{L}_c(\pi/2) = -2\frac{\sqrt{1-\mathcal{D}^2}}{1-\mathcal{D}\mathcal{V}}(|\alpha\rangle\langle\beta|+|\beta\rangle\langle\alpha|). \quad (27)$$

Quantum Fisher information is saturated by projecting the spatial degree of freedom onto the eigenstates of $\hat{L}_c(\theta)$ (refs 42,43) given explicitly for $\theta=\pi/2$ by $|\pm\rangle=(|\alpha\rangle\pm|\beta\rangle)/\sqrt{2}$. These states are non-trivial superpositions of photon pairs prepared in combinations of R, L spatial modes at two different output ports of the interferometer.

Experimental details. In the experiment we used a photon pair source based on the II-type SPDC process in a 5-mm-long periodically poled KTP crystal (Raicol Crystals) pumped with 8 mW of 400 nm light from a continuous wave diode laser. The produced pairs are transmitted through a 3 nm full-width at half maximum interference filter, carefully synchronized in time using a delay line and spatially filtered using the single-mode fibre. The gaussian-like spatial modes of the photons after the fibre have a flat phase and the half-width $\sigma=122\,\mu$m at $1/e$ height for the intensity distribution measured at the position of the camera system. The residual spectral distinguishability of the photons was determined from the depth of the Hong-Ou-Mandel dip scanned using the delay line and measured with standard avalanche photodiodes for the HWP orientation corresponding to $\theta=\pi/2$.

Our camera system[38,39] begins with an image intensifier (Hamamatsu V7090D) where each detected photon that induces a photoelectron emission produces a macroscopic charge avalanche resulting in a bright flash at the output phosphor screen. The flashes are subsequently imaged with a relay lens onto a fast, low-noise $6.5\times6.5\,\mu$m pixel size sCMOS sensor (Andor Zyla) and recorded as $\sim$25-px gaussian wide spots, which can be easily discriminated from the low-noise background. The central positions of the spots are retrieved from each captured frame with a subpixel resolution by a real-time software algorithm that provides full information about transverse coordinates of each registered coincidence event as illustrated in the inset of Fig. 4. For the sake of simplicity, we consider only the coordinate x in the horizontal plane of the setup and integrate the signals in the vertical direction. A cylindrical lens with $f=30$ mm in front of the detector was used to reduce the vertical size of the image, producing effectively a 700×22 px stripe, which significantly decreases frame readout time and allows us to reach 7 kHz collection rate of frames with exposure time 30 ns each. The overall quantum efficiency of the camera system is 23%.

Three-photon scheme. A straightforward but tedious calculation shows that for a three-photon input state of the form $\left(\hat{a}^\dagger_{1R}\right)^2\left[\sqrt{\mathcal{V}}\left(\sqrt{\mathcal{D}}\hat{b}^\dagger_{1R}+\sqrt{1-\mathcal{D}}\hat{b}^\dagger_{1L}\right)\right.$ $\left.+\sqrt{1-\mathcal{V}}\left(\sqrt{\mathcal{D}}\hat{b}^\dagger_{2R}+\sqrt{1-\mathcal{D}}\hat{b}^\dagger_{2L}\right)\right]|\mathrm{vac}\rangle$ spatially resolved distributions for events 21 and 12 when the photons are split between the interferometer output ports into two and one are given by

$$\begin{aligned}p_{21}(x_1,x_2;x'|\theta) &= \mathcal{V}\left|\frac{1}{2}\sin\theta\sin\frac{\theta}{2}(u(x_1)v(x_2)\right.\\ &\left.+u(x_2)v(x_1))u(x') - \cos^3\frac{\theta}{2}u(x_1)u(x_2)v(x')\right|^2\\ &+(1-\mathcal{V})\left[\frac{1}{4}\sin^2\theta\sin^2\frac{\theta}{2}(|u(x_1)|^2|v(x_2)|^2+|v(x_1)|^2|u(x_2)|^2)\right.\\ &\left.\times|u(x')|^2+\cos^6\frac{\theta}{2}|u(x_1)|^2|u(x_2)|^2|v(x')|^2\right]\end{aligned} \quad (28)$$

and

$$\begin{aligned}p_{12}(x;x'_1,x'_2|\theta) &= \mathcal{V}\left|\sin^3\frac{\theta}{2}v(x)u(x'_1)u(x'_2)\right.\\ &\left.-\frac{1}{2}\sin\theta\cos\frac{\theta}{2}u(x)\left(u(x'_1)v(x'_2)+u(x'_2)v(x'_1)\right)\right|^2\\ &+(1-\mathcal{V})\left[\frac{1}{4}\sin^2\theta\cos^2\frac{\theta}{2}|u(x)|^2(|u(x'_1)|^2|v(x'_2)|^2\right.\\ &\left.+|u(x'_2)|^2|v(x'_1)|^2+\sin^6\frac{\theta}{2}|v(x)|^2|u(x'_1)|^2|u(x'_2)|^2\right]\end{aligned} \quad (29)$$

where we have made use of equation (10). Fisher information taking into account spatially resolved detection of events 21 and 12 is given by

$$\begin{aligned}F^{(3)}(\theta) &= \frac{1}{p_{30}(\theta)}\left(\frac{dp_{30}}{d\theta}\right)^2 + \frac{1}{p_{03}(\theta)}\left(\frac{dp_{03}}{d\theta}\right)^2\\ &+\int\frac{dx_1dx_2dx'}{p_{21}(x_1,x_2;x'|\theta)}\left(\frac{\partial}{\partial\theta}p_{21}(x_1,x_2;x'|\theta)\right)^2\\ &+\int\frac{dxdx'_1dx'_2}{p_{12}(x;x'_1,x'_2|\theta)}\left(\frac{\partial}{\partial\theta}p_{12}(x;x'_1,x'_2|\theta)\right)^2.\end{aligned} \quad (30)$$

Assuming Gaussian spatial modes introduced in equation (12), the above expression was optimized over the displacement d for $\theta_0=2\arctan(1/\sqrt{2})$ and $\mathcal{V}=93\%$. The relative uncertainty of a phase estimate for the obtained value $d=1.45\sigma$ has been depicted for the entire range $0\leq\theta\leq\pi$ in Fig. 6c.

References

1. Brida, G., Genovese, M. & Ruo Berchera, I. Experimental realization of sub-shot-noise quantum imaging. *Nat. Photon.* **4,** 227–230 (2010).
2. Ono, T., Okamoto, R. & Takeuchi, S. An entanglement-enhanced microscope. *Nat. Commun.* **4,** 2426 (2013).
3. Schwartz, O. *et al.* Superresolution microscopy with quantum emitters. *Nano Lett.* **13,** 5832–5836 (2013).
4. Tsang, M. Quantum imaging beyond the diffraction limit by optical centroid measurements. *Phys. Rev. Lett.* **102,** 253601 (2009).
5. Wolfgramm, F., Vitelli, C., Beduini, F. A., Godbout, N. & Mitchell, M. W. Entanglement-enhanced probing of a delicate material system. *Nat. Photon.* **7,** 28–32 (2012).
6. Crespi, A. *et al.* Measuring protein concentration with entangled photons. *Appl. Phys. Lett.* **100,** 233704 (2012).
7. LIGO Collaboration. Enhanced sensitivity of the LIGO gravitational wave detector by using squeezed states of light. *Nat. Photon.* **7,** 613–619 (2013).
8. Caves, C. M. Quantum-mechanical noise in an interferometer. *Phys. Rev. D* **23,** 1693–1708 (1981).
9. Yurke, B., McCall, S. L. & Klauder, J. R. SU(2) and SU(1,1) interferometers. *Phys. Rev. A* **33,** 4033–4054 (1986).
10. Bollinger, J. J., Itano, W. M., Wineland, D. J. & Heinzen, D. J. Optimal frequency measurements with maximally correlated states. *Phys. Rev. A* **54,** R4649–R4652 (1996).
11. Boto, A. N. *et al.* Quantum interferometric optical lithography: exploiting entanglement to beat the diffraction limit. *Phys. Rev. Lett.* **85,** 2733–2736 (2000).
12. Giovannetti, V., Lloyd, S. & Maccone, L. Quantum metrology. *Phys. Rev. Lett.* **96,** 10401 (2006).
13. Paris, M. G. A. Quantum estimation for quantum technology. *Int. J. Quantum Inf.* **7,** 125 (2009).
14. Banaszek, K., Demkowicz-Dobrzański, R. & Walmsley, I. A. Quantum states made to measure. *Nat. Photon.* **3,** 673–676 (2009).
15. Maccone, L. & Giovannetti, V. Quantum metrology: Beauty and the noisy beast. *Nat. Phys.* **7,** 376–377 (2011).
16. Tóth, G. & Apellaniz, I. Quantum metrology from a quantum information science perspective. *J. Phys. A* **47,** 424006 (2014).
17. Demkowicz-Dobrzański, R., Jarzyna, M. & Kołodyński, J. Quantum limits in optical interferometry. *Prog. Optics* **60,** 345–435 (2015).
18. Spring, J. B. *et al.* Boson sampling on a photonic chip. *Science* **339,** 798–801 (2013).

19. Spagnolo, N. *et al.* Experimental validation of photonic boson sampling. *Nat. Photon.* **8,** 615–620 (2014).
20. Tichy, M. C., Mayer, K., Buchleitner, A. & Mølmer, K. Stringent and efficient assessment of boson-sampling devices. *Phys. Rev. Lett.* **113,** 020502 (2014).
21. Motes, K. R., Gilchrist, A., Dowling, J. P. & Rohde, P. P. Scalable boson sampling with time-bin encoding using a loop-based architecture. *Phys. Rev. Lett.* **113,** 120501 (2014).
22. Rarity, J. G. *et al.* Two-photon interference in a Mach-Zehnder interferometer. *Phys. Rev. Lett.* **65,** 1348–1351 (1990).
23. Huelga, S. F. *et al.* Improvement of frequency standards with quantum entanglement. *Phys. Rev. Lett.* **79,** 3865–3868 (1997).
24. Dorner, U. *et al.* Optimal quantum phase estimation. *Phys. Rev. Lett.* **102,** 40403 (2009).
25. Knysh, S., Smelyanskiy, V. N. & Durkin, G. A. Scaling laws for precision in quantum interferometry and the bifurcation landscape of the optimal state. *Phys. Rev. A* **83,** 21804 (2011).
26. Jiang, K. *et al.* Strategies for choosing path-entangled number states for optimal robust quantum-optical metrology in the presence of loss. *Phys. Rev. A* **86,** 013826 (2012).
27. Datta, A. *et al.* Quantum metrology with imperfect states and detectors. *Phys. Rev. A* **83,** 63836 (2011).
28. Escher, B. M., de Matos Filho, R. L. & Davidovich, L. General framework for estimating the ultimate precision limit in noisy quantum-enhanced metrology. *Nat. Phys.* **7,** 406–411 (2011).
29. Demkowicz-Dobrzański, R., Guta, M. & Kołodyński, J. The elusive Heisenberg limit in quantum enhanced metrology. *Nat. Commun.* **3,** 1063 (2012).
30. Peeters, W., Renema, J. & van Exter, M. Engineering of two-photon spatial quantum correlations behind a double slit. *Phys. Rev. A* **79,** 043817 (2009).
31. Shin, H., Chan, K. W. C., Chang, H. J. & Boyd, R. W. Quantum spatial superresolution by optical centroid measurements. *Phys. Rev. Lett.* **107,** 083603 (2011).
32. Rozema, L. A. *et al.* Scalable spatial superresolution using entangled photons. *Phys. Rev. Lett.* **112,** 223602 (2014).
33. Abouraddy, A. F., Nasr, M. B., Saleh, B. E. A., Sergienko, A. V. & Teich, M. C. Demonstration of the complementarity of one- and two-photon interference. *Phys. Rev. A* **63,** 63803 (2001).
34. Edgar, M. P. *et al.* Imaging high-dimensional spatial entanglement with a camera. *Nat. Commun.* **3,** 984 (2012).
35. Moreau, P.-A., Devaux, F. & Lantz, E. Einstein-Podolsky-Rosen paradox in twin images. *Phys. Rev. Lett.* **113,** 160401 (2014).
36. Lemos, G. B. *et al.* Quantum imaging with undetected photons. *Nature* **512,** 409–412 (2014).
37. Morris, P. A., Aspden, R. S., Bell, J. E. C., Boyd, R. W. & Padgett, M. J. Imaging with a small number of photons. *Nat. Commun.* **6,** 5913 (2015).
38. Chrapkiewicz, R., Wasilewski, W. & Banaszek, K. High-fidelity spatially resolved multiphoton counting for quantum imaging applications. *Opt. Lett.* **39,** 5090–5093 (2014).
39. Jachura, M. & Chrapkiewicz, R. Shot-by-shot imaging of Hong-Ou-Mandel interference with an intensified sCMOS camera. *Opt. Lett.* **40,** 1540–1543 (2015).
40. Kay, S. M. *Fundamentals of Statistical Signal Processing: Estimation Theory* (Prentice Hall, 1993).
41. Demkowicz-Dobrzański, R., Banaszek, K. & Schnabel, R. Fundamental quantum interferometry bound for the squeezed-light-enhanced gravitational wave detector GEO 600. *Phys. Rev. A* **88,** 41802 (2013).
42. Helstrom, C. W. *Quantum Detection and Estimation Theory* (Academic press, 1976).
43. Braunstein, S. L. & Caves, C. M. Statistical distance and the geometry of quantum states. *Phys. Rev. Lett.* **72,** 3439–3443 (1994).
44. Nagata, T. L., Okamoto, R., O'Brien, J. L., Sasaki, K. & Takeuchi, S. Beating the standard quantum limit with four-entangled photons. *Science* **316,** 726–729 (2007).
45. Holland, M. J. & Burnett, K. Interferometric detection of optical phase shifts at the Heisenberg limit. *Phys. Rev. Lett.* **71,** 1355–1358 (1993).
46. Jarzyna, M. & Demkowicz-Dobrzański, R. True precision limits in quantum metrology. *New J. Phys.* **17,** 013010 (2015).
47. Afek, I., Ambar, O. & Silberberg, Y. High-NOON states by mixing quantum and classical light. *Science* **328,** 879–881 (2010).
48. Knill, E., Laflamme, R. & Milburn, G. J. A scheme for efficient quantum computation with linear optics. *Nature* **409,** 46–52 (2001).
49. Yao, X. C. *et al.* Experimental demonstration of topological error correction. *Nature* **482,** 489–494 (2012).

Acknowledgements

We acknowledge inspiring discussion with M.G. Raymer and I.A. Walmsley. This project was financed by the National Science Centre, No. DEC-2013/09/N/ST2/02229, DEC-2011/03/D/ST2/01941, Polish Ministry of Science and Higher Education Iuventus Plus program for years 2015-2017, No. 0088/IP3/2015/73, European Commission under the FP7 projects SIQS (Grant Agreement No. 600645) and PhoQuS@UW (Grant Agreement No. 316244) co-financed by the Polish Ministry of Science and Higher Education. R.C. was supported by Foundation for Polish Science (FNP).

Author contributions

M.J., R.C. and W.W. designed and performed the experiment. R.D.D. developed theory and analysed the experimental data. K.B. conceived the scheme and interpreted results. All authors contributed to the writing of the manuscript.

Additional information

Competing financial interests: The authors declare no competing financial interests.

Permissions

All chapters in this book were first published in NC, by Nature Publishing Group; hereby published with permission under the Creative Commons Attribution License or equivalent. Every chapter published in this book has been scrutinized by our experts. Their significance has been extensively debated. The topics covered herein carry significant findings which will fuel the growth of the discipline. They may even be implemented as practical applications or may be referred to as a beginning point for another development.

The contributors of this book come from diverse backgrounds, making this book a truly international effort. This book will bring forth new frontiers with its revolutionizing research information and detailed analysis of the nascent developments around the world.

We would like to thank all the contributing authors for lending their expertise to make the book truly unique. They have played a crucial role in the development of this book. Without their invaluable contributions this book wouldn't have been possible. They have made vital efforts to compile up to date information on the varied aspects of this subject to make this book a valuable addition to the collection of many professionals and students.

This book was conceptualized with the vision of imparting up-to-date information and advanced data in this field. To ensure the same, a matchless editorial board was set up. Every individual on the board went through rigorous rounds of assessment to prove their worth. After which they invested a large part of their time researching and compiling the most relevant data for our readers.

The editorial board has been involved in producing this book since its inception. They have spent rigorous hours researching and exploring the diverse topics which have resulted in the successful publishing of this book. They have passed on their knowledge of decades through this book. To expedite this challenging task, the publisher supported the team at every step. A small team of assistant editors was also appointed to further simplify the editing procedure and attain best results for the readers.

Apart from the editorial board, the designing team has also invested a significant amount of their time in understanding the subject and creating the most relevant covers. They scrutinized every image to scout for the most suitable representation of the subject and create an appropriate cover for the book.

The publishing team has been an ardent support to the editorial, designing and production team. Their endless efforts to recruit the best for this project, has resulted in the accomplishment of this book. They are a veteran in the field of academics and their pool of knowledge is as vast as their experience in printing. Their expertise and guidance has proved useful at every step. Their uncompromising quality standards have made this book an exceptional effort. Their encouragement from time to time has been an inspiration for everyone.

The publisher and the editorial board hope that this book will prove to be a valuable piece of knowledge for researchers, students, practitioners and scholars across the globe.

List of Contributors

Thomas Louail
Institut de Physique Théorique, CEA-CNRS (URA 2306), Orme-des-Merisiers Batiment 774, F-91191 Paris, France
Géographie-Cités, CNRS-Paris 1-Paris 7 (UMR 8504), 13 rue du four, FR-75006 Paris, France

José J. Ramasco and Maxime Lenormand
IFISC, Instituto de Física Interdisciplinar y Sistemas Complejos (CSIC-UIB), Campus Universitat de les Illes Balears, Palma de Mallorca E-07122, Spain

Miguel Picornell, Oliva García Cantú and Ricardo Herranz
Nommon Solutions and Technologies, calle Cañas 8, Madrid E-28043, Spain

Enrique Frias-Martinez
Telefonica Research, Madrid E-28050, Spain

Marc Barthelemy
Institut de Physique Théorique, CEA-CNRS (URA 2306), Orme-des-Merisiers Batiment 774, F-91191 Paris, France
Centre d'Analyse et de Mathématique Sociales, EHESS-CNRS (UMR 8557), 190-198 avenue de France, FR-75013 Paris, France

Daniela Abdelkhalek and Roman Schnabel
Institut für Laserphysik, Universität Hamburg, Hamburg 22761, Germany
Institut für Gravitationsphysik, Leibniz Universität Hannover and Max-Planck-Institut für Gravitationsphysik (Albert-Einstein-Institut), Hannover 30167, Germany

Mareike Syllwasschy
Institut für Gravitationsphysik, Leibniz Universität Hannover and Max-Planck-Institut für Gravitationsphysik (Albert-Einstein-Institut), Hannover 30167, Germany

Nicolas J. Cerf
Quantum Information and Communication, Ecole Polytechnique de Bruxelles, CP 165, Université libre de Bruxelles, Brussels 1050, Belgium

Jaromír Fiurášek
Department of Optics, Palacký University, Olomouc 77146, Czech Republic

Johannes Knebel, Markus F. Weber and Erwin Frey
Arnold Sommerfeld Center for Theoretical Physics and Center for NanoScience, Department of Physics, Ludwig-Maximilians-Universität München, Theresienstraβe 37, 80333 München, Germany

Torben Krüger
IST Austria, Am Campus 1, 3400 Klosterneuburg, Austria

Xiaogang Qiang, Kanin Aungskunsiri, Jeremy L. O'Brien and Jonathan C.F. Matthews
Centre for Quantum Photonics, H.H. Wills Physics Laboratory and Department of Electrical and Electronic Engineering, University of Bristol, Bristol BS8 1UB, UK

Thomas Loke and Jingbo B. Wang
School of Physics, The University of Western Australia, Crawley, WA 6009, Australia

Ashley Montanaro
School of Mathematics, University of Bristol, Bristol BS8 1TW, UK

Xiaoqi Zhou
Centre for Quantum Photonics, H.H. Wills Physics Laboratory and Department of Electrical and Electronic Engineering, University of Bristol, Bristol BS8 1UB, UK
State Key Laboratory of Optoelectronic Materials and Technologies and School of Physics, Sun Yat-sen University, Guangzhou 510275, China

M. Naghiloo, N. Foroozani, D. Tan and A. Jadbabaie
Department of Physics, Washington University, St Louis, Missouri 63130, USA

K. W. Murch
Department of Physics, Washington University, St Louis, Missouri 63130, USA
Institute for Materials Science and Engineering, St Louis, Missouri 63130, USA

M. Kira
Department of Physics, Philipps-University Marburg, Renthof 5, D-35032 Marburg, Germany

M. Böttcher and S. Heinze
Institute of Theoretical Physics and Astrophysics, University of Kiel, Leibnizstrasse 15, 24098 Kiel, Germany

B. Dupé
Institute of Theoretical Physics and Astrophysics, University of Kiel, Leibnizstrasse 15, 24098 Kiel, Germany
Peter Grünberg Institut (PGI-1) and Institute for Advanced Simulation (IAS-1), Forschungszentrum Jülich and JARA, 52425 Jülich, Germany

S. Blügel and G. Bihlmayer
Peter Grünberg Institut (PGI-1) and Institute for Advanced Simulation (IAS-1), Forschungszentrum Jülich and JARA, 52425 Jülich, Germany

Tobias Graβ and David Raventós
ICFO-Institut de Ciències Fotòniques, The Barcelona Institute of Science and Technology, Avinguda Carl-Friedrich Gauss 2, Castelldefels 08860, Spain

Bruno Juliá-Díaz
ICFO-Institut de Ciències Fotòniques, The Barcelona Institute of Science and Technology, Avinguda Carl-Friedrich Gauss 2, Castelldefels 08860, Spain
Departament de Fìsica Quàntica i Astrofìsica, Facultat de Fìsica, Universitat de Barcelona, Barcelona 08028, Spain
Institut de Ciències del Cosmos, Universitat de Barcelona, ICCUB, Martí i Franquès 1, Barcelona 08028, Spain

Christian Gogolin
ICFO-Institut de Ciències Fotòniques, The Barcelona Institute of Science and Technology, Avinguda Carl-Friedrich Gauss 2, Castelldefels 08860, Spain
Max-Planck-Institut für Quantenoptik, Hans-Kopfermann-Straβe 1, Garching 85748, Germany

Maciej Lewenstein
ICFO-Institut de Ciències Fotòniques, The Barcelona Institute of Science and Technology, Avinguda Carl-Friedrich Gauss 2, Castelldefels 08860, Spain
ICREA-Instituciö Catalana de Recerca i Estudis Avanc¸ats, Lluís Companys 23, Barcelona 08010, Spain

Michael D. Mazurek and Kevin J. Resch
Institute for Quantum Computing, University of Waterloo, Waterloo, Ontario Canada N2L 3G1
Department of Physics & Astronomy, University of Waterloo, Waterloo, Ontario Canada N2L 3G1

Matthew F. Pusey and Robert W. Spekkens
Perimeter Institute for Theoretical Physics, 31 Caroline Street North, Waterloo, Ontario Canada N2L 2Y5

Ravi Kunjwal
Optics & Quantum Information Group, The Institute of Mathematical Sciences, C.I.T Campus, Taramani, Chennai 600 113, India

Amin Doostmohammadi, Michael F. Adamer, Sumesh P. Thampi and Julia M. Yeomans
The Rudolf Peierls Centre for Theoretical Physics, 1 Keble Road, Oxford OX1 3NP, UK

Yige Chen
Department of Physics, University of Toronto, Toronto, Ontario M5S 1A7, Canada

Yuan-Ming Lu
Department of Physics, University of California, Berkeley, California 94720, USA
Materials Sciences Division, Lawrence Berkeley National Laboratory, Berkeley, California 94720, USA

Hae-Young Kee
Department of Physics, University of Toronto, Toronto, Ontario M5S 1A7, Canada
Canadian Institute for Advanced Research, Toronto, Ontario M5G 1Z8, Canada

Joel Yuen-Zhou
Department of Chemistry and Biochemistry, University of California–San Diego, La Jolla, California 92093, USA

Semion K. Saikin
Department of Chemistry and Chemical Biology, Harvard University, Cambridge, Massachusetts 02138, USA
Department of Physics, Kazan Federal University, Kazan 420008, Russian Federation

Tony Zhu
Department of Physics, Massachusetts Institute of Technology, Cambridge, Massachusetts 02139, USA
Center for Excitonics, Research Laboratory of Electronics, Massachusetts Institute of Technology, Cambridge, Massachusetts 02139, USA

Mehmet C. Onbasli and Caroline A. Ross
Department of Materials Science and Engineering, Massachusetts Institute of Technology, Cambridge, Massachusetts 02139, USA

Marc A. Baldo and Vladimir Bulovic
Center for Excitonics, Research Laboratory of Electronics, Massachusetts Institute of Technology, Cambridge, Massachusetts 02139, USA
Department of Electrical Engineering and Computer Science, Massachusetts Institute of Technology, Cambridge, Massachusetts 02139, USA

Antonio Majdandzic, Chester Curme and H. Eugene Stanley
Center for Polymer Studies and Department of Physics, Boston University, 590 Commonwealth Avenue, Boston, Massachusetts 02215, USA

Lidia A. Braunstein
Center for Polymer Studies and Department of Physics, Boston University, 590 Commonwealth Avenue, Boston, Massachusetts 02215, USA
Instituto de Investigaciones Físicas de Mar del Plata (IFIMAR), Physics Department, Universidad Nacional de Mar del Plata-CONICET, Funes 3350, 7600 Mar del Plata, Argentina

Irena Vodenska
Center for Polymer Studies and Department of Physics, Boston University, 590 Commonwealth Avenue, Boston, Massachusetts 02215, USA
Administrative Sciences Department, Metropolitan College, Boston University, Boston, Massachusetts 02215 USA

Sary Levy-Carciente
Center for Polymer Studies and Department of Physics, Boston University, 590 Commonwealth Avenue, Boston, Massachusetts 02215, USA
Economics and Social Sciences Faculty, Central University of Venezuela, 1040 Caracas, Venezuela

Shlomo Havlin
Center for Polymer Studies and Department of Physics, Boston University, 590 Commonwealth Avenue, Boston, Massachusetts 02215, USA
Department of Physics, Bar-Ilan University, 52900 Ramat-Gan, Israel

Alan Hastings
Department of Environmental Science and Policy, University of California, Davis, California 95616, USA

Andrew E. Noble
Department of Environmental Science and Policy, University of California, Davis, California 95616, USA
Department of Physics, University of Massachusetts, Amherst, Massachusetts 01003, USA

Jonathan Machta
Department of Physics, University of Massachusetts, Amherst, Massachusetts 01003, USA Santa Fe Institute, 1399 Hyde Park Road, Santa Fe, New Mexico 87501, USA

Owen Benton, L. D. C. Jaubert, Han Yan and Nic Shannon
Okinawa Institute of Science and Technology Graduate University, Onna-son, Okinawa 904-0495, Japan

Manlio De Domenico and Alexandre Arenas
Departament d'Enginyeria Informática i Matemátiques, Universitat Rovira I Virgili, Avda Paisos Catalans 26, Tarragona 43007, Spain

Vincenzo Nicosia
School of Mathematical Sciences, Queen Mary University of London, London E1 4NS, UK

Vito Latora
School of Mathematical Sciences, Queen Mary University of London, London E1 4NS, UK
Dipartimento di Fisica ed Astronomia, Università di Catania and INFN, Catania I-95123, Italy

Michele Campisi
NEST, Scuola Normale Superiore & Istituto Nanoscienze-CNR, Pisa I-56126, Italy

Rosario Fazio
NEST, Scuola Normale Superiore & Istituto Nanoscienze-CNR, Pisa I-56126, Italy
ICTP, Strada Costiera 11, Trieste 34151, Italy

Manuel Mielenz, Henning Kalis, Matthias Wittemer, Frederick Hakelberg and Ulrich Warring
Albert-Ludwigs-Universität Freiburg, Physikalisches Institut, Hermann-Herder-Strasse 3, Freiburg 79104, Germany

Roman Schmied
Department of Physics, University of Basel, Klingelbergstrasse 82, Basel 4056, Switzerland

Matthew Blain and Peter Maunz
Sandia National Laboratories, PO Box 5800 Albuquerque, New Mexico 87185-1082, USA

David L. Moehring
Sandia National Laboratories, PO Box 5800 Albuquerque, New Mexico 87185-1082, USA
The Intelligence Advanced Research Projects Activity, College Park, MD, USA

Dietrich Leibfried
Time and Frequency Division, National Institute of Standards and Technology, 325 Broadway, Boulder, Colorado 80305, USA

Tobias Schaetz
Albert-Ludwigs-Universität Freiburg, Physikalisches Institut, Hermann-Herder-Strasse 3, Freiburg 79104, Germany
Albert-Ludwigs-Universität Freiburg, Freiburg Institute for Advanced Studies, Albertstr. 19, 79104 Freiburg, Germany

Yelena Guryanova, Sandu Popescu, Anthony J. Short and Paul Skrzypczyk
H.H. Wills Physics Laboratory, University of Bristol, Tyndall Avenue, Bristol BS8 1TL, UK

Ralph Silva
H.H. Wills Physics Laboratory, University of Bristol, Tyndall Avenue, Bristol BS8 1TL, UK
Département de Physique Théorique, Université de Genève, 1211 Genève, Switzerland

Leopoldo R. Gómez, Nicolás A. García and Daniel A. Vega
Department of Physics, Universidad Nacional del Sur – IFISUR – CONICET, 8000 Bahía Blanca, Argentina

Vincenzo Vitelli
Instituut-Lorentz, Universiteit Leiden, 2300 RA Leiden, The Netherlands

José Lorenzana
Institute for Complex Systems, Consiglio Nazionale delle Ricerche, I-00185 Rome, Italy
Physics Department, University of Rome La Sapienza, I-00185 Rome, Italy

Filippo Caruso
LENS, QSTAR & Dipartimento di Fisica e Astronomia, Università di Firenze, via Nello Carrara 1, I-50019 Sesto Fiorentino, Italy

Andrea Crespi and Roberto Osellame
Istituto di Fotonica e Nanotecnologie, Consiglio Nazionale delle Ricerche (IFN-CNR), Piazza Leonardo da Vinci 32, I-20133 Milano, Italy
Dipartimento di Fisica, Politecnico di Milano, Piazza Leonardo da Vinci 32, I-20133 Milano, Italy

Anna Gabriella Ciriolo
Dipartimento di Fisica, Politecnico di Milano, Piazza Leonardo da Vinci 32, I-20133 Milano, Italy

Fabio Sciarrino
Dipartimento di Fisica, Sapienza Università di Roma, Piazzale Aldo Moro 5, I-00185 Roma, Italy

Demid Sychev
Russian Quantum Center, 100 Novaya Street, Skolkovo, Moscow 143025, Russia

Alexander E. Ulanov
Russian Quantum Center, 100 Novaya Street, Skolkovo, Moscow 143025, Russia
Moscow Institute of Physics and Technology, Dolgoprudny 141700, Russia
P.N. Lebedev Physics Institute, Leninskiy prospect 53, Moscow 119991, Russia

Ilya A. Fedorov
Russian Quantum Center, 100 Novaya Street, Skolkovo, Moscow 143025, Russia
P.N. Lebedev Physics Institute, Leninskiy prospect 53, Moscow 119991, Russia

Philippe Grangier
Laboratoire Charles Fabry, Institut d'Optique Graduate School, CNRS, Université Paris-Saclay, Palaiseau 91127, France

A. I. Lvovsky
Russian Quantum Center, 100 Novaya Street, Skolkovo, Moscow 143025, Russia
P.N. Lebedev Physics Institute, Leninskiy prospect 53, Moscow 119991, Russia
Institute for Quantum Science and Technology, University of Calgary, Calgary, Alberta, Canada T2N 1N4

Nicole Yunger Halpern
Institute for Quantum Information and Matter, Division of Physics, Mathematics, and Astronomy, California Institute of Technology, Pasadena, California 91125, USA

Philippe Faist
Institute for Theoretical Physics, Department of Physics, ETH Zürich, 8093 Zürich, Switzerland

Jonathan Oppenheim
Department of Physics and Astronomy, University College London, Gower Street, London WC1E 6BT, UK

Andreas Winter
Departament de Fìsica, Grup d'Informació Quántica, Universitat Autònoma de Barcelona, ES-08193 Barcelona, Spain
ICREA, Pg. Lluís Companys 23, 08010 Barcelona, Spain

Filippo Cardano, Maria Maffei, Bruno Piccirillo and Enrico Santamato
Dipartimento di Fisica, Universita` di Napoli Federico II, Complesso Universitario di Monte Sant'Angelo, via Cintia, Napoli 80126, Italy

Francesco Massa
Dipartimento di Fisica, Università di Napoli Federico II, Complesso Universitario di Monte Sant'Angelo, via Cintia, Napoli 80126, Italy
Faculty of Physics, University of Vienna, Boltzmanngasse 5, Vienna 1090, Austria

Corrado de Lisio, Giulio De Filippis and Vittorio Cataudella
Dipartimento di Fisica, Università di Napoli Federico II, Complesso Universitario di Monte Sant'Angelo, via Cintia, Napoli 80126, Italy
CNR-SPIN, Complesso Universitario di Monte Sant'Angelo, Via Cintia, Napoli 80126, Italy

Lorenzo Marrucci
Dipartimento di Fisica, Università di Napoli Federico II, Complesso Universitario di Monte Sant'Angelo, via Cintia, Napoli 80126, Italy
CNR-ISASI, Via Campi Flegrei 34, Pozzuoli (NA) 80078, Italy

Patrick J. Coles, Eric M. Metodiev and Norbert Lütkenhaus
Department of Physics and Astronomy, Institute for Quantum Computing, University ofWaterloo,Waterloo, Ontario, Canada N2L3G1

K. Balzer and M. Eckstein
Max Planck Research Department for Structural Dynamics, University of Hamburg-CFEL, 22761 Hamburg, Germany

J. H. Mentink
Max Planck Research Department for Structural Dynamics, University of Hamburg-CFEL, 22761 Hamburg, Germany
Radboud University, Institute for Molecules and Materials, Heyendaalseweg 135, 6525 AJ Nijmegen, The Netherlands

H. Krull and G. S. Uhrig
Lehrstuhl für Theoretische Physik I, Technische Univerität Dortmund, Otto-Hahn Strasse 4, D-44221 Dortmund, Germany

N. Bittner, D. Manske and A.P. Schnyder
Max-Planck-Institut für Festkörperforschung, Heisenbergstrasse 1, D-70569 Stuttgart, Germany

Ido Kaminer, Yichen Shen, Ognjen Ilic, Josué J. López, John D. Joannopoulos and Marin Soljačić
Department of Physics, Massachusetts Institute of Technology, 77 Massachusetts Avenue, Cambridge, Massachusetts 02139, USA

Yaniv Tenenbaum Katan
Physics Department and Solid State Institute, Technion, Haifa 32000, Israel

Hrvoje Buljan
Department of Physics, University of Zagreb, Bijenička c. 32, 10000 Zagreb, Croatia

Liang Jie Wong
Singapore Institute of Manufacturing Technology, 2 Fusionopolis Way, Innovis, Singapore 138634, Singapore

Michał Jachura, Radosław Chrapkiewicz, Rafał Demkowicz-Dobrzański, Wojciech Wasilewski and Konrad Banaszek
Faculty of Physics, University of Warsaw, Pasteura 5, 02-093 Warsaw, Poland

Index